T0235966

Lecture Notes in Artificial Intelligence 10363

Subseries of Lecture Notes in Computer Science

More information about this series at http://www.springer.com/series/1244

De-Shuang Huang · Abir Hussain
Kyungsook Han · M. Michael Gromiha (Eds.)

Intelligent Computing Methodologies

13th International Conference, ICIC 2017
Liverpool, UK, August 7–10, 2017
Proceedings, Part III

 Springer

Editors
De-Shuang Huang
Tongji University
Shanghai
China

Kyungsook Han
Inha University
Incheon
Korea (Republic of)

Abir Hussain
Liverpool John Moores University
Liverpool
UK

M. Michael Gromiha
Indian Institute of Technology Madras
Chennai
India

ISSN 0302-9743   ISSN 1611-3349   (electronic)
Lecture Notes in Artificial Intelligence
ISBN 978-3-319-63314-5   ISBN 978-3-319-63315-2   (eBook)
DOI 10.1007/978-3-319-63315-2

Library of Congress Control Number: 2017946068

LNCS Sublibrary: SL7 – Artificial Intelligence

Printed on acid-free paper

This Springer imprint is published by Springer Nature
The registered company is Springer International Publishing AG
The registered company address is: Gewerbestrasse 11, 6330 Cham, Switzerland

Preface

The International Conference on Intelligent Computing (ICIC) was started to provide an annual forum dedicated to the emerging and challenging topics in artificial intelligence, machine learning, pattern recognition, bioinformatics, and computational biology. It aims to bring together researchers and practitioners from both academia and industry to share ideas, problems, and solutions related to the multifaceted aspects of intelligent computing.

ICIC 2017, held in Liverpool, UK, August 7–10, 2017, constituted the 13th International Conference on Intelligent Computing. It built upon the success of ICIC 2016, ICIC 2015, ICIC 2014, ICIC 2013, ICIC 2012, ICIC 2011, ICIC 2010, ICIC 2009, ICIC 2008, ICIC 2007, ICIC 2006, and ICIC 2005 that were held in Lanzhou, Fuzhou, Taiyuan, Nanning, Huangshan, Zhengzhou, Changsha, China, Ulsan, Korea, Shanghai, Qingdao, Kunming, and Hefei, China, respectively.

This year, the conference concentrated mainly on the theories and methodologies as well as the emerging applications of intelligent computing. Its aim was to unify the picture of contemporary intelligent computing techniques as an integral concept that highlights the trends in advanced computational intelligence and bridges theoretical research with applications. Therefore, the theme for this conference was "Advanced Intelligent Computing Technology and Applications." Papers focused on this theme were solicited, addressing theories, methodologies, and applications in science and technology.

ICIC 2017 received 612 submissions from 21 countries and regions. All papers went through a rigorous peer-review procedure and each paper received at least three review reports. Based on the review reports, the Program Committee finally selected 212 high-quality papers for presentation at ICIC 2017, included in three volumes of proceedings published by Springer: two volumes of *Lecture Notes in Computer Science* (LNCS) and one volume of *Lecture Notes in Artificial Intelligence* (LNAI).

This volume of *Lecture Notes in Artificial Intelligence* (LNAI) includes 67 papers.

The organizers of ICIC 2017, including Tongji University and Liverpool John Moores University, UK, made an enormous effort to ensure the success of the conference. We hereby would like to thank the members of the Program Committee and the referees for their collective effort in reviewing and soliciting the papers. We would like to thank Alfred Hofmann, executive editor at Springer, for his frank and helpful advice and guidance throughout and for his continuous support in publishing the proceedings. In particular, we would like to thank all the authors for contributing their papers. Without the high-quality submissions from the authors, the success of the conference would not have been possible. Finally, we are especially grateful to the IEEE Computational Intelligence Society, the International Neural Network Society, and the National Science Foundation of China for their sponsorship.

May 2017

De-Shuang Huang
Abir Hussain
Kyungsook Han
M. Michael Gromiha

ICIC 2017 Organization

General Co-chairs

De-Shuang Huang, China
Abir Hussain, UK

Program Committee Co-chairs

Kang-Hyun Jo, Korea
M. Michael Gromiha, India

Organizing Committee Co-chairs

Dhiya Al-Jumeily, UK
Tom Dowson, UK

Award Committee Co-chairs

Vitoantonio Bevilacqua, Italy
Phalguni Gupta, India

Tutorial Co-chairs

Juan Carlos Figueroa, Colombia
Valeriya Gribova, Russia

Publication Co-chairs

Kyungsook Han, Korea
Laurent Heutte, France

Workshop/Special Session Chair

Wenzheng Bao, China

Special Issue Chair

Ling Wang, China

International Liaison Chair

Prashan Premaratne, Australia

Publicity Co-chairs

Chun-Hou Zheng, China
Jair Cervantes Canales, Mexico
Huiyu Zhou, China

Exhibition Chair

Lin Zhu, China

Program Committee

Khalid Aamir	Yannis Goulermas	Chunmei Liu
Mohd Helmy Abd Wahab	Michael Gromiha	Shuo Liu
Abbas Amini	Fei Han	Xingwen Liu
Vasily Aristarkhov	Kyungsook Han	Xiwei Liu
Waqas Haider	Tianyong Hao	Yunxia Liu
Khan Bangyal	Wei-Chiang Hong	Ahmad Lotfi
Shuui Bi	Yuexian Hou	Jungang Lou
Hongmin Cai	Saiful Islam	Yonggang Lu
Jair Cervantes	Chuleerat Jaruskulchai	Yingqin Luo
Pei-Chann Chang	Kang-Hyun Jo	Jinwen Ma
Chen Chen	Dah-Jing Jwo	Xiandong Meng
Shih-Hsin Chen	Seeja K.R.	Kang Ning
Weidong Chen	Sungshin Kim	Ben Niu
Wen-Sheng Chen	Yong-Guk Kim	Gaoxiang Ouyang
Xiyuan Chen	Yoshinori Kuno	Francesco Pappalardo
Jieren Cheng	Takashi Kuremoto	Young B. Park
Michal Choras	Xuguang Lan	Eros Pasero
Angelo Ciaramella	Xinyi Le	Marzio Pennisi
Jose Alfredo Costa	Choong Ho Lee	Prashan Premaratne
Guojun Dai	Xiujuan Lei	Yuhua Qian
Yanrui Ding	Bo Li	Daowen Qiu
Ji-Xiang Du	Guoliang Li	Jiangning Song
Pufeng Du	Kang Li	Stefano Squartini
Jianbo Fan	Ming Li	Zhixi Su
Paul Fergus	Qiaotian Li	Shiliang Sun
Juan Carlos	Shuai Li	Zhan-Li Sun
Figueroa-Garcia	Honghuang Lin	Jijun Tang
Liang Gao	Bin Liu	Joaquin Torres-Sospedra
Dunwei Gong	Bingqiang Liu	Antonio Uva

Bing Wang
Jim Jing-Yan Wang
Shitong Wang
Xuesong Wang
Yong Wang
Yuanzhe Wang
Ze Wang
Wei Wei
Ka-Chun Wong
Hongjie Wu

QingXiang Wu
Yan Wu
Junfeng Xia
Shunren Xia
Xinzheng Xu
Wen Yu
Junqi Zhang
Rui Zhang
Shihua Zhang
Xiang Zhang

Zhihua Zhang
Dongbin Zhao
Xiaoguang Zhao
Huiru Zheng
Huiyu Zhou
Yongquan Zhou
Shanfeng Zhu
Quan Zou
Le Zhang

Additional Reviewers

Honghong Cheng
Jieting Wang
Xinyan Liang
Feijiang Li
Furong Lu
Mohd Farhan Md Fudzee
Husniza Husni
Rozaida Ghazali
Zarul Zaaba
Shuzlina Abdul-Rahman
Sasalak Tongkaw
Wan Hussain Wan Ishak
Mohamad Farhan
    Mohamad Mohsin
Masoud Mohammadian
Mario Diván
Nureize Arbaiy
Mohd Shamrie Sainin
Francesco Masulli
Nooraini Yusoff
Aida Mustapha
Nur Azzah Abu Bakar
Anang Hudaya Muhamad
    Amin
Sazalinsyah Razali
Norita Md Norwawi
Nasir Sulaiman
Ru-Ze Liang
Wan Hasrulnizzam
    Wan Mahmood
Azizul Azhar Ramli

Qian Guo
Mingjie Tang
Fang Jin
Muhammad Imran
Barnypok
Weizhi Li
Ri Hanafi
Jieyi Zhao
Zhichao Jiang
Y-H. Taguchi
Chenhao Xu
Shaofu Yang
Jian Xiao
Haoyong Yu
Prabakaran R.
Anusuya S.
A. Mary Thangakani
Fatih Adiguzel
Shanwen Zhu
Yuheng Wang
Xixun Zhu
Ambuj Srivastava
Xing He
Nagarajan Raju
Kumar Yugandhar
Sakthivel Ramasamy
Anoosha Paruchuri
Sherlyn
Ramakrishnan
Chandrasekaran
Akila Ranjith

Harihar Balasubramanian
Vimala A.
Farheen Siddiqui
Sameena Naaz
Parul Agarwal
Soniya Balram
Jaya Sudha
Deepa Anand
Yun Xue
Santo Motta
Nengfu Xie
Lei Liu
Nengfu Xie
Xiangyuan Lan
Rushi Lan
Jie Qin
Tomasz Andrysiak
Rafal Kozik
Adam Schmidt
Zhiqiang Liu
Liang Xiaobo
Zhiwei Feng
Xia Li
Chunxia Zhang
Jiaoyun Yang
Yi Xiong
Wenjun Shen
Xihao Hu
Xiangli Zhang
Xiaoyong Bian
Yan Xu

Lu-Chen Weng
Guan Xiao
Jianchao Fan
Ka-Chun Wong
Fei Wang
Beeno Cheong
Wei-Jie Yu
Meng Zhang
Jiao Zhang
Zhouhua Peng
Shankai Yan
Yueming Lyu
Xiaoping Wang
Cheng Liu
Jiecong Lin
Shan Gao
Xiu-Jun Gong
Lijun Quan
Haiou Li
Junyi Chen
Cheng Chen
Ukyo Liu
Weizhi Li
Xi Yuga
Bo Chen
Binbin Pan
Yaran Chen
Xiangtao Li
Fei Guo
Leyi Wei
Yungang Xu
Bin Qin
Yuanpeng Zhang
Wenlong Hang
Huo Xuan
Feng Cong
Hongguo Zhao
Xiongtao Zhang
Li Liu
Baohua Wang
Jianwei Yang
Naveed Anwer Butt
Seung Hoan Choi
Farid Garcia-Lamont
Sergio Ruiz

Josue Vicente Cervantes
  Bazan
Abd Ur Rehman
Ta Zhou
Wentao Fan
Xin Liu
Qing Lei
Ziyi Chen
Yewang Chen
Yan Chen
Jialin Peng
Sihai Yang
Hong-Bo Zhang
Xingguang Pan
Biqi Wang
Arturo
Lisbeth Rodríguez
Asdrubal Lopez-Chau
Xiaoli Li
Yuriy Orlov
Ken Wing-Kin Sung
Zhou Hufeng
Gonghong Wei
De-Shuang Huang
Dongbo Bu
Dariusz Plewczyński
Jialiang Yang
Musa Mhlanga
Shanshan Tuo
Dawei Li
Chee Hong Wong
Zhang Zhizhuo
Xiang Jinhai
Wada Youichiro
Shuai Cheng Li
Xuequn Shang
Ralf Jauch
Limsoon Wong
Edwin Cheung
Filippo Castiglione
Ping Zhang
Hong Zhu
Tianming Liang
Jiancong Fan
Guanying Wang

Shumei Zhang
Raymond Bond
Ginestra Bianconi
Antonino Staiano
Dong Shan
Xiangying Jiang
Hongjie Jia
Xiaopeng Hua
Sixiao Wei
Shumei Zhang
Francesco Camastra
Weikuan Jia
Huajuan Huang
Hui Li
Xinzheng Xu
Yan Qin
Jiayin Zhou
Xin Gao
Antonio Maratea
Wu Qingxiang
Xiaowei Zhang
Cunlu Xu
Yong Lin
Haizhou Wu
Ruxin Zhao
Alessio Ferone
Zheheng Jiang
Kun Zhang
Fei Chu
Chunyu Yang
Wei Dai
Gang Li
Guo Yanen
Hao Xin
Zhang Zanchao
Yonggang Wang
Yajun Zhang
Guimin Lin
Pengfei Li
Antony Lam
Hironobu Fujiyoshi
Yoshinori Kobayashi
Jin-Soo Kim
Erchao Li
Miao Rong

Yiping Liu
Nobutaka Shimada
Hae-Chul Choi
Irfan Mehmood
Yuyan Han
Jianhua Zhang
Biao Xu
Meirong Chen
Kazunori Onoguchi
Gongjing Chen
Yuntao Wei
Rina Su
Lu Xingjia
Casimiro Aday
    Curbelo Montanez
Mohammed Khalaf
Atsushi Yamashita
Hotaka Takizawa
Yasushi Mae
Hisato Fukuda
Shaohua Li
Keight Robert
Raghad Al-Shabandar
Xie Zhijun
Abir Hussain
Morihiro Hayashida
Ziding Zhang
Mengyao Li
Shen Chong
Wei Wang
Basma Abdulaimma
Hoshang Kolivand
Ala Al Kafri
Abbas, Rounaq
Kaihui Bian
Jiangning Song
Haya Alaskar
Francis, Hulya
Chuang Wu
Xinhua Tang
Hong Zhang
Sheng Zou
Cuci Quinstina
Bingbo Cui
Hongjie Wu
Wenrui Zhao

Xuan Wang
Junjun Jiang
Qiyan Sun
Zhuangguo Miao
Yuan Xu
Dengxin Dai
Zhiwu Huang
Xuetao Zhang
Shaoyi Du
Jihua Zhu
Bingxiang Xu
Zhixuan Wei
Hanfu Wang
Prashan Premaratne
Alessandro Naddeo
Jun Zhou
Bing Feng
Carlo Bianca
Raúl Montoliu
German Martín
    Mendoza-Silva
Ximo Torres
Bulent Ayhan
Qiang Liu
Xinle Liu
Jair Cervantes
Junming Zhang
Linting Guan
Guodong Zhao
Tao Yang
Zhenmin Zhang
Tao Wu
Yong Chen
Chien-Lung Chan
Liang-Chih Yu
Chin-Sheng Yang
Chin-Yuan Fan
Si-Woong Lee
Zhiwei Ji
Ke Zeng
Filipe Saraiva
Mario Malcangi
Francesco Ferracuti
José Alfredo Costa
Jacob Schrum
Shangxuan Tian

Yitian Zhao
Yonghuai Liu
Julien Leroy
Pierre Marighetto
Jheng-Long Wu
Falcon Rafael
Haytham Fayek
Yong Xu
Kozou Abdellah
Damiano Rossetti
Daniele Ferretti
Chengdong Li
Zhen Ni
Pavel Osinenko
Bakkiyaraj Ashok
Hector Menendez
Pandiri Venkatesh
Davendra Donald
Ming-Wei Li
Leo Chen
Xiaoli Wei
Chien-Yuan Lai
Yongquan Zhou
Ma Xiaotu
Yannan Bin
Wei-Chiang Hong
Seongho Kim
Xiang Zhang
Geethan Mendiz
Brendan Halloran
Kai Xu
Di Zhang
Sheng Ding
Jun Li
Lei Wang
Jin Gu
Wang, Xiaowo
Hao Lin
Yao Nie
Vincenzo Randazzo
Yujie Cai
Derek Wang
Gang Li
Fuhai Li
Junfeng Luo
Wei Lan

Chaowang Lan
Lasker Ershad Ali
Shuyi Zhang
Pengbo Bo
Abhijit Kundu
Jingkai Yang
Shamim Reza
Lee Kai Wah
Anass Nouri
Arridhana Ciptadi
Lun Li
Cheeyong Kim
Helen Hong
Guangchun Gao
Weilin Deng
Taeho Jo
Jiazhou Chen
Enhong Zhuo
Carlo Bianca
Gai-Ge Wang
Kwanggi Kim
Jiangling Song
Xiandong Xu
Guohui Zhang
Long Wen
Kunkun Peng
Qiqi Duan
Hui Wang
Shiying Sun
Yu Yongjia
Jianbo Fan
Lishuang Shen
Liming Xie
Geethan Mendiz
Ying Bi
Hong Wang
Juntao Liu
Yang Li
Enfeng Qi
Xiangyu Liu
Jia Liu
Hou Yingnan
Weihua Deng
Chuanfeng Li
Fuyi Li
Tomasz Talaska

Nathan Cannon
Bo Gao
Ting Yu
Kang Li
Li Zhang
Xingjia Lu
Qingfeng Li
Yong-Guk Kim
Kunikazu Kobayashi
Takashi Kuremoto
Shingo Mabu
Takaomi Hirata
Tuozhong Yao
Lvzhou Li
Fuchun Liu
Shenggen Zheng
Jingkai Yang
Chen Li
Sungshin Kim
Hansoo Lee
Jungwon Yu
Yongsheng Dong
Zhihua Cui
Duyu Liu
Timothy L. Bailey
Hongda Mao
Cai Xiao
Xin Bai
Qishui Zhong
Yansong Deng
Giansalvo Cirrincione
Shi Zhenwei
Shuhui Wang
Junning Gao
Ziye Wang
Hui Li
Chao Wu
Fuyuan Cao
Rui Hong
Wahyono Wahyono
Rafal Kozik
Yihua Zhou
Wenyan Wang
Juan Figueroa
Artem Lenskiy
Qiang Yan

Wenlong Xu
Yi Gu
Yi Kou
Chenhui Qiu
Tianyu Yang
Xiaoyi Yu
Rongjing Hu
Yuanyuan Liu
Zhang Haitao
Libing Shen
Dongliang Yu
Yanwu Zeng
Yuanyuan Wang
Giulia Russo
Emilio Mastriani
Di Tang
Yan Jiang
Hong-Guan Liu
Liyao Ma
Panpan Du
Hongguan Liu
Chenbin Liu
Dangdang Shao
Yunze Yang
Yang Chen
Yang Li
Heye Zhang
Austin Brockmeier
Hui Li
Xi Yang
Yan Wang
Lin Bai
Laksono Kurnianggoro
Ajmal Shahbaz
Yan Zhang
Xiaoyang Wang
Bo Wang
Sajjad Ahmed
Ke Zeng
Liang Mao
Yuan You
Qiuyang Liu
Shaojie He
Zehui Cao
Zhongpu Xia
Bin Ye

Zhi-Yu Shen
Alexander Filonenko
Zhenhu Liang
Jing Li
Yang Yang
Wenjun Xu
Yongjia Yu
Jyotsna Wassan
Vibha Patel
Saiful Islam
Ekram Khan

Angelo Ciaramella
Junyu Chen
Ziang Dong
Jingjing Fei
Meng Lei
Xuanfang Fei
Bing Zeng
Taifeng Li
Yan Wu
Paul Fergus
Jiulun Cai

Jingying Huang
Francesco Pappalardo
Xiaoguang Zhao
Qingfang Meng
Mohamed Alloghani
Ruihao Li
Zhengyu Yang
Xin Xie
Yifan Wu
Hong Zeng

Contents – Part III

Intelligent Agent and Web Applications

Fuzzy Theory and Algorithms

Supervised Learning

Unsupervised Learning

Kernel Methods and Supporting Vector Machines

Knowledge Discovery and Data Mining

Natural Language Processing and Computational Linguistics

Advances of Soft Computing: Algorithms and Its Applications - Rozaida Ghazali

Advances in Swarm Intelligence Algorithm

Computational Intelligence and Security for Image Applications in Social Network

Biomedical Image Analysis

Information Security

Machine Learning

Intelligent Computing in Robotics

An Adaptive Position Synchronization Controller Using Orthogonal Neural Network for 3-DOF Planar Parallel Manipulators

Quang Dan Le[1], Hee-Jun Kang[2]([✉]), and Tien Dung Le[3]

[1] Graduate School of Electrical Engineering,
University of Ulsan, Ulsan, South Korea
ledanmt@gmail.com
[2] School of Electrical Engineering, University of Ulsan, Ulsan, South Korea
hjkang@ulsan.ac.kr
[3] The University of Danang - University of Science & Technology,
Danang, Vietnam
ltdung@dut.udn.vn

Abstract. This paper proposes an adaptive position synchronization controller using orthogonal neural network for 3-DOF planar parallel manipulators. The controller is designed based on the combination of computed torque method with position synchronization technique and orthogonal neural network. By using the orthogonal neural network with online turning gains can overcome the drawbacks of the traditional feedforward neural network such as initial values of weights, number of processing elements, slow convergence speed and the difficulty of choosing learning rate. To evaluate the effectiveness of the proposed control strategy, simulations were conducted by using the combination of SimMechanics and Solidworks. The tracking control results of the parallel manipulators were significantly improved in comparison with the performance when applying non-synchronization controllers.

Keywords: Planar parallel manipulator · Position synchronization controller · Adaptive controller · An orthogonal neural network · Online self-turning

1 Introduction

The past decade has seen a steady increase of parallel manipulators in industrial production due to their advantages over serial manipulators such as high velocity, high accuracy, low moving inertia and strong structure. However, the parallel manipulators gradually reveal weaknesses because of their closed-chain structures. For example, they have complex kinematic, dynamic models and the limited workspace even with many singularities inside. Therefore, the parallel manipulator control needs more advanced technologies.

For parallel manipulator, the motion of end-effector based on the cooperation motion of each sub-manipulator with one another is called synchronization. The technical synchronization is based on cross-coupling error of each sub-manipulator with one another. This is different from the conditional non-synchronized controller

© Springer International Publishing AG 2017
D.-S. Huang et al. (Eds.): ICIC 2017, Part III, LNAI 10363, pp. 3–14, 2017.
DOI: 10.1007/978-3-319-63315-2_1

that concerns only the convergence of each joint tracking error. The cross-error in controller was firstly defined by Koren [1] for CNC machine tool. Then, Koren and his cooperators applied this technique in mobile robot [2]. Upon the definition of cross-error, Lu Ren et al. defined a new cross-error for parallel manipulator [3, 4]. With inspiration synchronization, Liu and Chopra [5] and Rodriguez-Angeles and Nijmeijer [6] applied this technique in multi manipulator cooperated.

The adaptive methods for robot control are based on intelligent technique using neural network that increase wide variety. The Feedforward Neural Networks (FNN) used in adaptive control for robot [7–10] achieve high accuracy and can deal with uncertainties. Authors in [11, 12] combined fuzzy and neural network in control to compromise each other. The majority of these researchers used intelligence control in control systems. But the classical FNN have some drawbacks such as local minimum, initial values of weight, number of processing elements, slow convergence speed and the difficulty of choosing learning rate. In order to solve convergence speed problem, Chen and Manry replaced sigmoid active function by polynomial function [13]. Enrique Castillo and his cooperators used sensitivity analysis technique [14], while Karayiannis and Venetsanopoulos using the fast delta rule algorithm [15]. Yam et al. proposed linear algebraic method in [16] and Wang et al. presented the extreme learning machines in [17]. To resolve the remaining problems in conventional FNN, Yang and Tseng [18] presented Orthogonal Neural Network (ONN) based on feed-forward network and using polynomial function. This neural network handles most of the drawbacks in traditional FNN. Sher et al. [19] also showed comparison between the training method and the numerical method of ONN in function approximation. Stanisa and his group applied successfully ONN in anti-lock braking system and modeling of dynamical systems [20–22]. In addition, ONN was proved to have good results in robot controller by Timmis et al. [23] and Wang [24].

In this paper, an adaptive Position Synchronization Controller Using Orthogonal Neural Network (PSC-ONN) for 3-DOF planar parallel manipulator was presented. In this method, the PSC-ONN consists of computed torque control, error synchronization and ONN, The PSC-ONN deals with the existence of the uncertainty in robotic systems and obtain the higher tracking performance. In addition, the 3-DOF planar parallel manipulator simulation model was established using combination Solidworks and SimMechanics to verify effectiveness of the controller proposed.

The remainder of this paper is arranged as follows. Section 2 describes the dynamics of 3-DOF planar parallel manipulator. Section 3 presents the position synchronization controller. In Sect. 4 the combination of the PSC and the ONN is described for the proposed controller. Simulation results are shown in Sect. 5 and finally, some remarks are addressed in Sect. 6.

2 The 3-DOF Planar Parallel Manipulators

2.1 The Geometry of the 3-DOF Planar Parallel Manipulators

The 3-DOF planar parallel manipulator was depicted in Fig. 1. It includes three active joints, six passive joints and the end-effector. The link lengths of the parallel manipulator

$l_1 = A_iB_i$, $l_2 = B_iC_i$, $(i = 1, 2, 3)$, the end-effector $C_1C_2C_3$ is equilateral triangle with $l_3 = C_iP$ which is radius of the circle circumscribing the three vertices.

In Fig. 1 denotes: $\theta_a = [\theta_{a1}, \theta_{a2}, \theta_{a3}]^T$ is the vector of active joints angular, $\theta_p = [\theta_{p1}, \theta_{p2}, \theta_{p3}]^T$ is the vector of three important passive joints angular and $P = [x_P, y_P, \phi_P]^T$ is the vector of position of end-effector.

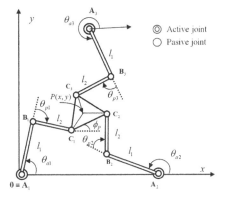

Fig. 1. The kinematics planar parallel manipulator 3-DOF

2.2 Dynamic Models of 3 DOF Planar Parallel Manipulators

Dynamic models of 3 DOF planar parallel manipulators are established using virtually cut and Lagrange-D'alamber principle. The virtually cut showed in Fig. 2, we derive the Lagrangian equation of the open-chain system and compute the joints torques needed to generate a given motion which satisfies the loop constraints.

Dynamic models of each planar serial 2 DOF manipulator and end-effector given by Largane without fiction and external distribution follow:

$$\frac{d}{dt}\left(\frac{\partial L_i}{\partial \dot\theta_i}\right) - \frac{\partial L_i}{\partial \theta_i} = \tau_{\theta_i} \quad (1)$$

$$\frac{d}{dt}\left(\frac{\partial L_i}{\partial \dot X}\right) - \frac{\partial L_i}{\partial X} = \tau_X \quad (2)$$

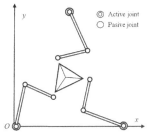

Fig. 2. The open-chain systems of planar parallel manipulator

where L_i the Lagrangian function for each serial manipulator $(i = 1,2,3)$.

$\theta_i = [\theta_{ai}, \theta_{pi}]^T$ is the angle joints vector. $\tau_{\theta_i} = [\tau_{ai}, \tau_{pi}]^T$ is the joint torque vector. $X = [x_P, y_P, \phi_P]^T$ is the position of the end-effector on Cartesian. $\tau_X = [\tau_{xP}, \tau_{yP}, \tau_{\phi P}]^T$ is the end-effector torque vector in each direction.

From (1) and (2), using virtual displacement principle we have:

$$\left(\frac{d}{dt}\left(\frac{\partial L}{\partial \dot\theta_a}\right) - \frac{\partial L}{\partial \theta_a} - \tau_a\right)\delta\theta_a + \left(\frac{d}{dt}\left(\frac{\partial L}{\partial \dot\theta_p}\right) - \frac{\partial L}{\partial \theta_p} - \tau_p\right)\delta\theta_p$$

$$+ \left(\frac{d}{dt}\left(\frac{\partial L}{\partial \dot X_P}\right) - \frac{\partial L}{\partial X_P} - \tau_X\right)\delta X_P = 0 \quad (3)$$

Rewrite the Eq. (3) with matrix form and actuators only have at active joints we obtain:

$$\left[\frac{d}{dt}\left(\frac{\partial L}{\partial \dot{\theta}_a}\right) - \frac{\partial L}{\partial \theta_a}, \frac{d}{dt}\left(\frac{\partial L}{\partial \dot{\theta}_p}\right) - \frac{\partial L}{\partial \theta_p}, \frac{d}{dt}\left(\frac{\partial L}{\partial \dot{X}_P}\right) - \frac{\partial L}{\partial X_P}\right]\left[I, \frac{\partial \theta_p}{\partial \theta_a}, \frac{\partial X_P}{\partial \theta_a}\right]^T = \tau_a \quad (4)$$

Equation 4 can be rewritten:

$$W^T \tau = \tau_a \quad (5)$$

where $\tau = [\tau_a, \tau_p, \tau_X]^T \in \Re^{9 \times 1}$ is torque of robot system. $W = [I, \partial \theta_p / \partial \theta_a, \partial X_p / \partial \theta_a]^T \in \Re^{9 \times 3}$ is vector prismatic joint and velocity active joints. This is the similar to the work of Nakamura in [25].

The dynamic of the open-chain system can be arranged in the form

$$M_t(q)\ddot{q} + C_t(q, \dot{q})\dot{q} = \tau \quad (6)$$

where $q = [\theta_a, \theta_p, X_P]^T$ is the vector of joints and end-effector's position. $M_t(q) \in \Re^{9 \times 9}$ is the inertial matrix. $C_t(q, \dot{q}) \in \Re^{9 \times 9}$ is the Coriolis matrix.

Substituting (5) and (6) we have

$$W^T (M_t(q)\ddot{q} + C_t(q, \dot{q})\dot{q}) = \tau_a \quad (7)$$

From kinematic constraints in (5) we have:

$$\dot{q} = [I, \partial \theta_p / \partial \theta_a, \partial X_P / \partial \theta_a]^T \dot{\theta}_a = W \dot{\theta}_a \quad (8)$$

After derivative both side of (8), it leads to:

$$\ddot{q} = \dot{W} \dot{\theta}_a + W \ddot{\theta}_a \quad (9)$$

Now, by substituting (8) and (9) into (7) we obtain the dynamic model of the 3-DOF planar parallel manipulator in active joints space:

$$M \ddot{\theta}_a + C \dot{\theta}_a = \tau_a \quad (10)$$

where $M = W^T M_t \dot{W} \in \Re^{3 \times 3}$ is the inertial matrix of robot. $C = W^T M_t \dot{W} + W^T C_t W \in \Re^{3 \times 3}$ is the Croliolis matrix.

The Eq. (10) is dynamic model of robot 3-DOF planar parallel manipulator without friction and external disturbances in actuator at three active joints.

3   The Conventional Synchronization Controller

3.1   Problem Statements of the Conventional CTC

The dynamic model (10) did not contain the friction and external disturbances. However in real system robot, the uncertainties term always exists. General equation of robot can be written

$$M\ddot{\theta}_a + C\dot{\theta}_a + F_a + f = \tau_a \tag{11}$$

where F_a is friction at active joints. f is a load disturbance.

The computed torque controller (CTC) is used to control the robot which has dynamic model described by (10) follow a desired trajectory. The structure of the CTC is designed as:

$$\tau_a = M(q)\left[\ddot{\theta}_d + K_e(\theta_d - \theta) + K_v(\dot{\theta}_d - \dot{\theta})\right] + C(\theta, \dot{\theta})\dot{\theta} \tag{12}$$

where $\theta_d \in \Re^{3\times1}$ is the desired manipulator trajectory, $K_e \in \Re^{3\times3}$ and $K_v \in \Re^{3\times3}$ are the proportional gain matrices.

The computed torque controller in (12) have drawbacks in real application such as requirement of an exact dynamic model of robot which usually impossible, lack of robustness to structured and unstructured uncertainties. In addition, the conventional CTC only concerns the position error of each joints and information from other joints is neglected while the uncertainties at each joint is different so they will make time convergence at each joint is different.

This paper thus aims to design an adaptive control scheme for uncertain robot planar parallel manipulator based on CTC, synchronization error, and orthogonal neural network such that the system output will follow the desired trajectories.

3.2   The Position Synchronization Controller

In this part, the position synchronization controller (PSC) based on synchronization error and cross-coupling error is presented. The position tracking error of each active joints is defined:

$$e_i(t) = \theta_i^d(t) - \theta_i(t), \quad i = 1, 2, \ldots, n \tag{13}$$

where $\theta_i^d(t)$ denotes the desired position of the each active joint. In the position synchronization, not only the position error $e_i(t) \to 0$ but also aimed to regulate motion relationship among multiple active joints during the tracking $e_1(t) = e_2(t) = \ldots = e_n(t)$.

The synchronization error are defined [2] of a subset of all possible pairs of two active joints from total of n joints and this case is three active joints in the following $\varepsilon_1(t) = e_1(t) - e_2(t)$, $\varepsilon_2(t) = e_2(t) - e_3(t)$, $\varepsilon_3(t) = e_3(t) - e_1(t)$.

The goal of synchronization error can be achieved if $\varepsilon_i(t) = 0$ for all active joints. Unlike conditional controller non-synchronize that concerns the convergence of position tracking error, in synchronize controller consider error between each other active joints. Specifically, the controller is to control $e_i(t) \to 0$ and at the same time.

The cross-coupling error was defined in [4]

$$e_i^* = e_i + \beta \int_0^t (\varepsilon_i - \varepsilon_{i-1}) dw \tag{14}$$

where β is positive constant. w is variable from time zero.

In this paper, the position synchronization controller is defined:

$$\tau_a = M(q)\left[\ddot{\theta}_d + K_e e + K_v \dot{e}\right] + C(\theta, \dot{\theta})\dot{\theta} + K_s s(t) + K_c r(t) \tag{15}$$

where K_s is positive matrix. ε is synchronization error matrix. $s(t) = \varepsilon_i - \varepsilon_{n+1-i}$. $r(t) = e_i^* + \Gamma \dot{e}_i^*$.

The friction at each active joint is $F_i = 0.4\,\mathrm{sign}\,(\dot{\theta}_{ai}) + 0.5\dot{\theta}_{ai}$. The results using CTC and the position synchronization controller (PSC) showed on Fig. 3. We can see, the error when using PSC achieves smaller error than of CTC. Figure 3 was also presented how errors reduced when using the principle of synchronization errors. The error at each joint promote each other convergence to zero at same time.

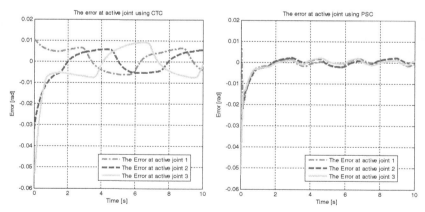

a) The error of active joints when using CTC   b) The error of active joints when using PSC

Fig. 3. The result of tracking errors in the cases of using CTC and PSC

4 The Proposed Adaptive Synchronization Controller

4.1 The Orthogonal Neural Network

In this part, the orthogonal neural network was based on FNN and polynomial for active function [18, 19] is showed. According to theory of orthogonal function, an arbitrary function $f(x), f : [a, b] \rightarrow \Re$ will have an orthogonal polynomial.

$$F_n(x) = w_1\phi_1(x) + w_2\phi_2(x) + \ldots + w_n\phi_n(x) \tag{16}$$

Such that

$$\lim_{n \to \infty} \int_a^b (f(x) - F_n(x))^2 dx = 0 \tag{17}$$

where

$$\int_a^b \phi_i(x)\phi_j(x)dx = \begin{cases} 0 & i \neq j \\ A_i & i = j \end{cases} \tag{18}$$

$$w_i = \int_a^b f(x)\phi_i(x)dx/A_i \quad i = 1, 2, \ldots, n \tag{19}$$

$\{\phi_1(x), \phi_2(x), \ldots\}$ is an orthogonal set.

In case there is a function with m variables. Orthogonal function set will be $\{\Phi_1(x), \Phi_2(x), \ldots\}$. Each of orthogonal function is defined as

$$\Phi_i(X) = \phi_{1i}(x_1)\phi_{2i}(x_2)\ldots\phi_{mi}(x_m) \tag{20}$$

where $X = [x_1, x_2, \ldots, x_n]^T$ is input vector.

There are orthogonal functions such as Fourier series, Bessel function, Legendre, and Chebyshev polynomial.

Orthogonal neural network based on feedforward network with one hide layer showed in Fig. 4 by Tseng and Chen. In Fig. 4, (a) is multi input-one input and (b) is multi input-multi output. The connection between input layer and hide layer with the weight is one and the bias is zero. The $\phi_i (i = 0, 1, \ldots, n)$ is orthogonal function. The weight between is w_{ij} (with i is number orthogonal function and j is number output). The output of ONN can be showed below:

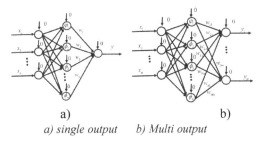

a)   b)

a) single output   b) Multi output

Fig. 4. Orthogonal neural network.

$$y = F(X) = \sum_{i=1}^{n} w_i(t)\Phi_i(X) = W^T(t)\Phi \qquad (21)$$

4.2 The Proposes Adaptive Synchronization Controller

According to the nice result of the Eq. (15), the propose controller adding ONN for estimating uncertainties for the planar parallel manipulator is expressed as follow:

$$\tau_a = M(q)\left[\ddot{\theta}_d + K_e(e) + K_v(\dot{e})\right] + C(\theta, \dot{\theta})\dot{\theta} + K_s s(t) + K_c r(t) + f_{ONN} \qquad (22)$$

where f_{ONN} is output of ONN. The ONN which is based on feedforward network has three layers. The block diagram of the proposed adaptive position synchronization controller using orthogonal neural network is illustrated in Fig. 5.

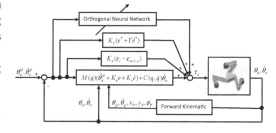

In the Eq. (22), the sliding function is defined as:

$$s = \dot{e} + \Gamma e + \Lambda\varepsilon \qquad (23)$$

Fig. 5. The block diagram of the proposed controller

The Input Layer: The input vector of the ONN is denoted by

$$X = [x_1, x_2, \ldots, x_9]^T = [e_1, \dot{e}_1, \varepsilon_1, e_2, \dot{e}_2, \varepsilon_2, e_3, \dot{e}_3, \varepsilon_3]^T \qquad (24)$$

where $e_i = \theta_{ai}^d - \theta_{ai}, \dot{e}_i = \dot{\theta}_{ai}^d - \dot{\theta}_{ai}, \varepsilon_i = e_i - e_{i+1}, i = 1, 2, 3$. The input must be inside $[-1, 1]$. The limitation of input inside [a,b] we can be transformed as:

$$t_i = \frac{2}{b-a}x_i - \frac{b-1}{b-a}, \; t_i \in [-1, 1] \qquad (25)$$

The Hidden Layer: The hidden layer using Chebyshev polynomial of the first kind for orthogonal function expressed follow:

$$\phi_i(x) = \frac{(-2)^i i!}{(2i)!}\sqrt{1-x^2}\frac{d^i}{dx^i}(1-x^2)^{i-1/2} \; i = 0,1,2,\ldots,n \qquad (26)$$

The equation can be rewritten $\phi_0(x) = 1, \; \phi_1(x) = x, \; \phi_2(x) = 2x^2 - 1, \; \ldots, \phi_{n+1}(x) = 2x\phi_n(x) - \phi_{n-1}(x)$. In this paper, $n = 10$ is chosen.

The Output Layer: The weight matrix connection between the hidden with output layer is showed by:

$$W = [w_1, w_2, \ldots, w_n]^T \qquad (27)$$

The output of ONN expressed by:

$$y = W^T \Phi \qquad (28)$$

where $\Phi = \{\Phi_1(X), \Phi_2(X), \ldots, \Phi_n(X)\}$ with $\Phi_i(X)$ showed on (20).

In the proposed controller, the weights of the ONN are adjusted based on the gradient descent method. Although in this case is multiple outputs showed in Fig. 4b. However, each of single-output neural networks has its independent weights, their respective weights can be separately trained. The cost function given by

$$E = \frac{1}{2}e^2 = \frac{1}{2}(y - \hat{y})^2 \in \Re \qquad (29)$$

where e is the learning error. $\hat{y}$ is the output of ONN. y is the actual output.

The ONN's weight update law for the instantaneous gradient descent algorithm is given as $\dot{W} = \delta e \Phi$ which δ is the learning rate. e is the learning error.

5   The Result Simulation

To illustrate the effectiveness of the adaptive position synchronization controller using ONN for 3-DOF planar parallel manipulator in this paper, simulation studies were conducted on Matlab-Simulink and mechanical model of the parallel manipulator was built on Solidworks and exported to SimMechainics. By this way, the mechanical model is almost the same with the real model and the forward dynamics don't need to be used. The parallel robot manipulator has full parameter exported from Solidworks designed.

The parameter of 3-DOF planar parallel manipulator were presented in Table 1, where l_i is length of link i^{th}, l_{ci} is distance from the joint to the center of mass for link i^{th}, m_i is mass of link i^{th}, and I_i is inertial of link i^{th} (i = 1,2,3). The gains of the conventional CTC were chosen as $K_e = 20 \times I_{3\times3}$, $K_v = 50 \times I_{3\times3}$ in which $I_{3\times3}$ is the identity matrix with dimension equal to 3×3. The learning rate of FNN is $\eta = \mu = 0.2$ and the number of hide neurons is $n = 10$. The synchronization rate $\beta = 0.2$, $K_s = 220 \times I_{3\times3}$, $K_c = 0.8 \times I_{3\times3}$, $\Gamma = 0.4 \times I_{3\times3}$.

Table 1. Parameter of the manipulator

Links	l_i (m)	L_{ci} (m)	m_i (kg)	I_i (kg.m^2)
Active link	$l_1 = 0.4$	0.2	5.118	9139×10^{-6}
Passive link	$l_2 = 0.6$	0.3	7.39	26763×10^{-6}
End-effector	$l_3 = 0.2$	–	1.84	3170×10^{-6}

In ONN, the learning rate is $\eta = 0.2$ and $\Lambda = 0.5 \times I_{3 \times 3}$. The number of Chebyshev polyniminal was chosen $n = 10$.

The input trajectories of end-effector given: $x(t) = 0.49 + 0.03 \cos(\pi(t)/3)$, $y(t) = 0.37 + 0.03 \sin(\pi(t)/3)$, $\phi_P(t) = \pi/12$. After computed inverse kinematic, the desired angular is given at each active joints of planar parallel manipulator $\theta_{ai}^d = [\theta_{a1}^d, \theta_{a2}^d, \theta_{a3}^d]^T$. The uncertainties are given by addition the friction each active joints as the following $f_{ai} = 0.4 sign(\dot{\theta}_{ai}) + 0.5\dot{\theta}_{ai}$, $i = 1,2,3$.

In Figs. 6, 7, 8 and 9, the performance of our proposed controller was compared with adaptive CTC using FNN (CTC-FNN) and adaptive PSC using FNN (PSC-FNN). Firstly, the feature of synchronization was clearly shown in Figs. 6, 7 and 8. The convergence of error at joint one includes two steps. (1) The error at joint one equals to the error at other ones. (2) The error at each joint approaches to zero together. As the results can be seen in Figs. 7, 8 and 9, PSC-FNN and PSC-ONN showed the same position error trend, while CTC-FNN has the reverse position error trend. The reason is that due to the constraints of joint two and joint three as shown in Figs. 7 and 8. The position error of joint one using PSC-FNN and PSC-ONN has to change adaptively to the error of joint two and joint three. However, although PSC-ONN has the convergence slower than PSC-FNN and CTC-FNN in the first 2 s, it approaches to zero faster after that. The results of tracking trajectory were presented in Fig. 9.

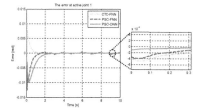

Fig. 6. The error of active joint 1

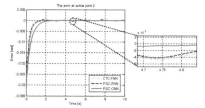

Fig. 7. The error of active joint 2

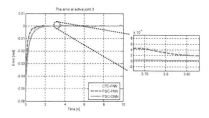

Fig. 8. The error of active joint 3

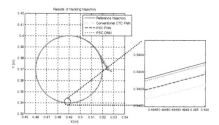

Fig. 9. The result of tracking trajectory

6   The Conclusion

In this paper, an adaptive position synchronization controller using orthogonal neural network for 3-DOF planar parallel manipulators was presented. The proposed controller is composed of the computed torque control method with position synchronization control algorithm and orthogonal neural network. The orthogonal neural network was trained online during the trajectory tracking control based on slide mode function and online learning method to deal with the existence of the uncertainties. To illustrate the effectiveness of the proposed strategy, simulations were conducted in order to compare the proposed controller with adaptive computed torque controller using traditional FNN and adaptive position synchronization controller using traditional FNN. The simulation results shows that the joint errors converges quickly to zero almost at the same time and the proposed controller has the better tracking performance than CTC and PSC control algorithm using traditional FNN.

Acknowledgement. This research was supported by Basic Science Research Program through the National Research Foundation of Korea (NRF) funded by the Ministry of Education (NRF-2016R1D1A3B03930496).

References

1. Koren, Y.: Cross-coupled biaxial computer control for manufacturing systems. J. Dyn. Syst. Meas. Control **102**, 265 (1980)
2. Feng, L., Koren, Y., Borenstein, J.: Cross-coupling motion controller for mobile robot. IEEE Control Syst. **13**, 35–43 (1993)
3. Ren, L., Mills, J.K., Sun, D.: Controller design applied to planar parallel manipulator for trajectory tracking robot. In: Proceeding of the 2005 IEEE International Conference on Robotics and Automation (2005). doi:10.1109/ROBOT.2005.1570242
4. Ren, L., Mills, J.K., Sun, D.: Experimental comparison of control approaches on trajectory tracking control of a 3 DOF parallel robot. IEEE Trans. Control Syst. Technol. **15**(5), 982–988 (2007)
5. Liu, Y.-C., Chopra, N.: Controlled synchronization of heterogeneous robotic manipulators in the task space. IEEE Trans. Robot. **28**, 268–275 (2012)
6. Rodriguez-Angeles, A., Nijmeijer, H.: Mutual synchronization of robot via estimated state feedback: a cooperative approach. IEEE Trans. Control Syst. Technol. **12**(4), 542–554 (2004)
7. Adaptive robot control using neural network: Sad, M., Bigras, P., Dessaint, L.-A., Al-Haddad, K. IEEE Trans. Ind. Electron. **41**, 176–181 (1994)
8. He, W., Chen, Y., Yin, Z.: Adaptive neural network control of an uncertain robot with full-state constraints. IEEE Trans. Cybern. **46**, 620–629 (2016)
9. Li, X., Cheah, C.C.: Adaptive neural network control of robot based on a unified objective bound. IEEE Trans. Control Syst. Technol. **22**, 1032–1043 (2014)
10. Le, Q.D., Kang, H.-J., Le, T.D.: Adaptive extended computed torque control of 3 DOF planar parallel manipulators using neural network and error compensator. In: Huang, D.-S., Han, K., Hussain, A. (eds.) ICIC 2016. LNCS, vol. 9773, pp. 437–448. Springer, Cham (2016). doi:10.1007/978-3-319-42297-8_41

11. Gao, Y., Er, M.J., Yang, S.: Adaptive control of robot manipulators using fuzzy neural networks. IEEE Trans. Ind. Electron. **48**, 1274–1278 (2001)
12. Chen, C.-S.: Dynamic structure neural fuzzy networks for robust adaptive control of robot manipulators. IEEE Trans. Ind. Electron. **55**, 3402–3414 (2008)
13. Chen, M.S., Manry, M.T.: Conventional modeling of the multilayer perception using polynominal basis functions. IEEE Trans. Neural Netw. **4**, 14–16 (1992)
14. Castillo, E., Guijarro-Berdinas, B., Fontenla-Romero, O., Alonso-Betanzos, A.: A verry fast learning method for neural networks based on sensitivity analysis. J. Mach. Learn. Res. **7**, 1159–1182 (2006)
15. Karayiannis, N.B., Venetsanopoulos, A.N.: Fast learning algorithm for neural networks. IEEE Trans. Circ. Syst. **39**(7), 453–474 (1992)
16. Yam, Y.F., Chow, T.W.S., Leung, C.T.: A new method in determining the initial weights of feedforward neural networks. Neuralcomputing **16**, 23–32 (1997)
17. Wang, S., Chung, F.-L., Wang, J., Wu, J.: A fast learning method for feedforward neural networks. Neuralcomputing **149**, 295–307 (2015)
18. Yang, S.-S., Tseng, C.-S.: An orthogonal neural network for function approximation. IEEE Trans. Syst. Man Cybern. **26**(5), 779–785 (1996)
19. Sher, C.F., Tseng, C.-S., Chen, C.-S.: Properties and performance of orthogonal neural network in function approximation. Int. J. Intell. Syst. **16**, 1377–1392 (2001)
20. Peric, S.L., Antic, D.S., Milovanovic, M.B., Mitic, D.B., Milojkovic, M.T., Nikolic, S.S.: Quasi-sliding mode control with orhtogonal endocrine neural network-based estimator applied in anti-lock braking system. IEEE/ASME Trans. Mechatron. **21**(2), 754–764 (2016)
21. Milojkovic, M.T., Antic, D.S., Milovanovic, M.B., Nikolic, S.S., Peric, S.L., Almawlawe, M.: Modeling of dynamic systems using othogonal endocrine adaptive neural-fuzzy inference systems. J. Dyn. Syst. Meas. Control **137**(9), DS-15-1098 (2015)
22. Milojkovic, M.T., Nikolic, S.S., Dankovic, B., Antic, D., Jovanovic, Z.: Modeling of dynamical systems based on almost orthogonal polynominal. Math. Comput. Model. Dyn. Syst. **16**(2), 133–144 (2010)
23. Timmis, J., Neal, M., Thorniley, J.: An adaptive neuro-endocrine system for robotic systems. In: IEEE Workshop on Robotic Intelligence in Informationally Structured Space, pp. 129–136 (2009). doi:10.1109/RIISS.2009.4937917
24. Wang, P.: Control of robot manipulators based on legendre orthogonal neural network. Appl. Mech. Mater. **427–429**, 1089–1092 (2013)
25. Nakamura, Y., Ghodoussi, M.: Dynamics computation of close-link robot mechanisms with no redundant and redundant actuators. IEEE Trans. Robot. Autom. **5**, 294–302 (1989)

Navigation of Mobile Robot Using Type-2 Fuzzy System

Rahib H. Abiyev$^{(\boxtimes)}$, Besime Erin, and Ali Denker

Department of Computer Engineering, Near East University,
P.O. Box 670 Lefkosa, North Cyprus, Mersin-10, Turkey
rahib.abiyev@neu.edu.tr

Abstract. One of the important problems of robotics is the navigation of mobile robots in uncertain environments that are densely cluttered with obstacles. The control of robots using the traditional control algorithms is not satisfactory as far as the navigational accuracy and the distance and time to reach the goal are concerned, when the robot is in a complicated surrounding. One of alternative and efficient ways of constructing a control system that explicitly deal with uncertainty is the use of fuzzy systems approach. The paper is devoted to navigation of mobile robot using type-2 fuzzy system. The design principle of navigation algorithm using type-2 fuzzy system is presented. The fuzzy knowledge base that describes the relation between the input- current angle and distance signals and output signals that determine the robot turn angle is developed. The control rules of navigation and inference engine operations have been described. The comparative simulation results of robot navigation system demonstrate the advantage of the fuzzy navigation algorithm.

Keywords: Robot navigation · Type-2 fuzzy systems · Fuzzy obstacle avoidance

1 Introduction

The robot navigation problem is very wide and complex. The environments where robot moves may vary from static areas with fixed obstacles, to fast-changing dynamic areas with many moving obstacles. The environment is uncertain if no prior information available about the obstacles. In such environment, the basic aim is to safely move and reach the prescribed destination point by finding the shortest path without collision of the obstacles existing on the road of the robot [1, 2]. The problem is the design of a fast and efficient algorithm that will lead mobile robots to the destination point. The navigation algorithm should determine continuous motion from one configuration to the other for the given the initial and final configurations of a mobile robot and find such a motion if one exists.

Recently various methodologies have been purposed for mobile robot navigation [3–10]. The most widely used are Artificial Potential Fields (APF) [4], Vector Field Histogram (VFH) [5], VFH+ [6], local navigation [7], fuzzy navigation [8–11] techniques. Others are Dynamic Window Approach [12], Rule-Based Methods [13], agoraphilic [14], Rapidly-exploring Random Trees (RRTs) [15], RRT smooth [16], A star [17].

© Springer International Publishing AG 2017
D.-S. Huang et al. (Eds.): ICIC 2017, Part III, LNAI 10363, pp. 15–26, 2017.
DOI: 10.1007/978-3-319-63315-2_2

Artificial potential field method of obstacle avoidance is based on the repulsive potential field around the obstacle and an attractive potential field around the goal to force away or to attract the robot. Various version of APF has been developed for various environments. VFH uses the concept of potential field and two-dimensional Cartesian grid as a world model. But VFH is timely expensive. The used conventional path planning algorithms are basically time-consuming and have difficulties in navigation of an uncertain environment. Finding the shortest path to reach a goal is one of important problem in robot navigation. Different algorithms have been used for finding shortest path in different environments. The comparisons of some path finding algorithms are given in [18–22].

Fuzzy logic is one of the efficient techniques for navigation of mobile robots in uncertain environments. Fuzzy logic can be used to model human perception process and deal with uncertainties in the control process. Usually, fuzzy logic based control is based on fuzzy if-then rules that describe the locations and relative positions of mobile robot and obstacles. The developed fuzzy rules are basically direction based rules and they describe the action of a mobile robot in different situations. A layered goal-oriented motion planning strategy using fuzzy logic is developed in [9] for navigation of a mobile robot in an unknown environment. Due to simplicity and capability for real-time implementation of the fuzzy navigation system, it is implemented on a real mobile robot, Koala. The design of fuzzy obstacle avoidance for robot navigation is presented in [10, 11]. An obstacle avoidance algorithm based on the fuzzy matching of obstacle environment is presented in [24]. Fuzzy rules and procedure to perform fuzzy navigation for behaviour based robot navigation is presented in [11, 25]. As a result of using fuzzy logic, the number of rules to be determined is reduced and the design of the robotic controller is simplified. The evolutionary algorithms are also used for tuning of the fuzzy system in the path motion controller. In [26] evolutionary programming is used for designing fuzzy logic path planner and motion controller.

As shown, a number of techniques are applied to the control and navigation of mobile robots. Fuzzy navigation algorithms are based on the usage of fuzzy knowledge bases. Mobile robots are usually navigating in changing and unstructured environments characterizing with a large amount of uncertainties. Sometimes the type-1 fuzzy systems cannot handle such kind of uncertainties. In such cases, type-2 fuzzy sets are used to handle uncertainties and increase the performance of navigation system. The uncertainties in type-2 fuzzy systems can arise from different sources. The types of uncertainties are given in [27]. Because the membership functions of type-2 fuzzy systems are themselves fuzzy, they provide a powerful framework to represent and handle such types of uncertainties. In this paper, a mobile robot navigation using a type-2 fuzzy rule-based algorithm for obstacle avoidance is proposed.

The type-2 fuzzy sets introduced by Zadeh. Mendel and Karnik have further developed the theory of type-2 fuzzy sets [27, 28]. The theoretical background of interval type-2 fuzzy system and its design principles are described in [27–29]. In literature, various applications of type-2 fuzzy logic systems can be seen; such as robot control [30], prediction of hot strip mill temperature [31], for identification and control nonlinear system [32–37], credit rating [35].

In the paper type-2 fuzzy system is presented for navigation of mobile robot. The distance and the angle variables are applied for designing navigation system. The paper is organized as follows. Section 2 represents the design of fuzzy rule base for obstacle avoidance using type-2 fuzzy sets. Section 3 represents the type-2 fuzzy inference mechanism for if-then rule base describing the obstacle avoidance. Section 4 represents simulation results of robot navigation system using different techniques. Section 5 gives the conclusions of the paper.

2   Obstacle Avoidance Using Type-2 Fuzzy System

In robot navigation, the fuzzy system focuses on the goal reaching, while keeping avoidance of obstacles. An example of scenarios that demonstrates obstacle avoidance is shown in Fig. 1. Here, the mobile robot uses the distance measure in order to detect the presence of the obstacle. After detection of the obstacle, the local sensor detects the left and right boundaries of the obstacle in the local sensing region. The sensor region is determined by the user on the base of characteristics of the sensor used. It is impossible to know the shapes and sizes of all the obstacles outside the sensing region. During avoidance of an obstacle, the boundary of the obstacle is further enlarged to ensure the safety of the robot. This expanded boundary is called the "safe boundary". This ensures the safety of the robot during avoidance of obstacles. Each obstacle will have two different boundaries, the "real boundary" and the "safe boundary", as shown in Fig. 1. The robot detects the real boundary of obstacles. Then the safe boundaries of obstacles are determined by controller calculations performed within the navigation algorithm.

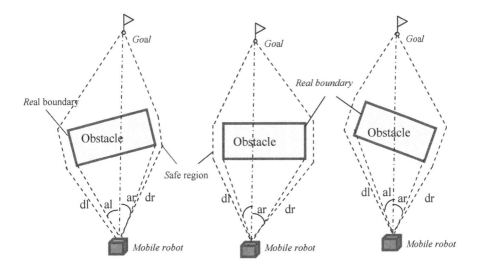

Fig. 1. Obstacle avoidance

The obstacle avoidance is carried out using knowledge base. This knowledge base includes If-Then rules that describe how the robot should avoid the obstacles.

The basic idea is to avoid the obstacle using shortest path and follow to the goal. In Fig. 1 three different scenarios are given in order to demonstrate turn angle of the mobile robot. Each scenario can be represented by one If-Then rule. Using different scenarios the rule base of the mobile robot is constructed. During designing of the rule base, the input variables are the left and right angle and distance variables, output variable is the turn angle of the robot. The left (al) and right angles (ar) are determined using the line connecting the robot and goal and the line connecting the robot and the real boundaries (left and right) of the robot. The left and right distances are determined as the distance between robot and left and right boundary points of the robot. Figure 2 describes the fuzzy If-Then rules for obstacle avoidance of mobile robot. In the rule base, the linguistic terms are used to represent the values of input and output parameters.

Rule 1: If ar =S and dr=M and al = M and dl=L then ta = PS
Rule 2: If ar=M and dr=M and al= S and dl=S then ta = NS
Rule 3: If ar=M and dr=S and al= S and dl=L then ta = NS
...
Rule n: If ar= VL and dr=VL and al= VL and dl=L then ta = NVL

Fig. 2. Fragment of rule base

Here Z, VS, S, M, L, VL are linguistic terms and denote zero, very small, medium, large and very large, NVL, NL, NM, NS, Z, PS, PM, PL, PVL are negative very large, negative large, negative medium, negative small, zero, positive small, positive medium, positive large, positive very large.

Using the rule base and current values of input variables, the value of turn angle (ta) of the robot is determined. After determination of turn angle on the output of rule base, the safe angle is added to this turning angle in order to find the real turn angle of the robot for the avoidance of obstacle. Safe obstacle boundary is determined for the left and right side of the robot. The determination of the value of the input variables is performed as follows. At first, the distance between robot and obstacle are determined by sensors. At the same time by inspecting left and right borders of the obstacle from the target direction the left and right distances between the robot and left and right boundaries of the robot are determined. After detecting the left and right distances between robot and obstacle, the corresponding left αl and right αr angles are determined. Then the safe angle characterising the safe region is used for final detection of left and right angles. Based on the left and right angles (al and ar) and also left and right distances (dl and dr) and If-Then rule base a corresponding decision on turn angle of the robot are made by the robot. The basic strategy is to select the direction that has the smallest distance from the robot to the goal. The calculated new turn angle is used to avoid an obstacle that will be:

$$ta(k) = F(\alpha l, \alpha r, dl, dr) \tag{1}$$

dl are dr left and right angles and left and right distances correspondingly.

In the rule base the values of input and output variables are represented using type-2 fuzzy sets. The type-2 fuzzy inference is applied for calculating output turn angle.

3 Type-2 Fuzzy System

The design of type-2 fuzzy model for the navigation of mobile robot is considered. As mentioned using angle and distance input variables and rule base the value of turn angle is determined. For these variables the term sets are determined. These values are determined using the highest and lowest value of the parameter and divided the value into three parts. Figure 2 describes the fuzzy values defined for the angle, distance and turn angle parameters. For simplicity, in the rule base, the input and output variables are scaled. The data can be easily transformed to the required range of the variables.

As mentioned, a type-2 fuzzy logic system can handle imprecise and uncertainty data to produce complex decision outcomes and minimize the effects of uncertainties. Type-2 fuzzy systems are applied for solution of different engineering problems [27–37]. In this paper, the type-2 fuzzy sets are applied for the navigation of robot.

The knowledge base given in Fig. 2 uses fuzzy IF-THEN rules including type-2 fuzzy values of the parameters in the antecedent and consequent parts of the rules. In the paper, the triangle type type-2 fuzzy sets are used to represent the fuzzy values of the parameters. In triangle type-2 fuzzy sets uncertainties can be associated to the mean. In the paper, the multi-input single output fuzzy rules are used. The type-2 TSK fuzzy rules used in this paper has the following form.

$$\text{IF } x_1 \text{ is } \tilde{A}_{1j} \text{ and } x_2 \text{ is } \tilde{A}_{2j} \text{ and} \ldots \text{and } x_m \text{ is } \tilde{A}_{mj} \text{ THEN } y_j \text{ is } \tilde{B}_j \qquad (2)$$

where x_1, x_2, ...,x_m are the input variables, y_j are the output variables, $\tilde{A}_{ij}$ is type-2 interval fuzzy membership functions of the antecedent part assigned for the j-th rule of the i-th input, $\tilde{B}_j$ is type-2 interval fuzzy membership functions of the consequent part for j-th rule. In the paper, triangle membership functions are used for $\tilde{A}_{ij}$ and $\tilde{B}_j$.

The type-1 is extended to an interval type-2 fuzzy system by adding uncertainties in both antecedent and consequent parts of each rule. For each input i and rule j triangular membership functions (MF) with uncertain mean are used to represent fuzzy values of the parameters. Figure 3 depicts the membership functions with uncertain mean, used in the antecedent and consequent parts of the fuzzy rules.

As we know triangular membership function has the following formula.

$$\mu_A(x) = \begin{cases} 0, & x < l \\ \frac{x-l}{c-l}, & l \leq x < c \\ \frac{r-x}{r-c}, & c < x \leq r \\ 0, & x > r \end{cases} \qquad (3)$$

where l, r, c are left, right and centre parts of triangle. respectively.

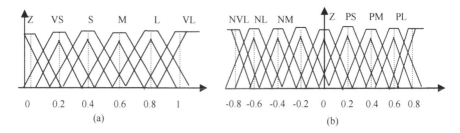

Fig. 3. Type-2 interval membership functions for - (a) input and (b) output variables

We apply this formula for designing type-2 membership functions and obtaining the upper and lower MFs. The lower $\underline{\mu}_{\tilde{F}}(x)$ and upper $\overline{\mu}_{\tilde{F}}(x)$ membership values are determined using (3).

In this paper, each membership function of the antecedent and consequent parts are represented by an upper and a lower membership functions. They are denoted as $\overline{\mu}(x)$ and $\underline{\mu}(x)$, or $\overline{A}(x)$ and $\underline{A}(x)$.

$$\mu_{\tilde{A}_k^i}(x_k) = [\underline{\mu}_{\tilde{A}_k^i}(x_k), \overline{\mu}_{\tilde{A}_k^i}(x_k)] = [\underline{\mu}^i, \overline{\mu}^i] \tag{4}$$

The inference engine can be implemented using "min" or "prod" t-norms operations. In this paper, "min" operation is chosen to calculate the firing strengths of the rules. As shown below the "min" operation is denoted as *. The firing strength of the jth rule is an interval type-2 fuzzy set determined by the left most and its rightmost points, which are calculated as follows:

$$\underline{f} = \underline{\mu}_{\tilde{A}_1}(x_1) * \underline{\mu}_{\tilde{A}_2}(x_2) * \ldots * \underline{\mu}_{\tilde{A}_n}(x_n); \quad \overline{f} = \overline{\mu}_{\tilde{A}_1}(x_1) * \overline{\mu}_{\tilde{A}_2}(x_2) * \ldots * \overline{\mu}_{\tilde{A}_n}(x_n) \tag{5}$$

where * is t-norm prod operator. After computing firing strength type reduction and defuzzification operations are performed. The type-reduction generates type-1 fuzzy set output. Defuzzifier uses this output and converts to a crisp number. Defuzzification provides mathematical formulas for the inner and outer bound sets which can be used to approximate the type-reduced set.

If we use the centre of sets type reduction then we need to compute the centroid of every consequent set, then computing a weighted average of these centroids.

$$Y = [y_l, y_r] = \int_{y^1} \ldots \int_{y^M} \int_{f^1} \ldots \int_{f^M} 1 \bigg/ \frac{\sum_{i=1}^{M} f^i y^i}{\sum_{i=1}^{M} f^i} \tag{6}$$

Here Y is interval set that is determined by y_l and y_r, f^i is $[\underline{f}^i, \overline{f}^i]$, $y^i = [y_l^i, y_r^i]$ is centroid of the type-2 interval fuzzy set in consequent part.

Karnik and Mendel [27, 28] have shown that the two end points y_l and y_r depend on mixture of $\underline{f}_i$ and $\bar{f}_i$ values.

$$
\begin{aligned}
y_l &= y_l(\bar{f}^1, \ldots, \bar{f}^M, \underline{f}^{L+1}, \ldots, \underline{f}^M, y_l^1, \ldots, y_l^M); \\
y_r &= y_r(\underline{f}^1, \ldots, \underline{f}^R, \bar{f}^{R+1}, \ldots, \bar{f}^M, y_r^1, \ldots, y_r^M);
\end{aligned}
\tag{7}
$$

Here $\bar{f}_j$ and $\underline{f}_j$ are determined using (4). Karnik and Mendel developed special iterative procedure [27, 28] for computing the values of y_l and y_r.

$$
y_l = \frac{\sum_{i=1}^{M} f_i y_i^l}{\sum_{i=1}^{M} f_i}; \quad y_r = \frac{\sum_{i=1}^{M} f_i y_i^r}{\sum_{i=1}^{M} f_i}
\tag{8}
$$

Where $f^i \in F^i = [\underline{f}_i, \bar{f}_i]$, $y_i = [y_i^l, y_i^r]$. Karnik-Merndel have developed algorithm that find switch points L and R and compute two end points yl and yr of type-reduced set.

$$
y_l = \frac{\sum_{i=1}^{L} \bar{f}_i y_i^l + \sum_{i=L+1}^{M} \underline{f}_i y_i^l}{\sum_{i=1}^{L} \bar{f}_i + \sum_{i=L+1}^{M} \underline{f}_i}; \quad y_r = \frac{\sum_{i=1}^{R} \underline{f}_i y_i^r + \sum_{i=R+1}^{M} \bar{f}_i y_i^r}{\sum_{i=1}^{R} \underline{f}_i + \sum_{i=R+1}^{M} \bar{f}_i}
\tag{9}
$$

Here switch points that can be calculated using Karnik-Merndel algorithm. The crisp outputs in defuzzification layer can be computed as follows:

$$
y = \frac{y_l + y_r}{2}
\tag{10}
$$

Here y is defuzzified crisp output. Although the Karnik-Merndel algorithm is time consuming but it is efficient for the design of type-2 fuzzy logic system.

4   Simulation Results

Robot navigation software that includes the implementations of four algorithms is designed. The simulation of Potential Field Method (PFM), Local Navigation (LN), Vector Field Histogram Plus (VFH+) and also type-2 Fuzzy Navigation (T2FN) are performed. In simulation using the GUI interface, different obstacles having rectangular and circular form can be drawn. Start and goal positions of the robot can be specified. Then, a path planning algorithm is selected and a navigation of the mobile robot is performed. In all of the software runs below with algorithms, the gridSize has been taken as 1.0 and line length has been taken as 1. The number of points generated indicates the total number of points required to reach the goal with the selected line length. That is, the array that contains the path points has this number of members.

Figure 4 is a typical run of the Robot Navigation software with obstacles, using type-2 Fuzzy navigation. As can be seen from the figure, all obstacles were avoided successfully by the robot. The fuzzy algorithm avoided the obstacles by making decisions when the robot has approached the obstacles. Rule base with fuzzy inference mechanism is applied to find turning angle of the robot. As it is seen from the figure robot uses safe region during avoidance of obstacles.

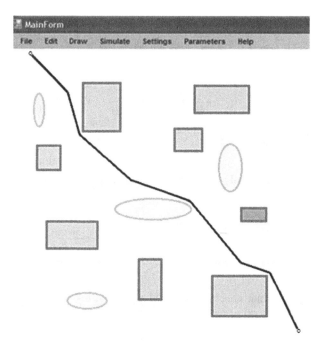

Fig. 4. Simulation result of type-2 fuzzy navigation algorithm

In next stage, we test fuzzy navigation algorithm using two different rule bases (RB): RB using angle and distance measures (left and right angels and distances) and RB using distance measures (left and right distances). In our simulation, RB that use angle and distance measures have given better results than other RB. Figure 5 demonstrate graphical simulation results. Table 1 demonstrates test results of FN algorithm with two different rule bases.

Another interesting simulation result is given in Fig. 6. Here, three obstacles are used with all four algorithms: Potential Field Method (PFM), Local Navigation (LN), Vector Field Histogram Plus (VFH+) and type-2 Fuzzy Navigation (T2FN). The obstacles are scattered around the path that the robot is expected to follow. Table 2 gives the statistics for Fig. 6 in terms of time to reach goal and distance measures. Here the number of generated point can be accepted as the length of the path) distance measure). The best result was obtained with the type-2 fuzzy navigation algorithm. The results in terms of time taken to reach the goal are different compared to the case shown in Fig. 6. It is interesting to note that there are differences in the time taken to reach the

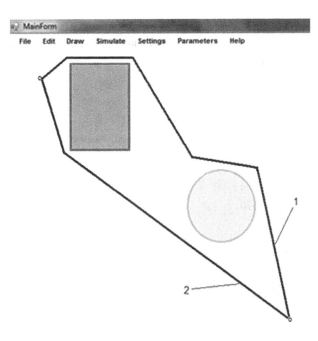

Fig. 5. Simulation result of type-2 fuzzy navigation algorithm using two different RBs: 1-RB using distance measures, 2-RB using angle and distance measures

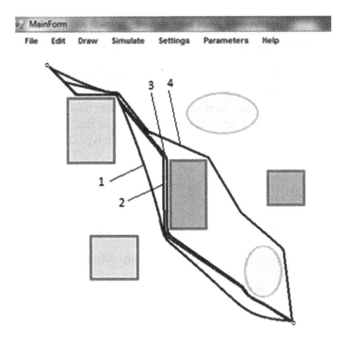

Fig. 6. Obstacle avoidance: 1 – APF, 2 – VFH, 3 – LN, 4 – T2FN.

Table 1. Statistics for Fig. 5

	Time taken to reach goal (ms)	Number of points
KB with distance measures	360	720
KB with angle and distance measures	380	674

Table 2. Statistics for Fig. 6

Algorithm	Time taken to reach goal (ms)	Number of points
PFM	21885	5304
VFH+	5599	5415
LN	398	5624
Fuzzy	370	5260

goal and the number of points generated by the fuzzy algorithm and other algorithms. The superiority of the T2FN algorithm is probably because of its resemblance to the intelligent human reasoning and decision-making process.

5 Conclusions

The type-2 fuzzy system is presented for mobile robot navigation in the presence of obstacles. The knowledge based is designed for obstacle avoidance. For simulation of type-2 fuzzy navigation and other above-mentioned navigation algorithms, a software package using C# have been developed. The workspace consisted of drawing different obstacles and robot starting and goal points on the screen. Various parameters of the algorithms can be adjusted manually in order to compare the advantages and disadvantages of the algorithms. Simulation of obstacle avoidance using type-2 fuzzy system has been performed. Comparison of simulation results of different algorithms has shown that the algorithm having shortest path among other navigation algorithms was the type-2 fuzzy navigation. The obtained results show that the algorithm having shortest path among other navigation algorithms was the type-2 fuzzy navigation.

References

1. Siegwart, R., Nourbakhsh, R.I.: Introduction to Autonomous Mobile Robots. MIT Press, Cambridge (2004)
2. Choset, H., et al.: Principles of Robot Motion, Theory. Algorithms and Implementations. MIT Press, Cambridge (2005)
3. Khatib, O.: Real-time obstacle avoidance for manipulators and mobile robots. In: Proceedings of IEEE International Conference on Robotics and Automation, St. Louis, MO, pp. 500–505 (1985)
4. Borenstein, J., Koren, Y.: Real-time obstacle avoidance for fast mobile robots. IEEE Trans. Syst. Man Cybern. **19**, 1179–1187 (1989)

5. Borenstein, J., Koren, Y.: The vector field histogram-fast obstacle avoidance for mobile robots. IREE J. Robot. Autom. **7**(3), 278–288 (1991)
6. Wlrich, I., Borenstein, J.: VFH+: reliable obstacle avoidance for fast mobile robots. In: Proceedings of the IEEE International Conference on Robotics and Automation (1998)
7. Barber, R., Salichs, M.A.: A new human based architecture for intelligent autonomous robots. In: The 4th IFAC Symposium on Intelligent Autonomous Vehicles, Sapporo, Japan, pp. 85–90 (2001)
8. Wu, C.J.: A learning fuzzy algorithm for motion planning of mobile robots. J. Intell. Rob. Syst. **11**, 209–221 (1995)
9. Yang, X., Moallem, M., Rajni, V.P.: A layered goal-oriented fuzzy motion planning strategy for mobile robot navigation. IEEE Cybern. **35**(6), 1214–1224 (2005)
10. Abiyev, R., Ibrahim, D., Erin, B.: Navigation of mobile robots in the presence of obstacles. Adv. Softw. Eng. **41**(10–11), 1179–1186 (2010)
11. Abiyev, R.H., Günsel, I., Akkaya, N., Aytac, E., Çağman, A., Abizada, S.: Robot soccer control using behaviour trees and fuzzy logic. In: 12th International Conference on Application of Fuzzy Systems and Soft Computing, ICAFS 2016. (Book Series: Procedia Comput. Sci. **102**, 477–484 (2016))
12. Fox, D., Burgard, W., Thrun, S.: The dynamic window approach to collision avoidance. IEEE Robot. Autom. **4**(1), 23–33 (1997)
13. Ehjimura, K.: Motion Planning in Dgnarrtic Environments. Springer, Heidelberg (1991)
14. Ibrahim, M.Y.: Mobile robot navigation in a cluttered environment using free space attraction "agoraphilic" algorithm. In: Proceedings of the 9th International Conference on Computers and Industrial Engineering, vol. 1, pp. 377–382 (2002)
15. LaValle, S.M., Kuffner, J.J.: Randomized kinodynamic planning. Int. J. Robot. Res. **2**(5), 378–400 (2001)
16. Abiyev, R.H., Akkaya, N., Aytac, E., Ibrahim, D.: Behaviour tree based control for efficient navigation of holonomic robots. Int. J. Robot. Autom. **29**(1), 44–57 (2014)
17. Yan, Z., Zhao, Y., Hou, S., Zhang, H., Zheng, Y.: Obstacle avoidance for unmanned undersea vehicle in unknown unstructured environment. Math. Prob. Eng. (2013)
18. Abiyev, R., Ibrahim, D., Erin, B.: EDURobot: an educational computer simulation programm for navigation of mobile robots in the presence of obstacles. Int. J. Eng. Educ. **26** (1), 18–29 (2010)
19. Erin, B., Abiyev, R., Ibrahim, D.: Teaching robot navigation in the presence of obstacles using a computer simulation program. Procedia – Soc. Behav. Sci. **2**(2), 565–571 (2010). Elsevier
20. Abiyev, R.H., Akkaya, N., Aytac, E., Günsel, I., Çağman, A.: Improved path-finding algorithm for robot soccers. J. Autom. Control Eng. **3**(5), 398–402 (2015)
21. Abiyev, R.H., Akkaya, N., Aytac, E.: Control of soccer robots using behaviour trees. In: ASCC, Istanbul, June 2013
22. Abiyev, R.H., Akkaya, N., Aytac, E.: Navigation of mobile robot in dynamic environments. In: IEEE International Conference on Computer Science and Automation Engineering, CSAE 2012, Zhangjiajie, China, pp. 480–484 (2012)
23. Abiyev, R.H., Bektas, S., Akkaya, N., Aytac, E.: Behaviour tree based control of holonomic robots. In: Recent Advances in Mathematical Methods, Intelligent Systems and Materials, WSEAS Conference, Limasol, Cyprus, pp. 54–59 (2013)
24. Guo, H., Cao, C., Yang, J., Zhang, O.: Research on obstacle-avoidance control algorithm of lower limbs rehabilitation robot based on fuzzy control. In: 6th International Conference on Fuzzy Systems and Knowledge Discovery, p. 151 (2009)

25. Shirinzadeh, B., Parasuraman, S., Ganapathy, V.: Fuzzy decision mechanism combined with neuro-fuzzy controller for behavior based robot navigation. In: The 29th Annual Conference of the IEEE (2003)
26. Min, B.-C., Lee, M.-S., Kim, D.: Fuzzy logic path planner and motion controller by evolutionary programming for mobile robots. Int. J. Fuzzy Syst. **11**(3), 154–163 (2009)
27. Mendel, J.M.: Uncertain Rule-Based Fuzzy Logic System: Introduction and New Directions. Prentice Hall, Upper Saddle River (2001)
28. Karnik, N.N., Mendel, J.M., Liang, Q.: Type-2 fuzzy logic systems. IEEE Trans. Fuzzy Syst. **7**, 643–658 (1999)
29. Wu, H., Mendel, J.: Uncertainty bounds and their use in the design of interval type-2 fuzzy logic systems. IEEE Trans. Fuzzy Syst. **10**, 622–639 (2002)
30. Hagras, H.: A hierarchical type-2 fuzzy logic control architecture for autonomous mobile robots. IEEE Trans. Fuzzy Syst. **12**(4), 524–539 (2004)
31. Castillo, O., Melin, P.: Intelligent systems with interval type-2 fuzzy logic. Int. J. Innov. Comput. Inf. Control **4**(2), 771–783 (2008)
32. Lin, Y.-C., Lee, C.-H.: System identification and adaptive filter using a novel fuzzy neuro system. Int. J. Comput. Cogn. **5**(1), 15–26 (2007)
33. Abiyev, R.H., Kaynak, O.: Type-2 fuzzy neural structure for identification and control of time-varying plants. IEEE Trans. Ind. Electron. **57**(12), 4147–4159 (2010)
34. Abiyev, R.H., Kaynak, O., Alshanableh, T., Mamedov, F.: A type-2 neuro-fuzzy system based on clustering and gradient techniques applied to system identification and channel equalization. Appl. Soft Comput. **11**(1), 1396–1406 (2011)
35. Abiyev, R.H.: Credit rating using type-2 fuzzy neural networks. Math. Probl. Eng. **2014** (2014). Hindawi Publications
36. Abiyev, R.H., Kaynak, O., Kayacan, E.: A type-2 fuzzy wavelet neural network for system identification and control. J. Franklin Inst. Eng. Appl. Math. **350**(7), 1658–1685 (2013)
37. Kayacan, E., Oniz, Y., Aras, A.C., Kaynak, O., Abiyev, R.: A servo system control with time-varying and nonlinear load conditions using type-2 TSK fuzzy neural system. Appl. Soft Comput. **11**(8), 5735–5744 (2011)

Intelligent Computing in Computer Vision

An Effective EM-PND Based Integrated Approach for NRSFM with Small Size Sequences

Xia Chen[1], Zhan-Li Sun[1(✉)], Shang Li[2], Tao Shen[1], and Chao Zheng[1]

[1] School of Electrical Engineering and Automation,
Anhui University, Hefei, China
zhlsun2006@126.com
[2] Department of Communication Technology,
Electronic Information Engineering College, Suzhou Vocational University,
Suzhou, China

Abstract. The performance of non-rigid structure from motion (NRSFM) generally deteriorates when the image sequence is small. In this paper, an effective approach is proposed to deal with NRSFM with small size sequences based on the Expectation and Maximization-Procrustean Normal Distribution (EM-PND) algorithm. In the proposed method, the sub-sequences are first extracted from the original small size sequence. Further, some weaker estimators are constructed by inputting the sub-sequences to the EM-PND algorithm. Finally, the 3Dstructures of the sequences are estimated by integrating the outputs of these weaker estimators. Experimental results on several widely used sequences demonstrate the effectiveness and feasibility of the proposed algorithm.

Keywords: Non-rigid structure from motion · Small size sequence · Weaker estimator

1 Introduction

The aim of non-rigid structure from motion (NRSFM) is to recover the time varying 3D coordinates of a deformable object by using a set of 2D feature points of an image sequence [3, 6]. As one of the hottest issues in computer vision, NRSFM has been studied extensively over the past three decades [1]. However, in the absence of any prior knowledge on 3D shape deformation, reconstructing non-rigid structure from motion is still a challenging and underconstrained problem.

The standard matrix factorization approach is a classical approach of structure from motion [2]. In the factorization algorithm, the observation matrix is decomposed as the product of a motion matrix and a shape basis matrix. There are two main types of matrix factorization approach. For the first type, all the estimated 3D shapes are constrained to lie within a linear space spanned by K 3D shape bases.

The other dual concept was proposed in [3] that models independent 3D point trajectories instead of 3D shapes. For the former, the large majority of works [4, 5]

© Springer International Publishing AG 2017
D.-S. Huang et al. (Eds.): ICIC 2017, Part III, LNAI 10363, pp. 29–36, 2017.
DOI: 10.1007/978-3-319-63315-2_3

assumed that the number of basis shapes is a prior information. Under this theory, Torresani et al. [6] deemed the shape deformation satisfied a Gaussian prior and the optimum solution to this problem can be given by using Expectation-Maximization (EM). According to the later concept, recent research works generally using DCT as an object independent basis due to its excellent energy compaction for highly correlated data [7–9]. Based on the nuclear minimization method, Dai et al. [10] proposed Block Matrix Method (BMM) to recover both the camera motion matrix and the non-rigid shape matrix from image measurement matrix W. However, it is too difficult to choose the right number of 3D shape bases or DCT bases (3D point trajectories), which is greatly affects the reconstruction performance, and a unified way is actually still absent.

In order to deal with the lack of new constraints (priors) on the number of bases, a novel NRSFM with Procrustean normal distribution (EM-PND) is proposed in [11] by considering NRSFM as an alignment problem and present an additional constraint obtained from the generalized Procrustes analysis (GPA) [12, 13] to each rotation matrix and it does not require any rank constraint. This idea makes the NRSFM problem more tractable.

However, in real situations, the quantity of superior images can be got hardly in many cases, the face images in a monitor system for example. In this paper, a sub-sequence based integrated algorithm is proposed to deal with the problem of NRSFM with small sample sequence. In the proposed method, each sample of original small size image sequence is estimated in order. Further, we choose a frame as test frame, extracting training frames from the original sequence sequentially after rule out the test frame. Then, we can obtain the input of EM-PND combining the extracted frames with the test frame, which we call the sub-sequences. According to integrated learning theory, an estimation of 3D shapes can be received from a weaker estimator constituted by the sub-sequences and estimation process of EM-PND. Finally, the z-coordinates corresponding 2D points obtained by multiple weaker estimators are integrated through two methods. Experimental results on several widely used image sequences prove the effectiveness and feasibility of the proposed method.

The remainder of the paper is organized as follows. In Sect. 2, we present the proposed EM-PND-based integrated approach. Experimental results and the related discussions are given in Sect. 3, and our concluding remarks are presented in Sect. 4.

2   Methodology

Figure 1 shows the flowchart of the proposed EM-PND-based integrated approach algorithm. There are three steps in the proposed method: extract sub-sequences from the original sequence, construct the weaker estimators by means of the EM-PND algorithm, compute the final estimation results by utilizing the outputs of weaker estimators. A detailed description of these three steps is presented in the following subsections.

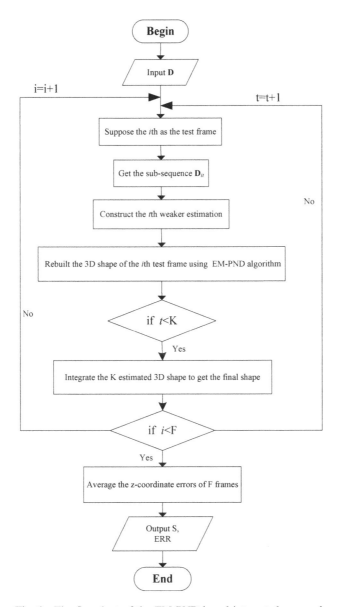

Fig. 1. The flowchart of the EM-PND-based integrated approach.

2.1   Sub-sequence Extraction

The first step of the proposed method is to produce the inputs of weaker estimators. Let $[x_{i,j}, y_{i,j}]^T$ $(i = 1, 2, \cdots, F; j = 1, 2. \cdots, n)$ denote the 2D projection of the jth 3D feature point observed on the ith frame. The n 2D point tracks of F frames can be represented as a $2F \times n$ observation matrix **D**, i.e.

$$\mathbf{D} = \begin{bmatrix} x_{1,1} & x_{1,2} & \cdots & x_{1,n} \\ y_{1,1} & y_{1,2} & \cdots & y_{1,n} \\ \vdots & \vdots & \ddots & \vdots \\ x_{F,1} & x_{F,2} & \cdots & x_{F,n} \\ y_{F,1} & y_{F,2} & \cdots & y_{F,n} \end{bmatrix} \tag{1}$$

In the proposed method, the 3D coordinates of feature points are estimated one frame by one frame. Assume that the ith frame is the one to be estimated, its 2D observation $\mathbf{D}$ is a $2 \times n$ matrix,

$$\mathbf{D}_i = \begin{bmatrix} x_{i,1} & x_{i,2} & \cdots & x_{i,n} \\ y_{i,1} & y_{i,2} & \cdots & y_{i,n} \end{bmatrix}. \tag{2}$$

Except for $\mathbf{D}_i \in \Re^{2 \times n}$, the remained $\mathbf{D}_t \in \Re^{2(F-1) \times n}$ ($t = 1, 2, \cdots, i-1,$ $i+1, \cdots, F$) is a sequence with $K = F - 1$ frames. Then, a sub-sequence can be built by choosing $K - 1$ observations from $\mathbf{D}_t$. The sub-sequence and the ith frame are merged to be used as the input of weaker estimator. Similarly, K sets of sub-sequence are extracted successively.

2.2   EM-PND Based Weaker Estimator

In the proposed method, the EM-PND algorithm [11] is adopted as the weaker estimator due to its good performance. Let $\mathbf{X}_i \in \Re^{3 \times n}$ denote the 3D shapes to be estimated for the ith sample, the objective function of generalized Procrustes analysis (GPA) can be formulated as,

$$\min_{s_i, \mathbf{R}_i, t_i, \bar{\mathbf{X}}} \quad \sum_{i=1}^{T} \| s_i \mathbf{R}_i \mathbf{X}_i + t_i - \bar{\mathbf{X}} \|, \tag{3}$$
$$\text{subject to} \quad \mathbf{R}_i \mathbf{R}_i^T = \mathbf{I}, f(s) = 1$$

where $s_i \in \Re^{1 \times 1}$, $\mathbf{R}_i \in \Re^{3 \times 3}$, and $t_i \in \Re^{3 \times 1}$ are the scale factor, rotation and translation, respectively. The solution of EM-PND can be obtained by maximizing the following log-likelihood function via the EM algorithm,

$$\sum \log(p(\mathbf{D}_i, \mathbf{X}_i | \sigma, s_i, \mathbf{R}_i, \bar{\mathbf{X}}, \Sigma)) = \sum [\log(p(\mathbf{D}_i | \sigma)) + \log(p(\mathbf{X}_i | s_i, \mathbf{R}_i, \bar{\mathbf{X}}, \Sigma))] \tag{4}$$

where σ denotes the standard deviation of Gaussian noise in $\mathbf{D}_i$.

2.3   Integration of Weaker Estimator

As mentioned earlier, for the ith test frame, we can get one 3D coordinate $\mathbf{S}_i^t$ from each sub-sequence $\mathbf{D}_i^t$, where $t = 1, 2, \cdots, K, i = 1, 2, \cdots, F$. In this paper, we proposed two methods to integrate the results obtained by K weaker estimators, called Method1 and Method2. Let $w_{i,t}$ be the weight of tth weaker estimator, then the final estimated 3D shape $\mathbf{S}_i^t$ test image is computed as

$$\mathbf{S}_i = \sum_{t=1}^{K} w_{i,t} \mathbf{S}_i^t, \tag{5}$$

where

$$w_{i,t} = \frac{\mathbf{S}_i^t}{\sum_{t=1}^{K} \mathbf{S}_i^t}, \tag{6}$$

The above method called Method1. Owing to the fact that it is difficult to determine the weight coefficient, the other approach, Method2, is to divide K weaker estimators into several intervals with different sizes. The integrated results of the tth test image is the mean of the highest probability among those intervals, and it can be calculated as

$$\mathbf{S}_i = \frac{1}{N} \sum_{t=1}^{N} \mathbf{S}_i^t, \tag{7}$$

where N is the number of factors in the highest probability interval.

3   Experimental Results

3.1   Experiments

In this section, we adopt three widely used motion sequences, containing dance, yoga and pick-up [11], to compare the performance of our proposed method and EM-PND. To evaluate the estimation accuracy, the average of z-coordinate error can be seen as a performance index, i.e.,

$$e\left(z, \hat{z}\right) = \frac{1}{F} \sum_{i=1}^{F} \left| \left(z_i - \hat{z}_i\right) \right|, \tag{8}$$

where z_i and $\hat{z}_i$ are the true z-coordinates and the estimated z-coordinates of ith frame, respectively. As the problem mentioned is the 3D reconstruction of small size sequence, we randomly select 10 frames from each sequence and used as the experimental data, i.e. $F = 10$. The corresponding numbers of frames (NF) and the numbers of 3D feature point tracks (NP) for those sequences have been given in Table 1.

To evaluate the performance of our proposed integrated algorithms, we compare our two methods with the original method on 3 widely used datasets. As the problem mentioned above, we first select 10 frames from the dance sequence and use the z-coordinates errors as the experimental results. In our experiments, the length of

Table 1. Three databases used in experiment.

Datasets	NF	NP
Dance	10	75
Yoga	10	41
Pick-up	10	40

sub-sequences and the number of weaker estimators are both set 9. Compared to [11], we calculate the average z-coordinates errors and the corresponding standard deviation on every dataset. Then, we evaluate the z-coordinates errors produced by our sub-sequence based integrated algorithm (denoted as Method1 and Method2). Finally, the experimental result is easy to be obtained.

Table 2 shows the z-coordinate errors of two algorithms on the dance sequence, where the frame number 1 to 10 represents the 1th to 10th frame. The final two lines display the corresponding mean (μ) and standard deviation (δ) of the 10 frames. From Table 2 we can see our algorithm get relatively high reconstructed accuracy than those of PND. Moreover, the standard deviations of our integrated methods are lower than those of the original method, which means our proposed methods have better robustness.

Table 2. The z-coordinate errors, and the corresponding mean (μ) and standard deviation (δ), of two algorithms on the dance sequence when the sample size is 10.

Frame number	PND	Method1	Method2
1	0.1327	0.1833	0.1358
2	0.1329	0.1586	0.1131
3	0.1312	0.1369	0.1346
4	0.1312	0.1249	0.1304
5	0.1348	0.1725	0.1359
6	0.1412	0.1684	0.1407
7	0.1490	0.1846	0.1475
8	0.1629	0.1629	0.1415
9	0.1853	0.2580	0.1830
10	0.2094	0.3751	0.1778
μ	0.1511	0.0925	0.1460
δ	0.0269	0.0734	0.0188

Further, five equally-sized groups of ten consecutive images are randomly selected from three different databases. Table 3 tabulate the mean and standard deviation ($\mu \pm \delta$) of the z-coordinates errors for three datasets. It is observed that the proposed methods achieve a better performance than the original PND method.

Table 3. The mean and standard deviation $(\mu + \delta)$ between the true and estimated z coordinates when choose randomly 10 consecutive frames from three different datasets.

Datasets	Trial	PND	Method1	Method2
Dance	1	0.1511 ± 0.0269	0.1925 ± 0.0734	$\mathbf{0.1460 \pm 0.0188}$
	2	0.1797 ± 0.0162	0.1799 ± 0.0166	$\mathbf{0.1711 \pm 0.0139}$
	3	0.1239 ± 0.0336	$\mathbf{0.1085 \pm 0.0200}$	0.1245 ± 0.0354
	4	0.1365 ± 0.0086	0.1331 ± 0.0087	$\mathbf{0.1329 \pm 0.0070}$
	5	0.1869 ± 0.0037	$\mathbf{0.1796 \pm 0.0075}$	0.1849 ± 0.0043
Yoga	1	0.1121 ± 0.0228	0.1661 ± 0.0525	$\mathbf{0.1118 \pm 0.0202}$
	2	0.1393 ± 0.0422	0.1273 ± 0.0371	$\mathbf{0.1019 \pm 0.0283}$
	3	0.1050 ± 0.0254	0.1568 ± 0.0285	$\mathbf{0.1013 \pm 0.0183}$
	4	0.1648 ± 0.0569	0.1725 ± 0.0272	$\mathbf{0.1453 \pm 0.0478}$
	5	0.1401 ± 0.0362	0.1424 ± 0.0373	$\mathbf{0.1375 \pm 0.0345}$
Pickup	1	0.1531 ± 0.0119	0.1914 ± 0.0354	$\mathbf{0.1512 \pm 0.0095}$
	2	0.0905 ± 0.0047	0.0961 ± 0.0070	$\mathbf{0.0891 \pm 0.0052}$
	3	0.1770 ± 0.0151	0.1760 ± 0.0153	$\mathbf{0.1721 \pm 0.0133}$
	4	0.0985 ± 0.0040	0.0982 ± 0.0038	$\mathbf{0.0956 \pm 0.0040}$
	5	0.0718 ± 0.0152	0.0715 ± 0.0151	$\mathbf{0.0696 \pm 0.0138}$

3.2 Discussion

Our proposed algorithm can be divided into three steps: choosing 9 sub-sequences for each weaker estimator; calculating the 3D shapes of each weak estimator and integrating these estimators as the final results. In this paper, two methods have been given to integrate 9 weaker estimators with each other. A simple linearity weighted method in the limited data on z-coordinate is presented, named Method1. And Method2 is the average of these estimation intensive. It's possible to try a better weighted average, but how to determined the weight is a difficult problem. Future work on Method1 will overcome this challenge to further improve the algorithm accuracy.

4 Conclusion

In this paper, a integrated-based solution to NRSFM with small size sequence is proposed. Compared to the original algorithm, the proposed method has a higher estimation accuracy and robustness. The experimental results on three widely used datasets have verified the effectiveness and feasibility of the proposed method.

Acknowledgement. The work was supported by a grant from National Natural Science Foundation of China (No. 61370109), a key project of support program for outstanding young talents of Anhui province university (No. gxyqZD2016013), a grant of science and technology program to strengthen police force (No. 1604d0802019), and a grant for academic and technical leaders and candidates of Anhui province (No. 2016H090).

References

1. Hartley, R., Zisserman, A.: Multiple View Geometry in Computer Vision. Cambridge University Press, Cambridge (2003)
2. Tomasi, C., Kanade, T.: Shape and motion from image streams under orthography: a factorization method. Int. J. Comput. Vis. **9**(2), 137–154 (1992)
3. Bregler, C., Hertzmann, A., Biermann, H.: Recovering non-rigid 3D shape from image streams. In: IEEE Conference on Computer Vision and Pattern Recognition, pp. 690–696 (2000)
4. Akhter, I., Sheikh, Y., Khan, S.: Computational prediction models for early detection of risk of cardiovascular events using mass spectrometry data. In: IEEE Conference on Computer Vision and Pattern Recognition, pp. 1534–1541 (2009)
5. Hamsici, O.C., Gotardo, P.F.U., Martinez, A.M.: Learning spatially-smooth mappings in non-rigid structure from motion. In: Fitzgibbon, A., Lazebnik, S., Perona, P., Sato, Y., Schmid, C. (eds.) ECCV 2012. LNCS, vol. 7575, pp. 260–273. Springer, Heidelberg (2012). doi:10.1007/978-3-642-33765-9_19
6. Torresani, L., Hertzmann, A., Bregler, C.: Nonrigid structure-from-motion: estimating shape and motion with hierarchical priors. IEEE Trans. Pattern Anal. Mach. Intell. **30**(5), 878–892 (2008)
7. Akhter, I., Sheikh, Y., Khan, S., et al.: Nonrigid structure from motion in trajectory space. In: IEEE Transactions on Pattern Analysis and Machine Intelligence, pp. 41–48 (2009)
8. Akhter, I., Sheikh, Y., Khan, S., et al.: Trajectory space: a dual representation for nonrigid structure from motion. IEEE Trans. Pattern Anal. Mach. Intell. **33**(7), 1442–1456 (2011)
9. Gotardo, P.F.U., Martinez, A.M.: Non-rigid structure from motion with complementary rank-3 spaces. In: IEEE Conference on Computer Vision and Pattern Recognition, pp. 3065–3072 (2011)
10. Dai, Y., Li, H., He, M.: A simple prior-free method for non-rigid structure-from-motion factorization. Int. J. Comput. Vis. **107**(2), 101–122 (2014)
11. Lee, M., Cho, J., Choi, C.H., et al.: Procrustean normal distribution for non-rigid structure from motion. In: IEEE Conference on Computer Vision and Pattern Recognition, pp. 1280–1287 (2013)
12. Gower, J.C.: Generalized procrustes analysis. Psychometrika **40**(1), 33–51 (1975)
13. Zelditch, M.L., Swiderski, D.L., Sheets, H.D.: Geometric Morphometrics for Biologists: A Primer. Academic Press, Cambridge (2012)

Single Laser Bidirectional Sensing for Robotic Wheelchair Step Detection and Measurement

Shamim Al Mamun$^{(\boxtimes)}$, Antony Lam, Yoshinori Kobayashi,
and Yoshinori Kuno

Graduate School of Science and Engineering, Saitama University, Saitama, Japan
{shamim,antonylam,kobayashi,
kuno}@cv.ics.saitama-u.ac.jp

Abstract. Research interest in robotic wheelchairs is driven in part by their potential for improving the independence and quality-of-life of persons with disabilities and the elderly. Moreover, smart wheelchair systems aim to reduce the workload of the caregiver. In this paper, we propose a novel technique for 3D sensing of the terrain using a conventional Laser Range Finder (LRF). We mounted this sensing system onto our new six-wheeled robotic step-climbing wheelchair and propose a new step measurement technique using the histogram distribution of the laser data. We successfully measure the height of stair steps in a railway station. Our step measurement technique for the wheelchair also enables the wheelchair to autonomously board a bus. Our experiments show the effectiveness and its applicability to real world robotic wheelchair navigation.

Keywords: Step detection · Histogram · LRF

1 Introduction and Motivation

In recent years, many people around the world have exhibited mobility impairments due to aging problems. According to the World Bank overview on disabilities [1], about 15% of the world's populations experience significant disabilities. Currently the needs of many individuals with disabilities can be satisfied with powered wheelchairs. However, many disabled people have trouble when driving powered wheelchairs using standard joysticks or remotes. A clinical study [2] shows that 9%–10% of people with severe disabilities have this issue. To recover their mobility, researchers have come up with robotic concepts for wheelchairs that can autonomously perform their functions. Indeed, several assistive wheelchairs possessing user-friendly interfaces and/or autonomous functions for meeting the needs of the disabled and an aging society have been proposed [3–5].

Robotic wheelchairs should be able to assist their users in avoiding obstacles for collision-free travel, going to pre-designated places, and maneuvering through doorways or other narrow and crowded premises. In addition, such smart wheelchairs would be more helpful if they could follow their user's caregiver autonomously; by detecting the registered person in the system hence reducing the workload of the caregiver [6, 7].

However, most robotic wheelchairs are currently used in indoor environments to navigate through doorways or corridors while avoiding obstacles. Nevertheless, robotic

© Springer International Publishing AG 2017
D.-S. Huang et al. (Eds.): ICIC 2017, Part III, LNAI 10363, pp. 37–47, 2017.
DOI: 10.1007/978-3-319-63315-2_4

wheelchairs should provide navigational assistance in urban areas like roads and rail or bus stations. Our previous work [8] was to detect any given terrain's smoothness and recognize categories of the terrain like muddy, watery or rough places. The primary concern in developing outdoor navigating robotic wheelchairs is the safe navigation through a given path. Therefore, it is important to detect the types of terrain while traversing urban outdoor pathways. We found that our smart wheelchair successfully detected four classes of terrain with a success rate of 97% which helps maintain smooth outdoor navigation. However, there are many more important and difficult barriers along a given path that a given robotic wheelchair should have to detect for safe and smooth autonomous operation. In addition, smart wheelchairs intended to run autonomously in environments designed for humans must also be able to climb up steps in order to say, board a bus or train. In this paper, we develop a new mechanism for robotic wheelchairs in collaboration with Toyota Motor Corporation and the University of Tokyo that can detect steps and have the capability to move on stair steps.

We note that most robotic wheelchairs use one or more 2D laser devices and vision systems to avoid hazards like narrow passes, slopes, stairs etc. [9]. In particular, much of the previous work concerns the use of 2D Laser Range Finder (LRF) sensors for obstacle detection. However, using multiple sensors or any moving mechanism for sensing results in a more complex system thus increasing cost and the need for maintenance. But if one were to naively use a single laser, it would not provide enough information to properly sense the environment. In this paper, we propose a novel concept for sensing with a single 2D laser sensor that can obtain horizontal and vertical scan data simultaneously. We show that our setup is highly effective for finding steps on stairs or buses, which allows for smooth climbing with our newly designed type of robotic wheelchair.

As a prerequisite, the wheelchair has to be able to acquire accurate measurements of the steps on buses and stairs on the ground floor of a building or railway station. Regarding this, researchers are working on building accurate models for overcoming these difficulties [10]. They rely on 3D laser range data for finding a given stair's planar surfaces from the observed point cloud to reconstruct models of staircases. 3D mapping [11, 12] shows that 3D sensors can be used for terrain mapping and obstacle detection for smooth navigation. Obwald et al. [13] presented an approach to autonomous stair climbing with humanoids given a known 3D model of the whole staircase using a 3D laser. For accurate and faster detection of obstacles on the path of the robot, researchers are using 3D sensing devices on the top of their systems. At the same time, some researchers are finding ways to get accurate measurements of environments with the help of 2D LRF to lower system cost. In [14, 15] the authors presented some 3D sensing systems using 2D LRFs. In those systems, multiple LRF sensors installed in different directions were mounted on a rotatable unit to sense environmental hazards or for terrain classification. Fujita and Kondo [16] developed a sensing system using an arm-type sensor movable unit, which can sense in horizontal and vertical spaces to model stairs and gaps. The authors showed that their robot could measure the step size of stairs fairly well. However, they had a 5.8% error in calculating the step height and width. In [17], researchers fused depth images and laser readings for detecting and climbing stairs using a mobile robot. Vision based wheelchairs are also used mainly for navigation and positioning of the wheelchair in outdoor environments [18, 19].

In this paper, we propose a new robotic wheelchair in collaboration with Toyota Motor Corporation and the University of Tokyo [20], which can measure and climb stairs, escalators or board buses. Here, we also discuss a vision-assisted system to give navigational information to the user by mapping the laser data onto video images so that for example, people with partial visual impaired can benefit from seeing what the robotic wheelchair senses. The visual display of the laser data to the user can also be beneficial when driving the wheelchair in manual mode. We achieve all this with a novel approach where we use a single 2D sensor that is capable of sensing the outdoor environment, simultaneously in the horizontal and vertical directions.

2 Robotic Wheelchair System

Figure 1 shows the mechanical design of the wheelchair, developed by Toyota Motor Corporation and the University of Tokyo. We incorporate our new outdoor environment sensing system on this new type of six-wheeled robotic wheelchair equipped with four hydraulic suspensions for lifting itself. Figure 1 shows that the front wheel and middle wheel are attached to the suspension. When the wheelchair finds a step to move on, its control system lifts the front wheel onto it, while the rear and middle wheel balance it. Once the front wheel is able to maintain a good grip on the upper step, the middle wheel is lifted up and moved forward while balancing itself with the help of the rear wheel. The schematic workflow in Fig. 2 describes the operation of these wheels. Moreover, the new robotic wheelchair is equipped with data sensing devices in addition to its control unit. The wheelchair comes equipped with a standard webcam (iBuffalo) with our novel 2D laser range setup using a HOKUYO UTM-30LX unit installed at 75 cm above the ground plane. Detailed measurements of bus steps or stairs can then be obtained using the laser sensor so that information can be mapped onto a video display that is linked with the webcam, for the user's benefit. With the measurement of the heights of steps on buses or stairs from the ground level, the wheelchair can then be autonomously commanded to climb the sensed height if necessary.

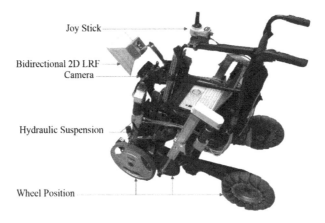

Fig. 1. Robotic wheelchair

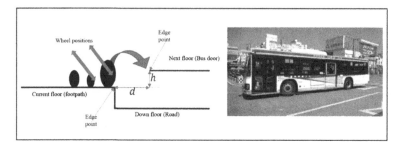

Fig. 2. Schematic diagram of wheelchair wheel positions for boarding a bus.

3   Data Acquisition Method

For smooth navigational support of our new wheelchair, we propose a novel approach for data sensing using a single 2D laser sensor and undistorted mirrors. Mounting this setup, we can scan both the horizontal and vertical data along the wheelchair's path.

3.1   2D Sensor Using Undistorted Mirrors

Conventionally 2D laser sensor has wide angle horizontal field of view. We use a 2D LRF (HOKUYO) that has a horizontal angle view field of 270°, with a measuring interval angle of 0.25°. A single scan from the LRF captures 1080 values in 25 ms, representing a viewing angle of 270° but we do not need to scan all 270° in the horizontal field of view. However, we do need some vertical scan data to detect steps. Thus, we divided the total view angle into three parts with a 90° view angle each by using three undistorted reflecting mirrors. Figure 3 shows the data acquisition apparatus, equipped with a HOKUYO UTM-30LX and three undistorted reflecting mirrors. The mirrors are placed at 45° angles around the 2D laser device.

The collected data from the single LRF and three mirror setup is divided into three different segments where each mirror's associated data block consists of 360 values. The vertical scan data comes from the left and right side of the laser and the horizontal scan comes from the top of it. From the LRF's scan, the first block of 360 values corresponds with the left side mirror within the $0\,to\,90^0$, the middle block of 360 values are the angles between $90^0\,to\,180^0$, and the last block is from the right mirror corresponding to the $180^0\,to\,270^0$ angles. The data that comes from boundaries between blocks are not reliable due to adjustments of the mirror. Therefore, we took values from the middle 20^0 of each block. Figures 3 and 4 show the schematic view of mirror arrangement and corresponding scan lines of our novel technique for laser based data collection. In this orientation of the laser, the top data comes from $xy - plane$, where $(x,y) \in \{360 < i < 720\}$; the right data and left data comes from $yz - plane$, where $(y,z) \in \{0 < i\,360\,and\,1080 > i > 720\}$ respectively.

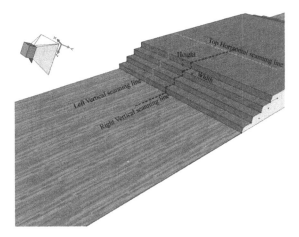

Fig. 3. Schematic diagram of our sensing system.

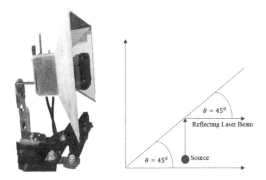

Fig. 4. Sensing apparatus using HOKUYO UTM-30LX with reflecting mirror.

3.2   Coordinate Transformation from Laser Data to Camera Image Plane

Mapping the LRF data on the image plane would be important for displaying information on what the wheelchair senses to the users. For example, people with partial visual impairment could benefit from seeing the data on paths in front of them and could act accordingly while seated on the wheelchair. The staircases or bus's door captured by the LRF has distance information, while the camera has the optical information projected onto the camera plane. Thus the LRF data needs to be appropriately mapped to the camera's coordinate system. In this paper, we divided conventional horizontal scanning LRF data into three different segments, each representing different 2D planes. Therefore, the coordinates of each block need to be transformed accordingly to the camera coordinates. We use the approach described in [21] for each block of our LRF data to map onto the image plane and for completeness, the approach summarized here.

Here, the LRF coordinates system of each block is denoted $f(x_t, y_t, z_t)$, the coordinates of the camera image plane is $c(u, v)$, and world coordinate system is $W(x_c, y_c, z_c)$. The most widely used camera model is the typical pinhole camera. The equation of the model is as shown in (1):

$$
\begin{bmatrix} u \\ v \\ 1 \end{bmatrix} = \frac{f}{z_c} \begin{bmatrix} k_u & 0 & \frac{u_0}{f} \\ 0 & k_v & \frac{v_0}{f} \\ 0 & 0 & \frac{1}{f} \end{bmatrix} \begin{bmatrix} x_c \\ y_c \\ z_c \end{bmatrix} = \frac{1}{z_c} \begin{bmatrix} f_u & 0 & u_0 \\ 0 & f_v & v_0 \\ 0 & 0 & 1 \end{bmatrix} \begin{bmatrix} x_c \\ y_c \\ z_c \end{bmatrix} \tag{1}
$$

where u, v represents the coordinates of the image plane and x_c, y_c, z_c are the world are coordinates of the steps for stairs or buses. Here, the camera coordinate system is also considered. The scale factors k_u, k_v along the axes of the pixel coordinates are defined as the adjacent pixels' physical distance ratios in the horizontal and vertical directions, respectively in the images. Moreover, the coordinates of the image center u_0, v_0 are in pixels, with f denoting the focal length. The orthonormal rotation matrix $[O] = f(\emptyset_1, \emptyset_2, \emptyset_3)$ and translation vector $[T] = (t_1, t_2, t_3)$ are combined for transforming from the LRF distance data points to the camera coordinates, and we obtain conversion Eq. (2) according to coordinate transformation rules:

$$
\begin{bmatrix} x_c \\ y_c \\ z_c \end{bmatrix} = \begin{bmatrix} \omega_{11} & \omega_{12} & \omega_{13} \\ \omega_{21} & \omega_{22} & \omega_{23} \\ \omega_{31} & \omega_{32} & \omega_{33} \end{bmatrix} \begin{bmatrix} x_t \\ y_t \\ z_t \end{bmatrix} + \begin{bmatrix} t_1 \\ t_2 \\ t_3 \end{bmatrix} \tag{2}
$$

Therefore, substituting Eq. (2) into Eq. (1), the laser point's coordinates are transformed into the camera image coordinates as,

$$
\begin{bmatrix} z_c u \\ z_c v \\ z_c \end{bmatrix} = \begin{bmatrix} f_u & 0 & u_0 \\ 0 & f_v & v_0 \\ 0 & 0 & 1 \end{bmatrix} \cdot \begin{bmatrix} \omega_{11} & \omega_{12} & \omega_{13} \\ \omega_{21} & \omega_{22} & \omega_{23} \\ \omega_{31} & \omega_{32} & \omega_{33} \end{bmatrix} \cdot \begin{bmatrix} x_t \\ y_t \\ z_t \end{bmatrix} + \begin{bmatrix} f_u & 0 & u_0 \\ 0 & f_v & v_0 \\ 0 & 0 & 1 \end{bmatrix} \cdot \begin{bmatrix} t_1 \\ t_2 \\ t_3 \end{bmatrix} \tag{3}
$$

According to Eq. (3), the world coordinates are $z_c = \omega_{31} x_t + \omega_{32} y_t + \omega_{33} z_t + t_3$, so the points of the LRF mapping onto the camera images is successfully completed.

3.3   Positioning the Laser Data onto the Camera Image Plane

Figure 4 briefly explains the process of data plotting onto the camera image plane. For our new wheelchair system, we employed it for displaying the calibrated image for the wheelchair user's comfort and assistance in manual navigation. ThLRF data is expressed in the polar coordinates of two different planes, namely the $xy - plane$ for the top mirror distance values and the $yz - plane$ for the left and right mirror distance values. The laser distance data of the top mirror are then converted to 3D coordinates as,

$$
\begin{cases} x_i = d_{dist}.cos\theta \\ y_i = d_{dist}.cos\theta \\ z_i = 0 \end{cases}
$$

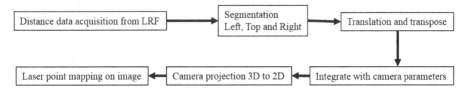

Fig. 5. Schematic flowchart of laser data mapping on to the image.

The data from the mirrors corresponding to the $yz-plane$ are converted in a similar manner. The extrinsic parameters are used in the experiments as follows:

$$\begin{bmatrix} x_{pos} \\ y_{pos} \\ z_{pos} \end{bmatrix} = \begin{bmatrix} 19.29 \\ 10.84 \\ 22.60 \end{bmatrix}$$

In addition, the camera intrinsic parameters are defined as in Eq. 4,

$$k_c = \begin{bmatrix} fc_1 & \alpha_c * fc_1 & cc_1 \\ 0 & fc_2 & cc_2 \\ 0 & 0 & 1 \end{bmatrix} \tag{4}$$

where the experimental values of camera matrix, $k_c = [0.213, -0.0455, 0.0019, -0.0002]$ and the focal length is, $fc = [310.5370, 310.5219]$; the principal point is, $cc = [307.0099, 226.8956]$ and the skew coefficient, $\alpha_c = 0$.

As the LRF is placed in the middle of the mirror triangle, the single laser sensor acts as the equivalent of three lasers where the left and right side reflections from the mirrors are denoted as vertical scanning and the top reflection is denoted as the horizontal scanning LRF. In addition, the top reflecting data comes from the $0.5z_i$ direction, and the left and right reflecting data are from $-0.5x_i$ and $0.5x_i$ directions in the world coordinate system respectively. The normal vector of the top part is in the positive z directions and left and right parts are in negative and positive x directions respectively. After translation and transposing of our laser data according the camera parameters for the three segments, we plotted our data on to the image plane. The real time fusion results for the staircases are given in Fig. 6.

Fig. 6. Real time fusion results from video images.

4   Step Measuring Methodology

Our proposed system will detect and measure the stairs or elevation gap for the bus door from roadside. Figure 7 illustrates the overall methodology of measuring the step size of the stairs.

Fig. 7. Methodology of finding the step size of the bus or stair

All acquired data from the 2D laser had outliers removed so that we could accurately measure the step size of stairs or the elevation gaps of bus doors. In order to remove any undesirable artifacts, we use the median values and standard deviation of the data series. We also note, our bidirectional sensing device is titled towards the ground at 45^0 and 100 cm away from the stair or bus. We calculated the slope of the laser data to find whether it is positive or negative. A positive tangent indicates a planar surface and negative slope indicates steps. Moreover, for stairs the slope of the steps changes in intervals. As mentioned earlier, our sensing laser gives two vertical scan lines for the stairs. The left and right side data values are almost the same, as expected. For the vertical positional data, we get two distributions for every step of a given stairway in the $y - direction$ and the $z - direction$. Here, y and z indicate the horizontal and vertical spaces of the data plot respectively. $z - values$ give the height of the stairs where as $y - values$ give the width of the stairs respectively. The standard deviation and mean of each direction were used for finding the outliers. Then mode of the distribution of the values was applied to find the peak points and lowest point. The distance of those two point was then used to determine the height of the steps. For finding the mode, we used 1 cm interval in each direction.

5   Experiment Result

Several experiments were employed to confirm the effectiveness of our new sensing system for staircases and buses. The results for these experiments show that the proposed system is useful for sensing and measuring the step size of staircases and bus floor heights and widths from the ground.

5.1   Upward Stairs

Figure 5 shows an overview of the experiment's environment. In this experiment, we consider the wheelchair to be located at the bottom of the staircase if it is 100 cm away from the first step. Each stair step is 26 cm wide and 17 cm high. From a stationary position, the wheelchair scans the data from the laser's left and right vertical mirrors. The top sensing mirror gives the distance of the wheelchair from stairs.

Table 1. Measurement of the steps (in cm) and error percentage of the setup in Fig. 4.

System calculation	Actual values	% of error HPM
17.2	17.0	−1.2
26.4	26.0	−1.6
17.3	17.0	−1.7
26.3	26.0	−1.1
17.4	17.0	−2.2
25.6	26.0	1.6
16.4	17.0	3.3
26.6	26.0	−2.3

Table 1 shows the result with compare to ground truth values and experimental values of the stair height and width. It is an accurate measurement of the stairs suitable for our wheelchair to determine how it should move the stairs. Figure 8 shows the data plot of the left vertical data from our experiment. Our sensing laser gives two vertical scan lines for the stairs. The left and right side data values are almost same, as expected.

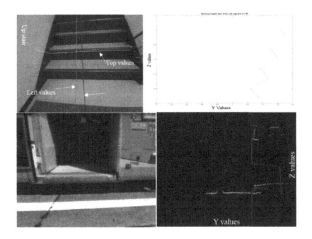

Fig. 8. Upward stair data and height of the bus floor measured from the red vertical lines. (Color figure online)

5.2   Bus Floor

In Fig. 7, the vertical red line shows the height of the bus floor and the blue line shows the bus ground and body data distributions. We calculated the histogram distribution and the difference between peaks of the broken line. Our system calculated the value of the bus floor height to be 29.34 cm while the ground truth height is 30 cm. With this measurement, the new wheelchair can then climb onto the bus using the new wheel positioning system.

6   Conclusion

This paper proposed a novel 3D sensing system using a triangular shaped apparatus with undistorted reflecting mirrors for vertical and horizontal sensing of the floor. This sensing system is able to accurately measure bus or staircase steps for the purpose of climbing them. In contrast, such measurements are difficult with a single conventional 2D laser. Moreover, the experiment setup of this sensing system for our new six-wheeled wheelchair shows the effectiveness of the overall system for users to move about in outdoor terrains and onto steps such as those of stairways in railways stations or to board buses. Our next research goal is to make the wheelchair autonomously climb on steps by using the proposed sensing system.

Acknowledgement. This work was supported by the Saitama Prefecture Leading-edge Industry Design Project and JSPS KAKENHI Grant Number 26240038 and in collaboration with Dr. Tomoyuki Takahata and Professor Iaso Shimoyama, at The University of Tokyo and Toyota Motor Corporation.

References

1. Overview of Disability, 21 September 2016. http://www.worldbank.org/en/topic/disability/overview. Accessed 22 Feb 2017
2. Fehr, L., Langbein, W.E., Skaar, S.B.: Adequacy of power wheelchair control interfaces for persons with severe disabilities: a clinical survey. J. Rehabil. Res. Dev. **37**(3), 353–360 (2000)
3. Kuno, Y., Shimada, N., Shirai, Y.: Look where you're going: a robotic wheelchair based on the integration of human and environmental observations. IEEE Robot. Autom. **10**(1), 26–34 (2003)
4. Min, J., Lee, K., Lim, S., Kwon, D.: Human friendly interfaces of wheelchair robotic system for handicapped persons. In: Proceedings of International Conference on Intelligent Robots and Systems (IROS), vol. 2, pp. 1505–1510 (2011)
5. Satoh, Y., Sakaue, K.: An omnidirectional stereo vision-based smart wheelchair. J. Image Video Process. **2007**, 87646 (2007)
6. Suzuki, R., Yamada, T., Arai, M., Sato, Y., Kobayashi, Y., Kuno, Y.: Multiple robotic wheelchair system considering group communication. In: International Conference on Visual Computing (ISVC 2014), vol. 8887, no. Part I, pp. 805–814 (2014)

7. Kobayashi, Y., Kinpara, Y., Takano, E., Kuno, Y., Yamazaki, K., Yamazaki, A.: Robotic wheelchair moving with caregiver collaboratively. In: Huang, D.-S., Gan, Y., Gupta, P., Gromiha, M.M. (eds.) ICIC 2011. LNCS, vol. 6839, pp. 523–532. Springer, Heidelberg (2012). doi:10.1007/978-3-642-25944-9_68

8. Mamun, S.A., Suzuki, R., Lam, A., Kobayashi, Y., Kuno, Y.: Terrain recognition for smart wheelchair. In: Huang, D.-S., Han, K., Hussain, A. (eds.) ICIC 2016. LNCS, vol. 9773, pp. 461–470. Springer, Cham (2016). doi:10.1007/978-3-319-42297-8_43

9. Lv, J., Kobayashi, Y., Hoshino, Y., Emaru, T.: Slope detection based on orthogonal assumption. In: IEEE/SICE International Symposium on system Integration (SII) (2012)

10. Hwang, Y., Lee, J.: Robust 3D map building for a mobile robot moving on the floor. In: IEEE International Conference on Advanced Intelligent Mechatronics (AIM) (2015)

11. Yi, C., Jeong, S., Cho, J.: Map representation for robots. Smart Comput. Rev. 2(1), 18–27 (2012)

12. Nemoto, Z., Takemura, H., Mizoguchi, H.: Development of small-sized omnidirectional laser range scanner and its application to 3D background difference. In: Proceedings of IEEE 33rd Annual Conference Industrial Electronics Society (IECON 2007), pp. 2284–2289 (2007)

13. Obwald, S., Görög, A., Hornung, A., Bennewitz, M.: Autonomous climbing of spiral staircases with humanoids. In: Proceedings of the IEEE/RSJ International Conference on Intelligent Robots and Systems (IROS) (2011)

14. Poppinga, J., Birk, A., Pathak, K.: Hough based terrain classification for real time detection of drivable ground. J. Field Robot. 25(1–2), 67–88 (2008)

15. Sheh, R., Kadous, M., Sammut, C., Hengst, B.: Extracting terrain features from range images for autonomous random step field traversal. In: Proceedings of IEEE International Workshop on Safety, Security and Rescue Robotics (2007)

16. Fujita, T., Kondo, Y.: Robot Arms (2011). ISBN 978-953-307-160-2. (Book Edited by Satoru Goto)

17. Hsiao, C.W., Chien, Y.H., Wang, W.Y., Li, I.H., Chen, M.C., Su, S.F.: Wall following and continuously stair climbing systems for a tracked robot. In: 2015 IEEE 12th International Conference on Networking, Sensing and Control, Taipei, pp. 371–375 (2015)

18. Shen, J., Xu, B., Pei, M., Jia, Y.: A low-cost tele-presence wheelchair system. In: 2016 IEEE/RSJ International Conference on Intelligent Robots and Systems (IROS), pp. 2452–2457 (2016)

19. Carbonara, S., Guaragnella, C.: Efficient stairs detection algorithm assisted navigation for vision impaired people. In: 2014 IEEE International Symposium on Innovations in Intelligent Systems and Applications (INISTA) Proceedings, pp. 313–318 (2014)

20. Ishikawa, M., et al.: Travel Device. Patent Number: WO2016/006248 A1 (2016)

21. Shin, Y., Park, J., Bae, J., Baeg, M.: A study on reliability enhancement for laser and camera calibration. Int. J. Control Autom. Syst. 10(1), 109–116 (2012)

Detecting Inner Emotions from Video Based Heart Rate Sensing

Keya Das[✉], Sarwar Ali, Koyo Otsu, Hisato Fukuda, Antony Lam,
Yoshinori Kobayashi, and Yoshinori Kuno

Graduate School of Science and Engineering, Saitama University, Saitama, Japan
{keya0612, sarwar_ali, otsu, fukuda, antonylam, yoshinori,
kuno}@cv.ics.saitama-u.ac.jp

Abstract. Recognizing human emotion by computer vision is an interesting and challenging problem. In particular, the reading of inner emotions, has received limited attention. In this paper, we use a remote video-based heart rate sensing technique to obtain physiological data that provides an indication of a person's inner emotions. This method allows for contactless estimates of heart rate data while the subject is watching emotionally stimulating video clips. We also compare against a wearable heart rate sensor to validate the usefulness of the proposed remote heart rate reading framework. We found that the reading of heart rates of a subject effectively detects the inner emotional reactions of human subjects while they were watching funny and horror videos—despite little to no facial expressions at times. These findings are validated from the reading of heart rates for 40 subjects with our vision-based method compared against conventional wearable sensors. We also find that the change in heart rate along with emotionally stimulating content is statistically significant and our remote sensor is well correlated with the wearable contact sensor.

Keywords: Heart rate · Video PPG · Wearable sensor

1 Introduction

Emotion recognition is a challenging task with many applications such as human robot interaction, movie marketing, and more. In this paper, we make the distinction between external and inner emotions. External emotions are readily visible through facial expressions and there is a wealth of work on this topic. For example, Cohen et al. [1] proposed a method for facial expression recognition from video. They introduced a Tree-Augmented-Naive Bayes (TAN) classifier that learns the dependencies between facial features and they provide an algorithm for finding the best TAN structure. In Zhang et al. [2] they propose a method of facial expression recognition based on local binary patterns (LBP) and local Fisher discriminant analysis (LFDA). On the other hand, there has been less attention on inner emotion recognition. Despite this, there do exist proposed techniques for detecting inner emotions in the psychophysiology literature. Psychophysiology attempts to discern human emotions by studying of the interrelationships between the physiological and psychological aspects of behavior. John Stern defined psychophysiology as "any research in which the dependent variable

© Springer International Publishing AG 2017
D.-S. Huang et al. (Eds.): ICIC 2017, Part III, LNAI 10363, pp. 48–57, 2017.
DOI: 10.1007/978-3-319-63315-2_5

(the subject's response) is a physiological measure and the independent variable (the factor manipulated by the experimenter) a behavioral one" [3]. Physiological parameters include electrodermal activity, heart rate, skin temperature and respiration rate because these variables respond to signals from the autonomic nervous system, which is not under conscious control [4]. Thus many researchers on inner emotions focus on using different physiological channels. Kim et al. [5] reported an emotion recognition system with 78.4% and 61.8% accuracy for the recognition of 3 and 4 classes of emotions using an electrocardiogram (ECG), skin temperature variation, and electrodermal activity. Zong and Chetouani [6] used 25 features from ECG, electromyogram, skin conductivity (SC) and respiration changes by the Hilbert-Huang transform to obtain 76% accuracy for 4 classes. Picard et al. [7] used 40 features from heart rate, muscle tension, temperature, and SC to get 81% recognition accuracy on 8 classes. Chanel et al. [8] obtained 80% recognition accuracy on 3 classes using an electroencephalogram (EEG). These researchers reinforce the finding that physiological changes primarily respond to emotion.

However, to realize an actual application system, we need a method that does not require any wearable attachments or devices. We consider heart rate (HR) in this paper and its connection to emotional states. However, conventional methods for measuring HR such as ECG or photoplethysmography (PPG) using optical sensors require the sensor to make physical contact with the person and this can be inconvenient and uncomfortable for the person. Fortunately, in controlled conditions it has been shown that it is possible to use a conventional camera to remotely detect small changes in skin color due to a person's cardiac pulse [13].

Thus, in our paper, we use a vision based video method. In particular, we choose one that is robust to motion and illumination changes [14]. This vision video-based method takes into account hemoglobin effects on skin appearance along with the effects of illumination changes and selects good local regions to use in cardiac signal extraction.

To validate the use of remote video HR estimation for inner emotion detection, we also compare against a wearable sensor. We show that our remote heart rate sensing framework, which uses only a conventional camera, performs favorably in comparison. In addition, our proposed approach has the benefit that commercial cameras are readily available so no special sensors are needed. For example, webcams, surveillance cameras, and cellphone cameras could be used. With the ability to see HR without contact sensors, we present a convenient system for detecting inner emotions.

2 Estimating Heart Rates from Video

In recent years, there have been interesting developments in estimating HR from video such as observing very small movements of the head. Balakrishnan et al. [20], exploited subtle head oscillations that accompany the cardiac cycle to extract information about cardiac activity from videos. In addition to providing an unobtrusive way of measuring HR, the method can be used to extract other clinically useful information about cardiac activity, such as the subtle changes in the length of heart beats that are associated with the health of the autonomic nervous system. However, the most common approach is still to consider minute changes in skin color. This approach to reading HR is based on the well-known PPG technique whereby a pulse oximeter makes contact with skin, illuminates it, and the changes in light absorption due to cardiac pulse is observed. In the case of remote video based PPG, it has been found that consumer grade cameras (e.g. Canon Powershot) can capture small changes in light absorption (color) that correspond well with the cardiac pulse [13]. Later on, this work was extended to account for more realistic imaging conditions [12, 14–17]. These approaches have different pros and cons. In our work, we choose to employ the method by Lam and Kuno [14] because it was designed to address the issue of changing illumination (a problem that can adversely affect color based HR estimation). This is because our work here involves estimating people's emotional responses to video stimuli on a screen.

For completeness, we briefly summarize the method of Lam and Kuno [14]. The basic model employed assumes skin to consist of two components, hemoglobin, where light absorption is influenced by cardiac activity, and pigments not immediately influenced by cardiac activity such as melanin. Formally, this model expresses the pixel value for a single channel camera of a given point I on skin at time t as

$$I(t) = a_m a_l(t) \int R_m(\lambda) L(\lambda, t) C(\lambda) d\lambda + a_h(t) a_l(t) \int R_h(\lambda, t) L(\lambda, t) C(\lambda) d\lambda \qquad (1)$$

where $R_m(\lambda)$ and $R_h(\lambda)$ are the normalized reflectance spectra of the melanin and hemoglobin pigments at wavelength λ, respectively, $L(\lambda,t)$ is the normalized light source spectrum at wavelength λ and time t. Similarly, $C(\lambda)$ is the camera's spectral response. The a terms are for scaling the different spectra inside the integrations. For example, $a_l(t)$ scales the light spectrum $L(\lambda,t)$ at time t and $a_h(t)$ scales $R_h(\lambda,t)$ at time t. The integrations sum over the range of wavelengths λ the camera response $C(\lambda)$ is sensitive to.

We can see from Eq. 1 that if the light source spectrum (e.g. from a video display) changes colors over time, this affects the amount of light at different wavelengths. This in turn, is reflected by the melanin and hemoglobin pigments differently. Thus the observed color changes in skin can be thought of as a mixture of two signals. It is well-known that techniques such as Independent Components Analysis (ICA) can be used to perform linear blind source separation (BSS) of signals. Thus these findings suggest we should be able to take two points on a person's face, obtain the pixel traces of those points, and treating them as signals, apply ICA to separate the melanin and hemoglobin influenced color changes. With the hemoglobin part of the color changes determined, we should be able to calculate the HR.

However, Lam and Kuno showed that not all pairs of skin surface points could be subjected to ICA for linear BSS. For example, if two skin points were illuminated by different colors, the melanin at the different skin points could reflect very different colors and thus linear BSS would not be applicable. However, provided the two points are illuminated by the same color spectrum (even at different brightness) and the ratios of melanin to hemoglobin between the two skin points are different, ICA can be used to effectively separate out the hemoglobin portion of the signal.

Since appropriate skin point pairs are not known a priori, they opted to randomly test using ICA on different pairs of points of the face and observe the histogram of estimated HRs. Then the most common HR in the histogram was used as the final estimate of the HR. The intuition is that randomly chosen point pairs that to satisfy the conditions for linear BSS should consistently give the same HR estimate from the separated hemoglobin portion of the signal. On the other hand, point pairs that violate the conditions for linear BSS should give less consistent HR estimates. As a result, performing a majority vote on the many determined HRs from random point pairs should give a robust estimate for a single final HR. See Fig. 1 for the basic flow of the algorithm.

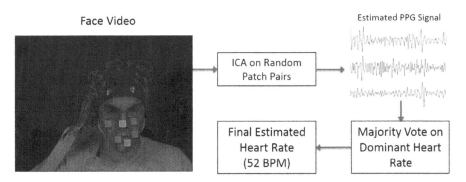

Fig. 1. Basic flow of the video PPG algorithm for estimating heart rate

With a robust algorithm for estimating HR in the presence of illumination changes (referred to as the Vision Based Video Method, VBVM), we proceed to determine whether it can be used for determining the emotional states of people watching real-world videos. This is done in comparison to a commonly available wearable sensor, namely the Fitbit.

3 Experiment

For this study, 40 undergraduate and postgraduate engineering students from Saitama University were recruited for data collection. The participant's ages ranged from 20 to 25 years old and there were 25 males and 15 females. They were all of Asian descent and in sound health. The experiment consisted of three parts. In the first part, we measured the human subject's resting HR for 1 min. Then using a Sony 4 K FDR-AX30 camera, we recorded 29 FPS videos of each subject's face for 4 min while the subject watched a funny video. After that, we recorded the subject for 4 min while he/she watched a horror video clip. In all cases, we determined the average HR for each minute of the viewing sessions.

Figure 2 shows the experiment's setup during HR measurement while watching funny and horror clips on a laptop.

We should briefly discuss our choice in using the FitBit sensor. Concerning commercial wearable sensors, there are many on the market. Generally, companies

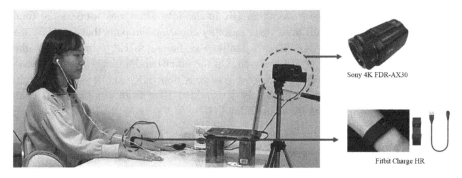

Fig. 2. Experiment's setup for measuring heart rate during watching video clips.

display advertising for these kinds of products and depict them as beneficial, user friendly, and accurate but it is important of consider the research literature to objectively determine their usefulness. Kaewkannate and Kim [18] in their paper summarizes and compares wearable fitness devices. In addition, Evenson et al. [19] proposed summarizing the evidence for validity and reliability of the most popular consumer-wearable activity trackers. Among a variety of trackers on the market, approximately 3.3 million were sold between April 2013 to March 2014, with 96% of the sensors made by Fitbit (67%), Jawbone (18%), and Nike (11%) [19]. In their meta-study, they reported that the FitBit was often found to have high reliability for monitoring steps, energy expenditure, and sleep in some FitBit models but that none were reported for the Jawbone. Due to the better understood performance characteristics of the FitBit for various monitoring tasks, we also chose to use the FitBit. In addition, internal studies by FitBit report their trackers are 95–97% accurate.

4   Results and Discussion

4.1   Relation Between Vision Based Video Method and Wearable Sensor

Here we show an analysis of Vision Based Video Method (VBVM) HR estimation and wearable sensor HR from funny and horror video clips. We compare the HR data from VBVM and the wearable sensor (Fitbit). We found that the two methods give us similar results from two emotional situations. In fact, the two methods had HR estimates with a correlation of 0.90 (Fig. 3). Figure 4 also shows the mean HRs (over the entire viewing session for each subject) for the VBVM and wearable sensor are consistent across different subjects. Thus, the VBVM provides comparable performance to the wearable sensor.

4.2   Emotion Changes from Heart Rate

We collected all subjects' HR data using VBVM and the wearable sensor (Fitbit). We observed them while they watched video clips (funny and horror) and compared their

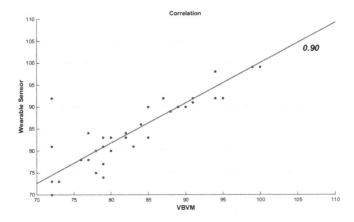

Fig. 3. Correlation of the VBVM and wearable sensor

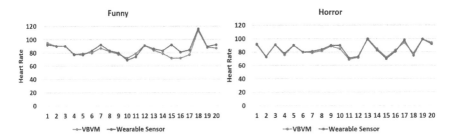

Fig. 4. Sample mean heart rates for 20 subjects from the viewing sessions.

resting HRs with their HRs while viewing the videos (taken as the maximum between the two to three minute points in each viewing session because these were the most emotionally stimulating parts).

In addition, Fig. 5 shows t-test results for various cases. These tests compare the HRs during emotional stimulation relative to the mean resting HR (69 ± 5.78 BPM) of the subjects. It can be seen that all the p-values of these tests show clear statistical significance indicating that when emotionally stimulated, people do experience increased HRs. Both the VBVM and wearable sensor pick up these changes in HRs reliably.

Figure 5 shows the bar graphs of the resting, funny, and horror cases for both VBVM and the wearable sensor data. From the diagram, we see that the HR increases by about 22%–26% and for the funny video case and 18%–19% for the horror video case. The HR for the funny video case is typically higher than the horror video case for both VBVM and the wearable sensor.

4.3   Demonstration on Multiple Subjects

We now present a demonstration of our system measuring the HR of multiple human subjects in our lab. In this demonstration, we used a funny video clip in English. We

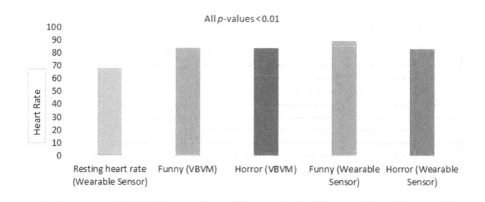

Fig. 5. Average HR bar graphs of the resting, funny, and horror cases.

selected three subjects for the demonstration. We then recorded the three subjects watching the video clip (Fig. 6). The video was then used to estimate their HRs over 12 min and the plot of these HRs can be seen in Fig. 7. The average HRs for subjects A, B, and C were estimated at 74, 77, and 70 respectively, which is within the normal range of HRs. We can also see from Fig. 7, that in some parts of the viewing session, the HRs increased and decreased at the same times. However, since these subjects had varying levels of English comprehension, we believe there were some differences in their appreciation of the humor, which would result in some differences in HR changes over the session between the subjects. In future work, we will consider videos in languages more suited to the particular human subjects being tested.

We also note that Subject B (the middle person in Fig. 6) did not show much change in facial expressions. However, we see from Fig. 7, that his HR had the most

Fig. 6. Measuring heart rate during viewing of video clips for three different subjects. From left to right, subjects A, B, and C.

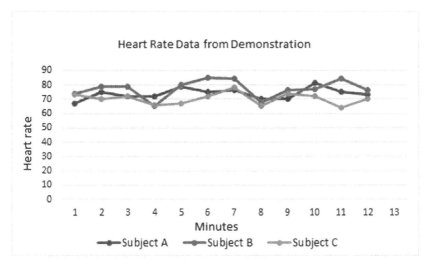

Fig. 7. Comparison between heart rate data for three subjects

variability. Subject A and Subject C showed smiling faces but their HRs actually had less changes. From this demonstration, we can see an example of how facial expressions may not always indicate genuine emotions. But physiological responses such as HR can reveal a person's hidden emotional responses.

5   Conclusion and Future Work

In this paper, we collected HR data for three emotional states (normal resting, funny, and horror) and performed a statistical analysis of the results. The HR data received using the VBVM is strongly correlated with the wearable sensor ground truth data. Moreover, our experiment's results show that the HRs for the funny clips increase approximately 24.04% and there is a 19.06% rise for horror clips. We were able to detect emotional state changes from HR.

Our method has some limitations like extreme changes in illumination and rapid motion that affect accuracy. In the future, we will resolve these problems to improve our system's performance.

Another interesting line of future work is that we will develop an approach to automatically decide whether a person has experienced some kind of inner emotional change. The current work here establishes the ground work by statistically verifying the feasibility of the basic sensors used in our framework but we currently have no automated way to decide if someone is reacting to emotional stimuli. (For example, a Support Vector Machine or other machine learning technique could be used to take in various HR data and classify people's reactions.)

Moreover, we will account for additional emotional states like sadness, anger, and frustration. Differentiating between the various types of emotions will likely require some form of detailed subtle facial expression analysis but detecting changes in HR would still be needed to indicate the presence of emotional stimulation as a first step.

We also plan to continue our work to evaluating multiple people at the same time (as is done in the demonstration at the end of the paper). This would be particularly interesting as we could then observe the emotional reactions of audience members in movie screenings. This would be useful for pilot screenings so that movie producers could adjust content to improve the quality of movies. An interesting idea might even be to have a system that could adaptively alter the movie's content based on the sensed emotions in order to enhance the emotional experience of watching a movie.

In addition, we will also investigate using the cardiac PPG signal itself and computing metrics such as HR variability, which is known to also be a good indicator of emotional change. It would be interesting to see if observing the continuous changes in the cardiac PPG signal itself could provide even more information about a person's emotional state.

Acknowledgments. This work was supported by JSPS KAKENHI Grant Number 26540131 and the Tateishi Science and Technology Foundation.

References

1. Cohen, I., Sebe, N., Garg, A., Chen, L.S., Huang, T.S.: Facial expression recognition from video sequences. In: Proceedings of 2002 IEEE International Conference on Multimedia and Expo, ICME 2002 (2002)
2. Zhang, S., Zhao, X., Lei, B.: Facial expression recognition based on local binary patters and local fisher discriminant analysis. In: PMC (2011)
3. Stern, R.M., Ray, W.J., Quigley, K.S.: Psychophysiological Recording, 2nd edn. Oxford University Press, New York (2001)
4. Stern, J.A.: Toward a Definition of Psychophysiology, vol. 1, no. 1. Version of Record online: 30 January 2007. http://onlinelibrary.wiley.com/doi/10.1111/j.1469-8986.1964. tb02626.x/abstract
5. Kim, K.H., Bang, S.W., Kim, S.R.: Emotion recognition system using short-term monitoring of physiological signals. Med. Biol. Eng. Comput. **42**, 419–427 (2004)
6. Zong, C., Chetouani, M.: Hilbert-huang transform based physiological signals analysis for emotion recognition. In: IEEE International Symposium on Signal Processing and Information Technology, pp. 334–339 (2009)
7. Picard, R.W., Vyzas, E., Healey, J.: Toward machine emotional intelligence: analysis of affective physiological state. IEEE Trans. Pattern Anal. Mach. Intell. **23**(10), 1176–1189 (2001)
8. Chanel, G., Kierkels, J.J.M., Soleymani, M., Pun, T.: Short-term emotion assessment in a recall paradigm. Int. J. Hum.-Comput. Stud. **67**, 607–627 (2009)
9. Li, X., Chen, J., Zhao, G., Pietkainen, M.: Remote heart rate measurement from face videos under realistic situations. In: Proceedings of the 2014 IEEE Conference on Computer Vision and Pattern Recognition, CVPR 2014, pp. 4264–4271 (2014)
10. Monkaresi, H., Hussain, M.S., Calvo, R.A.: Using remote heart rate measurement for affect detection. In: Florida Artificial Intelligence Research Society Conference, the Twenty-Seventh International Flairs Conference, pp. 119–123 (2014)

11. Wu, H.-Y., Rubinstein, M., Shih, E., Guttag, J., Durand, F., Freeman, W.T.: Eulerian video magnification for revealing subtle changes in the world. ACM Trans. (2012)
12. Kwon, S., Kim, H., Park, K.S.: Validation of heart rate extraction using video imaging on a built-in camera system of a smartphone. In: IEEE Engineering in Medicine and Biology Society (EMBC), pp. 2174–2177, August 2012
13. Verkruysse, W., Svaasand, L.O., Nelson, J.S.: Remote plethysmographic imaging using ambient light. Opt. Expr. **16**, 21434–21445 (2008)
14. Lam, A., Kuno, Y.: Robust heart rate measurement from video using select random patches. In: ICCV 2015, pp. 3640–3648 (2015)
15. Poh, M.-Z., McDuff, D., Picard, R.: Advancements in noncontact, multiparameter physio-logical measurements using a webcam. IEEE Trans. Biomed. Eng. **58**(1), 7–11 (2011)
16. Li, X., Chen, J., Zhao, G., Pietikainen, M.: Remote heart rate measurement from face videos under realistic situations. In: IEEE Computer Vision and Pattern Recognition (CVPR), pp. 4264–4271 (2014)
17. Tulyakov, S., Alameda-Pineda, X., Ricci, E., Yin, L., Cohn, J.F., Sebe, N.: Self-adaptive matrix completion for heart rate estimation from face videos under realistic conditions. In: The IEEE Conference on Computer Vision and Pattern Recognition (CVPR) (2016)
18. Kaewkannate, K., Kim, S.: A comparison of wearable fitness devices. BMC Public Health **16**(1), 433 (2016)
19. Evenson, K.R., Goto, M.M., Furberg, R.D.: Systematic review of the validity and reliability of consumer-wearable activity trackers. Int. J. Behav. Nutr. Phys. Act. **12**(1), 159 (2015)
20. Balakrishnan, G., Durand, F., Guttag, J.: Detecting pulse from head motions in video. In: IEEE Computer Vision and Pattern Recognition (CVPR) (2013)

Agricultural Pests Tracking and Identification in Video Surveillance Based on Deep Learning

Xi Cheng, You-Hua Zhang, Yun-Zhi Wu, and Yi Yue[(✉)]

School of Information and Computer,
Anhui Agricultural University, Hefei, China
{opteroncx, zhangyh, wuyzh, yyyue}@ahau.edu.cn

Abstract. Agricultural pests can cause serious damage to crops and need to be identified during the agricultural pest prevention and control process. In comparison with the low-speed and inefficient artificial identification method, it is important to develop a fast and reliable method for identifying agricultural pests based on computer vision. Aiming at the problem of agricultural pest identification in complex farmland environment, a recognition method through deep learning is proposed. The method could recognize and track the agricultural pests in surveillance videos of farmlands by using deep convolutional neural network and Faster R-CNN models. Compared with the traditional machine learning methods, this method has higher recognition accuracy in high background noise, and it can still effectively recognize agricultural pests with protective colorations. Therefore, compared with the current agricultural pest static-image recognition method, this method has a higher practical value and can be put into the actual agricultural production environment with the agricultural networking technology.

Keywords: Agricultural pests · Video recognition · Convolutional neural networks · Deep learning · VGG16 · Faster R-CNN

1 Introduction

China is one of the largest agricultural producer in the world and also a high incidence area of agricultural pest problems. The agricultural pests could cause a great damage of crops every year. Without professional knowledge, people could hardly identify the pests and the erroneous use of pesticides tends to cause secondary damage to areas affected by pests. In recent years, computer technology has developed rapidly and deepened into the field of agricultural production. With the development of agricultural network technology, the image and video in the farmland environment become easier to obtain, so the development of a computer vision technology based pest image and video recognition system is of great significance to improve the efficiency of agricultural pest controlling.

Deep learning technology have shown great results in image recognition works, this research used the deep convolutional neural networks to realize the task of pest image classification and build a system for pest detection in surveillance videos through Faster R-CNN. The system is robust enough to deal with the videos with complex background

© Springer International Publishing AG 2017
D.-S. Huang et al. (Eds.): ICIC 2017, Part III, LNAI 10363, pp. 58–70, 2017.
DOI: 10.1007/978-3-319-63315-2_6

and could even detect and track the pest with a protective coloration correctly. Also it could directly process the videos took from the farmland without any feature preprocessing works. Compared with other state of the art pest identification systems for still pest images, our system for pest video is more advanced and valuable for agricultural development.

2 Related Works

Agricultural pest recognition has been intensively studied in recent years, including some methods based on computer vision. Larios et al. [1] developed a system that used concatenated feature histogram method to classify stonefly larvae. Zhu and Zhang [2] proposed an insect classify method by analyzing the color histogram and Gray-Level Co-occurrence Matrices of wing images. Wang et al. [3] designed an automatic insect identification system. In their system, artificial neural networks and support vector machine are used as pattern recognition methods to classify the insects. Faithpraise et al. [4] proposed a pest detection system based on a combination of k-means cluster and correspondence filter. Xia et al. [5] used the watershed algorithm to separate the pest images from the background and then extract the color characteristics of the pests by Mahalanobis distance to classify the pest images captured by the mobile and embedded devices. Wang et al. [6] proposed a local Chan-Vese model to do the image segmentation tasks. Xie et al. [7] developed an insect recognition system with advanced multiple task sparse representation and multiple-kernel learning. Maharlooei et al. [8] developed an image based system that could identify and count soybean aphids in a green house.

However, the pest recognition methods above are largely dependent on the pest features selected artificially and the features always affect a lot on the model performance. In addition, in the actual agricultural pest control work, only identify the static agricultural pest image cannot meet the real needs. The detection of insect pests in the video collected by the agricultural networking system is more close to the identification requirements in the actual prevention and control process. However, the research on pest video identification is relatively few.

3 Methodology

3.1 Deep Convolutional Neural Networks

The history of the convolutional neural networks could date back to 1959, Hubel and Wiesel [9] discovered a hierarchical structure from the cat's visual cortical cells, and that inspired the Japanese scholar Fukushima to propose a model called Neocognitron [10] in 1980. Then LeCun et al. [11, 12] developed a modern structure of the convolutional neural networks. Neural networks were proved to be strong image classifiers and many researches [13, 14] were done to improve their performance. However due to the limitations of computer performance, the deep neural network model with more hidden layers is difficult to train at the early stage. In recent years, with the continuous improvement of computer performance and the introduction of GPU computing technology [15], the deep learning technology has been fully developed. In the ILSVRC2012, Krizhevsky proposed a deep convolutional neural network model named Alexnet [16] and is significantly better than other machine learning methods the field of image recognition and has aroused widespread focus. To increase the recognition accuracy, two deeper models named VGG16 [17] and GoogLeNet [18] were proposed in 2014. Then come the deep residual network, which enabled the depth of the neural network to be deeper, appeared to ResNet [19] model as the representative of the ultra-high depth convolution neural networks.

The basic structure of convolutional neural networks is shown in the following Fig. 1, which mainly includes convolution layers, pooling layers and full connected layers, and the head is usually a softmax classifier.

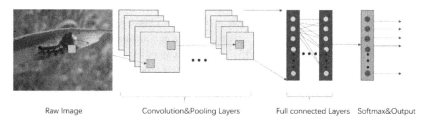

Fig. 1. The basic structure of convolutional neural networks

3.2 Agricultural Pest Identification Model

In the experiments of agricultural pest identification, two kinds of linear convolutional neural networks with different structures and depths were used, which were modified Alexnet and VGG16. In the study, the two deep neural networks were fine-tuned and adjusted the number of neurons in the full connected layer so that it can handle the number of categories corresponding to the dataset.

Alexnet is a convolutional neural network model with 8-layer depth. It has good recognition performance in image recognition. The basic structure of Alexnet is shown in the following Fig. 2. It contains five convolution layers, three pooling layers and two full connected layers, then the softmax classifier and output.

The VGG16 model is a 16-layer convolutional neural network, the basic structure is shown in Fig. 3, which is different to Alexnet, each group contains multiple convolution layer, and in each convolution layer, the convolution kernel size is also relatively small.

In the convolutional neural networks, the main working process of the convolution layers can be regarded as the filter sliding on the feature map of a certain layer, doing Hadamard product with the value of each pixel on the image. The process can be expressed by the following formula.

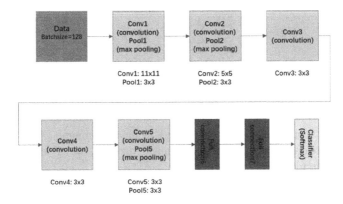

Fig. 2. Structure of Alexnet

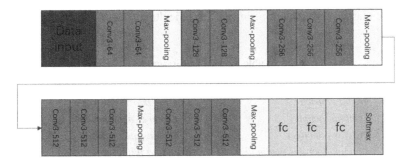

Fig. 3. Structure of VGG16

$$x_j^l = f\left(\sum_{i \in M_j} x_i^{l-1} * k_{ij}^l + b_j^l\right) \tag{1}$$

b_j^l is the bias, $f(.)$ is the activation function. The activation function is usually sigmoid or tanh in common neural networks. However, in some deeper structures, Relu is usually used as an activation function in order to suppress gradient vanish or gradient explosion.

The pooling layer is a down-sampling operation. Usually there are two pooling method called max pooling and average pooling. Where the max pooling strategy retains the maximum value of one of the sliding windows, while the latter holds the average of the pixels in the sliding window.

The classifier on the top uses the softmax classifier to output a probability vector whose loss function is its cross entropy with the actual class vector. The loss function can be expressed by the following equation.

$$J(\theta) = -\frac{1}{N}\left[\sum_{i=1}^{N}\sum_{i=1}^{C} 1\{y_i = t\}log\frac{e^{\theta_t^T x_i}}{\sum_{l=1}^{T}e^{\theta_l^T x_i}}\right] \qquad (2)$$

Where $1\{.\}$ is an indicative function, the value is 1 only when the i-th image is correctly classified. The $\frac{1}{\sum_{l=1}^{T}e^{\theta_l^T x_i}}$ indicates that the probability distribution is normalized.

At the full connected layer, compared with the common backpropagation (BP) neural networks, it has a strategy named dropout, which can suppress the over-fitting of the model to a certain extent.

3.3   Experiment

In the experiment, the image set mainly contains agricultural pests that are common in China. The classes and number of images of agricultural pests is shown in Table 1.

Table 1. Pest and images

Pest	Images	Pest	Images
Aelia sibirica	55	Dolerus tritici	55
Cifuna locuples	55	Gonepteryx amintha	55
Cletus punctiger	55	Pentfaleus major	55
Cnaphalocrocis medinalis	55	Erthesina fullo	55
Colaphellus bowvingi	55	Tettigella viridis	55
Total	550		

The image set in the experiment was obtained from Xie's research [6]. Figure 4 gives an example of these agricultural pests.

Fig. 4. Examples of 10 classes of pests

3.3.1   Experiment Setup

The software and hardware environment of the experiment are shown in Table 2. In the experiment, Caffe is used as the deep learning framework to fine tuning the models and train and test. We also used the GPU to accelerate the optimization speed.

Table 2.   Experiment equipment

Hardwear	Softwear
CPU: Intel Core i7 7700	Ubuntu14.04
RAM: 16 GB DDR4	CUDA8.0 + CUDNN5.1
GPU: NVIDIA GTX1070 (8 GB)	Caffe + Python2.7.12 + Matlab R2014a

3.3.2   Feature Extraction

Convolutional neural networks can automatically extract the features of agricultural pests in the input images. Figure 5 shows the original input image of a *Cletus punctiger*. Figure 6 shows the visualized image of the Conv3 layers of VGG16, Fig. 7 shows the features of pooling layers 1 to 3 of VGG16.

Fig. 5.   Input image of the *C. punctiger*

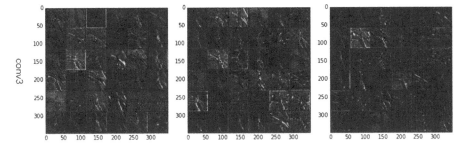

Fig. 6.   Features of Conv3

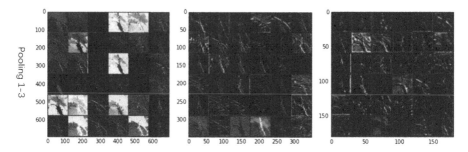

Fig. 7. Features of pooling 1–3

In these visualized feature maps, the main content of the image is separated from the background and we can easily find the outline of the *C. punctiger* is activated and the feature is extracted by the convolutional neural networks.

3.3.3 Comparison

In reality, images obtained from farmland environments usually have large background noise and too much pretreatment and noise reduction could lead to a large time consumption. In addition, the feature preprocessing means for static images does not apply to the video and is less applicable in the practical situations. So it is important to study the recognition performance of different models in the harsh environment with background noise.

In the experiment we choose 40 images of these pest images for training and 15 images for testing the model. All these images did not make any noise reduction on the background. Only the matrix of the RGB channel of the image is centered on the preprocessing operation, the process could be expressed by the following equation.

$$x^* = x - \mu \tag{3}$$

Where μ is the mean value on the training set and x is the value of a pixel on a color channel.

The recognition accuracy and loss functions of Alexnet and VGG16 on the image set used in the experiment are shown in Figs. 8 and 9.

In the backpropagation neural network model used as a control, in order to control the model size, the input image dimension of each channel is adjusted to 32×32 and expanded into a one-dimensional vector of length 3072, which is input to a traditional BP neural network with only one hidden layer, the size of hidden neuron is also 3072, respectively, using Sigmoid and Relu as the activation function. On the top of the networks we used Softmax classifier to output pest identification probability, the other control group we used linear support vector machine to classify the pests. Table 3 shows the recognition performance of different recognition models in pest images with complex background.

The experimental results show that in the complex background, since no feature pre-processing was performed on the input images, the recognition accuracy of the traditional BP neural networks and the support vector machine is not ideal. The results

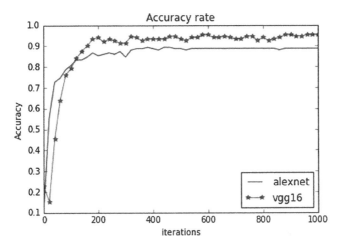

Fig. 8. Alexnet and VGG16 accuracy

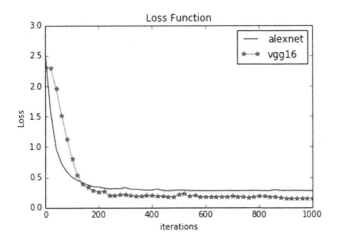

Fig. 9. Alexnet and VGG16 loss function

Table 3. Accuracy of models

Models	Accuracy
SVM	25.33%
BP neural network (sigmoid)	16.67%
BP neural network (relu)	27.33%
Alexnet	86.67%
VGG16	95.33%

also show that the deep CNN's recognition ability of pest images is much higher than that of traditional BP neural networks and linear support vector machine since they have the ability to extract features automatically. In the experiment Alexnet has the accuracy rate of 86.67%, while the deeper VGG16 model has a higher accuracy rate of 95.33%.

3.4   Pest Tracking

In practice, clear images of static pests are often difficult to obtain, but it is simply to get pest video directly from the agricultural network monitoring system. In addition, the identification system needs not only the recognition ability but also the ability to track and locate the moving pest targets.

3.4.1   Faster R-CNN

The convolutional neural networks are not limited to do only classification tasks. Models with target detection and tracking function were developed. In 2014 Girshick et al. [20] proposed the R-CNN model, which added a header to process the location information of the target in the image on the basis of the deep convolutional neural network model which generally performs the classification tasks. However, the R-CNN model can only handle fixed-size input images. To solve this problem, He et al. [21] added the spatial pyramid pool layer to the last convolution layer of R-CNN, and constructed SPP-net model. Then in 2015 Girshick [22] further proposed the Fast R-CNN model which can train and detect targets faster. In order to deal with real-time video processing, Ren et al. [23] proposed the Faster R-CNN model, which adds the RPN network structure on the basis of Fast R-CNN, so that the proposal selection and convolution operation of the area can be carried out at the same time, significantly increased the speed of object detection. The above-mentioned deep convolutional neural network for the target detection task has been applied in the field of human face tracking [24], vehicle identification [25], pedestrian detection [26], and instance search [27]. These applications of deep CNNs have been proved to have excellent tracking and recognition performance.

3.4.2   Video Processing System

In the process of dealing with the pest videos, the first step is to split the video frame by frame and input these frames to the pest target detection system, then the frame merge system transfers the detection result into video. The workflow is shown in Fig. 10.

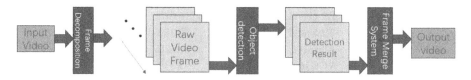

Fig. 10. Video detection system workflow

3.4.3 Fine-Tuning Faster R-CNN

In the experiment, the Faster R-CNN model was used to track the agricultural pests. In this model, the input image is convoluted and pooled to extract the features, and the pest proposals in the images is selected by the region proposal networks (RPN). Finally, the target recognition probability is output by the softmax classification head and combined with the position information. The work process is shown in Fig. 11.

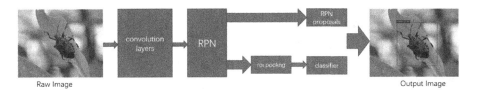

Raw Image   Output Image

Fig. 11. Faster R-CNN model for pest detection

In order to improve the training speed of the Faster R-CNN model for agricultural pest detection, fine tuning was used in the experiment. In the previous study, VGG16 was found to be higher in the accuracy rate of pest identification, so the VGG16 model was still used in the classification of Faster R-CNN. In training, a VGG16 model, which is pre-trained on the ImageNet dataset, is first fine-tuned to initialize the weight of all convolution layers before the full connected layers. Training using the learning rate of 0.001, and using alt-opt method to train the model in two main stages, each stage iterates 8000 times.

3.4.4 Detection Result

In the experiment we tested videos of two different insects named *Erthesina fullo* and *Gonepteryx amintha*. The total frame of *E. fullo* video is 1569 while the other video of *G. amintha* has 1845 frames. The video detection result obtained by the video recognition system is shown in Figs. 12 and 13. The output clearly shows the location information and the label of probability of the insects.

In the experiment, the NVIDIA GTX1070 GPU is used to identify the video and the recognition speed is about 10FPS to 11FPS. The experimental results show that the pest targets can be identified and tracked in video with the protective colorations and the motion blur state. When the video of *E. fullo* is analyzed, the frame loss rate is only 0.83%, which was caused by failing to detect the targets, and all the targets in frames of the second video were successfully detected.

Fig. 12. *E. fullo* detection results

Fig. 13. *G. amintha* detection results

4 Conclusion

In this study, a new recognition method based on deep learning is introduced to recognize the agricultural pests. The agricultural pests in the surveillance video in the agricultural networking are identified and tracked by the convolutional neural networks and the Faster RCNN model. The experimental results show that this method has a good ability to identify agricultural pests in farmland with complex background noise, and is superior to traditional BP neural network and other machine learning classification algorithms. In addition, this method has the ability of video processing, comparing with static pest image recognition methods, it is closer to the actual needs. However, due to the depth of deep neural network model, the scale of the parameters increased significantly compared to the traditional neural networks, so the hardware requirements for training is much higher. Future research needs to continue optimize the hardware resource occupancy and improve the detection speed in order to achieve smoother real-time detection capabilities.

Acknowledgements. This research was supported by the Anhui Agricultural University High-level Scientific Research Foundation for the introduction of talent (yj2016-4). National Natural Science Foundation of China (No. 31671589). 2017 National Undergraduate Training Programs for Innovation and Entrepreneurship (No. 201710364041).

References

1. Larios, N., Deng, H., Zhang, W., et al.: Automated insect identification through concatenated histograms of local appearance features: feature vector generation and region detection for deformable objects. Mach. Vis. Appl. **19**(2), 105–123 (2008)
2. Zhu, L.Q., Zhang, Z.: Auto-classification of insect images based on color histogram and GLCM. In: 2010 Seventh International Conference on Fuzzy Systems and Knowledge Discovery (FSKD), vol. 6, pp. 2589–2593. IEEE (2010)
3. Wang, J., Lin, C., Ji, L., et al.: A new automatic identification system of insect images at the order level. Knowl.-Based Syst. **33**, 102–110 (2012)
4. Faithpraise, F., Birch, P., Young, R., et al.: Automatic plant pest detection and recognition using k-means clustering algorithm and correspondence filters. Int. J. Adv. Biotechnol. Res. **4**(2), 189–199 (2013)
5. Xia, C., Chon, T.S., Ren, Z., et al.: Automatic identification and counting of small size pests in greenhouse conditions with low computational cost. Ecol. Inform. **29**, 139–146 (2015)
6. Wang, X.-F., Huang, D.S., Xu, H.: An efficient local Chan-Vese model for image segmentation. Pattern Recogn. **43**(3), 603–618 (2010)
7. Xie, C., Zhang, J., Li, R., et al.: Automatic classification for field crop insects via multiple-task sparse representation and multiple-kernel learning. Comput. Electron. Agric. **119**, 123–132 (2015)
8. Maharlooei, M., Sivarajan, S., Bajwa, S.G., et al.: Detection of soybean aphids in a greenhouse using an image processing technique. Comput. Electron. Agric. **132**, 63–70 (2017)

9. Hubel, D.H., Wiesel, T.N.: Republication of the Journal of Physiology (1959) 148, 574–591: receptive fields of single neurones in the cat's striate cortex. 1959. J. Physiol. **587**(12), 2721–2732 (2009)

10. Fukushima, K.: Neocognitron: a self-organizing neural network model for a mechanism of pattern recognition unaffected by shift in position. Biol. Cybern. **36**(4), 193–202 (1980)

11. LeCun, Y., Boser, B., Denker, J., et al.: Backpropagation applied to handwritten zip code recognition. Neural Comput. **1**(4), 541–551 (1989)

12. LeCun, Y., Bottou, L., Bengio, Y., et al.: Gradient-based learning applied to document recognition. Proc. IEEE **86**(11), 2278–2324 (1998)

13. Huang, D.S., Du, J.-X.: A constructive hybrid structure optimization methodology for radial basis probabilistic neural networks. IEEE Trans. Neural Netw. **19**(12), 2099–2115 (2008)

14. Zhao, Z.-Q., Huang, D.S., Sun, B.-Y.: Human face recognition based on multiple features using neural networks committee. Pattern Recogn. Lett. **25**(12), 1351–1358 (2004)

15. Coates, A., Baumstarck, P., Le, Q., et al.: Scalable learning for object detection with GPU hardware. In: IEEE/RSJ International Conference on Intelligent Robots and Systems, pp. 4287–4293. IEEE Press (2009)

16. Krizhevsky, A., Sutskever, I., Hinton, G.E.: ImageNet classification with deep convolutional neural networks. In: Advances in Neural Information Processing Systems, vol. 25, no. 2 (2012)

17. Simonyan, K., Zisserman, A.: Very deep convolutional networks for large-scale image recognition (2014). arXiv preprint arXiv:1409.1556

18. Szegedy, C., Liu, W., Jia, Y., et al.: Going deeper with convolutions. In: Proceedings of the IEEE Conference on Computer Vision and Pattern Recognition, pp. 1–9 (2015)

19. He, K., Zhang, X., Ren, S., et al.: Deep residual learning for image recognition. In: Proceedings of the IEEE Conference on Computer Vision and Pattern Recognition, pp. 770–778 (2016)

20. Girshick, R., Donahue, J., Darrell, T., et al.: Rich feature hierarchies for accurate object detection and semantic segmentation. In: Proceedings of the IEEE Conference on Computer Vision and Pattern Recognition, pp. 580–587 (2014)

21. He, K., Zhang, X., Ren, S., et al.: Spatial pyramid pooling in deep convolutional networks for visual recognition. IEEE Trans. Pattern Anal. Mach. Intell. **37**(9), 1904–1916 (2014)

22. Girshick, R.: Fast R-CNN. In: Proceedings of the IEEE International Conference on Computer Vision, pp. 1440–1448 (2015)

23. Ren, S., He, K., Girshick, R., et al.: Faster R-CNN: towards real-time object detection with region proposal networks. In: Advances in Neural Information Processing Systems, pp. 91–99 (2015)

24. Sun, X., Wu, P., Hoi, S.C.H.: Face detection using deep learning: an improved faster R-CNN approach (2017). arXiv preprint arXiv:1701.08289

25. Wang, L., Lu, Y., Wang, H., et al.: Evolving boxes for fast vehicle detection (2017). arXiv preprint arXiv:1702.00254

26. Stewart, R., Andriluka, M., Ng, A.Y.: End-to-end people detection in crowded scenes. In: Proceedings of the IEEE Conference on Computer Vision and Pattern Recognition, pp. 2325–2333 (2016)

27. Salvador, A., Giró-i-Nieto, X., Marqués, F., et al.: Faster R-CNN features for instance search. In: Proceedings of the IEEE Conference on Computer Vision and Pattern Recognition Workshops, pp. 9–16 (2016)

Accuracy Enhancement of the Viola-Jones Algorithm for Thermal Face Detection

Arwa M. Basbrain[1,2(✉)], John Q. Gan[1], and Adrian Clark[1]

[1] School of Computer Science and Electronic Engineering,
University of Essex, Colchester, UK
{amabas,jqgan,alien}@essex.ac.uk
[2] Faculty of Computing and Information Technology,
King Abdul-Aziz University, Jeddah, Kingdom of Saudi Arabia

Abstract. Face detection is the first step for many facial analysis applications and has been extensively researched in the visible spectrum. While significant progress has been made in the field of face detection in the visible spectrum, the performance of current face detection methods in the thermal infrared spectrum is far from perfect and unable to cope with real-time applications. As the Viola-Jones algorithm has become a common method of face detection, this paper aims to improve the performance of the Viola-Jones algorithm in the thermal spectrum for detecting faces with or without eyeglasses. A performance comparison has been made of three different features, HOG, LBP, and Haar-like, to find the most suitable one for face detection from thermal images. Additionally, to accelerate the detection speed, a pre-processing stage is added in both training and detecting phases. Two pre-processing methods have been tested and compared, together with the three features. It is found that the proposed process for performance enhancement gave higher detection accuracy (95%) than the Viola-Jones method (90%) and doubled the detection speed as well.

Keywords: Thermal image · Face detection · Viola-Jones algorithm

1 Introduction

Human facial analysis is an active research area due to its wide variety of potential applications, such as face recognition, emotion recognition and human-computer interaction. Recently, there has been an increased interest in facial analysis in the thermal infrared spectrum, because it combats some drawbacks of the visible spectrum and provides a higher level of liveness detection such as face temperature, emotions and health state [1, 2].

Face detection is the first step towards automatic facial analysis. Considerable work has been done on developing face detection methods in the visible spectrum, whereas face detection in the thermal spectrum has received less attention. To detect faces from thermal images, some related issues should be addressed. For example, detecting faces with eyeglasses is a challenging issue and heat-emitting objects may cause false positive face detection. Reese et al. [1] compared Viola-Jones [2] algorithm with Gabor feature extraction and classification [3] and the non-training method Projection Profile

© Springer International Publishing AG 2017
D.-S. Huang et al. (Eds.): ICIC 2017, Part III, LNAI 10363, pp. 71–82, 2017.
DOI: 10.1007/978-3-319-63315-2_7

Analysis [4] for face detection from both thermal and visible images. Reese's study concluded that the Viola-Jones algorithm achieved better performance, with average accuracy of 74.8% and average speed of 0.03 s per image, and suggested that more work could be done to improve the accuracy of the Viola-Jones algorithm for face detection from thermal infrared images. This paper proposes a thermal face detecting process which can meet the requirements of real-time applications. The proposed process significantly improves the accuracy of the Viola-Jones algorithm to detect faces from thermal images. The potential benefits of the proposed process have been investigated. Firstly, the performances of two pre-processing methods, Otsu's method [8] and Gradient Magnitude, were compared. Secondly, the performances of LBP features [5] and HOG descriptors [6] were compared with the performance of the Haar-like features that are originally utilized by the Viola-Jones algorithm. The Viola-Jones algorithm and three types of features are described in the following section. The rest of this paper is organised as follows. The proposed methods are introduced in Sect. 3. Experimental setup and databases used are described in Sect. 4. Experimental results and discussion are presented in Sect. 5. Section 6 draws conclusions.

2 Related Work

Since thermal cameras can capture face skin temperature, many algorithms for thermal facial analysis simply use thresholding to separate areas of interest from background. Cheong et al. [5], Mekyska et al. [6], and Sumathi et al. [7] convert a thermal image into a gray image and binarise it using the Otsu's method [8], followed by identifying the location of the head region using the global minimum point from the horizontal projection of the image. Trujillo et al. [9] and Wong et al. [10] proposed non-training face detection algorithms by using the head curve geometry. Zheng [4] used the projection profile analysis algorithm for thermal face detection. To separate areas of interest from the background, Zheng used region growing segmentation, which is slow and lowers the accuracy slightly. Reese et al.'s experiments [1] showed that learning based methods (Viola-Jones and Gabor) are able to detect faces from both thermal and visible images and the Viola-Jones algorithm is the best for face detection in the thermal spectrum.

2.1 Viola-Jones Method

Since Viola and Jones' face detector [2] is the most popular, the state of the art for face detection, it is adopted to detect faces from thermal images as the baseline method. Their contribution is three-fold. In order to achieve fast calculation with high accuracy, the first contribution is a simple and efficient classifier to choose a small number of critical visual features from a large set of potential features, by using the AdaBoost learning algorithm. The second is a method for combining classifiers in a cascade that allows background regions of the image to be easily discarded while spending more computation time on promising face-like regions. The last is the introduction of the "integral image", which allows Haar-like features to be computed very quickly. A brief review of Haar-like, HOG and LBP features is presented below.

2.2   Haar-like Features

The basic idea of Haar-like features is to make use of the difference between summed pixel values of rectangular image regions. A rectangle with white and gray areas is moved over the original image to calculate the difference of the sum of pixel values within the gray area and the sum of pixel values in the white area. Features from rectangles that have one white area and one gray area are called 2-rectangle features. Viola-Jones [2] defined 3-rectangle features and 4-rectangle features as shown in Fig. 1. The Haar-like features indicate certain characteristics of a particular area of the image such as the existence or absence of edges or changes in texture.

Fig. 1. *n*-rectangular features [2]. (A) and (B): $n = 2$; (C): $n = 3$; (D): $n = 4$.

2.3   Local Binary Patterns (LBP) Features

LBP was first proposed as a gray level invariant texture primitive [11]. LBP features describe each pixel by its relative gray level to its neighboring pixels. Each centre pixel is represented as a binary string, by comparing its gray value with the values of the eight neighborhood pixels. If the value of the centre pixel is greater than the neighbours' values, its value is set to zero, otherwise to one. The combination of the ones and zeros of the eight neighbours' values are represented as an 8-bit binary number, resulting in 28 distinct values for the binary pattern.

2.4   Histograms of Oriented Gradient (HOG) Descriptors

In the context of human detection, Dalal and Triggs [12] proposed image descriptors that describe the local object appearance and shape by computing the dense grid of histograms of oriented gradients. Their method divides an image into blocks of various sizes, where each block consists of a number of cells. A local 1-D centered orientation histogram of gradients is calculated from the gradient orientations of sample pixels within a cell. Depending on the values found in the gradient, each pixel within the cell casts a weighted vote into the orientation histogram. Each histogram splits the gradient angle range into a pre-defined number of bins.

3   Methods

The proposed thermal face detecting process consists of training and detecting phases, where each phase consists of three stages, as shown in Fig. 2.

In the training phase, training images are divided into two groups to create positive and negative samples. A positive image contains a single face, whereas a negative

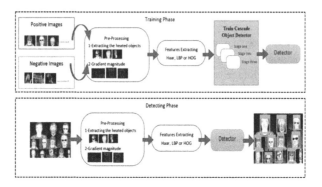

Fig. 2. Thermal face detecting process

image contains no face. In the first stage, both positive and negative samples are pre-processed by using gradient magnitude or Otsu's methods [8]. The features are extracted from the pre-processed samples in the second stage. Both positive and negative features are used to train the cascade face detectors in the final stage.

In the detecting phase, testing images are pre-processed by using the same pre-processing method as in the training phase. In the second stage, features are extracted from the testing image which is scanned by sub-windows of different sizes to find the face. In the third stage, the extracted features are evaluated per scanned sub-window by the cascade face detector. The detector rejects non-face sub-windows and detects sub-windows containing a face. If multiple sub-windows occur around each face in the scanned image, the detected sub-windows are combined to convert the overlapping detections into a single detection to return one last detection per face.

3.1   Pre-processing

Due to the loss of facial feature properties and appearance features in thermal images, we propose using the gradient magnitude method to enhance the texture and the edge boundary of the face in thermal images. Also, we suggest object extraction as a pre-processing step to increase detection accuracy.

3.1.1   Gradient Magnitude
Gradient magnitude is utilized to extract useful information from images, such as edges. The pixels with large gradient magnitude values become possible edge pixels, as shown in Fig. 3. The Gradient magnitude can also be used for feature and texture matching.

3.1.2   Object Extraction
It is essential in image processing to extract objects from their background, by selecting appropriate threshold values of gray level. Otsu [8] proposed a method for selecting an optimal threshold by utilizing the discriminant criterion to maximize the separability of the gray level classes. Due to its efficiency, most state-of-the-art methods for thermal

Fig. 3. Samples of gradient magnitude images by using different color maps

image processing commonly use the Otsu's method for pre-processing. In order to extract heated objects from their background, the global threshold value is used to convert a thermal image to a binary image. Then each pixel in the original image is multiplied by its correspondent pixel in the binary image, as shown in Fig. 4.

Fig. 4. Samples of extracted heated objects in thermal images by applying the Otsu's method

3.2 Feature Extraction and Classification

Due to its efficiency, Haar-like features [2] have been commonly used in face detection. The Viola-Jones algorithm utilises Haar-like features rather than the pixels directly. We propose using LBP features alternatively because they work well for representing fine-scale textures and it is often used to detect faces. Also, we suggest using HOG features which are useful for capturing the overall shape of an object.

The Viola-Jones algorithm uses Adaboost training to select the most effective features. However, there are still a huge number of features to be calculated. The classifiers are organised in the form of a cascade to avoid worthless calculation. As shown in Fig. 2, the cascade classifier consists of multiple stages, in which multiple simple classifiers are used. If a sub-window fails on any weak classifier, it will be excluded as it contains no face. Only the sub-windows, which have reached the final stage, would be considered as having faces.

4 Experiments

In this section, we first describe the thermal facial image databases used in the training phase. Thereafter, we examine the effect of different parameters on the face detection performance on a validation dataset. Finally, we compare the performance of different face detectors on a testing dataset.

4.1   Databases

Two thermal facial image databases, NVIE [13] and I.Vi.T.E [14], were adopted in our experiments. The Natural Visible and Infrared facial Expression (NVIE) database contains both posed and spontaneous expressions of 215 subjects (157 males and 58 females) with three different illumination directions: left, right, and front. The subjects were required to pose with seven different expressions, wearing glasses and without them. To record infrared videos, an infrared camera capturing 25 frames per second, with resolution 320 × 240 and band wave 8–14 mm, was used. The camera was placed 0.75 m in front of the subject. The naturalistic database of thermal emotion, which was named Italian Visible-Thermal Emotion (I.Vi.T.E.), consists of spontaneous expressions (one image per second in .PNG format) of 40 subjects (Italian under-graduate students aged 22 to 28 years). The thermal image resolution is 160 × 120 pixels.

In order to create negative samples, 50 thermal videos were downloaded from YouTube and converted to 13082 frames of thermal images, of which 10,000 images were used to train all detectors whilst 3082 images were used for testing.

4.2   Experiment Design

A dataset of 10,021 thermal infrared frontal face images from the NVIE database was constructed from posed images and the first frame of the spontaneous expression image sequences. The faces were manually extracted from images by placing a bounding box around each face just underneath the chin and about half-way between the hairline and the eyebrows. The Training Image Labeler app [15] in MATLAB was used for extracting the faces from images. No further alignment or resizing was done. The NVIE dataset was split into three subsets: training, validation, and testing. The training and validation subsets, consisting of 3530 thermal images each, were used to conduct two-fold cross-validation for parameter selection. The remaining 2961 thermal images from the NVIE database were used as the testing subset. To test the ability to detect more than one face in an image, a 12-In-One dataset was created from the NVIE database. It contains 180 images, which were selected from the thermal spontaneous expressions database. Each image comprises twelve randomly selected faces. Samples of the 12-In-One dataset are shown in Fig. 5.

The third dataset is the I.Vi.T.E. database. Since it contains video frames with spontaneous expressions of each subject, some frames are not suitable for testing because a large part of the face is covered or the whole face is missing as shown in

Fig. 5. Two samples of the 12-In-One dataset which are randomly selected from the NVIE database

Fig. 6. Some 3583 images were screened out. The rectangular regions of interest were specified semi-manually for 31217 images. The primary ROIs were determined by using three trained detectors; then these were combined to specify the final rectangular region. All images were manually checked and then the Training Image Labeler app [15] in MATLAB was used for misdirecting faces in 624 images.

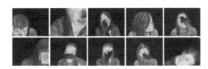

Fig. 6. Samples of separated thermal images from the I.Vi.T.E. database where the face does not fully appear

4.3   Criterion for Calculating True/False Positive Rate

Motivated by the work of Wang et al. [16], face detection rate is used to measure the effectiveness of each detector by calculating the displacement from the automatically located rectangular of the target face to the true (manually annotated) rectangles, defined as the overlap ratio between them. To compute the ratio, the area of intersection between rectangle A and rectangle B is divided by the area of the union of the two, as shown in Fig. 7. The value of the overlap ratio is between 0 and 1, where 1 implies a perfect overlap and 0 implies no overlap. For true positive cases, the minimum acceptance value of the overlap ratio was set to 0.5, whereas the cases are regarded as false positives if the overlap ratio is less than 0.5.

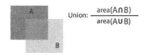

Fig. 7. The overlap ratio between rectangle A and rectangle B

5   Results and Discussion

In order to demonstrate the performance of the proposed methods, Haar-like features were compared with LBP and HOG features for face detection. Before starting the comparisons, the holdout cross-validation was adopted to find the optimal values for the parameters of each pre-processing and feature selection method. There is a trade-off between the training parameters: the number of cascade stages, the false positive and true positive rates are required to be set at each stage. In order to find optimal values for these parameters, several detectors were trained on different combinations of them. The detectors with the highest mean accuracy on the validation datasets were selected among the others for final testing.

Figure 8a–c give the ROC curves comparing the performances of nine detectors tested on the NVIE testing dataset. To create the ROC curve, the threshold value for merging detected boxes around a face was adjusted from 7 to 1. This value was used to perform the merging operation, where there are multiple detections around a face. The detections were grouped and then merged to produce one bounding box around the face if they meet the merging threshold value. When the merging threshold is set to 0, all detections were returned separately without performing a thresholding or merging operation. This will increase both detection and false positive rates while adjusting the merging threshold value to ∞ will yield a detection rate of 0 and a false positive rate of 0. To compute the true positive and false positive rates, the number of true positive detections should be divided by the total number of all faces in all the images, while for calculating false positive rate the number of false positive detections is divided by the total number of sub-windows scanned in all the images [2].

Figure 8a illustrates the Viola-Jones method's performance when using Haar-like, HOG, and LBP features respectively without any pre-processing phase, while Fig. 8b and c provide similar comparisons, but with the proposed pre-processing phase. Figure 8b and c show the performance of the three types of features when using the gradient magnitude method and Otsu's method respectively in the pre-processing phase. The results indicate that the Viola-Jones method using LBP features achieved the highest accuracy in comparison with Haar-like and HOG features when detecting faces from thermal images. In addition, using the Otsu's method in the pre-processing phase reduced the false positive rate. However, using HOG features reduced the accuracy with or without pre-processing.

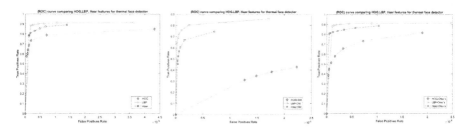

Fig. 8. (a) ROC curves comparing the performance of the Viola-Jones algorithm on thermal images when using different features (HOG, LBP, and Haar-like), without pre-processing. (b) ROC curves comparing the performance of the Viola-Jones algorithm on thermal images when using different features and applying Gradient Magnitude (GM) method for pre-processing. (c) ROC curves comparing the performance of the Viola-Jones algorithm on thermal images when using different features and applying Otsu's method for pre-processing.

Figure 8a and c show that there is little difference in the accuracy of the Viola-Jones method when using LBP features with or without a pre-processing phase. However, the pre-processing phase (Otsu's method) speeds up the detection time to be almost twice as fast, because the speed of the detector is related to the number of features evaluated per scanned sub-window. By applying the Otsu's method in thermal images, most of the features vanish in non-heat-emitting regions. Thus, the first stage of the detection process discards a vast majority of the sub-windows, so that they are not evaluated in subsequent stages. This increases the detection speed and reduces the false positive rate at the same time.

The McNemar's statistical test [17] was used to determine the significance of the results. The null hypothesis assumes that there is no statistical difference in performance of the Viola-Jones method when using LBP or Haar-like features. Based on the calculation of McNemar's statistic Eq. (1), the null hypothesis can be rejected with an error probability of 0.05 if $|Z| > 1.96$, which indicates that the differences in performance are statistically significant. N_{sf} is the number of occurrences when the first algorithm succeeds and the second algorithm fails and N_{fs} is the opposite. If $N_{sf} + N_{fs} > 20$, the statistic is reliable and Z can be converted into a probability using tables [17].

$$Z = \frac{|N_{sf} - N_{fs}| - 1}{\sqrt{N_{sf} + N_{fs}}} \tag{1}$$

The Z-value and the related parameters $(N_{ss}, N_{sf}, N_{fs}, N_{ff})$ on the NVIE database as well as the I.Vi.T.E. database are shown in Tables 1 and 2 respectively. The tables are split into three parts. The first part shows the results of McNemar's test for LBP features without pre-processing phase versus HOG and Haar-like features: without pre-processing phase in the first column and with pre-processing phases (Gradient Magnitude-GM, Otsu methods) in the second and third columns respectively. Like the first part, the second and third parts show the results of McNemar's test for LBP features with pre-processing phases (GM and Otsu) versus other features. The last column displays the comparison of McNemar's test results for the different pre-processing phases with LBP features. The Haar-like features without pre-processing phase represents the standard Viola-Jones algorithm trained to detect faces from thermal images.

According to the McNemar's test, LBP features outperformed other features in all cases except when using the Gradient Magnitude method in pre-processing phase. Since some geometrical and appearance features are lost in thermal images, HOG features have the lowest accuracy in comparison with LBP and Haar-like features in face detection from thermal images. Also, the differences between the gray values of pixels in the same heated area, such as the face, are quite small in the thermal image. Each LBP feature represents the exact differences of each pixel with the eight neighborhood pixels whereas each Haar-like feature represents differences of the value of white and gray rectangles which combine more than one pixel. Therefore, LBP features give a more accurate representation of the heated areas in the thermal image than Haar-like features. The performance of the Viola-Jones algorithm using LBP features is significantly better than using Haar-like features in thermal images. The Z-value of LBP vs. LBP-Otsu's in Table 1 (0.20) indicates that there is no significant difference between the two algorithms on the NVIE database, while the Z-value (10.18) of LBP vs. LBP-Otsu's in Table 2 shows that the difference between them is significant on the I.Vi.T.E. database.

Table 1. Comparison of LBP features with HOG and Haar-like features on the NVIE

LBP VS	HOG	Haar-like	HOG GM	Haar-like GM	HOG Otsu's	Haar-like Otsu's	LBP-GM
N_{ss}	2531	2636	2189	2210	2436	2621	2545
N_{sf}	200	95	542	521	295	110	186
N_{fs}	4	13	17	2	2	7	22
N_{ff}	226	217	213	228	228	223	208
Z-value	13.65	7.79	22.16	22.65	16.94	9.42	11.30
LBP-GM VS	HOG	Haar-like	HOG GM	Haar-like GM	HOG Otsu's	Haar-like Otsu's	LBP Otsu's
N_{ss}	2445	2527	2114	2193	2377	2464	2545
N_{sf}	122	40	453	374	190	103	22
N_{fs}	90	122	92	19	61	164	188
N_{ff}	304	272	302	375	333	230	206
Z-value	2.13	6.36	15.42	17.86	8.08	3.67	11.39
LBP-Otsu's VS	HOG	Haar-like	HOG GM	Haar-like GM	HOG Otsu's	Haar-like Otsu's	LBP
N_{ss}	2532	2639	2191	2208	2437	2623	2719
N_{sf}	201	94	542	525	296	110	14
N_{fs}	3	10	15	4	1	5	12
N_{ff}	225	218	213	224	227	223	216
Z-value	13.79	8.14	22.29	22.61	17.06	9.700	0.20

Table 2. Comparison of LBP features with HOG and Haar-like features on the I.Vi.T.E.

LBP VS	HOG	Haar-like	HOG GM	Haar-like GM	HOG Otsu's	Haar-like Otsu's	LBP-GM
N_{ss}	28306	27488	24244	23762	26023	26177	25509
N_{sf}	1567	2385	5629	6111	3850	3696	4364
N_{fs}	484	579	649	388	378	444	400
N_{ff}	860	765	695	956	966	900	944
Z-value	23.89	33.15	62.84	70.98	53.38	50.53	57.42
LBP-GM VS	HOG	Haar-like	HOG GM	Haar-like GM	HOG Otsu's	Haar-like Otsu's	LBP Otsu's
N_{ss}	25180	24146	21203	22260	23595	23630	25422
N_{sf}	729	1763	4706	3649	2314	2279	487
N_{fs}	3610	3921	3690	1890	2806	2991	4075
N_{ff}	1698	1387	1618	3418	2502	2317	1233
Z-value	43.72	28.61	11.08	23.62	6.86	9.79	53.11
LBP-Otsu's VS	HOG	Haar-like	HOG GM	Haar-like GM	HOG-Otsu's	Haar-like Otsu's	LBP
N_{ss}	28023	27204	23958	23682	25873	25999	29007
N_{sf}	1474	2293	5539	5815	3624	3498	490
N_{fs}	767	863	935	468	528	622	866
N_{ff}	953	857	785	1252	1192	1098	854
Z-value	14.91	25.44	57.21	67.44	48.03	44.79	10.18

Table 3 shows the speed of the detectors. It can be seen that the LBP-Otsu's detector has the highest speed in comparison with other detectors; it can process an image in about 0.0088 s odn a 2.70 GHz Intel® i7 processor.

Table 3. Comparisons of LBP, HOG and Haar-like features with the two pre-processing methods on NVIE, I.Vi.T.E, and 12-In-One databases.

	NVIE database		I.Vi.T.E. database		12-In-One dataset		Speed
	TPR	FPR	TPR	FPR	TPR	FPR	S/Image
HOG	0.86	4.17E-06	0.92	4.42E-05	0.90	4.98E-07	0.0283
LBP	0.92	3.50E-06	0.96	7.59E-05	0.97	1.07E-06	0.0132
Haar-like	0.89	1.31E-06	0.90	9.13E-05	0.93	1.09E-06	0.0190
HOG-GM	0.75	6.64E-05	0.80	1.05E-03	0.70	2.55E-05	0.0313
LBP-GM	0.87	1.64E-05	0.83	1.94E-04	0.93	2.84E-06	0.0093
Haar-like-GM	0.75	6.87E-06	0.77	1.20E-04	0.87	8.26E-07	0.0101
HOG-Otsu's	0.82	1.80E-06	0.85	4.87E-05	0.80	7.39E-08	0.0271
LBP-Otsu's	0.92	2.34E-06	0.94	8.34E-05	0.98	5.50E-07	0.0088
Haar-like-Otsu's	0.89	9.90E-07	0.85	4.53E-05	0.92	3.39E-07	0.0096

6 Conclusions

This paper presents an efficient and effective process to improve the performance of the Viola-Jones algorithm for face detection from thermal images, with or without eyeglasses. To test whether the difference in performance between the proposed process and the standard Viola-Jones algorithm is statistically significant, the McNemar's test was employed. The results of the test demonstrate that the LBP features outperformed other features significantly in most cases and applying Otsu's method in the pre-processing phase reduced the false positive rate in face detection from thermal images. The proposed enhancement process reduced the detection time of the Viola-Jones algorithm by roughly a factor of two while retaining high detection accuracy.

References

1. Reese, K., Zheng, Y., Elmaghraby, A.: A comparison of face detection algorithms in visible and thermal spectrums. In: International Conference on Advances in Computer Science and Application (2012)
2. Viola, P., Jones, M.: Robust real-time face detection. Int. J. Comput. Vis. **57**(2), 137–154 (2004)
3. Gupta, B.S.G., Tiwari, A.: Face detection using gabor feature extraction and artificial neural networks. In: Proceedings of ISCET, pp. 18–23 (2010)
4. Zheng, Y.: Face detection and eyeglasses detection for thermal face recognition. In: Proceedings of SPIE 8300 (2012)
5. Cheong, Y.K., Yap, V.V., Nisar, H.: A novel face detection algorithm using thermal imaging. In: IEEE Symposium on Computer Applications and Industrial Electronics (ISCAIE) (2014)
6. Mekyska, J., Espinosa-Duro, V., Faundez-Zanuy, M.: Face segmentation: a comparison between visible and thermal images. In: IEEE International Carnahan Conference on Security Technology (ICCST) (2010)

7. Sumathi, C.P., Santhanam, T., Mahadevi, M.: Automatic facial expression analysis: a survey. Int. J. Comput. Sci. Eng. Surv. **3**(6), 47–59 (2012)
8. Otsu, N.: A threshold selection method from gray-level histograms. IEEE Trans. Syst. Man Cybern. **9**(1), 62–66 (1979)
9. Trujillo, L., et al.: Automatic feature localization in thermal images for facial expression recognition. In: IEEE Computer Society Conference on Computer Vision and Pattern Recognition (2005)
10. Wong, W.K., et al.: Face detection in thermal imaging using head curve geometry. In: The 5th International Congress on Image and Signal Processing (CISP) (2012)
11. Ojala, T., Pietikainen, M., Maenpaa, T.: Multiresolution gray-scale and rotation invariant texture classification with local binary patterns. IEEE Trans. Pattern Anal. Mach. Intell. **24** (7), 971–987 (2002)
12. Dalal, N., Triggs, B.: Histograms of oriented gradients for human detection. In: IEEE Computer Society Conference on Computer Vision and Pattern Recognition (2005)
13. Wang, S., et al.: A natural visible and infrared facial expression database for expression recognition and emotion inference. IEEE Trans. Multimed. **12**(7), 682–691 (2010)
14. Esposito, A., Capuano, V., Mekyska, J., Faundez-Zanuy, M.: A naturalistic database of thermal emotional facial expressions and effects of induced emotions on memory. In: Esposito, A., Esposito, A.M., Vinciarelli, A., Hoffmann, R., Müller, V.C. (eds.) Cognitive Behavioural Systems. LNCS, vol. 7403, pp. 158–173. Springer, Heidelberg (2012). doi:10. 1007/978-3-642-34584-5_12
15. Training Image Labeler (2014). http://uk.mathworks.com/help/vision/ref/ trainingimagelabeler-app.html
16. Wang, S., et al.: Eye localization from thermal infrared images. Pattern Recogn. **46**(10), 2613–2621 (2013)
17. Kanwal, N., Bostanci, E., Clark, A.F.: Evaluation method, dataset size or dataset content: how to evaluate algorithms for image matching? J. Math. Imaging Vis. **55**(3), 378–400 (2016)

Leaf Categorization Methods for Plant Identification

Asdrúbal López-Chau[1]([⊠]), Rafael Rojas-Hernández[1],
Farid García Lamont[2], Valentín Trujillo-Mora[1],
Lisbeth Rodriguez-Mazahua[3], and Jair Cervantes[2]

[1] Universidad Autónoma del Estado de México,
Centro Universitario UAEM Zumpango, Zumpango, Mexico
{alchau, rrojashe}@uaemex.mx
[2] Universidad Autónoma del Estado de México,
Centro Universitario UAEM Texcoco, Texcoco, Mexico
[3] Division of Research and Postgraduate Studies,
Instituto Tecnológico de Orizaba, Orizaba, Mexico

Abstract. In most of classic plant identification methods a dichotomous or multi-access key is used to compare characteristics of leaves. Some questions about if the analyzed leaves are lobed, unlobed, simple or compound need to be answered to identify plants successfully. However, very little attention has been paid to make an automatic distinction of leaves using such features. In this paper we first explore if incorporating prior knowledge about leaves (categorizing between lobed simple leaves, and the unlobed simple ones) has an effect on the performance of six classification methods. According to the results of experiments with more than 1,900 images of leaves from Flavia data set, we found that it is statically significant the relationship between such categorization and the improvement of the performances of the classifiers tested. Therefore, we propose two novel methods to automatically differentiate between lobed simple leaves, and the unlobed simple ones. The proposals are invariant to rotation, and achieve correct prediction rates greater than 98%.

Keywords: Leaf features · McNemar test · Plant identification

1 Introduction

Plant identification is a challenging issue which has aroused researchers' attention in recent years. Classic plant identification methods are based on observing specific features of leaves to categorize leaves. However, in the literature very little attention has been paid to make an automatic distinction between different types of leaves to improve plant identification. This difference is important because most of the classic methods for plant identification use dichotomous or multi-access keys that compare characteristics of the leaves, asking if they are lobed, unlobed, simple or compound, among others features.

In this paper, we first analyze if the relationship between the knowledge about the type of leaf (unlobed simple or lobed simple) and the classification accuracy of

© Springer International Publishing AG 2017
D.-S. Huang et al. (Eds.): ICIC 2017, Part III, LNAI 10363, pp. 83–94, 2017.
DOI: 10.1007/978-3-319-63315-2_8

classification methods is significant statistically. As a first approach to explore the relationship, we carried out the following experiment: we extract basic leaf features, and use them with standard classification methods. Then, we add the type of leaf as a feature and test again the same methods. In both cases, the classification accuracies were measured, and then compared applying the McNemar test. According to the results of the experiments, using the type of leave as a binary feature has a positive impact on the performance of the methods tested, such impact is statistically significant. To be fair, only basic features were extracted from leaves because we are interested in observing the effect of another basic binary leaf feature. Therefore, we propose two novel methods to categorize leaves. The first method presented uses concentric circles to detect the changes of color. The second method uses convex hulls. These methods do not vary neither to scale nor to rotation.

The rest of the paper is organized as follows. Section 2 describes basic leaf features for plant identification and gives a brief review about main works related to extraction of leaf features. Section 3 shows the results of the exploration on the effect of incorporate previous knowledge about the type of leaf to classification methods for plant identification. We present two methods to identify lobed simple leaves in Sect. 4, then Sect. 5 shows experiments and results. Finally, last section of this paper presents conclusions and future works.

2 Related Works

Leaf features are extracted from images previously processed. Then, leaf features are encoded as a set of numbers (vectors) or nominal values, also known as feature descriptors.

Most of leaf features can be categorized into the following six main types [1]. *Geometric*: defined as sets of points that form points, lines, etc.; *Morphological*: related to form and structure of a leaf; *Texture*: these descriptors characterize image textures or regions; *Color*: based on RGB image and its 3 channels; *Shape*: contour of leaves has to be taken into account to describe the structure. Current work is mostly focused on this type of descriptors; *Vein network*: leaf veins are analyzed to extract specific characteristics; *Others*: image descriptors borrowed from computer vision to describe leaves, such as Fourier descriptors, SIFT, and border detectors or filters, for example Gabor. Shape is the most popular feature in literature on plant identification [2], among the six types of descriptors explained.

In [3], authors propose "shape-defining feature" (SDF), by using slopes and distances between two consecutive points. The shape of a leaf along with its fine serrations is retrieved using this method. In order to compute SDF, they draw a total of 400 lines (vertical and horizontal) over the image of a leaf, and then detect the endpoints of these lines. The larger the number of lines, finer is the detail of serrations. For the classification authors use a Neural Network along with AR, CH, Ec and Roundness. A drawback of the method presented in [3] is that the number of features is large (800 features per leaf), compared with the number of images per leaf in data sets.

Shape context (SC) descriptor, proposed in 2000 by Belongie and Malik [4], is used to compute shape correspondences and similarities between two images. Based on SC,

Zhi et al. [5] proposed "Arc Length Shape Contexts" (ARC-SC). This descriptor is composed of two parts: the sum of Euclidean distances between adjacent points, and the angle between two pixels on the silhouette of leaf. Minimum cost of matches between all ARC-SC of a leaf and the extracted from training set is computed to identify a plant.

Other descriptor that uses leaf shape (specifically, points on the border of the leaf) is Multi scale Distance Matrix (MDM), introduced by Hu et al. [6]. The first step to build MDM is to create a symmetric matrix D in R^{nxn}, whose entry $d_{i,j}$ is the distance between points x_i and x_j, both on the border of the leaf. Then, dimensionality reduction is applied, retaining only unrepeated elements in D.

MDM descriptor is invariant to rotation, scaling and translation; however, to apply MDM, the shape of leaf must be stable, i.e., without noise. Different from the previous methods, Gwo et al. [7] do not use all points on the border, but retain only few ones, compared to other methods. The selection of feature points is realized by comparing distances.

Several methods use distances from a reference points to the border of leaf. Hajjdiab and Al Maskari [8] use the centroid of image as reference. They take 32 points chosen circularly, at equally spaced angles. Shen et al. [9] compute a centroid considering only the points located on the border of the leaf. Then, they subsample the border, obtaining 36 points. This type of methods require the detection of the silhouette of the leaf from clean images.

Kala et al. [10] use the border of leaf in a different way to other works. They compute a sinuosity measure, which expresses the meandering of a curve. An issue of the sinuosity measure is that it requires the silhouette of the leaf to be differentiable and this measure is not rotation invariant.

Texture of leaves has also been used to identify plants. In [11, 12], authors combined shape and texture of a leaf to identify plants. For shape analysis, Beghin et al. [11] extract the contour signature from leaf, and then compute the dissimilarities between all leaves in data set using the Jeffrey-divergence measure. Meanwhile, Chaki et al. [12] apply curvelet transform coefficients together with invariant moments. The method for texture feature extraction presented in [11] uses Sobel directions histogram. Chaki et al. [12] use Gabor filter (GF) and gray level co-occurrence matrix (GLCM).

In [13], authors propose a combination of morphological and geometric features of a leaf. They remove irrelevant features using a fuzzy surface selection method. Few remaining features are used with a Neural Network. Classification is performed quickly by using this simple scheme. However, extraction of features is computationally costly. That method was tested with only four species of plants, all of them have simple leaves.

One of the least used features for plant identification is color. Most of methods for the same purpose work with binary images. de M. Sá Junior et al. [14] use a gravitational approach, which produces success rate above 90%; however, their method requires a manual selection of texture windows and orientation of leaf. A general problem with color features, is that many factors have to be taken into account, for example, illumination conditions, maturity of plant, diseases and environment [14, 15].

3   Studying the Effect of Adding Prior Knowledge About the Type of Leaf on Classification

To explore if the prior knowledge (type of leaf) has an effect of the performance of classification methods, we executed the statistical test of McNemar. It tests consistency in responses across two variables. McNemar test recognizes that some instances will move from incorrectly predicted to correctly predicted and others from correctly predicted to incorrectly predicted just randomly. If the prior knowledge is having no effect on performance of a classification method, the number of instances which move from incorrectly predicted to correctly predicted should be about equal to those who move in the other direction.

Six basic leaf features were extracted from images leaves. We first built and tested six different classifiers with these characteristics of leaves. Then, we manually added a binary leaf feature, assigning a value of true for lobed simple leaves with smooth margins, and a value of false for the rest of the leaves. The six classifiers were trained and tested again. The classification accuracies were measured in both cases. The number of instances correctly/incorrectly predicted before and after adding the binary leaf features were counted to create the contingency tables.

3.1   Materials

One of the most widely used data set for testing plant identification systems is Flavia. It is publicly available at http://flavia.sourceforge.net, this set contains 1,907 color images of 32 different species of plants. These images have a dimension of 1,600 × 1,200 pixels.

In general, leaves can be classified according to their blade (simple or compound), edge (smooth, dentate, etc.), petiole (petiolated or sessile), shape of blade, etc. Among these categories, simple, compound, unlobed and lobed are very common in dichotomous keys. For simple leaves, the leaf blade is a single, continuous unit. For compound leaves the blade is divided into two or more leaflets arising from the petiole. Figure 1 shows an example of simple leaf and compound leaf. In this case, it is really easy to categorize these leaves. However, in many other cases this categorization it is really complicated. This is because there are many subtypes of leaves. For example, simple leaves can be unlobed or lobed. For unlobed leaves, the blade is completely undivided. Lobed leaves have projections off the midrib with individual inside veins. Figure 2 shows two examples of simple leaves which are very different from the simple leaf presented in Fig. 1.

Fig. 1. Example of a simple leaf (left), and a compound leaf (right).

Fig. 2. Examples of lobed simple leaf (left) and simple leaf (right).

3.2   First Experiment

We manually identified the type of leaf and added a label (binary attribute) called *Lobed* to each leaf. The value of this attribute was set to *true* for the images of leaves of classes C6, C8, C28 and C32 (lobed simple leaves with smooth margins). For the rest of the leaves the value of the attribute was set to *false* (unlobed simple leaves). This attribute is to explore if using the type of leaf can improve the performance of classification methods for automatic plant identification.

As we said before, only basic features were selected, because we are interested in observing the effect of a basic binary leaf feature (Lobed). We use six basic leaf features to extract from each one of the preprocessed images. The features extracted to each leaf are the following:

1. Aspect Ratio (**AR**): The ratio of length of the major axis to that of the minor axis of the leaf.
2. Area convex hull (**A$_{CH}$**): The area of the smallest convex set that contains a leaf.
3. Leaf area (**A**): The number of pixels forming the leaf.
4. Diameter (**D**): The longest distance between any two points in the leaf.
5. Area convexity (**A$_C$**): The ratio of the leaf area to convex area.
6. Leaf perimeter (**P**): The number of pixels at the margin of the leaf.

The classification methods tested were the following: Decision tree C4.5, k-nearest neighbors (KNN), Random forest, Multiclass classifier, Neural network (NN), Naive Bayes (NB) and Random Tree.

We created two data sets, *Data set 1* and *Data set 2*, which contain the features extracted from images of leaves. The first data set does not contain information about the type of leaf, whereas the second does contain this information (prior knowledge). Table 1 shows a summary of attributes in each data set.

We use 10-fold cross validation in the experiments. Table 2 summarizes the classification accuracy achieved by each classification method. The best performances are in bold. We observed that performances of classifiers are lower than those reported in the literature. However, in our experiments we only considered six basic leaf

Table 1. Attributes in each data set for experiment 1.

Dataset	AR	A$_{CH}$	A	D	A$_C$	P	Lobed
Data set 1	Yes	Yes	Yes	Yes	Yes	Yes	No
Data set 2	Yes	Yes	Yes	Yes	Yes	Yes	Yes

Table 2. Effect prior knowledge (type of leaf as a binary attribute) on six classification methods.

Classification method	Classification accuracy (%) Data set 1	Classification accuracy (%) Data set 2	p-value
Decision tree C4.5	58.78	**60.15**	0.0305
KNN (k = 1)	64.13	**64.24**	0.0478
Random forest	65.44	**66.60**	0.0216
Multiclass classifier	68.38	**71.95**	0.0025
Naïve Bayes	55.06	**56.58**	0.0296
Random tree	56.63	**57.63**	0.0380

features. This number is lesser and simpler than the used in many other works [12]. Our goal is to compare basic leaf features with the type of leaf, as we consider it a basic leaf feature too.

In order to validate if the improvement in the performance of classifiers is statistically significant, we apply the McNemar test. The p-values achieved are shown in the last column of Table 2. Although the improvement of performances is slight, the p-values values suggest that the difference of frequencies observed in instances correctly classified before and after adding the attribute is not due to randomness. Based on these results, in next Section we propose two methods to detect lobed simple leaves.

4 Proposed Methods to Categorize Leaves

4.1 Method Based on Concentric Circles

The first method that we propose in this paper utilizes concentric circles. By using a preprocessed binary image L of a leaf, we detect changes of color (black to white) along a curve that crosses the image of a leaf. These changes are produced by the leaf or by its leaflets.

First, we create a number of concentric circles over the leaf. Figure 3 shows how to compute the center and the radio of those circles. Then, we count color changes on the trajectory of each circle. To avoid counting noisy pixels, we only consider it a color change when there are at least K pixels of the same color once color variation has been rendered. We empirically determined that a value of K = 10 works for most cases.

As a result, we obtained a vector V with A components (A is the number of concentric circles). Each component of V comprises the number of changes of color (from black to white) minus one.

Figure 4 shows an example of two leaves with eight concentric circles over them. One of the leaves is unlobed simple and the other one is lobed simple. For the first leaf V = [0 0 0 1 1 1 1 1], for the second leaf V = [0 0 0 0 3 3 3 5]. Value 0 means that corresponding circle only touches the foreground (black), and never crosses through the background (white), i.e. there are not changes of color along the contour of the circle. The greater the value of a component V, the greater the number of times the contour of the corresponding circle detects changes from black to white.

Input: L: a binary image of a leaf; A: Number of
     concentric cicles
Output: Center and radios of A concentric circles
1 Compute the center of mass $C = (c_x, c_y)$ of L using;

$$c_x = \frac{1}{N} \sum_{i,j} x_i \ s.t. \ I(x_i, y_j) \in L$$

$$c_y = \frac{1}{N} \sum_{i,j} y_i \ s.t. \ I(x_i, y_j) \in L.$$

N is the area of L (see Table 1).

2 Compute the convex hull CH of L;
3 Compute the average Euclidean distance d_{CHav}: the
   mean of all distances from C to the vertexes of CH;
4 The radius of greatest circle is d_{CHav}. The radius of
   the smallest circle is d_{CHav}/A. The rest of
   concentric circles (A-2) are equally spaced between
   maximum and minimum radios;
5 **Return** Center and radios;

Fig. 3. Pseudocode of algorithm to generate concentric circles on a leaf.

In order to discriminate lobed simple leaves from unlobed simple leaves, we add
the number of components of V with a value greater than two, this sum is named S.
Value two is based on the observation that lobed leaves have, in general, at least three
leaflets, which produce three color changes. Then the criterion shown in Eq. (1) is
applied:

$$Lobed = \begin{cases} \textit{true} \text{ if } S \geq 2 \\ \textit{false} \text{ otherwise} \end{cases} \tag{1}$$

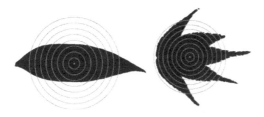

Fig. 4. Concentric circles generated on unlobed simple leaf (left), and lobed simple leaf (right).

The center of mass $C(c_x, c_y)$, convex hull CH and the average distance d_{CHav} are all of them invariant to the orientation of the image. Any circle with radius R centered in C will pass exactly through the same points even if the leaf is rotated, therefore the algorithm proposed in this subsection is invariant to rotation.

4.2   Method Based on Convex Hulls

According to the results presented in Sect. 3.2, incorporation of prior knowledge about the type of leaf (as a feature) improves the performance of classification methods for plant identification. Therefore, a second method to identify lobed simple and unlobed simple leaves was developed based on the concept of convex hull.

Given a binary image L of a leaf, our second method computes the difference between the convex hull of L and the binary image L. Thereafter, the connected components in the resulting image are identified, and the area (number of pixels) of each component is computed. The underlying idea is that the area of the connected components in lobed simple leaves is greater than the areas of the connected components in unlobed simple leaves; an example of this is shown in Fig. 5. The procedure that implements this part of our method is presented in Fig. 6.

Fig. 5. Examples of the difference (white areas) of convex hull and leaf.

Input: Image of a leaf L
Output: Areas of connected components
1 $A_L \leftarrow$ Obtain area of L
2 $C \leftarrow$ extract the contour of L
3 $CH \leftarrow$ Compute the convex hull of C
4 $A_{CH} \leftarrow$ Obtain area of CH
5 $R_L \leftarrow$ Compute $A_{CH} - A_L$
6 $C_c \leftarrow$ Detect connected components of R
7 $H \leftarrow \emptyset$
8 **for** $c \in C_c$ **do**
9     $\lfloor$ $H \leftarrow H \cup$ Area of c
10 **return** H

Fig. 6. Pseudocode of algorithm to compute residual areas.

In general, the set of areas computed by applying the algorithm shown in Fig. 6 has different cardinality for each image. Therefore, we only retain the greatest ten areas. This produces a numeric vector in $\mathbb{R}^{10}$, which is used to train a Random Forest Classifier. The class of each instance is the label that we manually set for lobed leaves, and that is explained in Subsect. 3.2.

In next Section, we present the results of experiments to measure the performance of our two methods.

5 Evaluation of the Proposed Methods

We measured the performance of our proposals to identify lobed simple leaves. Both methods were tested with images of Flavia data set. In all our experiments, we did not rotate or scale any image.

Because in the literature there are not features specifically designed to identify lobed leaves, we do not compare the obtained results with others methods. Instead, we measure accuracy, specificity and sensitivity of the two introduced methods.

Henceforth, the method based on circles will be referred as M_{Circ}, whereas the method based on convex hull will be referred as $M_{ConvexH}$.

In order to measure the performance of M_{Circ} and $M_{ConvexH}$, we use the whole Flavia data set. The confusion matrices obtained are presented in Tables 3 and 4. The positive cases correspond to lobed simple leaves, whereas the negative cases are the unlobed simple ones. Based on these matrices, the following measures are obtained:

Table 3. Confusion matrix for the method based on circles M_{Circ}.

Real class	Prediction	
	Lobed = false	Lobed = true
Lobed = false	1,629	64
Lobed = true	6	208

Table 4. Confusion matrix for the method based on convex hulls $M_{ConvexH}$.

Real class	Prediction	
	Lobed = false	Lobed = true
Lobed = false	1,672	21
Lobed = true	0	214

- **Accuracy**: the proportion of the total number of predictions (positive and negative) that were correct.
- **Sensitivity or Recall**: the proportion of actual lobed simple leaves which are correctly identified.
- **Specificity**: the proportion of actual unlobed simple leaves which are correctly identified.

It can be observed in Tables 3 and 4 that most of the prediction errors are committed in actual unlobed simple leaves, which are incorrectly identified as lobed ones. The method M_{Circ} produces some errors in the identification of lobed simple leaves, whereas $M_{ConvexH}$ does not commit this error in this type of leaves.

Comparing the performances of both methods (Table 5), it is possible to claim proposal with the highest performance is $M_{ConvexH}$. One more time, we tested the performances of the six classification methods using the outcomes of $M_{ConvexH}$. Table 6 shows the classification accuracies obtained. The performances are quite similar – although slightly lower - to those show in Table 2. This is could be due to the method $M_{ConvexH}$ does not identify the 100% of leaves correctly.

Table 5. Performances of proposed methods.

Real class	Accuracy (%)	Recall	Specificity
M_{Circ}	96.33	0.9720	0.9963
$M_{ConvexH}$	98.89	1.000	0.9876

Table 6. Performance of classification method using $M_{ConvexH}$

Classification method	Classification accuracy (%) Data set 2
Decision tree C4.5	**60.01**
KNN (k = 1)	**64.14**
Random forest	**66.61**
Multiclass classifier	**71.87**
Naïve Bayes	**56.45**
Random tree	**57.61**

It can be seen in Tables 2 and 6 that the method with best performance is Multiclass classifier. It transforms a multiclass problem into several two-class ones, each one of these problems is solved with logistic regression. The second best method is Random forest, which uses a number of decision trees to solve the multiclass problem. These two classification methods are more suitable for plant identification with our methods.

6   Conclusions and Future Work

Many classic plant identification methods use dichotomous keys that take into account specific features of leaves, such as lobed, unlobed, simple or compound. However, state-of-the-art methods are not oriented to detect these leaf features. In this paper, we firstly explore if adding the type of leaf (distinguishing between lobed simple leaves and unlobed simple ones) as a basic binary feature can improve the performance of six classification methods. We found that incorporating this previous knowledge is beneficial for classifier, although the improvement is slight.

Motivated by the results obtained, we designed two new methods to discriminate automatically between unlobed simple and lobed simple leaves. The first method detects changes from black-to-white (and vice versa) in binary images. The second method uses the differences of areas between convex hull and the leaf.

Both methods were tested with color images from Flavia data set. The correct prediction rate is above 96% for the method based on circles, and greater than 98% for the method based on convex hull. These methods are invariant to rotation of images.

Adding the prior knowledge about the type of the leave for creating a complete plant identification system is out of the scope of the proposals presented in this paper; however, preliminary results of experiments with Flavia data set have shown our methods can help to improve the performance of such type of systems, although at this moment the improvement achieved is not statistically significant yet. This is due to our methods only discriminate between unlobed simple and lobed simple leaves, the proportion between these types of leaves is 8:1 in Flavia data set. The number of lobed leaves is very small compared to the number of unlobed leaves, Currently, we are working in an improved version of our methods, to discriminate between more types of leaves.

References

1. Pahikkala, T., Kari, K., Mattila, H.: Classification of plant species from images of overlapping leaves. Comput. Electron. Agric. **118**, 186–192 (2015). doi:10.1016/j.compag. 2015.09.003
2. Jamil, N., Hussin, N.A.C., Nordin, S., Awang, K.: Automatic plant identification: is shape the key feature? Procedia Comput. Sci. **76**, 436–442 (2015). doi:10.1016/j.procs.2015.12. 287
3. Aakif, A., Khan, M.F.: Automatic classification of plants based on their leaves. Biosyst. Eng. **139**, 66–75 (2015). doi:10.1016/j.biosystemseng.2015.08.003
4. Belongie, S., Malik, J.: Matching with shape contexts. In: IEEE Work on Proceedings of the Content-Based Access Image Video Libraries, pp. 20–26 (2000)
5. Zhi, Z.-D., Hu, R.-X., Wang, X.-F.: A new weighted ARC-SC approach for leaf image recognition. In: Huang, D.-S., Ma, J., Jo, K.-H., Gromiha, M.M. (eds.) ICIC 2012. LNCS, vol. 7390, pp. 503–509. Springer, Heidelberg (2012). doi:10.1007/978-3-642-31576-3_64
6. Hu, R., Jia, W., Ling, H., Huang, D.: Multiscale distance matrix for fast plant leaf recognition. IEEE Trans. Image Process. **21**, 4667–4672 (2012)
7. Gwo, C.Y., Wei, C.H., Li, Y.: Rotary matching of edge features for leaf recognition. Comput. Electron. Agric. **91**, 124–134 (2013). doi:10.1016/j.compag.2012.12.005
8. Hajjdiab, H., Al Maskari, I.: Plant species recognition using leaf contours. In: 2011 IEEE International Conference on Imaging Systems and Techniques (IST), pp. 306–309 (2011)
9. Shen, Y., Zhou, C., Lin, K.: Leaf image retrieval using a shape based method. In: Li, D., Wang, B. (eds.) AIAI 2005. ITIFIP, vol. 187, pp. 711–719. Springer, Boston (2005). doi:10. 1007/0-387-29295-0_77
10. Kala, J.R., Viriri, S., Moodley, D.: Sinuosity coefficients for leaf shape characterisation. In: Pillay, N., Engelbrecht, A.P., Abraham, A., du Plessis, M.C., Snášel, V., Muda, A.K. (eds.) Advances in Nature and Biologically Inspired Computing. AISC, vol. 419, pp. 141–150. Springer, Cham (2016). doi:10.1007/978-3-319-27400-3_13

11. Beghin, T., Cope, J.S., Remagnino, P., Barman, S.: Shape and texture based plant leaf classification. In: Blanc-Talon, J., Bone, D., Philips, W., Popescu, D., Scheunders, P. (eds.) ACIVS 2010. LNCS, vol. 6475, pp. 345–353. Springer, Heidelberg (2010). doi:10.1007/978-3-642-17691-3_32

12. Chaki, J., Parekh, R., Bhattacharya, S.: Plant leaf recognition using texture and shape features with neural classifiers. Pattern Recogn. Lett. **58**, 61–68 (2015). doi:10.1016/j.patrec.2015.02.010

13. Tzionas, P., Papadakis, S.E., Manolakis, D.: Plant leaves classification based on morphological features and a fuzzy surface selection technique. In: Fifth International Conference on Technology and Automation, Thessaloniki, Greece, pp. 365–370 (2005)

14. de M. Sá Junior, J.J., Backes, A.R., Cortez, P.C.: Plant leaf classification using color on a gravitational approach. In: Wilson, R., Hancock, E., Bors, A., Smith, W. (eds.) CAIP 2013. LNCS, vol. 8048, pp. 258–265. Springer, Heidelberg (2013). doi:10.1007/978-3-642-40246-3_32

15. McCarthy, C.L., Hancock, N.H., Raine, S.R.: Applied machine vision of plants: a review with implications for field deployment in automated farming operations. Intell. Serv. Robot. **3**, 209–217 (2010). doi:10.1007/s11370-010-0075-2

An Integrated Learning Framework for Pedestrian Tracking

Taihong Xiao and Jinwen Ma[(✉)]

Department of Information Science,
School of Mathematical Sciences and LMAM, Peking University,
Beijing 100871, China
xiaotaihong@pku.edu.cn, jwma@math.pku.edu.cn

Abstract. Pedestrian tracking has been arguably addressed as a special topic beyond general object tracking. Although many learning or data driven object trackers as well as recent deep learning object trackers have shown excellent performance for general object tracking, they have limited success on pedestrian tracking because there exist three major challenges emerging from pedestrian tracking such as vast variations of human bodies, distraction from similar persons and complete occlusion. In this paper, we propose an integrated learning framework for pedestrian tracking to overcome these problems. It is demonstrated by the experimental results on the SVD-B dataset that our proposed framework can achieve competitive results in comparison with state-of-the-art object trackers under the evaluation of the precision and success rate as well as fps.

Keywords: Pedestrian tracking · Deep re-id feature · Online learning · R-CNN · Switching detection · Precision and success plots

1 Introduction

Pedestrian tracking is a special kind of object tracking task. Although there are lots of general object tracking algorithms which can achieve good performance on different object tracking such as TLD [1], KCF [2], SAMF [3], Staple [4], Struck [5], C-COT [6], MDNet [7], FCNT [8], CF2 [9] and SCT [10], they have limited success in pedestrian tracking due to three challenges. First, the appearance, pose and gestures can change very greatly in the same video. For example, a person can be walking on the road, riding a bicycle, holding a book, etc. Actually, it is rather difficult when a person who is facing towards the camera at this moment turns around with his back to the camera after several frames. Second, similar persons can lead to a distraction in real-life surveillance videos. As we know from general object tracking datasets such as VOT2013 and VOT2015, it is quite easy to discriminate the single target object from the background, because there is no other object of the same kind in videos. However, there may be many similar people appearing in the real-life surveillance camera, which may render distraction to general object trackers. Third, as a more complicated situation, complete occlusion occurs more frequently in real-life surveillance videos than in common benchmark datasets.

All these challenges motivate us to come up with a new framework to address the above challenges in pedestrian tracking. In this paper, we present an *Integrated*

© Springer International Publishing AG 2017
D.-S. Huang et al. (Eds.): ICIC 2017, Part III, LNAI 10363, pp. 95–106, 2017.
DOI: 10.1007/978-3-319-63315-2_9

Learning Framework for Pedestrian Tracking (ILFPT) that overcomes the three difficult problems: vast variation of human bodies, distraction from similar persons and complete occlusion, being fatal to most general object trackers. Our main contributions are three folds: (1) we propose an effective framework specific to pedestrian tracking problem, where many general object tracking algorithms do not perform well; (2) we demonstrate that the deep re-id features could be used in pedestrian tracking problems. This is unprecedented to our knowledge, since no publications had investigated on using re-id features for tracking problems; (3) we establish a mechanism of switching detection frames and non-detection frames, which is a simple yet powerful way towards high runtime efficiency and better performance. Remarkably, the experiments on SVD-B dataset demonstrate that ILFPT achieves competitive results in comparison with state-of-the-art general object trackers on the precision and success rate for evaluation. Meanwhile, our method has a test-time speed of 16.429 frame per second (fps), which is faster than most deep-leaning approaches and conventional learning approaches.

The rest of the paper is organized as follows: Sect. 2 outlines the related work. Section 3 describes our ILFPT model. Section 4 demonstrates the experimental results in comparison with state-of-the-art trackers and further investigates the effectiveness of the deep re-id feature and our sampling scheme. Section 5 draws our conclusions.

2   Related Work

2.1   Visual Object Tracking

As a fundamental problem in computer vision, visual object tracking has been studied for decades. Great progress has been made in the past few years [8, 11]. Most general object tracking algorithms fall into either generative or discriminative approaches. Generative methods describe the target appearance using certain generative model and search for the target regions that fit the model best. Various generative target appearance modeling algorithms have been established such as sparse representation [12], density estimation [13], and incremental subspace learning [14]. In contrast, discriminative methods aim to build a model that distinguishes the target objects from the background. Typically, they learn and update the classifiers online, such as multiple instance learning [15], P-N learning [1], online boosting [16], structured output SVMs [5], etc. As deep neural networks (DNNs) have recently shown their outstanding feature representation ability in a wild range of vision tasks such as image classification, object detection and image segmentation, many deep learning based tracking algorithms come into our eyes and achieve better performance in general object tracking tasks, such as MDNet [7], FCNT [8], CF2 [9] and SCT [10], etc.

2.2   Person Re-identification

Closely related to detection and tracking, person re-identification aims to establish a correspondence between some images of a person taken from different cameras. A typical re-id system has two basic functions: capturing the unique feature descriptor of the person and then comparing the descriptors of two different appearances to infer either a match or a non-match. The difficulties lie in the appearance changes across different cameras or even within frames of the same camera. Because we are lack of spatio-temporal relationships between different persons, it is hard to compare person descriptors captured in different locations, time instants, and over different durations. Most of the current approaches fall into two either contextual methods or non-contextual methods. Contextual methods rely on external contextual information either for pruning the correspondences or extracting the features for re-id, such as, topology of cameras [17], camera calibration [18]. Non-contextual methods including SDALF [19], CPS [20], RDC [21] and RankSVM [22] focus more on designing visual descriptors to characterize the appearance of one person.

3   The ILPFT Model

An overview of our proposed Integrated Learning Framework for Pedestrian Tracking (ILFPT) model is shown in Fig. 1. All the details are introduced in the following subsections.

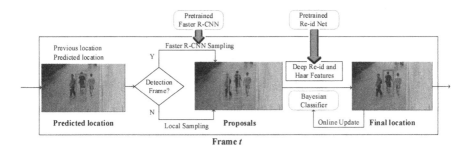

Fig. 1. The pipeline of the ILFPT model.

3.1   Offline Training

To detect the pedestrians in each frame, we employ Faster R-CNN [23] as our detector. In fact, Faster R-CNN consists of two components: a fully convolutional Region Proposal Network (RPN) for locating candidate regions and a downstream Fast R-CNN [24] classifier. Faster R-CNN achieves the state-of-the-art object detection accuracy on PASCAL VOC 2007, 2012, and MS COCO datasets. So we retrain the network by feeding with pedestrian training data and transform it into a pedestrian detector.

Another network that needs to be offline trained is the re-id network. There are many famous pedestrian re-id datasets that can be used for training a re-id network such as CUHK01 [25], CUHK03 [26], PRID [27] and VIPeR [28]. However, the quality, resolution and sample distribution vary greatly among these datasets. Xiao et al. [29] proposed a domain guided dropout method for training the re-id network across different domains simultaneously so solve this problem. We adopt this re-id network and extract the output following fc7 layer as the deep re-id feature.

3.2 Sampling

Since the quality of samples is crucial to our online learning and tracking process, we employ a strategy to refine our sampling. Specifically, we make a prediction about the location and speed of the target before we select samples. Assume that we are tracking on frame t, the speed of the target in the current frame can be estimated by

$$v_x^t = \rho v_x^t + (1 - \rho)\Delta_x^{t-1}, v_y^t = \rho v_y^{t-1} + (1-\rho)\Delta_y^{t-1} \tag{1}$$

where v_x^{t-1} and v_x^{t-1} are respectively the horizontal and vertical speed in frame $t - 1$, $\rho \in [0, 1]$ is the momentum factor that controls the weights of the previous speed and $\Delta_x^{t-1} = x^{t-1} - x^{t-2}, \Delta_y^{t-1} = y^{t-1} - y^{t-2}$. Therefore, the center of tracking window in frame t can be predicted as

$$x^t = x^{t-1} + v_x^t, \ y^t = y^{t-1} + v_y^t \tag{2}$$

As long as we obtain a coarse estimation about the target position, we can select samples more accurately. To facilitate the online learning process, we propose two kinds of sampling methods to select candidate bounding boxes over the current frame.

1. **Local sampling generates** samples locally. It takes samples within a fixed search size based on our prediction about the target location (x^t, y^t) computed by (2). In frame t, we select bounding boxes whose centers are no more than in-radius distance away from the predicted target center as positive samples, that is, $\left\{ (x,y) \middle| \sqrt{(x-x^t)^2 + (y-y^t)^2} \leq \text{inrad} \right\}$. As for negative samples, the distances of their centers from the predicted bounding box are greater than in-radius and no more than out-radius, that is, $\{(x, y) \mid \text{inrad} < \sqrt{(x-x^t)^2 + (y-y^t)^2} \leq \text{outrad}\}$.

2. **Faster R-CNN sampling** generates pedestrian samples $\mathcal{F} = \{F_i\}$ globally in the whole image using the pretrained faster R-CNN detection network. And then we select those samples whose centers are no more than an inrad radius distance away from the predicted location, i.e. $\left\{ (x,y) \in \mathcal{F} \middle| \sqrt{(x-x^t)^2 + (y-y^t)^2} \leq \text{inrad} \right\}$.

To better explain our switching mechanism, we first introduce the concepts of detection frames and non-detection frames. Detection frames consist of periodical ones and feed-back ones. As shown in Fig. 2, we set the period parameter $\tau = 4$ in advance, then one periodical frame comes in every 4 frames. On the other hand, a frame is defined as a feed-back frame if no sample is generated or selected by our sampling methods in its previous frame. In this case, no candidate is sampled in the 4-th and 6-th

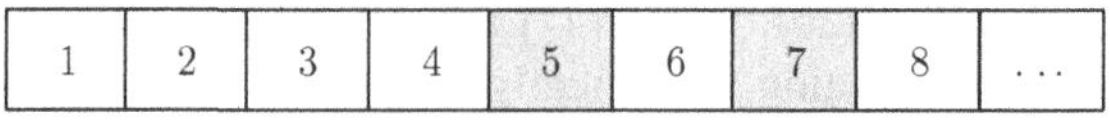

| 1 | 2 | 3 | 4 | 5 | 6 | 7 | 8 | ... |

Fig. 2. A series of frames represented as boxes in which the blue and yellow frames are respectively periodical and feed-back frames, while the white boxes are non-detection frames. (Color figure online)

frames, therefore, the 5-th and 7-th frames become feed-back frames. Oppositely, the rest frames are called non-detection frames.

Faster R-CNN sampling is activated only when a detection frame comes. Therefore, the detection network periodically validates our predictions on these periodical frames. Besides, when certain complete occlusion occurs, the next frame becomes a feed-back frame because there is no appropriate candidate. Then, the detection network helps search for the target until the target is found or reappears again. In this way, the complete occlusion can be effectively addressed.

Faster R-CNN sampling plays an important role in adapting the bounding box size to the target. The pedestrian detection network is pretrained offline, thus capable of selecting appropriate bounding boxes with high confidence. By combining the two sampling methods together, we can adjust the size of bounding box dynamically so as to improve the accuracy.

Moreover, the switching scheme helps improve the tracking speed. Faster R-CNN sampling relies on the deep neural network, and thus takes more time than local sampling inevitably. So, we organically combine two sampling methods together for the balance between quality and time. It is demonstrated to be efficient in our experiments (See Table 1).

Table 1. The comparative results among different trackers on the average **precision scores** and **success scores** with different attributes and **fps (speed)** values, where the best and second best results are denoted in red and *green* colors, respectively.

Trackers	Precision Scores					Success Scores					FPS
	SV	OCC	BC	DEF	IPR	SV	OCC	BC	DEF	IPR	
KCF	0.278	0.293	0.376	0.257	0.330	0.402	0.461	0.453	0.461	0.447	**116.610**
SAMF	0.568	0.465	0.507	*0.443*	*0.528*	0.597	0.576	0.542	0.617	0.583	4.170
Staple	**0.670**	0.508	0.573	0.424	0.525	**0.687**	0.606	0.592	0.633	0.647	9.307
Struck	0.239	0.292	0.353	0.256	0.275	0.413	0.478	0.469	0.488	0.458	1.524
TLD	0.275	0.188	0.270	0.208	0.233	0.363	0.287	0.337	0.328	0.300	10.201
C-COT	0.601	0.529	0.628	0.415	0.463	0.670	0.619	0.599	*0.684*	*0.652*	0.547
MDNet	*0.606*	*0.601*	**0.690**	0.435	0.496	*0.683*	**0.681**	**0.659**	0.676	**0.664**	0.664
FCNT	0.380	0.318	0.358	0.222	0.307	0.438	0.438	0.443	0.385	0.422	1.954
CF2	0.282	0.324	0.386	0.283	0.342	0.407	0.486	0.479	0.484	0.467	1.206
SCT	0.241	0.271	0.352	0.242	0.280	0.375	0.457	0.433	0.455	0.432	11.579
Ours	0.588	**0.645**	*0.675*	**0.615**	**0.605**	0.622	*0.673*	*0.658*	**0.685**	0.647	*16.429*

3.3   Online Learning Model

A positive sample set $\mathcal{S}^+$ and a negative sample set $\mathcal{S}^-$ are initialized for storing new pedestrian patterns and updating online model. For each candidate sample c, we obtain a compressed low-dimensional feature $\mathbf{v} = (v_1, v_2, \ldots, v_n)^{\mathrm{T}} \in \mathbb{R}^n$. Assuming that all the elements in $\mathbf{v}$ are independently distributed and the prior probability is equally distributed, i.e. $p(y = 1) = p(y = 0)$, we have

$$R(\mathbf{v}) = \log \frac{p(y=1|\mathbf{v})}{p(y=0|\mathbf{v})} = \log \frac{p(\mathbf{v}|y=1)p(y=1)}{p(\mathbf{v}|y=0)p(y=0)}$$

$$= \log \frac{\prod_{i=1}^{n} p(v_i|y=1)p(y=1)}{\prod_{i=1}^{n} p(v_i|y=0)p(y=0)} = \sum_{i=1}^{n} \log \frac{p(v_i|y=1)}{p(v_i|y=0)} \tag{3}$$

The conditional probabilities $p(v_i|y=1)$ and $p(v_i|y=0)$ in the classifier with $R(\mathbf{v})$ are assumed to be Gaussian distributed with four parameters $\left(\mu_i^1, \sigma_i^1, \mu_i^0, \sigma_i^0\right)$, that is,

$$p(v_i|y=1) = \mathcal{N}\left(\mu_i^1, \sigma_i^1\right), p(v_i|y=0) = \mathcal{N}\left(\mu_i^0, \sigma_i^0\right) \tag{4}$$

According to the principle of maximal likelihood estimation, we use the following update rules for parameters:

$$\sigma_i^j \leftarrow \sqrt{\lambda\left(\sigma_i^j\right)^2 + (1-\lambda)\left(\tilde{\sigma}_i^j\right)^2 + \lambda(1-\lambda)\left(\mu_i^j - \tilde{\mu}_i^j\right)^2}$$

$$\mu_i^j \leftarrow \lambda\mu_i^j + (1-\lambda)\tilde{\mu}_i^j \tag{5}$$

$$(i = 1, 2, \ldots, n; j = 0, 1)$$

where $\lambda \in (0, 1)$ is the inertial factor that controls the updating speed and

$$\tilde{\mu}_i^j = \frac{1}{|S_j|}\sum_{k \in S_j} v_i(k), \quad \tilde{\sigma}_i^j = \sqrt{\frac{1}{|S_j|}\sum_{k \in S_j}\left(v_i(k) - \tilde{\mu}_i^j\right)^2} \tag{6}$$

where $v_i(k)$ denotes the i-th component of feature of k-th sample in the storage sample set. It should be noted that in (6) $\tilde{\mu}_i^j$ and $\tilde{\sigma}_i^j$ are computed on all the samples in the corresponding storage sample set, i.e., $S_1 = \mathcal{S}^+$ or $S_0 = \mathcal{S}^-$. So, these parameters get renewed each time as the store samples get updated.

We employ two pairs of classifiers respectively for two kinds of feature, since dimensions of different features are not the same. In detection frames, we use Faster R-CNN sampling to get candidates and then compute $R(\mathbf{v})$ using the deep re-id feature. Whereas in other frames, we use local sampling to get candidates and then compute $R(\mathbf{v})$ using the Haar feature.

4 Experimental Results

4.1 Dataset and Setup

Training and test data come from SVD-B dataset[1] where all videos are collected from 14 cameras in a university campus. The training data consists of 14 videos, each of which lasts 15 min. Only 22362 frames of them are annotated with one or more

---

[1] The source code and dataset are available at http://github.com/prinsphield/ILFPT.

pedestrians. Besides, there are 20 video sequences provided for test. All the image frames have a resolution of 1280×720.

Our proposed pedestrian tracking framework is implemented in MATLAB based on the wrapper of Caffe framework [30]. All the experiments in this paper are done on a server with 8 2.4 GHz CPUs and a GTX 1080 GPU. The well-trained re-id net is directly employed for extracting the deep re-id feature. The storage capacity of S^+ and S^- is set to 200. The period parameter τ is 10. The search window size is set to be 50. The learning rate λ and inertia factor ρ are set to be 0.86 and 0.79, respectively. The classification threshold θ is set to be -60.

4.2 Evaluation Methodology

As for the performance criteria, we choose the mean center location error (CLE) and Pascal VOC overlap ratio (VOR) [31]. CLE reflects the difference between the center of tracked result and the ground truth, while VOR is a measure of coincidence defined as $\text{VOR} = \frac{\text{Area}(G \cap T)}{\text{Area}(G \cup T)}$, where G is the ground truth bounding box and T is the tracking bounding box. In the CLE method, a frame is considered correctly tracked if the center distance between the predicted target and the ground truth is below a threshold. In fact, a precision curve can clearly show the percentage of correctly tracked frames for a range of all the thresholds. For comparison, all the comparative trackers can be ranked according to the precision scores at a typical threshold of 20 pixels. In the VOR method, a frame is considered successfully tracked if the VOR is greater than a threshold. A success curve can show the percentage of successfully tracked frames for all the thresholds ranging from 0 to 1. The area under the curve (AUC) of each success plot can be used to rank these comparative tracking algorithms. All the evaluation results are based on one pass evaluation (OPE) [31].

4.3 Comparison with State-of-the-Art Trackers

As shown in Fig. 3, we plot both precision and success curves of our proposed pedestrian tracker as well as some top rank trackers including non-deep-learning trackers, TLD [1], KCF [2], SAMF [3], Staple [4], Struck [5], C-COT [6] and deep-learning trackers, MDNet [7], FCNT [8], CF2 [9] and SCT [10]. Our proposed pedestrian tracker achieves outstanding performance compared with the other state-of-the-art trackers in terms of average precision score, success score and fps.

To facilitate better analysis on the tracking performance, we further evaluate performance of all the trackers with respect to 5 attributes: scale variation (SV), occlusion (OCC), background cluttered (BC), deformation (DEF) and in-plane rotation (IPR). Table 1 demonstrates that our proposed pedestrian tracker can well handle a variety of challenging factors and consistently outperform state-of-the-art methods in some attributes. Figure 4 demonstrates that our methods can efficiently deal with the vast variations of human bodies, distraction from similar persons and complete occlusion. Moreover, our pedestrian tracker achieves higher runtime efficiency compared with other methods especially involved with DNNs, though using deep neural networks.

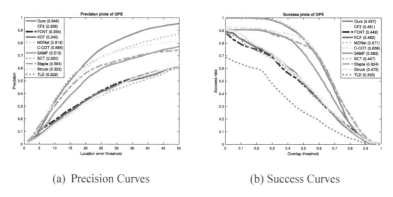

(a) Precision Curves   (b) Success Curves

Fig. 3. The precision and success plots of the ILFPT and comparative methods on SVD-B in which the denoted numbers for the methods are the corresponding precision scores at the error threshold of 20 pixels in (a) and AUC values in (b), respectively.

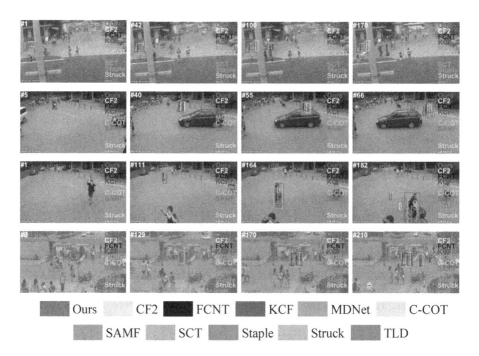

Fig. 4. The screenshots of tracking results on some challenging sequences (video01, video02, video03, video19). Frame number is displayed at top left corner. Our algorithm can address the problem of complete occlusion (video01), pose changes (video02), scale variation (video03), distractions from similar person and objects (video19).

4.4   Effectiveness of Deep Re-id Feature

To further investigate the effectiveness of the deep re-id feature on pedestrian tracking, we present the performances of different variants of the proposed approach by substituting the deep re-id feature with the other features (Fig. 5). We find out that the combination of the Haar and deep re-id features achieves the best performance. The Haar and Color features are computationally efficient, but lack in representing the high-level information about pedestrian. VGG16 network [32] is now wildly used for all kinds of visual tasks. But the speed would become unbearable if we directly use its final output as our feature. So, we extract the output just after pool3 layer in VGG16 network, and then reduce its dimension into 256, 1024 and 4096 via random projection as our VGG16-256, VGG16-1024 and VGG16-4096 features respectively. The overall performance of VGG16 feature is inferior to our deep re-id feature, since deep re-id feature catches the difference among persons while VGG16 feature captures the difference among general objects. This further demonstrates that our proposed re-id feature is effective in pedestrian tracking.

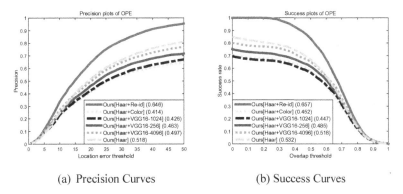

(a) Precision Curves   (b) Success Curves

Fig. 5. The precision and success plots with different features in which the denoted numbers for the features are the corresponding precision scores at the error threshold of 20 pixels in (a) and AUC values in (b), respectively.

4.5   The Effectiveness of the Sampling Scheme

We finally discuss the mechanism of switching detection frames and non-detection frames. The Faster R-CNN sampling and deep re-id feature are bundled while local sampling and Haar feature are bundled. They are two branches in our pipeline. We explore the effectiveness of two sampling methods by changing the period parameter τ. We plot the precision scores (@20 pixels) and success scores at different τ (See Fig. 6).

When $\tau = 1$, the Faster R-CNN sampling is employed in each frame, whereas the local sampling is not used. This point reflects the tracking ability of offline pretrained detector. When τ is large enough, the plot point reflects almost the tracking ability of Haar feature in our online learning model, since the pretrained detection network is rarely used. From Fig. 6, the best case is when $\tau = 10$, since two classifiers are

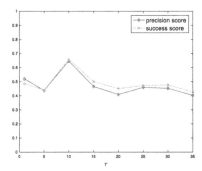

Fig. 6. The precision and success scores at different τ.

effectively combined. The parameter τ is independent of the speed of the moving pedestrian, because our framework works in all test videos, including some videos where the target pedestrian is riding a bike, which is faster than walking pedestrians. So we can conclude that our switching scheme is effective.

5   Conclusion

We have proposed an integrated framework for pedestrian tracking. Our framework learns pedestrian representation by offline pretraining and captures the target-specific information through certain efficient online learning during tracking. It actually achieves outstanding performance on SVD-B video datasets in comparison with some state-of-the-art tracking algorithms. The combination of Haar and deep re-id feature is experimentally proved to be better than the other commonly used features in the pedestrian tracking task. Our sampling scheme is demonstrated to be effective in both performance and runtime efficiency.

Acknowledgments. This work was supported by the Natural Science Foundation of China for Grant 61171138.

References

1. Kalal, Z., Mikolajczyk, K., Matas, J.: Tracking-learning-detection. IEEE Trans. Pattern Anal. Mach. Intell. **34**(7), 1409–1422 (2012)
2. Henriques, J.F., Caseiro, R., Martins, P., Batista, J.: High-speed tracking with kernelized correlation filters. IEEE Trans. Pattern Anal. Mach. Intell. **37**(3), 583–596 (2015)
3. Li, Y., Zhu, J.: A scale adaptive kernel correlation filter tracker with feature integration. In: Agapito, L., Bronstein, M.M., Rother, C. (eds.) ECCV 2014. LNCS, vol. 8926, pp. 254–265. Springer, Cham (2015). doi:10.1007/978-3-319-16181-5_18
4. Bertinetto, L., Valmadre, J., Golodetz, S., Miksik, O., Toor, P.H.: Staple: complementary learners for real-time tracking. In: Proceedings of the IEEE Conference on Computer Vision and Pattern Recognition, pp. 1401–1409 (2016)

5. Hare, S., Saffari, A., Torr, P.H.: Struck: structured output tracking with kernels. In: International Conference on Computer Vision, pp. 263–270. IEEE (2011)
6. Danelljan, M., Robinson, A., Shahbaz Khan, F., Felsberg, M.: Beyond correlation filters: learning continuous convolution operators for visual tracking. In: Leibe, B., Matas, J., Sebe, N., Welling, M. (eds.) ECCV 2016. LNCS, vol. 9909, pp. 472–488. Springer, Cham (2016). doi:10.1007/978-3-319-46454-1_29
7. Nam, H., Han, B.: Learning multi-domain convolutional neural networks for visual tracking, arXiv preprint arXiv:1510.07945 (2015)
8. Wang, L., Ouyang, W., Wang, X., Lu, H.: Visual tracking with fully convolutional networks. In: International Conference on Computer Vision (2015)
9. Ma, C., Huang, J., Yang, X., Yang, M.: Hierarchical convolutional features for visual tracking. In: Proceedings of the IEEE International Conference on Computer Vision, pp. 3074–3082 (2015)
10. Choi, J., Jin Chang, H., Jeong, J., Demiris, Y., Young Choi, J.: Visual tracking using attention-modulated disintegration and integration. In: Proceedings of the IEEE Conference on Computer Vision and Pattern Recognition, pp. 4321–4330 (2016)
11. Gao, J., Ling, H., Hu, W., Xing, J.: Transfer learning based visual tracking with gaussian processes regression. In: Fleet, D., Pajdla, T., Schiele, B., Tuytelaars, T. (eds.) ECCV 2014. LNCS, vol. 8691, pp. 188–203. Springer, Cham (2014). doi:10.1007/978-3-319-10578-9_13
12. Mei, X., Ling, H.: Robust visual tracking using ℓ_1 minimization. In: International Conference on Computer Vision, pp. 1436–1443. IEEE (2009)
13. Han, B., Comaniciu, D., Zhu, Y., Davis, L.S.: Sequential kernel density approximation and its application to real-time visual tracking. IEEE Trans. Pattern Anal. Mach. Intell. **30**(7), 1186–1197 (2008)
14. Ross, D.A., Lim, J., Lin, R.S., Yang, M.H.: Incremental learning for robust visual tracking. Int. J. Comput. Vis. **77**(1–3), 125–141 (2008)
15. Babenko, B., Yang, M.H., Belongie, S.: Robust object tracking with online multiple instance learning. IEEE Trans. Pattern Anal. Mach. Intell. **33**(8), 1619–1632 (2011)
16. Grabner, H., Grabner, M., Bischof, H.: Real-time tracking via on-line boosting. In: British Machine Vision Conference, vol. 1, p. 6 (2006)
17. Loy, C.C., Xiang, T., Gong, S.: Time-delayed correlation analysis for multi-camera activity understanding. Int. J. Comput. Vis. **90**(1), 106–129 (2010)
18. Lantagne, M., Parizeau, M., Bergevin, R.: VIP: vision tool for comparing images of people. In: Vision Interface, vol. 2 (2003)
19. Bazzani, L., Cristani, M., Murino, V.: Symmetry-driven accumulation of local features for human characterization and re-identification. Comput. Vis. Image Underst. **117**(2), 130–144 (2013)
20. Cheng, D.S., Cristani, M., Stoppa, M., Bazzani, M., Murino, V.: Custom pictorial structures for re-identification. In: British Machine Vision Conference, vol. 1, p. 6 (2011)
21. Zheng, W., Gong, S., Xiang, T.: Reidentification by relative distance comparison. IEEE Trans. Pattern Anal. Mach. Intell. **35**(3), 653–668 (2013)
22. Prosser, B., Zheng, W., Gong, S., Xiang, T., Mary, Q.: Person re-identification by support vector ranking. In: British Machine Vision Conference, vol. 2, p. 6 (2010)
23. Ren, S., He, K., Girshick, R., Sun, J.: Faster R-CNN: towards real-time object detection with region proposal networks. In: Advances in Neural Information Processing Systems, pp. 91–99 (2015)
24. Girshick, R.: Fast R-CNN. In: Proceedings of the IEEE International Conference on Computer Vision, pp. 1440–1448 (2015)
25. Li, W., Wang, X.: Locally aligned feature transforms across views. In: Computer Vision and Pattern Recognition, pp. 3594–3601 (2013)

26. Li, W., Zhao, R., Xiao, T., Wang, X.: DeepReID: deep filter pairing neural network for person re-identification. In: Computer Vision and Pattern Recognition, pp. 152–159 (2014)

27. Hirzer, M., Beleznai, C., Roth, P.M., Bischof, H.: Person re-identification by descriptive and discriminative classification. In: Heyden, A., Kahl, F. (eds.) SCIA 2011. LNCS, vol. 6688, pp. 91–102. Springer, Heidelberg (2011). doi:10.1007/978-3-642-21227-7_9

28. Gray, D., Brennan, S., Tao, H.: Evaluating appearance models for recognition, reacquisition, and tracking. In: Proceedings of IEEE International Workshop on Performance Evaluation for Tracking and Surveillance (PETS), vol. 3. Citeseer (2007)

29. Xiao, T., Li, H., Ouyang, W., Wang, X.: Learning deep feature representations with domain guided dropout for person re-identification. In: Computer Vision and Pattern Recognition (2016)

30. Jia, Y., Shelhamer, E., Donahue, J., Karayev, S., Long, J., Girshick, R., Guadarrama, S., Darrell, T.: Caffe: convolutional architecture for fast feature embedding. In: Proceedings of the 22nd ACM International Conference on Multimedia, pp. 675–678. ACM (2014)

31. Wu, Y., Lim, J., Yang, M.H.: Online object tracking: a benchmark. In: Computer Vision and Pattern Recognition, pp. 2411–2418 (2013)

32. Simonyan, K., Zisserman, A.: Very deep convolutional networks for large-scale image recognition (2014). arXiv preprint arXiv:1409.1556

Lumbar Spine Discs Labeling Using Axial View MRI Based on the Pixels Coordinate and Gray Level Features

Ala S. Al Kafri[1(✉)], Sud Sudirman[1], Abir J. Hussain[1], Paul Fergus[1],
Dhiya Al-Jumeily[1], Hiba Al Smadi[1], Mohammed Khalaf[1],
Mohammed Al-Jumaily[2], Wasfi Al-Rashdan[3],
Mohammad Bashtawi[3], and Jamila Mustafina[4]

[1] Faculty of Engineering and Technology, Liverpool John Moores University,
Byrom Street, Liverpool L3 3AF, UK
a.s.alkafri@2015.ljmu.ac.uk, {s.sudirman,a.hussain,
p.fergus,d.aljumeily}@ljmu.ac.uk,
H.B.Alsmadi@2016.ljmu.ac.uk,
m.i.khalaf@2014.ljmu.ac.uk
[2] Dr. Sulaiman Al Habib Hospital, Dubai Healthcare City, Dubai, UAE
maljumaily@yahoo.fr
[3] Irbid Speciality Hospital, Irbid, Jordan
drwasfil@hotmail.com, mohdbakhiet@yahoo.com
[4] Kazan Federal University, 18 Kremlyovskaya str,
Kazan 420008, Russian Federation
DNMustafina@kpfu.ru

Abstract. Disc herniation is a major reason for lower back pain (LBP), a health issue that affects a very high proportion of the UK population and is costing the UK government over £1.3 million per day in health care cost. Currently, the process to diagnose the cause of LBP involves examining a large number of Magnetic Resonance Images (MRI) but this process is both expensive in terms time and effort. Automatic labeling of lumbar disc pixels in the MRI to detect the herniation area will reduce the time to diagnose and detect the cause of LBP by the physicians. In this paper, we present a method for automatic labeling of the lumbar spine disc pixels in axial view MRI using pixels locations and gray level as features. Clinical MRIs are used for the training and testing of the method. The pixel classification accuracy and the quality of the reconstructed disc images are used as the main performance indicators for our method. Our experiments show that high level of classification accuracy of 91.1% and 98.9% can be achieved using Weighted KNN and Fine Gaussian SVM classifiers respectively.

Keywords: LBP · MRI · Lumbar spine disc · Disc herniation

1 Introduction

Low back is pain considered as the second most popular illness after the common cold. More than half of the world population were affected by the lower back pain once in their lives [1]. In the UK, the figure is higher with about sixty to eighty percent of its

© Springer International Publishing AG 2017
D.-S. Huang et al. (Eds.): ICIC 2017, Part III, LNAI 10363, pp. 107–116, 2017.
DOI: 10.1007/978-3-319-63315-2_10

population will suffer from back pain once in their lives [2]. As a result back pain cost the UK government £1.3 million per day [3]. One of the main techniques for diagnosing the cause of lower back pain is through MRI examination by a radiologist. It has been reported that the number of MRI examination in 2014 has increased in the UK by 11.3% to reach 2.61 million in compared with the number of imaging tests in 2013 [4]. However, the number of radiologists in the UK is insufficient for the clinical demand made by the radiology services as there is an increase of 26% for the MRI requests in comparison to only 3% increase in the consultant radiology workforce [5]. The gap between the increasing rate of the radiology services compared to the number of radiologists is very wide which justifies the need of automating the diagnosing process which usually includes two steps that start with labeling the intervertebral discs area then diagnosing the disc abnormality. The focus of this paper is on labeling and localizing the disc area in the lumbar spine using the axial view MR images. In our previous work [6] we have developed a framework for detecting the disc herniation in the lumbar spine which requires labeling the disc area to be able to detect the herniation.

Currently diagnosing the lower back pain is done by a visual observation and analysis of the lumbar spine MR images and this process could take up much of a physician time and effort. Moreover, it can increase the possibility of misdiagnosis. In some other disease, a computer aided diagnosing systems (CAD) were developed to help the physician in the diagnosing process as an example of these systems, CAD system for detecting colonic polyp, CAD system for detecting breast cancer in mammography and CAD system for detecting prostate cancer using MR images [7–10]. Meanwhile, there are many works used the neural network and machine learning to improve the diagnosis process for different diseases [11, 12]. Unfortunately, this type of CAD does not exist to diagnose the back pain unlike other diseases as the CAD systems are used in the diagnosing process. At the same time, there is a pressing need for this type of application to help radiologists and orthopedists in their tasks. This makes one of the motivations of this research. Such an application will need to employ sophisticated algorithms to overcome a number of technical challenges due to the wide range of imaging characteristics and resolutions [13] as well as technical limitations in detecting and highlighting areas of interest.

2   Related Research

Image analysis and comparison are performed by means of classifying its features. This is done by comparing the features from the test image in question with those from training data. A brute force approach for comparing two sets of image features would compare every feature in one set to every feature in the other and keep track of the "best so far" match. This results in a heavy computational complexity in the order of O (N^2) where N is the number of features in each image. A number of algorithms have been proposed to improve the computational complexity, including the popular kd-tree technique [14]. This technique uses exact nearest neighbour search and works very well for low dimensional data but quickly loses its effectiveness as dimensionality increases. The popularity of the kd-tree technique has seen a number of proposals to further improve the algorithm including [15, 16]. Jiang *et al.* [17] proposed a visualization and quantitative analysis framework using image segmentation technique to derive six features that are extracted from patients MR images, which have a close relationship with Lumbar Disc Herniation score [17]. The six features are the distribution of the protruded disc, the ratio between the protruded part and the dural sacs, and its relative signal intensity. Alomari *et al.* [18, 19] proposed a probabilistic model for automatic herniation detection work by combining the appearance and shape features of the lumbar intervertebral discs. The technique models the shape depending on both the T1-weighted and T2-weighted co-registered sagittal views for building a 2D feature image. The disc shape feature is modelled using Active Shape Model algorithm while the appearance is modelled using the normalised pixel intensity. These feature-pairs are then classified using Gibbs-based classifier. The paper reported that 91% accuracy is achieved in detecting the herniation. More details about the related research are available in our previous publication [6].

3   Proposed Approach

The main goal of for our work is to detect the disc herniation automatically. To reach this goal we need firstly to automatically label the disc area via machine learning and artificial intelligence. In our experiment, we will concentrate on the gray level for each pixel in the area of interest of the lumbar spine MRI and the coordinate value for that pixel.

3.1   Feature Definition

Our feature set covers the coordinate value for each pixel in the disc area in addition to the intensity value for that pixel. Those features will be used to predict the disc area in the axial view MRI.

Training Data. One of the main steps in using the machine learning is preparing suitable training data to train the classifiers. In our experiments, the training data is MRIs that contain the truth value, i.e., correctly labelled disc and non-disc area. This training data is developed manually by labelling each disc image to create a mask image. An example of such mask images with the corresponding MRI can be seen in Fig. 1. The intensity level for the disc on the mask is used for this comparison. The disc area on the mask filled in black colour with intensity level 0 and the remaining area of the mask image have the white colour with intensity level 255. The comparison will work pixel by pixel. For the same disc, each two pixels with the same coordinate in the MRI and the mask will be compared. If the pixel on the original MR image matches a pixel with intensity 255 the output will be zero indicating a non-disc area and if the original image pixel matches a pixel with intensity zero in the mask then the output will be one indicating a disc area. At the end of this comparison, an extra column of data will be produced indicating the disc area which will be used for system training process.

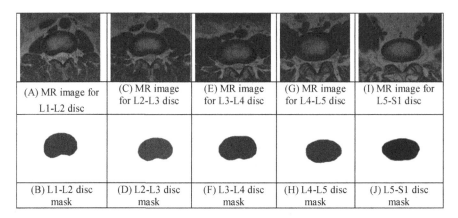

Fig. 1. Discs MRI and the developed mask for each disc.

System Classifier. There are a number of algorithms that could be used in the knowledge-based/artificial intelligence system. Our approach is to experiment with a number of classifiers and perform the training and classification process. The best classifier will be selected based on the achieved accuracy (high true positive and false negative rates as well as low false positive and true negative rates). To illustrate the training and classification process, the training will use the truth data which have been developed using a contrast weighted MR images as discussed earlier. Contrast weighted images are used to emphasise different types of tissues within the same MR images. The trained system will then be able to produce labelled images of the affected areas if a disc herniation is detected in the input MR images.

3.2   Performance Measurement

In order to evaluate the capability of our classifiers, we introduce a framework of performance measures, posed in conjunction with a series of comparator trials. In the comparator trials, we use models that are purposely selected from different theoretical classes. We reason that to be justified, our test classifiers should significantly outperform both linear classes of models and also simplistic models such as K-nearest neighbour (weak learners). Additionally, to demonstrate that the data we use as input contains true dependency, we compare the performance of all of the models presented with a random guessing baseline, showing that uninformed decisions are insufficient to produce significant results. Additionally, since generalisation is the goal of our classifier, we apply the described performance framework to the training (70% of the sample data), testing (20% of the sample data) and validation (10% of the data). A comparison of results between the reconstructed images from the classifiers with the original images is used to provide an indication of the classifier performance as described in Sect. 4.3. To furnish the classifier responses with objective measures of performance, we utilise the scalar metrics method described below.

Scalar Metrics. To characterise the capability of the classifiers simulated in our experimental trials, we use a number of scalar measures, each of which provides a different summary of the deviation between our classifier outputs and the corresponding ground truth values. Such statistical measures are necessary, since complete discrimination between sets of outcomes is often unrealistic, prompting a trade-off between the types of errors committed. A listing of the measures and their derivations is presented in Table 1. We use accuracy which represents the most general correct classification proportion, grouping correct and incorrectly classified outcomes without reference to the underlying error types for each classifier then we depend on other statistical measures.

Table 1. Performance metrics.

Metric	Abbreviation	Computation	Range
Area under curve	AUC	$0 <= area(ROC) <= 1$	[0,1]
Sensitivity	SEN	TP/(TP+FN)	[0,1]
Specificity	SPEC	TN/(TN + FP)	[0,1]
Accuracy	ACC	(TP + TN) /(TP + FN + TN + FP)	[0,1]

TP = True Positive Count, TN = True Negative Count
FP = False Positive Count, FN = False Negative Count

4 Experiment Result and Analysis

To evaluate the capability of our proposed classifiers, we conducted a series of empirical simulations, using the extracted feature from the MRI as our sample. The models used in our experiment are listed in Table 2.

Table 2. First and second experiment results based on the evaluation of 12 models

Model	The first experiment accuracy	The second experiment accuracy
Complex tree [20, 21]	98.4	94.0
Medium tree [20, 21]	98.3	92.0
Fine KNN [22, 23]	98.4	89.1
Medium KNN [22, 23]	98.9	91.1
Coarse KNN [22, 23]	98.6	90.7
Cosine KNN [22, 23]	93.4	88.7
Cubic KNN [22, 23]	98.9	91.0
Weighted KNN [22, 23]	**98.9**	90.8
Quadratic SVM [24]	98.8	90.6
Medium Gaussian SVM [24]	98.7	90.8
Fine Gaussian SVM [24]	98.8	**91.1**
Ensemble RUSBoosted trees [25, 26]	98.0	84.1

Keys: SVM Support Vector Machine, KNN K-Nearest Neighbour
Note: The selected classifier is highlighted in bold in the table

4.1 Feature Origination

Three features are extracted from the selected five MR images and the produced masks for 101 patients using Matlab as defined in Sect. 3.1. Two experiments have been performed using two data samples. In the first set of experiments, we train the system using one patient data by selecting a random sample containing 7500 pixels from the five discs and use this sample in the system training. Table 2 shows the list of classifiers and their result whereas, Fig. 3 shows the reconstructed output from the Weighted KNN model as a model with the highest accuracy. In the second set of experiments, we train the system by using 7000 pixels as a random sample from seven patients with 1000 pixel data selected from each one of them. Figure 4 shows the reconstructed output from SVM Fine Gaussian as a model with the highest accuracy among the benchmarked models. The trials presented in this work were conducted using a common dataset partitioning scheme. We reserved 70% of the data for model training and the remaining 30% was divided 20 for testing and 10% for validation. The results of our experiment are divided into two parts: the first part is the classifier training result and this will be explained in Sect. 4.2 whereas the second part is the reconstructed image of predicted data from the classifier that have the highest accuracy and this explained in Sect. 4.3.

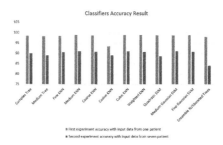

Fig. 2. The classifiers for the two experiments and the accuracy for each of them.

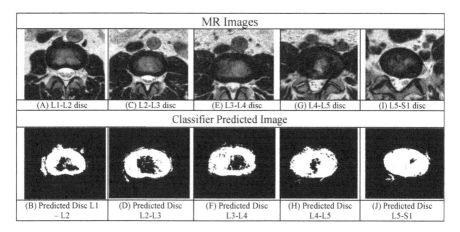

Fig. 3. The original and predicted images by weighted KNN

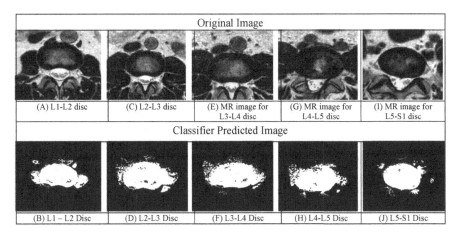

Fig. 4. The original and predicted images by SVM fine Gaussian

4.2   Classifier Training Result

The results from our experimental procedure are presented and organised for each respective classifier for the first experiment in Table 2 when the data source was one patient details with 7500 pixels for the first experiment whereas, the second experiment the data source used seven patient details with 7000 pixels. We then proceed to present our evaluation of the classifiers according to the accuracy. We have experimented with all the 23 available classifiers in Matlab. Only the best 12 classifiers with the highest accuracy have been discussed. Figure 2 shows the classifiers for the two experiments and the achieved accuracy for each one of them. In our experiments, we concentrate on the prediction of the disc area hence a positive result is defined as disc pixel prediction whereas a negative result is defined as non-disc pixel prediction. True prediction means the prediction matches the correct class whereas false prediction means otherwise.

In our experiments, we concentrate on classifiers' accuracy as the quality of the constructed images depends on the value of the accuracy compared to the other performance measures. The best classifier for the first experiment is the Weighted KNN, Medium KNN and Cubic KNN classifiers with accuracy = 98.9%. Whereas, for the second experiment SVM Fine Gaussian classifier and Medium KNN achieve the same accuracy which is 91.1%. However, Weighted KNN is the selected classifier in the first and SVM Fine Gaussian classifier for the second experiments respectively as they got the highest true positive rate. The selected classifiers will be used in subsequent steps and analysis.

4.3   Image Reconstruction

In this section, we have reconstructed "re-visualized" the images based on the classifiers result as discussed in Sect. 4.2. From the first experiment, we took the result of the Weighted KNN classifier for each x and y coordinate as input to Matlab program which converts this matrix to image as shown in Fig. 3. On the other hand, the result of the

SVM Fine Gaussian classifier has been selected in the second experiment and used as an input with x and y coordinate to the Matlab code to reconstruct the output image. Figure 4 shows the predicted image for SVM Fine Gaussian classifier.

5 Discussion and Conclusion

Overall, it can be seen from the two experiments that the highest accuracy for the first experiment using one patient data is 98.9% which is higher than the highest accuracy in the second experiment using seven patients' data which reached 91.1%. Looking at the detail, the Medium, Cubic and Weighted KNN classifiers achieved the highest accuracy of 98.9% and as a second factor for selecting one of these classifiers we looked at the true positive rate for all of them to find a slight difference between them as medium and Cubic KNN classifiers got 93% while Weighted KNN got 94% and this classifier has been selected for the rest of work. In the second experiment, both the Fine Gaussian SVM and Medium KNN achieved 91.1% accuracy rate but the ROC for the true positive rate for Fine Gaussian SVM was 62% and 60% for Medium KNN. Hence the Fine Gaussian SVM classifier was selected for the rest of the work. The second part of the experiment is to apply the selected trained classifiers to label discs pixels of new patient MRIs. The same image features are calculated from these new MRIs to classify the pixels and corresponding disc masks are produced by reconstructing the labeled image from the classified pixels.

The reconstructed images in Figs. 3 and 4 show the discs area in the lumbar spine for the same patient. In general, the predicted images from the Weighted KNN classifier has better accuracy than SVM Fine Gaussian predicted images and we relate that to the difference in the classifier accuracy. However, the reconstructed images from both classifiers contain some deformations when compared with the original image for that disc. This is true even for those that have a very high accuracy rate (98.9%) using the Weighted KNN. It is very important to have a very good quality mask from these experiments as this mask will be used to detect the disc herniation as mentioned earlier. Further improvement in the quality of the reconstructed label images can be achieved by (1) balancing the ratio between the two number of pixels that belong to both classes during the training process, (2) using feature sets that take into account interrelationship between adjacent pixels intensities, and (3) applying some pre-processing and post-processing algorithms to remove noise before and after the reconstruction process. It is our expectation that the combination of these three steps should improve the quality of the reconstruction label images and make them more suitable for the disc herniation detection process that follows.

References

1. Koh, J., Alomari, R.S., Chaudhary, V., Dhillon, G.: Lumbar spinal stenosis CAD from clinical MRM and MRI based on inter-and intra-context features with a two-level classifie. In: SPIE Medical Imaging, vol. 7963, pp. 796304-1–796304-8 (2011)

2. Waddell, G., Burton, A.K.: Occupational health guidelines for the management of low back pain at work: evidence review. Occup. Med. (Chic. Ill.) **51**(2), 124–135 (2001)
3. Gordon, R., Bloxham, S.: A systematic review of the effects of exercise and physical activity on non-specific chronic low back pain. Healthc. (Basel, Switz.) **4**(2), 22 (2016)
4. Dixon, S.: Diagnostic imaging dataset annual statistical release 2013/14, pp. 1–27 (2014)
5. Royal College of Radiologists: 2015 Clinical Radiology UK Workforce Census (2016)
6. Al Kafri, A.S., Sudirman, S., Hussain, A.J., Fergus, P., Al-Jumeily, D., Al-Jumaily, M., Al-Askar, H.: A framework on a computer assisted and systematic methodology for detection of chronic lower back pain using artificial intelligence and computer graphics technologies. In: Huang, D.-S., Bevilacqua, V., Premaratne, P. (eds.) ICIC 2016. LNCS, vol. 9771, pp. 843–854. Springer, Cham (2016). doi:10.1007/978-3-319-42291-6_83
7. Al-Jumeily, D., Iram, S., Vialatte, F.-B., Fergus, P., Hussain, A.: A novel method of early diagnosis of Alzheimer's disease based on EEG signals. Sci. World J. **2015**, 931387 (2015)
8. Al-Jumeily, D., Hussain, A., Fergus, P.: Using adaptive neural networks to provide self-healing autonomic software. Int. J. Space-Based Situat. Comput. **5**, 129–140 (2015)
9. Taher, F., Werghi, N., Al-Ahmad, H.: Computer aided diagnosis system for early lung cancer detection. In: 2015 International Conference on Systems, Signals and Image Processing (IWSSIP), pp. 5–8 (2015)
10. Alomari, R.S., Corso, J.J., Chaudhary, V., Dhillon, G.: Automatic diagnosis of lumbar disc herniation with shape and appearance features from MRI. Prog. Biomed. Opt. Imaging **11**, 76241A (2010)
11. Hussain, A.J., Fergus, P., Al-Askar, H., Al-Jumeily, D., Jager, F.: Dynamic neural network architecture inspired by the immune algorithm to predict preterm deliveries in pregnant women. Neurocomputing **151**(P3), 963–974 (2015)
12. Khalaf, M., Hussain, A.J., Keight, R., Al-Jumeily, D., Fergus, P., Keenan, R., Tso, P.: Machine learning approaches to the application of disease modifying therapy for sickle cell using classification models. Neurocomputing **228**(February 2016), 154–164 (2016)
13. Lootus, M., Kadir, T., Zisserman, A.: Vertebrae detection and labelling in lumbar MR images. In: Yao, J., Klinder, T., Li, S. (eds.) Computational Methods and Clinical Applications for Spine Imaging. LNCVB, vol. 17, pp. 219–230. Springer, Cham (2014). doi:10.1007/978-3-319-07269-2_19
14. Freidman, J.H., Bentley, J.L., Finkel, R.A.: An algorithm for finding best matches in logarithmic expected time. ACM Trans. Math. Softw. **3**(3), 209–226 (1977)
15. Arya, S., Mount, D.M., Netanyahu, N.S., Silverman, R., Wu, A.Y.: An optimal algorithm for approximate nearest neighbor searching in fixed dimensions. In: Proceedings of 5th ACM-SIAM Symposium Discrete Algorithms, vol. 1, no. 212, pp. 573–582 (1994)
16. Beis, J.S., Lowe, D.G.: Shape indexing using approximate nearest-neighbour search in high-dimensional spaces. In: Proceedings of IEEE Computer Society Conference on Computer Vision and Pattern Recognition, pp. 1000–1006 (1997)
17. Jiang, H., Qi, W., Liao, Q., Zhao, H., Lei, W., Guo, L., Lu, H.: Quantitative evaluation of lumbar disc herniation based on MRI image. In: Yoshida, H., Sakas, G., Linguraru, M.G. (eds.) ABD-MICCAI 2011. LNCS, vol. 7029, pp. 91–98. Springer, Heidelberg (2012). doi:10.1007/978-3-642-28557-8_12
18. Alomari, R.S., Corso, J.J., Chaudhary, V., Dhillon, G.: Automatic diagnosis of lumbar disc herniation with shape and appearance features from MRI. In: SPIE Medical Imaging, p. 76241A (2010)
19. Alomari, R.S., Corso, J.J., Chaudhary, V., Dhillon, G.: Lumbar spine disc herniation diagnosis with a joint shape model. In: Yao, J., Klinder, T., Li, S. (eds.) Computational Methods and Clinical Applications for Spine Imaging. LNCVB, vol. 17, pp. 87–98. Springer, Cham (2014). doi:10.1007/978-3-319-07269-2_8

20. Alpaydin, E.: Introduction to Machine Learning. MIT Press, Cambridge (2014)
21. Tan, P.-N., Steinbach, M., Kumar, V.: Classification: basic concepts, decision trees. Introd. Data Min. **67**(17), 145–205 (2006)
22. Wu, X., Kumar, V., Ross, Q.J., Ghosh, J., Yang, Q., Motoda, H., McLachlan, G.J., Ng, A., Liu, B., Yu, P.S., Zhou, Z.H., Steinbach, M., Hand, D.J., Steinberg, D.: Top 10 algorithms in data mining. Knowl. Inf. Syst. **14**(1), 1–37 (2008)
23. Bhatia, N., Author, C.: Survey of nearest neighbor techniques. (IJCSIS) Int. J. Comput. Sci. Inf. Secur. **8**(2), 302–305 (2010)
24. Aggarwal, C.C. (ed.): Data Classification: Algorithms and Applications. CRC Press, Boca Raton (2014)
25. Seiffert, C., Khoshgoftaar, T.M., Van Hulse, J., Napolitano, A.: RUSBoost: improving classification performance when training data is skewed. In: 2008 19th International Conferences on Pattern Recognition, no. March 2016, pp. 8–11 (2008)
26. Seiffert, C., Khoshgoftaar, T.M., Van Hulse, J., Napolitano, A.: RUSBoost: a hybrid approach to alleviating class imbalance. IEEE Trans. Syst. Man Cybern. Part A Syst. Hum. **40**(1), 185–197 (2010)

Speeding Up Dilated Convolution Based Pedestrian Detection with Tensor Decomposition

Yan Wu$^{(\boxtimes)}$, Wei Jiang, Jiqian Li, and Tao Yang

College of Electronics and Information Engineering, Tongji University,
Shanghai 201804, China
yanwu@tongji.edu.cn

Abstract. Researches show in the test phase of Convolutional Neural Network (CNN), the most time cost occurred in convolutional layers, while the most memory cost occurred in the fully connected layers. With the rapid development of the pedestrian methods, which are based on deep learning, the performance is going better and better. Especially using resemble models, pedestrian detection can get a more excellent performance. However, the performance is improved by the increase in parameters and slow of speed in price. Meanwhile, for some specific tasks, such as driverless cars, due to the limitations of hardware facilities, it is impossible to use these methods on them. In this paper, we applied the tensor decomposition to pedestrian detection task, in order to accelerate the whole pedestrian detection processing. Experiments show even though the decomposition brings some of the rise of miss rate (MR), the saved memory and time indicates the efficiency of our method.

Keywords: Pedestrian detection · Tensor decomposition · Deep model acceleration · Dilated convolution

1 Introduction

In driverless cars and advanced driver assistant system, we usually use the camera to detect the pedestrian in close distant, and use the millimeter wave radar to detect the pedestrian in remote distant. Due to the own characteristics of millimeter wave radar, the detection precision is constraint, our researches mainly focus on the camera based pedestrian detection. According to the obtained video or picture information, we use the image processing and computer vision related algorithms to determine whether there are pedestrians and find the location and size of pedestrians, which can be further used into the driverless car decision making or pedestrian re-identification. Researches on pedestrian detection can provide a strong guarantee for the safety of pedestrians, which has important research significance and application value.

At present, there are three mainstream pedestrian detection methods, which are the motion information based ones, the template based ones and the machine learning based ones respectively [1]. Among them, the machine learning based ones have the most extensive research and application, due to its better performance and speed. The

D.-S. Huang et al. (Eds.): ICIC 2017, Part III, LNAI 10363, pp. 117–127, 2017.
DOI: 10.1007/978-3-319-63315-2_11

classical machine learning based pedestrian detection methods include HOG+SVM [2], LBP [3], DPM [4] and other deep learning models [5–10]. With the rapidly development of deep learning based object detection, the deep learning based pedestrian detection is becoming more and more inevitable. The detection precision is becoming higher and higher, which greatly promote the development of pedestrian detection. However, the complex model, hundreds of millions of parameters and the slow calculation processing lead to its difficult to achieve real-time demands.

Tensor decomposition can be considered to be a high order generalization of matrix decomposition in multiple linear algebra. Since the matrix decomposition has a wide use in many areas such as dimension reduction processing, coefficient data padding and implicit relation mining, as the high order extension of it, tensor decomposition also has good compression performance. Therefore, we apply the tensor decomposition to the best performed deep learning based pedestrian detection model, in order to shrink down the model and accelerate the calculation.

The rest of this paper is organized as follows: Sect. 2 introduces the related work of this paper briefly and Sect. 3 presents our main methods. The quantitative experiments are shown in Sect. 4 and the conclusions and future work are discussed in Sect. 5.

2   Related Work

With the increase of precision in deep learning based pedestrian detection, many compression algorithms for deep models emerge. These algorithms can be divided into three categories: knowledge distilling [11–13], pruning and coding [14], and decomposition [15–17]. In the compression of deep models, decomposition is aimed to extract the important features from dense matrix. The more commonly used is the eigenvalue decomposition, singular value decomposition and tensor decomposition. Since most of the parameters in Convolutional Neural Networks (CNNs) can be seen as tensor, many recent works have achieved significant performance on CNN acceleration with tensor decomposition. Rigamonti et al. [15] proposed a low-rank decomposition based convolutional decompression in 2013, they separated the 2D or 3D filters into a small number of separable ones. The method delivered a very substantial speed-up at no loss in performance. Soon afterward, Jaderberg et al. [16] described a more efficient scheme which approximates the 4D kernel tensor to the combination of two 3D kernel tensor. They also introduce the local fine tuning into the calculation, despite it is eventually proved to be ineffective. Meanwhile, Denton et al. [17] based on the CANDECOMP/ PARAFAC (CP decomposition) proposed a scheme to obtain the kernel tensor decomposition using bilateral clustering. The bilateral clustering considers the two non-spatial dimensions as subgroups and decreases the effective rank in CP decomposition. The most related to the Denton's, Lebedev et al. [18] used the fine-tuned CP decomposition to speed up the convolutional neural networks. They directly applied the CP decomposition into the fully convolutional kernel tensors and used the non-linear least square into CP decomposition instead of the greedy calculation. To the best of our knowledge, the knowledge distilling based method [13] is the only one which has been used in deep learning based pedestrian detection. However, as with other knowledge distilling methods, it is also limited by training samples and training time.

With the rapid development of R-CNN [19] based object detection, many pedestrian detection models based on CNN came into being. Based on Faster R-CNN [20], Cai et al. [9] adopted features of different layers in different ways, according to the different features of different layers. For instance, the low layer features of conv-3 can be sensitive to the small objects because of its detail information, while the high layer features of conv-5 will be more precise for the large objects. Designing different scale object detectors for different layers, the multi scale object detection can be solved. Zhang et al. proposed RPN+BF [10], which successfully apply the Faster R-CNN on pedestrian detection in improving the resolution and mining the hard example. In our previous work [5], we applied the dilated convolution which was most popularly used in semantic segmentation into pedestrian detection with RPN and boosted decision trees and experiments showed that the high resolution feature map is the key to solve the pedestrian detection problem. Later, Du et al. proposed Fused DNN [21], which adopt a soft rejection with multi networks in order to optimize the candidate boxes extracted by SSD [22] detectors. Along with enforcement of the introduction of pixel-wise semantic segmentation information, Fused DNN is achieving the state-of-the-art in Caltech Pedestrian Detection Benchmark [23]. Most of these methods adopted fine-tuning the pre-trained classification models on ImageNet [24] or other large data sets, which always consist of a large number of convolutions. Even though the Miss Rate (MR) is becoming lower and lower, the detection time is trapped into a bottleneck. Taking the Fused DNN for an example, processing one image will take about 2.48 s, which cannot meet the demand to a large extent.

In this paper, we try to apply the CP decomposition to our previous work [5] to make a time and precision tradeoff. Similar with the Ref. [18], we apply the CP decomposition into the fully convolutional kernel tensors and use the non-linear least square into CP decomposition. With the fine tuning of whole network, we got a significant speed up for the detection task with only minor MR rise.

3   Method

Our method consists of three processions: a layer-wise decomposition of each layer in VGG16, take the accelerated model as the pre-trained model for pedestrian detection and fine tuning the whole network only if the decomposition introduces large MR increase.

3.1   Tensor Decomposition

The concept of tensor was first proposed by mathematicians at the end of 19th century, and it was widely concerned after the use of tensor language in Einstein's related work on general relativity. The essence of tensor is a multidimensional array in which the elements can be regarded as containing multiple indicators, which correspond to the order of the tensor. Tensor decomposition essentially can be considered to be a high order generalization of matrix decomposition in multiple linear algebra, and there are two main types of tensor decomposition: Tucker decomposition and CP decomposition.

In this paper, we focus on the acceleration of convolutional operation using CP decomposition, so the CP decomposition of a tensor will be explained here.

For a d dimensional and size of $n_1 * n_2 * \ldots * n_d$ array A, the CP decomposition can be expressed as follows:

$$A(i_1, i_2, \ldots, i_d) = \sum_{r=1}^{R} A_1(i_1, r) \ldots A_d(i_d, r) \tag{1}$$

where the smallest possible r is called the rank of the tensor. The advantage of this decomposition is we only need to store $(n_1 + n_2 + \ldots + n_d)R$ elements instead of a whole tensor with $n_1 n_2 \ldots n_d$ elements.

In this paper, we choose the TensorLab [25], which was developed by Sorber et al. in 2014, to execute the CP decomposition. We use the Gauss-Newton approximation method to minimize L2 norm. By this way, the minimal variance optimization algorithm is better than the greedy search for the best ran-optimization method to get the best R.

3.2   Dilated Convolution Based Pedestrian Detection

In our previous work Ref. [5], we proposed a pedestrian detection framework, which consisted of three components: region proposal network, dilated convolution and boosted decision trees, where region proposal network was adopted to obtain boxes and scores for samples, dilated convolution was adopted to obtain high resolution features, and boosted decision trees can mine hard examples of pedestrian detection. Among them, the region proposal network and the dilated convolution are based on convolution operations. We first use the pedestrian dataset to fine-tune the pre-trained Faster R-CNN model, which has been proved very successful in object detection. The pre-trained Faster R-CNN model uses VGG16 [26] as its basic model, which consists of 5 groups of convolutional layers and 3 fully connected layers. Even though each kernel in VGG16 has the same size of 3 * 3, the parameters of VGG16 still have reached 138 million and the pre-trained model has reached over 550 MB. The simple network structure and the detail parameters can be seen in Fig. 1 and Table 1.

Despite the dilated convolution has discard the 4th pooling layer, with the different dilation based features stitching, it still retains most of the VGG16 parameters. The large amount of parameters and convolutional calculations make the detection model large and the whole detection processing slow.

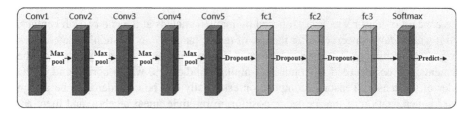

Fig. 1. Network Structure of VGG16

Table 1. The detail network parameters of VGG16

Input	Conv3-64	Conv3-128	Conv3-256	Conv3-512	Conv3-512	FC-4096	FC-4096	FC-1000	Softmax
	Conv3-64	Conv3-128	Conv3-256	Conv3-512	Conv3-512				
			Conv3-256	Conv3-512	Conv3-512				

In this paper, based on the model of our previous work [5], we introduce the tensor decomposition into it to accelerate the model to make a time and memory tradeoff.

3.3 Apply the Tensor Decomposition on Dilated Convolution Based Pedestrian Detection

The application of tensor decomposition on dilated convolution based pedestrian detection is mainly concentrated on the convolutional kernels' decomposition. Here we use the method proposed in [17] to process the decomposition. Taking a tensor with the size of X * Y * S as the input, we can obtain an output tensor with the size of $(X - d + 1) * (Y - d + 1) * T$ through the following linear function:

$$V(s, y, t) = \sum_{i=x-\delta}^{x+\delta} \sum_{j=y-\delta}^{y+\delta} \sum_{s=1}^{S} K(i - x + \delta, j - y + \delta, s, t) U(i, j, s) \qquad (2)$$

where $K(., ., ., .)$ is a 4 order kernel tensor with the size of $d * d * S * T$, among which the first two dimensions are corresponding to the spatial dimension and the third dimension is corresponding to the input channels, while the fourth dimension is corresponding to the output channels. The height and width of the kernel tensor is marked as d, and δ is half of the width.

Then the CP decomposition of 4 order kernel tensor can be expressed as follows:

$$K(i, j, s, t) = \sum_{i=1}^{R} K^x(i - x + \delta, r) K^y(j - y + \delta, r) K^s(s, r) K^t(t, r) \qquad (3)$$

where $K^x(., .) K^y(., r) K^s(., .) K^t(., .)$ are four 2 order tensors with the size of $d * R, d * R, S * R$ and $T * R$ respectively.

Put the formula 3 into 2 and perform the corresponding permutation and grouping summation, we can obtain the approximated formula of convolutional calculation as follows:

$$V(s, y, t) = \sum_{r=1}^{R} K^t(t, r) \left(\sum_{i=x-\delta}^{x+\delta} K^x(i - x + \delta, r) \right.$$

$$\left. \left(\sum_{j=y-\delta}^{y+\delta} K^y(j - y + \delta, r) \left(\sum_{s=1}^{S} K^s(s, r) U(i, j, s) \right) \right) \right) \qquad (4)$$

Based on formula 4, we can calculate the output tensor $V(.)$ from the input tensor $U(.)$ through 4 small convolutional calculations:

$$U^s(i,j,r) = \sum_{s=1}^{S} K^s(s,r)U(i,j,s) \tag{5}$$

$$U^{sy}(i,y,r) = \sum_{j=y-\delta}^{y+\delta} K^y(j-y+\delta,r)U^s(i,j,r) \tag{6}$$

$$U^{syx}(x,y,r) = \sum_{i=x-\delta}^{x+\delta} K^x(i-x+\delta)U^{sy}(i,y,r) \tag{7}$$

$$V(x,y,t) = \sum_{r=1}^{R} K^t(t,r)U^{syx}(x,y,r) \tag{8}$$

where $U^s(.)U^{sy}(.)$ and $U^{syx}(.)$ are middle tensors. From $U(.)$ to $U^s(.)$ and from $U^{syx}(.)$ to $V(.)$, we use the same 1 * 1 convolutional process with "Network in Network" [27], which can efficiently make the pixel-wise linear reconstruction of the input feature maps. From $U^s(.)$ to $U^{sy}(.)$ and from $U^{sy}(.)$ to $U^{syx}(.)$, we use the normal convolutional operation, which only flat one of the spatial dimensions of the convolution kernel to form a relatively small convolutional kernel. This process can be visualized as shown below, where R indicates the rank of the kernel tensor.

In this paper, we finished the decomposition of the whole network through the layer-wise decomposition as described in Fig. 2, and the following layers' decomposition procession is as the same.

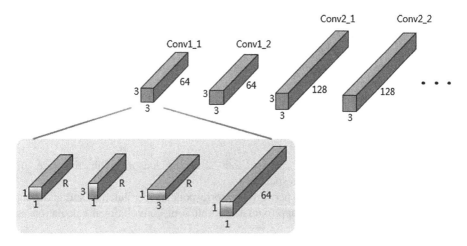

Fig. 2. Tensor decomposition procession demo of the conv1_1 kernel tensor in VGG16

4   Experiment

In this section, we will evaluate the proposed method on the model compression of VGG16 and the dilated convolution based pedestrian detection model proposed in [5] from three aspects, which are the degree of compression of model, pedestrian detection time and pedestrian detection miss rate.

4.1   Model Compression

In this subsection, we apply the tensor decomposition to the 13 convolutional layers' kernels of VGG16 layer wisely. That is to say, we first decompose the conv1_1 using the CP decomposition to obtain an accelerated model. Then take the accelerated model as the original model, we decompose the conv1_2 as above and so on. Finally, we have all of the convolution kernels decompressed and obtained a final model, which only has the size of 391.4 MB, the decomposition process can be shown as Table 2.

We can see from Table 2, our tensor decomposition based method decompose the VGG16 from 553.4 MB to 394.1 MB, which indicates that our method performs well in model compression.

Table 2. The size of model after tensor decomposition in each layer of VGG16

Model	Size
Original model	553.4 MB
Conv1_1	553.3 MB
Conv1_2	553.2 MB
Conv2_1	552.9 MB
Conv2_2	550.8 MB
Conv3_1	548.3 MB
Conv3_2	541.5 MB
Conv3_3	538.6 MB
Conv4_1	522.7 MB
Conv4_2	497.9 MB
Conv4_3	469.3 MB
Conv5_1	442.5 MB
Conv5_2	412.7 MB
Conv5_3	394.1 MB

4.2   Caltech Benchmark Pedestrian Detection Dataset

The same as the previous algorithm in previous section, we calculate the MR after the decomposition of each layer. For the layers which greatly impact on MR after decomposition, we adopt the fine tuning with the Caltech Pedestrian Detection Benchmark on the whole network. The Caltech Pedestrian Detection Benchmark is proposed by Dollar et al., which is captured by car camera and has the resolution 640 * 480. The whole dataset takes about 10 h, whose the frame rate is 30 frames per second, and is organized by set00 ~ set10, among which set00 ~ set05 are training set and set06 ~ set10 are testing set. We use the Pitors toolbox [28] to assist in evaluation, and choose the 'Reasonable' setting as evaluation standard. Since we totally used the decomposed convolution kernel instead of the original convolution kernel, the convolution of the transmission operation may bring noise, which may cause the rise of MR.

Figure 3 shows the impact of tensor decomposition on Caltech Pedestrian Detection Benchmark, where Dilated Convolution indicates the detection results of [5], while the tensor decomposition indicates our proposed method in this paper. Meanwhile, we also recorded the run time of our model at the test time. Compared with the dilated convolution based pedestrian detection algorithm in our previous work [5], the results are shown in Table 3. The running environment is Ubuntu 14.10 system, Intel Core i7 processor, the largest memory of Titan X GPU is 16G.

From Fig. 3 and Table 3, we can discover that even though our method brings minor MR increase, our method reduced the time complexity and spatial complexity of the pedestrian detection model, which highly indicates the effectiveness of our algorithm.

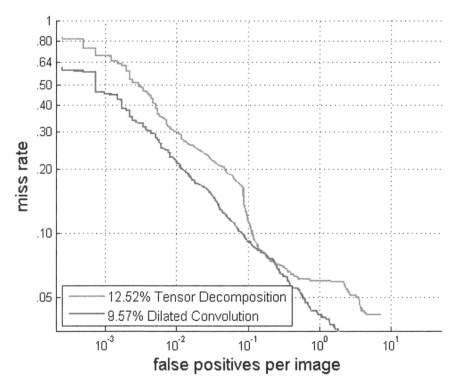

Fig. 3. Comparisons on the Caltech set (legends indicate MR; Reasonable)

For further verification, we randomly select some samples from the testing set of Caltech Pedestrian Detection Benchmark to analyze our algorithm qualitatively. As is shown in Fig. 4, the three columns are set07_V007_I00329.jpg, set07_V001_I00449. jpg and set07_V000_I01829.jpg respectively, and the four rows are original images, dilated convolution based method [5], our detection results and the ground truth. We can see our algorithm has detected most of the pedestrians, while there are still some missed inspections.

Table 3. Test time comparisons

Method	Test time (s/img)
Dilated convolution	0.56 s
Tensor decomposition	0.32 s

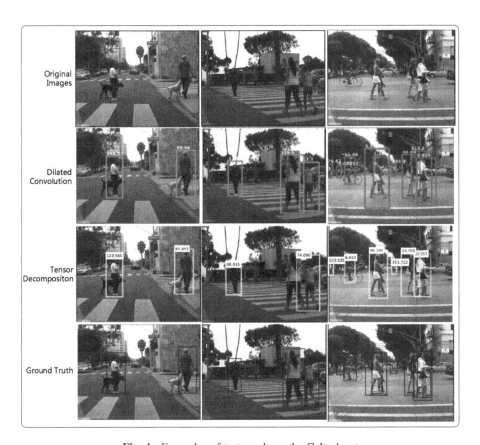

Fig. 4. Examples of test result on the Caltech set

5 Conclusion

In this paper, taking into account the excellent characteristics of tensor decomposition and the defect of most of the detection models, we apply tensor decomposition into the acceleration of pedestrian detection models. We have conducted some experiments to compare the performance with and without tensor decomposition. The results demonstrate that tensor decomposition performs well in the acceleration of pedestrian detection models, which may make a little impact on the performance.

As described above, some methods for model compression such as FitNets, the final small network outperforms the original big network. In the future we are interested in how to use parameter fine tuning or bring in some new method, to achieve better performance.

References

1. Yao, Q., An, S.Q., Yao, L.: Algorithms of pedestrian detection and tracking based on three frame different method and mean-shift algorithm. Comput. Eng. Des. **35**(1), 223–227 (2014). (in Chinese)
2. Dalal, N., Triggs, B.: Histograms of oriented gradients for human detection. In: IEEE Computer Society Conference on Computer Vision and Pattern Recognition, vol. 1, pp. 886–893. IEEE Computer Society (2005)
3. Xia, D., Sun, H., Shen, Z.: Real-time infrared pedestrian detection based on multi-block LBP. In: 2010 International Conference on Computer Application and System Modeling (ICCASM), vol. 12, pp. V12-139–V12-142. IEEE (2010)
4. Felzenszwalb, P., Mcallester, D., Ramanan, D.: A discriminatively trained, multiscale, deformable part model. In: IEEE Computer Society Conference on Computer Vision and Pattern Recognition, pp. 1–8. DBLP (2008)
5. Li, J., Wu, Y., Zhao, J., et al.: Pedestrian detection with dilated convolution, region proposal network and boosted decision trees. In: The 2017 International Joint Conference on Neural Networks (IJCNN 2017), pp. 4052–4057 (2017)
6. Yang, B., Yan, J., Lei, Z., et al.: Convolutional channel features. In: Proceedings of the IEEE International Conference on Computer Vision, pp. 82–90 (2015)
7. Tian, Y., Luo, P., Wang, X., et al.: Deep learning strong parts for pedestrian detection. In: Proceedings of the IEEE International Conference on Computer Vision, pp. 1904–1912 (2015)
8. Cai, Z., Saberian, M., Vasconcelos, N.: Learning complexity-aware cascades for deep pedestrian detection. In: Proceedings of the IEEE International Conference on Computer Vision, pp. 3361–3369 (2015)
9. Cai, Z., Fan, Q., Feris, R.S., et al.: A unified multi-scale deep convolutional neural network for fast object detection. In: European Conference on Computer Vision, pp. 354–370. Springer International Publishing, Heidelberg (2016)
10. Zhang, L., Lin, L., Liang, X., et al.: Is faster R-CNN doing well for pedestrian detection? In: European Conference on Computer Vision, pp. 443–457 (2016)
11. Hinton, G., Vinyals, O., Dean, J.: Distilling the knowledge in a neural network. Comput. Sci. **14**(7), 38–39 (2015)
12. Romero, A., Ballas, N., Kahou, S.E., et al.: FitNets: hints for thin deep nets. arXiv preprint arXiv:1412.6550 (2014)
13. Shen, J., Vesdapunt, N., Boddeti, V.N., et al.: In Teacher We Trust: Learning Compressed Models for Pedestrian Detection. arXiv preprint arXiv:1612.00478 (2016)
14. Han, S., Mao, H., Dally, W.J.: Deep compression: compressing deep neural networks with pruning, trained quantization and Huffman coding. Fiber **56**(4), 3–7 (2015)
15. Rigamonti, R., Sironi, A., Lepetit, V., et al.: Learning separable filters. In: IEEE Conference on Computer Vision and Pattern Recognition, pp. 2754–2761. IEEE Computer Society (2013)

16. Jaderberg, M., Vedaldi, A., Zisserman, A.: Speeding up convolutional neural networks with low rank expansions. arXiv preprint arXiv:1405.3866 (2014)
17. Denton, E., Zaremba, W., Bruna, J., et al.: Exploiting linear structure within convolutional networks for efficient evaluation. In: Advances in Neural Information Processing Systems, pp. 1269–1277 (2014)
18. Lebedev, V., Ganin, Y., Rakhuba, M., et al.: Speeding-up convolutional neural networks using fine-tuned CP-decomposition. In: Proceedings of the International Conference Learn, Represent (2015)
19. Girshick, R., Donahue, J., Darrell, T., et al.: Rich feature hierarchies for accurate object detection and semantic segmentation. In: Computer Vision and Pattern Recognition, pp. 580–587. IEEE (2014)
20. Ren, S., He, K., Girshick, R., Sun, J.: Faster R-CNN: towards realtime object detection with region proposal networks. In: Advances in Neural Information Processing Systems, pp. 91–99 (2015)
21. Du, X., El-Khamy, M., Lee, J., et al.: Fused DNN: a deep neural network fusion approach to fast and robust pedestrian detection. arXiv preprint arXiv:1610.03466 (2016)
22. Liu, W., Anguelov, D., Erhan, D., et al.: SSD: Single Shot MultiBox Detector. arXiv preprint arXiv:1512.02325 (2015)
23. Dollar, P., Wojek, C., Schiele, B., et al.: Pedestrian detection: an evaluation of the state of the art. IEEE Trans. Pattern Anal. Mach. Intell. **34**(4), 743–761 (2012)
24. Deng, J., Dong, W., Socher, R., et al.: Imagenet: a large-scale hierarchical image database. In: Computer Vision and Pattern Recognition, pp. 248–255 (2009)
25. Sorber, L., Van, B.M., De, L.: Tensorlab v2.0 (2014). http://tensorlab.net
26. Simonyan, K., Zisserman, A.: Very deep convolutional networks for large-scale image recognition. In: ICLR (2015)
27. Lin, M., Chen, Q., Yan, S.: Network in network. arXiv preprint arXiv:1312.4400 (2013)
28. Piotr, D.: Piotr's Computer Vision Matalab Toolbox (PMT). https://github.com/pdollar/toolbox

A Robust Method for Multimodal Image Registration Based on Vector Field Consensus

Xinmei Wang, Xianhui Liu$^{(\boxtimes)}$, Yufei Chen, and Zhiping Zhou

CAD Research Center, Tongji University, Shanghai 201804, China
lxh@tongji.edu.cn

Abstract. Popular registration methods can be applying into multimodal images, such as Harris-PIIFD, SURF-RPM, GMM, GDB-ICP and so on. There exist some challenges in existing multimodal image registration techniques: (1) They fail to register image pairs with some significantly different content, illumination and texture changes; (2) They fail to register image pairs with too small overlapping or too much noise. To address these problem, this paper improves the multimodal registration by contribute a novel robust framework SURF-PIIFD-BBF-VFC (SPBV). The SURF-PIIFD method can provide enough repeatable and reliable local features; the bilateral matching method and vector field consensus (VFC) can establish robust point correspondences of two point sets. For evaluation, we compare the performance of the proposed SPBV with two existing methods Harris-PIIFD and SURF-RPM on two multimodal data sets. The results indicate that our SPBV method outperforms the existing methods and it is robust to low quality and small overlapping multimodal images.

Keywords: Multimodal registration · SURF-PIIFD · Bilateral matching · Vector field consensus

1 Introduction

Image registration is the process of matching and superimposing two or more images acquired by different sensors (image capture equipment), different time or different conditions (illumination, viewpoint angle). Image registration can be applicable in many field, for instance computer vision, image processing, remote sensing image analysis and medical image analysis. The purpose of image registration is to align two or more images in a same coordinate system so as to obtain more data information. Image registration include three research direction: spatial registration, temporal registration, multimodal registration [1]. Just as their name implies, spatial registration and temporal registration images are captured by the same sensor with different perspectives or at different times, multimodal registration images are captured by different sensors.

In this paper, we propose a new framework for multimodal image registration and apply it into multimodal retinal image analysis. Note that during the last decades, various registration algorithms have been proposed in these fields. In the following, we will first introduce some prior work related to multimodal image registration, and then propose our own SPBV framework.

© Springer International Publishing AG 2017
D.-S. Huang et al. (Eds.): ICIC 2017, Part III, LNAI 10363, pp. 128–139, 2017.
DOI: 10.1007/978-3-319-63315-2_12

1.1 Prior Work

A large number of methods have been presented for multimodal image registration. We can classify these existing registration techniques into three groups: area-based methods, feature-based methods, hybrid methods [2] considering both of them. Area-based methods compare image intensity information of image pairs, and match them use a similarity metric. Existing similarity measures of multimodal image registration contain cross correlation [3], mutual information [4], phase correlation [5], fourier methods [6] and entropy correlation coefficient [7]. Maximize similarity value to find suitable parameters of transformation mode function. However, these methods have many problems when they apply into multimodal image registration: (1) They fail to register image pairs with some significantly different content, illumination and texture changes; (2) They fail to register image pairs with high dimension transformation, such as high-dimensional nonrigid transformations; (3) They fail to register image pairs with too small overlaps; (4) The huge searching space and similarity computation using the entire content undoubtedly increasing the computational complexity.

Feature-based methods are intuitive and simple which relying on detecting and matching distinctive features between image pairs. As to achieve mismatch removal and enough correct matching, the features must be highly salient and distinctive. The features can be presented by geometrical entities at different level, for example regions, contours, lines and points [8]. In the following paper, we will discuss point-based methods in the application of multimodal image registration. Compared with area-based methods, feature-based methods are more robust in the case of image distortions and illumination changes. However, the sparsity of the features may lead to some unavoidable mismatches: (1) Considering the case of occlusion, if outliers occupy a large proportion, it will match only a small portion of the initial features; (2) Considering the case of the image have some repeated patterns, the matching criteria must be flexible if ambiguity happens, for example if the closest neighbor is sufficiently close to the second-closest neighbor, the matched point pair should be abandoned.

Hybrid methods combine both area-based methods and feature-based methods together to improve the general performance. For example, Chen et al. [9] proposed a partial intensity invariant feature description (PIIFD) for multimodal image registration, the idea is to extract structural contour in the surrounding area of each Harris corner. The idea is based on two assumptions, one is that in multimodal images the matched image pairs have similar anatomical structure at corresponding locations; another is that in multimodal images the gradient orientation point to the same or opposite directions at corresponding locations. However, the PIIFD method can not detect scale invariant and repeatable interesting points. Ghassabi et al. [10] presented an uniformed and robust scale invariant method (UR-SIFT) instead of Harris corner, improved the robustness of PIIFD and can obtained efficient repeatable interesting points.

Now, popular multimodal registration methods are mostly based on local feature descriptor. There are two steps, one is that locate local feature points by specific detector, another is that match point pairs by different matching rules. Lowe [11] proposed a popular descriptor scale invariant feature transform (SIFT), which assign each point a main orientation to achieve rotate invariant and scale invariant. However,

SIFT can be used very well in monomodal image, and it can't provide sufficient key points for multimodal image of high dimension transformation. Many methods have been proposed to improve the performance of SIFT in application of multimodal image registration, as mentioned earlier PIIFD and UR-SIFT. Bay et al. [12] proposed a speed up robust feature method (SURF), it used integral image instead for convolution. SURF based on the strengths of SIFT and Haar wavelet, it can provide robust, repeatable and distinctive features, yet it can increasing computation speed greatly. Liu et al. [13] proposed a feature-guided Gaussian mixture model (GMM) to implement mismatches removal and the transformation mode parameters estimation. One point set be represented by a GMM, coincide it with another point set by constrained the centers and features of Gaussian densities. However, the GMM model is sensitive to outliers. Yang et al. [14] presented a general dual bootstrap iterative closest point method (GDB-ICP). The algorithm mapping the face or corner points to enlarge the area around initial matches iteratively. It is robust even only one correct initial match pair exists. However, the GDB-ICP method will perform badly if initial points distribution are significantly affected by noise.

1.2 Problem Statement and Proposed Method SPBV

The performance of the above mentioned methods is poor when multimodal images are severely affected by distortion or noise. The SIFT and SURF methods can perform well when deal with poor quality images, but they are only limited to the monomodal images. Generally speaking, there still exist some challenges in multimodal image registration: (1) How to detect and describe feature points in multimodal images; (2) How to remove mismatch points in initial matches; (3) How to estimate the transformation mode parameters.

To address these problem, this paper improves the multimodal registration by contribute a novel robust framework SPBV. The feature points detector and descriptor SURF-PIIFD can provide enough repeatable and reliable local features for multimodal images. And use the bilateral matching method based on best bin first (BBF) [15] technique to achieve initial feature matching. We achieve mismatches removal by vector field consensus (VFC) [16] to establish robust point correspondences of two point sets. Finally, we choose appropriate transform mode for image fusion.

The rest of this paper is organized as follows: Sect. 2 presents our new registration framework specifically. Section 3 addresses our own experiment data sets and evaluation criterion, and then evaluate different methods' performance to prove the superiority of our method in detail. Section 4 concludes the paper.

2 Proposed Registration Framework

Our method hybridize both area-based and feature-based method in the application of multimodal image registration. The improved multimodal image registration framework comprises the following six steps, as is shown in Fig. 1.

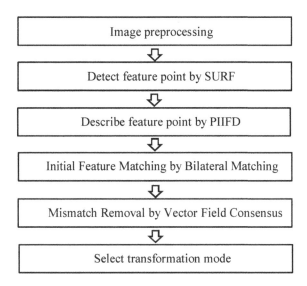

Fig. 1. The improved multimodal image registration framework SPBV consists of six steps.

In the following sections, we will introduce the improvements of our multimodal image registration framework in detail.

2.1 Image Preprocessing

In our improved framework, we need to do three image preprocessing steps before feature points extraction: (1) Resize images to a fixed size (about 500 * 500 pixels), as to eliminate scale differences in some extent; (2) Select green channel from the RGB input images; (3) Scale the intensities of images to the full 8 bits intensity range [0, 255].

2.2 SURF Feature Points Detector

Harris and Stephens [17] proposed a widely used corner detector called Harris corner detector in 1988. This method can generate corner points of rotation, illumination, and translation invariant, however Harris corners don't have scale invariant property. SIFT method can provide scale invariant points but it can't provide sufficient key points for multimodal image of high dimension transformation. Compared with other methods, SURF method can be computed more faster and can provide more repeatable feature points. To be clear that we only use SURF method to locate feature points, because of its superior performance in low time cost and high repeatability. We will introduce how to detect feature points by SURF in detail.

SURF method use Hessian matrix theory [18], integral images [19], box filter and scale theory. For a certain point $X = (x, y)$ in an image I at a definite scale σ, the Hessian matrix $H_L(X, \sigma)$ is computed as follows:

$$H_L(X, \sigma) = \begin{bmatrix} L_{xx}(X, \sigma) & L_{xy}(X, \sigma) \\ L_{xy}(X, \sigma) & L_{yy}(X, \sigma) \end{bmatrix} \tag{1}$$

where $L(X, \sigma) = G(X, \sigma) * I(X)$ means the convolution of image I with Gaussian kernel modal, L_{xx}, L_{xy}, L_{yy} means the Gaussian second order derivatives convolution.

SIFT method uses Difference of Gaussian (DOG) [20] to approximate Laplacian of Gaussian (LOG) operator to decrease time consumption. Similarly, SURF uses box filers to approximate the second order derivative model of Gaussian. Using box filters and integral images to speed up computing process and finally get approximate value D_{xx}, D_{yy}, D_{xy}. Because the use of approximate value, we need to balance the correlation in the expression. For example, we can approximate the $\sigma = 1.2$ Gaussian second order derivatives by the 9×9 box filters, which means the minimum scale. And then compute the normalized scale template ratio,

$$\frac{|L_{xy}(1.2)|_F |D_{xx}(9)|_F}{|L_{xx}(1.2)|_F |D_{xy}(9)|_F} \simeq 0.9 \tag{2}$$

where $|x|_F$ means the Frobenius norm. We can get the $\det(H(L, \sigma))$ directly,

$$\det(H_{approx}) = D_{xx}D_{yy} - (0.9D_{xy})^2 \tag{3}$$

Furthermore, SURF method uses box filters of different sizes to build the image pyramid. Contrary to SIFT method, SURF construct a inverted pyramid by keeping the image invariant and expanding the size of template gradually. Subsequently, we build the whole scale space step by step. We set the initial scale layer size of 9×9, and set the first layer of every octave $9 \times 9, 15 \times 15, 21 \times 21, 27 \times 27$ etc. In each octave we increase the filter size doubly, such as going from 6 to 12 etc.

Finally, in the $3 \times 3 \times 3$ neighborhood we can find feature points by a non-maximum suppression. Feature points detected by the above method are extreme points of discrete space, we can fitting the two dimensional function to locate points accurately and simultaneously remove low contrast and unstable edge response points. For more information about the SURF method, please see [12].

2.3   PIIFD Feature Descriptor

As we can know, PIIFD descriptor is one of the most excellent descriptor in the application of multimodal image registration. PIIFD method can detect rotation invariant interesting points, furthermore the descriptor is partially invariant to affine, intensity and viewpoint.

The idea is based on the assumption that the matched image regions have similar anatomical structure at corresponding locations in multimodal images. Because in multimodal images at corresponding locations the gradient orientations may be same or opposite and the gradient magnitudes maybe change greatly, we have to do two operations to guarantee the property of partial intensity invariant: (1) Degrade gradient orientation histogram from 16 bins to 8 bins by accumulate opposite gradient orientation, namely ranged orientation from $[0°, 360°]$ to $[0°, 180°]$; (2) Normalize the gradient magnitudes from 1 to 0, namely the minimum 20% is set to 0, the second minimum 20% is set to 0.25 and the maximum 20% is set to 1.

To achieve the property of rotation invariant, we will calculate main orientations to feature points after we detect them by SURF. As in multimodal images at corresponding locations the gradient orientations may be opposite, they will be canceled if only averaged gradients in an corner point neighborhood window. Differ from SIFT, we use a continuous average squared gradients method [21] instead of discrete orientation histogram method to keep the opposite direction. For double the angle and square the length, polar coordinates is introduced to convert the gradient vectors $[G_x, G_y]^T$ in Cartesian coordinates, which is $[G_\rho, G_\varphi]^T$. And the squared gradient vectors $[G_{s,x}, G_{s,y}]^T$ is calculated:

$$\begin{bmatrix} G_{s,x} \\ G_{s,y} \end{bmatrix} = \begin{bmatrix} G_\rho^2 \cos 2G\varphi \\ G_\rho^2 \sin 2G\varphi \end{bmatrix} = \begin{bmatrix} G_x^2 - G_y^2 \\ 2G_x Gy \end{bmatrix} \tag{4}$$

And then, the average squared gradient $[\overline{G_{s,x}}, \overline{G_{s,y}}]^T$ can be found in the image:

$$\begin{bmatrix} \overline{G_{s,x}} \\ \overline{G_{s,y}} \end{bmatrix} = \begin{bmatrix} \sum_W G_{s,x} \\ \sum_W G_{s,y} \end{bmatrix} = \begin{bmatrix} \sum_W G_x^2 - G_y^2 \\ \sum_W 2G_x Gy \end{bmatrix} = \begin{bmatrix} G_{xx} - G_{yy} \\ 2G_{xy} \end{bmatrix} \tag{5}$$

Next, the main orientation ϕ ranged from 0 to π can be presented by:

$$\phi = \frac{1}{2} \begin{cases} \tan^{-1}(\overline{G_{s,y}}/\overline{G_{s,x}}) + \pi, & \overline{G_{s,x}} \geq 0 \\ \tan^{-1}(\overline{G_{s,y}}/\overline{G_{s,x}}) + 2\pi, & \overline{G_{s,x}} < 0 \cap \overline{G_{s,y}} \geq 0 \\ \tan^{-1}(\overline{G_{s,y}}/\overline{G_{s,x}}), & \overline{G_{s,x}} < 0 \cap \overline{G_{s,y}} < 0 \end{cases} \tag{6}$$

Therefore, we can assign each feature point $p(x, y)$ a main orientation $\phi(x, y)$ by above steps. Similar to SIFT, extracting the descriptions in the local neighborhood of these feature points, the big square is composed by 4×4 small squares and each square denote an 8 bins orientation. Finally the PIIFD description is a $4 \times 4 \times 8 = 128$ dimension and is normalized to a unit length.

2.4 Initial Feature Matching by Bilateral Matching Based on BBF

We use the bilateral matching method based on best bin first (BBF) technique to achieve initial feature matching. For two images I1 and I2, corresponding point set descriptors vectors are U and V. According to the idea of BBF, we use euclidean distance to measure the similarity of point pairs, such as two point is initial matched only if the distance ratio is smaller than a threshold τ (set the τ value to 0.9 in our experiment), the ratio is defined as the closest distance to the second-closest distance.

On the thought of bilateral matching, we can briefly do the BBF technique from U to V and from U to V to achieve one-to-one correspondence.

2.5   Mismatch Removal by Vector Field Consensus

A vector field is a mapping function which defined by a vector function, it can assign a vector y to each location x to satisfy $y = f(x)$. Vector field consensus method (VFC) is a robust vector field interpolation technique which can learn the entire smooth vector field from sparse samples with noises and outliers. Further more, VFC is so robust that it even can tolerate 90% mismatches. We will give a detailed introduction to the VFC method which based on the representer theorem for Tikhonov regularization in reproducing kernel Hilbert space (RKHS) [22] as follows.

Firstly we must do some preprocessing operations: (1) Normalize each initial match (u_n, v_n) to $(\hat{u}_n, \hat{v}_n)$; (2) Convert it to a vector field sample (x_n, y_n), where $x_n = \hat{u}_n, y_n = \hat{v}_n - \hat{u}_n$ means that the output is the displacement of the corresponding point in another image. The sample pair $S = \{(x_n, y_n) \in \mathcal{X} \times \mathcal{Y}\}_{n=1}^{N}$ which composed by both inliers and outliers, where $\mathcal{X} \subseteq \mathbb{R}^P, \mathcal{Y} \subseteq \mathbb{R}^P$ are input and output space. VFC is aimed at learning a mapping f function to fit the inliers well, where $f : \mathcal{X} \to \mathcal{Y}, f \in \mathcal{H}$ and $\mathcal{H}$ represents a RKHS. We assume that the distribution of inliers is Gaussian with standard uniform deviation σ and zero mean μ, the distribution of outliers is uniform $1/a$. The parameter γ means the percentage of inliers in the samples. We set a parameter $z_n \in \{0, 1\}$, where 1 means that the inliers and 0 means the outliers. Thus we can construct the likelihood function for the mixture model:

$$p(Y|X, \theta) = \prod_{n=1}^{N} p(y_n, z_n | x_n, \theta) = \prod_{n=1}^{N} \left(\frac{\gamma}{(2\pi\sigma^2)^{D/2}} e^{-\frac{\|y_n - f(x_n)\|}{2\sigma^2}} + \frac{1-\gamma}{a} \right) \quad (7)$$

where $\theta = \{f, \sigma^2, \gamma\}$, $X_N \times P = (x_1, x_2, \ldots, x_N)^T$ and $Y_N \times D = (y_1, y_2, \ldots, y_N)^T$. Define the prior of f:

$$p(f) \propto e^{-\frac{\lambda}{2}\|f\|_{\mathcal{H}}^2} \quad (8)$$

where $\mathcal{H}$ is the norm of the RKHS and λ the regularization parameter. According to the maximum posterior $p_n = p(z_n = 1|x_n, y_n, \theta^{old})$ probability (MAP) theory $p(\theta|X, Y) \propto p(Y|X, \theta)p(f)$ and $q^* = \arg\max_{\theta} p(Y|X, \theta)p(f)$, where θ^* is the true estimate value of θ.

Using the EM algorithm, we can get $p_n = p(z_n = 1|x_n, y_n, \theta^{old})$, $\sigma 2$ and γ. Using the representer theorem and matrix kernel, we can get f:

$$f(x) = \sum_{n=1}^{N} \Gamma(x, x_n) c_n \quad (9)$$

where Γ is a matrix kernel and we can get c_n via $(\Gamma + \lambda\sigma^2 P^{-1})C = Y$. Finally, once the EM algorithm converges, we can get vector field and the consistent set.

2.6 Transformation Mode

Considering that there is no significant transformation in the multimodal images and the surface of retina is nearly spherical, we choose affine and second-order polynomial transformations. If we have more than two matches then we will apply the affine transformation, and if we have more than five matches then we will apply the second-order polynomial transformation. Finally we use the least square method to acquire the transformation function parameters.

3 Experiments and Result

In this section, we evaluate the proposed SPBV method and compare it with SURF-RPM and Harris-PIIFD on 70 pairs of multimodal retinal images, and all parameter settings are suggested by the authors. We implement all experiments in Matlab R2013a and test them on an CPU I5 3.20 GHz desktop computer with 8 GB of RAM.

3.1 Data Description

There is no recognized public benchmark data set in the application of multimodal image registration. We collect two groups multimodal image data sets to evaluate the registration performance of above three techniques. Both the two data sets comprise two modalities images: Red-Free (RF) and Fundus Auto-Fluorescene (FAF) [23]. The first data set contains 45 pairs images which is collected from previous related papers [9, 10] and their test sets, these images sizes range from 520 * 380 to 2400 * 2000 pixels. The second data set contains 25 pairs images which is collected from the internet, these images sizes approximately 500 * 500 pixels. The minimum overlap ratio of the multimodal retinal image pairs in the two data sets is about 20%.

3.2 Evaluation Criterion

It is very important to choose fair and reliable criterion to measure the performance of above registration methods. In this paper, we choose two criterions to evaluate the performance of these above three methods.

The first one is successful rate which mean whether the registration method acquire enough matching point pairs, for example three point pairs for affine transformation and six point pairs for quadratic transformation. In our paper, we choose six point pairs as the successful criterion.

The second one is the root of mean square error (RMSE), the median error (MEE) and the maximum error (MAE) computed by ground truth. Let (x_r, y_r) and (x_s, y_s) be the corresponding point pairs in the model and reference images. Because there is no available public multimodal image data set and ground truth, we obtain ground truth by manual: (1) select at least six matched point pairs (a_r, b_r) and (a_s, b_s) manually using MATLAB R2013a and using function 'cpcorr' to get the precise point locations. Note that these points should be distributed uniformly. Setting these corresponding point pairs to be manual ground truth. (2) via the manual ground truth to generate the quadratic transformation parameters by compute $(a_r, b_r) = A(a_s, b_s)$. (3) via the forward spacial transformation to get the transformed points $(x_t, y_t) = A(x_s, y_s)$. (4) calculate Euclidean distances between the transformed points (x_t, y_t) and the corresponding points (x_r, y_r) in the reference points. In this paper, we choose 12 point pairs manually and calculate RMSE as follows:

$$
\text{RMSE} = \sqrt{\frac{\sum\limits_{k=1}^{n} (x_r^k - x_t^k)^2 + (y_r^k - y_t^k)^2}{n}} \tag{10}
$$

where $n = 12$. In this ground truth measure section, we can use RMSE, MEE and MAE values to evaluate the performance of registration methods. We classify the registration results as three groups: acceptable (MAE $\leq$ 10 pixels and MEE $\leq$ 1.5 pixels), inaccuracy (MAE $\leq$ 10 pixels and MEE $>$ 1.5 pixels) and incorrect (MAE $>$ 10 pixels).

3.3   Experiment Results

Table 1 shows success rate of Harris-PIIFD, SURF-RPM and SPBV in the two multimodal data sets, namely whether these methods obtain six matching point pairs. Through experiments, our SPBV method is proved to be the highest success rate among these three methods. Our method can generate enough matching point pairs even in the low equality images or small overlapping. Harris-PIIFD perform badly when faced low equality or small overlapping images.

Table 1. Success rate for all three methods.

Criterion	Harris-PIIFD	SURF-RPM	SPBV
Success rate (%)	62.86	84.29	91.43

We classify the registration results as three groups: acceptable (MAE $\leq$ 10 pixels and MEE $\leq$ 1.5 pixels), inaccuracy (MAE $\leq$ 10 pixels and MEE $>$ 1.5 pixels) and incorrect (MAE $>$ 10 pixels). Using our SPBV approach, some registration results of final matching are shown in Fig. 2(a)–(c) which on behalf of these three groups.

We evaluate the accuracy of Harris-PIIFD, SURF-RPM and our proposed SPBV by RMSE and MEE as shown in Table 2. Our SPBV method performs better than other methods in accuracy with low values both in three groups and overall results.

However, as shown in Table 3 our SPBV method consumes more time than Harris-PIIFD method. On the other hand, SURF-RPM consumes more time than Harris-PIIFD and our method. So in consideration of both Tables 2 and 3, our SPBV method gets balance between computational complexity and performance.

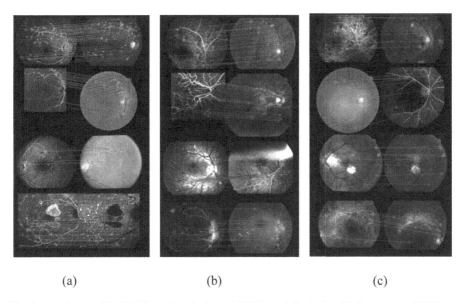

(a)   (b)   (c)

Fig. 2. (a) Acceptable (MAE $\leq$ 10 pixels and MEE $\leq$ 1.5 pixels); (b) inaccuracy (MAE $\leq$ 10 pixels and MEE > 1.5 pixels); (c) incorrect (MAE > 10 pixels).

Table 2. RMSE and MEE for all three methods.

Groups	Criterion	Harris-PIIFD	SURF-RPM	SPBV
Acceptable	RMSE	2.43 $\pm$ 0.57	2.25 $\pm$ 0.55	1.73 $\pm$ 0.68
Acceptable	MEE	1.33 $\pm$ 0.22	1.44 $\pm$ 0.25	0.95 $\pm$ 0.27
Inaccuracy	RMSE	4.46 $\pm$ 1.69	4.07 $\pm$ 0.76	4.10 $\pm$ 1.31
Inaccuracy	MEE	3.56 $\pm$ 1.36	3.05 $\pm$ 1.38	2.96 $\pm$ 0.71
Incorrect	RMSE	31.41 $\pm$ 23.12	22.20 $\pm$ 19.08	13.35 $\pm$ 14.98
Incorrect	MEE	17.62 $\pm$ 13.57	16.29 $\pm$ 14.57	6.02 $\pm$ 6.42
Overall	RMSE	25.65 $\pm$ 23.12	14.60 $\pm$ 13.29	6.56 $\pm$ 4.40
Overall	MEE	12.59 $\pm$ 11.47	8.74 $\pm$ 10.59	4.84 $\pm$ 2.46

Table 3. Runtime for all three methods.

Criterion	Harris-PIIFD	SURF-RPM	SPBV
Runtime (s)	819.88	2289.71	1926.21

4   Conclusion

In this work, we proposed a simple and robust framework SPBV to achieve multimodal images registration. The experiment results demonstrate that our proposed SPBV method outperforms the existing methods in accuracy when testing on our two multimodal data sets. Our method can be used to register images of large initial mismatching or rotation change and it is robust to register low quality or small overlapping multimodal images.

Further work is to apply the SPBV to other area, not only in the multimodal retinal images. In the experiment, parameter is setting by manual, other future work will pay more attention to parameter automatization.

Acknowledgement. This work was supported by the Shanghai Innovation Action Project of Science and Technology (15DZ1101202) and the National Key Technology Support Program of China (No. 2015BAF17B00).

References

1. Laliberté, F., Gagnon, L., Sheng, Y.: Registration and fusion of retinal images-an evaluation study. IEEE Trans. Med. Imaging **22**(5), 661–673 (2003)
2. Chanwimaluang, T., Fan, G., Fransen, S.R.: Hybrid retinal image registration. IEEE Trans. Inf. Technol. Biomed. **10**(1), 129–142 (2006)
3. Cideciyan, A.V.: Registration of ocular fundus images: an algorithm using cross-correlation of triple invariant image descriptors. IEEE Eng. Med. Biol. Mag. **14**(1), 52–58 (1995)
4. Legg, P.A., Rosin, P.L., Marshall, D., et al.: Improving accuracy and efficiency of mutual information for multi-modal retinal image registration using adaptive probability density estimation. Comput. Med. Imaging Graph. **37**(7), 597–606 (2013)
5. Kolar, R., Harabis, V., Odstrcilik, J.: Hybrid retinal image registration using phase correlation. Imaging Sci. J. **61**(4), 369–384 (2013)
6. Ma, J., Zhou, H., Zhao, J., et al.: Robust feature matching for remote sensing image registration via locally linear transforming. IEEE Trans. Geosci. Remote Sens. **53**(12), 6469–6481 (2015)
7. Studholme, C., Hawkes, D.J., Hill, D.L.: Normalized entropy measure for multimodality image alignment. In: Medical Imaging 1998, International Society for Optics and Photonics, pp. 132–143 (1998)
8. Brown, L.G.: A survey of image registration techniques. ACM Comput. Surv. (CSUR) **24**(4), 325–376 (1992)
9. Chen, J., Tian, J., Lee, N., et al.: A partial intensity invariant feature descriptor for multimodal retinal image registration. IEEE Trans. Biomed. Eng. **57**(7), 1707–1718 (2010)
10. Ghassabi, Z., Shanbehzadeh, J., Sedaghat, A., et al.: An efficient approach for robust multimodal retinal image registration based on UR-SIFT features and PIIFD descriptors. EURASIP J. Image Video Process. **2013**(1), 1–16 (2013). doi:10.1186/1687-5281-2013-25
11. Lowe, D.G.: Distinctive image features from scale-invariant keypoints. Int. J. Comput. Vis. **60**(2), 91–110 (2004)
12. Bay, H., Tuytelaars, T., Gool, L.: SURF: speeded up robust features. In: Leonardis, A., Bischof, H., Pinz, A. (eds.) ECCV 2006. LNCS, vol. 3951, pp. 404–417. Springer, Heidelberg (2006). doi:10.1007/11744023_32
13. Liu, C., Ma, J., Ma, Y., et al.: Retinal image registration via feature-guided Gaussian mixture model. JOSA A **33**(7), 1267–1276 (2016)
14. Yang, G., Stewart, C.V., Sofka, M., et al.: Registration of challenging image pairs: initialization, estimation, and decision. IEEE Trans. Pattern Anal. Mach. Intell. **29**(11) (2007)

15. Beis, J.S., Lowe, D.G.: Shape indexing using approximate nearest-neighbour search in high-dimensional spaces. In: Proceedings 1997 IEEE Computer Society Conference on Computer Vision and Pattern Recognition, pp. 1000–1006. IEEE, (1997)
16. Ma, J., Zhao, J., Tian, J., et al.: Robust point matching via vector field consensus. IEEE Trans. Image Proc. **23**(4), 1706–1721 (2014)
17. Harris, C., Stephens, M.: A combined corner and edge detector. In: Alvey Vision Conference, vol. 15, no. 50 (1988). 10.5244
18. Lindeberg, T.: Feature detection with automatic scale selection. Int. J. Comput. Vis. **30**(2), 79–116 (1998)
19. Viola, P., Jones, M.: Rapid object detection using a boosted cascade of simple features. In: Proceedings of the 2001 IEEE Computer Society Conference on Computer Vision and Pattern Recognition, 2001, CVPR 2001, vol. 1, p. I-I. IEEE (2001)
20. Lowe, D.G.: Object recognition from local scale-invariant features. In: The Proceedings of the Seventh IEEE International Conference on Computer vision, vol. 2, pp. 1150–1157. IEEE (1999)
21. Bazen, A.M., Gerez, S.H.: Systematic methods for the computation of the directional fields and singular points of fingerprints. IEEE Trans. Pattern Anal. Mach. Intell. **24**(7), 905–919 (2002)
22. Aronszajn, N.: Theory of reproducing kernels. Trans. Am. Math. Soc. **68**(3), 337–404 (1950)
23. Quellec, G., Lamard, M., Cazuguel, G., et al.: Automated assessment of diabetic retinopathy severity using content-based image retrieval in multimodal fundus photographs. Invest. Ophthalmol. Vis. Sci. **52**(11), 8342–8348 (2011)

Intelligent Computing in
Communication Networks

A New Indoor Location Method Based on Real-Time Motion and Sectional Compressive Sensing

Yichun Li[(⊠)] and Ningkang Jiang[(⊠)]

East China Normal University,
No. 3663 North Zhongshan Road, Putuo, Shanghai, China
378820235@qq.com, nkjiang@sei.ecnu.edu.cn

Abstract. This paper presents a sectional algorithm for indoor location using wireless sensor networks. This algorithm uses the motion regularity of target to compute the next motion area quickly and apply the pre-processed compressive sensing method to that area, which reduce the location problem to a sparse signal reconstruction problem. Then we carry out the proposed algorithm on the motion of next time turn by turn, such procedure is able to locate with fewer data collection, wireless links and wireless nodes as well as raise the accuracy of location. The simulation results show that the proposed algorithm of dynamic motion based compressive sensing sectional location method has a good performance.

Keywords: Real-time motion · RSSI · Wireless · Compressive sensing · Indoor location

1 Introduction

Currently, there are two main location methods, one is based on ranging algorithm, and the other is algorithm without distance measurement. Ranging algorithm obtains the relation between the location information of unknown target node and wireless nodes, then determine the location by algorithms based on triangulation or maximum likelihood estimation [1]. On the contrary, the algorithm does not need the distance and angle information, which is able to get node location according to the network connectivity and other information [2].

Traditional real time target locating methods have to collect, store and process huge amount of data and traditional compressive sensing locating methods demand complicated calculation and a useful observation matrix. Many compressive sensing based methods are proposed to solve the problems. [3] proposed a compressive sensing method to recover sparse signals from a small number of noisy measurements by solving an l1-minimization problem. [4] proposed a block-sparsity-based locating method in wireless sensor networks based on OMP.

Although the hardware deployment and signal acquisition of RSSI distance estimation method [5] cost less time and space, the indoor environment is more complex, and the RSSI value is greatly affected by the environment. It is necessary to deploy the

© Springer International Publishing AG 2017
D.-S. Huang et al. (Eds.): ICIC 2017, Part III, LNAI 10363, pp. 143–155, 2017.
DOI: 10.1007/978-3-319-63315-2_13

beacon nodes in a wide range of directions in the environment to improve the precision and accuracy of RSSI measurement.

In order to reduce the cost and raise the performance of indoor locating, we proposed an indoor locating algorithm PPCSGPSSR- LACA. The algorithm is carried out in an 8 m × 8 m room, where the wireless nodes are distributed uniformly as beacon nodes. We use the sensor to do actual motion sampling and then use the median filtering to process the sample data, from which we estimate the approximate locating area for signal reconstruction. The first part of the algorithm is called LACA, which uses a low cost received signal strength indication(RSSI) distance measurement technology and the characteristic of low velocity motion under indoor environment to predict the location area. Then we use this area to wake up the sleeping nodes to participate in the locating. The most important part of PPCSGPSSR- LACA is the sparse signal reconstruction method called PPCSGPSSR. We pre-process the change of RSSI value data, which is the basic of this part. In the simulation experiment, we validate the algorithm in three aspects: performance, accuracy and energy saving, and the experimental result is encouraging.

2   Related Works

We made research in several categories, such as location system model, shadowing model and filtering model.

2.1   Locating Model Construction

In this paper, the basic locating model is described in Fig. 1, all the wireless nodes evenly distributed around the locating area. The whole area is divided into the N × N grid, each grid can be regarded as a pixel and the central point of each grid is the location of this pixel. If a link is blocked by the object, the RSSI measured value of this link will be definitely different from the RSSI measured value when the link is not

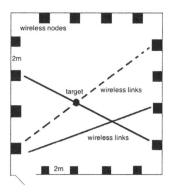

Fig. 1. Location model

blocked. By searching the target in the areas of which the RSSI values change most, we can estimate approximate locating area of the target.

2.2 Shadowing Model

Displayed equations or formulas are centered and set on a separate line (with an extra line or half-line space above and below). Displayed expressions should be numbered for reference. Shadow Model is a mathematical model based on the characteristic that in the process of wireless signal propagation, the signal intensity decreases with the increase of distance. Currently, the most common theory model in the wireless transmission is Shadowing model [6].

$$R_i t = G_i + P_i - L_i. \tag{1}$$

$$P = P_0 + 10 n \lg \frac{d}{d_0} + \xi. \tag{2}$$

$R_i t$ is the RSSI value of the ith wireless link where G_i is the node acceptance increment, L_i is the signal attenuation value at the distance of 1 m. d_0 is the reference distance, P_0 is the signal strength received at d_0, d is the distance from the current position to the wireless signal source. P is the wireless signal strength of target to be measured received at current location, ξ is the shadow factor, of which the average is 0 and the mean square error is σ_{dB} normal random variable, n is the path loss index.

2.3 Filtering Model

Median filtering is able to effectively suppress the noise based on order statistical theory. The basic principle of Median filtering is that the value of a point in a sequence of numbers is replaced by the median value of each point in a neighborhood of that point, which makes the value of the surrounding close to the true value and estimate the isolated noise. In this paper, the median filtering is used to sample multiple RSSI values for a given position and its vicinity and we can obtain the numerical data that is close to the true state as the result of computation by filtering.

3   Construction of Locating Area

According to the shadow model, the target pixels in the locating area are the pixels that contribute more to the change of RSSI ΔR. That is to say, if we want to find out those pixels, we only need to compute and vote the change of every pixel Δp_i. In order to reduce the locating time, we proposed a method based on motion prediction model to predict the location range of search area and reduce the locating range, which saves the search time.

3.1   Single State Motion Location Model

When we can only collect few history pixel points data on the movement, we can construct the prediction model through single state motion model based on triangular locating and real time moving states. The velocity can be defined as follows:

$$v_t = v_0 + \int_0^t a_t dt. \tag{3}$$

where v_0 is the initial velocity. The distance of the moving direction at time t can be defined as follows:

$$S = \int_0^t v_t dt. \tag{4}$$

Since the movement of human body has the randomness, we tend to catch the instantaneous acceleration of motion (a_{xt}, a_{yt}) in continuous time. The characteristic of low velocity movement splits the motion into several quite short periods of time Δt. In every Δt, we suppose that the initial velocity is $\left(\sum_{t=0}^n a_{xt}\Delta t, \sum_{t=1}^n a_{yt}\Delta t\right)$ and the acceleration is (a_{xt}, a_{yt}). The displacements computed as follows:

$$S_t = v_{t0}T + \frac{1}{2}a_t T^2. \tag{5}$$

$$S = \sum_{t=0}^n S_t. \tag{6}$$

$$\alpha = \arctan\left(\frac{S_x}{S_y}\right). \tag{7}$$

we estimate the approximate location of the object according to the Figs. 2 and 3.

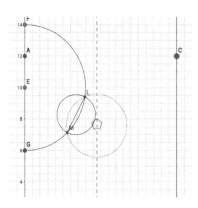

Fig. 2. Single state motion tracking

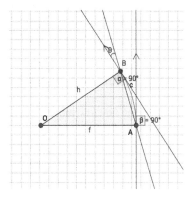

Fig. 3. History states motion tracking

We can estimate the final locating candidate pixel set $\mathbb{S}_1$, which is composed of the pixels in the circle with diameter ML.

3.2 History States Motion Location Model

For objects with regular moving, suppose that we have known the two history position pixels A and B on the track, we can calculate the track radius R by position A and B. First we obtain the track radius R_{AB} of object from Fig. 3 as follows:

$$R_{AB} = \frac{l_{AB}}{2 \sin \theta}. \tag{8}$$

where $l_{AB} = \sqrt{(x_A - x_B)^2 + (y_A - y_B)^2}$; θ is the angle of vector $\overrightarrow{AB}$ and v_B, then we have:

$$R_{AB} = \frac{l_{AB}}{2} \times \sqrt{1 + \left(\frac{1 + k_{AB}k_B}{k_{AB} - k_B}\right)^2}, \tag{9}$$

where $k_{AB} = \frac{y_B - y_A}{x_B - x_A}$; $k_{AB} = \tan \theta_B$, θ_B is the direction of moving object at point B.

According to the Eq. (8), we calculate the track radius in this Δt period, then we choose m times by Eq. (9) to obtain the corresponding radius set $S_R = \{R_1, R_2, \cdots, R_m\}$. Now we let P be the probability of the S_R where $R \leq \lambda r$. P^1 is the threshold used to judge the types of moving track. If $P \geq P^1$, the track is an arc, otherwise, it is a linear track. In the linear track motion prediction, let (x_1, y_1) be the initial pixel of moving object, v_1 be the velocity, θ_1 be the moving direction, t be the timestamp and the linear moving track is defined as follows:

$$\begin{cases} x = x_1 + v_1 t \cos \theta_1 \\ y = y_1 + v_1 t \sin \theta_1 \end{cases}. \tag{10}$$

In Arc track motion prediction, we choose the radiuses that satisfy the condition $R_1 \leq \lambda r$ from the set S_R and construct the new radius set $S_R^1 = \{R_1^1, R_2^1, \cdots, R_n^1\}$. The arc moving track radius R of moving object with irregular moving is defined as follows:

$$R = \frac{1}{n} \sum_{i=1}^{n} R_i^1 \tag{11}$$

Let A (x_A, y_A) be the initial pixel of moving object, B (x_B, y_B) be the history pixel, θ_A is the moving direction angle at pixel A, B be the track radius and we can define the coordinates of two pixels as follows according to Fig. 4:

$$O \begin{cases} x_0 = x_A - R \sin \theta_A \\ y_0 = y_A + R \cos \theta_A \end{cases} \quad O' \begin{cases} x_{O'} = x_A - R \sin \theta_A \\ y_{O'} = y_A - R \cos \theta_A \end{cases} \tag{12}$$

According to the Fig. 4, let l_{BO} be the distance between B and O, $l_{BO'}$ be the distance between B and O'. If $l_{BO} > l_{BO'}$, the center of the circle track should be O', otherwise, it should be O.

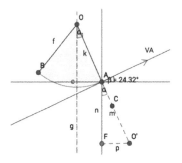

Fig. 4. History states motion center

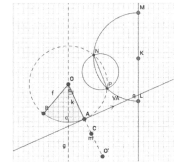

Fig. 5. History states motion tracking

In the experimental environment, we adopt the obtained states of moving object into the history track prediction model and get the final result, which contains the track type τ, track radius R, moving direction angle α of moving object and center of track O. Then we estimate the approximate location of the moving object according to the Fig. 5.

We can estimate the final locating candidate pixel set $\mathbb{S}_2$, which is composed of the pixels in the circle with diameter NP.

3.3   LACA Algorithm

The (LACA) locating area construction algorithm is described in the Algorithm 1. τ is the timestamp, the hardware address of beacon node is B_{addr} and the signal strength of beacon node is recorded as s B_{strength}. We use T to denote the signal acquisition of sensing device and use m to denote the times of sampling for tracking. The set of instantaneous acceleration of two vertical directions in the moving plane is $S_a = \{(a_{x_1}, a_{x_2}), (a_{x_2}, a_{x_2}), \ldots, (a_{x_m}, a_{x_m})\}$. The set of RSSI value in the acquisition is $R = \{r_1, r_2, \ldots, r_m\}$ and the set of history locating pixel is $S_B = \{(x_{B_1}, y_{B_1}), (x_{B_2}, y_{B_2}), \ldots, (x_{B_m}, y_{B_m})\}$. Then we use $A(x_A, y_A)$ to denote the current locating pixel and use t_i to denote the interval between the runs of Algorithm 1. The max velocity of moving object V_m. The direction angle set of $\vec{B}A = \{\theta_1, \theta_2, \ldots, \theta_m\}$., the positive adjustment coefficient is λ and the average probability of arc type is P^1.

4 Compressive Sensing Model

Based on a series of pre-process of measured RSSI value, we proposed and construct the compressive sensing model and PPCSGPSSR-LACA algorithm.

4.1 RSSI Ranging Model

Based on the Eq. (2), we defined the RSSI measured value $R_i t$ of ith wireless link as follows:

$$R_i t = -L_i - 10\beta \lg d_i - S_i(t) + v_i(t). \tag{13}$$

where L_i is the signal attenuation at the distance of 1 m, β is the path attenuation factor, d_i is the distance from the current position to the wireless signal source, $S_i(t)$ is the signal attenuation caused by environmental disturbance and $v_i(t)$ is the environmental noise. At the time t_1 and t_2, the change ΔR_i of RSSI measured value is

$$\Delta R_i = R_i(t_1) - R_i(t_2). \tag{14}$$

$$\Delta R_i = S_i(t_2) - S_i(t_1) + v_i(t_1) - v_i(t_2). \tag{15}$$

Thus we can see that, ΔR_i mainly determined by the shadow attenuation and the shadow attenuation can be approximated as the sum of each pixel attenuation as follows:

$$\Delta R_i = \sum_{j=1}^{N} \lambda_1 w_{ij} \Delta p_j + \lambda_2 v_i. \tag{16}$$

where N is the number of pixels in the area, Δp_j is the change of RSSI measured value in pixel j, w_{ij} is the contribution of pixel j to the change of link i. In this paper, we choose ellipse model [7] to describe the w_{ij}.

Algorithm 1: LACA algorithm

Input : τ, strength, T, m,
$S_o\{(a_{x_1},a_{y_1}),(a_{x_2},a_{y_2}),\ldots,(a_{x_m},a_{y_m})\}$
$\{r_1,r_2,\ldots,r_m\}$
$S_B = \{(x_{B_1},y_{B_1}),(x_{B_2},y_{B_2}),\ldots,(x_{B_m},y_{B_m})\}$
$A (x_A,y_A)$, $BA=\{\theta_1,\theta_2,\ldots,\theta_m\}$
λ, P^1, t_1, V_m

Output: the set of pixels K

1 $S_R := \emptyset$, $K := \emptyset$;
2 for $j = 1$ and $j < m$ do
3 | $R_j :=$ computed by Eq.(18);
4 | $S_R := S_R \cup \{R_j\}$;
5 end
6 arcCount := 0, area$_t$:= $\emptyset$;
7 for $R \in S_R$ do
8 | if $R \leq \lambda V_m \times T$ then
9 | | arcCount := arcCount + 1;
10 | end
11 end
12 trackType := arcCount/m $\geq P^1$? arc : linear;
13 if $S_B == \emptyset$ or trackType $\neq$ arc then
14 | $S_B := S_1 \cup S_2 \cup \ldots \cup S_m$; resultSet := $\emptyset$;
15 | for $S \in S_B$ do
16 | | for data $\in S$ do
17 | | | computed $S_{r_{x-y}}$ by Eq.(8);
18 | | | computed S_x by Eq.(14);
19 | | | computed S_y by Eq.(14);
20 | | end
21 | | displacement of S $(S_x,S_y) := (avg(S_x),avg(S_y))$;
22 | | $K_i := (x_i,y_i)$ according to Fig.2(l);
23 | | resultSet := resultSet $\cup \{K_i\}$;
24 | end
25 | area$_t$:= max(circle with the center K_iK_j);
26 else
27 | $R := S_R$, o := (x_0,y_0);
28 | computed O by Eq.(21);
29 | computed O' by Eq.(22);
30 | if $l_{BO} > l_{BO'}$ then
31 | | o := O'
32 | else
33 | | o := O
34 | end
35 | computed the NP according to the Fig.3(τ);
36 | area$_t$:= circle with the center NP;
37 end
38 for pixeli $\in N$ do
39 | if $i \in$ area$_t$ then
40 | | $K := K \cup i$;
41 | end
42 end
43 return K;

The RSSI value is variable according to the experiment, we can express the variables in the form of matrix as follows according to Eq. (16):

$$\Delta R = \lambda_1 W \Delta p + \lambda_2 v. \tag{17}$$

Where ΔR is a $M \times 1$ vector, representing the change of RSSI value, $\Delta R = [\Delta R_1, \Delta R_2, \cdots, \Delta R_M]^T$, v is the $M \times 1$ environment noise vector. Δp is the $N \times 1$ reconstructed result signal vector, $\Delta p = [\Delta p_1, \Delta p_2, \cdots, \Delta p_N]^T$. W is the weight contribution to the corresponding link of each pixel and ij are the coordinates of pixels.

4.2   Median Filtering Model

According to the median filtering model, we collect several groups of RSSI value and estimate the result by linear regression using filtering algorithm. Finally, we compute the parameter A and n that are close to the truth. The mean value filtering model to process the motion data as follows.

$$RSSI = \frac{1}{n} \sum_{i=1}^{n} RSSI_i \tag{18}$$

$$a = \frac{1}{n} \sum_{i=1}^{n} a_i \tag{19}$$

After obtaining the data for the first time, we group the data of several continuous time slot and adopt the mean process to the data of new groups.

4.3   Orthogonal Preprocess Compressive Sensing Model

In order to reconstruct the sparse signals by a few sample signals based on compressive sensing theory [8], we need to ensure that the signals should have sparsity under certain condition. Suppose that there are K targets in the whole locating area, if one of the targets is in the j pixel grid, the value of Δp_j is non-zero. Otherwise, the value of Δp_j is zero. we can easily get that the number of targets in the area equals the number of non-zero elements in Δp. If the number of non-zero elements in Δp is much smaller than the number of pixels, then Δp is sparse and we can reconstruct is by signal reconstruction algorithm [9]. The reconstruction can be defined based on ℓ_1-norm function as follows:

$$\Delta \hat{p} = argmin \|\Delta p\|_1, \quad s.t. \quad \|\Delta R - W \Delta p\|_2 < v. \tag{20}$$

The condition it must satisfy is very hard and can't be satisfied commonly. Thus, we tend to preprocess the RSSI observation vector rather than using ΔR and W directly to reconstruct Δp. Using P and Z to pre-process the RSSI observation vector, the result of $\Delta \tilde{R}$ is as follows:

$$\Delta \tilde{R} = PZR. \tag{21}$$

where $P = ZW^{-1}$, $Z = orth(W^T)^T$ and according to the Eq. (17), we can rewrite the Eq. (21) as follows:

$$\Delta \tilde{R} = ZW^{-1} \Delta R = ZW^{-1W} \Delta p + \lambda_2 PW^{-1} v = \lambda_1 Z \Delta P + \lambda_2 \tilde{v} \tag{22}$$

Then we proposed the PPCSGPSSR (preprocessed compressive sensing gradient pursuit sparse signal reconstruction) algorithm to solve the sparse signal reconstruction problem.

4.4 PPCSGPSSR Algorithm

After obtaining the pre-processed $\Delta\tilde{R}$ and locating area K according to the Algorithm1, we proposed PPCSGPSSR(preprocessed compressive sensing gradient pursuit sparse signal reconstruction)algorithm to reconstruct the source signals. The principal of the algorithm is that it updates the search direction by gradient of each iteration, thus each iteration can make the objective function to be optimized gradually decreased. We tend to find the pixel j that minimize the value of $\|y - \varphi\Phi\|_2$, where Φ is a N × 1 vector that has only one non-zero element. In this paper, we compute the orthogonal weight matrix φ according to the relationship between the wireless links and pixel location in the location environment. Then we update the vector y by new vector $y - \varphi\Phi$. If there is no non-zero element in the whole iteration, then we end the algorithm to find the max $\Delta\hat{P}$. The PPCSGPSSR Algorithm is shown in Algorithm 2.

Algorithm 2: PPCSGPSSR Algorithm

Input : pre-processed RSSI mesured value vector $\Delta\tilde{R}$
orthogonal weight matrix Z
Approximate locating area K computed by LACA Algorithm
Output: N × 1 current time reconstruction of signals $\Delta\tilde{p}$

1 $y := \Delta\tilde{R}$;
2 $\varphi := Z$;
3 $\phi := [\phi_1, \phi_2, ..., \phi_N]^T$;
4 **for** $\phi_i \in \phi$ **do**
5 | $\phi_i \leftarrow 0$;
6 **end**
7 $i = 0$;
8 **while** *true* **do**
9 | **for** pixel $i \in K$ **do**
10 | | $tmp := [0, 0, ..., \phi_i, ..., 0, 0]^T$;
11 | | $(i, \phi_i) \leftarrow argmin_{i \in K} \|y - \varphi tmp\|_2$;
12 | | **if** $\phi_i == 0$ **then**
13 | | | break;
14 | | **end**
15 | | $\Delta\hat{p} \leftarrow \Delta\hat{p} + \phi_i$;
16 | | $\phi = [0, 0, ..., \phi_i, ..., 0, 0]^T$;
17 | | $y \leftarrow y - \varphi\phi$;
18 | | $\phi_i \leftarrow 1$;
19 | **end**
20 | **for** $\phi_i \in \phi$ **do**
21 | | **if** $\phi_i == 0$ **then**
22 | | | break;
23 | | **end**
24 | | **return** max$\Delta\tilde{p}$;
25 | **end**
26 **end**
27 **return** $\Delta\tilde{p}$;

4.5 PPCSGPSSR-LACA Algorithm

Next, we adopt PPCSGPSSR algorithm to LACA algorithm. In the approximate area computed by LACA algorithm, we compare the change of each pixel p_i in that area effectively and find out the pixel that contributes most. The whole algorithm PPCSGPSSR- LACA is described as Algorithm 3.

Algorithm 3: PPCSGPSSR-LACA Algorithm

Input : real time motion data set M at time t
         pre-processed RSSI mesured value vector $\Delta \tilde{R}$
         orthogonal weight matrix Z
Output: coordinate(x, y)
1 **while** *true* **do**
2     $K :=$
     $LACA(M.T, M.S_a, M.S_B, M.A(x_A, y_B), M.\acute{B}A, M.t_i, M.V_m, M.\lambda, M.P^1);$
3     $\Delta \tilde{p} := PPCSGPSSR(\Delta \tilde{R}, Z, K);$
4   **if** $\Delta \tilde{p} \neq null$ **then**
5       $i := i$ of $\Delta \tilde{p};$
6   **return** center location of $i(x, y);$
7   **end**
8 **end**

---

5 Experiment

In this paper, we place the routers in a small room with four walls and a door. There are 16 wireless nodes in the whole area. The internal between nodes is 1 m and the experiment is carried out in an 8 m × 8 m room. The receiving device is an Android terminal with a wireless receiver and a motion acceleration sensor supporting the 802.11 protocol. The number of pixel points is 64 and the size of pixel point is $2.0 \times 2.0\,m^2$. We take 50 wireless links as samples to measure performance of different locating algorithm and we compare the final locating effect of different locating method under 100 wireless links.

We compare the proposed algorithm PPCSGPSSR- LACA with ML, L1, OMP and IGMP algorithm. First, we tend to obtain the wireless signal strength blocked by barrier, the result is shown in the Fig. 6. The locating effect is shown in Fig. 7, we catch the locating result and the supposed location of target at time t_1 and t_2.

The result is indicative of the greater locating effect using PPCSGPSSR- LACA algorithm. We sum up the mean square error of different locating method under 50 wireless links as Fig. 8. The locating error of proposed method PPCSGPSSR- LACA is far much lower than others when we don't have enough wireless links data as is shown in Fig. 9. The computational complexity comparisons of different methods are shown in the Table 1.

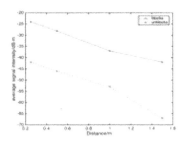

Fig. 6. Signal strength

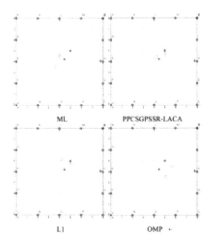

Fig. 7. Effects of locating under different method

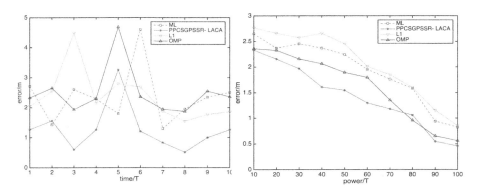

Fig. 8. Mean square error   **Fig. 9.** Certain signal power

Table 1. The computational complexity of methods.

Algorithm	Avg. computational complexity(s)
ML	0.054
PPCSGPSSR-LACA	0.043
OMP	0.056
L1	0.119

6 Conclusion

This paper proposed a sectional compressive sensing locating method based on real time motion area. This method aims to achieve the accurate locating with the measurement of fewer wireless links to some degree. It is based on shadow model as well as formulated sparsity approaches. In order to reduce the cost of locating, we design the new representation of sparse observation matrix which is easy to implement and low correlation with the sparse observation matrix. At the same time, we carry out the PPCSGPSSR algorithm in a small area that computed by LACA algorithm according to the motion of next time, which make up the shortcut that usually caused by the traditional locating where the wireless measurements are fewer and consumption of energy is higher. In the experiment, we verified the effectiveness of the algorithm according to the simulation experiment.

Although the PPCSGPSSR- LACA algorithm achieves a more accurate effect as well as saving the energy, it still costs time to process and store the motion data. Therefore, we will further study how to use fewer motion data to apply to the LACA algorithm and wake up fewer wireless nodes to participate in the reconstruction of sparse signals to recuse the time for calculating.

References

1. Zhu, M.-H., Zhang, H.-Q.: Research on model of indoor distance measurement based on RSSI. Transducer Microsyst. Technol. 19–22 (2010)
2. Liu, X.-D., He, W., Tian, Z.-S.: The improvement of RSS-based location fingerprint technology for cellular networks. In: International Conference on (CSSS) 2012, pp. 1267–1270 (2012)
3. Benkic, K., Malajner, M., Planinsic, P., et al.: Using RSSI value for distance estimation in wireless sensor networks based on ZigBee. In: 15th International Conference on Systems, Signals and Image Processing, IWSSIP 2008, pp. 303–306. IEEE (2008)
4. Feng, C., Au, W.S.A., Valaee, S., et al.: Received-signal-strength-based indoor positioning using compressive sensing. IEEE Trans. Mob. Comput. **11**(12), 1983–1993 (2012)
5. Bay, A., Carrera, D., Fosson, S.M., Fragneto, P., Grella, M., Ravazzi, C., Magli, E.: Block-sparsity based location in wireless sensor networks. EURASIP J. Wirel. Commun. Netw. 1–15 (2015)
6. Patwari, N., Agrawal, P.: Effects of correlated shadowing: connectivity, location, and RF tomography. In: Proceedings of the 7th International Conference on Information Processing in Sensor Networks, pp. 82–93. IEEE Computer Society, April 2008
7. Zhou, J., Chu, K.M.-K., Ng, J.K.-Y.: Providing location services within a radio cellular network using ellipse propagation model. In: 19th Advanced Information Networking and Applications (AINA 2005) (AINA papers), vol. 1, pp. 559–564 (2005)
8. Wang, J., Gao, Q., Zhang, X., et al.: Device-free location with wireless networks based on compressive sensing. IET Commun. **6**(15), 2395–2403 (2012)
9. Zhang, B., Cheng, X., Zhang, N., Cui, Y., Li, Y., Liang, Q.: Sparse target counting and location in sensor networks based on compressive sensing. In: Proceedings IEEE INFOCOM 2011, pp. 2255–2263. IEEE, April 2011

An Efficient Allocation Mechanism for Crowdsourcing Tasks with Minimum Execution Time

Xiaocan Wu[1], Danlei Huang[1], Yu-E Sun[2,5(✉)], Xiaofei Bu[3], Yu Xin[4], and He Huang[1,5]

[1] School of Computer Science and Technology, Soochow University, Suzhou, China
[2] School of Urban Rail Transportation, Soochow University, Suzhou, China
sunye12@suda.edu.cn
[3] College of Software, Shenyang Normal University, Shenyang, China
[4] Beijing Institute of Remote Sensing Information, Beijing, China
[5] Suzhou Institute for Advanced Study, University of Science and Technology of China, Suzhou, China

Abstract. Crowdsourcing is used to leverage external crowds to perform specialized tasks quickly and inexpensively. In the application of crowdsourcing, one task may often include many ordered steps. Based on the different requirements (*e.g.* workers' skills and *etc.*) of steps, the task requester may divide the task into many sub-tasks, and publish the subtasks in the crowdsourcing system. Moreover, service requesters usually want to finish their submitted tasks as immediately as possible. However, there has been no allocation mechanism with consideration of crowdsourcing tasks with precedence constraints and minimization of the total execution time simultaneously. To tackle this challenge, we consider the precedence constraints among tasks and design an efficient task allocation mechanism for the crowdsourcing system. In this work, we first introduce the crowdsourcing system model and formulate the task allocation problem. After proving that the studied problem is NP-hard, we propose an approximation algorithm that can minimize the total execution time of all the tasks. Then, we conduct extensive simulations to evaluate the performance of the proposed algorithm, and the simulation results show that the proposed algorithm has good approximate optimal ratios under different parameter settings.

Keywords: Crowdsourcing · Task allocation · Execution time minimization · Precedence constraint

1 Introduction

With the proliferation of information and social web technologies, crowdsourcing has emerged as a promising paradigm in recent years which can leverage the external crowds to perform specialized work quickly and inexpensively [4]. Many commercial crowdsourcing-based websites (*e.g.* Amazon Mechanical Turk [13], CrowdFlower [1],

© Springer International Publishing AG 2017
D.-S. Huang et al. (Eds.): ICIC 2017, Part III, LNAI 10363, pp. 156–167, 2017.
DOI: 10.1007/978-3-319-63315-2_14

MicroWorkers [2], Yahoo! Answers [3], and *etc.*) allow the requesters to publicize plenty of heterogeneous tasks to the platform, and harnessing the crowd intelligence from a large pool of workers all around the world [15]. Thus, crowdsourcing can also be seemed as the outsourcing of task requests to a crowd of individuals through the specific platform. Considering the full potential of crowdsourcing, many companies and researchers have proposed numerous applications which aim to maximize the benefit of the requester. One of the most well-known crowdsourcing application is the "crowdsensing", which can leverage the power of large crowds to complete the complicated sensing tasks by using their smartphones or other mobile devices at a lower cost [8]. For instance, Rana *et al.* [11] implements an Ear-Phone system for monitoring the environmental noise pollution in urban areas through crowdsourcing data collection. Therefore, it is clear that the idea of crowdsourcing can be designed for many purposes. Except for the data collection in, the idea of crowdsourcing is also widely applied in scientific discovery [10], transcribing text, designing logos, and *etc.* [12]. Generally, users on a crowdsourcing platform can be categorized into two types, which are called "requester" and "worker" respectively. The requester is the one who aims at carrying out computationally hard tasks through the websites, and the workers (*a.k.a.* crowds) are focusing on solving the small units of work in return for monetary reward.

In real world, large companies or the non-profit organizations often act as the task requesters. The publicized tasks sometimes are computationally hard, and these tasks need to be achieved by many workers with different skill levels. Thus, the task allocation issue is playing an important role in mechanism design. Nowadays, most of the crowdsourcing platforms consider that the skilled crowds and specialized work, so it is necessary to design an efficient sub-tasks allocation mechanism [5, 7–9, 14]. When allocating sub-tasks to a large pool of online workers, the platform may face some challenges. First, the workers may have different skills since their different backgrounds. Second, most of the tasks have multiple steps in realistic world, and these steps often have precedence constraints. The precedence constraint can be interpreted as one task can only be started after some steps have been achieved. Usually, the task requester will divide the tasks with multiple steps into many sub-tasks, then publish the sub-tasks in the crowdsourcing system. [6] considers the precedence constraints of specialized tasks and designs a crowdsourcing task allocation mechanism. Moreover, a multi-step task of a company may be a project, and companies usually want to minimize the execution time of their project. However, there is no allocation mechanism for crowdsourcing tasks with precedence constraints can minimize the total execution time of all the tasks.

In this paper, we address the challenges by studying the crowdsourcing task allocation issue for tasks with precedence constraints, and aim to design an efficient task allocation algorithm which can minimize the total execution time of all the tasks. Unfortunately, the studied problem is proved to be NP-hard. Thus, we need to judiciously design an approximation algorithm for this problem. The proposed task allocation algorithm includes four parts: task level division, final task set construction, allocation priority sequence construction and the task allocation. We first divide the task level based on their precedence constraints. The total execution time of tasks is equal to the finishing time of the last task in the task set. Define the tasks that not in the conditional task set of other tasks as final tasks. Obviously, the total execution time of tasks is determined by the expected finishing time of final tasks. Therefore, we construct the final task set in the second part. Then, we construct an allocation priority sequence according to the expected finishing time of tasks. When we allocate tasks to the workers in the fourth part, the tasks with lower order in the allocation priority sequence have higher priority to allocate to workers than the tasks with high order. The main contribution of this work is listed as follows:

- We design an efficient task allocation algorithm for tasks with precedence constraints. As far as we know, this is the first work which considers the precedence constraints of tasks and the proposed algorithm can minimize the total execution time of all the tasks.
- Define the approximate optimal ratio as the ratio of the execution time of our algorithm and the optimal one. We conduct extensive simulations to evaluate the performance of the proposed algorithm, and the results show that our algorithm has good approximate optimal ratios under different parameter settings.

2 Preliminaries

In this section, we first introduce the system model. After that we formally formulate the task allocation problem. And in the rest part of this section, we prove the studied problem is NP-hard.

2.1 System Model

There are three roles in the studied crowdsourcing system: a service requester, a crowdsourcing platform and many crowdsourcing workers. To make each allocated tasks easy enough for workers to finish, the service requester tends to divide tasks into pieces of indivisible tasks. Afterward, the service requester submits these tasks to the crowdsourcing platform. The platform makes all of that a one-time release. The workers, offering services to the platform, choose tasks which they are interested in and submit their decisions to the platform. The structure of the proposed crowdsourcing system is described in Fig. 1. Because of the priorities between tasks and the distinct choices among workers, the platform has to consider these constraints. According to the constraints set on the tasks, the platform allocates them to workers. Each task is only performed by one worker. In the end, after receiving notification, workers complete tasks in the deadline and return the answer to the crowdsourcing platform. The platform sends the answer to the service requester, who determines whether the answer meets requirements or not. After the answer satisfies the demand, the crowdsourcing worker is paid accordingly. And the crowdsourcing platform also gets a part of reward with commission.

We denote the set of tasks which released in platform by $\mathcal{T}$ We assume there are m tasks, denoted by $\mathcal{T} = \{t_1, t_2, t_3, \ldots, t_m\}$ Each worker is denoted as ω_i. Let $\mathcal{W}$ expresses the set of all crowdsourcing workers, $\mathcal{W} = \{\omega_i | 1 \leq i \leq n\}$. The task t_j which is posted on the platform is presented as $t_j = \{C_j, h_j, p_j\}$, where C_j denotes the conditional task

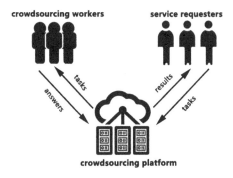

Fig. 1. The structure of the crowdsourcing system

set, h_j is the performing time needed to complete the task t_j, p_j represents the payment for the worker who finished the task t_j with qualified answer. Due to the priorities among tasks, only when all the tasks in the conditional task set C_j have been completed, can the task t_j be assigned to a worker to perform. However, if the conditional task set $C_j = \phi$ (*i.e.* the task t_j has no conditional task), the task t_j can be allocated by platform at any time. For the sake of simplicity, we temporarily suppose that each worker spends the identical amount of time on the same task, and h_j is a reasonable assessment towards the execution time of task t_j. The symbol T_i is a set of the requested tasks that worker ω_i is interested in performing. If the platform finally decides to allocate the task t_j for the worker ω_i to finish, the task t_j must be included in T_i.

2.2 Problem Formulation

Now, we formally formulate our task allocation problem.

After the workers have submitted their requests, the crowdsourcing platform decides how to properly assign the tasks in the task set to individuals in the workers collection. The goal of this work is to minimize the final finishing time of all the tasks. Since there are priorities among tasks and the interested tasks of workers are different, the platform has to take these constraints into account and makes an appropriate decision on the allocation of the tasks. Let $y_j = \{0, 1\}$ represents whether the task t_j is finished. If the task t_j is a completed task, $y_j = 1$; otherwise, $y_j = 0$ We use the $a_j = \{0, 1\}$ to denote if the task t_j is permitted to be allocated, and it is conspicuous that $a_j = \Pi_{k \in C_j} y_k$. If task t_j satisfies the constraint $a_j = 1$, then the task t_j can be allocated to a worker. We further use $x_{i,j} = \{0, 1\}$ to indicate whether the platform allocates the task t_j to the worker ω_i. If t_j is assigned to the worker ω_i, $x_{i,j} = 1$, otherwise, $x_{i,j} = 0$. $pr_{j,k}$ is used to represent whether task t_j is prior to task t_k to be allocated. If $pr_{j,k} = 1$, it means that task t_j should be allocated before task t_k. Suppose s_j is the expected beginning time of task t_j, (*i.e.* the earliest time that the platform can assign task t_j to a worker). Obviously, each s_j satisfies the constraint that $s_j \geq \max\limits_{t_k \in \mathcal{T}, t_k \neq t_j} pr_{k,j}(s_k x_i, k + h_k)$.

After all the workers have submitted their request task sets, the platform will determine the outcome of a suitable assignment. Define the total time that the workers finish all the tasks in the task set as the execution time of a task set. Then, the goal of the this work is to minimize the execution time of the released task set $\mathcal{T}$. *i.e.*

$$\min \quad \max_{t_j \in \mathcal{T}}(s_j + h_j)$$

s.t. some constraints

Theorem 1. *The crowdsourcing task allocation problem with the objective of minimizing the execution time of the total publicized tasks is NP-hard.*

Proof. Consider a simple case of our problem, where there is no priority constraints among the tasks and each worker is interested in all the tasks, (*i.e.* task t_j's conditional task set $C_j = \phi$ and worker ω_i's submitted tasks $T_i = \mathcal{T}$). Then, we can reduce this simple case to the multiprocessor scheduling problem. The tasks in our problem are the jobs in the multiprocessor scheduling problem, and the workers in our problem can be viewed as the processors in the multiprocessor scheduling problem. Then, the goal of our problem is equivalent to find the minimum possible time required to schedule all jobs on the processors such that there is none overlap. As is known to all, the multiprocessor scheduling problem is NP-hard. Since the simple case of our problem can be reduced to the multiprocessor scheduling problem, the problem studied in this work is also NP-hard. This finished our proof.

3   Algorithm Design

We have proved that the task allocation problem with minimum execution time is NP-hard, and there is no efficient algorithm can be used to solve this problem directly. Thus, we design an approximation algorithm with polynomial-time to solve this problem. There are four parts in our algorithm: task level division, final task set construction, allocation priority sequence construction and task allocation.

3.1   Task Level Division

We use level to denote the priority relationship among tasks. In our model, tasks can be allocated to workers only when all the tasks in their conditional task set have been finished. Thus, we constrict that the level of all the tasks in C_j should be less than k if task t_j's level is k. We use L_j to denote the level of t_j. Obviously, each task t_j with $C_j = \phi$ is in the lowest level (*i.e.* $L_j = 1$). The algorithm of how to divide the level of tasks is as shown in Algorithm 1.

In Algorithm 1, we first make a copy for each conditional task set C_j, which is de-noted as C_j'. Initially the current task level $k = 1$. Our task level division algorithm runs in an iterative way. In each iteration, we scan all the tasks in temporary task set $\mathcal{T}'$. When t_j is scanned, we check whether the temporary conditional set $C_j' = \phi$ or not. If

$C_j' = \phi$, we set the task level of is equal to k (*i.e.* set $L_j = k$). After all the tasks in $\mathcal{T}'$ have been scanned, we will delete the tasks with level k from the temporary conditional set of other tasks and delete t_j from the temporary task set $\mathcal{T}'$. Finally, we set $k = k + 1$, and start the next iteration until the task set $\mathcal{T}' = \phi$.

By confirming the task level, we are more clear about the relationship of tasks. And we will find its necessity when construct the allocation priority sequence.

Algorithm 1 Task level division.

Input:
    the task set $\mathcal{T}$
Output:
    $\mathcal{L} = \{L_j\}_{t_j \in \mathcal{T}}$
  1: Set $\mathcal{T}' = \mathcal{T}$;
  2: **for** each t_j in $\mathcal{T}$ **do**
  3:    Set $C_j' = C_j$;
  4: Set $k = 1$;
  5: **while** $\mathcal{T}' \neq \phi$ **do**
  6:    **for** each task $t_j \in \mathcal{T}'$ **do**
  7:      **if** $C_j' = \phi$ **then**
  8:        Set $L_j = k$;
  9:    **for** each task $t_j \in \mathcal{T}'$ **do**
10:      **for** each task $t_q \in C_j'$ **do**
11:        **if** $L_q = k$ **then**
12:          Delete t_q from set C_j';
13:       **if** $L_j = k$ **then**
14:         Delete t_j from set $\mathcal{T}'$;
15:    $k + +$;
16: **return** $\mathcal{L} = \{L_j\}_{t_j \in \mathcal{T}}$;

3.2 Final Task Set Construction

Definition 1. *Define the tasks that don't exist in any other tasks' conditional task set as final tasks. There exists at least one final task t_j in $\mathcal{T}$. If t_j is a final task, then it satisfied that for any $t_k \in (\mathcal{T}\{t_j\})$, $t_j \notin C_k$ stands.*

Since the total execution time of all the tasks is bounded by final tasks, we first construct the final task set before we sort the allocation priority of tasks. By taking use of the characteristics of the final task, we first suppose that each task in task set $\mathcal{T}$ is the final task. Then we check our suppositions about each task. If the task have been inspected exists in a conditional task set for any other tasks, we have to redress the supposition. If the task does not exist in the conditional task set for any other tasks, it proves that our hypothesis is true and the task is a final task. The details are as shown in Algorithm 2.

As is the feature that a final task has, we can know that if all the final tasks in $\mathcal{T}$ have been finished, then all the tasks which are not the final tasks in $\mathcal{T}$ are also completed. In another word, if all the final tasks in $\mathcal{T}$ have been finished, all the tasks in $\mathcal{T}$ have been done.

Algorithm 2 Final task set construction

Input:
    the task set $\mathcal{T}$
Output:
    the final task set $\mathcal{F}$
1: Set $\mathcal{F} = \mathcal{T}$;
2: **for** each task $t_j \in \mathcal{F}$ **do**
3:    **for** each task $t_k \in \mathcal{T}$ **do**
4:       **for** each task $t_q \in C_k$ **do**
5:          **if** $t_j = t_q$ **then**
6:             Delete task t_j from the final task set $\mathcal{F}$;
7: **return** the final task set $\mathcal{F}$;

3.3   Allocation Priority Sequence Construction

Our optimization objective is bounded by the final tasks with maximum expected finishing time. Thus, we need to compute the expected finishing time of all the final tasks when we allocate tasks to workers. To achieve the designed goal, we will sort the tasks with their expected finishing time and construct an allocation priority sequence.

Definition 2. Task Sequence: *A task sequence is sequence of tasks which satisfies the l-th task in the sequence is in the conditional task set of the l + 1-th task and this sequence ends in a final task.*

Since the tasks can only be performed one by one in the task sequence, the expected finishing time of a task sequence is equal to the expected finishing time of the final task in the sequence.

Definition 3. The Critical Task Sequence: *The critical task sequence is defined as the task sequence with maximum expected finishing time.*

Our allocation priority sequence construction algorithm runs in iteration. In each iteration, we will first find the critical task sequence of task set $\mathcal{T}$, and then put the task with lowest level in the critical task sequence into our allocation priority sequence. The details are as shown in Algorithm 3.

Suppose f_j is the order of task t_j in our allocation priority sequence. In each iteration, we aim to find a task sequence and its value is greater than any other task sequences. Note that the value of a task sequence is equal to the expected finishing time of the final task in the sequence. To get the expected finishing time of the final tasks, we need to know the expected finishing time of the tasks in their conditional task set. Thus, Algorithm 3 first computes the expected finishing time of each task in $\mathcal{T}$. Let h_j^e

be the expected finishing time of task t_j. Obviously, $h_j^e = h_j$ when $L_j = 1$. When $L_j \geq 2$, $h_j^e = h_j + \max\{h_p^e\}_{t_p \in C_j}$. Notice that we compute the expected finishing time of tasks from low level to high level, and the levels of tasks in C_j are lower than t_j. Thus, $\{h_j^e\}_{t_p \in C_j}$ are known when we compute h_j^e. We also use c_j^t to record the task with maximum expected finishing time, since c_j^t can help us to construct the critical task sequence.

After computing the expected finishing time of all the final tasks, we can find the task t_q with maximum expected finishing time, and construct the critical task sequence of task t_q with the help of recorded c_j^t. We suppose t_p is the task with lowest level in the critical task sequence of final task t_q. Then, we put t_p into the allocation priority sequence by setting $f_p = l$. After that, we delete t_p from task set $\mathcal{T}$ and all the conditional task sets of other tasks. If t_p is a final task, we also need to delete t_p from the final task set $\mathcal{F}$. Finally, the next iteration begins until $\mathcal{T} = \phi$.

Algorithm 3 Allocation priority sequence construction.

Input:
    the task set $\mathcal{T}$, the final task set $\mathcal{F}$
Output:
    the allocation priority sequence $\{f_j\}_{t_j \in \mathcal{T}}$
1: Set $l = 1$;
2: **While** $\mathcal{T} \neq \phi$ **do**
3:     **for** $k = 1$ to $\max\{L_j\}_{t_j \in \mathcal{T}}$ **do**
4:       **for** each task in $\mathcal{T}$ **do**
5:         **if** $L_j = k$ **then**
6:           Compute the expected finishing time of t_j, and use h_j^e to denote;
7:           Set c_j^t is the task with maximum expected finishing time among all tasks in C_j;
8:     Find the final task $t_q \in \mathcal{F}$ with maximum expected finishing time;
9:     Construct the critical task sequence of task t_q;
10:    Find the task t_p with the lowest level in critical task sequence of task t_q;
11:    Set $f_p = l$;
12:    Delete t_p from $\mathcal{T}$ and all the conditional task sets of other tasks;
13:    **If** $t_p \in \mathcal{F}$ **then**
14:       Delete t_p from $\mathcal{F}$;
15:    Set $l = l + 1$;

3.4   Task Allocation

Our task allocation algorithm mainly includes four steps. The details are as follows:

Step 1: Sort the tasks in $\mathcal{T}$ according to the allocation priority sequence.
Step 2: Sort the workers in $\mathcal{W}$ in increasing order according to the number of tasks in their requested task sets.

Step 3: Scan the tasks in the sorted list one by one. When t_j is scanned, the platform scans the workers in the sorted list one by one until finds a worker who are interested in task t_j. When worker w_i is scanned, the platform checks if $t_j \in T_i$ stands. If $t_j \in T_i$, the platform allocates t_j to w_i, deletes t_j from T and deletes w_i from W. In the case of there is no task can be allocated to workers, all the tasks and workers remain in T and W should wait for new worker arrives or some of the allocated tasks finished.

Step 4: When an allocated task finished (*i.e.* worker w_i finishes task t_j), the plat-form first adds w_i in the sorted worker list, then runs step 3 to check if there are some tasks can be allocated to workers already and continue allocates tasks to workers until $T = \phi$.

4   Simulation

4.1   Simulation Setting

Let *Alg(Opt)* be the execution time of task set T under the allocation of our algorithm (the optimal allocation algorithm). Then, **the approximate optimal ratio** is the ratio of *Alg* and *Opt*.

In order to show the performance of our proposed algorithm, especially about the performance of the approximate optimal ratio, we vary the number of the released tasks. The number of the involved workers and the level of the tasks set T with the symbol of m, n, l. More in details, $m = |T|$, $n = |W|$ and $l = \max\{L_j\}_{t_j \in T}$. To show the efficiency of our proposed algorithm, we also compare the performance of our algorithm with another algorithm under the same circumstance.

The number of total tasks in each level is uniformly distributed in $[m/l - 3, m/l + 3]$. For any task t_j in task set T, it has attributes of execution time h_j and conditional task set C_j, and the size of C_j is distributed in $[1, 3]$ at random. In Fig. 2(a), (b) and (d), the parameter h_j appears to be follow $\mathcal{N}(10, 3)$ normal distribution. And in Fig. 2(c) simulation, the parameter h_j can also be $[1, 19]$ uniformly distributed. Each worker submits a set of tasks that he or she is interested in performing, and the size of T_i is randomly generated in $[3, 5]$ or $[5, 8]$ in different experiments. In each case of $< m, n, l >$, we generate 2000 instances and take the average value them. The average value is the outcome of the case finally.

4.2   Simulation Results

We first vary the number of involved workers n from 50 to 110 while the number of released task m is fixed to 200. The level of task set l is set to be 6. We get the outcome of these cases and plot these result in Fig. 2(a). Obviously, when the number of involved workers increases, the approximate optimal ratio decreases. As the theoretically optimal value of the allocation for the task set T is only in connection with the size and structure of the task set T. The theoretically optimal value is fixed no matter how the involved tasks change. However, when the number of involved workers

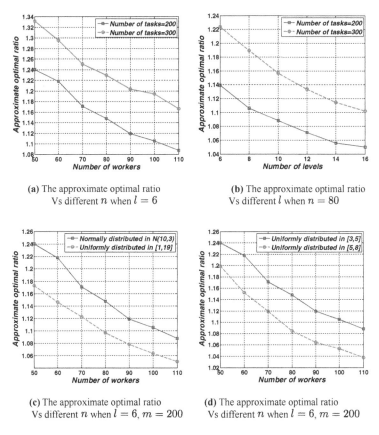

(a) The approximate optimal ratio
Vs different n when $l = 6$

(b) The approximate optimal ratio
Vs different l when $n = 80$

(c) The approximate optimal ratio
Vs different n when $l = 6$, $m = 200$

(d) The approximate optimal ratio
Vs different n when $l = 6$, $m = 200$

Fig. 2. The approximate optimal ratio of the proposed algorithm

increases, the tasks are more likely to be allocated once they are available, which would make the execution time of the task set $\mathcal{T}$ decrease. Then, the approximate optimal ratio decreases. Therefore, the approximate optimal ratio decreases as the number of involved workers increases. We also set m to be 300, making the simulation and plot the results in Fig. 2(a). The results are consistent with the case of $m = 200$.

We also make the simulation to show the influence on the level of the task set $\mathcal{T}$. We vary the number of levels l from 6 to 16 while the number of released tasks m is fixed to 200, 300, and the number of workers n is set to be 80. We get the outcome of these cases and plot these results in Fig. 2(b). Apparently, the approximate optimal ratio will decrease if the levels of task set increases. It is because more levels of task set make the number of tasks in each task less, and the tasks are more likely to be allocated once they are available, which would make the execution time of the task set $\mathcal{T}$ more closer to the theoretically optimal value of the allocation for the task set. To sum up, the approximate optimal ratio decreases as the level of task set increases.

In Fig. 2(c), we vary the number of involved workers n from 50 to 110 while the number of released task m is fixed to 200. The level of task set l is set to be 6. The number of each worker's submitted tasks $|T_i|$ is uniformly distributed in $[3, 5]$. We compare the approximate optimal ratio when the number of each task's conditional tasks $|C_j|$ is distributed in two different ways. As the theoretically optimal value tends to be larger when the number of each task's conditional tasks $|C_j|$ is uniformly distributed in $[1, 19]$. Besides, the difference between the theoretically optimal value and the outcome of the proposed algorithm is finite in a range, no matter how $|C_j|$ is distributed. Then, we can find the approximate optimal ratio is larger when $|C_j|$ is uniformly distributed in $[1, 19]$ which is consistent with the results in Fig. 2(c).

In Fig. 2(d), we vary the number of involved workers n from 50 to 110 while the number of released task m is fixed to 200. The level of task set l is set to be 6. The number of each task's conditional tasks $|C_j|$ is normally distributed in $\mathcal{N}(10, 3)$. Then, we compare the results under the circumstances that the number of each worker's submitted tasks $|T_i|$ is uniformly distributed in $[3, 5]$ or $[5, 8]$. Apparently, after the analysis of Fig. 2(a), we can easily get that the approximate optimal ratio is larger when $|C_j|$ is uniformly distributed in $[5, 8]$, and the outcome of our simulation accords with our assumption.

5 Conclusion

Considering the precedence constraints of tasks, we design an efficient task allocation algorithm for crowdsourcing systems which can minimize the total execution time of all the tasks. The designed algorithm first divides the level of tasks and finds all the final tasks. Then, we construct an allocation priority sequence according to the expected finishing time of tasks. Finally, we allocate the tasks to workers based on the constructed allocation priority sequence. Our simulation results verify the efficiency of the designed algorithm.

Acknowledgements. The research of authors is partially supported by National Natural Science Foundation of China (NSFC) under Grant No. 61572342, No. 61672369, Natural Science Foundation of Jiangsu Province under Grant No. BK20151240, No. BK20161258, China Postdoctoral Science Foundation under Grant No. 2015M580470, No. 2016M591920. Postdoctoral Science Foundation of Jiangsu Province under Grant No. 1501130B, and No. 1501078A. Any opinions, findings, conclusions, or recommendations expressed in this paper are those of author(s) and do not necessarily reflect the views of the funding agencies.

References

1. Crowdflower. http://crowdflower.com
2. Microworkers. http://microworkers.com
3. Adamic, L., Zhang, J., Bakshy, E., Ackerman, M.: Knowledge sharing and yahoo answers: everyone knows something. In: Proceedings of the 17th International Conference on World Wide Web (WWW), pp. 665–674 (2008)

4. Bi, R., Zheng, X., Tan, G.: Optimal assignment for deadline aware tasks in the crowdsourcing. In: Proceedings of the 2016 IEEE International Conferences on Big Data and Cloud Computing (BDCloud), Social Computing and Networking (SocialCom), Sustainable Computing and Communications (SustainCom) (BDCloud-SocialCom-SustainCom), pp. 178–184 (2016)
5. Boutsis, I., Kalogeraki, V.: On task assignment for real-time reliable crowdsourcing. In: Proceedings of the IEEE ICDCS 2014, pp. 1–10 (2014)
6. Chatterjee, A., Borokhovich, M., Varshney, L.R., Vishwanath, S.: Efficient and flexible crowdsourcing of specialized tasks with precedence constraints. In: Proceedings of the IEEE INFOCOM 2016, pp. 1–9 (2016)
7. Goel, G., Nikzad, A., Singla, A.: Mechanism design for crowdsourcing markets with heterogeneous tasks. In: Proceedings of the Second AAAI Conference on Human Computation and Crowdsourcing, pp. 77–86 (2014)
8. He, S., Shin, D.-H., Zhang, J., Chen, J.: Toward optimal allocation of location dependent tasks in crowdsensing. In: Proceedings of the IEEE INFOCOM 2014, pp. 745–753 (2014)
9. Jin, H., Su, L., Chen, D., Nahrstedt, K., Xu, J.: Quality of information aware incentive mechanisms for mobile crowdsensing systems. In: Proceedings of the ACM MobiHoc 2015, pp. 167–176 (2015)
10. Khatib, F., Cooper, S., Tyka, M.D., Xu, K., Makedon, I., Popović, Z., Baker, D.: Algorithm discovery by protein folding game players. Proc. Nat. Acad. Sci. **108**(47), 18949–18953 (2011)
11. Rana, R.K., Chou, C.T., Kanhere, S.S., Bulusu, N., Hu, W.: Ear-phone: an end-to-end participatory urban noise mapping system. In: Proceedings of the ACM/IEEE IPSN 2010, pp. 105–116 (2010)
12. Tran-Thanh, L., Huynh, T.D., Rosenfeld, A., Ramchurn, S.D., Jennings, N.R.: Crowdsourcing complex workflows under budget constraints. In: Proceedings of the AAAI 2015, pp. 1298–1304 (2015)
13. Amazon Mechanical Turk: Amazon mechanical turk (2012). Accessed 17 Aug 2012
14. Xu, W., Huang, H., Sun, Y.-E, Li, F., Zhu, Y., Zhang, S.: DATA: a double auction based task assignment mechanism in crowdsourcing systems. In: Proceedings of the 8th International ICST Conference on Communications and Networking in China (CHINACOM 2013), pp. 172–177 (2013)
15. Zhang, Y., Van der Schaar, M.: Reputation-based incentive protocols in crowdsourcing applications. In: Proceedings of the IEEE INFOCOM 2012, pp. 2140–2148 (2012)

Research on Link Layer Topology Discovery Algorithm Based on Dynamic Programming

Yunjia Li[1] and Zheng Yao[2(✉)]

[1] The Department of English Teaching, Hunan Normal University,
Changsha 410081, China
liyj@hunnu.edu.cn
[2] The Department of Computer Teaching, Hunan Normal University,
Changsha 410081, China
790975278@qq.com

Abstract. This paper proposes a link layer topology discovery algorithm based on dynamic programming. The basic idea of this algorithm is to express the topology of the switch and the switch, the switch and the host in the form of tree. According to the principle of multi-stage decision process, we set to construct a single order tree for the whole network topology using the address forwarding table (AFT) port. The results of theoretical analysis and practical application show that the topology discovery algorithm has been greatly improved in terms of efficiency, accuracy and effectiveness.

Keywords: Dynamic programming · Topology discovery · Address forwarding table · Single order tree

1 Introduction

Domestic and international researchers have put forward a lot of methods [1–3] to find the third layer of network, that is, the network layer topology. The three layer network topology [4, 5] can't provide all kinds of topology information needed for local area network management and maintenance. In 2000, IETF launched MIB (management information base) [6], trying to solve the discovery problem about the network layer topology. But as there is no mechanism for determining how to get these MIB objects, the automatic discovery on the second layer network (link layer) still needs more research on topology.

Richard Black et al. proposed a method of detecting packet based in work [7], whose basic idea is to set up a proxy process on each host, some packets, and adapts the network settings in mixed mode (in this mode, the card can receive all the data in all network segment); then, it determines the connection between the equipment according to the data packet received by each host. This method can judge the connections between switches as well as connections between switch and hosts. The disadvantage is that it is necessary to set up a proxy process on each host device, which is unlikely for a larger network. Breitbart et al. from Baer laboratory proposed a physical network topology discovery method based on switch address forwarding table AFT (Address Forwarding Table) [8]. The core of the algorithm is to determine the switch port direct

© Springer International Publishing AG 2017
D.-S. Huang et al. (Eds.): ICIC 2017, Part III, LNAI 10363, pp. 168–177, 2017.
DOI: 10.1007/978-3-319-63315-2_15

theorem: if and only if two ports within the same subnet address forwarding entries in the intersection is empty, and the concentrated and contains all the switches in a subnet address entry, then a port belonging to the two switches is connected. In order to ensure the integrity of the address delivering table, Breitbart designed a mping program and modified the address of the target based on the original ping program, and added MAC address of target switch address forwarding table, so as to ensure the integrity of the address table. Zheng [3, 9–11] gives a set of theorems, determining links between switches by the upstream and downstream ports; Sun et al. [12, 13] proposed to judge the links between switches by the network node's immediate relations and collateral relations, which can construct forwarding network topology by incomplete address.

Dynamic programming [14] is a branch of operations research, and it is a optimized mathematical method to solve the decision process (Decision Process). At the beginning of the 1950s, when American mathematician R.E. Bellman et al. worked on the optimization problem in the study of multistage decision process, they proposed the famous principle of optimality, transited the multi stage process into series of single stage problems one by one, created the new optimization problem – dynamic programming. In 1957 he published his famous book "Dynamic Programming", the first work in the field.

This paper proposes a new link layer topology discovery algorithm. Compared with the existing algorithms: (1) it proposes a method for constructing topological structure based on dynamic programming. Dynamic programming is algorithm strategy through dividing problem instances into smaller and similar sub problems, and storing solutions to solve and avoid the repeated calculation of sub problems, and to solve the optimization problem. The algorithm build a single order tree using the address forwarding table (AFT) port set from the end of the network nodes until the whole network topology. Compared with the existing algorithms, the time complexity of the algorithm is greatly reduced, and the derivation method is easier to understand; (2) this paper proposes a kind of tree-structure deductive method, which can traverse all the vertices and edges by the deductive method; (3) the algorithm in this paper supports the expression of the link layer topology relation in the form of multiple sub-trees; (4) it proposed that the discovery of the topology relation of the switch requires not only the address forwarding table but also the spanning tree. Because the address forwarding table is only connected to the MAC address and the port information terminal virtual port number of the other port, the virtual port mapping to the local actual port must be obtained by the mapping relationship to generate the tree (BRIDGE-MIB: dot1dBase PortTable: 1.3.6.1.2.1.17.1.4).

2 Link Layer Topology Discovery Algorithm

2.1 Algorithm Base

This article proposes the definition method of connection relation and MAC address set and deduce on this basis.

Definition 1. For the same subnet: if the i device is a switch as S_i, the j port of switch S_i port as S_{ij}; if the i device is a host, it is represented as H_i.

Definition 2. For any port S_{ij} of switch, P_{ij} switch means set of the rest of the switch port and the hosts learned through the port S_{ij} address forwarding table, P_{ij} called as switch reachable path set.

Definition 3. For any port S_{ij} of switch, "L $<S_{ij}, S_{kl}>$" represents that the switch S_i connects the switch S_k port S_{kl} through the port S_{ij}; "L $<S_{ij}, H_k>$ "indicates that the switch S_i connected with the host H_k through port S_{ij}.

Definition 4. For any port S_{ij} of switch, if the leaned port MAC address has only port S_{kl} MAC switch S_k address of another switch, then the switch is called edge switches, the corresponding switch connection L $<S_{kl}>$ S_{ij} is called the edge line.

Definition 5. Set V represents the set of all port reachable paths P_{ij}, i.e., $\{P_{11}, P_{12}, ..., P_{1m}, P_{21}, P_{22}, ..., P_{2n}, P_{i1}, P_{i2}, ..., P_{ik}, ...\}$.

Definition 6. Set E represents the set of all connections of the switch.

2.2   Algorithm Framework

The undirected graph connected without a loop is called tree represented by G = <V, E>. A topological graph can be represented by one or more trees. "n" for vertex tree in "G", "m" for the number of edges in "G", then n = m + 1. According to the characteristics of tree, it is known that: the vertex of degree 1 is called a leaf, and the vertex with a degree greater than 1 is called a branch. If a branch point has only one edge connected to the other branch points, the vertices of the other edges are all leaves, which is called the branching points. In this paper, the definition of the evolution of the tree is shown in Fig. 1: delete all the leaves and the edges connected with the leaves of the undirected graph G_1 to get the undirected graph G_2. Conclusion is as below:

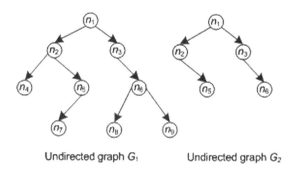

Undirected graph G_1 Undirected graph G_2

Fig. 1. Evolution process of tree

(1) All the evolutionary branching points become leaves
     Prove: Due to the evolution of branch points with only one edge to the other branch points, the rest are connected with the leaves. Thus, after removing the edges attached to the leaves, the number of vertices of the evolved branch points also becomes 1, evolving into leaves.

(2) The resulting sub-graph is still tree

Prove:

(a) connectivity: suppose to delete all the leaves n_i and the edge m_j connected to the leaf, the sub-graph becomes non-connected. Then n_i must be connected to other nodes, and n_i itself is a contradiction of leaves. Thus, the subgraph is connected;

(b) there is no loop: as no new edges are formed as no new edge is added during the deletion. It can be concluded that the sub-graph is connected and there is no loop.

(3) Recursive call in turn, make ergodic set from the end of the network node without repeat, and delete the current traversal of the node and the set of nodes from the spanning tree set the tree set is empty.

2.3   Algorithm Steps

(1) Go through the forwarding tables of all switch ports in the subnet, and the MAC address of the switch is collected as a set of switch ports P_{ij};

(2) Construct all the obtained P_{ij} into the spanning tree set V;

(3) Find the connection between the switch and the host: remove the set P_{ij} {H_k} or P_{ij} {H_k,..., H_l} of switch containing only the host MAC address, if it is P_{ij} {H_k}, then the switch port S_{ij} is connected to the host H_k, delete set P_{ij}, and remove the H_k in the rest of the set of spanning trees, and add connection <S_{ij}, H_k>; if it is {P_{ij},..., H_l}, then the host H_k,..., H_l is connected to the switch by HUB, delete the set P_{ij}, and remove the rest of the spanning tree set in the host H_k,..., H_l, adding the connections <S_{ij}, H_k>...<S_{ij}, H_l>.

(4) Found the edge switch, edge connection:

(a) Take exchange unary relation set P_{ij} {S_{kl}}, we can know that the switch S_i and switch S_k port are directly connected through the port S_{ij} via the theorem; S_k is the edge switch, <S_{ij}, S_{kl}> is the edge connection;

(b) Delete set P_{ij}; delete all elements in the spanning tree set V of the edge switch S_k; delete all the sets in all ports on the edge switch S_k in the spanning tree;

(c) Add the connection relationship of the switch <S_{ij}, S_{kl}>;

(d) Cycle process the single set of switches in the next round until the single set of the round is processed.

(5) Follow step (4) to continue the processing of the spanning tree collection until the spanning tree is empty, then algorithm program exits.

2.4   Algorithm Analysis

This paper proposes a typical algorithm of topology discovery of connection topology. In the network, there are 7 switches, 3 hosts, and 2 hosts are connected through HUB switches, 1 hosts connected with switch directly. Topology connection graph is shown in Fig. 2.

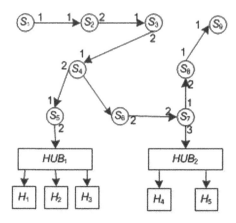

Fig. 2. Topology connection graph

Step 1. Generating spanning trees. The spanning tree set V from the address forwarding table is below:

$P_{11}\{S_{21}, S_{31}, S_{41}, S_{51}, S_{61}, S_{72}, S_{82}, S_{91}, H_1, H_2, H_3, H_4, H_5\}$
$P_{21}\{S_{11}\}$
$P_{22}\{S_{31}, S_{41}, S_{51}, S_{61}, S_{72}, S_{82}, S_{91}, H_1, H_2, H_3, H_4, H_5\}$
$P_{31}\{S_{11}, S_{22}\}$
$P_{32}\{S_{41}, S_{51}, S_{61}, S_{72}, S_{82}, S_{91}, H_1, H_2, H_3, H_4, H_5\}$
$P_{41}\{S_{11}, S_{22}, S_{32}\}$
$P_{42}\{S_{51}, H_1, H_2, H_3\}$
$P_{43}\{S_{61}, S_{72}, S_{82}, S_{91}, H_4, H_5\}$
$P_{51}\{S_{11}, S_{22}, S_{32}, S_{42}, S_{61}, S_{72}, S_{82}, S_{91}, H_4, H_5\}$
$P_{52}\{H_1, H_2, H_3\}$
$P_{61}\{S_{11}, S_{22}, S_{32}, S_{43}, H_1, H_2, H_3\}$
$P_{62}\{S_{72}, S_{82}, S_{91}, H_4, H_5\}$
$P_{71}\{S_{82}, S_{91}\}$
$P_{72}\{S_{11}, S_{22}, S_{32}, S_{43}, S_{51}, S_{62}, H_1, H_2, H_3\}$
$P_{73}\{H_4, H_5\}$
$P_{81}\{S_{91}\}$
$P_{82}\{S_{11}, S_{22}, S_{32}, S_{43}, S_{51}, S_{62}, S_{71}, H_1, H_2, H_3, H_4, H_5\}$
$P_{91}\{S_{11}, S_{22}, S_{32}, S_{43}, S_{51}, S_{62}, S_{71}, S_{81}, H_1, H_2, H_3, H_4, H_5\}$

Step 2. Finding the connection between the switch and the host. Take the switch contains only the host MAC address set $P_{52}\{H_1, H_2, H_3\}$, $P_{73}\{H_4, H_5\}$, and delete the set P_{52}, P_{73}; remove H_1, H_2, H_3, H_4, H_5 from the rest sets; add connection $<S_{52}, H_1>$, $<S_{52}, H_2>$, $<S_{52}, H_3>$, $<S_{73}, H_4>$, $<S_{73}, H_5>$. After search of the round, the connection relationship E: $<S_{52}, H_1>$, $<S_{52}, H_2>$, $<S_{52}, H_3>$, $<S_{73}, H_4>$, $<S_{73}, H_5>$.

After the search of this round, the topology is shown in Fig. 3.

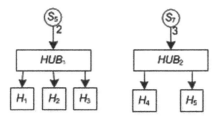

Fig. 3. Topology from step 2

After the search of this round, the spanning tree to be processed is as below:

$P_{11}\{S_{21}, S_{31}, S_{41}, S_{51}, S_{61}, S_{72}, S_{82}, S_{91}\}$
$P_{21}\{S_{11}\}$
$P_{22}\{S_{31}, S_{41}, S_{51}, S_{61}, S_{72}, S_{82}, S_{91}\}$
$P_{31}\{S_{11}, S_{22}\}$
$P_{32}\{S_{41}, S_{51}, S_{61}, S_{72}, S_{82}, S_{91}\}$
$P_{41}\{S_{11}, S_{22}, S_{32}\}$
$P_{42}\{S_{51}\}$
$P_{43}\{S_{61}, S_{72}, S_{82}, S_{91}\}$
$P_{51}\{S_{11}, S_{22}, S_{32}, S_{42}, S_{61}, S_{72}, S_{82}, S_{91}\}$
$P_{61}\{S_{11}, S_{22}, S_{32}, S_{43}\}$
$P_{62}\{S_{72}, S_{82}, S_{91}\}$
$P_{71}\{S_{82}, S_{91}\}$
$P_{72}\{S_{11}, S_{22}, S_{32}, S_{43}, S_{51}, S_{62}\}$
$P_{81}\{S_{91}\}$
$P_{82}\{S_{11}, S_{22}, S_{32}, S_{43}, S_{51}, S_{62}, S_{71}\}$
$P_{91}\{S_{11}, S_{22}, S_{32}, S_{43}, S_{51}, S_{62}, S_{71}, S_{81}\}$

Step 3. The edge switch is found in the first round and the set is $P_{21}\{S_{11}\}$, $P_{42}\{S_{51}\}$, $P_{81}\{S_{91}\}$; delete the set P_{21}, P_{81}, and remove the sets of all ports S_1, S_5, S_9 and all the ports from the spanning tree; add connection $<S_{21}, S_{11}>$, $<S_{42}, S_{51}>$, $<S_{81}, S_{91}>$. After the round search, the connection E:

$<S_{52}, H_1>$, $<S_{52}, H_2>$, $<S_{52}, H_3>$, $<S_{73}, H_4>$, $<S_{73}, H_5>$, $<S_{21}, S_{11}>$, $<S_{42}, S_{51}>$, $<S_{81}, S_{91}>$.

After the round search, the topology is shown in Fig. 4.
After the search of this round, the spanning tree to be processed is as below:

$P_{22}\{S_{31}, S_{41}, S_{61}, S_{72}, S_{82}\}$
$P_{31}\{S_{22}\}$
$P_{32}\{S_{41}, S_{61}, S_{72}, S_{82}\}$
$P_{41}\{S_{22}, S_{32}\}$
$P_{43}\{S_{61}, S_{72}, S_{82}\}$
$P_{61}\{S_{22}, S_{32}, S_{43}\}$
$P_{62}\{S_{72}, S_{82}\}$
$P_{71}\{S_{82}\}$

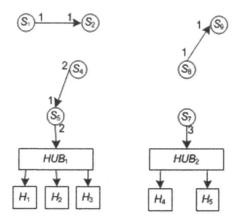

Fig. 4. Topology diagram from step 3

$P_{72}\{S_{22}, S_{32}, S_{43}, S_{62}\}$
$P_{82}\{S_{22}, S_{32}, S_{43}, S_{62}, S_{71}\}$

Step 4. In the second round, the edge switch is found, and the connection set is $P_{31}\{S_{22}\}$, $P_{71}\{S_{82}\}$; delete the set P_{31}, P_{71}, and remove the sets of all ports S_2, S_8 and all the ports from the spanning tree; add connection $<S_{31}, S_{22}>$, $<S_{71}, S_{82}>$. After this round, the connection set E: $<S_{52}, H_1>$, $<S_{52}, H_2>$, $<S_{52}, H_3>$, $<S_{73}, H_4>$, $<S_{73}, H_5>$, $<S_{21}, S_{11}>$, $<S_{42}, S_{51}>$, $<S_{81}, S_{91}>$, $<S_{31}, S_{22}>$, $<S_{71}, S_{82}>$.

After the search of this round, the topology is shown in Fig. 5.

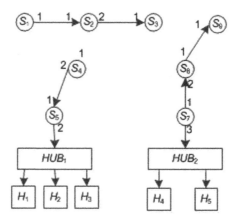

Fig. 5. Topology diagram from step 4

After the search of this round, the spanning tree to be processed is as below:

$P_{32}\{S_{41}, S_{61}, S_{72}\}$
$P_{41}\{S_{32}\}$
$P_{43}\{S_{61}, S_{72}\}$
$P_{61}\{S_{32}, S_{43}\}$
$P_{62}\{S_{72}\}$
$P_{72}\{S_{32}, S_{43}, S_{62}\}$

Step 5. In the third round, the edge switch is found, and the connection set is $P_{41}\{S_{32}\}$, $P_{62}\{S_{72}\}$; delete the set P_{41}, P_{62}, and remove the sets of all ports S_3, S_7 and all the ports from the spanning tree; add connection $<S_{41}, S_{32}>$, $<S_{62}, S_{72}>$. After this round, the connection set E: $<S_{52}, H_1>$, $<S_{52}, H_2>$, $<S_{52}, H_3>$, $<S_{73}, H_4>$, $<S_{73}, H_5>$, $<S_{21}, S_{11}>$, $<S_{42}, S_{51}>$, $<S_{81}, S_{91}>$, $<S_{31}, S_{22}>$, $<S_{71}, S_{82}>$, $<S_{41}, S_{32}>$, $<S_{62}, S_{72}>$. After the search of this round, the spanning tree to be processed is shown in Fig. 6.

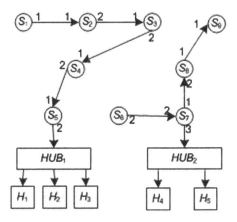

Fig. 6. Topology diagram from step 5

After the search of this round, the spanning tree to be processed is as below:

$P_{43}\{S_{61}\}$
$P_{61}\{S_{43}\}$

Step 6. In the fourth round, the edge switch is found, and the connection set is $P_{43}\{S_{61}\}$, $P_{61}\{S_{43}\}$; delete the set P_{43}, P_{61}, and remove the sets of all ports S_6, S_4 and all the ports from the spanning tree; add connection $<S_{43}, S_{61}>$, $<S_{61}, S_{43}>$. As the set $<S_{43}, S_{61}>$, $<S_{61}, S_{43}>$ is the same connection, it's not necessary to repeat again. After this round, the connection set E: $<S_{52}, H_1>$, $<S_{52}, H_2>$, $<S_{52}, H_3>$, $<S_{73}, H_4>$, $<S_{73}, H_5>$, $<S_{21}, S_{11}>$, $<S_{42}, S_{51}>$, $<S_{81}, S_{91}>$, $<S_{31}, S_{22}>$, $<S_{71}, S_{82}>$, $<S_{41}, S_{32}>$, $<S_{62}, S_{72}>$, $<S_{43}, S_{61}>$. After the search of this round, the spanning tree to be processed is shown in Fig. 7.

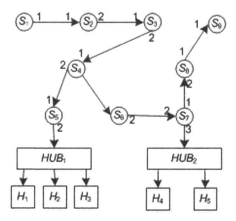

Fig. 7. Topology diagram from step 6

3 Algorithm Performance Test

In order to test the efficiency of the system topology discovery algorithm, we can use the Hewlett-Packard product OpenView to compare with it. Using only one topology discovery engine to run on a host, the topology discovery can completed in 3 min and 11 s. Run the HP OpenView network node manager on the host, and it requires 16 min and 24 s to found the entire network. We can know that the system is much faster than the HP OpenView network node manager. In addition, we also have an efficient analysis of several common topology discovery algorithms, as shown in Table 1.

Table 1. The contrast of physical topology discovery algorithms

Algorithm	Hub	Complexity
DC	No support	$O(n2)$
Zheng.H	No support	$O(n2)$
MKR	Support	$O(n2)$
Skeleton path	Support	$O(n7)$
Skeleton tree	Support	$O(n3)$
Present algorithm	Support	$O(n)$

4 Concluding Remarks

Network topology discovery is important basis for people to know, understand and study the network behavior and the judgment of the connection relationship is a technical difficulty. This article studies the link layer discovery algorithm and proposes a link layer network topology discovery algorithm based on dynamic programming, to judge the relationship between switches and hosts. Compared with the existing algorithms, a new method is proposed to construct a single order tree from the end node of the network. The tree can be used to traverse all the vertices and edges. Compared to the present algorithms, the time complexity is greatly reduced, and the derivation method is easier to understand. Finally, the effectiveness of the algorithm is verified in actual environment.

We can't find a VLAN in exchange network topology based on the idea of this algorithm. In the following research, we will try to find exchange network topology in VLAN by the algorithm.

Acknowledgement. This research was financially supported by the Key Program of Hunan Science and Technology Foundation (2016SK2017).

References

1. Donnet, B., Friedman, T.: Internet topology discovery: a survey. IEEE Commun. Surv. Tutor. **9**(4), 56–69 (2007)
2. Laurent, B., Traian, M.: Integrated genetic algorithm and goal programming for network topology design problem with multiple objectives and multiple criteria. IEEE/ACM Trans. Netw. **16**(3), 680–690 (2008)
3. Breitbart, Y., Garofalakis, M., Jai, B.: Topylogy discovery in heterogeneous IP networks: the NetInventory system. IEEE/ACM Trans. Netw. **12**(3), 401–414 (2004)
4. Breitbart, Y., Carofalakis, M.C.: Topylogy discovery in heterogeneous IP networks. In: Proceedings of INFOCOM 2000, Tel Aviv, Israel (2000)
5. Jin, X., Tu, W., Chan, S.: Scalable and efficient end-to-end network topology inference. IEEE Trans. Parallel Distrib. Syst. **19**(6), 837–850 (2008)
6. Srisuresh, P., Egevang, K.: Traditional IP Network Address Translator (Traditional NAT). IETF RFC 3022 (2001)
7. Aboba, B., Dixon, W.: IPSec-NAT Compatibility Requirements. IETF RFC 3715 (2004). Kivinen, T., Volpe, V.: Negotiation of NAT- Traversal in the IKE. IETF RFC 3947 (2005), Huttunen, A., DiBurro, L.: UDP Encapsulation of IPSec Packets. IEFF RFC 3948 (2005)
8. Breitbart, Y., Garofalakis, M., Martin, C., et al.: Topology discovery in heterogeneous IP networks. In: IEEE International Conference on Computer Communication 2000 Proceedings, Tel Aviv, pp. 265–274, 1–3 March 2000
9. Zheng, H., Zhang, G.Q.: An algorithm for physical network topology discovery. J. Comput. Res. Dev. **39**(3), 264–268 (2002). (in Chinese with English abstract)
10. Lowekamp, B., O'Hallaron, D.R., Gross, T.R.: Topology discover for large ethernet networks. In: ACM Special Internet Group on Data Communication, pp. 237–248. ACM Press, New York (2000)
11. Kent, S., Atkinson, R.: Security Architecture for the Internet Protocol. IETF RFC 2401 (1998)
12. Sun, Y., Wu, Z., Shi, Z.: A method of topology discovery for switched ethernet based on address forwarding tables. J. Softw. **17**(12), 2565–2576 (2006)
13. Sun, Y., Shi, Z., Wu, Z.: Automatic discovery of physical topology in switched Ethernets. J. Comput. Res. Dev. **44**(2), 208–215 (2007)
14. Bellman, R.E.: Dynamic Programming. Princeton University Press, Princeton (1957)

Intelligent Control and Automation

Multiple-Crane Integrated Scheduling Problem with Flexible Group Decision in a Steelmaking Shop

Xie Xie[1](✉) and Yongyue Zheng[2]

[1] Key Laboratory of Manufacturing Industrial and Integrated Automation,
Shenyang University, Shenyang, Liaoning, China
xiexie8118@gmail.com
[2] Liaoning Institute of Standardization, Shenyang, Liaoning, China
zhengyongyue@163.com

Abstract. This paper abstracts a practical problem from the iron and steel enterprise, and researches a multiple crane integrated scheduling problem with flexible group decision (crane grouping scheduling for short). For this demonstrated NP-hard problem, we propose a heuristic algorithm based on some analyzed properties. For a restrict case, we analyze the worst-case performance. Further, for the general case, the average performance of the heuristic algorithm is computationally evaluated. The results show that the proposed heuristic algorithm is capable of generating good quality solutions.

1 Introduction

Steel is by far the most widely used metallic material and continues to be vitally important to our society. Since the 1990s, China's steel industry has undergone a substantial transformation and has made remarkable achievements. Typically, steel making operation planning in every iron and steel company involves ladle allocation, crane assignment, crane scheduling in coordination of the production and transportation process, and so on. In the practical of a steelmaking shop, crane often connects with the other transportation tools. All these operations impact each other. In particular, the scheduling of multiple-crane significantly influences the completion time of steelmaking process.

The melted steel in each ladle is transported firstly by trolleys and then operated by cranes which are mounted on the tracks over the trolley railway to the parallel converters. The whole process is considered as a loaded move. After a crane drops off the melted steel from a ladle, the crane moves empty ladle to trolley to perform the next loaded move. Usually an area in the steelmaking shop is served by several bridge cranes as illustrated in Fig. 1. The crane moves along the track over the area while its pickup device (hoist) can move along the crane bridge. In this way the hoist of the crane can reach any position in the area. The main objective of this process is to determine the sequence of loading operations so that the makespan of all required melted steel in ladles, that is, the latest completion time among all the ladles of the melted steel, is minimized.

© Springer International Publishing AG 2017
D.-S. Huang et al. (Eds.): ICIC 2017, Part III, LNAI 10363, pp. 181–187, 2017.
DOI: 10.1007/978-3-319-63315-2_16

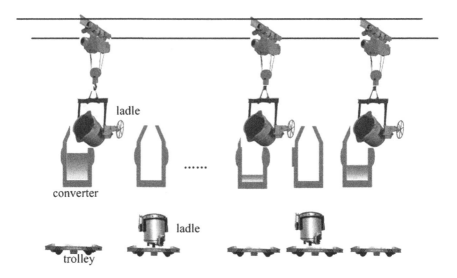

Fig. 1. Steel making process in an area of the steelmaking shop.

Most papers focused on crane scheduling problem arsing from the warehouse of the steelmaking shop. Zapfel and Wasner [1] first investigated a single crane scheduling problem in a distribution centre of steel coils to store incoming coils and retrieve coils required by customers. The problem was viewed as a job shop scheduling problem and formulated as a nonlinear integer programming model which is hard to solve. A local search based heuristic was proposed and tested through computation. Rei et al. [2] considered a single crane scheduling problem to store and retrieve steel items with known arrival and retrieval dates to minimize the number of crane movements. A simulation based heuristic was proposed to solve the problem. Tang et al. [3] studied a single crane scheduling problem in a warehouse to determine the shuffling positions and crane movements so that all required coils are retrieved in the shortest possible time. They formulated a mixed integer linear programming model and implemented a sequential solution approach. A dynamic programming algorithm was proposed for optimally solving a restricted case, and heuristic algorithm is proposed for the general case and its worst-case performance is analyzed. For this problem, Xie et al. [4] proposed a genetic algorithm and further analyzed its worst case performance. Xie et al. [5] extended this problem to the multiple crane scheduling problem. They modeled the problem as a MILP, proposed a genetic algorithm and analyzed its performance from the worst-case point of view. Tang et al. [6] studied the plate shuffling problems and the coil shuffling problems by a single crane. A greedy heuristic was proposed to solve the general problems and its worst-case performances on both problems are analyzed. A tabu search (TS) method with a tabu list of variable length was proposed to further improve the heuristic solutions.

Moreover, the problem studied here concerns material handling activities in steel production plant. Tanizaki et al. [7] considered steelmaking plant scheduling problem with crane transportation without waiting times constraint and hot consumption consideration. Dohn and Clausen [8] decomposed and modeled a slab yard planning and a crane scheduling problem. They proposed a two-stage heuristics for the studied problem. Xie et al. [9] focused on a hub reentrant shop scheduling problem in the actual packing production line of the iron and steel industry. They showed that the problem is NP-hard in the strong sense. Some properties were derived and further a heuristic algorithm was proposed with the demonstrated worst performance ratio. Sun et al. [10] introduced a mixed-timed petri-net modeling to minimize the makespan of the whole steelmaking and continuous casting process. For batch annealing process, Moon and Hrymak [11] and Liu et al. [12] mainly focused on the production process but hardly considering the crane scheduling. For studying this process, Tang et al. [13] formulated a mixed integer linear programming (MILP) model by considering both machine and crane positions. They showed that the problem is NP-hard in the strong sense and constructed a two-phase algorithm.

So far these papers only studied each crane scheduling separately without considering crane flexible group decision. In this paper, we study a multiple crane integrated scheduling problem with flexible group decision and aim to minimize the makespan of a given set of all required melted steel in ladles. The non-crossing constraints between the cranes must be satisfied owing to the common track on which these cranes can move. The remainder of this paper is organized as follows. In Sect. 2, we describe our problem in detail and analyze some properties. Section 3 proposes a heuristic algorithm and analyzes the quality for a restricted case by the worst case performance point of view. Computational results are reported in Sect. 4 comparing the heuristic solutions with the lower bound. Some concluding remarks are contained in Sect. 5.

2   Problem Description

Throughout the paper, we study the scheduling of cranes with non-crossing constraints to minimize the makespan of all required melted steel in ladles. Suppose that the ladles from left to right is numbered by 1, 2, …, n, respectively served by m bridge cranes (see Fig. 1). Only one crane can handle one ladle till completion. Throughout the paper, picking up of a required ladle and moving it to its designated converter will be called a crane operation. In this operation, a ladle concerned needs to be lifted up from its trolley, moved to a converter position and then dropped the melted steel off from the ladle. The whole process is considered as a full loaded move for the crane. After the crane drops off the full ladle, the crane can move the empty ladle its trolley. Then the crane empty move another trolley to perform the next full loaded move. For a given set of required ladles, we explore the multiple crane integrated scheduling problem to determine the sequence for full loaded moves for each crane. The objective is to minimize the *makespan*, i.e., the time by which the dropping off of all required ladles is completed.

Let A be the set of all positions in the area. Each position $q \in A$ can be identified uniquely by its row-column-height coordinates (r_q, c_q, h_q). In the rest of the paper we may refer to a position using its number or its coordinates, whichever being more convenient. Considering that a position can also be represented by its coordinates, each move time can be denoted in two ways: $t_{oq} \equiv t_{r_o c_o h_o, r_q c_q h_q}$, $e_{oq} \equiv e_{r_o c_o h_o, r_q c_q h_q}$.

Let the operation time of the j th ladle be $p_j > 0$ and for convenience, assume that the cranes are numbered by 1, 2, ..., m, respectively (see Fig. 1). Note that each operation time consists of the crane loaded move or empty move, and the time for lifting up and dropping off. Our problem is therefore finding a sequence to determine an exact time s_j for each ladle j when it starts. Owing to the crane non-crossing constraints, a scheme is feasible if and only if for any two ladles j, j' with $j < j'$, either their operation windows do not overlap, i.e., $[s_j, s_j + p_j) \cap [s_{j'}, s_{j'} + p_{j'}) = \emptyset$ or they are assigned to cranes with $\sigma(j) < \sigma(j')$. With such constraints, a simple observation can follow.

Proposition 1. At any moment while the ith crane is operating the jth ladle, neither a crane i' with $i' > i$ can operate a ladle j' with $j' \leq j$, nor a crane i'' with $i'' < i$ can operate a ladle j'' with $j'' \geq j$.

For simplicity, let $\sum_{j=1}^{n} p_j = P$ and $p_{\max} = \max_{j=1,2,\cdots,n} \{p_j\}$. Further, let C^A and C^* be the *makespan* generated by an algorithm A and the optimal *makespan*, respectively. Then, similar to the parallel machine scheduling problem, we have the following proposition.

Proposition 2. The lower bound LB on the optimal *makespan*

$$C^* \geq LB = \max\{\frac{P}{m}, p_{\max}\}$$

Knowing that even without considering the flexible group decision, the studied problem have been demonstrated strongly NP-hard if $m \geq 2$ by Lim et al. [14]. Thus, it will probably not be possible to solve instances of realistic size by an exact procedure inacceptable time. For this reason it is appropriate to use a heuristic, which will not necessarily find an optimal solution, but at least a reasonable one in acceptable time. Here we introduce a heuristic algorithm.

3   Heuristic Algorithm and Its Performance Analysis

In the section, we propose an efficient heuristic solution method. The algorithm partitions the ladles from the leftmost to the rightmost into m subsets and processes each subset independently by one of the m cranes, on the condition that each subset is not too large. The algorithm starts with an empty sequence for each crane or a fixed assignment of a ladle if the crane is currently executing a ladle. By each assignment of a ladle to a crane, a new, so-called partial solution is generated.

In the presentation of the heuristic algorithm below, the groups $\Omega_1, \Omega_2, \ldots \ldots, \Omega_m$ are m disjoint subsets during the solution process. Let $\sigma = (1, 2, \ldots, n)$ be a given schedule of the n ladles which is specified by the description: the n ladles are divided into m disjoint subsets. (r_{m0}, c_{m0}, h_{m0}) is also used as a dynamic parameter. It is used to represent the initial position of the crane m at the beginning and then used to represent the current position of the crane at any stage. The procedure of the *Heuristic* is as follows.

Step 1. If $\Omega_1 = \emptyset$, assign one by one from the leftmost ladle to the first group with $\Omega_1 = \{j : \sum_{j=1}^{k} p_j \geq \max\{\frac{P}{m}, p_{\max}\}\}$ which is operated by the leftmost crane.

Step 2. Similarly, if $\Omega_2 = \emptyset$, assign one by one from the $k+1th$ ladle to the second group with $\Omega_2 = \{j : \sum_{j=k+1}^{k'} p_j \geq \max\{\frac{P}{m}, p_{\max}\}\}$ operated by the second leftmost crane, until each crane has been divided into a group.

Step 3. In each group, choose a ladle j with minimal $(e_{r_{m0}c_{m0}h_{m0},r_jc_jh_j} + t_{r_qc_qh_q,r_jc_jh_j})$ where $t_{r_qc_qh_q,r_jc_jh_j}$ is the time for moving ladle j to the nearest empty position q. In case of a tie, priority is given to the left ladle. Any further tie is broken by choosing the ladle with minimal $e_{r_{m0}c_{m0}h_{m0},r_jc_jh_j}$. Create a sequence by scheduling every ladle to its assigned crane as soon as possible, that means starting at time 0 from the first ladle to the last one, once crane becomes available.

Step 4. If $\Omega_1 \cup \Omega_2 \cup \ldots \cup \Omega_m = \emptyset$, stop. Otherwise, go to step 3.

Partitioning the ladles in steps 1 and 2 take at most $O(n)$ time. Crane scheduling procedure in step 3 takes at most $O(mnA)$ time. Therefore, the computational complexity of the heuristic algorithm is $O(mn^2A)$. In the following section, we will analyze the computational performance of the heuristic algorithm.

Since our algorithm assigns a subset of ladles integrally to an available crane, the *makespan* must be determined by one of the subsets. More precisely, it is attained by the subset with the maximum total processing time. The following observation further proves that the proposed algorithm can be bounded within $2 - 2/m + 1$ if there are not too many subsets and the total processing time of each subset is not too large.

Proposition 3. If our heuristic algorithm can divide at most m group, and the overall processing time of each group is no more than $C^H \leq \max\{2P/m+1, p_{\max}\}$, combining it with the fact that $C^* \geq \max\{P/m, p_{\max}\}$, then $C^H \leq 2(1 - 1/m + 1)C^*$.

4   Computational Results

In order to assess the performance of the proposed heuristics and gain some insights into the problem structure, we implemented our algorithms in C# 5.0 and had them solve a series of test instances emulating real-world data from an iron and steel company. All tests were run on an x64 PC with an Intel Core i7-3770 3.4 GHz CPU and 8,192 MB of RAM. To show the performance of the proposed algorithm, 500 instances were randomly generated for the algorithm. The test problems are generated randomly by considering the following parameters:

For each instance, the full loaded move time followed a uniform discrete distribution in the range [30, 180]. The empty move time followed a uniform discrete distribution in the range [10, 60]. The number (n) of ladles and the number (m) of cranes that ranged in [6, 30] and [3, 6], respectively. The crane traveling speeds for loaded and empty moves are $v = 2$ and $\lambda = 4$. The time for lifting up and dropping off a ladle is $\mu = 2.5$. Now a series of computational experiments are conducted to examine the average performance of this algorithm. The quality is measured by its relative deviation from the LB, $(C_{\max} - LB)/LB * 100\%$. The average error ratio (Avg.ER) and the maximum error ratio (Max.ER) measured over the derived lower bound of the *makespan* are used for the performance test. The results of the evaluation are reported in Table 1 where the optimal value is estimated by the lower bounds given in proposition 2. To test the effect of the time for lifting up and dropping off that is not related to distance, we doubled its value and solved the problem instances again.

Table 1. Average optimality gaps of the lower bounds with respect to the heuristic algorithm

		$v = 2,\ \lambda = 4,\ \mu = 2.5$			$v = 2,\ \lambda = 4,\ \mu = 5$		
		Max.ER	Avg.ER	Avg.CPU	Max.ER	Avg.ER	Avg.CPU
$m = 3$	$n = 6$	22.463	18.702	0.000	31.586	27.378	0.000
	$n = 10$	21.599	19.041	0.000	31.659	26.718	0.000
	$n = 20$	21.653	19.214	0.002	31.604	26.814	0.003
	$n = 30$	21.736	19.050	1.206	31.451	27.375	1.310
$m = 4$	$n = 6$	22.686	18.590	0.000	31.292	26.390	0.000
	$n = 10$	22.287	18.112	0.000	31.319	26.382	0.000
	$n = 20$	21.909	18.302	0.003	31.096	26.501	0.002
	$n = 30$	21.621	18.454	1.211	31.301	26.389	1.323
$m = 5$	$n = 6$	22.914	18.513	0.000	31.295	26.598	0.000
	$n = 10$	22.133	18.176	0.000	31.141	26.236	0.000
	$n = 20$	22.471	17.880	0.002	31.113	26.314	0.003
	$n = 30$	22.160	18.012	1.300	31.175	26.183	1.421
$m = 6$	$n = 6$	22.639	18.201	0.000	31.121	26.460	0.000
	$n = 10$	21.738	17.996	0.000	31.001	26.161	0.000
	$n = 20$	22.299	17.829	0.002	31.092	26.153	0.003
	$n = 30$	22.391	17.963	1.536	30.949	26.118	**1.621**

These computational results demonstrate that the algorithm is capable of generating near-optimal solutions at most 1.621 s of CPU time. The proposed heuristic algorithm can solve any of the test instances almost instantly and the solution quality is not far from optimal. As seen in the table, the error ratio appears in a decreasing trend as the value of the ladle number increases. One of its reasons may be that the more ladles, the less waiting time for steel making. When the lift-up and drop-off time becomes larger, the ARDs for all methods are increases. This can be explained that increasing μ increases the constant part of the objective function and so the relative deviation from the *LB* becomes smaller. Therefore, it will be more important to solve the problem effectively in cases where the crane needs little time to lift up and drop off ladles and spends most of the time on moving ladles.

5   Conclusions

The paper takes the iron and steel-making process in the iron and steel enterprises as background, considers multiple-crane integrated scheduling problem with flexible group decision, the objective function is to minimize the latest completion time among all ladles of the melted steel. We analyze the property of the studied problem and propose a heuristic algorithm. For a restrict case, the worst case of the proposed heuristic algorithm has been proved. Computational experiments were carried out on instances randomly generated based on practical data. The results illustrate that the proposed heuristic algorithms can generate robust and acceptable solutions quickly.

Acknowledgements. This research is supported by National Natural Science Foundation of China (Grant Nos. 71672117 and 71201104), and Liaoning Province Natural Science Foundation (Grant No. 201602526).

References

1. Zapfel, G., Wasner, M.: Warehouse sequencing in the steel supply chain as a generalized job shop model. Int. J. Prod. Econ. **104**, 482–501 (2006)
2. Rei, R.J., Kubo, M., Pedroso, J.P.: Simulation-based optimization for steel stacking. In: Proceedings of 2nd International Conference on Modelling, Computation and Optimization, Metz, France, pp. 254–263(2008)
3. Tang, L.X., Xie, X., Liu, J.Y.: Crane scheduling in a warehouse storing steel coils. IIE Trans. **46**, 267–282 (2014)
4. Xie, X., Zheng, Y.Y., Li, Y.P.: Genetic algorithm and its performance analysis for scheduling a single crane. Discret. Dyn. Nat. Soc. (2015). http://dx.doi.org/10.1155/2015/618436
5. Xie, X., Zheng, Y.Y., Li, Y.P.: Multi-crane scheduling in steel coil warehouse. Expert Syst. Appl. **41**, 2874–2885 (2014)
6. Tang, L.X., Zhao, R., Liu, J.Y.: Models and algorithms for shuffling problems in steel plants. Nav. Res. Logist. **59**, 502–524 (2012)
7. Tanizaki, T., Tamura, T., Sakai, H., Takahashi, Y.: A heuristic scheduling algorithm for steel making process with crane handling. J. Oper. Res. Soc. Jpn. **49**, 188–201 (2006)
8. Dohn, A., Clausen, J.: Optimising the slab yard planning and crane scheduling problem using a two-stage heuristic. Int. J. Prod. Res. **48**, 4585–4608 (2010)
9. Xie, X., Tang, L.X., Li, Y.P.: Scheduling of a hub-reentrant job shop to minimize makespan. Int. J. Adv. Manuf. Technol. **56**, 743–753 (2011)
10. Sun, L.L., Liu, W., Chai, T.Y.: Crane scheduling of steel-making and continuous casting process using the mixed-timed petri net modelling via CPLEX optimization. In: Proceedings of 18th IFAC World Congress, Milano, Italy, pp. 9482–9487 (2011)
11. Moon, S., Hrymak, A.N.: Scheduling of the batch annealing process–deterministic case. Comput. Chem. Eng. **23**, 1193–1208 (1999)
12. Liu, Q.L., Wang, W., Zhan, H.R.: Optimal scheduling method for bell-type batch annealing shop and its application. Control Eng. Pract. **13**, 1315–1325 (2005)
13. Tang, L.X.: Scheduling of a single crane in batch annealing process. Comput. Oper. Res. **36**, 2853–2865 (2009)
14. Lim, A., Rodrigues, B.: A m-parallel crane scheduling problem with a non-crossing constraint. Nav. Res. Logist. **54**, 115–127 (2007)

Fault Diagnosis of Internal Combustion Engine Using Empirical Mode Decomposition and Artificial Neural Networks

Md. Shiblee[1](✉), Sandeep K. Yadav[2], and B. Chandra[3]

[1] Department of Computer Engineering, KKU, Abha, Kingdom of Saudi Arabia
[2] Department of Electrical Engineering, IIT Jodhpur, Jodhpur, India
[3] Sprinklr (India) and (Adjunct) IT School, IIT Delhi, New Delhi, India

Abstract. In this paper, a novel approach has been proposed for fault diagnosis of internal combustion (IC) engine using Empirical Mode Decomposition (EMD) and Neural Network. Live signals from the engines were collected with and without faults by using four sensors. The vibration signals measured from the large number of faulty engines were decomposed into a number of Intrinsic Mode Functions (IMFs). Each IMF corresponds to a specific range of the frequency component embedded in the vibration signal. This paper proposes the use of EMD technique for finding IMFs. The Cumulative Mode Function (CMF) was chosen rather than IMFs since all the IMFs are not useful to reveal the vibration signal characteristics due to the effect of noise. Statistical parameters like shape factor, crest factor etc. of the envelope spectrum of CMF were investigated as an indicator for the presence of faults. These statistical parameters are used in turn for classification of faults using Neural Networks. Resilient Propagation which is a rapidly converging neural network algorithm is used for classification of faults. The accuracy obtained by using EMD-ANN technique effectively in IC engine diagnosis for various faults is more than 85% with each sensor. By using a majority voting approach 96% accuracy has been achieved in fault classification.

Keywords: Fault diagnosis · Empirical mode decomposition · Intrinsic mode function · Cumulative mode function · Neural network · Resilient Propagation · Classification

1 Introduction

Faults in internal combustion engine usually have very simple origins. However, if undetected, they compound and manifest in different forms, making it difficult to pin-point the original cause. An effective diagnosis system should therefore, warn the operator as soon as the first signs of malfunctioning begin to appear, so that corrective action can be taken. In condition monitoring and fault diagnosis of internal combustion engine, the Acoustic emission and Vibration signals are used, because they always carry the dynamic information of the mechanical system. Traditionally, Acoustic and Vibration signals are mainly analyzed using signal processing techniques, such as the Fourier transform (FT), short time Fourier transform (STFT), symmetrized dot pattern

© Springer International Publishing AG 2017
D.-S. Huang et al. (Eds.): ICIC 2017, Part III, LNAI 10363, pp. 188–199, 2017.
DOI: 10.1007/978-3-319-63315-2_17

(SDP) and Wigner-Ville distribution [1, 2]. However, these methods only provide limited performance for the analysis of vibration signals of internal combustion engines, as they are non-stationary and consist of many transient components [3]. In recent years, wavelet transforms [4–6], has been investigated for its applicability in feature extraction from non-stationary, transient signals. However, the result of wavelet transform depends on the choice of the wavelet basis function. Any signal characters that correlate well with the shape of the wavelet basis function yield coefficients of high value, while all other features will be masked or completely ignored [7]. In the recent years, Huang et al. [8] proposed a new method for analyzing nonlinear and non-stationary signals. This method decomposes a signal using empirical mode decomposition (EMD) into a finite sum of components known as intrinsic mode functions (IMFs). This paper explores the utility of empirical mode decomposition as an effective tool for feature extraction of internal combustion engine fault diagnosis. The instantaneous frequencies and amplitudes can be extracted by applying the Hilbert transform (HT) on the IMFs. As we are analyzing the live signals and these signals are having a lot of ambient noise. Sometimes noise will increase the EMD error so greatly that the IMFs are distorted seriously and fail to represent the actual modes in the signal. We have selected the most representative IMFs based on, correlation and energy criterion. Cumulative Mode Function (CMF) is the aggregation of the most representative IMFs is proposed. We have extracted the envelope of CMF by applying Hilbert transform and then computed the Time domain and Spectral domain features from the envelope. These statistical features were used for fault classification using neural network. Resilient Back-propagation algorithm was used with feed-forward neural network as a classifier [9]. To eliminate the harmful effect of the magnitudes of partial derivative, the Resilient Propagation algorithm is used to train the neural network [9]. The live vibration signals were acquired under four conditions which include 3 fault states and the normal operating state. In our study, we have placed four vibration sensor (Accelerometer) at different locations of the IC engine. We have analyzed the data for each vibration sensor output using the proposed EMD-ANN technique and finally adopted the majority voting scheme for the final decision.

The rest of the paper is organized as follows. In Sect. 2, we discuss the proposed technique for feature extraction and classification. Section 3 gives and overview of types of fault and experimental setup. In Sect. 4, performance of the proposed method has been tested on real life data sets. Conclusions are given in Sect. 5.

2 Proposed Technique

The Proposed method comprises of the various stages. The first stage is meant for finding the Intrinsic Mode Functions (IMFs). In this paper, we have used EMD method for finding the IMFs. The second stage comprises of finding the most representative IMFs which resembles the actual signal to a large extent. The criteria used are namely energy and correlation for this purpose.

The third stage involves finding the Cumulative Mode Function (CMF). In the fourth stage we have extracted the envelope of CMF using Hilbert Transform. In the Fifth stage Time-domain and Spectral-domain features of the extracted envelope has

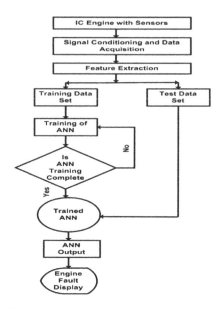

Fig. 1. Flow diagram of proposed technique

been computed. In the last stage neural network has been used for classification. The whole process is depicted in the following Fig. 1, and details are given in the following subsections.

2.1 EMD for Finding IMFs

For internal combustion engine fault diagnosis, vibration signals are decomposed into IMFs using Empirical Mode Decomposition. Each IMF extracted from the vibration signal serves as a filter and is self-adaptive in that it is directly originated from the signal and defined by inherent characteristic time scale.

Empirical Mode Decomposition is an emerging technique used for signal decomposition. It is derived from the signal itself, and based on following assumption [10–12].

(1) The signal has at least two extrema, i.e., one maximum and one minimum.
(2) The characteristic timescale is clearly defined by the time lapse between alternations of local maxima and local minima.
(3) If the signal has no extrema, but contains inflection points, then it can be differentiated one or more times to reveal the extrema.

A signal satisfying the above assumptions is decomposed into a number of IMFs, with each IMF being independent of others, due to the different characteristics time scales of the EMD. For example, the first IMF represents the shortest time scale of the signal, corresponding to the signal components in the highest frequency range.

Conversely, the last IMF is associated with longest time scale of the signal, and contains frequency components in the lowest frequency range.

The procedure for extracting the IMFs from the signal $x(t)$, the shifting process is implemented as follows. First, identify all the local maxima and minima of the signal, the upper and lower envelopes are generated through curve fitting. The mean values of the upper and lower envelopes of the signal $m_{11}(t)$ are calculated as

$$m_{11}(t) = \left(x_{up}(t) + x_{low}(t)\right)/2 \qquad (1)$$

where $x_{up}(t)$ and $x_{low}(t)$ are the upper and lower envelopes of the signal, respectively. Accordingly the difference between the signal $x(t)$ and the mean of the envelopes of the signal $m_{11}(t)$, which is denoted as $g_{11}(t)$, is given by

$$g_{11}(t) = x(t) - m_{11}(t) \qquad (2)$$

Due to the approximation nature of the curve fitting method, $g_{11}(t)$ has to be further processed (by treating $g_{11}(t)$ as the signal itself and repeating the process continually) until it satisfies the following two conditions.

(1) The number of extrema and the number of zero-crossing are either equal to each other or differ by at most one.
(2) At any point, the mean value between the envelope defined by local maxima and local minima is zero.

Through the iteration process, the difference between the signal and the mean envelope values, which is denoted as $g_{1k}(t)$, is obtained as

$$g_{1k}(t) = g_{1(k-1)}(t) - m_{1k}(t) \qquad (3)$$

where $g_{1k}(t)$ is the mean envelope value after the kth iteration, and $g_{1(k-1)}(t)$ is the difference between the signal and the mean envelope values at the $(k-1)^{th}$ iteration. The function $g_{1k}(t)$ is then defined as the first IMF and expressed as

$$c_1(t) = g_{1k}(t) \qquad (4)$$

After subtracting $c_1(t)$ from the original signal $x(t)$, the residue is obtained as

$$r_1(t) = x(t) - c_1(t) \qquad (5)$$

Subsequently, the residue $r_1(t)$ can be treated as the new signal, and the above illustrated iteration process is repeated to extract the rest of IMFs inherent to the signal $x(t)$ as

$$r_1(t) - c_2(t) = r_2(t)$$
$$\vdots \qquad\qquad (6)$$
$$r_{(n-1)}(t) - c_n(t) = r_n(t)$$

The signal decomposition is terminated when $m(t)$ becomes a monotonic function, from which no further IMFs can be extracted. By substituting (6) into (5), the signal $x(t)$ is decomposed into a number of intrinsic mode functions, that are constituent components of the signal. As a result, the signal $x(t)$ can be expressed as

$$x(t) = \sum_{j=1}^{n} c_j(t) + r_n(t) \tag{7}$$

Where $c_j(t)$ is the jth intrinsic mode function, and $m(t)$ id the residue of the signal decomposition. Equation (7) provides a complete description of Empirical Mode Decomposition.

2.2   Most Representative IMFs Selection

To ensure that from all the IMFs produced during the EMD process, the one that correlates with the vibration signal generated by the different fault in internal combustion engine are chosen and its corresponding frequency range is subsequently considered as the representative filter band. Two criteria on the most representative IMF selection were chosen, as discussed below:

2.2.1   Energy-Based Selection

Signal strength can be measured by evaluating the energy content in the signal. Thus, this parameter can be used to characterize the signal. The energy content in all the IMFs of a signal produced by EMD process is given by:

$$E_{x(t)} = \sum_{j=1}^{n} E_{c_j(t)} \tag{8}$$

where n is the number of IMFs and $E_{c_j(t)}$ is the amount of energy contained in the jth IMF, and it is calculated as

$$E_{c_j(t)} = \sum_{t=0}^{T} \left| c_j(t) \right|^2 \tag{9}$$

where T is total time duration of the signal being analyzed. Given the nature of data sampling, each IMF component $c_j(t)$ has a discrete value, thus the energy contained in the jth IMF can be estimated from the sampled data as

$$E_{c_j(l)} = \sum_{l=0}^{N} \left| c_j(l) \right|^2 \tag{10}$$

Where N is the number of data samples. The energy based criteria for IMF selection is based on the IMF containing the highest amount of energy among all the IMFs extracted from the signal should be chosen as the representative IMF [12].

2.2.2 Correlation-Based Selection

The correlation between two signals describes the similarity to each other. As IMFs are extracted from the internal combustion engine vibration signal, they are inherently correlated with each other. The degree of similarity between the IMF and the original signal can be measured in terms of a correlation coefficient, defined as [13]

$$\rho_{x(t)c_j(t)} = \frac{C_{x(t)c_j(t)}}{\rho_{x(t)}\rho_{c_j(t)}} \tag{11}$$

where $\rho_{x(t)}$ and $\rho_{c_j(t)}$ are the standard deviations of the signal $x(t)$ and the jth IMF $c_j(t)$, and $C_{x(t)c_j(t)}$ is the covariance of $x(t)$ and $c_j(t)$. In practice, the signal $x(t)$ and IMF $c_j(t)$ are sampled as discrete values; thus the correlation coefficient is estimated from the sampled data as

$$\hat{\rho}_{xc_j} = \frac{\sum\limits_{i=1}^{n} (x(i) - \bar{x})(c_j(i) - \overline{c_j})}{\left[\sum\limits_{i=1}^{n} (x(i) - \bar{x})^2 \sum\limits_{i=1}^{n} (c_j(i) - \overline{c_j})^2\right]^{1/2}} \tag{12}$$

where N is the number of data samples, and $\hat{x}$ and $\hat{c}_j$ are the mean values of $x(i)$ and $c_j(i)$, respectively. If an IMF contains high frequency resonant components, the correlation strength between that IMF and the signal will be relatively high. Thus, the IMF that has the highest correlation coefficient with signal should be chosen as the most representative IMF.

2.2.3 Cumulative Mode Function (CMF)

In the present study of internal combustion engine fault diagnosis, the vibration signal of each fault is decomposed using EMD and first 5 most representative IMFs selected using the above mentioned selection criteria. By using cumulative mode function (CMF), we will combine them to obtain CMF as:

$$z = c_1 + c_2 + c_3 + c_4 + c_5 \tag{13}$$

The first five most representative IMFs may be the consecutive IMFs or may not be the consecutive IMFs. In our internal combustion engine vibration signal analysis for different faults order of these IMFs comes different. Hilbert transform is studied in the present research for envelope extraction [14]. Analytically, performing the Hilbert transform on a signal, e.g., $z(t)$, is equivalent to formulating a corresponding analytic signal whose real and imaginary parts are the original signal $z(t)$ itself and its Hilbert transform, respectively. Such an analytic signal is expressed as

$$B(t) = z(t) + j\hat{z}(t) \tag{14}$$

with $\hat{z}(t)$ indicating the Hilbert transform of the signal $z(t)$ and expressed as

$$\hat{z}(t) = \frac{1}{\pi} \int \frac{z(\tau)}{t - \tau} d\tau \tag{15}$$

The modulus of such an analytic signal represents the original signal's envelope, defined as

$$e(t) = abs\left(B(t)\right) = \sqrt{z(t)^2 + \hat{z}(t)^2} \tag{16}$$

2.2.4 Feature Extraction

Time-domain and Spectral-domain features of envelope defined in Eq. (16), are computed in this paper which are in turn used for classification.

2.1.4.1 Time-Domain Statistical Features. In order to obtain time-domain statistical features, that effectively represent the IC engine characteristic in time-domain, 9 major features can be extracted from the envelope of the CMF [15]. These features are given in Table 1.

Table 1. Statistical features

Feature number	Time domain features	Feature number	Spectral domain features
1. Absolute mean	$\|\bar{e}\| = \frac{1}{N}\sum_{i=1}^{N}\|e_i\|$	10	$S_{10} = \frac{\int_0^{B_f} E(f)df}{\int_0^{F_s/2} E(f)df}$
2. Maximum peak value	$e_p = \max(e)$	11	$S_{11} = \frac{\int_{B_f}^{2B_f} E(f)df}{\int_0^{F_s/2} E(f)df}$
3. Root mean square	$e_{rms} = \sqrt{\frac{1}{N}\sum_{i=1}^{N} e_i^2}$	12	$S_{12} = \frac{\int_{2B_f}^{3B_f} E(f)df}{\int_0^{F_s/2} E(f)df}$
4. Square root value	$e_r = \left(\frac{1}{N}\sum_{i=1}^{N}\sqrt{e_i}\right)^2$	13	$S_{13} = \frac{\int_{3B_f}^{4B_f} E(f)df}{\int_0^{F_s/2} E(f)df}$
5. Variance	$var = \frac{1}{N-1}\sum_{i=1}^{N}(e_i - \bar{e})^2$	14	$S_{14} = \frac{\int_{4B_f}^{5B_f} E(f)df}{\int_0^{F_s/2} E(f)df}$
6. Kurtosis	$\beta = \frac{1}{N}\sum_{i=1}^{N} e_i^4$	15	$S_{15} = \frac{\int_{5B_f}^{6B_f} E(f)df}{\int_0^{F_s/2} E(f)df}$
7. Crest factor	$C = \frac{\|e\|_{peak}}{e_{rms}}$	16	$S_{16} = \frac{\int_{6B_f}^{7B_f} E(f)df}{\int_0^{F_s/2} E(f)df}$
8. Shape factor	$S_f = \frac{e_{rms}}{\|\bar{e}\|}$	17	$S_{17} = \frac{\int_{7B_f}^{8B_f} E(f)df}{\int_0^{F_s/2} E(f)df}$
9. Standard deviation	$e_{std} = \sqrt{\frac{1}{N}\sum_{i=1}^{N}(e_i - \bar{e})^2}$		

2.1.4.2 Spectral-Domain Statistical Features. The time-domain statistical features generally reflect the pattern of time-domain signal that only contains the typical frequency band. However, these features are not sufficient as the IC engine vibration signals are not-stationary in nature. We can find some typical frequency bands that could reflect the effective behavior of signals for different class of faults [16].

We can transform time-domain signal $e(t)$ to frequency-domain $E(F)$. The frequency spectra of the spectral domain signal are divided into 8 equal bands with bandwidth $B_f (B_f = (F_s/2)/8$, where F_s is the sampling frequency). The spectral energy on each band is calculated as the sum of the frequency amplitudes and the relative spectral energies are defined as the ratio of the spectral energy on each band to the total energy value in the spectral domain. These 8 relative spectral energies are taken as the spectral domain statistical features to represent the IC engine fault patterns. Mathematically, these features are given in Table 1.

2.2.5   Classification Using Neural Network
Statistical features extracted from CMF are used as inputs of neural network for classification of faults. Descriptions of extracted features are given in the Table 1.

The multilayer feed-forward network with resilient back-propagation algorithm is used to train the multilayer feed-forward neural network [17].

3   Experiment Setup

In order to evaluate the proposed EMD and Neural Network based method for fault diagnosis, a single-cylinder internal combustion engine testing test-rig has been used to measure the vibration signals of the engine. The experimental setup contained a large number of single-cylinder internal combustion engines, an optical encoder for speed measurement, PCB general purpose industrial ceramic shear ICP accelerometer (603C01), Data Acquisition module (NI cDAQ-9172). In order to verify the effectiveness of the proposed technique, three different faults were seeded in the engine. In view of recording the data for these seeded faults four vibration sensor (accelerometer) were placed at four different locations of the engine. LabVIEW 8.0 was used as a platform to communicate with the cDAQ-9172 hardware. Using VI's in LabVIEW, an easy to use menu driven user interface was used to record the engine vibration samples. Engine is operated in the 2500 rpm speed. The sampling rate of the data acquisition system was 50 kHz. The seeded faults are described below:

3.1   Cam Chain Fault

The Cam Chain is the element within the engine which transfers the drive from the Crank Shaft to the Cam Shaft. The Cam Chain rides along two riders. The tension on Cam Chain is adjusted by pushing the slider inwards or outwards. This can be done using the tension adjuster, which is accessible. Whenever Cam Chain is under tension, it produces the Cam Chain noise. All Cam Chain noise faults were seeded by varying tension on the Cam Chain using the tension adjuster. The vibration signal for Cam

Chain fault was recorded and analyzed using the EMD based technique. From which a total of 13 IMFs were extracted. All the extracted IMFs arranged according to their respective magnitude in energy (Joule) and correlation coefficient, as shown in Fig. 2. The first five most representative IMFs with the highest energy and highest correlation coefficient are selected by the developed algorithm. Figure 4, shows the Cam Chain Fault vibration signal and its five most representative IMFs. It is evident that the five most representative IMFs for Cam Chain Fault are IMF1, IMF2, IMF3, IMF4, IMF8.

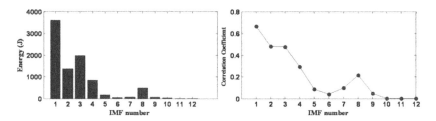

Fig. 2. Energy and correlation coefficient of each IMF for Cam Chain fault.

3.2   Tappet Fault

Tappet fault appears whenever the tappet clearance is too high. Under ideal settings the inlet tappet and outlet tappet clearance kept close to 0.07/0.08 mm. Any deviation from these set values results in generation of tappet fault. To seed tappet fault the tappet clearance of both inlet and outlet ports were disturbed from their ideal settings. The vibration signal of the engine with tappet fault is decomposed using EMD and the most representative IMFs are selected using above mentioned technique. Figure 3, shows the IMFs with the highest energy and correlation coefficient. It is observed that the five most representative IMFs for the Tappet fault are IMF1, IMF2, IMF3, IMF4, IMF8.

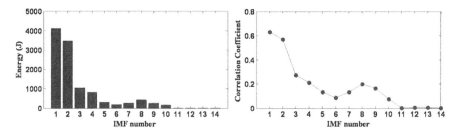

Fig. 3. Energy and correlation coefficient of each IMF for Tappet fault.

3.3   Primary Gear Damage (PGD) Fault and Healthy Engine

The gear assembly is located with the crank case. It comprises the set of drive gear and driven gear assemblies. Any abnormality in these gear assembly in the from of tooth damage, tooth to tooth error, eccentric/inclined bore, results in typical impact kind of sound. The Primary Gear damage fault was seeded by individually introducing defects in the drive gear and the driven gear. Figure 4, shows the IMFs with the highest energy and correlation coefficient. It is observed that the five most representative IMFs for the Primary Gear damage fault are IMF1, IMF2, IMF3, IMF7, IMF8. Figure 5, shows the IMFs with the highest energy and correlation coefficient for Healthy engine. It is observed that the five most representative IMFs for the Healthy engine are IMF1, IMF2, IMF3, IMF4, IMF7.

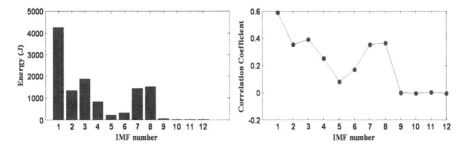

Fig. 4. Energy and correlation coefficient of each IMF for PGD fault.

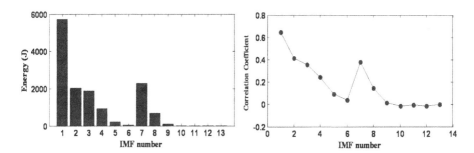

Fig. 5. Energy and correlation coefficient of each IMF for Healthy engine.

4   Results

This section presents the classification results for faults of IC engine using EMD-ANN technique. 765 vibration signals were collected for different faults. Time-domain and Spectral - domain features (as discussed in section-II) were extracted out of which features of 665 vibration signals were used for training the Neural Network and features of remaining 100 were used for testing. Twenty neurons were taken in the hidden layer and training was performed till 2000 iterations using Resilient Propagation algorithm. Classification accuracy corresponding to all four sensors is shown in Table 2. As summarized in Table 2, all the testing results have a performance of more than 85% for all the positions and various faults. For overall classification we have chosen the majority voting scheme (choosing at least 3 out of 4 outputs), and for this

Table 2. Classification performances

Fault category	Sensor-1	Sensor-2	Sensor-3	Sensor-4	Majority voting
Good	89.47	90.00	95.00	100	95.00
CCN	92.85	68.5	100	73.68	94.73
PGD	89.47	83.33	85.71	92.85	97.61
Tappet	90.00	94.73	100	78.94	95.00

scheme accuracy is more than 96%. The experimental results show that the EMD-ANN technique can be used effectively in IC engine diagnosis for various faults through measurement of engine vibration signals.

5 Conclusions

The Empirical Mode Decomposition provides a new method for time-frequency analysis and have got the great attention in various areas. In this paper an EMD-ANN based technique has been proposed for fault diagnosis of single cylinder four stroke IC engine. Experimental investigations are carried out to evaluate the proposed technique for online fault diagnosis of single cylinder four stroke IC engine at one of the leading automobile two-wheeler manufacturing company in India. Analyzed results show that the proposed technique is effective for the online fault diagnosis. From a more general point of view, we anticipate that the proposed EMD-ANN would become an appealing tool in non-stationary vibration analysis for diagnosis of localized faults of other mechanical systems.

Acknowledgment. The authors would like to express their gratitude to King Khalid University, Saudi Arabia for providing administrative and technical support 20070174.

References

1. Li, H., Zhang, Y.P.: Bearing faults diagnosis based on EMD and WIgner-Ville distribution. In: Proceedings of 6th World Congress on Intelligent Control and Automation, Dalian, China, pp. 5447–5451 (2006)
2. Shibata, K., Takashashi, A., Shirai, T.: Fault diagnosis of rotating machinery through visualization of sound signals. Mech. Syst. Sig. Process. **14**, 229–241 (2000)
3. Zheng, G.T., Leng, A.Y.T.: Internal combustion engine noise analysis with time-frequency distribution. J. Eng. Gas Turbines Power **124**, 645–649 (2002)
4. Wang, C., Gao, R.: Wavelet transform with spectral post-processing for enhanced feature extraction. IEEE Trans. Instrum. Meas. **52**(4), 1296–1301 (2003)
5. Frank, P.M.: On-line fault detection in uncertain nonlinear systems using diagnostic observers: a survey. Int. J. Syst. Sci. **25**, 2129–2154 (1994)
6. Wang, W., Jian, A.: A smart sensing unit for vibration measurement and monitoring. IEEE/ASME Trans. Mechatron. 1–9 (2009, accepted for future publication)
7. Loutridis, S.J.: Damage detection in gear system using empirical mode decomposition. Eng. Struct. **26**, 1833–1841 (2004)

8. Huang, N.E., Shen, Z., Long, S.R., et al.: The empirical mode decomposition and the Hilbert spectrum for nonlinear non-stationary time series analysis. In: Proceedings of Royal Society, London, vol. 454, pp. 903–995 (1998)
9. Riedmiller, M., Braun, H.: A direct adaptive method for faster back propagation learning: the RPROP algorithm. In: Proceedings of International Conference on Neural Networks, San Francisco (1993)
10. Huang, N.E., Shen, Z., Long, S.R.: A new view of nonlinear water waves: the Hilbert spectrum. Annu. Rev. Fluid Mech. **31**, 417–457 (1998)
11. Huang, N.E., Wu, M., Long, S.R., et al.: A confidence limit for the empirical mode decomposition and Hilbert spectrum analysis. In: Proceedings of Royal Society of London, Series A, vol. 459, pp. 2317–2345 (2003)
12. Yan, R., Gao, R.X.: Rotary machine health diagnosis based on empirical mode decomposition. J. Vib. Acoust. **130**, 1–12 (2008)
13. Bendat, J.S., Piersol, A.G.: Randaom Data: Analysis and Measurement Procedures, 3rd edn. Wiley, New York (2001)
14. Hahn, S.L.: Hilbert Transform in Signal Processing. Artech House Inc., Norwood (1996)
15. Chen, K., Lee, C.: Machine Condition Monitoring and Fault Diagnosis Technology. Beijing Science and Technology Press, Beijing (1991)
16. Zhu, Z.K., Feng, Z.H., Kong, F.R.: Cyclostationarity analysis for gearbox condition monitoring: approaches and effectiveness. Mech. Syst. Sig. Process. **19**, 467–482 (2005)
17. Haykin, S.: Neural Networks: A comprehensive Foundation. Pearson Education, Singapore (2003)

Intelligent Data Fusion

Distributed Attack Prevention Using Dempster-Shafer Theory of Evidence

Áine MacDermott[✉], Qi Shi, and Kashif Kifayat

School of Computer Science, Liverpool John Moores University,
Liverpool L3 3AF, UK
a.mac-dermott@2008.ljmu.ac.uk,
{q.shi,k.kifayat}@ljmu.ac.uk

Abstract. This paper details a robust collaborative intrusion detection methodology for detecting attacks within a Cloud federation. It is a proactive model and the responsibility for managing the elements of the Cloud is distributed among several monitoring nodes. Since there are a wide range of elements to manage, complexity grows proportionally with the size of the Cloud, so a suitable communication and monitoring hierarchy is adopted. Our architecture consists of four major entities: the Cloud Broker, the monitoring nodes, the local coordinator (Super Nodes), and the global coordinator (Command and Control server - C2). Utilising monitoring nodes into our architecture enhances the performance and response time, yet achieves higher accuracy and a broader spectrum of protection. For collaborative intrusion detection, we use the Dempster Shafer theory of evidence via the role of the Cloud Broker. Dempster Shafer executes as a main fusion node, with the role to collect and fuse the information provided by the monitors, taking the final decision regarding a possible attack.

Keywords: Intrusion detection · Cloud computing · Security · Collaboration · Dempster Shafer · Fusion algorithm · Autonomous systems

1 Introduction

Adoption of Cloud technologies allows critical infrastructure to benefit from dynamic resource allocation for managing unpredictable load peaks. Given the public awareness of critical infrastructure and their importance, there needs to be an assurance that these systems are built to function in a secure manner. Appropriate security procedures have to be selected when developing such systems and documented accordingly. Most existing technologies and methodologies for developing secure applications only explore security requirements in either critical infrastructure or Cloud Computing. Individual methodologies and techniques or standards may even only support a subset of specific critical infrastructure requirements. Requirement based security issues can be quite different for these applications and for common IT Cloud applications but need to be considered in combination for the given context.

Automation has become an indispensable part of service provision and has increased exponentially as demand for digital services and interconnectivity has

© Springer International Publishing AG 2017
D.-S. Huang et al. (Eds.): ICIC 2017, Part III, LNAI 10363, pp. 203–212, 2017.
DOI: 10.1007/978-3-319-63315-2_18

increased. The reliance on these systems has resulted in ICT playing a key role in the provisioning of services that critical infrastructures deliver to the general population. Disruptions in one part of an infrastructure may propagate throughout the system and have cascading effects on other sectors (Ten et al. 2010). Critical infrastructure protection relates to application processes, electronic systems, and information stored and processed by such systems.

The concern is that critical IT resources and information in Cloud systems may be vulnerable to cyber attacks or unauthorised access. The primary security concerns with Cloud environments pertain to security, availability, and performance. Many attacks are designed to block users from accessing services and providers from delivering services, i.e. Denial of Service (DoS) or Distributed Denial of Service (DDoS). Service providers may face significant penalties due to their inability to deliver services to customers in accordance with regulatory requirements and Service Level Agreements (SLA) (Rak et al. 2012). DDoS attacks are a serious and growing problem for corporate and government services conducting their business via the Internet. Resource management to prevent DDoS attacks is receiving attention, as the Infrastructure as a Service (IaaS) architecture, effectively 'supports' the attacker. When the Cloud system observes the high workload on the flooded service, it is likely the Cloud federation (which is the practice of interconnecting the Cloud computing environments of two or more service providers for the purpose of load balancing traffic and accommodating spikes in demand) will start providing more computational power in order to cope with it.

Traditional network monitoring schemes are not scalable to high speed networks such as Cloud networks, let alone Cloud federations. It is clear that an Intrusion Detection System (IDS) alone cannot protect the Cloud environment from attack. If an IDS is deployed in each Cloud Computing region, but without any cooperation and communication, it may easily suffer from single point of failure attack. The Cloud environment could not support services continually, as it is not always easy for the victim to determine that is being attacked, or where the attack is originating from. A new and novel approach to the aforementioned problem is required, that is, providing Security as a Service in a Cloud federation. Our solution encompasses the following methodological attributes:

- We represent Cloud Service Providers (CSPs) within a Cloud federation as interconnected domains.
- Once a Belief is generated that an attack is underway (b_a), this is sent to a Command and Control server (C2). The C2 queries the Cloud Broker, and the Broker checks the value against its stored values (as it may not have been published yet), and invokes a global poll procedure in which other C2s within the other domains are queried.
- The Cloud Broker coordinates attack responses, both within the domain itself, and with other domains, and is facilitating inter-domain cooperation.
  - Dempster-Shafer (D-S) is used to fuse the generated beliefs and make a system wide decision. This cooperation between CSPs ensures that the scalable defence required against DDoS attacks is in an efficient manner; aiming to improve the overall resilience of the interconnected infrastructure.

- The development of a collaborative intrusion detection heuristic based on D-S theory of evidence, and the inclusion of confidence values for improved accuracy.
  - We are improving the decision making precision and accuracy for autonomous information sharing in a federated Cloud environment via a two stage fusion process.

2 Background

The Cloud Computing paradigm is increasingly being adopted in critical sectors such as energy, transport, and finance. Deploying high assurance services in the Cloud increases cyber security concerns, as successful attacks could lead to outages of key services that have high socioeconomic implications. This exposes these infrastructures to cyber risks and results in demand for protection against cyber-attacks, even more than traditional systems. Critical security issues include: data integrity; user confidentiality; availability of data; and trust among entities. Securing applications and services provided in the Cloud against cyber attacks is hard to achieve due to the complexity, heterogeneity, and dynamic nature of such systems.

Site management and monitoring has improved for critical infrastructure facilities as they have become more progressively connected to the Internet. The added convenience of connectivity, however, has turned the once-limited attack surface of these industries into a fertile landscape for cyber-attacks. Due to the potentially high profile effects of attacks to critical infrastructure systems, these industries have become even more attractive targets for cybercriminals (Trend Micro Incorporated 2015). The sensitive nature of critical infrastructure services deems their protection critical, and their services hereof. This is predominantly caused by the inadequacies and limitations of current security protection measures which fail to cope with the sheer size and vast dynamic nature of the Cloud environment.

Attacks and failures are inevitable; therefore, it is important to develop approaches to understand the Cloud environment under attack. The current lack of collaboration among different components within a cloud federation, or among different providers, for detection or prevention of attacks is the focus of our work (MacDermott et al. 2015). Our research focuses on maintaining the availability of the data, as previously described, the service in question could be financial, organisational, or on demand. Protecting the Cloud environment from DDoS attacks is imperative as these attacks can threaten the availability of Cloud functionalities.

3 Security as a Service

Detecting intrusion patterns in the Cloud environment involves looking for behavioural changes. This process could involve signature based detection for DoS attacks, which must be robust against noise data, and false positives and false negatives produced. Anomaly detection is an approach for detecting behavioural changes as these schemes often aggregate normal behaviour through their modelling of normal versus abnormal traffic. The main requirement of our solution is to provide protection for critical infrastructure services being hosted in the Cloud environment through novel intrusion detection techniques (MacDermott et al. 2015).

Monitoring is a core function of any integrated network and service management platform. Cloud computing makes monitoring an even more complex infrastructure support function, since it includes multiple physical and virtualized resources and because it spans several layers of the Cloud software stack, from IaaS to PaaS, and SaaS. Using our Security as a Service method, collaborative intrusion detection is possible in a federated Cloud environment.

The system uses a Cloud Broker to propagate information to the C2 entities in each CSP domain – this is in the form of Black lists and White lists. Monitoring nodes are used to observe changes or suspicious activities in local domains. These values are stored in a Grey list of ambiguous observations. For pre-emptive warning, Beliefs are generated and assigned to all subsets of possible outcomes based on the trigger.

We assign Beliefs to the outcome in the form of, random attack variable x would have a basic probability assignment of { }, {attack}, {no attack}, and {attack, no attack}. { } represents an empty subset with a value of 0, whereas {attack, no attack} represents uncertainty, i.e. it could be either. An advantage of using D-S theory of evidence to fuse Beliefs is that the algorithm can start from an uncertain state and allow the observed evidences form in each of the subnets gradually. D-S utilises orthogonal sum to combine the evidences. We define the belief functions, describing the belief in a hypothesis A, as $\mathrm{Bel}_1(A), \mathrm{Bel}_2(A)$; then the belief function after the combination is defined as:

$$\mathrm{Bel}(A) = \mathrm{Bel}_1(A) \oplus \mathrm{Bel}_2(A)$$

The mass function after the combination can be described as:

$$m(A) = K^{-1}. \sum_{A_i \cap B_i = A} m_1(A_i) m_2(B_j)$$

Where K is called Orthogonal Coefficient, and it is defined as:

$$K = \sum_{A_i \cap B_i \neq \emptyset} m_1(A_i) m_2(B_j)$$

3.1   Implementation

Collaborative security between CSPs in a Cloud federation can offer holistic security to those in this scheme. Information sharing in this approach is automated which we conceive to be an important aspect of our approach. Dividing the system into domains makes the system more scalable, and Belief generation and sharing of threat information could be used as a warning of an imminent attack. Previous work of ours (MacDermott et al. 2015) details our simulations using Riverbed Modeler 18.0 and convey how attacks could propagate throughout a Cloud federation. At this stage of the simulation, the main purposes were to analyse the role a Broker could have with autonomous sharing of information; the role of a single monitoring entity on the entire federation vs the C2s monitoring their own sub domains; and how an attack within a Cloud federation could affect the interdependent services present.

Next, we are showing the actions to be taken in the simulation, from the point where an intrusion is believed to have been detected. The integration of the decisions coming from different IDSs has emerged as a technique that could strengthen the final decision. Sensor fusion can be defined as the process of collecting information from multiple and possibly heterogeneous sources and combining them to obtain a more descriptive, intuitive and meaningful result (Thomas and Narayanaswamy 2011). Related work in the field of sensor fusion has been carried out mainly with methods such as probability theory, evidence theory, voting fusion theory, fuzzy logic theory, or neural network in order to aggregate information.

Our implementation of our D-S for collaborative intrusion detection is in C#, and focuses on demonstrating the application of the fusion algorithm in an autonomous information sharing scheme. For proof of concept we are using a lower amount of entities to convey how the communication occurs and the information would be exchanged within the infrastructure; future work would involve expanding this solution to cope with a larger scale. Firstly, an IP address is entered into the program and the value is compared to the Black list and White list to see if the values are present (Fig. 1).

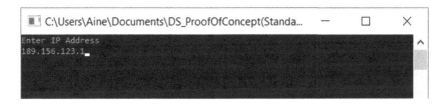

Fig. 1. Checking list values to determine if IP present

When compared against the lists, if the IP address is in the Black list then the user is 'Blocked' – source code for this is conveyed in Fig. 2.

```
//if the IP is in the black list (Key will be true)
if (BlackValue.Key)
{
    Console.WriteLine("***BLACK***");
    //block IP
    return "Blocked User";
```

Fig. 2. Blocked user key return

If the entered IP address is present on the White list then the user is 'Permitted Access', as illustrated in Fig. 3. The console outputs the other values from the White list, and this is also a separate file than can be viewed.

```
Allow access, user on White List

 White 192.168.22.14 White 192.168.66.12 White 192.168.222.74 White 192.168.98.62 White 192.168.837
836 White 192.168.78.73 White 192.168.1.1 White 156.156.126.12 White 1925.16587.1568.1568 White 192
1647.1346.132 White 123456.123.123123.123 White 123789.123789.123789.123789 White 321.654.987.321 W
ite 123456.123456789.1324586798.12 White 1234563.1.1.1 White 1236.12654.15646512.156 White 111.111.
.1 White 222.222.2.2 White 1.1.1.1 White 7.7.7.7 White 8.8.8.8 White 1.11.1.1 White 154.1325.156.45
White 154.447.145.122 White 35.56.4522.254 White 123456.123456.11.11 White 123456.12354.15645.55 Wh
te 123456.456496.5646.4864 White 456.486.486.54 White 654.54.5454.45 White 15645.4864.546.546 White
465.486.486.84 White 456.456.255.55 White 456.456.456.12 White 45.45.12.23 White 44.55.66.22 White
```

Fig. 3. Value on white list

If the entered IP address is not present on either list, the value is stored in the Grey list and assigned a threat value which we use to form the Belief. Hypothesis sets based on all values between 0 and 1 are included within the program, as well as mass values and plausibility functions.

Figure 4 shows the threat value ranges used, and the ability to increase/decrease the associated risk due to occurrences on the list is also an option. Increased occurrences could cause the risk score to increase, e.g. beginning on the white list, moving to the grey list, but then being promoted to the black list. For a value over 70 this would trigger a Belief generation and the associated hypothesis values output. Figure 5 is an example of a threat score of 80 and the associated hypothesis set generated, and Fig. 6 shows a score of 60.

```
namespace DS_ProofOfConcept
{
    2 references
    class ListChecking
    {

        //values decice what returned percentage equals which tier
        public int HIGHTHREAT = 70;
        public int MIDTHREAT = 40;
        public int LOWTHREAT = 20;

        //amount to increase threat level on grey list if threat detected
        //and in which tier
        public int highTeirIncreaseAmount = 50;
        public int midTeirIncreaseAmount = 20;
        public int lowTeirIncreaseAmount = 10;
        public int belowLowTierReduction = 5;
```

Fig. 4. Threat value ranges

```
***No List***


Threat Score  80
-----------------------------------------------------------------
---Hypothesis-||--Mass--||---Belief--||-Plausibility||
-----Attack --||-- 0.8  ||   0.8     ||     0.9     ||
---No Attack -||-- 0.1  ||   0.1     ||     0.2     ||
-----Either --||-- 0.1  ||   1.0     ||     1.0     ||
-----------------------------------------------------------------
```

Fig. 5. Example hypothesis set generation for a threat score of 80

```
Threat Score   60
------------------------------------------------------------------
---Hypothesis-||--Mass--||---Belief--||-Plausibility||
-----Attack --||-- 0.6  ||   0.6     ||     0.7     ||
---No Attack -||-- 0.3  ||   0.3     ||     0.4     ||
-----Either --||-- 0.1  ||   1.0     ||     1.0     ||
------------------------------------------------------------------
```

Fig. 6. A belief generation of 0.6

This value is sent to the Broker and compared against the Black and White lists, as the information may not have been propagated to the C2s within the federation. The Broker then queries the adjacent monitoring entities and requests they generate a Belief based on the original value.

In the example of a Cloud federation, the Broker takes three belief values and fuses them together to make a system wide decision. The values would then be updated to the lists (White or Black) and calculations of these combinations are as follows:

Belief Combination of Two Values – $\boldsymbol{Bel(A) = Bel_1(A) \oplus Bel_2(A)}$

(m_1) we have belief that the proposition is true for just state Attack is m_1({Attack}) = 0.8) and similarly m_1({No Attack}) = 0.1 with m_1({Either}) = 0.1.

Then we take another assessment m_2 with m_2 ({Attack}) = 0.6, m_2({No Attack}) = 0.3 with m_2({Either}) = 0.1.

The joint mass function would be $m_{1,2}$(A) = (1/1−K) m_1({Either}) m_2({Either}) with K = m_1({Attack}) m_2({No Attack}) + m_1({No Attack}) m_2({Attack}) = 0.8 * 0.3 + 0.1 * 0.6 = 0.30

$$\text{So}\, m_{1,2}(\text{A}) = (1/1 - 0.30) * 0.1 * 0.1 = 0.007$$

Belief Combination of Three Values - $\boldsymbol{Bel(A) = Bel_1(A) \oplus Bel_2(A) \oplus Bel_3(A)}$

(m_1) we have belief that the proposition is true for state Attack which is 0.3 (i.e. m_1({Attack}) = 0.8) and similarly m_1({No Attack}) = 0.1 with m_1({Either}) = 0.1.

Then we take another assessment m_2 with m_2 ({Attack}) = 0.6, m_2({No Attack}) = 0.3 with m_2({Either}) = 0.1.

m_3 associated values include {Attack}) = 0.0, m_3({No Attack}) = 0.5 with m_3({Either}) = 0.1.

The joint mass function would be $m_{1,2,3}$(A) = (1/1−K) m_1({Either}) m_2({Either}) m_3({Either}) with: K = m_1({Attack}) m_2({No Attack}) + m_1({No Attack}) m_2({Attack}) + m_1({Attack}) m_3({No Attack}) + m_1({No Attack}) m_3({Attack}) + m_2({Attack}) m_3({No Attack}) + m_2({No Attack}) m_3({Attack}) = 0.8 * 0.3 + 0.1 * 0.6 + 0.8 * 0.5 + 0.1 * 0.0 + 0.6 * 0.5 + 0.3 * 0.0 = 1

$$\text{So}\, m_{1,2,3}(\text{A}) = (1/1 - 1) * 0.1 * 0.1 * 0.5 = \text{n/a} - \text{cannot divide by 0}$$

Belief Combination of Three Values - $\boldsymbol{Bel(A) = Bel_1(A) \oplus Bel_2(A) \oplus Bel_3(A)}$

(m_1) we have belief that the proposition is true for state Attack which is 0.3 (i.e. m_1({Attack}) = 0.8) and similarly m_1({No Attack}) = 0.1 with m1({Either}) = 0.1.

Then we take another assessment m_2 with $m_2(\{\text{Attack}\}) = 0.6$, $m_2(\{\text{No Attack}\}) = 0.3$ with $m_2(\{\text{Either}\}) = 0.1$.

m_3 associated values include $\{\text{Attack}\}) = 0.9$, $m_3(\{\text{No Attack}\}) = 0.05$ with $m_3(\{\text{Either}\}) = 0.05$.

The joint mass function would be $m_{1,2,3}(A) = (1/1-K) \, m_1(\{\text{Either}\}) \, m_2(\{\text{Either}\}) \, m_3(\{\text{Either}\})$ with: $K = m_1(\{\text{Attack}\}) \, m_2(\{\text{No Attack}\}) + m_1(\{\text{No Attack}\}) \, m_2(\{\text{Attack}\}) + m_1(\{\text{Attack}\}) \, m_3(\{\text{No Attack}\}) + m_1(\{\text{No Attack}\}) \, m_3(\{\text{Attack}\}) + m_2(\{\text{Attack}\}) \, m_3(\{\text{No Attack}\}) + m_2(\{\text{No Attack}\}) \, m_3(\{\text{Attack}\}) = 0.8 * 0.3 + 0.1 * 0.6 + 0.8 * 0.05 + 0.1 * 0.9 + 0.6 * 0.05 + 0.3 * 0.09 = 0.73$

$$\text{So } m_{1,2,3}(A) = (1/1 - 0.73) * 0.1 * 0.1 * 0.5 = 0.01851$$

4 Evaluation

The use of D-S rule is mathematically possible only if m^a and m^b are not conflicting, i.e. if there is a focal element of m^a and a focal element z of m^b satisfying $(y \cap z) \neq \emptyset$. Merging two belief masses with the conjunctive rule defined above produces a sub-additive belief probability assignment, meaning that the sum of belief masses on focal elements can be less than one, in which case it is assumed that the missing or complement belief mass gets assigned to the empty set. If desirable, the normality assumption $m(\, /0) = 0$ can be recovered by dividing each belief mass by a normalization coefficient (Josang and Pope 2012). This rule is associative, and the normalisation in D-S redistributes conflicting belief masses to non-conflicting ones, and tends to eliminate any conflicting characteristics in the resulting belief mass distribution. This rule of combination can be applied to avoid this particular problem by allowing all conflicting belief masses to be allocated to the empty set.

When performing the belief calculations by two values the returned result is quite surprising. When comparing two high belief generations the assumption is that the combined belief value would also be a high number, however it is a lower value. The correlation between high belief values and low fused outputs suggests that the lower the fused output the higher the risk. The same is understood for two fused low belief values generate a high fused output, which would be a low risk. It is not clear if this is due to our calculations but these metrics have been compared on numerous belief fusions and this is a similar occurrence. The mass value must be between 0 and 1 but not inclusive as this seems to skew the calculations, e.g. using a value of 0 would render the combination calculation $(m_{1,2}(A))$ uncalculatable as you cannot divide by 0 which would be a pertinent value. A coefficient value of 1 would leave the combination calculation having to divide by 0 $(1/1-1)$ which is an impossible calculation. Also, having a coefficient of 0 would give a negative risk output, which is also an unusable value.

Implementing our Security as a Service solution, the below limitations of D-S can be evident:

- Associative – for rule combination, the order of the information in the aggregated evidences does not impact the result. A non-associative combination is necessary for many cases.
- Non-weighted – rule combination implies we trust all evidences equally. However, in reality, our trust on different evidences may differ, which means we should consider various factors for each evidence.

We have demonstrated how D-S can provide collaborative intrusion detection, however there may be cases where the decision may be inaccurate, and if a domain under attack generation the Belief of origin then it would still need to take action against the condition. D-S when applied in an autonomous collaborative environment should apply a weight of confidence when the belief generation occurs. If CSPs collaboratively vote no attack, but one CSP is adamant it is being attacked, there should be a way to overrule the decision based on the strength of the associated trust or confidence value. The algorithm should be extended further to take this into consideration, and we propose a two stage collaborative detection process for conflicting decisions.

Two stage D-S fusion for conflicting decisions is an option for solving this issue. Post Belief generation processing is needed for application to this area to facilitate information exchange for defence. Via the inclusion of confidence values the accuracy of decisions can be improved. Protecting the local services of the CSP but proactively warning others of the potential threat. If the fused decision is "No Attack" but the Belief of origin has a high confidence value, then the domain of origin would take action against the suspect observation, but send the Belief value ($B_A(IP$, *timestamp*, *ConfidenceValue*) to the Broker to store in its local Grey list. Should an adjacent CSP query the Broker regarding the suspect IP in the future, it has the information from the origin CSP.

5 Conclusion

We have presented our Security as a Service solution, a novel platform for the protection of infrastructure services in a federated Cloud environment. D-S theory is extended to meet the domain management needs and to facilitate autonomous sharing of information. The novel contributions of this project are that it provides the means by which DDoS attacks are detected within a cloud federation, so as to enable an early propagated response to block the attack, particularly by the interdependent CSPs within the federation. This is effectively inter-domain cooperation as these CSPs will cooperate with each other to offer holistic security, and add to the defence in depth. D-S is used to facilitate this autonomous sharing of information, and to fuse the generated beliefs to form a system wide decision. This cooperation between CSPs ensures the scalable defence required against DDoS attacks is in an efficient and cost effective way. Protecting the federated cloud against cyber-attacks is a vital concern, due to the potential for significant economic consequences. The effects of attacks can span from the loss of data, to the potential isolation of parts of the federation. Our simulations offer proof of concept, and deem the applicability of D-S to this area promising but still with evident limitations.

Acknowledgements. The work reported in this paper is partly supported under the Newton Research Collaboration Programme by the Royal Academy of Engineering.

References

Josang, A., Pope, S.: Dempster's rule as seen by little coloured balls. Comput. Intell. **28**(4), 453–474 (2012)

MacDermott, Á., Shi, Q., Kifayat, K.: Collaborative intrusion detection in a federated Cloud environment using the Dempster-Shafer theory of evidence. In: European Conference on Information Warfare and Security, ECCWS (2015)

MacDermott, Á., Shi, Q., Kifayat, K.: Collaborative intrusion detection in federated cloud environments. J. Comput. Sci. Appl. Big Data Anal. Intell. Syst. **3**(3A), 10–20 (2015)

Rak, M., Ficco, M., Luna, J., Ghani, H., Suri, N., Panica, S., Petcu, D.: Security issues in cloud federations. In: Achieving Federated and Self-Manageable Cloud Infrastructures: Theory and Practice, pp. 176–194 (2012). http://doi.org/10.4018/978-1-4666-1631-8.ch010

Ten, C.W., Manimaran, G., Liu, C.C.: Cybersecurity for critical infrastructures: attack and defense modeling. IEEE Trans. Syst. Man Cybern. Part A: Syst. Hum. **40**(4), 853–865 (2010). http://ieeexplore.ieee.org/xpls/abs_all.jsp?arnumber=5477189

Thomas, C., Narayanaswamy, B.: Sensor fusion for enhancement in intrusion detection. In: Sensor Fusion - Foundation and Applications, pp. 61–76 (2011)

Trend Micro Incorporated.: Report on Cybersecurity and Critical Infrastructure in the Americas (2015)

Xiao, Z., Xiao, Y.: Security and privacy in cloud computing. IEEE Commun. Surv. Tutor. **15**(2), 843–859 (2013)

Intelligent Agent and Web Applications

A Code Quality Metrics Model for React-Based Web Applications

Yiwei Lin[1], Min Li[2], Chen Yang[1], and Changqing Yin[1(✉)]

[1] College of Software Engineering, Tongji University, Shanghai 201804, China
yin_cq@qq.com
[2] Fudan University, Shanghai 200433, China

Abstract. With the increasing of website user experiences, the interactive UI effect adds the complexity of web applications, which results in the development of Web front-end frameworks. Nevertheless, the existing code quality metrics models are not able to measure the code quality of Web front-end frameworks effectively. On the other hand, it is also difficult to present a quality metrics model that can evaluate the software frameworks comprehensively, so we should focus on a specific measurement object in order to improve accuracy of metrics. On this groundwork, we present a code quality metrics model for Web front-end frameworks based on React. In this model, we put forward 16 metric units, according to the characteristics of JavaScript and React through experiments. Then we determine the quantitative rules for each metric unit. At long last, we generate a set of scores to measure Web applications. In the part of experiment, firstly we select two projects as our benchmarks, then we choose a real project to evaluate the code qualify and compare the results with people's evaluation of this project. The experiment results imply that this model is reliable and has some directing significance. The primary contributions of this paper are proposing a new code quality metrics model for JavaScript and React, which is a new research direction in the area of code quality metrics. By this model, we can monitor code quality and enhance development efficiency.

Keywords: React · Web front-end language · Code quality metrics model · Quantitative rules

1 Introduction

Nowadays the demand for user experiences of website is growing progressively and the scope of Web technology has not been restricted to a website and it can be applied to all platforms, which is called "Develop Once, Deploy Everywhere". Due to the plethora of applications that JavaScript serves and the variety of programming needs, JavaScript Frameworks (JFs) have been developed in order to facilitate the work of Web programmers [1].

Nevertheless, the wide use of frameworks may also bring many troubles, such as the complexity of software structure, the question of low-quality code and so on. In society to solve these problems, it is important to hold an effective quality assessment for the front-end framework, which can help developers to review code and improve

© Springer International Publishing AG 2017
D.-S. Huang et al. (Eds.): ICIC 2017, Part III, LNAI 10363, pp. 215–226, 2017.
DOI: 10.1007/978-3-319-63315-2_19

software quality. Unfortunately, there are few studies for Web front-end technology in the area of software quality metrics.

In the area of front-end development, JavaScript is extremely popular in the area of Web applications. Meanwhile, as a popular framework, React [2] has attracted increasing attention by its lightweight, innovativeness and excellent maintainability. Before our study, we have reviewed many papers relevant to Web front-end frameworks, and find that only a few researches have been managed to develop new measurement metrics for JavaScript or React. So we focus on the code quality metrics model for web applications. In this paper, we choose React and JavaScript as the measurement objects, and present a software quality metrics model that is targeted at Web front-end framework. Using this model, we choose a real project to measure the code qualify for it. In conclusion, this experiment result verifies the reliability of this model.

In summary, this work makes the following main contributions:

- We firstly propose a domain-specific code quality metrics model for web application based on React, which is not offered in the area of software quality measurement yet.
- We narrow the scope of quality metrics and present a specific model because it is overwhelmingly difficult to present a quality metrics model that can evaluate all the software frameworks using in Web front-end.
- We propose a list of metric units to measure code quality of Web applications, collected from various characteristics of JavaScript and React.
- We present customized quantization rules for each metric unit and apply the result to grade Web applications based on React.
- We evaluate the effectiveness and reliability of our model in measuring code quality of Web applications.

The principal challenge in our paper is the selection of metric units and quantization rules. Because there are only a few researches that propose metrics for JavaScript frameworks, we should list all the possible metric units at first, and then select the representational metric units in them. After the selection of metric units, another challenge is to determine the quantization rule for each metric unit. In this paper, we analyze and determine different quantization rules for each metric unit.

This paper proceeds as follows: After introducing the related works in front-end software quality metrics model, we present a new quality metrics model for Web front-end software in Sect. 3. In Sect. 3.1, we present the 16 metric units, according to the characteristics of JavaScript and React. Then in Sect. 3.2 we present the quantitative rules for these metric units and show the quantitative evaluation using by these quantitative rules. Section 4 is devoted to introduce the experiment for this model and show the experiment results in Sect. 5.

2   Related Works

As this study concentrates on the front-end software metrics for Web front-end technology, this section presents studies on this topic. The main foundation of this work is static analysis technology for Web front-end framework.

Many researches [3–7] in this field are about the basic quality metrics model of traditional software. Huang and Yang [3] present the specific research process of code quality metric using static analysis. They integrate Ant with other open source static testing tools and develop a comprehensive evaluation of Java code quality that supports to evaluate in six different aspects. In Sun et al.'s research [4], they improve and merge the Logiscope and McCabe software metric system, and selected 16 metric units to establish a scoring function model. Our research is based on this process of developing the quality metrics model.

In terms of Web software quality metric, Olsina and Rossi [8] presents the method of applying the quantitative assessment strategies to evaluate the quality of Web applications earlier, which is called WebQEM. This tool is used to refine the usability, practicability, reliability and efficiency of software and generate metric units for each aspect. Then it defines the standard preference function for each metric unit and gets the final assessment results. The model in this paper is widely used to measure Web Applications quality in 2002, but it is not suitable for the highly framed Web Applications nowadays.

JavaScript is increasingly employed to develop Web applications, notwithstanding, merely a few researches have been managed to develop new measurement techniques and metrics for JavaScript. In the paper of Sanjay [9], they propose a new complexity quality metrics model that can evaluate the code quality of JavaScript. In Andreas's paper [1], they use the common software metrics, including scale metrics, complexity metrics and maintainability metrics, to measure six common JavaScript frameworks (JSF). Lastly, they compare the measurement results and generate the best JavaScript framework.

In the end, we will shortly introduce React, which is our measurement object in this paper. React [2] originated from the internal project of Facebook primarily, and ·it is mainly applied for the development of big data applications using a predictable and declarative way to build the Web user interface. In React, all the elements that can be seen in the interface can be packaged as a Component (C), and each component contains its own state and the state (props) passed by the parent component.

3   Code Quality Metrics Model

3.1   Metric Units

Metrics are a mapping of things in the real world to standard mathematical indicators, which is often presented when we have a deep understanding of one thing [3]. Like other things in the real world, software is a kind of production, which can be delivered and applied. So it should have many certain quality characteristics and we can present metric units by measuring the quality characteristics of software [4]. Metric units can reflect the common characteristics of codes with higher quality, and these common features can be extracted as metric units to measure other codes. Based on the existing research on the quantitative measurement method for traditional software application code quality, this paper summarizes the metric units by extracting the characteristics of JavaScript and React framework, which are presented in Table 1.

Table 1. Overview of metrics units

Category	Metric unit	Abbreviation	Quantization rule
JavaScript	Code indentation ratio	JSCIR	Logarithmic quantization
	Code annotation ratio	JSCAR	Normal quantization
	Variables and functions correct naming ratio	JSVFCNR	Logarithmic quantization
	Internal code replication ratio	JSICRR	Linear quantization
React components	Component line of code	RCLOC	Normal quantization
	Component dependent graph leaf node number	RCDGLNN	Normal quantization
	Component dependent graph depth	RCDGD	Normal quantization
	Component Functions average cyclomatic complexity	RCFACC	Normal quantization
	Component coupling factors	RCCF	Normal quantization
React states	Component self-state ratio	RCSSR	Normal quantization
	Parent component transition state ratio	RPCTSR	Normal quantization
	State utilization ratio	RSUR	Linear quantization
	State transmissibility ratio	RSTR	Normal quantization
	State transition weight	RSTW	Normal quantization
	State call basic function ratio	RSCBFR	Normal quantization
	State required ratio	RSRR	Normal quantization

As shown in Table 1, we divide the metric units into three categories: JavaScript, React Component and React States (including the variables and functions).

In the category of JavaScript, we pour attention to the code standards such as the code indention and the complexity of comments. These metric units are common metrics indicators in most of the metrics model and they can represent the readability of code. Readability refers to the ease with which a human reader can comprehend the purpose, control flow, and operation of source code. It involves the aspects of quality above, including portability, usability and most importantly maintainability [9], which is an indispensable indicator in the metrics model.

In the category of React Component, we focus on the component and the relationship between components in React. React uses components to encapsulate the interface module and the entire interface is regarded as a combination of components. The development process of React is to split, optimize and establish the component tree constantly. The relationships between components in React can reflect the degree of coupling and cohesion in an application, which are the common metrics indicators in the metrics model.

In the category of React State, we pay attention to State, another basic conception in React. In React, there are two types of data that control a component: props and state. Props are set by the parent and they are fixed throughout the lifecycle of a component. For data that is going to change, we have to use state.

State can reflect the degree of coupling in an application, state places extra emphasis on the communication data flow between parallel components, which don't have the relationship of composability and dependency. According to this situation, we measure the degree of coupling and cohesion in two aspects.

In order to understand the specific meaning of each metric unit more clearly, we list the calculation of each metric unit in Table 2.

3.2   Quantitative Rules of Metric Units

In Table 1, we list the metric units and quantitative rules of them based on the characteristics of JavaScript and React, which could be summarized as the form of $y = f(x)$. In this formula, x is the value of each metric unit obtained by the calculation method in Table 2, and y is the score of metric unit calculated by $f(x)$. In order to facilitate the subsequence calculation, we select the appropriate quantity rule for each metric unit according to its characteristics. The process of selection is based on the previous research and our experiments. In this section, we will introduce the calculation method for each quantitative rule individually.

Linear Quantization. Linear quantization represents a proportional linear relationship between the quantized object and the quantization result, which is simpler and more intuitional. We can calculate it using Eq. (1).

$$y = a * x + b \tag{1}$$

Note that the values of a and b should ignore the discreteness of data in order to facilitate the subsequent calculation of scores. The value range of y is (0, 1), and when the value of x is larger, the ratio of y and x is closer to 1. In the above-mentioned metric units, JSICRR and RSUR should be quantized using linear quantization. Research

Table 2. Calculation method of metric units

Metric unit	Calculation method
JSCIR	$JSCIR = line_{cir} / line_{total}$ In a system, $line_{cir}$ are the lines of code with the right indentation format and $line_{total}$ are the total lines of code
JSCAR	$JSCAR = line_{note} / line_{total}$ In a system, $line_{note}$ are the lines of code with annotation and $line_{total}$ are the total lines of code
JSVFCNR	$JSVFCNR = VF_{cnr} / VF_{total}$ In a system, VF_{pnr} is the amount of variables and functions with the right naming conventions (including the right meaning of variables and functions should express) and VF_{total} is the total amount of variables and functions
JSICRR	$JSICRR = 1 - line_{copy} / line_{total}$ In a system, $line_{copy}$ are lines of code which are duplicated and $line_{total}$ are the total lines of code
RCLOC	$RCLOC = \left(\sum_{i=0}^{n} Z_i \right) / n \quad Z_i = \begin{cases} 1 & if \ line_{C_i} \geq 300 \\ 0 & if \ line_{C_i} < 300 \end{cases}$ In a system, n is the amount of components and $line_{C_i}$ are the lines of code in component i
RCDGLNN	$RCDGLNN = \left(\sum_{j=0}^{m} leaf_{CDG_j} \right) / m$ In a system, m is the amount of complete component dependent graphs and $leaf_{CDG_j}$ is the amount of leaf nodes in graph j
RCDGD	$RCDGD = \left(\sum_{j=0}^{m} deap_{CDG_j} \right) / m$ In a system, m is the amount of complete component dependent graphs and $deap_{CDG_j}$ is the depth of graph j
RCFACC	$RCFACC = \left(\sum_{i=0}^{n} \sum_{k=0}^{l} CC_k \right) / n$ In a system, n is the amount of components and l is the amount of functions in component i and CC_k is the cyclomatic complexity of function k
RCCF	$RCCF = \left(\sum_{i=1}^{n} C_i \right) / (n * (n - 1))$ In a system, n is the amount of components and C_i is the amount of dependencies between component i and other components
RCSSR	$RCSSR = \sum_{j=0}^{m} \max \left(CS_j / total \right)$ In a system, m is the amount of complete component dependent graphs and $\max \left(CS_j / total \right)$ is the maximum self-state ratio of components in graph j
RPCTSR	$RPCTSR = \sum_{j=0}^{m} \max \left(CP_j / total \right)$ In a system, m is the amount of complete component dependent graphs and $\max \left(CP_j \right)$ is the maximum parent component transition state ratio of components in graph j

(continued)

Table 2. (*continued*)

Metric unit	Calculation method
RSUR	$$\text{RSUR} = \left(\sum_{i=0}^{n} US_i \, / \, (TS_i + PS_i) \right) / n$$ In a system, n is the amount of components and US_i is the used state number of components i and TS_i is the self-state number of component i and PS_i is the number of states which are delivered from parent components to component i
RSTR	$$\text{RSTR} = \left(\sum_{i=0}^{n} PS_i \, / \, SS_i \right) / n$$ In a system, n is the amount of components and PS_i is the number of states which are delivered from parent component to component i and SS_i is the number of states which are delivered from component i to child components
RSTW	$$\text{RSTW} = \left(\sum_{i=0}^{n} (BPS_i * \alpha + CPS_i * \beta + APS_i * \gamma) \right) / n$$ In a system, n is the amount of components and BPS_i is the number of basic states which are delivered from parent components to component i and CPS_i is the number of complicated states which are delivered from parent components to component i and APS_i is the number of any type of states which are delivered from parent components to component i. In this formula, α, β, γ are weights which can indicate the complexity of state transition and in this paper we divide them into 1, 2 and 3
RSCBFR	$$\text{RSCBFR} = \left(\sum_{i=0}^{n} FB_i \, / \, FA_i \right) / n$$ In a system, n is the amount of components and FB_i is the ratio of calling public basic functions in component i and FA_i is the amount of member functions in component i
RSRR	$$\text{RSRR} = \left(\sum_{i=0}^{n} RS_i \, / \, AS_i \right) / n$$ In a system, n is the amount of components and RS_i is the number of states which must be delivered from parent components to component i and AS_i is the number of states which are delivered from parent components to component i

shows that with the increase of code replication rate, the effect is linearly increasing [10]. It implies that when the code replication ratio is lower, the score of the code indentation ratio is nearer to 1.

Logarithmic Quantization. Logarithmic quantization represents the logarithmic relationship between the quantized object and the quantization result. When the value of metric unit is larger, the growth rate of quantization result is slowed. We can calculate it using Eq. (2).

$$y = \log_a(x) + b \tag{2}$$

To be specific, JSCIR should be quantized using logarithmic quantization. When the value of x, which is the ratio of codes with the right indentation format, is small, the code will be hard to understand. If x increases constantly, most codes will have the right indentation and the readability of code will increase rapidly, which approaches the form of logarithmic functions.

Normal Quantization. Normal quantization is one of the most widely used quantization rules in our model, and it utilizes a normal distribution to score the value of metric unit. The calculation is shown in Eq. (3).

$$y = \frac{1}{e^{(x-\mu)^2}} \tag{3}$$

In formula (3), the value range of y is (0, 1) and μ represents the best value of the metric unit. That is to say, when x equals to μ, the value of y is 1, which is the optimal score of metric unit. The difference between x and μ represents the difference between the actual score and the optimal score.

For example, JSCAR should be quantized using normal quantization. JSCAR means the ratio of code with annotation. Research indicates that a project or framework should have a proper code annotation ratio. With plethoric code annotations, the code structure is more complicated and is hard to maintain. With less code annotations, it is difficult for developers to understand the code structure. The characteristics of code annotation are similar with normal quantization, which delivers a peak value at a proper key. So we should use normal quantization to measure this unit.

3.3   Metrics Result

Using different quantization rules, we can score each metric unit and obtain three basic score metrics indicators for React-based Web application code quality model. They are RWCQ_JS (the score of JavaScript code), RWCQ_C (the score of Component in React framework) and RWCQ_S (the score of State in React framework). And the calculation methods for them are shown in Eqs. (4), (5) and (6).

$$RWCQ_{JS} = \frac{f(JSCIR) + f(JSCAR) + f(JSVFCNR) + f(JSICRR)}{4} \tag{4}$$

$$RWCQ_C = \frac{f(RCLOC) + f(RCDGLNN) + f(RCDGD) + f(RCFACC) + f(RCCF)}{5} \tag{5}$$

$$RWCQ\_S = \frac{f(RCSSR) + f(RPCTSR) + f(RSUR) + f(RSTR) + f(RSTW) + f(RSCBFR) + f(RSRR)}{7} \tag{6}$$

At last, we can obtain the final code quality metrics model using Eq. (7).

$$RWCQ = RWCQ\_JS^{w_{JS}} * RWCQ\_C^{W_C} * RWCQ\_S^{w_S} \tag{7}$$

4   Experiments

4.1   Metric Unit

In this metrics model, it is indispensable to determine the parameters of quantization rules before evaluating a real project. For this reason, the solution proposed in this paper is to select some React-based projects that are widely used and popular on GitHub, which are called standard projects in this paper. Then we can evaluate this project and set the quantization parameters. The mean value of the result is taken as the final value of parameter, which provides the basis for the measurement of other projects. After determining the parameters, we should choose another React-based project as our experiment object to measure the code quality and verify the reliability of the quality metrics model.

After the process of research and comparison, we choose two projects with high attention on GitHub as our standard projects and one project as our experiment project. The standard and experimental projects are shown in Table 3.

Table 3. Overview of standard projects and experiment project

Project name	Scale	Description	Project type
Ant-design	1251	An enterprise-level UI design language	Standard project
Material_UI	35246	React components that implement Google's material design	Standard project
fofEasy	26807	An analysis and display platform for foundations	Experimental project

4.2   Quantization Tools

In this paper, we choose SonarQube [11] and MeasureReact to gather the data in metric units and calculate the score of metric units. SonarQube is an open source platform for code quality management. It can support the quality metrics of more than 20 kinds of programming languages. With SonarQube, we can detect the quality of code from different dimensions comprehensively.

MeasureReact is a React metric acquisition tool we implemented in this paper. It can use the package of Babel [12] to convert the project code from ES6 to ES5 (Most of the projects based on React are developed using ES6) and generate the abstract syntax tree of ES5 code. We use this tool to extract experimental parameters in this paper.

4.3   Experiment Process

In this paper, the experiment process is divided into two parts. In part one, we measure the standard projects and determine the coefficient in formulas (1)–(3). After that, we measure the experimental project using the determined coefficient and obtain the score of each metric unit. One thing we should note is that we assign the score y of each metric unit in the standard project to 1 in order to determine the coefficient. Next we calculate the RWCQ_JS, RWCQ_C and RWCQ_S using formulas (4)–(6). Finally, we get the final score of the project using formula (7). The process of calculating is shown in Fig. 1.

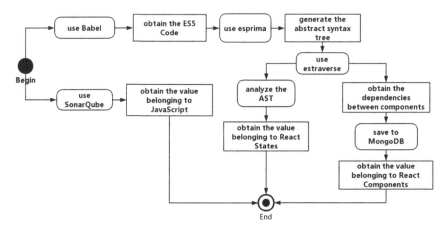

Fig. 1. The process of calculating the value of metric units.

In the process of calculating the value of metric units, firstly we use SonarQube to obtain the value of metric units belonging to the category of JavaScript. Secondly, we use Babel to covert project code from ES6 to ES5. Thirdly, we use esprima [13] to generate the abstract syntax tree using ES5 code. The abstract syntax tree can show the states and relationships between states of the framework and we can obtain the value of metric units belonging to the category of React States. Fourthly, we use estraverse [14] to extract the dependencies between components of the abstract syntax tree. Finally, we use MongoDB, which is a NoSQL document-oriented database program, to save the dependencies and calculate the value of metric units belonging to the category of React Components.

4.4   Experiment Results

We apply the experimental tools to determine the parameters of metric units of standard projects. Then we use the experimental tools to obtain the parameters of metric units of experiment project. Finally, we obtain the score of each metric unit using the quantization rules shown in Sect. 3.2. The experiment results are shown in Table 4.

According to Table 4, we put the scores into formulas (4), (5) and (6), and then we can obtain the results: RWCQ_JS $\approx$ 0.697, RWCQ_C $\approx$ 0.751, RWCQ_S $\approx$ 0.827.

According to the existing research in this field, the weight coefficients of RWCQ_JS, RWCQ_C and RWCQ_S are equal to 0.33. Using formula (7), the final score RWCQ of fofEasy is 0.776. Because the value range of RWCQ is $(0, 1)$, we can

Table 4. Parameters and scores of metric units of fofEasy

Metric unit	Ant-design	Material_UI	Parameter	Value of metric unit	Score
JSCIR	a = 2, b = 1	a = 2, b = 1	a = 2, b = 1	1	0.98
JSCAR	$\mu = 0.0187$	$\mu = 0.0194$	$\mu = 0.019$	0.053	0.998
JSVFCNR	a = 2, b = 1	a = 2, b = 1	a = 2, b = 1	0.64	0.549
JSICRR	a = 1, b = 0	a = 1, b = 0	a = 1, b = 0	0.24	0.241
RCLOC	$\mu = 0.7167$	$\mu = 0.533$	$\mu = 0.62$	0.7	0.979
RCDGLNN	$\mu = 4.1$	$\mu = 3.2$	$\mu = 3.65$	4.22	0.723
RCDGD	$\mu = 2.26$	$\mu = 1.41$	$\mu = 1.84$	1.35	0.787
RCFACC	$\mu = 1.0$	$\mu = 1.4$	$\mu = 1.2$	1.8	0.698
RCCF	$\mu = 2.72$	$\mu = 4.38$	$\mu = 3.55$	4.37	0.51
RCSSR	$\mu = 1$	$\mu = 1$	$\mu = 1$	0.89	0.988
RPCTSR	$\mu = 0.857$	$\mu = 1$	$\mu = 0.93$	1	0.995
RSUR	a = 1, b = 0	a = 1, b = 0	a = 1, b = 0	0.634	0.648
RSTR	$\mu = 0.61$	$\mu = 0.74$	$\mu = 0.68$	0.426	0.938
RSTW	$\mu = 5.4$	$\mu = 6.28$	$\mu = 8.68$	7.8	0.461
RSCBFR	$\mu = 0.21$	$\mu = 0.16$	$\mu = 0.19$	0.05	0.98
RSRR	$\mu = 0.82$	$\mu = 0.87$	$\mu = 0.84$	0.33	0.771

divide the code quality into five levels: A is $[0.8, 1)$, B is $[0.6, 0.8)$, C is $[0.4, 0.6)$, D is $[0.2, 0.4)$ and E is $(0, 0.2)$. So the level of fofEasy is B, and we think it is in accordance with the actual evaluation.

5   Conclusions

This paper summarizes and proposes a new code quality metrics model for React-based Web application according to the understanding of JavaScript and React. In this paper, we put forward 16 metric units based on the characteristics of JavaScript and React. Then we use different quantization rule to score each metric unit and obtain three basic score metrics indicators. In the end, we set out the final score of metrics model and verify the accuracy and reliability of the experimental results. From this model, readers can benefit from the method of narrowing quality metrics scope and selecting metric units. Second, we present customized quantization rules for each metric unit based on the characteristics and evaluate the effectiveness and reliability of this model. At last, we expect our model to encourage further research on fast-developing front-end frameworks in the area of code quality measurement.

References

1. Gizas, A., Christodoulou, S.P., Papatheodorou, T.S.: Comparative evaluation of JavaScript frameworks. In: WWW 2012 Companion, Proceedings of 21st International Conference on World Wide Web, pp. 513–514 (2012)

2. React Homepage. https://hulufei.gitbooks.io/react-tutorial/content/introduction.html
3. Huang, P., Yang, M.: Research and application of code quality static measurement. Comput. Eng. Appl. **47**(23), 61–64 (2011)
4. Sun, M., Song, X., Chao, Y.: Research on software program code quality measurement technology. Comput. Eng. Des. **27**(2), 325–327 (2006)
5. McIntosh, S., Kamei, Y., Adams, B., Hassan, A.E.: An empirical study of the impact of modern code review practices on software quality. Empir. Softw. Eng. **21**(5), 2146–2189 (2016)
6. Chaparro, O., Bavota, G., Marcus, A., Di Penta, M.: On the impact of refactoring operations on code quality metrics. In: 2014 IEEE International Conference on Software Maintenance and Evolution (ICSME), pp. 456–460 (2014)
7. Seffah, A., Donyaee, M., Kline, R.B., Padda, H.K.: Usability measurement and metrics: a consolidated model. Softw. Qual. J. **14**(2), 159–178 (2006)
8. Olsina, L., Rossi, G.: Measuring web application quality with WebQEM. IEEE Multimedia **9**(4), 20–29 (2002)
9. Misra, S., Cafer, F.: Estimating quality of JavaScript. Int. Arab J. Inf. Technol. **9**(6), 535–543 (2012)
10. Hotta, K., Sasaki, Y., Sano, Y.: An empirical study on the impact of duplicate code. Adv. Softw. Eng. **2012**, 5 (2012)
11. SonarQube Homepage. https://www.sonarqube.org/
12. Babel Homepage. http://babeljs.io/
13. Esprima Homepage. http://esprima.org/
14. Estravese Homepage. https://github.com/estools/estraverse
15. Yue, C., Wang, H.: Characterizing insecure JavaScript practices on the web. In: WWW 2009, Proceedings of 18th International Conference on World Wide Web, pp. 961–970 (2009)
16. Ant-Design Homepage. https://github.com/ant-design/ant-design
17. Material_UI Hompage. http://www.material-ui.com
18. Hristoski, I.S., Mitrevski, P.J.: Evaluation of business-oriented performance metrics in e-Commerce using web-based simulation. Emerg. Res. Solut. ICT (ERSICT) **1**(1), 1–16 (2016)
19. Rodriguez, J.M., Mateos, C., Zunino, A.: Assisting developers to build high-quality code-first web service APIs. Web Eng. **14**(3&4), 251–285 (2015)

An Interaction-Centric Approach to Support Coordination in IoT-Based Enterprise Systems

Djamel Benmerzoug[1(✉)], Fouzi Lezzar[1], and Ilham Kitouni[2]

[1] LIRE Laboratory, Faculty of NTIC, University Constantine 2,
Constantine, Algeria
{djamel.benmerzoug, fouzi.lezzar}@univ-constantine2.dz
[2] MISC Laboratory, Faculty of NTIC, University Constantine 2,
Constantine, Algeria
ilham.kitouni@univ-constantine2.dz

Abstract. Internet of Things (IoT) becomes a reality for many reasons: low powers processors, improvements in wireless communication technologies and electronic devices. This will initiate new business opportunities in providing these novel applications and services, which integrate efficiently IoT services into enterprise applications. The work presented here proposes an Agent-based approach for an effective integration of the IoT in enterprise systems. The proposed approach defines a new meta-model to describe this integration. Also, the paper presents an agent-based system architecture whose main goal is to address and tackle interoperability challenges in the context of IoT environments. Also, it solves the interoperability issues between heterogeneous Cloud services environments by offering a harmonized API.

Keywords: Business process modeling · Enterprise systems · Internet of Things (IoT) · Multi-agent system

1 Introduction

The Internet of Things (IoT) is the most promising area which penetrates the advantages of Wireless Sensor and Actuator Networks and Pervasive Computing domains. Different applications of IoT have been developed and researchers of IoT well identified the opportunities, problems, challenges and the technology standards used in IoT such as Radio-Frequency IDentification (RFID) tags, sensors, actuators, mobile phones, etc. [1]. Modern enterprises systems will lead to a further integration of real-world entities into Internet applications. That means that smart devices will be the fundamental building blocks for the future enterprises systems.

In an IoT-based enterprise system, business processes will depend heavily on the coordination of heterogeneous smart devices, both among themselves and with business systems. This coordination requires a uniform description format that facilitates the design, customization, and composition. In this context, Agent Interaction Protocols (AiP) are a useful way for structuring communicative interaction among business partners, by organizing messages into relevant contexts and providing a common guide to the all parts.

© Springer International Publishing AG 2017
D.-S. Huang et al. (Eds.): ICIC 2017, Part III, LNAI 10363, pp. 227–238, 2017.
DOI: 10.1007/978-3-319-63315-2_20

It was noted in [8] that AiP are appropriate approaches to define and manage collaborative processes in B2B relationships where the autonomy of participants is preserved. The idea of adopting AiP for managing collaborative processes was first introduced and proposed in our previous work [9]. Whereas in [10], we demonstrated the practicability of our approach by embedding it in a Web services language for specifying business protocols, which conducive to reuse, refinement and aggregation of our business protocols.

To address the collaboration and interaction issues in modern enterprises, it is normal to turn our attention to IoT, which gives an immediate access to information about physical objects and leads to innovative services with high efficiency and productivity. This paper is of two fold, the first fold covers the integration of the IoT and AiP paradigm with business processes. The proposed approach defines a new meta-model to describe this integration. The second fold of this paper presents an agent-based architecture for the development of IoT-based enterprise system. Specifically, a Broker-based architecture whose main goal is to address and tackle interoperability challenges in the context of IoT environments.

The remainder of the paper is organized as follows: Sect. 2 overviews some related work. In Sect. 3, we propose a domain model for IoT-based business application. Section 4 presents an agent-based architecture in a nutshell introducing the main modules and the core functionality. In Sect. 5, we describe the AiP as a high abstraction level in the modelling of enterprise application. Finally, Sect. 6 concludes this paper and presents future directions.

2   Related Work

Multiagent systems are a very active area of research and development. In fact, several researchers are working at the intersection of agents and modern enterprise systems.

For example, [3] proposes an agent-based middleware which permits to create, deploy, execute, orchestrate, and manage services over highly dynamic and radically distributed IoT environments. The platform provides some intelligent workflow tools for service providers, by autonomously creating, compositing and deploying IoT services at run-time in order to meet the demands of user-personalized services, and seamlessly integrates the service enablers container service engine into the hybrid service exposure.

In [6], an agent-oriented approach to IoT networks modeling is proposed by exploiting the ACOSO model. Then, agent-modelled IoT networks of different scales are simulated through the Omnet++ simulation platform, with the goal of analyzing issues and bottlenecks at communication level.

In [5, 7], authors have presented the design of an architecture that can strongly couple the envisioned IoT infrastructure with enterprise services. The proposed approach is event-based and its interaction with external entities as well as internal communication is based on web services. Theoretically each component could be hosted in different locations. As an example, parts of the architecture could be running in the cloud offering high-performance services, while parts interacting locally with the devices could run even on embedded systems located at the device layer.

When examining the above cited approaches we observe a recurrent theme. They do not allow for easy extensibility or customization options. Better ways are necessary for IoT service consumers to orchestrate a cohesive service and provision Cloud stack services that range across networking, computing, storage resources, and applications from diverse service providers.

Our approach is based on the idea of reducing the complexity involved when developing IoT-based application. In our work, the knowledge obtained during services composition is stored, shared, and reused. In fact, we proposed a basis for a theoretical approach for reusing and aggregating existing protocols to create a new desired business application.

Agent-oriented software methodologies aim to apply software engineering principles in the agent context e.g. Tropos, AMCIS, Amoeba, and Gaia. Tropos [14] and AMCIS [15] differ from these in that they include an early requirements stage in the process. Amoeba [16] is a methodology for business processes that is based on business protocols. Protocols capture the business meaning of interactions among autonomous parties via commitments. Gaia [17] differs from others in that it describes roles in the software system being developed and identifies processes in which they are involved as well as safety and liveness conditions for the processes. It incorporates protocols under the interaction model and can be used with commitment protocols.

Our approach differs from these in that it is aimed at achieving protocol re-usability by separation of protocols and business rules. It advocates and enables reuse of protocols as building blocks of business processes. Protocols can be composed and refined to yield more robust protocols.

3 Domain Model for Enterprise Application Based on Agent and IoT

Figure 1 shows our conceptual model for the treatment of enterprises applications collaboration based on IoT. The proposed model is based on our previous work on advanced enterprise systems [9, 10] and those of the IoT domain [20, 22].

A public business process is the aggregation of the private processes and/or Web services participating in it. Let us notice that private processes are considered as the set of processes of the enterprise itself and they are managed in an autonomous way to serve local needs. The public processes span organizational boundaries. They belong to the enterprises involved in a B2B relationship and have to be agreed and jointly managed by the partners.

In the context of B2B relationships, an interaction protocol is a specification of the allowed interactions between two or more participant business partners. These interactions represent public business processes that enterprises agreed on the collaboration. In our case the public process is specified by interaction protocols, which are used by the agents to enact the public process at run time.

The most generic IoT scenario can be identified as that of a generic Agent needing to interact with a Physical Entity in the physical world. Physical Entities are represented in the digital world by a Virtual Entity. An IoT Object (or a thing) is the

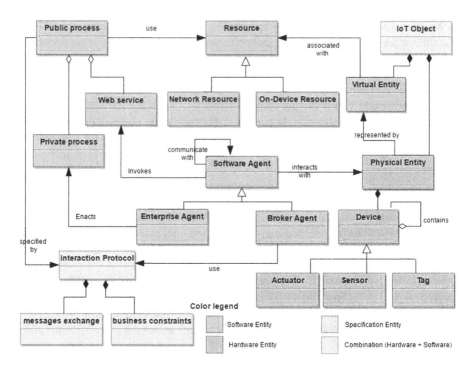

Fig. 1. Conceptual model: enterprise application based on AiP and IoT

composition of a Virtual Entity and the Physical Entity it is associated to, in order to highlight the fact that these two concepts belong together.

In the physical environment, Physical Entity is the composition of one or more Devices. This can be achieved by using Devices of the same class, as in the case of certain similar kinds of body-area network nodes, or by using Devices of different classes, as in the case of an RFID tag and reader.

In addition, it's important to realize that the Device entity will often perform in a bi-directional manner with the "IoT Object" at the edge of the network either acting as input devices (Sensors) or output devices (actuators). Besides input and output devices, Tags are used to identify Physical Entities, to which the Tags are usually physically attached. The identification process is called "reading", and it is carried out by specific Sensor devices, which are usually called readers. The primary purpose of Tags is to facilitate and increase the accuracy of the identification process. This process can be optical, as in the case of barcodes and QR codes, or it can be RF-based, as in the case of RFID.

Public business process (which is the aggregation of the private processes and/or Web services participating in it) consumes resources. Resources are software components which used in the actuation on Physical Entities. There is a distinction between On-Device Resources and Network Resources. As the name indicates, On-Device Resources are hosted on Devices. They include executable code for accessing, processing, and storing Sensor information, as well as code for controlling Actuators. On the other hand, Network Resources are Resources available somewhere in the network, e.g., back-end or cloud-based databases. A Virtual Entity can also be associated with Resources that enable interaction with the Physical Entity that the Virtual Entity represents.

4   An Agent Based Architecture for IoT-Based Business Applications

As shown in Fig. 2, future enterprise resources will be service-oriented. As such, new functionality will be introduced by combining services in a cross-layer form, i.e. services relying on the enterprise system, on the network itself and at device level will be combined. In addition, sophisticated services can be created at any layer taking into account and based only on the provided functionality of other entities that can be provided as a service. In parallel, Agent-to-Agent communication will allow to optimally exploit the functionality of a given device. As stated in [5], it is clear that we move away from isolated stand-alone hardware and software solutions towards more cooperative models.

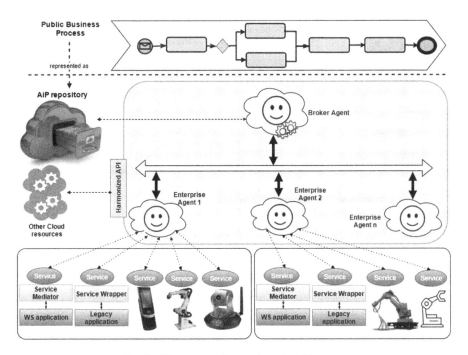

Fig. 2. The proposed agent-based architecture

Consequently, our main contribution is an agent-based middleware layer, between the Cloud service offerings and the enterprises systems. As shown in Fig. 2, our architecture defines two types of agent, namely, the Enterprise Agent representing an individual enterprise, and the Broker Agent, which facilitates the Cloud based application developers in searching for, deploying and governing their business applications on the Cloud offerings that best match their needs.

4.1 Description of Agents

To ensure an effective integration of the IoT resources, the Enterprise Agent exposes real-world devices with embedded software to standard IT systems by making them accessible in a service-oriented way. Our agent-based architecture hides the heterogeneity of hardware, software, data formats and communication protocols that is present in today's embedded systems. As shown in Fig. 3, the following layers can be distinguished: IoT Objects Register, IoT Objects Management, Application Monitoring, AiP Composition and Monitoring, Agent-to-Agent Communication, Harmonized API, Semantic Model and GRC toolkit.

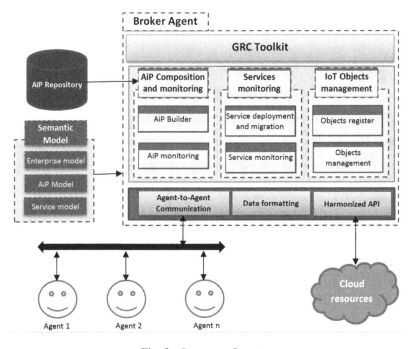

Fig. 3. Structure of agents

IoT Objects Register: Contains list of IoT devices and different equipment that collect data and send it to the server throw the middleware, to avoid any external and unwanted attempts to connect to the server or to the gateway.

IoT Objects Management: This module is used to manage objects connects to the system. User can add, remove and configure devices remotely.

Application Monitoring: Supports the efficient deployment and governance of applications. The developers can manage the life-cycle of their applications as a homogenized way independently of the specific platform offering the application is deployed.

AiP Composition and Monitoring: Orchestrate AiP and control the access to them. It receives requests to resolve requirements from applications developers. Then, it handles the requests via their associated AiP. It also provides operations for monitoring interaction (i.e., creating and deleting instances).

Agent-to-Agent Communication Module: Contains all the processes required to handle agent to agent communication, such as: reception, filtering, and translation of incoming messages, and formulation and sending of the outgoing messages. Agent to agent communication occurs via FIPA Agent Communication Language [18], where XML is used for the description of the content of the message.

Harmonized API: Provides the necessary tools, which enable the management of applications across different Cloud offerings.

Protocols Repository: Maintains repository of commonly used and generic protocols. It facilitates the reuse of a variety of well-defined, well-understood and validated protocols. It provides the AiP specification that describes the functionality, input and output of protocols. An AiP is described by the requirement it resolves, and the parameters of the requirement correspond to the input of the AiP. The AiP output is a set of parameters that results from resolving the requirement.

Semantic Model: Is the backbone of the architecture and spans the entire architecture, resolving interaction conflicts by providing a common basis for publishing and searching different services offerings.

GRC Toolkit: The GRC is an integrated approach to managing Governance, Risk and Compliance, enabling enterprises to manage risk and regulatory issues across the organization. It provides a set of essential services and functional components that encompass various areas of risk and compliance management including operational risk, policy and compliance management, and control management. This module is described in detail in the next section.

5 Supporting Collaborative Enterprise Systems Using AiP

The enterprise information system of the future will comprise of unbounded numbers and combinations of service eco-systems, which are network-structured, software-intensive, geographically dispersed, and have a global reach [4]. A service eco-system is a system of systems [21], which depends on distributed control, cooperation, influence, cascade effects, orchestration, and other emergent behaviours as primary compositional mechanisms to achieve its purpose.

To address these challenges, our approach is based on the notion of AiP through which we specify complex services and break the complexity of running systems by recognising larger chunks that have a meaning in the application domain. This notion of AiP, which is inspired by work of Multiagent Systems (MAS), supports the modelling of composite services as entities whose business logic involves a number of interactions among more elementary service components.

5.1 Agent Interaction Protocols

Let P be a set of interaction protocols and let $R = \{r_1, r_2, \ldots, r_n\} : (n > 1)$ be a set of elementary service components (private process and/or Web service) in interaction. $CA \subseteq FIPA\ ACL$ *communicative act* is a set of standardized communication actions. All constraints and Post/Pre-conditions over communication actions are specified in *Option* (e.g. Synchronous/Asynchronous message, constraints on message, timed constraint and deadlines, etc…).

Let $M = R \times R \times CA \times Option$ be a non-empty set of primitive messages exchanged at the time of Agent or process communication. A Primitive Message is $pm = (Id_{send}, Id_{recv}, C, Opt)$. pm is defined by the following grammar:

$$pm ::= pm \mid pm_i \circledast pm_j \text{ where} : pm \in M \text{ and } \circledast \in \{xor, \wedge, \vee\} \text{ and for all}$$
$$i, j : i \neq j \wedge pm_i.Id_{send} = pm_j.Id_{send}$$

Formally an AiP is defined over a set P of interaction protocols as follow:

Definition 1. An Agent Interaction Protocol is a tuple $AiP = (ID, R, M, f_M)$ where ID is an identifier of the interaction protocol in P and $f_M \subseteq R \times R$ is a flow relation.

Developing effective protocols to be executed by autonomous partners is challenging. Similar to protocols in traditional systems, AiP in open and Web-based settings need to be specified rigorously so that business partners can interact successfully. For this reason, we have developed a method for AiP design and verification [2]. Our method motivates the use of AiP based on AUML/WSBPEL, where pre- and post-conditions, rules, guards are specified in OCL (Object Constraint Language).

AUML (Agent UML) notation [19] is a UML profile dedicated to agents trying to simplify the transition from software engineering to multi-agent system engineering. In other hand, WSBPEL [12] (Business Process Execution Language for Web Services) is a de-facto standard for describing Web services composition. In our context, WSBPEL was used as a specification language for expressing the interaction protocols of the multi-agent system [8].

Driven by the motivation of reuse, we would like to treat protocols as modular components, potentially composed into additional protocols, and applied in a variety of business processes. By maintaining repositories of commonly used, generic, and modular protocols, we can facilitate the reuse of a variety of well-defined, well-understood, and validated protocols.

5.2   AiP Composition for More Complex Business Services

Because protocols address different business goals, they often need to be composed to be put to good use. For example, an enterprise that is interested in selling books could focus on this protocol while outsourcing other protocols such as payment and shipment.

The composition of two or more AiP generates a new protocol providing both the original individual behavioral logic and a new collaborative behavior for carrying out a new composite task. This means that existing protocols are able to cooperate although the cooperation was not designed in advance.

Definition 2 (Composite Protocol). A Composite Protocol (CP) is a tuple $CP = (P, OP, Input, Output, P_{init}, P_{fin})$ where:

- P is a non empty set of basic protocols,
- Op is a non empty set of operators, $OP \subseteq (P \times P)$,
- Input, Output are a set of the elements required (produced) by the composite protocol CP,
- P_{init} is non empty set of initial protocols, $P_{init} \in P$ and
- P_{fin} is non empty set of final protocols, $P_{fin} \in P$.

Protocol Contract

The proposed approach provides the underpinnings of aggregation abstractions for protocols. To achieve this goal, we require an agreement between AiP in the form of a shared contract. A contract describes the details of a protocol (participants in the protocol, produced elements, required elements, constraints,...) in a way that meets the mutual understandings and expectations of two or more protocols. Introducing the contract notion gives us a mechanism that can be used to achieve a meaningful composition.

Definition 3 (Protocol Contract). A contract C, is a collection of elements that are common across two protocols. It represents the mutually agreed upon protocol schema elements that are expected and offered by the two protocols.

1. Let us denote by P_k^{in}, P_k^{out} the set of elements required (produced) by the protocol P_k where $P_k^{in} = \{x_i, i \geq 1\}$ and $P_k^{out} = \{y_j, j \geq 1\}$
2. Let us define a function Φ called a contract-mapping, that maps a set of P_k^{out} elements (produced by the protocol P_k) and a set of P_r^{in} (consumed by the protocol P_r), $\Phi = \{\exists y_i \in P_k^{out} \wedge \exists x_j \in P_r^{in} | (xi, yj)\}$, which means that the protocol Pr consumes the element y_j provided by the protocol P_k, and $C = (P_k^{out})_{\Phi}(C) \cup (P_r^{in})_{\Phi}(C)$.

Protocols Composability Relationship

The composability relationship means that AiP can be joined together to create a new protocol that performs more sophisticated applications. That composition can then served as a protocol itself. We can classify the nature of protocols composition on the dependence degree between them. We may thus distinguish between two kinds of composability relationship:

Definition 4 (Partial composability). Two protocols P_k and P_r meet the *Partial composability* relationship iff $\exists x_i \in (P_r^{in})_\Phi(C)$, $\exists y_j \in (P_k^{out})_\Phi(C) \mid (x_i, y_j)$, which means that the protocol P_r can be executed after the protocol P_k but it must wait until all its "required elements" will be provided by others protocols.

Definition 5 (Full composability). Two protocols P_k and P_r are called *Full composability* iff $\forall x_i \in (P_r^{in})_\Phi(C)$, $\exists y_j \in (P_k^{out})_\Phi(C) \mid (x_i, y_j)$, which means that the protocol P_r must be (immediately) executed after the protocol P_k because *all* its "required elements" are offered by the protocol P_k.

We note here that the partial (full) composability relationship is not commutative. So, if a protocol P_k has a partial (full) composability relationship with a protocol P_r, it does not mean that P_r has a partial (full) composability relationship with P_k.

Proposition. Let $P = \{P_1, P_2, \ldots, P_m\}$ a set of AiP. The set P constitutes a meaningful composition if:[1]

$$\forall Pk \in (P - P_{init}^1); \ \forall x_i \in (P_k^{in})_\Phi(C); \ \exists y_j \in (P_r^{out})_\Phi(C) \mid (x_i, y_j), \text{ where } r,$$
$$k \in [1, m] \text{ and } r \neq k.$$

This proposition states that, if we have a set of protocols $P = \{P_1, P_2, \ldots, P_m\}$ where all their "required elements" are offered (by other protocols), this means that all the protocols of P can be executed.

6   Conclusion and Future Work

In this paper we presented an agent based approach for integrating enterprise application with IoT and its smart devices. First, we presented the idea of AiP, which is a useful way for structuring interaction among business partners. Second, we proposed a basis for a theoretical approach for aggregating protocols to create a new desired business application. This research is among the earliest efforts, to the best of the author' knowledge, in adopting an AiP-based approach for supporting business applications based on IoT. Furthermore, an agent-based architecture that supports our theoretical approach is developed. The main goal of the proposed architecture is to address and tackle interoperability challenges in the context of IoT environments. Also, it solves the interoperability issues between heterogeneous Cloud services environments by offering a harmonized API.

The design and implementation was done with scalability in mind. However, we plan to conduct extended scalability and performance tests in the future. A special focus will be on performance tests when running services on resource-constrained devices.

---

[1] $P_{init} \in P$ and represent the initial protocols, which their "required elements" are provided by external events.

References

1. Porkodi, R., Bhuvaneswari, V.: The internet of things (IoT) applications and communication enabling technology standards: an overview. In: IEEE International Conference on Intelligent Computing Applications (2014)
2. Benmerzoug, D.: An agent-based approach for hybrid multi-cloud applications. Scalable Comput.: Pract. Exp. **14**(2), 95–109 (2013)
3. Wang, J., Zhu, Q., Ma, Y.: An agent-based hybrid service delivery for coordinating internet of things and 3rd party service providers. J. Netw. Comput. Appl. **36**(6), 1684–1695 (2013)
4. Taher, Y., Nguyen, D.K., Lelli, F., van den Heuvel, W.-J., Papazoglou, M.: On engineering cloud applications - state of the art, shortcomings analysis, and approach. Scalable Comput.: Pract. Exp. **13**(3), 215–231 (2012)
5. Spiess, P., Karnouskos, S., Guinard, D., Savio, D., Baecker, O., de Souza, L.M.S., Trifa, V.: SOA-based integration of the internet of things in enterprise services. In: ICWS 2009, pp. 968–975 (2009)
6. Fortino, G., Russo, W., Savaglio, C.: Agent-oriented modeling and simulation of IoT networks. In: FedCSIS 2016, pp. 1449–1452 (2016)
7. Guinard, D., Trifa, V., Karnouskos, S., Spiess, P., Savio, D.: Interacting with the SOA-based internet of things: discovery, query, selection, and on-demand provisioning of web services. IEEE Trans. Serv. Comput. **3**(3), 223–235 (2010)
8. Benmerzoug, D., Boufaida, M., Kordon, F.: A specification and validation approach for business process integration based on web services and agents. In: Proceedings of 5th International Workshop on Modelling, Simulation, Verification and Validation of Enterprise Information Systems, MSVVEIS-2007, Conjunction with ICEIS 2007, pp. 163–168. NSTIIC Press (2007)
9. Benmerzoug, D., Kordon, F., Boufaida, M.: Formalisation and veri cation of interaction protocols for business process integration: a Petri net approach. Int. J. Simul. Process. Model. **4**(3–4), 195–204 (2008)
10. Benmerzoug, D.: Towards AiP as a service: an agent based approach for outsourcing business processes to cloud computing services. Int. J. Inf. Syst. Serv. Sect. **7**(2), 1–17 (2015)
11. Van Hoecke, S., Waterbley, T., Devos, J., Deneut, T., De Gelas, J.: Efficient management of hybrid clouds. In: Second International Conference on Cloud Computing, GRIDs, and Virtualization, Rome, Italy, pp. 167–172, September 2011
12. IBM, Microsoft, SAP, and Systems Siebel: Business process execution language for web services version 1.1. Technical report (2003)
13. Garcia, O., Sim, K.M.: Agent-based cloud service composition. Appl. Intell. Int. J. Artif. Intell. Neural Netw. Complex Probl.-Solv. Technol. **22**(2) (2012)
14. Penserini, L., Kuflik, T., Busetta, P., Bresciani, P.: Agent-based organizational structures for ambient intelligence scenarios. Ambient Intell. Smart Environ. **2**(4), 409–433 (2010)
15. Benmerzoug, D., Boufaida, Z., Boufaida, M.: From the analysis of cooperation within organizational environments to the design of cooperative information systems: an agent-based approach. In: Meersman, R., Tari, Z., Corsaro, A. (eds.) OTM 2004. LNCS, vol. 3292, pp. 495–506. Springer, Heidelberg (2004). doi:10.1007/978-3-540-30470-8_63
16. Desai, N., Chopra, A.K., Singh, M.P.: Amoeba: a methodology for modeling and evolving cross-organizational business processes. J. ACM Trans. Softw. Eng. Methodol. **19**(2), 6 (2009)
17. Cernuzzi, L., Molesini, A., Omicini, A., Zambonelli, F.: Adaptable multi-agent systems: the case of the Gaia methodology. Int. J. Softw. Eng. Knowl. Eng. **21**(4), 491–521 (2011)

18. FIPA-ACL: FIPA Communicative Act Library Specification. Technical report, FIPA - Foundation for Intelligent Physical Agents (2001). http://www.pa.org/specs/XC00037/
19. Bauer, B., Odell, J.: UML 2.0 and agents: how to build agent-based systems with the new UML standard. Int. J. Eng. Appl. AI **18**(2), 141–157 (2005)
20. Haller, S., Serbanati, A., Bauer, M., Carrez, F.: A domain model for the internet of things. In: GreenCom/iThings/CPScom, pp. 411–417 (2013)
21. Fisher, D.A.: An emergent perspective on interoperation in systems of systems. Technical report, Software Engineering Institute, Carnegie Mellon University (2006)
22. Bauer, M., Bui, N., De Loof, J., Magerkurth, C., Nettsträter, A., Stefa, J., Walewski, J.W.: IoT reference model. In: Bassi, A., Bauer, M., Fiedler, M., Kramp, T., van Kranenburg, R., Lange, S., Meissner, S. (eds.) Enabling Things to Talk, pp. 113–162. Springer, Heidelberg (2013). doi:10.1007/978-3-642-40403-0_7

Fuzzy Theory and Algorithms

Fuzzy PID Controller for Reactive Power Accuracy and Circulating Current Suppression in Islanded Microgrid

Minh-Duc Pham and Hong-Hee Lee$^{(\boxtimes)}$

School of Electrical Engineering, University of Ulsan,
Ulsan 680-749, South Korea
minhducpham2009@gmail.com, hhlee@mail.ulsan.ac.kr

Abstract. In this paper, an intelligent control scheme based on fuzzy proportional-integral-derivative controller (FPIDC) is proposed for distributed generations in islanded microgrid. The proposed FPIDC method is composed of a close-loop of the virtual impedance to compensate the feeder mismatch between distributed generators (DGs). Together with feedback of the inaccurate reactive power, the uncertainty of proportional-integral-derivative (PID) parameters is removed by adaptive tuning. Therefore, the power sharing performance is improved and the circulating current between DGs is mitigated. The dynamic response of the microgrid system with the proposed FPIDC is much better than that of the conventional PID controller. The comparison and analysis of the proposed control with conventional control are carried out to evaluate the superiority of the proposed method.

Keywords: Islanded microgrid · Fuzzy logic control · Reactive power control · Secondary control

1 Introduction

Recently, the use of distributed generation (DG) units including renewable energy resources has been increased in power system, and the use of low efficient traditional power plants are anticipated to be decreased. Microgrids are becoming an important part to integrate DG and distributed energy storage system [1].

Figure 1 shows the basic structure of the microgrid. An indispensable function of microgrid is to achieve desired power management among DG units in islanding operation. When the static switch is open, the microgrid is isolated from the main grid and DGs have to work independently. Therefore, each DG has a responsibility to regulate not only the voltage but also the frequency to ensure the stability and autonomously power sharing feature of the microgrid.

For this purpose, DGs are generally operated in the voltage control mode (VCM) to maintain the constant voltage at the point of common (PCC) bus. In order to control DGs for power sharing, the droop control is commonly used [1]. The main advantage of droop control scheme is the plug and play ability which means each DG can be connected and disconnected freely without the need of high bandwidth communication

© Springer International Publishing AG 2017
D.-S. Huang et al. (Eds.): ICIC 2017, Part III, LNAI 10363, pp. 241–252, 2017.
DOI: 10.1007/978-3-319-63315-2_21

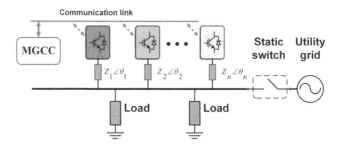

Fig. 1. Typical islanded microgrid configuration.

[2]. In addition, the control scheme can be easily extended to a number of DGs in microgrid without voltage and current information. Meanwhile, the droop control algorithm still has some disadvantages such as circulating current and inaccurate power sharing. Since the frequency of the microgrid is not affected by the feeder impedance mismatches, the real power sharing capabilities can be achieved successfully. However, the reactive power sharing is not accurate because of unequal feeder impedance in microgrid configuration.

There are many methods to increase power decoupling between active and reactive power. In [3], authors propose a transformation into the new virtual frame to decouple the active and reactive power. However, this method requires the exact value of the feeder impedance, which is hard to detect in practical applications. On the other hand, authors in [4] have modified the inverter output impedance to control the current and power sharing. This concept is robust to control directly the inverter output voltage by considering virtual impedance at the output of inverter. The virtual impedance is mainly calculated according to the error of reactive power sharing through the proportional-integral-derivative (PID) controllers. Therefore, the performance of the controller is significantly depended on PID parameters. As well known, the performance of the PID controller is affected by many non-ideal system parameters such as noise and disturbance. So, it is important to remove the effect of these parameter uncertainty, and fuzzy logic controller (FLC) is one of the advantageous solution to overcome the drawback of the conventional PID control method.

In this paper, we propose an enhanced fuzzy proportional-integral-derivative controller (FPIDC) to achieve accurate power sharing in islanded microgrid by the hybrid controller which combines the advantage of the nonlinear control of FLC with small steady state error of PID controller. The FPIDCs are already introduced in [5, 6]. However, the application of FPIDC to control virtual impedance in microgrid has not been presented up to now. Moreover, this paper presents a feedback compensation method based on FPIDC method to improve power sharing performance and circulating current rejection. The proposed method is guaranteed and verified by investigating the dynamic response and stability of DGs in islanded microgrid with MATLAB/SIMULINK.

2 Control Strategy for Single Phase Inverter

Figure 2 shows the single-phase inverter topology which is used in islanded microgrid. In Fig. 2, DC bus source is usually taken from photo voltaic source or energy storage system. The converter stage is a commonly full bridge two-level voltage source inverter (VSI). The output voltage of inverter with variable frequency and amplitude is regulated by modulating the converter stage. The LCL filter, which was introduced in [7, 8] is generally used to smooth the inverter output voltage.

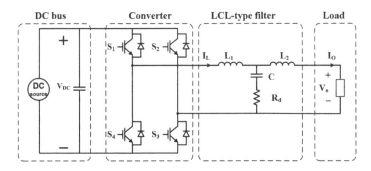

Fig. 2. Schematic diagram of the single phase stand-alone inverter.

2.1 Droop Control Theory

An crucial function of microgrid is autonomously sharing the power among DG units in islanding operation. From the relationship between active and reactive power with voltage amplitude and frequency, the power transfer through the line impedance can be expressed as

$$P = \frac{1}{R^2 + X^2} (R\,E^2 - REV\,\cos\,\delta + XEV\,\sin\,\delta) \tag{1}$$

$$Q = \frac{1}{R^2 + X^2} (XE^2 - XEV\,\cos\,\delta - REV\,\sin\,\delta) \tag{2}$$

where E and V are the amplitudes of inverter output voltage and PCC bus voltage, respectively, δ is the phase angle difference between E and V. R and X are the line resistance and line inductance of the feeder, respectively. Because the leakage inductance of the transformer and output filter inductor is much larger than the resistance, i.e., X>>R, the affect of R can be neglected [9]. In addition, the phase angle difference δ is normally small, and it is reasonable to assume $\sin(\delta) = \delta$ and $\cos(\delta) = 1$. Therefore, the Eqs. (1) and (2) can be rewritten as following:

$$\delta \approx (X.P)/(E.V) \tag{3}$$

$$E - V \approx (X.Q)/E \tag{4}$$

From the Eqs. (3) and (4), the active power P is proportional with the frequency, while the reactive power Q is influenced by the amplitude of the difference E-V, and the droop control concept is defined as

$$\omega = \omega_0 - m(P - P^*) \tag{5}$$

$$E = E_0 - n(Q - Q^*) \tag{6}$$

where ω_0 and E_0 are the nominal values of the output frequency and voltage, respectively, which are usually chosen equal to those of the main grid. m and n are the frequency and amplitude droop coefficients, respectively. In islanded microgrid, because the inverter determines the power sharing without the grid source, P^* and Q^* are set to zero. When the microgrid is transferred to the islanded mode, the output power of DG systems are changed according to their droop characteristics to supply power to the load.

2.2   Outer and Inner Loop Control

The following proportional-resonant (PR) voltage and current controllers are applied in this paper:

$$G_V(s) = K_{pV} + \frac{2.K_{rV}.\omega_{CV}.s}{s^2 + 2.\omega_{CV}.s + \omega_o^2} \tag{7}$$

$$G_I(s) = K_{pI} + \frac{2.K_{rI}.\omega_{CI}.s}{s^2 + 2.\omega_{CI}.s + \omega_o^2} \tag{8}$$

where K_{pV} and K_{pI} are the proportional gains, K_{rV} and K_{rI} are the resonant gains of the PR controller, respectively. ω_{CV} and ω_{CI} represent the voltage and current cut-off frequencies. Figure 3 shows the control diagram of outer and inner loop controller with PR controller. The double loop control is used to increase the phase margin, and to stabilize the system.

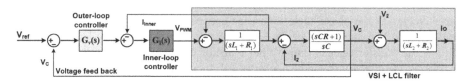

Fig. 3. Outer and inner loop control diagram.

In order to control the PR, the values of K_p, K_r and ω_C of two control loop should be optimized for the best results. It is important to note that the bandwidth of inner current control loop should be higher than the bandwidth of outer voltage control loop for stability and phase margin [10]. Typically, the current control loop bandwidth is selected 10 times higher than that of the voltage control loop.

3 Proposed Fuzzy Controller for Power Sharing Accuracy in Microgrid

The control method presented in previous section is suitable for inverter worked in islanded microgrid. However, the reactive power output of each inverter depends on the voltage amplitudes due to the mismatched voltage drop caused by unequal impedance [11]. To avoid the active and reactive power coupling and increase power sharing accuracy, the virtual impedance is applied in this paper. In order to obtain the virtual impedance, the PID controller can be used to increase the steady state performance. In islanded microgrid system, because system parameters are mainly non-linear and time varying values, it is necessary to tune the parameters to detect the virtual impedance value accuracy with PID controller. To solve this parameter uncertainty, the PID controller to find the virtual impedance is designed with the aid of the enhanced fuzzy logic control method.

3.1 Virtual Impedance Control

From Fig. 3, the capacitor voltage (V_C) is obtained from the close loop transfer function of the inner loop:

$$V_C = \frac{G_I G_V Z_C}{Z_C + Z_L + G_I + G_I G_V Z_C} V_{ref}(s) - \frac{Z_C(Z_L + G_I)}{Z_C + Z_L + G_I + G_I G_V Z_C} i_o(s) \qquad (9)$$

$$\triangleq G_{control}(s) V_{ref}(s) - Z_O(s) i_o(s) \qquad (10)$$

where $Z_L = sL_1 + R_1$ and $Z_C = (sCR + 1)/sC$.

In Eq. (10), the capacitor voltage (V_C) can be controlled by two elements, $G_{control}$ and Z_O: $G_{control}(s)$ is the voltage gain and $Z_O(s)$ is equivalent value of the inverter output impedance. Meanwhile, the output voltage reference is obtained by considering the virtual impedance at the output of inverter:

$$V_{ref}(s) = V_O^* - i_o(s) Z_d(s) \qquad (10)$$

where $Z_D(s)$ is the virtual output impedance and V_O^* is the inverter output voltage reference, which is calculated from the droop equation $V_O^* = E_{droop} \sin(\omega_{droop} t)$. To compensate the mismatched voltage drop, the proper value of the virtual impedance is adjusted by the external loop control through PID controller:

$$L_{Vir} = K_{p\_vir}(Q_{average} - Q_i) + K_{i\_vir} \int (Q_{average} - Q_i) dt + K_{D\_vir} d(Q_{average} - Q_i)/dt \qquad (11)$$

where $Q_{average}$ is average value of reactive power calculated from the total number of DGs, n:

$$Q_{average} = \left(\sum_{i=1}^{n} Q_i\right)/n \qquad (12)$$

The virtual impedance is calculated based on the output current feedback:

$$X_{L\_Vir} = I_{O\_\alpha}R_{vir} - I_{O\_\beta}L_{vir} \qquad (13)$$

Figure 4 shows the simplified model of the microgrid with two inverters including microgrid central controller (MGCC). From this model, each DG sends its reactive power information to MGCC through low bandwidth communication. Then, the MGCC determines the reactive power demand for each DG, and it is used to control virtual impedance by using PID controller. Because the mismatched impedance between DGs is compensated due to the accurate virtual impedance, the reactive power is shared equally, and the power decoupling performance are improved. Moreover, by modifying output impedance and compensating the voltage drop mismatch between DGs, the output voltage at the PCC becomes almost same, and the circulating current is significantly reduced.

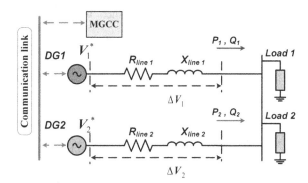

Fig. 4. Simplified model of the microgrid with two inverters.

3.2   Fuzzy Logic Controller Design

The FLC is not based on a mathematical model and is widely used to solve problems under uncertainty and high nonlinearities [12, 13]. To improve the performance and increase the dynamic response, there are some hybrid combination between conventional fuzzy and the most common controllers like PD, PI, or PID [14]. In this paper, the fuzzy PID controller is applied to increase the dynamic response and stability in islanded microgrid. The general continuous-time PID has the expression:

$$u(t) = K_p e(t) + K_I \int e(t)dt + K_D\dot{e}(t) \tag{14}$$

To find three parameters K_P, K_I, K_D of PID controller, the range of $[K_{Pmin}, K_{Pmax}]$, $[K_{Imin}, K_{Imax}]$, $[K_{Dmin}, K_{Dmax}]$ of each parameter should be defined first. In fuzzy controller, the range of the three parameters K_P', K_I', and K_D' are in $[0,1]$:

$$\begin{cases} 0 \le K_P' = \frac{K_P - K_{Pmin}}{K_{Pmax} - K_{Pmin}} \le 1 \\ 0 \le K_I' = \frac{K_I - K_{Imin}}{K_{Imax} - K_{Imin}} \le 1 \\ 0 \le K_D' = \frac{K_D - K_{Dmin}}{K_{Dmax} - K_{Dmin}} \le 1 \end{cases} \tag{15}$$

From (15), the gains for the PID controller are rewritten as following:

$$\begin{cases} K_P = (K_{Pmax} - K_{Pmin})K_P' + K_{Pmin} \\ K_I = (K_{Imax} - K_{Imin})K_I' + K_{Imin} \\ K_D = (K_{Dmax} - K_{Dmin})K_D' + K_{Dmin} \end{cases} \tag{16}$$

The main structure of the proposed FPIDC control scheme is shown in Fig. 5(a) and (b). In Fig. 5, the FPIDC consists of two input variables and three output variables. It should be noted that there is no saturation block because the range of PID gains are already limited with their minimum and maximum values.

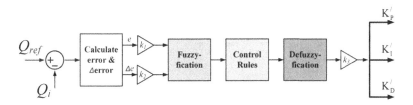

Fig. 5. Structure of the proposed FLC method

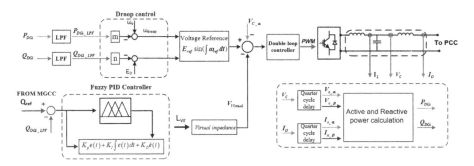

Fig. 6. Outer and inner loop control diagram.

The control block diagram of the DG is shown in Fig. 6. There are two input variables: the error (e) and the error change rate (de/dt). de/dt is usually replaced with the gradient of error (Δe) which is obtained by the difference between two sampling values. Then, the error (e) and the gradient of error (Δe) are calculated as follows:

$$e(k) = Q_{ref}(k) - Q_o(k) \tag{17}$$

$$\Delta e(k) = e(k) - e(k-1) \tag{18}$$

where k and $(k-1)$ are indicated the present and previous sampling signals.

The design of the proposed FPIDC algorithm is divided into three main parts: fuzzification, fuzzy rule and defuzzification. The asymmetrical triangular membership functions (MFs) are selected and there are five memberships for more precision as shown in Fig. 7. It is clear that the comprehensive range of fuzzy input and output variables, e(t) and $\Delta e(t)$, spreads symmetrically on both positive and negative sides. In FPIDC, there are five MFs for the input variables and the output variables, and the control rules are finally shown in Table 1.

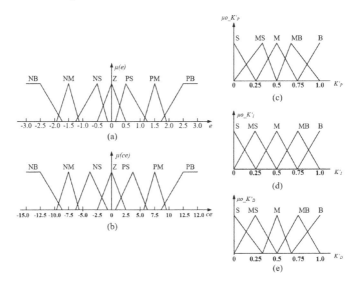

Fig. 7. Membership functions design for the input and output variable. (a) Input error e. (b) Input variable (Δe). (c) Output variable K'_P. (d) Output variable K'_I. (e) Output variable K'_D

Table 1. Fuzzy rule table for Fuzzy PID controller

e/Δe	NL	NS	ZR	PS	PL
NL	NL	NL	NL	NL	NL
NS	NM	NM	NS	NS	PS
ZR	NM	NS	ZR	PS	PM
PS	NS	NS	PS	PM	PM
PL	PL	PL	PL	PL	PL

4   Simulation Results

In order to verify the effectiveness of the proposed FPIDC method, some simulations have been carried out by using MATLAB/SIMULINK. The simulation parameters are given as follows: L_1 = 1.4mH, L_2 = 1.4mH, C = 20uF, R_d = 1.5 Ω, m = 0.012, n = 0.015, Vo = 110 V_{rms}/ 50 Hz, Z_{line1} = 0.1 + 0.39j Ω, Z_{line1} = 0.2 + 0.3j Ω, R_{load1} = R_{load2} = 10 Ω, L_{load1} = L_{load2} = 15mH.

Simulation is carried out with the microgrid in Fig. 4, where 2 DGs supply the power to 2 loads: load 1 and load 2. Figures 8 and 9 show the simulation results of DGs in islanded microgrid with the load 1 only. Each DG is operated by the droop controller at starting time to achieve the active power sharing, and the conventional PID controller is applied in Fig. 8 to improve the reactive power sharing at 2[s] while the FPIDC is applied in Fig. 9.

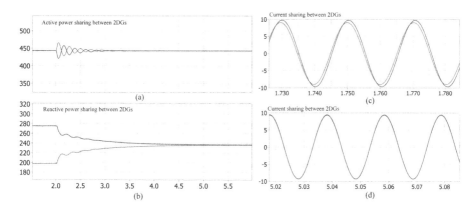

Fig. 8. Simulation with the conventional PID controller (DG1: blue line, DG2: red line). (a) Active power sharing between DGs. (b) Reactive power sharing between DGs. (c) Current sharing between 2DGs before the PID controller is activated. (d) Current sharing after the conventional PID is activated. (Color figure online)

In Figs. 8(a) and 9(a), the active power is shared equally regardless of the controller. However, due to the mismatched impedance, the reactive accurate power sharing is not satisfied in Figs. 8(b) and 9(b) before the controller is applied. When the controller is applied at 2 s, the reactive power sharing is accurately achieved after some transient interval. But, the dynamic performance with the FPIDC is faster than that of the conventional PID controller; the conventional PID controller takes about 1 s longer transient time than that of the proposed FPIDC. In addition, the conventional PID has more oscillation to reach the active power sharing, and is not smooth during the transient time in reactive power sharing compared with the FPIDC.

From Figs. 8(c) and 9(c), we can see that each DG currents has different magnitude and phase angle before 2 s. So, the circulating current appears between two DGs, which deteriorates the microgrid stability and quality. When the controller is applied at 2 s, the current sharing is balanced, which makes the circulating current becomes almost zero as shown in Figs. 8(d) and 9(d).

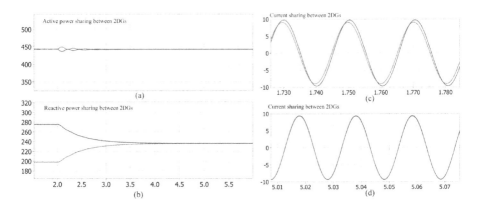

Fig. 9. Simulation with the proposed FPIDC controller (DG1: blue line, DG2: red line). (a) Active power sharing between DGs. (b) Reactive power sharing between DGs. (c) Current sharing between 2DGs before the proposed FPIDC is activated. (d) Current sharing after the proposed FPIDC is activated. (Color figure online)

Figures 10(a) and (b) show the dynamic responses and stabilities by connecting the load 2 in Fig. 4 at the time 6.6 s with the same condition in Figs. 8 and 9, respectively. Even though the conventional PID shows a good dynamic response, it turns into

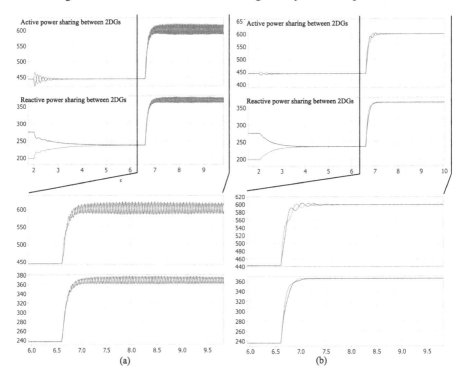

Fig. 10. The dynamic response comparison when the load step at 6.6 s (DG1: blue line, DG2: red line). (a) The conventional PID controller. (b) The proposed FPIDC. (Color figure online)

overshoot and oscillation as shown in Fig. 10(a) because PID parameters are not selected properly in the new operating condition. Moreover, because of the mutual interaction between parallel inverter, one DG variation induces the others variation, which makes whole islanded microgrid unstable. In contrast, the FPIDC has a very good response performance with no overshoot and the fast oscillation decay. In addition, 2 DGs smoothly reach a new equilibrium operating point to keep the accurate power sharing with the stable operation for islanded microgrid. If the membership function and rules are designed correctly, the optimal PID gains are adaptively obtained in spite of the system parameter variation, and it is evident from the simulated results that the performance of the FPIDC is better than the conventional PID controller.

5 Conclusion

This paper has presented an intelligent control method based on the fuzzy logic controller to improve the power sharing performance in islanded microgrid. With the proposed FPIDC scheme, the virtual impedance is adaptively tuned to share accurate reactive power and minimize the circulating current. Furthermore, the microgrid operating system becomes more stable in spite of the load variation. The dynamic performance of the proposed FPIDC is evaluated and compared with the conventional PID controller, and it is verified that the performance of the FPIDC is better than the conventional PID controller.

Acknowledgment. This work was supported by the National Research Foundation of Korea Grant funded by the Korean Government (NRF-2015R1D1A1A09058166).

References

1. Olivares, D.E., Mehrizi-Sani, A., Etemadi, A.H., Cañizares, C.A., Iravani, R., Kazerani, M., Hajimiragha, A.H., Gomis-Bellmunt, O., Saeedifard, M., Palma-Behnke, R., Jiménez-Estévez, G.A., Hatziargyriou, N.D.: Trends in microgrid control. IEEE Trans. Smart Grid **5**(4), 1905–1919 (2014)
2. Mahmoud, M.S., Hussain, S.A., Abido, M.A.: Modeling and control of microgrid: an overview. J. Franklin Inst. **351**(5), 2822–2859 (2014)
3. De Brabandere, K., Bolsens, B., Van den Keybus, J., Woyte, A., Driesen, J., Belmans, R.: A voltage and frequency droop control method for parallel inverters. IEEE Trans. Power Electron. **22**(4), 1107–1115 (2007)
4. Li, Y.W., Kao, C.N.: An accurate power control strategy for power-electronics-interfaced distributed generation units operating in a low-voltage multibus microgrid. IEEE Trans. Power Electron. **24**(12), 2977–2988 (2009)
5. He, S., Tan, S., Xu, F., Wang, P.: Fuzzy self-tuning of PID controllers. Fuzzy Sets Syst. **56**(1), 37–46 (1993)
6. Qiao, W.Z., Mizumoto, M.: PID type fuzzy controller and parameters adaptive method. Fuzzy Sets Syst. **78**(1), 23–35 (1996)
7. Loh, P.C., Holmes, D.G.: Analysis of multiloop control strategies for LC/CL/LCL-filtered voltage-source and current-source inverters. IEEE Trans. Ind. Appl. **41**(2), 644–654 (2005)

8. Tang, Y., Loh, P.C., Wang, P., Choo, F.H., Gao, F.: Exploring inherent damping characteristic of LCL-filters for three-phase grid-connected voltage source inverters. IEEE Trans. Power Electron. **27**(3), 1433–1443 (2012)
9. Guerrero, J.M., de Vicuna, L.G., Matas, J., Castilla, M., Miret, J.: A wireless controller to enhance dynamic performance of parallel inverters in distributed generation systems. IEEE Trans. Power Electron. **19**(5), 1205–1213 (2004)
10. Cha, H., Vu, T.K., Kim, J.E.: Design and control of proportional-resonant controller based Photovoltaic power conditioning system. In: 2009 IEEE Energy Conversion Congress and Exposition, pp. 2198–2205 (2009)
11. Vasquez, J.C., Guerrero, J.M., Luna, A., Rodriguez, P., Teodorescu, R.: Adaptive droop control applied to voltage-source inverters operating in grid-connected and islanded modes. IEEE Trans. Ind. Electron. **56**(10), 4088–4096 (2009)
12. Tran, Q.-H., Lee, H.-H.: A fuzzy logic controller for indirect matrix converter under abnormal input voltage conditions. In: Huang, D.-S., Han, K. (eds.) ICIC 2015. LNCS, vol. 9227, pp. 139–150. Springer, Cham (2015). doi:10.1007/978-3-319-22053-6_16
13. Buckley, J.J.: Fuzzy controller: further limit theorems for linear control rules. Fuzzy Sets Syst. **36**(2), 225–233 (1990)
14. Li, H.-X., Gatland, H.B.: Conventional fuzzy control and its enhancement. IEEE Trans. Syst. Man Cybern. Part B (Cybern.) **26**(5), 791–797 (1996)

Adaptive Control of DC-DC Converter Based on Hybrid Fuzzy PID Controller

Minh-Duc Pham and Hong-Hee Lee[(✉)]

School of Electrical Engineering, University of Ulsan,
Ulsan 680-749, South Korea
minhducpham2009@gmail.com, hhlee@mail.ulsan.ac.kr

Abstract. DC-DC converter is widely used in industrial applications with the aid of proportional-integral-derivative (PID) controller. The PID controller is commonly used because of its simple, efficiency and small steady state error. However, DC-DC converter is mainly nonlinear system due to its inherent switching operation, which makes the PID parameters very difficult to be found. In this paper, an intelligent PID controller based on a hybrid combination with the fuzzy logic controller (FLC) is proposed to control DC-DC converter. The proposed fuzzy based PID controller (FPIDC) is a close-loop control of the nonlinear PID parameters by measuring output voltage of the DC-DC converter, and it is designed effectively to reduce computation burden. By tuning the PID parameters continuously according to the load condition, the performance of the DC-DC converter is improved much better. The comparison between the proposed and conventional PID control method are investigated under various operating conditions. Simulation results and model analysis are carried in MATLAB/SIMULINK to prove the effectiveness of the proposed FPIDC method.

Keywords: DC-DC converter · Fuzzy logic controller (FLC) · Proportional-integral-derivative (PID) · Fuzzy based PID controller (FPIDC)

1 Introduction

DC-DC converter is widely used in industrial applications to supply DC power to the electric equipment [1]. Figure 1 shows the scheme of the DC-DC converter with the load in power system. In DC-DC converter system, the output voltage is typically adjusted by changing the duty cycle of the IGBT/MOSFET. Many parameters need to be considered in order to stabilize the DC-DC converter when the load or the input voltage is changed. Therefore, the effective design of the controllers is an important part in the real-time DC-DC converter system.

There are many controllers for industrial applications like proportional-integral (PI), proportional-derivative (PD), and proportional-integral-derivative (PID) controller. Among them, the PID controller is most versatile to control the system thanks to the small steady state error and a good dynamic performance [2]. To find the optimized parameters which are used for the PID controller, the Ziegler-Nichols frequency response method is usually used [3]. However, the system oscillation is inevitable

© Springer International Publishing AG 2017
D.-S. Huang et al. (Eds.): ICIC 2017, Part III, LNAI 10363, pp. 253–263, 2017.
DOI: 10.1007/978-3-319-63315-2_22

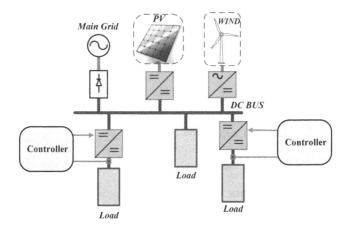

Fig. 1. Scheme of the DC electrical system with DC load

before getting into the steady state. And also, the traditional PID controller has still some problems to be solved such as the dynamic response and instability in the transient process. Furthermore, because the PID parameters are usually fixed suitable for a specific operating condition, its performance is not generally optimized and easily unstable in abnormal condition [4].

The traditional PID is combined with the fuzzy logic controller (FLC) to increase the performance and overcome the drawbacks of the PID controller in nonlinear system because the FLC leads to cope with the system uncertainties. The most important feature of FLC compared with the traditional tuning methods is that it is unnecessary to define the exact mathematic model about the plant [5]. However, to achieve a good response, the membership functions (MFs) and fuzzy rules are needed to be determined effectively. Moreover, it is hard to obtain a good dynamic performance for the system with only FLC [6]. Regarding to the PID based FLC, authors in [7] tried to improve the PID controller performance with the aid of optimized fuzzy rules and MFs. However, too many rules and MFs are used, which results in a large amount of computing time. In order to reduce the computing time, only three output variable MFs with the gauss and trap curve type was used to realize the PID controller [8]. However, three independent fuzzy controllers are applied separately, which makes the system bulky and hard to design.

In this paper, a hybrid combination of FLC based PID controller (FPIDC) is proposed with an optimized design of MFs and fuzzy rule to control DC-DC converter. The PID gains are continuously adjusted from the output voltage error and its derivative to make the transient smoothly and increase the dynamic performance of the DC-DC converter system. The proposed FPIDC consists of a close-loop control with the nonlinear PID parameters, and it is designed effectively to reduce computation burden with the optimized MFs and fuzzy rules by using PID gain transformation. The proposed FPIDC guarantees the dynamic response and stability of DC-DC converter in a wide-range operation and under abnormal condition. The efficiency of the proposed FPIDC is verified by simulation results using MATLAB/SIMULINK.

2   Control Strategy for DC-DC Converter

2.1   DC-DC Converter

DC-DC converter is an electromechanical device that converts the DC voltage from one level to another. The fundamental circuit of the DC-DC converter is shown in Fig. 2.

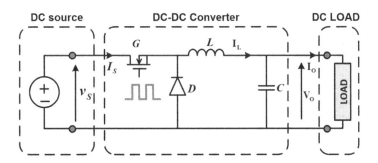

Fig. 2. DC-DC converter fundamental circuit

In order to obtain the system model, the state variables are defined as V_C and i_L and they are mapped as $i_L = x_1$, $V_C = x_2$. Then, the state space equation when the switch G is 'ON' state is expressed as

$$\begin{cases} \frac{di_L}{dt} = \frac{1}{L}(V_{in} - V_o) \\ \frac{dV_o}{dt} = \frac{1}{C}\left(i_L - \frac{V_o}{R}\right) \end{cases}, \ 0 < t < dT. \tag{1}$$

In case that the switch is 'OFF', the state space equation is given by

$$\begin{cases} \frac{di_L}{dt} = \frac{1}{L}(-V_o) \\ \frac{dV_o}{dt} = \frac{1}{C}\left(i_L - \frac{V_o}{R}\right) \end{cases}, \ dT < t < T. \tag{2}$$

By using the state space averaging method [8], the complete state space model of the DC-DC converter during 'ON' and 'OFF' state becomes

$$\begin{bmatrix} x_1' \\ x_2' \end{bmatrix} = \begin{bmatrix} 0 & -\frac{1}{L} \\ \frac{1}{C} & -\frac{1}{RC} \end{bmatrix} \begin{bmatrix} x_1 \\ x_2 \end{bmatrix} + \begin{bmatrix} \frac{d}{L} \\ 0 \end{bmatrix} V_{in}, \tag{3}$$

where d is the switching duty cycle of MOSFET/IGBT.

Meanwhile, the output state of V_C and i_L are given in (4):

$$\begin{bmatrix} y_1 \\ y_2 \end{bmatrix} = \begin{bmatrix} 1 & 0 \\ 0 & 1 \end{bmatrix} \begin{bmatrix} i_L \\ V_C \end{bmatrix} + \begin{bmatrix} 0 \\ 0 \end{bmatrix} V_{in} \tag{4}$$

Finally, the state space equation of the DC-DC converter is represented as following:

$$\begin{cases} \dot{x}(t) = A(t)x(t) + B(t)u(t) \\ y(t) = C(t)x(t) + D(t)u(t) \end{cases}, \tag{5}$$

where $A(t) = \begin{bmatrix} 0 & -\frac{1}{L} \\ \frac{1}{C} & -\frac{1}{RC} \end{bmatrix}$, $B(t) = \begin{bmatrix} \frac{d}{L} \\ 0 \end{bmatrix}$, $C(t) = \begin{bmatrix} 1 & 0 \\ 0 & 1 \end{bmatrix}$, and $D(t) = \begin{bmatrix} 0 \\ 0 \end{bmatrix}$ are the state matrix, input matrix, output matrix, and feedthrough matrix, respectively.

2.2 Traditional PID Controller

The PID controller is commonly used in industrial applications due to a small steady state error and a good dynamic performance. Figure 3 shows the block diagram of the DC-DC converter control system with the PID controller.

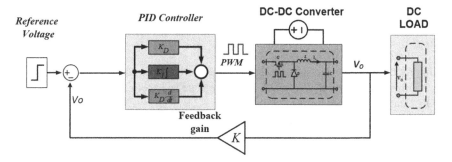

Fig. 3. Block diagram of the traditional PID controller

The traditional PID controller is expressed as (6):

$$G_{PID}(t) = K_P e(t) + K_I \int e(t)dt + K_D \frac{de(t)}{dt}, \tag{6}$$

where G_{PID} (s) is the PID output, K_P, K_I and K_D are the proportional gain, integral gain, and derivative gain, respectively. As well known, there are some criteria used to evaluate controller performance like rising time, settling time, steady state error, overshooting, etc. In order to satisfy the dynamic response of the system, many methods are proposed to find the effective values for the PID parameters. Ziegler-Nichols is the most favorite method to optimize the controller due to the open-loop tuning and its simplicity [3]. The objective of this method is to reach the target point with relatively small time delay, which is implemented with two parameters such as a horizontal axis θ and the slope a as shown in Fig. 4. With these two values, the parameters of the controller are determined according to Table 1.

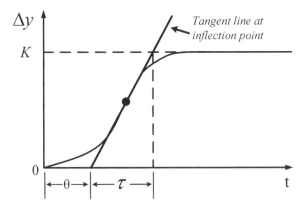

Fig. 4. Ziegler-Nichols parameters characteristic

Table 1. Ziegler-Nichols rules

Controller	K_P	$T_I = K_P/K_I$	$T_D = K_P/K_D$
P	τ/θ		
PI	$0.9\ \tau/\theta$	$10/3\ \theta$	
PID	$1.2\ \tau/\theta$	$2\ \theta$	$0.5\ \theta$

It should be noted that the gain value of Ziegler-Nichols is mainly based on experiment. Hence, the response is not always an optimized value and not exactly the sufficient value for the defined system. Moreover, the traditional PID controller has good performance only at particular operating range, and PID parameters have to be tuned again if the operating range is changed.

3   Proposed Fuzzy Based PID Controller

Typically, the performance of the PID controller depends on the sensitivity of the PID parameters [9]. So, in case of a parameter changing system, it is important to tune the PID parameters continuously. However, it is very difficult in practical applications. To overcome this difficulty, a hybrid combination with FLC and PID controller is proposed in this paper.

3.1   Fuzzy Based PID Controller

To find three parameters K_P, K_I, K_D of the PID controller, the ranges of $[K_{Pmin}, K_{Pmax}]$, $[K_{Imin}, K_{Imax}]$, $[K_{Dmin}, K_{Dmax}]$ for each parameter are defined as K_P', K_I', and K_D', respectively, in fuzzy controller. Then, K_P', K_I', and K_D' are defined as (7), and their ranges are in [0,1].

$$\begin{cases} K'_P = \dfrac{K_P - K_{Pmin}}{K_{Pmax} - K_{Pmin}} \\ K'_I = \dfrac{K_I - K_{Imin}}{K_{Imax} - K_{Imin}} \\ K'_D = \dfrac{K_D - K_{Dmin}}{K_{Dmax} - K_{Dmin}} \end{cases} \tag{7}$$

Therefore, the PID gains are expressed as

$$\begin{cases} K_P = (K_{Pmax} - K_{Pmin})K'_P + K_{Pmin} \\ K_I = (K_{Imax} - K_{Imin})K'_I + K_{Imin} \\ K_D = (K_{Dmax} - K_{Dmin})K'_D + K_{Dmin} \end{cases} \cdot \tag{8}$$

The Eq. (8) shows the relationship between the PID parameters K_P, K_I, K_D and new parameters K'_P, K'_I, and K'_D for FLC. From (8), because the PID parameters are always within the predefined their minimum and maximum values, the saturation block is not required in control diagram. Furthermore, the fuzzy rules can be optimized with the reduced MFs due to the PID gain transformation in (8), which reduces the computing time. Figure 5 shows the control block diagram of the FPID.

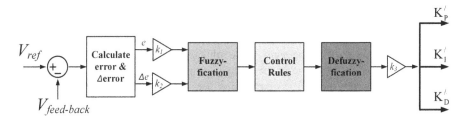

Fig. 5. The control block diagram of the FPIDC

The structure of the FLC is shown in Fig. 6. From Fig. 6, there are two input variables such as the error (e) and its changing rate (Δe), and they are defined as following:

$$e(k) = e_{REF}(k) - e_{feedback}(k) \tag{9}$$

$$\Delta e(k) = e(k) - e(k-1) \tag{10}$$

where k and (k−1) are represented by the current and the previous sampling time, respectively.

3.2   Fuzzy Controller Design

The design of the proposed FPIDC algorithm is divided into three main part: fuzzification, fuzzy rules, and defuzzification. Fuzzification is the first step in the fuzzy inferencing process. Whereby, the crisp inputs are measured by sensors and input to the

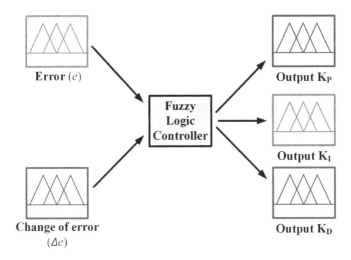

Fig. 6. Fuzzy logic controller structure

control system. Based on the membership function, all the crisp input values are transformed into the fuzzy inputs. For more accuracy and reduced computation, the asymmetrical triangular membership functions (MFs) are used. The Figs. 7(a) and (b) show the structure of MFs for a crisp input.

Similarly, the output MFs are presented in Figs. 7(c), (d) and (e). There are five MFs for input and output variables, and 25 fuzzy rules are obtained as shown in

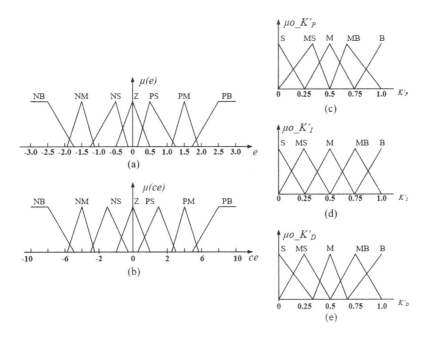

Fig. 7. Membership functions of the input and output variable.

Table 2. From the total control diagram shown in Fig. 8, there are three main parts: the proposed FPIDC, DC-DC converter, and DC load. As can be seen, the error between the feedback voltage and the reference voltage and its derivative are used for the FLC, which tunes the PID gains to control the DC-DC converter by adjusting the duty cycle.

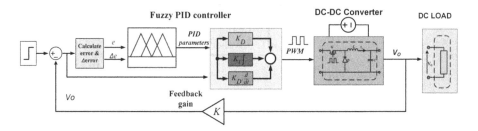

Fig. 8. Total block control diagram of the proposed controller

Table 2. Rule base for the proposed fuzzy PID controller

e/Δe	NL	NS	ZR	PS	PL
NL	NL	NL	NL	NL	NL
NS	NM	NM	NS	NS	PS
ZR	NM	NS	ZR	PS	PM
PS	NS	NS	PS	PM	PM
PL	PL	PL	PL	PL	PL

4   Simulation Results

In order to verify the effectiveness of the proposed FPIDC, some simulations have been carried out by using MATLAB/SIMULINK. The simulation parameters are given as follows: $V_{bus} = 300$ V, L = 1.5 mH, C = 100 uF, Vo = 180 V, $R_1 = 5$ Ω, $R_2 = 5$ Ω.

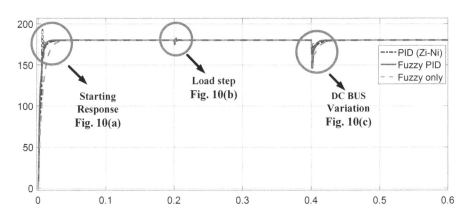

Fig. 9. The dynamic performance of the traditional PID controller, the FLC, and the proposed FPIDC for DC-DC converter with DC load.

The voltage controlled DC-DC converter is simulated based on three controllers: the traditional PID controller with Ziegler-Nichols method, the fuzzy logic controller, and the proposed control FPIDC. Normal DC load and DC motor are used as the output load of the DC-DC converter and the dynamic responses for all three controllers are shown in Fig. 9. The DC-DC converter starts to operate with the voltage reference 180 V when the DC load 1 is connected, as shown in Fig. 9. At t = 0.2 s, the Load 2 is connected, and DC bus voltage is suddenly decreased from 300 V to 250 V at t = 0.4 s. In order to investigate the dynamic performance in detail, Fig. 10 shows magnified waveforms arround the transient instants in Fig. 9. In Fig. 10(a), which shows the starting response, the PID controller designed by Ziegler-Nichols method

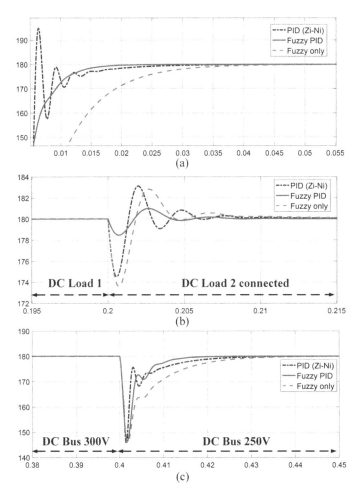

Fig. 10. (a) The starting response when DC Load 1 is connected. (b) System response when Load 2 is connected at 0.2 s. (c) System response when DC bus drop from 300 V to 250 V at 0.4 s.

has a very fast response at the first time, but it oscillates before the steady state because K_P is kept big from Ziegler-Nichols method for fast response with limitable oscillation. Even if the response of the FLC is smooth without overshoot, its transient response is slow compared to the proposed FPIDC since it is hard to describe perfectly the control constraints with only FLC which has a limit number of MFs and fuzzy rules.

At 0.2 s when DC load 2 is connected, the traditional PID controller and the FLC show larger oscillation and overshoot compared with the proposed FPIDC as shown in Fig. 10(b). In case of the input DC bus voltage drop, the proposed FPIDC shows the best recovery response as shown in Fig. 10(c). From Fig. 10, it is clear that the proposed FPIDC has fast response time with small oscillation compared to other methods.

Figure 11 shows the dynamic performance when DC motor is connected instead of normal DC load. The DC motor parameters are given as follows: Rated input voltage V_{motor} = 200 V, rated speed = 2000 rpm, back EMF constant K_e = 1.2 Vs/rad, viscous friction coefficient B_m = 0.01 Nms, torque constant K_t = 1.2 Nm/A. Armature resistance and inductance are 0.8 Ω and 0.016 H, respectively. In Fig. 11, DC motor speed changes from 1500 rpm to 2000 rpm at 1 s, and steps down from 2000 rpm to 1200 rpm at 2 s. From Fig. 11, the voltage applied to the armature is regulated to obtain the desired speed, and the proposed FPIDC shows superior performance compared to the traditional PID controller or the FLC with no overshoot and small transient time.

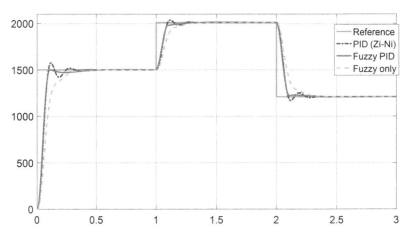

Fig. 11. Simulation with the DC motor

From the simulation, we can say that the proposed FPIDC is effective to control DC-DC converter and shows a good dynamic performance by adjusting PID gains adaptively.

5   Conclusion

This paper has presented an intelligent DC-DC converter control method based on the hybrid combination with the FLC and the PID controller. The proposed FPIDC can adjust its PID gains adaptively according to the load condition. Therefore, the performance of the DC-DC converter is improved in spite of the uncertainties and non-linear condition. Furthermore, the proposed FLC algorithm increases the control accuracy, and reduces the computation burden by using the optimized MFs and fuzzy rules with the PID transformation. Simulation results are carried out to evaluate the effectiveness of the proposed FPIDC, and it is verified that the performance of the proposed FPIDC is much better than the traditional PID controller and the FLC under the output load change and the input DC bus voltage variation.

Acknowledgment. This work was partly supported by the National Research Foundation of Korea Grant funded by the Korean Government (NRF-2015R1D1A1A09058166) and the Network-based Automation Research Center (NARC) funded by the Ministry of Trade, Industry & Energy.

References

1. Walker, G.R., Sernia, P.C.: Cascaded DC-DC converter connection of photovoltaic modules. IEEE Trans. Power Electron. **19**(4), 1130–1139 (2004)
2. Åström, K.J., Hägglund, T.: The future of PID control. Control Eng. Pract. **9**(11), 1163–1175 (2001)
3. Åström, K.J., Hägglund, T.: Revisiting the Ziegler-Nichols step response method for PID control. J. Process Control **14**(6), 635–650 (2004)
4. Datta, A., Ho, M., Bhattacharyya, S.P.: Structure and Synthesis of PID Controllers. Springer Science & Business Media, Heidelberg (2013)
5. Buckley, J.J.: Fuzzy controller: Further limit theorems for linear control rules. Fuzzy Sets Syst. **36**(2), 225–233 (1990)
6. Li, H.X., Gatland, H.B.: Enhanced methods of fuzzy logic control. In: Proceedings of 1995 IEEE International Conference on Fuzzy Systems, vol. 1, pp. 331–336 (1995)
7. Yuan, Y., Chang, C., Zhou, Z., Huang, X., Xu, Y.: Design of a single-input fuzzy PID controller based on genetic optimization scheme for DC-DC buck converter. In: 2015 International Symposium on Next-Generation Electronics (ISNE), pp. 1–4 (2015)
8. Dounis, A.I., Kofinas, P., Alafodimos, C., Tseles, D.: Adaptive fuzzy gain scheduling PID controller for maximum power point tracking of photovoltaic system. Renew. Energy **60**, 202–214 (2013)
9. Tran, Q.-H., Lee, H.-H.: A fuzzy logic controller for indirect matrix converter under abnormal input voltage conditions. In: Huang, D.-S., Han, K. (eds.) ICIC 2015. LNCS, vol. 9227, pp. 139–150. Springer, Cham (2015). doi:10.1007/978-3-319-22053-6_16

Fuzzy Uncertainty in Random Variable Generation: An α-Cut Approach

Christian Alfredo Varón-Gaviria, José Luis Barbosa-Fontecha, and Juan Carlos Figueroa-García$^{(\boxtimes)}$

Universidad Distrital Francisco José de Caldas, Bogotá, Colombia
chrisvg7593@outlook.es, jolubafo@yahoo.es,
jcfigueroag@udistrital.edu.co

Abstract. This paper presents a method for random variable generation based on α-cuts. The proposed method uses convex fuzzy numbers with single-element core and uniformly distributed random numbers to obtain random variables, mainly used in simulation models.

Keywords: Random variable generation · Convex fuzzy numbers

1 Introduction and Motivation

Fuzzy sets are mathematical tools to deal with uncertainty coming from human-like perceptions regarding words or concepts. While its application to fuzzy functions, optimization, differential equations etc. is wide, its application to simulation systems needs more definitions and methods. Many simulation problems lacks of statistical information to use classical simulation techniques, and sometimes the available information comes from experts, so fuzzy sets helps to represent their perceptions.

We focus on an efficient method to generate random variables regarding fuzzy sets where randomness is represented by a uniform random generator and fuzziness is defined by the shape of a fuzzy set. Its potential use in simulation of linguistic variables, human-like perceptions, or problems with no statistical information, is wide.

The paper is organized as follows: Sect. 1 is an Introductory section. Section 2 presents some basics about fuzzy sets; in Sect. 3, some concepts about random variable generation are provided. Section 4 presents an application example, and Sect. 5 presents the concluding remarks of the study.

2 Basics in Fuzzy Sets

Firstly, we establish basic notations. $\mathcal{P}(X)$ is the class of all crisp sets, and $\mathcal{F}(X)$ is the class of all fuzzy sets. A fuzzy set A is defined on an universe of discourse X and is

---

C.A. Varón-Gaviria and J.L. Barbosa-Fontecha are undergraduate students at the Universidad Distrital Francisco José de Caldas, Bogotá – Colombia.
J.C. Figueroa-García is Assistant Professor of the Universidad Distrital Francisco José de Caldas, Bogotá – Colombia.

© Springer International Publishing AG 2017
D.-S. Huang et al. (Eds.): ICIC 2017, Part III, LNAI 10363, pp. 264–273, 2017.
DOI: 10.1007/978-3-319-63315-2_23

characterized by a membership function $\mu_A(x)$ that takes values in the interval $[0,1]$, $A : X \to [0, 1]$. A fuzzy set A can be represented as a set of ordered pairs of an element x and its membership degree, $\mu_A(x)$, i.e.,

$$A = \{(x, \mu_A(x)) \,|\, x \in X\} \tag{1}$$

where $\mathcal{F}(\mathbb{R})$ is the class of all fuzzy sets.

The *core* of a fuzzy set $core(A)$ is the set of the elements with maximum membership, $core(A) = \{x \,|\, \mu_A(x) = 1\} \,\forall\, x \in X$. The α-cut of $\mu_A(x)$ namely $^\alpha A$ is the set of values with a membership degree equal or greatest than α, this means:

$$^\alpha A = \{x \,|\, \mu_A(x) \geq \alpha\} \,\forall\, x \in X, \tag{2}$$

$$^\alpha A \in \left[\inf_x {}^\alpha\mu_A(x), \; \sup_x {}^\alpha\mu_A(x) \right] = \left[A_\alpha^l, A_\alpha^r \right]. \tag{3}$$

A fuzzy number is then a convex fuzzy set defined as follows:

Definition 1 [Fuzzy Number]. *Let $A \in \mathcal{G}(\mathbb{R})$ where $\mathcal{G}(\mathbb{R}) \in \mathcal{F}(\mathbb{R})$ is the class of all normal, upper semicontinuous, and fuzzy convex sets. Then, A is a Fuzzy Number (FN) iff there exists a closed interval $[a, b] \neq 0$ such that*

$$\mu_A(x) = \begin{cases} 1 & for \quad x \in [a, b], \\ l(x) & for \quad x \in [-\infty, a], \\ r(x) & for \quad x \in [b, \infty], \end{cases} \tag{4}$$

where $l : (-\infty, a) \to [0, 1]$ is monotonic non-decreasing, continuous from the right, and $l(x) = 0$ for $x < \omega_1$, and $r : (b, \infty) \to [0, 1]$ is monotonic non-increasing, continuous from the left, and $r(x) = 0$ for $x > \omega_2$.

Figure 1 shows a triangular fuzzy set/number. Its universe of discourse is the set $x \in \mathbb{R}$, the *support of A, supp* (A) is the interval $x \in A_\alpha^l, A_\alpha^r$ and μ_A is a triangular function with parameters $A_\alpha^l, \bar{A}$ and A_α^r. α is the membership degree that an specific value x has regarding A and the dashed region is an α-*cut* done over A.

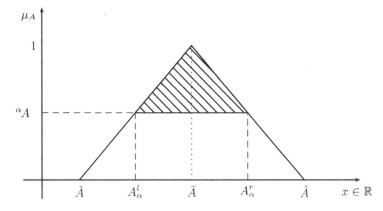

Fig. 1. Fuzzy set A

Note that any α-cut over a fuzzy number is monotonically increasing/decreasing, so for $\alpha_1 < \alpha_2, \alpha \in [0,1]$ then $^{\alpha_2}A \subseteq {}^{\alpha_1}A$ and $^{\alpha}A \subseteq supp\ (A)$, $\forall \alpha \in [0,1]$.

3 Fuzzy Uncertainty in Random Variable Generation

Some basic ideas about the use of fuzzy sets in simulation models come from control and dynamical systems analysis. Fishwick [1] proposed the use of fuzzy numbers in representation and simulation of qualitative models, and Hüllermeier [2] who applied fuzzy sets in simulation of dynamical systems. Other applications of fuzzy-based simulation (Monte Carlo methods, process simulation and fuzzy variable generation) were proposed by Suresh et al. [3], Zonouz and Miremadi [4], and Huang et al. [5].

A probability space is a triplet $(\Omega, \mathcal{D}, P)$ where Ω is the sample space, $\mathcal{D}$ is a set of probable events (a.k.a σ-algebra), and P is a probability measure. A random variable $X : \Omega \rightarrow \mathcal{D}$ is a measurable function $X(\omega)$ from the sample space to a σ-algebra, so if $\mathcal{D}$ is a topological space then it is called a Borel σ-algebra.

Random variable generation in probability spaces uses a random variable $Y \sim U[0,1]$ to obtain the inverse image of its cumulative probability function $F(X(\omega))$, this is $F^{-1}(X)$. Let $X \sim F(x)$ be the cdf $F(x)$ of X a monotonic increasing function, then if $Y = F(X);\ 0 \le y \le 1$, and assuming that F is invertible (in closed form or at least continuous over a Borel measurable interval) we have that $F(y) = P\{Y \le y\} = P\{F(X) \le y\}$, so $P\{X \le F^{-1}(y)\} = F(F^{-1}(Y)) = y$. This implies that $Y \sim U[0,1]$ since it is the image of X over P which leads to $F^{-1} : Y \rightarrow X$.

3.1   Fuzzy Random Variable Generation Using α-Cuts

Fuzzy sets come from human like information and help to represent words and concepts via a membership function. Basically, a fuzzy set can be also defined over similar spaces than probabilities, so we can also talk about the idea of a fuzzy space to then define a method for generating fuzzy random variables.

A fuzzy space is a triplet $(\Omega, \mathcal{D}, \mu_A)$ where Ω is the universe of discourse, $\mathcal{D}$ is a set of possible events (a.k.a σ-algebra), μ_A is a membership function, and A is label/concept/word. A fuzzy random variable $X : \Omega \rightarrow \mathcal{D}$ is a measurable function $X(\omega)$ from the sample space to a σ-algebra, so if $\mathcal{D}$ is a topological space then it is called a Borel σ-algebra. This way, what we propose is a method for generating random variables based on a membership function μ_A.

In this paper we consider a fuzzy number μ_A (see Definition 5) only the subclass $\mathcal{G}_1(\mathbb{R})$ of fuzzy sets with a singleton core, this is:

$$\mu_A(x) = \begin{cases} 1 & for \quad x = c, \\ l(x) & for \quad x \in [-\infty, c], \\ r(x) & for \quad x \in [c, \infty], \end{cases} \tag{5}$$

where $l : (-\infty, c) \rightarrow [0,1]$ is monotonic non-decreasing, continuous from the right, and $l(x) = 0$ for $x < \omega_1$, and $r : (c, \infty) \rightarrow [0,1]$ is monotonic non-increasing, continuous from the left, and $r(x) = 0$ for $x > \omega_2$.

Definition 2. *Let $\mu_A \in \mathcal{G}_1(\mathbb{R})$ be a normal, upper semicontinuous, single element core, and fuzzy convex set, then some of their properties include:*

- *Normalization:* $\max_x \{\mu_A(x)\} = 1$,
- *Convexity:* ${}^\alpha A = [A_\alpha^l, A_\alpha^r] \forall \alpha \in (0, 1]$,
- *Representation:* $A = \bigcup\limits_{\alpha \in [0,1]} \alpha \cdot {}^\alpha A$.

Now, we first define some concepts about the area of a fuzzy number.

Definition 3. *Let $\mu_A \in \mathcal{G}_1(\mathbb{R})$ be a fuzzy number, then its area Λ is as follows:*

$$\Lambda = \Lambda_1 + \Lambda_2 = \int_{x \in \mathbb{R}} l(x)dx + \int_{x \in \mathbb{R}} r(x)dx \tag{6}$$

Definition 4. *Let Λ_1, Λ_2 be the partial areas of $\mu_A \in \mathcal{G}_1(\mathbb{R})$. Then the normalized areas λ_1, λ_2 of $\mu_A \in \mathcal{G}_1(\mathbb{R})$ are defined as follows:*

$$\lambda_1 = \frac{\Lambda_1}{\Lambda}, \tag{7}$$

$$\lambda_2 = \frac{\Lambda_2}{\Lambda}, \tag{8}$$

$$\lambda_1 + \lambda_2 = 1. \tag{9}$$

Definition 5. *Let $\mu_A \in \mathcal{G}_1(\mathbb{R})$ be symmetrical, then the following properties hold:*

$$\lambda_1 = \lambda_2 = 0.5\lambda, \tag{10}$$

$$\lambda_1 = \lambda_2 = 0.5, \tag{11}$$

$$|A_\alpha^l - c| = |A_\alpha^r - c| \rightarrow A_\alpha^r = A_\alpha^l + 2(c - A_\alpha^l), \tag{12}$$

where c is the central value of μ_A.

The class $\mathcal{G}_1(\mathbb{R})$ of fuzzy numbers includes gaussian, triangular, exponential, etc. which are popular in decision making and simulation, and we can take advantage from its properties to efficiently generate fuzzy random variables, just like in the probablistic approach.

The goal of random variable generation is to use either $f^{-1}(x)$ or $F^{-1}(x)$ to return $X(\omega)$ using a random number $U[0,1]$ (see Devroye [6], and Law and Kelton [7]). As simple the method as easy-to-implement, so we use μ_A^{-1} and U to obtain $X(\omega)$.

Our proposal is simple: given the shape of μ_A, compute its partial areas λ_1, λ_2 then compute two uniform random numbers U_1, U_2, then use U_1 to compute ${}^\alpha A$ and U_2 to select which of the values A_α^l or A_α^r should be selected as random variable $X(\omega)$.

Thus, $Y = \alpha$ where $0 \le y \le 1$ where $F(Y) = {}^{\alpha}\mu_A(X)$ is invertible in closed form (or at least continuous over a Borel measurable interval), we can see that any α-cut $F(Y) = [A_\alpha^l, A_\alpha^r] = \{X \in F(Y)\}$, so $\{\alpha = F^{-1}(y)\} = F(F^{-1}(Y)) = y$. The proposal is shown in Algorithm 1.

Algorithm 1 α-cut based method

Require: $\mu_A \in \mathcal{G}_1(\mathbb{R})$ (see Eq. (5))
    Compute λ_1 and λ_2 using Definitions 3 and 4
    Compute ${}^{\alpha}A = [\check{A}_\alpha, \hat{A}_\alpha]$ using Eq. (3)
    Compute $U_1[0, 1]$ and $U_2[0, 1]$
    Set $\alpha = U_1[0, 1]$
    If $U_2 \le \lambda_1$ then $x = \check{A}_\alpha$, otherwise set $x = \hat{A}_\alpha$
    return α, x as the realization of $X(\omega)$ with membership α

---

4   Application Examples

Gaussian Symmetrical Fuzzy Set: As shown in Definition 5, symmetrical fuzzy sets lead to $\Lambda_1 = \Lambda_2$ so $\lambda_1 = \lambda_2 = 0.5$. Now, the exact equations to generate Gaussian fuzzy random variables (see Fig. 2) are shown next:

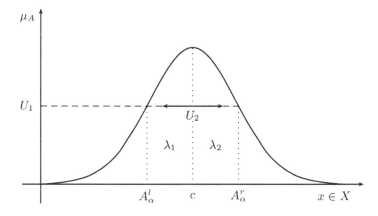

Fig. 2. Gaussian fuzzy set A

$$\mu_A(x) = e^{-\frac{1}{2}\left(\frac{x-c}{\delta}\right)^2} \ \forall\, x \in (-\infty, \infty),$$
$$A_\alpha^l = c - \sqrt{-2 \cdot \ln(\alpha)} \cdot \delta,$$
$$A_\alpha^r = c + \sqrt{-2 \cdot \ln(\alpha)} \cdot \delta.$$

Finally, the generation procedure for Gaussian fuzzy number given U_1, U_2, is:

$$X(\omega) = \begin{cases} c - \sqrt{-2 \cdot \ln(U_1)} \cdot \delta, & for \quad U_2 \leq 0.5, \\ c + \sqrt{-2 \cdot \ln(U_1)} \cdot \delta, & for \quad U_2 > 0.5. \end{cases} \tag{13}$$

Table 1. Simulated Gaussian variables x

i	U_1	U_2	x
1	0.086	0.985	8.323
2	0.145	0.554	7.948
3	0.255	0.747	7.480
4	0.828	0.996	5.922
5	0.712	0.207	3.764
6	0.275	0.229	2.590
7	0.052	0.767	8.648
8	0.157	0.586	7.886
9	0.721	0.262	3.786
10	0.130	0.667	8.028

Table 1 shows 10 simulated variables for $c = 5$ and $\delta = 1.5$.

For instance, x for $i = 3$ uses $U_{1,3} = \alpha = 0.255$, so ${}^{\alpha}A = [A_{\alpha}^l, A_{\alpha}^r] \rightarrow [2.520, 7.480]$, and as $U_{2,3} = 0.747 > 0.5$ then we set $A_{\alpha}^r = X(\omega) = 7.480$ as a random variable.

Non Symmetrical Triangular Example: A non symmetrical triangular random variable implies that $\Lambda_1 \neq \Lambda_2$ and consequently $\lambda_1 \neq \lambda_2$ (see Fig. 3). The equations to generate non symmetrical triangular fuzzy random variables are:

$$\mu_A(x) = \max(\min(\frac{x-a}{c-a}, \frac{b-x}{b-c}), 0) \ \forall \ x \in [a, b],$$

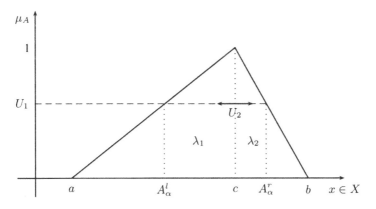

Fig. 3. Non symmetrical triangular fuzzy set A

$$A_\alpha^l = \alpha(c - a) + a,$$
$$A_\alpha^r = b - \alpha(b - c).$$

Its partial areas are computed as follows:

$$\Lambda = 0.5(b - a),$$
$$\Lambda_1 = \frac{c - a}{2},$$
$$\Lambda_2 = \frac{b - c}{2}.$$

Its normalized areas are:

$$\lambda_1 = \frac{c - a}{b - a},$$
$$\lambda_2 = \frac{b - c}{b - a},$$
$$\lambda_1 + \lambda_2 = 1.$$

Finally, the generation procedure for Gaussian fuzzy numbers given U_1, U_2, is:

$$X(\omega) = \begin{cases} U_1(c - a) + a, & \text{for} \quad U_2 \leq \frac{c-a}{b-a}, \\ b - U_1(b - c), & \text{for} \quad U_2 > \frac{c-a}{b-a}. \end{cases} \tag{14}$$

Table 2 shows 10 variables for $a = 2, c = 7, b = 9$. For instance, x for $i = 5$ uses $U_{1,5} = \alpha = 0.666$, so $\lambda_1 = 0.714, \lambda_2 = 0.286$, $^\alpha A = [A_\alpha^l, A_\alpha^r] \rightarrow [5.330, 7.668]$, and $U_{2,5} = 0.367 < 0.714$ then we set $A_\alpha^l = X(\omega) = 5.330$ as a random variable.

Table 2. Simulated triangular variables x

i	U_1	U_2	x
1	0.754	0.311	5.770
2	0.746	0.195	5.730
3	0.323	0.202	3.615
4	0.128	0.140	2.640
5	0.666	0.367	5.330
6	0.736	0.985	7.528
7	0.732	0.012	5.660
8	0.169	0.294	2.845
9	0.310	0.833	8.380
10	0.015	0.197	2.073

An Inventory Policy Example: Consider the well known single product inventory problem where the idea is to minimize the Total Inventory Cost (T_c):

$$T_c = d \cdot P_c + O_c \frac{d}{q} + I_c \frac{q}{2}$$

where $d \in \mathbb{R}$ is an estimate of the demand of the product, $q \in \mathbb{R}$ is the quantity to order of the product, $P_c \in \mathbb{R}$ is the unitary production cost, $O_c \in \mathbb{R}$ is the ordering cost, and $I_c \in \mathbb{R}$ is the unitary inventory cost.

In this problem it is assumed that d, P_c, O_c and I_c are deterministic parameters while q is a decision variable over we have to find a minimal inventory cost T_c (total cost). The optimal solution to this deterministic problem is:

$$q^* = \sqrt{\frac{2 \cdot O_c \cdot d}{I_c}} \tag{15}$$

If we do not have statistical data to estimate d, P_c, O_c and I_c, then the problem becomes a non-probabilistic problem where d, P_c, O_c and I_c can be estimated using experts opinions. Such information can be properly modeled using fuzzy sets.

Let us denote $\{D, O, I\} \in \mathcal{F}(\mathbb{R})$ as fuzzy estimations of demands, ordering and inventory costs of the problem. The solution of q^* can be seen as the fuzzy function based on the extension principle for fuzzy sets (see Bellman and Zadeh [8]):

$$Q(q^*) = \sup_{q^* = f(d, P_c, O_c)} \min\{D(d), O(O_c), I(I_c)\} \tag{16}$$

where Q is the set of optimal solutions q^* (defined in (15)).

We have simulated 5000 values of $Q(q^*)$ using $D = T(30, 55, 72), O = G(20, 2.5), I = G(5, 1.5)$, and Eqs. (13), (14). Aside from the optimal quantities q^*, we have computed the membership degree of every simulated q^* using Eq. (16). The obtained results are shown in Fig. 4.

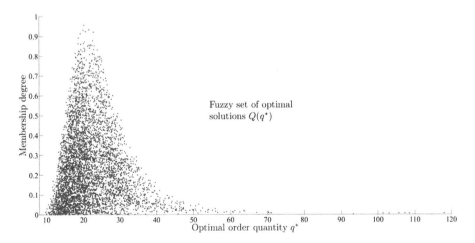

Fig. 4. Simulated set of optimal solutions

Note that the resultant set $Q(q^*)$ is also a fuzzy set obtained from fuzzy random variables. Some extreme values come from values of $D(d)$, $O(O_c)$, $I(I_c)$ near zero membership, so simulation helps to identify some extreme combination of parameters in order to see the system behavior in advance.

5 Concluding Remarks

We have proposed a simple method for generating fuzzy random variables using α-cuts. Using two uniform random numbers, we can compute the α-cut of any fuzzy number with single element core and then obtain $X(\omega)$. Our proposal can be used to generate random variables coming from some of the most used fuzzy numbers such as triangular, exponential, Gaussian, quadratic, etc.

The presented method is intended to be computationally efficient, so it provides closed form equations for generating fuzzy random variables. Its applicability in simulation models and decision making problems where there is no any available statistical information is wide.

Finally, the presented method works for some of the most important fuzzy numbers, and it opens the door to use experts opinions and perceptions into simulation models via fuzzy sets/numbers. It could be specially useful in cases where human-like information provides a base for decision making.

Further Topics

Random variable generation based on other kinds of fuzzy sets (non singleton core fuzzy sets), and representation of a variable using multi-label fuzzy sets are challenges to be covered in simulation models. Fields such as Markov stochastic processes (see Figueroa-García [9], and Figueroa-García et al. [10, 11]) or Type-2 fuzzy sets (see Figueroa-García et al. [12–14]) are potential applications of fuzzy simulation.

References

1. Fishwick, P.A.: Fuzzy simulation: Specifying and identifying qualitative models. Int. J. Gen Syst **19**(3), 295–316 (1991)
2. Hüllermeier, E.: A fuzzy simulation method. In: International Symposium on Soft Computing. (1996)
3. Suresh, P., Babar, A., Venkat-Raj, V.: Uncertainty in fault tree analysis: a fuzzy approach. Fuzzy Sets Syst. **83**, 135–141 (1996)
4. Zonouz, S.A., Miremadi, S.G.: A fuzzy-Monte Carlo simulation approach for fault tree analysis. In: RAMS 2006 Annual Reliability and Maintainability Symposium, pp. 428–433 (2006)
5. Huang, S.Y., Chen, X., Li, Y.P., Liu, T.: A fuzzy based simulation method for modelling hydrological processes under uncertainty. Hydrol. Process. **24**(25), 3718–3732 (2010)
6. Devroye, L.: Non-uniform Random Variate Generation. Springer-Verlag New York Inc., New York (1986)
7. Law, A., Kelton, D.: Simulation Modeling and Analysis. Mc Graw Hill, New York (2000)

8. Bellman, R.E., Zadeh, L.A.: Decision-making in a fuzzy environment. Manag. Sci. **17**(1), 141–164 (1970)
9. Figueroa-García, J.C.: Interval type-2 fuzzy Markov chains. In: Sadeghian, A., Mendel, J. M., Tahayori, H. (eds.) Advances in Type-2 Fuzzy Sets and Systems, vol. 301, pp. 49–64. Springer, Heidelberg (2013). doi:10.1007/978-1-4614-6666-6_4
10. Figueroa-García, J.C., Kalenatic, D., Lopéz, C.A.: A simulation study on fuzzy Markov chains. Commun. Comput. Inf. Sci. **15**(1), 109–117 (2008)
11. Kalenatic, D., Figueroa-García, Juan C., Lopez, C.A.: Scalarization of type-1 fuzzy Markov chains. In: Huang, D.-S., Zhao, Z., Bevilacqua, V., Figueroa, J.C. (eds.) ICIC 2010. LNCS, vol. 6215, pp. 110–117. Springer, Heidelberg (2010). doi:10.1007/978-3-642-14922-1_15
12. Figueroa-García, J.C., Chalco-Cano, Y., Román-Flores, H.: Distance measures for interval type-2 fuzzy numbers. Discrete Appl. Math. **197**(1), 93–102 (2015)
13. Figueroa-García, J.C., Hernández-Pérez, G.J.: On the computation of the distance between interval type-2 fuzzy numbers using α-cuts. In: 2014 Annual Meeting of the North American Fuzzy Information Processing Society (NAFIPS), pp. 1–6. IEEE (2014)
14. Figueroa-García, J.C., Hernández-Pérez, G.J., Chalco-Cano, Y.: On computing the footprint of uncertainty of an interval type-2 fuzzy set as uncertainty measure. Commun. Comput. Inf. Sci. **657**(1), 247–257 (2016)

Path Planning of Bionic Robotic Fish Based on BK Products of Fuzzy Relation

Yuntian Shi$^{(\boxtimes)}$ and Wei Pan

Fujian Key Lab of the Brain-Like Intelligent Systems, Xiamen University,
Xiamen, FJ 361005, China
sytshanli@163.com

Abstract. In this paper, a heuristic search technology is presented for the path planning of a bionic robotic fish equipped with single beam sonar. When the bionic robotic fish is navigating, the scanning range of the sonar is divided into 5 parts, and each sub-part is a candidate course, while the middle part as the current course. With the fuzzy relation between sub-parts of sonar and real time environment attributes as the core concept, the triangle sub-product relationship put forward by Bandler and Kohout reveals the relationship between the various parts of the sonar, and a sub-part of sonar is selected as the inheritance course of bionic fish. The simulation scene and the pool scene are set to validate the effect of the planning strategy, as a result of which it can conclude that the planning strategy based on BK triangle sub-product can effectively realize path planning of bionic robotic fish.

Keywords: BK triangle sub-product · Sonar · Bionic robotic fish · Path planning

1 Introduction

In recent years, autonomous underwater vehicles (AUV) are becoming the subject of the study of marine robotics, due to their commercial and military potential. In order to make the AUV complete the task of underwater exploration and underwater operation, it is of great significance to design an AUV with autonomous obstacle avoidance. The typical methods of obstacle avoidance include the method artificial potential field, fuzzy design method and neural network and so on. In 1994, Khatib O proposed the method of artificial potential field [1], using the target of AUV navigation process as attractive force, and the obstacles may encountered during the journey as a repulsive force, to select the course according to the force of the resultant force of them, but there may be a point of zero force. In terms of zero force, Johann B and Yoram K proposed the improved method [2]. DeMuth G, Springsteen S introduced the speed concept, and proposed a dynamic obstacle algorithm, using artificial intelligence method to train the obstacle graphics, and achieved dynamic path planning [3]. In addition, the product method based on fuzzy relation proposed by Bandler and Kohout is gradually used in autonomous heuristic obstacle avoidance of AUV [4, 5].

When AUV navigates, it must effectively avoid obstacles to protect the safety of fish. It can detect obstacles in the navigation path by being equipped with single beam

© Springer International Publishing AG 2017
D.-S. Huang et al. (Eds.): ICIC 2017, Part III, LNAI 10363, pp. 274–285, 2017.
DOI: 10.1007/978-3-319-63315-2_24

sonar, and the sonar scanning range is divided into several sub-part to represent the continuous candidate courses of AUV, while the center part is the current course of AUV, and AUV must choose the best from the candidate courses to avoid obstacles. Because the prior knowledge and information of the water environment are usually incomplete, uncertain and approximate, so the fuzzy logic control can be used to control the course [6]. In this paper, the best inheritance course of AUV is obtained by analyzing the basic inference rules and membership functions, and using the BK product method [7, 8].

The paper designed the bionic fish based on the National Natural Science Foundation of China (No. 60975084), used single beam sonar to collect the environmental information, and applied a heuristic search technique for the path planning strategy of bionic robotic fish. The first part of this article introduces the platform of bionic robotic fish and the structure of fish body; the second part designs obstacle avoidance algorithm and introduces the principle of the algorithm; the third part tests various simulation scenes, as well as the actual pool scene; and the final part is the conclusion.

2 Platform of Bionic Robotic Fish

2.1 Architecture of Bionic Robotic Fish

The bionic robotic fish with the shape design of imitating carp, and the length of 96 cm. There are equipped with two servo motors, one of which drives the tail fin, to control the swing amplitude and frequency of the handle, while the other motor controls the pectoral fins of bionic robotic fish, to achieve functions of moving up and down by swing in different angles. Its body part includes a core control module and an information collection module, while the former includes embedded Odroid, attitude sensor, compass, power management and pressure sensor, and the latter contains GPS, Wifi, camera, single beam sonar, the specific structure of which is shown in Fig. 1

Fig. 1. Appearance of bionic robotic fish

2.2 Single Beam Sonar

In this paper, a single beam sonar is used to obtain the information of the water environment, and the acoustic information is used to know whether there are obstacles on the navigation path. The single beam sonar is shown in Fig. 2, and when the sonar acquires the environmental information, the obvious color is shown in Fig. 2(c), indicating the obstruction information. The main mechanical parameters of single beam sonar are

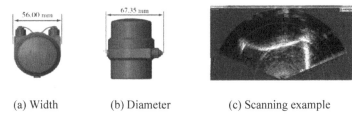

(a) Width   (b) Diameter   (c) Scanning example

Fig. 2. Appearance and scanning process of sonar

Table 1. Main mechanical parameters of sonar

Parameter name	Parameter values
Maximum peripheral dimension	56 mm * 68 mm * 79 mm
Weight in air	324 g
Maximum scanning angle	360°
Weight in water	180 g
Maximum scanning distance	75 m
Minimum resolution	7.5 mm

shown in Table 1, from which it can be seen that the sonar is small in size, light in weight and high in resolution, and can be used to adjust the scanning distance and scanning angle freely, meeting the requirements of the current aquatic environment.

3   Planning Strategy Based on BK Triangle Sub-product

3.1   Fuzzy Relation Proposed by Bandler and Kohout

Assume that the binary relation R is a relation matrix from the set X to the set Y, the binary relation S is a relation matrix from the set Y to the set Z, then the $R@S$ can represent the binary relation from the set X to the set Z by using a rule. This rule is BK-products [9, 10] proposed by Bandler and Kohout. In particular, it has the square product, the triangle super-product and the triangle sub-product [10], which are defined in detail as shown in Table 2.

Table 2. Definition of BK product

Product type	Symbol	Multivalued logic expression
Square product	$\square$	$(R\square S)_{ik} = \frac{1}{n}\sum_{j=1}^{n}(R_{ij} \leftrightarrow S_{jk})$
Triangle super-product	$\triangleright$	$(R \triangleright S)_{ik} = \frac{1}{n}\sum_{j=1}^{n}(R_{ij} \leftarrow S_{jk})$
Triangle sub-product	$\triangleleft$	$(R \triangleleft S)_{ik} = \frac{1}{n}\sum_{j=1}^{n}(R_{ij} \rightarrow S_{jk})$

In this paper, the multiple valued logic operator "→" existing in Table 2, uses Lukasiewicz fuzzy implication operation [11], $a{\rightarrow}b = \min(1,1 - a + b)$. If X is an object set, and Y is an attribute set, then the element R_{xy} of the relation set R represents the degree that the object x contains the attribute y; Conversely, the element R_{yx} of R^T that transposed from R represents the degree that the attribute y acts on the object x, and then the relation $(R@R^T)_{ij}$(@ represents $\sqsubset$, $\triangleright$, $\triangleleft$) of BK-products can represent the relation degree of between object x_i and object x_j on the same attribute.

3.2 Obstacle Avoidance

It is necessary to avoid obstacles in the path to ensure the safety of fish body, in the process from the beginning point to the target point in the water environment. For a specific navigation process, assume that the fish navigates from the starting point S to the target point T, detect the environmental obstacle information by the single beam sonar, and once the obstacle is detected, the bionic fish needs to planning the path again. The study assumes that sonar scanning range is divided into m parts, as shown in Fig. 3, each of which is as the candidate course of the bionic fish, and they constitute the candidate set $X=\{x_1,x_2,\ldots,x_m\}$, in which x_k represents the current course of the bionic robotic fish, $k = \lfloor(m + 1)/2\rfloor$.When an obstacle is detected by the sonar, it is possible to identify the sub-part of the sonar with obstacles, and the bionic fish must adjust the course to avoid the obstacle. Therefore, this paper proposes a heuristic strategy, which uses the fuzzy relation of BK triangle sub-product to avoid obstacle.

There are some factors that influence the navigation of the bionic robotic fish in the real water, such as the safety degree, the deviation distance between the current course and the target course, and the influence of the water flow. These attributes can be a set $Y=\{y_1,y_2,\ldots,y_n\}$, where n represents the number of attributes, each of which can act on any part of the sonar. According to the candidate set X and the attribute set Y, the fuzzy membership function is used to obtain the influence degree of the attribute y_i on the sonar sub part x_i, so as to obtain the relation matrix R (formula 1). In the decision making process of obstacle avoidance, it uses fuzzy relation and BK triangle sub-product, to calculate the relation matrix R and its transpose R^T to obtain the fuzzy relation matrix T (formula 2). The element t_{ij} of matrix T represents the association degree between the element x_i and x_j of the candidate set on the attribute set Y. In order

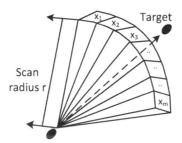

Fig. 3. Sonar blocks

to clarify the fuzzy relation matrix, it uses alpha-cut [12], $\alpha \in (0,1)$, if $R_{ij} \geq \alpha$, then $R_{ij} = 1$; otherwise, $R_{ij} = 0$, to simplify the fuzzy relation matrix T (formula 3).

$$R = X \times Y = \begin{Bmatrix} r_{11} & r_{12} & \cdots & r_{1n} \\ r_{21} & r_{22} & \cdots & r_{2n} \\ \cdots & \cdots & \cdots & \cdots \\ y_1 & y_2 & \cdots & y_n \\ r_{m1} & r_{m2} & \cdots & r_{mn} \end{Bmatrix} \begin{matrix} x_1 \\ x_2 \\ \cdots \\ x_m \end{matrix} \tag{1}$$

$$T = R \triangleleft R^T = \begin{Bmatrix} t_{11} & t_{12} & \cdots & t_{1m} \\ t_{21} & t_{22} & \cdots & t_{2m} \\ \cdots & \cdots & \cdots & \cdots \\ x_1 & x_2 & \cdots & x_n \\ t_{m1} & t_{m2} & \cdots & t_{mm} \end{Bmatrix} \begin{matrix} x_1 \\ x_2 \\ \cdots \\ x_m \end{matrix} \tag{2}$$

$$R_\alpha = \alpha\_cut(T,\alpha) = \begin{Bmatrix} a_{11} & a_{12} & \cdots & a_{1m} \\ a_{21} & a_{22} & \cdots & a_{2m} \\ \cdots & \cdots & \cdots & \cdots \\ x_1 & x_2 & \cdots & x_m \\ a_{m1} & a_{m2} & \cdots & a_{mm} \end{Bmatrix} \begin{matrix} x_1 \\ x_2 \\ \cdots \\ x_m \end{matrix} \tag{3}$$

Finally, the relationship among the elements is obtained by using Hasse diagram. Hasse diagram describes the partial order relation among the elements in the candidate set X, that is to directly obtain the advantages and disadvantages of each candidate course in the attribute set Y by Hasse diagram. Therefore, the fish should choose the top node of Hasse diagram as the inheritance course when navigating.

3.3   Concrete Examples

The scanning distance of the single beam sonar is 5 m, and the scanning range is about 60°, which is divided into 5 parts to form a candidate course set of $X = \{x_1,x_2,x_3,x_4,x_5\}$, using x_3 as the current course of bionic robotic fish. Because of the complex information of the real environment, the environmental information obtained by sonar is often incomplete and uncertain, so it uses fuzzy strategy when navigating. The safety degree and the offset degree are used as fuzzy control quantity to constitute the attribute set of $Y = \{y_1,y_2\}$. The safety degree means that the sub section xi of the sonar can be the next adjustment point of the course, when detecting the obstacle, and the offset degree indicates the deviation of $x_i (i = 1,2,3,4,5)$ from the target course. In this paper, the safety level is divided into three levels: dangerous, safe, very safe; the offset is divided into three levels: close, far, very far. Their membership functions are given by formulas 4 and 5, and the curves are shown in Fig. 4.

$$U_{\text{safety}}(x) = 1 - \frac{1}{1 + e^{5-x}} \tag{4}$$

$$U_{\text{offset}}(y) = 1 - \frac{1}{1 + e^{y-5}} \tag{5}$$

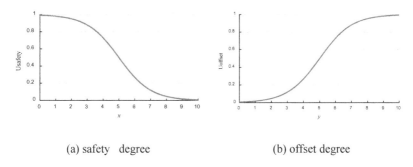

(a) safety degree   (b) offset degree

Fig. 4. Membership function curve

The fuzzy rule base can be set according to their membership functions of the safety degree and the offset degree, as shown in Tables 3 and 4.

Table 3. Fuzzy rule base of safety degree attributes

Degree	Description	x
Dangerous	If there are obstacles at x_i	5
Safe	If there are obstacles at one side of x_i	3
Very safe	If there is no obstacle on both sides of x_i	1

Table 4. Fuzzy rule base of deviation degree attributes

Degree	Description	y
Dangerous	If the distance between x_i and the target course is close	5
Safe	If the distance between x_i and the target course is far	3
Very safe	If the distance between x_i and the target course is very far	1

Specific case: there are obstacles at x_1, x_2, x_3 by scanning in the process of bionic robotic fish, and the target course is x_3, for the details as shown in Fig. 5.

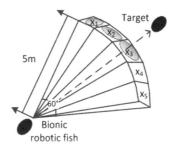

Fig. 5. Sonar model in the navigating process

The member value table can be obtained according to the membership function and fuzzy rule base of the attributes of the safety degree and the deviation degree, as shown in Tables 5 and 6.

Table 5. Membership function value of safety degree

Sonar sub	Safety degree attribute	Membership function value
x_1	Very safe	0.9820
x_2	Safe	0.8808
x_3	Dangerous	0.5000
x_4	Dangerous	0.5000
x_5	Dangerous	0.5000

Table 6. Membership function value of offset degree

Sonar sub	Offset degree attribute	Membership function value
x_1	Very far	0.0180
x_2	Far	0.1192
x_3	Close	0.5000
x_4	Far	0.1192
x_5	Very far	0.0180

The relation matrix R can be obtained according to the membership function value, (formula 6), and use the BK triangle sub-product to get the product relation between R and RT (formula 7), select $\alpha = 0.85(\alpha \in (0,1))$ and get a concise relation (formula 8).

$$R = \begin{Bmatrix} 0.9820 & 0.0180 \\ 0.8808 & 0.1192 \\ 0.5000 & 0.5000 \\ 0.5000 & 0.1192 \\ 0.5000 & 0.0180 \end{Bmatrix} \tag{6}$$

$$T = R \triangleleft R^T = \begin{Bmatrix} 1.0000 & 0.9494 & 0.7590 & 0.7590 & 0.7590 \\ 0.9494 & 1.0000 & 1.0000 & 0.8096 & 0.7590 \\ 0.7590 & 0.8096 & 1.0000 & 0.8096 & 0.7590 \\ 0.9494 & 1.0000 & 1.0000 & 1.0000 & 0.9494 \\ 1.0000 & 1.0000 & 1.0000 & 1.0000 & 1.0000 \end{Bmatrix} \tag{7}$$

$$T_{\alpha=0.85} = \begin{Bmatrix} 1 & 1 & 0 & 0 & 0 \\ 1 & 1 & 0 & 0 & 0 \\ 0 & 0 & 1 & 0 & 0 \\ 1 & 1 & 1 & 1 & 1 \\ 1 & 1 & 1 & 1 & 1 \end{Bmatrix} \tag{8}$$

And then get Hasse diagram from the condensed relation T_α, as shown in Fig. 6.

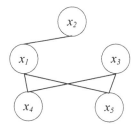

Fig. 6. Hasse diagram

Therefore, the bionic robotic fish should choose the x_2 direction as the next adjustment direction.

4 Test

In order to verify the performance of the heuristic path search strategy proposed in this paper, variety of simulation scenes are set to compare the time and path length of the heuristic search strategy and the A* search strategy. And in order to reduce the comparison error, the positions of the starting point and the target point are uniform in each scene.

4.1 Simulation Test

In order to test the performance of the algorithm, 10 different scenes are set. In the same size of space, it respectively sets the start point and the target point, and different sizes of the obstacles, to compare the heuristic search strategy between A* search algorithm and BK triangle sub-product (Table 7).

From Fig. 7, it can be found that in the same size of space, in terms of the scenes with different obstacle distribution, A*algorithm and BK triangle sub-product can all

Table 7. Path search results

Scene	Obstacle	Required time	
		A*(sec)	BK triangle sub-product (sec)
1	NO	0.098	0.032
2	1×2	0.134	0.041
3	1×3	0.142	0.045
4	1×4	0.145	0.048
5	$1 \times 2, 1 \times 3$	0.334	0.054
6	$1 \times 3, 1 \times 4$	0.375	0.062
7	$1 \times 2, 1 \times 4$	0.426	0.068
8	$1 \times 2, 1 \times 3, 1 \times 4$	0.524	0.081
9	$2 \times 2, 2 \times 3, 2 \times 4$	0.687	0.092
10	$2 \times 2, 3 \times 3, 4 \times 4$	0.752	0.095

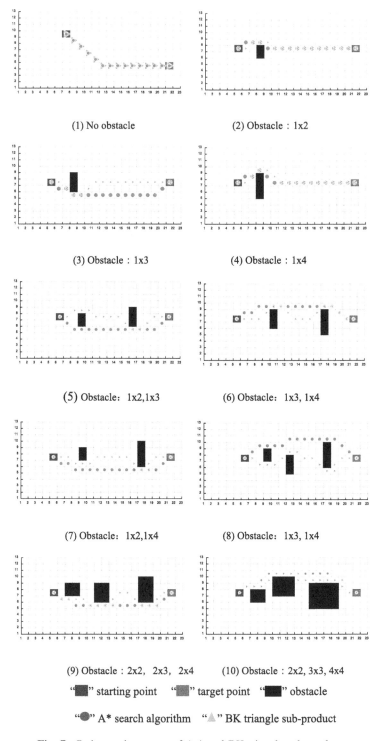

(1) No obstacle

(2) Obstacle : 1x2

(3) Obstacle : 1x3

(4) Obstacle : 1x4

(5) Obstacle：1x2,1x3

(6) Obstacle：1x3, 1x4

(7) Obstacle：1x2,1x4

(8) Obstacle：1x3, 1x4

(9) Obstacle : 2x2, 2x3, 2x4   (10) Obstacle : 2x2, 3x3, 4x4

" " starting point   " " target point   " " obstacle

" " A* search algorithm   " " BK triangle sub-product

Fig. 7. Path search process of A * and BK triangle sub-product

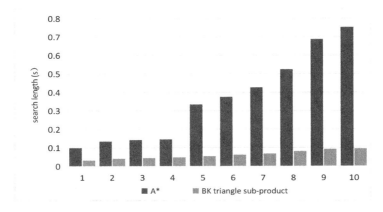

Fig. 8. The search duration of A* algorithm and BK triangle sub-product

achieve the process from the starting point to the target point, with the smaller path offset. From Fig. 8, it can be found BK triangle sub-product has the short search duration and the fast path achievement, while in the search process of A*, although it can reach the target point, it will deviate from the target point in the navigation process, and the security of the bionic robotic fish is not enough when encountering obstacles. In the search process of BK triangle sub-product, it almost does not deviate from the navigation target, obstacle, and can quickly adjust the decision when encountering obstacles. At the same time, it uses fuzzy decision-making in the calculation and selection of the next course, which is more suitable for the physical characteristics of bionic robotic fish.

4.2 Pool Test

In order to verify the validity of the actual algorithm, a pool test for path planning is carried out, and the operating effect of bionic robotic fish is observed through placing the obstacles with a certain volume in the predetermined area of the pool. The scene of the pool is shown in Fig. 9.

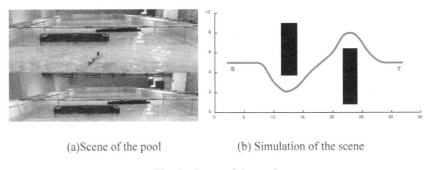

(a)Scene of the pool   (b) Simulation of the scene

Fig. 9. Scene of the pool

Two horizontal obstacles are placed in the center of the pool to test whether the bionic robotic fish can smoothly bypass the obstacles, the results of which show that during the voyage, the bionic robotic fish begins the journey from the starting point *S*, and can bypass the obstacles smoothly using the BK triangle sub-product algorithm once the transverse obstacle information is detected, to keep the stability of the fish to sail, and finally arrives at the target point *T* (Fig. 10).

Fig. 10. Obstacle avoidance of bionic robotic fish in the pool

It can be concluded that in the process of path planning of bionic robotic fish, once the single beam sonar detects obstacles, the use of heuristic search strategy proposed in this paper can adjust the direction of navigation of the robotic fish quickly and effectively, to realize the better path planning process.

5   Conclusion

The paper studied the platform of bionic robotic fish equipped with single beam sonar, used the sonar to scanning obstacle information, and divided the scanning range into several parts. It used the attributes of safety degree and offset property as the fuzzy input, when designing the strategy of path planning. When the fish encounters obstacles and needs to adjust the course, it will select the best inheritance course according to the heuristic search strategy of BK triangle sub-product. Finally, it can get the validity and superiority of the algorithm, through the simulation test and pool test.

References

1. Borenstein, J., Koren, Y.: Real-time obstacle avoidance for fast mobile robots in cluttered environments. In: Proceedings of IEEE International Conference on Robotics and Automation, vol. 1, pp. 572–577. IEEE Xplore (1990)
2. Johann, B., Yoram, K.: The vector field histogram-fast obstacle avoidance for mobile robotics. IEEE Trans. Robot. Autom. **7**(3), 278–288 (1991)
3. DeMuth, G., Springsteen, S.: Obstacle avoidance using neural networks. In: Proceedings of the Symposium on Autonomous Underwater Vehicle Technology, Piscat away, NJ, USA, pp. 213–215. IEEE (1990)
4. Antonelli, G., Chiaverini, S., Finotello, R., Schiavon, R.: Real-time path planning and obstacle avoidance for RAIS: an autonomous underwater vehicle. IEEE J. Oceanic Eng. **26**(2), 216–227 (2001)
5. Lee, Y.-I., Kim, Y.-G.: An intelligent navigation system for AUVs using fuzzy relational products. In: Joint 9th IFSA World Congress and 20th NAFIPS International Conference, 25–28 July 2001, vol. 2, pp. 709–714 (2001)
6. Lee, Y.-I., Kim, Y.-G., Kohout, J.: An intelligent collision avoidance system for AUVs using fuzzy relational products. Inf. Sci. **158**, 209–232 (2004). Elsevier, Amsterdam
7. Saffiotti, A.: The uses of fuzzy logic in autonomous robotic navigation. Soft. Comput. **1**, 180–197 (1997). Springer, Berlin
8. Bandler, W., Kohout, J.: Fuzzy relational products as a tool for analysis and synthesis of the behavior of complex natural and artificial system. In: Wang, S.K., Chang, P.P. (eds.) Fuzzy Sets: Theory and Application to Analysis and Information Systems, pp. 341–367. Plenum Press, New York (1980)
9. Kohout, J., Kim, E.: Reasoning with cognitive structures of agents, i: acquisition of rules for computational theory of perceptions by fuzzy relational products. In: Ruan, D., Kerre, E. (eds.) Fuzzy IF-THEN Rules in Computational Intelligence, pp. 161–188. Kluwer, Boston (2000)
10. Bandler, W., Keravnou, E.: Automatic documentary information retrieval by means of fuzzy relational products. In: Fuzzy Sets in Decision Analysis (1984)
11. Kohout, J., Kim, E.: The role of BK-products of relations in soft computing. Soft. Comput. **6**(2), 92–115 (2002)
12. Siler, W., Ying, H.: Fuzzy control theory: the linear case. Fuzzy Sets Syst. **33**(3), 275–290 (1989)

A Fuzzy Inference System to Scheduling Tasks in Queueing Systems

Eduyn Ramiro López-Santana[1]([⊠]), Carlos Franco[2],
and Juan Carlos Figueroa-Garcia[1]

[1] Universidad Distrital Francisco José de Caldas, Bogotá, Colombia
{erlopezs, jcfigueroag}@udistrital.edu.co
[2] Universidad del Rosario, Bogotá, Colombia
carlosa.franco@urosario.edu.co

Abstract. This paper studies the problem of scheduling customers or tasks in a queuing system. Generally the customers or a set of tasks in queuing system are attended according with different rules as round robin, equiprobable, shortest queue, among others. However, the condition of the system like the work in process, utilization and the length of queue is difficult to measure. We propose to use a fuzzy inference system in order to determine the status in the system depended of input variables like the length queue and the utilization. The experiment results shows an improvement in the performance measures compared with traditional scheduling policies.

Keywords: Fuzzy inference system · Scheduling · Queuing theory · Utilization

1 Introduction

Many real-world systems of manufacturing or services have situations where a set of tasks or customers must be waiting in order be transformed or receive a service, for instance the patients in a health services, the data in computing and communication services, the raw material in a production system, among the others. Generally, these process are complex when they have several steps, i.e., the customer service is not complete until this has been addressed in a sequential manner (or serial) in different stages, configuring around each one of them, queues whose requirements are measured in terms of response times, throughput, availability and security [1, 2].

To manage these systems, the decision making tools usually compare the efficiency of different configurations in terms of equipment, operators, storage areas, waiting areas, etc. and determine long-term decisions, for instance in capacity expansion [3]. Other features add more complexity like that feedback loops in the system, non-linearity, variability, product mixes, routing, equipment random failures and stochastic arrival times [4]. These systems can be considered as general queuing networks systems and there is not an analytical method to describe the solution of these systems [5].

In addition, when the decision's makers are faced to scheduling tasks, they has an uncertainty in order to define how long a queue is or a server are utilized. In this paper

© Springer International Publishing AG 2017
D.-S. Huang et al. (Eds.): ICIC 2017, Part III, LNAI 10363, pp. 286–297, 2017.
DOI: 10.1007/978-3-319-63315-2_25

we study the problem to scheduling tasks in a queuing systems considering the condition based systems in terms the queue's length, utilization and the cycle time involving the imprecision in their measurement process. We propose to use a fuzzy inference systems in order to determine the server to allocate a specific task according with queue's length and server's utilization.

Our solution is based in two aspects: the need to deal with complex systems, and to meet the growing demand for models that enable them to manage the uncertainty [6, 7]. The integration of traditional methods [8] and knowledge-based methods to build approximations through simple structures with multiples variables, non-linearity, uncertainty, among others. Recently, many researchers has been developed applications related with artificial intelligence [9], expert systems [8, 10] and intelligent agents [11, 12], where the common feature is the hybridization between different techniques.

The remainder of this paper is organized as follows: Sect. 2 presents a background and literature review of task scheduling in a queuing system. Section 3 describes the proposed fuzzy inference system. Section 4 shows an example of application of our proposed solution. Finally, Sect. 5 concludes this work and provides possible research directions.

2 Background and Literature Review

A queueing system consists of a stream of customers (humans, finished goods, messages) that arrive at a service facility, get served according to a given service discipline, and then depart [13]. The queues or waiting lines are an unavoidable component of modern life, for instance in grocery stores, banks, department stores, heath systems, amusement parks, movie theaters, among others. Queueing networks (QN) are configured when the service is completed by its step in different stages (or stations) where it is served in a sequential way. The complexity increases in QNs because scheduling several tasks in a QN with several stations consider more variables like capacity, routing probabilities, variability, blocking, reprocessing, among others [4].

2.1 Queuing Systems (QS)

Figure 1 shows the process in a QS. The customers requiring service are generated over time by an input source [14]. These customers enter the queueing system and join a queue. At certain times, a member of the queue is selected for service by some rule known as the queue discipline. The required service is then performed for the customer by the service mechanism, after which the customer leaves the QS.

A QS could be characterized in terms of Kendall's notation [15], whose encoding under the following structure:

$$1/2/3/4 \tag{1}$$

where 1 refers to the arrival process that can be Poisson (M), Deterministic (D) or general distribution different to Poisson (G); 2 is the service process that can be also M,

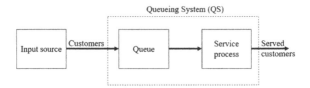

Fig. 1. The basic queuing system (source [14])

D o G; 3 represents the number of servers by stage of process in the network, which can be single (represented by 1) or multiple (represented by s); and 4 states the system's capacity, infinite when it is empty or a K to indicate the maximum queue's length.

The standard terminology and notation in QS consider as the state of system the number of customers in QS. The Queue length (Ql) is the number of customers waiting for service to begin or state of system minus number of customers being served.$P_n(t)$ denotes the probability of exactly n customers in the QS at time t, given number at time 0. s is the number of servers (parallel service channels) in the QS.λ_n is the mean arrival rate (expected number of arrivals per unit time) of new customers when n customers are in system and μ_n is the mean service rate for overall system (expected number of customers completing service per unit time) when n customers are in system.

When λ_n is a constant for all n, is denoted by λ. When the mean service rate per busy server is a constant for all $n \geq 1$, is denoted by μ. In this case, $\mu_n = s\mu$ when $n \geq s$, that is, when all s servers are busy. Also,$\rho = \lambda/s\mu$ is the utilization factor for the service facility, i.e., the expected fraction of time the individual servers are busy, because $\lambda/s\mu$ represents the fraction of the system's service capacity ($s\mu$) that is being utilized on the average by arriving customers (λ).

When a QS has recently begun operation, the state of the system (number of customers in the system) will be greatly affected by the initial state and by the time that has since elapsed. The system is said to be in a transient condition. However, after sufficient time has elapsed, the state of the system becomes essentially independent of the initial state and the elapsed time (except under unusual circumstances). The system has now essentially reached a steady-state condition, where the probability distribution of the state of the system remains the same (the steady-state or stationary distribution) over time. Queueing theory has tended to focus largely on the steady-state condition, partially because the transient case is more difficult analytically.

We assume that P_n is the probability of exactly n customers in queueing system. Then L is the expected number of customers in queueing system, it is computed by $\sum_{n=0}^{\infty} nP_n$, and L_q is the expected queue length (excludes customers being served), it is computed by $\sum_{n=s}^{\infty} (n - s)P_n$. In addition, W is the expected waiting time in system (includes service time) for each individual customer and W_q is the expected waiting time in queue (excludes service time) for each individual customer. Assume that λ_n is a constant for all n. It has been proved that in a steady-state queueing process,

$$L = \lambda W. \tag{2}$$

It is known as Little [4]. Furthermore, the same proof also shows that

$$L_q = \lambda W_q. \tag{3}$$

The Eqs. (1) and (2) are extremely important because they enable all four of the fundamental quantities L, W, L_q, and W_q to be immediately determined as soon as one is found analytically.

In QS including the so-called queueing networks (QN), arrivals and service times are presented in terms of probability distributions. Although these distributions are the main factors for the analysis of QNs, another factors must also be considered [16] like that characteristics of service stations, the configuration and routing protocols determine the flow of customers from one station to another, including the number of servers in each stage. Likewise, the arrival process by specifying the time between successive arrivals, number and destination of arrivals per time unit. Finally, the size of the waiting area for each station. When the size of the waiting area is limited, some of the transactions they avoid congestion at the previous station causing blockage in the following stations.

In a general sense, a QN must be defined in terms of arrival and service rates, and routing probabilities or proportion in which classes of customers are transferred sequentially from one service stage to another. Particularly, the routing probabilities induces feedback cycles that increase complexity in the understanding of this type of systems. Since the QNs is a system of nodes that interact, the operation of each node and the routing depend on what's happening along the network, given this dependence, can occur any or combination of synchronous or parallel processing of customers in multiple nodes, toggle the routing of transactions to avoid the congestion (or interference), speeding up or slowing down the rate of processing in the following nodes that can be idle or congested and the transactions are blocked from entry to a specific phase of the network when the phase is not capable to process more transactions. A key feature of the QNs is whether these are open, closed, or mixed. If the network permits the output of customers on the network then circular of default by the nodes required to complete the service, is considered an open network, otherwise it is considered a closed network. The network can receive different types of customers (multi-class) and can in turn be open for certain classes of customers and closed to other.

2.2   Review of Main Analytic and Approximated Techniques to QS

According with the QNs modeling purpose, the technical solution is selected taking into account the criterion of accuracy of the expected result with respect to the assumptions of the systems behavior. Baldwin et al. [17] proposes its classification in two types namely: exact and approximate. Within the analytical techniques are classified as exact: the Jackson Networks and BCMP networks. On the other hand, there are approximate techniques as: Mean Value Analysis (MVA) y Equilibrium Point Analysis. Table 1 presents the scope of the analysis techniques described above, by specifying the types of customer and network that can be modeled with precision (compared regard to the assumptions of each method). We include the technique called "Kingman's parametric decomposition" [4] considering their contribution for the modeling of flow times in QNs.

Table 1. Classification of analysis techniques scope of queueing networks

Analytical technique	Network topology	Customer type
Jackson Networks	Open, closed	Singleclass
BCMP (Type I, II, III, IV)	Open, closed, mixed	Multiclass
Kingman's parametric decomposition	Open, closed, mixed	Multiclass, multi-class with retry
Mean Value Analysis	Open, closed	Singleclass
Equilibrium Point Analysis (EPA)	Open, closed, mixed	Multiclass, multi-class with retry

Recently some applications have been developed in QS. Jain et al. [18] develop an iterative approach using MVA, for the prediction of performance in flexible manufacturing systems with multiple devices materials handling. They demonstrate improvements in the throughput, average time of service and average waiting time, with respect to the previous configuration of the materials handling devices; uses a neuro-fuzzy controller to compare the performance measures obtained with MVA demonstrating the consistency between the results of both techniques, giving the basis for the automation of the system using soft-computing. Cruz [19] examines the problem of maximizing the throughput in QNs with general service time, finding the reduction in the total number of waiting areas and the service rate, through genetic algorithms multi-objective to find a feasible solution to the need of improve the service given the natural conflict between the cost and throughput. Yang and Liu [20] develop a hybrid transfer function model that combines statistical analysis, simulation and analysis of queues, taking as input values the systems work rate, the performance variables throughput and work in process (WIP). However, the use of artificial intelligence to modelling QS is scarce, Azadeh et al. [21] has demonstrated optimize the modeling and simulation of QS, since under this scheme may include systems constraints and desired performance objectives, gaining flexibility and the ability to deal with the complexity and nonlinearity associated with the modeling of QNs.

3 Proposed Fuzzy Inference System (FIS) to Scheduling Tasks

Fuzzy inference system (FIS) is a making decision process that consists of formulating the mapping from a given input to an output using fuzzy logic. Figure 2 shows a basic architecture of a FIS, which has as the output the cycle time (W) and the inputs the queue's length (Ql) and the utilization (u). This intuition is based in Kingman equation given by:

$$W = VUT \tag{4}$$

where, V refers to a variability in the system, U is the utilization and T is the time. Likewise, the Little Law in Eq. (1) refer that the W depends the L. Thus, we have two

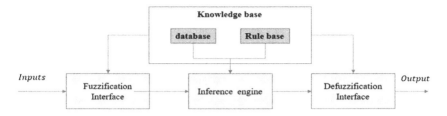

Fig. 2. Basic FIS architecture

equations to determine the cycle time W and we propose to combine both in a FIS system involving the imprecision in the definition of the queue's length and the server's utilization.

For the Fuzzification and Defuzzification Interfaces we need to define the membership function (MF) for the inputs and output. For the queue's length we define three linguistic labels: small, medium, and large. Also, for the server utilization we use three linguistic labels: low, medium, and high. Figure 3 shows an example of the MFs for utilization (see Fig. 3(a)) and queue's length (see Fig. 3(b)). These MFs must be defined according with the system behavior. Figure 3(c) presents the MF for the cycle time W, where we state three linguistic labels: small, medium and large.

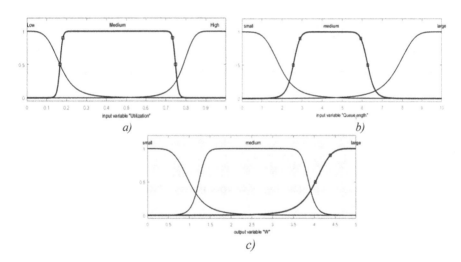

Fig. 3. Membership function of (a) utilization, (b) queue's length, and (c) cycle time W

Table 2 states the set of rules for the proposed FIS. For instance, if the utilization is medium and the queue's length is large the cycle time is Large. Figure 4 shows these rule based system in the graph (a) and its response surface in graph (b). As the utilization and queue's length increases the cycle time also increases. Our FIS is based on Mamdani inference [22, 23] because this method is intuitive, has widespread acceptance and is well suited to human input.

Table 2. Rules based system for the proposed FIS

Utilization\queue's length	Small	Medium	Large
Low	Small	Small	Large
Medium	Small	Medium	Large
High	Medium	Large	Large

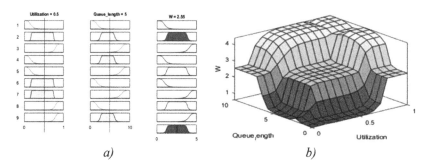

a)   b)

Fig. 4. Response of proposed FIS (*a*) rule base system, (*b*) response surface

This FIS system is evaluated in real time in a simulation process when a customer arrive to the queue, then a server l^* is determined as the server with the minimum value of W_l computed with the proposed FIS for all $l \in \{1, 2, .., s\}$ given as follows:

$$l^* = \text{argmin}_{l \in \{1,2,...,s\}} \{W_l\}. \tag{5}$$

4 Experiments

We develop a prototype of the QS in Matlab 2017 using the toolbox of SimEvents. Figure 5 presents the prototype for a system with 4 servers each one with a queue, the capacity is infinity, and the probability of rework is 20%. The discipline in the queue is FIFO (First in First Out).

4.1 Case of M/M/4 Without Rework

We consider an M/M/4 system. The inter-arrival and service time follow an exponential distributions with mean of 0.9 and 2 min, respectively. We don't consider the rework in this case. In order to compare the performance of our approach we consider a round robin scheduling policy that consists in allocate a server in sequential way and equiprobable policy that consists in allocate any server with a same probability. Figures 6 and 7 presents the results for the utilizations and queue's length for all servers, respectively. First, we consider a warm time of 100 min to transient condition. About the utilizations, all policies converge to 0.5 in average for all server, however the

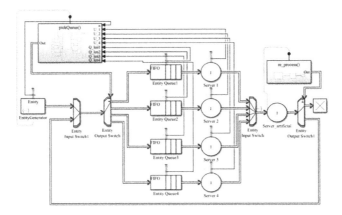

Fig. 5. Example of prototype for a QS with 4 servers with rework

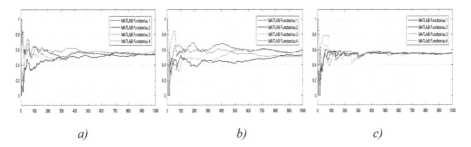

a)   *b)*   *c)*

Fig. 6. Results of utilizations case M/M/4 without rework for (*a*) round robin, (*b*) equiprobable and (*c*) FIS approach

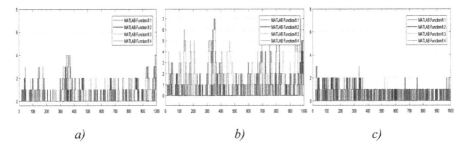

a)   *b)*   *c)*

Fig. 7. Results of queue's length case M/M/4 without rework for (*a*) round robin, (*b*) equiprobable and (*c*) FIS approach

FIS approach converge faster than round robin and equiprobable policies. Regarding with queue's length FIS approach obtain the shortest queue in the simulation horizon time and the largest queue is getting for equiprobable policy. These results allow to insight a rapid response of our FIS approach compared with the traditional policies.

4.2 Case of M/M/4 with Rework

We consider the same M/M/4 system but considering the rework with a probability of 20%. Figures 8 and 9 show the results for the utilizations and queue's length for all servers, respectively. About the utilizations the results is similar to previous experiment and converge to 0.6 in average for all servers due to the rework. Again, the FIS approach converge faster than round robin and equiprobable policies. In similar way, the queue's length if FIS approach get the shortest queue and the largest queue is getting for equiprobable policy. These results confirm the rapid response of our FIS approach compared with the traditional policies.

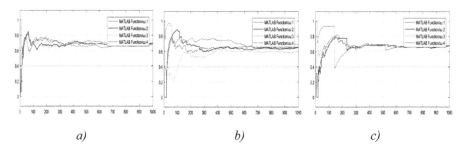

a)   *b)*   *c)*

Fig. 8. Results of utilizations case M/M/4 with rework for (*a*) round robin, (*b*) equiprobable and (*c*) FIS approach

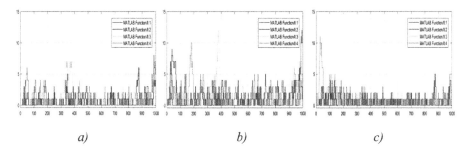

a)   *b)*   *c)*

Fig. 9. Results of queue's length case M/M/4 with rework for (*a*) round robin, (*b*) equiprobable and (*c*) FIS approach

4.3 Case G/G/4 with Rework

We consider the same system in the previous experiment but the inter-arrival time follows a uniform distributions between 0.5 and 1 min. The service time follows a uniform distribution between 1.5 and 3.0 min. Then, the system is G/G/4 with a rework probability of 20%.

Figures 10 and 11 show the results for the utilizations and queue's length for all servers, respectively. About the utilizations all converge to 0.87 in average for all

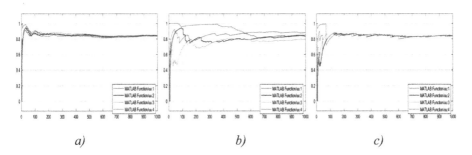

Fig. 10. Results of utilizations case G/G/4 with rework for (*a*) round robin, (*b*) equiprobable and *c*) FIS approach

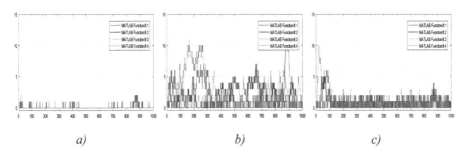

Fig. 11. Results of queue's length case G/G/4 with rework for (*a*) round robin, (*b*) equiprobable and (*c*) FIS approach

servers due to the rework. Again, the FIS approach converge faster than round robin and equiprobable policies. Respect to the queue's length, round robin policies gets the shortest queue and our FIS approach is lower than the queue in equiprobable policy. These results confirm the rapid response of our FIS approach compared with the traditional policies, but the queue length is not the best.

5 Conclusions

This paper review the problem of scheduling customers or tasks in a queuing system. Traditional scheduling policies attended the customers according with different rules as round robin, equiprobable, shortest queue, among others. We propose a FIS in order to determine the scheduling in a queuing system depended of input variables like the queue length and the server utilization.

Our results shows an improvement in the performance like utilization and queue length compared with round robin and equiprobable scheduling policies. Our FIS approach offers a rapid response compared with the traditional policies and involves information about the status of the system. In this sense, our FIS approach could be consider more information like breakdowns, variability, blocking, among others, and to provide a condition based scheduling policies for queening system.

This work generates possible future development lines, one of which is the validation in real-world case for example in heath service the task scheduling for authorizations, calls and transfers. In addition, we could be consider other variables as input for the FIS like breakdowns, variability and blocking. Finally, it is possible to design a multi-agent system that allow the load balancing of tasks in multi-class queueing networks systems and consider neuronal networks to determine the rules.

Acknowledgement. We would like to acknowledge to Centro de Investigaciones y Desarrollo Científico at Universidad Distrital Francisco José de Caldas (Colombia) by supporting partially under Grant No. 2-602-468-14. Last, but not least, the authors would like to thank the comments of the anonymous referees that significantly improved our paper.

References

1. Lopez-Santana, E.R., Méndez-Giraldo, G.A., Florez Becerra, G.F.: On the conceptual design of multi-agent system for load balancing using multi-class queueing networks. In: 2015 Workshop on Engineering Applications - International Congress on Engineering (WEA), pp. 1–7 (2015)
2. Cruz, F.R.B., Kendall, G., While, L., Duarte, A.R., Brito, N.L.C.: Throughput maximization of queueing networks with simultaneous minimization of service rates and buffers. Math. Probl. Eng. **2012**, 1–19 (2012)
3. Yang, F.: Neural network metamodeling for cycle time-throughput profiles in manufacturing. Eur. J. Oper. Res. **205**, 172–185 (2010)
4. Hopp, W.J., Spearman, M.L.: Factory Physics - Foundations of Manufacturing Management. Irwin/McGraw-Hill, New York (2011)
5. Rabta, B., Schodl, R., Reiner, G., Fichtinger, J.: A hybrid analysis method for multi-class queueing networks with multi-server nodes. Decis. Support Syst. **54**, 1541–1547 (2013)
6. Negi, D.S., Lee, E.S.: Analysis and simulation of fuzzy queues. Fuzzy Sets Syst. **46**, 321–330 (1992)
7. Zhang, H., Tam, C.M., Li, H.: Modeling uncertain activity duration by fuzzy number and discrete-event simulation. Eur. J. Oper. Res. **164**, 715–729 (2005)
8. López-Santana, E.R., Méndez-Giraldo, G.A.: A knowledge-based expert system for scheduling in services systems. In: Figueroa-García, J.C., López-Santana, E.R., Ferro-Escobar, R. (eds.) WEA 2016. CCIS, vol. 657, pp. 212–224. Springer, Cham (2016). doi:10.1007/978-3-319-50880-1_19
9. Rojek, I., Jagodziński, M.: Hybrid artificial intelligence system in constraint based scheduling of integrated manufacturing ERP systems. In: Corchado, E., Snášel, V., Abraham, A., Woźniak, M., Graña, M., Cho, S.-B. (eds.) Hybrid Artificial Intelligent Systems, pp. 229–240. Springer, Berlin Heidelberg (2012)
10. Álvarez, L., Caicedo, C., Malaver, M., Méndez, G.: Design of system expert prototype to scheduling in job-shop environment. Revista Científica **12**, 125–136 (2010). (in Spanish)
11. Pereira, I., Madureira, A.: Self-optimization module for scheduling using case-based reasoning. Appl. Soft Comput. **13**, 1419–1432 (2013)
12. Madureira, A., Pereira, I., Pereira, P., Abraham, A.: Negotiation mechanism for self-organized scheduling system with collective intelligence. Neurocomputing **132**, 97–110 (2014)
13. Ross, S.: Introduction to Probability Models. Academic Press, Cambridge (2006)

14. Hillier, F.S., Lieberman, G.J.: Introduction to operations research. McGraw-Hill Higher Education, New York (2010)
15. Kendall, D.G.: Stochastic processes occurring in the theory of queues and their analysis by the method of the imbedded Markov chain. Ann. Math. Stat. **24**, 338–354 (1953)
16. Gupta, D.: Queueing models for healthcare operations. In: Denton, B.T. (ed.) Handbook of Healthcare Operations Management, pp. 19–44. Springer, Heidelberg (2013). doi:10.1007/978-1-4614-5885-2_2
17. Baldwin, R.O., Davis IV, N.J., Midkiff, S.F., Kobza, J.E.: Queueing network analysis: concepts, terminology, and methods. J. Syst. Softw. **66**, 99–117 (2003)
18. Jain, M., Maheshwari, S., Baghel, K.P.S.: Queueing network modelling of flexible manufacturing system using mean value analysis. Appl. Math. Model. **32**, 700–711 (2008)
19. Cruz, F.R.B.: Optimizing the throughput, service rate, and buffer allocation in finite queueing networks. Electron. Notes Discrete Math. **35**, 163–168 (2009)
20. Yang, F., Liu, J.: Simulation-based transfer function modeling for transient analysis of general queueing systems. Eur. J. Oper. Res. **223**, 150–166 (2012)
21. Azadeh, A., Faiz, Z.S., Asadzadeh, S.M., Tavakkoli-Moghaddam, R.: An integrated artificial neural network-computer simulation for optimization of complex tandem queue systems. Math. Comput. Simul. **82**, 666–678 (2011)
22. Camastra, F., Ciaramella, A., Giovannelli, V., Lener, M., Rastelli, V., Staiano, A., Staiano, G., Starace, A.: A fuzzy decision system for genetically modified plant environmental risk assessment using Mamdani inference. Expert Syst. Appl. **42**, 1710–1716 (2015)
23. Alavi, N.: Quality determination of Mozafati dates using Mamdani fuzzy inference system. J. Saudi Soc. Agric. Sci. **12**, 137–142 (2013)

Campus Network Information Security Risk Assessment Based on FAHP and Matter Element Model

Fangfang Geng$^{(\boxtimes)}$ and Xiaolong Ruan

Henan University of Traditional Chinese Medicine, Zhengzhou
Henan Province, China
gfhactcm@126.com

Abstract. Based on the analysis of the main influencing factors of campus network information security, the risk evaluation index system of campus network information security is constructed. Meanwhile, a security risk assessment model of campus network information security based on FAHP and matter element model is proposed. It is used to complete the quantification of index, the determination of index weight and the calculation of correlation degree. The model can also be used to complete the campus network information security risk assessment, and then put forward the improvement measures. The results show that the model can be effectively applied to the campus network information security risk assessment. It provides a theoretical basis for improving campus network information security.

Keywords: FAHP · Matter element model · Campus network information · Risk assessment

1 Introduction

With the wide use of computer and the rapid development of network technology, information technology has entered the various walks of life. In order to improve the level of teaching, scientific research and management, colleges and universities have established campus network and application system. The construction of campus informatization marks another leap in the informatization level of our country. However, due to the openness and interconnection of the network, and the low level of the campus network users, the problem of campus network information security is becoming more and more serious. Security events such as hacking, malicious code, LAN attack, service attacks, worms, spam, power supply problem poses a threat to the service function of campus network information and reliable operation [1]. 2016 full year, due to power outages, campus network and information systems can not be accessed up to 12 h. The system could not be accessed up to 5 times by DOS attacks. The system could not be accessed up to 7 h due to network failure. The server could not provide the service up to 2 times due to air conditioning failure. Therefore, how to reduce the security threat to the campus network and application system, protect the campus network information security, so that the campus network can meet the needs

© Springer International Publishing AG 2017
D.-S. Huang et al. (Eds.): ICIC 2017, Part III, LNAI 10363, pp. 298–306, 2017.
DOI: 10.1007/978-3-319-63315-2_26

of school teaching, research and management development. This is the problem we need to solve. Therefore, FAHP and matter-element model are applied to the campus network information security assessment. The model can be used to determine the value of risk and provide some scientific basis and reference for the prevention and control of risk.

2 Construction of Campus Network Information Security Risk Index System

Campus network and its application system are constructed by the combination of complex hardware, software, physical environment, network environment and so on. It is mainly used to achieve interconnection, sharing and user services. The openness of the campus network and the complexity of the environment, the campus network information security are threatened.

Because of the complex structure of the campus network information system and the variety of equipment, it is necessary to take into account various factors to construct the safety risk assessment system. So it needs to be classified and summarized. Delphi expert and technical personnel consulting method is used to analyze and study the main safety risk index, and finally, the three level index system of campus network information security risk is constructed, as shown in Fig. 1.

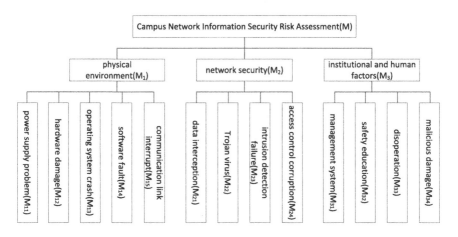

Fig. 1. Campus network information security risk assessment index system

The overall goal of the evaluation system is the campus network information security risk, as shown in Fig. 1. The factors that affect the total goal are three aspects: physical environment, network security, system and human factors. The security risk of the physical environment is the threat to the infrastructure of the campus network and application system. Security incidents include power supply problems, hardware damage, software failure, operating system crashes and communication link interruption. Any part of them will cause different degrees of threat to the campus network information. Network security risk refers to the user, the system authority is threatened. It includes many problems, such as illegal molecular data theft, computer Trojan virus infection caused by information security incidents, intrusion detection, access control failure caused damage to non intrusive authorized users. System and human factors are mainly from two aspects: first, the integrity of the management system; Second, the human factors. Due to the unclear security awareness, the administrator of the wrong operation, malicious damage and so on will be a threat to the campus network information, while causing a certain loss of property.

3 Algorithm Model

3.1 FAHP

Analytic hierarchy process (AHP) is a qualitative and quantitative system analysis method proposed by T.L. Satty in 1970s. Because of the combination of qualitative and quantitative analysis of the characteristics of various evaluation factors, as well as the advantages of system, flexibility and simplicity, AHP is welcomed by the risk assessor. However, the biggest problem is that a certain level of evaluation indicators (such as more than four), it is difficult to ensure the consistency of thinking. Therefore, combining fuzzy theory with analytic hierarchy process (AHP), fuzzy analytic hierarchy process (FAHP) is formed.

The basic steps of FAHP are as follows:

1. Determine the causal relationship between the various factors in the system, multi-level hierarchical structure model is established.
2. On the basis of the elements of the previous level, the elements of the same level are compared, as shown in Table 1. According to the evaluation scale, the relative importance degree is determined, and the fuzzy judgment matrix is established. For example, $a_{ii} = 0.5$ means that factors are as important as themselves. If $a_{ij} \in [0.1, 0.5)$, x_j is more important than x_i. If $a_{ij} \in [0.5, 0.9]$, x_i is more important than x_j.
3. The relative importance of each factor is determined by calculation.

Table 1. Scale and meaning

Scale	Meanings
0.5	Two elements are equally important
0.6	The first element is slightly more important than the latter
0.7	The first element is more important than the latter
0.8	The first element is much more important than the latter
0.9	The first element is extremely important than the latter
0.1–0.4	The comparison of the importance of the latter and the former

3.2   Matter Element Model

The theory of matter element analysis is a new subject which was founded by Professor Cai Wen in 1983. Matter element analysis is based on extension matter-element model, extension set and correlation function theory. Through the analysis, the complex problem is simplified into a formal, logical and quantitative mathematical model which is used to solve the related problems in natural and social sciences.

Matter element is the logical cell of matter element extension analysis. Matter element is also the basic element of describing things. The ordered triple R = {N,C,V} is used to represent one dimensional matter-element. N, C, V represents the characteristics of things, things and the amount of features respectively. In the matter element analysis, for a certain characteristic c of matter element, the range of the limit values of v is joint domain of c. A part of the characteristic section is called the classical domain. The matter element correlation function can describe quantitatively the degree that the elements of the evaluation object have certain attributes.

3.3   Campus Network Information Security Risk Assessment Based on FAHP and Matter Element Model

Aiming at the complexity of campus network information security evaluation index, FAHP and matter-element model are used to evaluate it. The task consists of the following steps.

1. Determination of evaluation index.
2. Determination of classical domain and joint domain
   According to the equipment parameters and operation and maintenance experience, the parameters of the normal operation of the campus network system are determined. Therefore, the evaluation index system of the system is established, which is the classical domain and the joint domain.
3. Calculation of correlation function
   The value of the correlation function (correlation degree) indicates that the value of each element is in line with the extent of the assessment. Based on the matter-element extension analysis theory, it is defined that the module of assessment level interval $X = [a, b]$ is $|X| = b - a$. The distance between any matter element characteristic factors x and assessment level interval X is $\rho(x, X)$.

$$\rho(x, X) = \left| x - \frac{a+b}{2} \right| - \frac{b-a}{2} \tag{1}$$

$K_j(x_i)$ indicates that the correlation degree between the ith evaluation matter-element factor and the level j.

$\rho(x_i, X_{oj})$ indicates that the distance between the ith evaluation matter-element factor and the classical domain of level j.

$\rho(x_i, X_p)$ indicates the distance between the ith matter-element factor and the joint domain.

$|X_{oj}|$ indicates the module of each classical domain.

$|X_p|$ indicates the module of each joint domain.

The formula of matter-element value index x_i about evaluating the rank correlation function is as follows.

$$
K_j(x_i) = \begin{cases} \dfrac{-\rho(x_i, X_{oj})}{|X_{oj}|} & x_i \in X_{oj} \\ \dfrac{\rho(x_i, X_{oj})}{\rho(x_i, X_p) - \rho(x_i, X_{oj})} & x_i \notin X_{oj} \end{cases} \tag{2}
$$

4. Comprehensive correlation evaluation

Based on the correlation between the value of each element of the matter element and the relative weight of each object factor, the correlation degree $K_j(R)$ of the whole system is obtained. The calculation method is as follows.

$$
K_j(R) = \sum_{i=1}^{n} \alpha_i K_j(x_i) \tag{3}
$$

α_i is the ith object element relative to the total target weight.

If $K_j = \max_{1 \leq j \leq m} K_j(R)$, where m is the number of evaluation grade, it belongs to the risk evaluation grade j.

4   Campus Network Information Security Risk Assessment Based on FAHP and Matter Element Model

Taking a campus network data as an example, according to the steps of information security risk assessment of FAHP and matter-element model, the information security of campus network is evaluated.

4.1   Determine the Weight of Evaluation Index

1. The hierarchical structure model of campus network information security risk assessment is established, as shown in Fig. 1. The first layer is the target layer, the second layer is the standard layer, and the third layer is the sub index layer.
2. Set up relation matri

    Due to the limitation of space, this paper only lists the relative importance matrix (M_1, M_2, M_3) of the target layer (M), as shown in Table 2, the standard layer - sub index layer fuzzy consistent matrix is no longer listed one by one.

Table 2. The relation matrix of target layer-criterion layer

M	M_1	M_2	M_3
M_1	0.5	0.6	0.9
M_2	0.4	0.5	0.8
M_3	0.1	0.2	0.5

3. Calculate the weight

In this paper, the relation matrix is transformed into a fuzzy consistency matrix, and then, on the basis of the single level sorting and the synthesis of the level, the comprehensive weight of the sub index layer is obtained (Table 3).

Table 3. Sub index weight

Serial number	Standard layer	Sub index layer	The weight of the standard layer	The weight of the target layer
1	M_1	M_{11}	0.1629	0.0759
2		M_{12}	0.2516	0.1173
3		M_{13}	0.0893	0.0416
4		M_{14}	0.3215	0.1496
5		M_{15}	0.1757	0.0819
6	M_2	M_{21}	0.1723	0.0608
7		M_{22}	0.3816	0.1347
8		M_{23}	0.2019	0.0713
9		M_{24}	0.2442	0.0862
10	M_3	M_{31}	0.3522	0.0637
11		M_{32}	0.2418	0.0437
12		M_{33}	0.2783	0.0503
13		M_{34}	0.1277	0.0231

4. Determine the joint domain and classical domain of risk assessment

Taking a campus network information security risk assessment as an example, the index system is shown in Fig. 1. In this paper, according to the running data of the university in 2016, the operation of the campus network and the operation of the business system, the indexes are quantified. Then according to the operation and maintenance experience and the actual operation of the indicators, the joint domain and classical domain of risk assessment are determined, as shown in Tables 4 and 5.

5. Determine the correlation degree between matter element and corresponding grade

Based on the evaluation of the correlation function calculation formula (2), the correlation degree of the matter-element parameter in the corresponding evaluation level is determined, as shown in Table 6.

6. Comprehensive correlation evaluation

According to the formula (3) and the weight of each element, the relative degree of the system relative to each level is calculated, as shown in Table 7.

Table 4. The classical domain of risk assessment

Serial number	Sub index layer	Evaluation parameter	Index value	Classical domain				
				Higher	High	Medium	Lower	Low
1	M_{11}	Reliability rate	99.73%	<0.990, 0.995>	<0.995, 0.998>	<0.998, 0.999>	<0.999, 0.9999>	<0.9999,1>
2	M_{12}	Usability	99.92%	<0.990, 0.995>	<0.995, 0.998>	<0.998, 0.999>	<0.999, 0.9999>	<0.9999,1>
3	M_{13}	Usability	99.98%	<0.990, 0.995>	<0.995, 0.998>	<0.998, 0.999>	<0.999, 0.9999>	<0.9999,1>
4	M_{14}	Usability	99.81%	<0.990, 0.995>	<0.995, 0.998>	<0.998, 0.999>	<0.999, 0.9999>	<0.9999,1>
5	M_{15}	Reliability rate	99.97%	<0.990, 0.995>	<0.995, 0.998>	<0.998, 0.999>	<0.999, 0.9999>	<0.9999,1>
6	M_{21}	Times per year	0.03	<0.8,1>	<0.5,0.8>	<0.2,0.5>	<0.1,0.2>	<0,0.1>
7	M_{22}	Times per year	0.19	<1,2>	<0.5,1>	<0.2,0.5>	<0.1,0.2>	<0,0.1>
8	M_{23}	Times per year	0.12	<4,5>	<3,4>	<2,3>	<1,2>	<0,1>
9	M_{24}	Times per year	1.30	<4,5>	<3,4>	<2,3>	<1,2>	<0,1>
10	M_{31}	Perfect degree	85.37%	<0.5,0.6>	<0.6,0.7>	<0.7,0.8>	<0.8,0.9>	<0.9,1>
11	M_{32}	Times per year	8.61	<6,7>	<7,8>	<8,10>	<10,11>	<11,12>
12	M_{33}	Times per year	0.28	<1,2>	<0.5,1>	<0.2,0.5>	<0.1,0.2>	<0,0.1>
13	M_{34}	Times per year	0.01	<0.8,1>	<0.5,0.8>	<0.2,0.5>	<0.1,0.2>	<0,0.1>

Table 5. The joint domain of risk assessment

Serial number	1	2	3	4	5	6	7
Sub index layer	M_{11}	M_{12}	M_{13}	M_{14}	M_{15}	M_{21}	M_{22}
Joint domain	<0.99,1>	<0.99,1>	<0.99,1>	<0.99,1>	<0.99,1>	<0,1>	<0,2>
Serial number	8	9	10	11	12	13	–
Sub index layer	M_{23}	M_{24}	M_{31}	M_{32}	M_{33}	M_{34}	–
Joint domain	<0,5>	<0,5>	<0.6,1>	<6,12>	<0,2>	<0,1>	–

Table 7 shows that the security risk level of campus network information security is lower. As can be seen from Table 6, the high risk factors are power supply problems and communication link interruption. The medium risk factors are software fault, safety education and misoperation. The lower risk factors are hardware failure, operating system crash, Trojan virus, intrusion detection failure, access control corruption, management system. The low risk factors are data theft and malicious damage.

Table 6. Correlation of matter element

Serial number	Sub index layer	The related degree value				
		Higher	High	Medium	Lower	Low
1	M_{11}	−0.1660	0.1128	−0.1541	−0.2982	−0.4126
2	M_{12}	−0.3941	−0.2653	−0.1098	0.0875	−0.1269
3	M_{13}	−0.6216	−0.3804	−0.2616	0.3216	−0.2009
4	M_{14}	−0.4122	−0.2300	0.0093	−0.2158	−0.4562
5	M_{15}	−0.2583	0.4272	−0.1208	−0.3677	−0.6212
6	M_{21}	−0.7127	−0.5129	−0.2504	−0.1903	0.0953
7	M_{22}	−0.1998	−0.1756	−0.0215	0.8648	−0.9001
8	M_{23}	−0.4957	−0.3500	−0.2844	0.2253	−0.1413
9	M_{24}	−0.2912	−0.2078	−0.1190	0.4012	−0.2745
10	M_{31}	−0.8993	−0.8554	−0.7214	0.4219	−0.6649
11	M_{32}	−0.2348	−0.1627	0.3451	−0.0256	−0.1843
12	M_{33}	−0.4005	−0.2102	0.2709	−0.4777	−0.5213
13	M_{34}	−0.9854	−0.9700	−0.9522	−0.9213	0.6155

Table 7. Comprehensive correlation degree of campus network information security

Assessment level	Higher	High	Medium	Lower	Low
The related degree value	−0.5610	−0.3222	−0.1469	−0.0926	−0.1547

Combined with the actual situation of the university, the university will carry on the power supply line overhaul every year. However, due to lack of UPS power supply, resulting in data center outage, further causing the business can not run normally. The construction of the city and the construction of the school also caused the interruption of communication lines. Therefore, these two factors do have a greater risk. A large number of software vulnerabilities lead to unavailability of business. In addition the school is a medical school, the user is not thorough understanding of information technology, but also caused a certain risk. However, the university has established a more comprehensive management system which provides a guarantee for information security to a certain extent.

4.2 Suggestions

It can be seen from the above, in order to improve the security of the campus network information, we need to do the following.

1. Reduce the maintenance time of the power supply system or increase the capacity of the UPS. Reduce the impact of power supply on the data center to ensure the normal operation of the business.
2. Regularly upgrade the software to reduce software errors.
3. Increase the training of the campus network users, making the campus network users can correctly access the campus network information.

4. Increase the construction of information security equipment, reducing the Trojan virus, data theft and other aspects of the loss.
5. Establish a sound room management system, business use process and information security management system.

5 Conclusion

By analyzing the influencing factors of campus network information security, the information security risk assessment index system is constructed, and the risk assessment method combining FAHP and matter - element model is put forward. Based on the operational data and expert experience of the indicators, the meta-elements are quantified and the classical domain and the joint domain are determined. The weight of each matter element is determined using the FAHP method. Finally, the security risk level is determined according to the association function of the matter-element model. Taking the campus network as an example, the evaluation model is successfully completed by the evaluation model. The results show that the evaluation model can effectively quantify and calculate the information security evaluation index of campus network, and draw a reasonable conclusion. At the same time, improvement measures of campus network information security operation are put forward. Therefore, to further protect the normal operation of the campus network business. The model provides a theoretical basis for ensuring the safe operation of campus network information.

References

1. Zhang, M.: Application of dynamic Bayesian network assessment method in university campus network. Netw. Secur. Technol. Appl. 11, 29–30 (2015)
2. Sun, J.W., Kang, Y.: Method for non-functional requirements evaluation of trusted software based on fuzzy analytic hierarchy process. Comput. Eng. Appl. 1–7 (2016)
3. Jiang, H., Zhang, Q.L.: Power quality evaluation of wind farm based on improved cloud matter - element model. Power Syst. Technol. 1, 205–210 (2014)
4. Li, C., Chen, J.H.: Application of the connecting number matterelement model in the civil engineering safety evaluation and risk forecast. J. Saf. Environ. 2, 71–75 (2016)

Supervised Learning

A Novel Semi-supervised Short Text Classification Algorithm Based on Fusion Similarity

Xiaohong Li$^{(\boxtimes)}$, Li Yan, Na Qin, and Hongyan Ran

College of Computer Science and Engineering, Northwest Normal University,
Lanzhou 730070, China
xiaohongli@nwnu.edu.cn

Abstract. A novel semi-supervised classification algorithm for short text based on fusion similarity is presented via analyzing of existing defects of short text classification algorithm. First of all, some words with the ability of indication of the category are extracted from the labeled dataset to construct a strong category features set. A valid fusion similarity measurement method is designed by combining cosine theorem and strong category features based similarity. Secondly, computing the mean value of the supervised information, and determining the virtual class center point of each class, and then finding the real class center point. Finally, we search those texts which have the highest similarity with each real class center in the unlabeled dataset, and give it the same class label with the real class center point. At the same time, we add it to the labeled collection, update the strong category features set and the similarity matrix. Repeat this process until all short texts have been labeled. Ultimately, experiments show that our method can significantly improve the efficiency of short text classification. The text of the most similarity with the center of the class.

Keywords: Semi-supervised classification · Short text · Strong category features · Fusion similarity · Class-center

1 Introduction

Semi-supervised classification learns from both labeled and unlabeled instances, the performance of text classifier could be improved effectively. At the same time, the field of semi-supervised classification has been very active in recent years, and many existing classification methods widely used in data analysis, text mining and so on. Recently, with the rise of social network services, a large number of short text data that carry a wealth of information appeared, such as network comments, news title and micro-blog. These data become a new information resource with great excavation value, and the classification requirement of them is also highlighted.

Different from traditional text, short text is usually of the following characteristics [1]: growing dynamically, lacking of information, short and huge. Sparse problem [2] is so seriously that the classification efficiency decreased when using those classical classification methods, such as KNN [3], Naive Bayes [4] and SVM [5]. Latterly, a number of classification algorithms for short text are proposed. Fan and Liu [6]

© Springer International Publishing AG 2017
D.-S. Huang et al. (Eds.): ICIC 2017, Part III, LNAI 10363, pp. 309–319, 2017.
DOI: 10.1007/978-3-319-63315-2_27

proposes a short text classification method of the combination of concept associativity and the semantic related concept sets according to the characteristics of Chinese short texts. Zhang and Liu [7] builds a semi-supervised classification model according to the bottleneck problem of annotation in dealing with large numbers of unlabeled samples. These two methods didn't take into account the problem of learning when the labeled data is small. Sriram et al. [8] adopts a short text classification algorithm based on the information of the author and the internal features of Tweets, which has achieved good classification results. However, as a result of manual search for category characteristics, so versatility is poor. The method proposed by Yao and Song [9] is that using LDA to construct a generative probabilistic model in order to classify short texts. Cai et al. [10] achieve short text classification by mean of using attribute selection algorithm and integration algorithm to build semi-supervised classification model. Methods mentioned above don't consider characteristic of poor theme and poor scalability.

We proposed a semi-supervised classification algorithm for short text based on fusion similarity and class center (Abbreviate SCFSC) according to analysis the characteristics and problems of the short text. First of all, the strong category features set is constructed, a valid fusion similarity calculation method is designed by combining cosine theorem and similarity based on strong category features collection. Then, we find each real class center by calculating the virtual class center of each class. In the next, the unlabeled documents which have the highest similarity with real class center are chosen to be added to the labeled collection. Finally, update the SCF collection, real class center and the similarity matrix. This process is repeated until all documents are labeled. The main contributions of this paper are summarized as follows:

(1) First, we construct a strong category features set as well as give the definition of strong category feature. In order to compare the similarity between short texts, a fusion similarity method is designed by combing Cosine theorem and SCF based similarity.
(2) We design a novel semi-supervised short text classification algorithm, as well as compute a real class center in each class, So it make K documents be labeled at the same time after the execution of the classification algorithm.
(3) We conduct experiments and compare with other kindred algorithms. The results show that our method can significantly improve the efficiency of classification.

The rest part of the paper is organized as follows. Extraction of SCFs is presented in Sect. 2. Sections 3 and 4 introduce fusion similarity method and detailed description of classification algorithm. Section 5 covers the experimental results and analysis. Finally, concluding remarks and future work are presented in Sect. 6.

2   Extraction of Strong Category Feature

In this section, we briefly introduce some related knowledge in two aspects: expected cross entropy, extraction method of strong category feature.

2.1 Expected Cross Entropy

Suppose that the set of labeled documents set is $D_L = \{d_1, d_2, ..., d_L\}$ set, D_L is also expressed as $D_L = C$, and $C = \{C_1, C_2, ..., C_K\}$, and let $D_U = \{d_{L+1}, d_{L+2}, ..., d_{L+U}\}$ $(U \gg L)$ denote the unlabeled data set. So the entire document collection is $D = D_L \cup D_U$.

Expected Cross Entropy [11] (*ECE*) is a feature selection method based on information theory, it reflects that the distance between the probability distribution of text category and the probability of text category under the condition of containing a certain feature.

In this paper, we consider that a word has ability of category indicative, furthermore, the ability of indicating is different in different categories. $ECE(t_i, C_r)$ of a word is just enough to measure this difference. Formula is defined as follows:

$$ECE(t_i, C_r) = P(t_i|C_r)P(C_r|t_i) \log_2 \frac{P(C_r|t_i)}{P(C_r)} \tag{1}$$

Where $P(C_r)$ represent the probability that the C_r class appears in the labeled set, $P(C_r|t_i)$ represent the probability that the short text containing the term t_i belongs to category C_r from formula (1) we can know, the greater the value of $P(C_r|t_i)$, the stronger the correlation between the characteristic word t_i and the category C_r. When the $P(C_r)$ is smaller, the term t_i is more indicative to the category C_r.

At the same time, we use $ECE(t_i, \bar{C}_r)$ to denote the indicative ability of the term t_i to the other classes:

$$ECE(t_i, \bar{C}_r) = \frac{\sum_{r \neq s} ECE(t_i, C_s)}{k - 1} \tag{2}$$

According to the formulas (1) and (2), the term t_i is more likely to be singled out for *SCF* when it has a high correlation with the certain category. On the other hand, t_i has a low correlation with the other categories. So we use $ECE'(t_i, C_r)$ denote the relevance of the feature t_i to the certain categories C_r in whole data set. Give the following formula:

$$ECE'(t_i, C_r) = \frac{ECE(t_i, C_r)}{ECE(t_i, \bar{C}_r) + 0.01} \tag{3}$$

2.2 Extraction of Strong Classification Features

We extract strong category feature which have class representation ability in the labeled dataset, and make them be used to guide the classification process of unlabeled short text as a priori knowledge. The following is the definition of strong category feature.

Definition (Strong Category Feature, abbreviate SCF) [12]: Assuming that the threshold filtering of feature is g, term t_i is defined as a **Strong Category Feature** if and only ECE' $(t_i, C_r) > g$.

As we know from the above definition, it is conditional whether or not a word is a *SCF*, one is if a word appears in the category, the other is how many times a word appears in the category. So we extract *SCF* based on the above two factors. Let $T = \{T_1,...T_r,...,T_K\}$ be a collection of SCFs for all category, where T_r is *SCF* set of the class C_r. The algorithm is as follows:

```
Program.1: ESCF(D_L, g):
Input: the labeled dataset D_L={C_1,C_2...,C_K},threshold g;
Output: strong category features set T={T_1,T_2,...,T_K}
1. Initialization: T=Φ, T_r=Φ, r=1;
2. while(r≤K)
   2.1 for d∈C_r do
       2.1.1 take the feature t_i∈d, calculate ECE'(t_i,C_r)
             according to the formula (3);
       2.1.2  if ECE'(t_i,C_r)≥g  then T_r=T_r∪t_i;
   2.2 T = T∪T_r;
   2.3 r++;
3. return T
```

3   Fusion Similarity Measurement Method

3.1   Similarity Based on Original Features

The vector of a piece of short text d_i can be represented as: $d_i = (w_{i1},...,w_{ij},...w_{in})$, where $w_{ij} = tf - idf(t_j, d_i)$ is weight of term t_j in d_i, in particular. Then the cosine similarity between d_i and d_j is as follows:

$$S_{\cos}(d_i, d_j) = \frac{d_i \bullet d_j}{|d_i| \times |d_j|} \quad (4)$$

3.2   Similarity Based on Strong Category Features

If the document d_i and d_j contain the common *SCF* in T_r, then we believe that they are more similar in class C_r.

$$S_{scf}^r(d_i, d_j) = \sum_{m=1}^{|T_r|} min\{f(d_i, t_m), f(d_j, t_m)\} \quad (5)$$

Suppose that two documents d_i and d_j are more similar in the category C_r if they contain more common *SCF*s in T_r.

$$S_{scf}^r(d_i, d_j)' = \frac{S_{scf}^r(d_i, d_j)}{|T_r| \max_{t_m \in T_r} \{f(d_i, t_m), f(d_j, t_m)\}} \quad (6)$$

The similarity between two documents d_i and d_j in the certain category is calculated via the formula (6), and then the similarity of them based on the *SCF* set in the whole dataset is defined as:

$$S_{scf}(d_i, d_j) = \max_{1 \le r \le K} \{S_{scf}^r(d_i, d_j)'\} \quad (7)$$

3.3 Fusion Similarity Combining Cosine Theorem and *SCF* Based Similarity

The above two section tell us, the cosine similarity ignores the indirect relation between two words in the labeled dataset, however these information is help to achieve correct classification of documents. While similarity based on *SCF* set could just be a complement for the cosine similarity, because it taken the categorical indication of a word into accounts. So we design a more effective similarity measurement method by integrating two similarity measures:

$$sim(d_i, d_j) = \alpha S_{cos}(d_i, d_j) + (1 - \alpha)S_{scf}(d_i, d_j) \quad (8)$$

Where α is the regulatory factor, $\alpha \in [0,1]$. The degree of similarity between short text d_i and d_j can be calculated by using the formula (8), the similarity matrix S is constructed, where the element $s_{ij} = sim(d_i, d_j)$:

$$S = \begin{bmatrix} s_{11} & s_{12} & \cdots & s_{1(L+U)} \\ \cdots & \cdots & \cdots & \cdots \\ s_{L1} & s_{L2} & \cdots & s_{L(L+U)} \\ \cdots & \cdots & \cdots & \cdots \\ s_{(L+U)1} & s_{(L+U)2} & \cdots & s_{(L+U)(L+U)} \end{bmatrix} \quad (9)$$

4 Class Center and Semi-supervised Classification Algorithm

The algorithm flow is in Fig. 1.

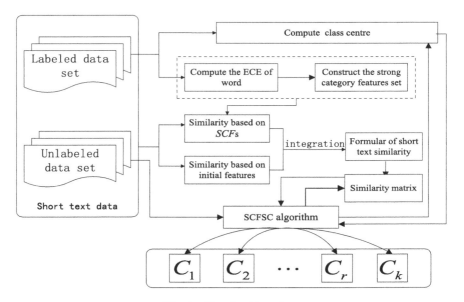

Fig. 1. The algorithm chart

4.1   Computing the Class Center

In general, a class center vector can be half of a category, so we compute each class center vector by the following steps. The class center vector of the class C_r represented by $V\_d(r)$, N_r is the number of documents that the class C_r contained.

$$V\_d(r) = \frac{1}{N_r}\left(\sum_{i=1}^{N_r} w_{i1}, \sum_{i=1}^{N_r} w_{i2}, \ldots, \sum_{i=1}^{N_r} w_{in}\right) \tag{10}$$

Finally, getting the real class center sets of known categories, denoted by Cen:

$$Cen = \{V\_d(1), \ldots, V\_d(r), \ldots, V\_d(K)\} \tag{11}$$

4.2   Algorithm Statement

The basic idea of the semi-supervised algorithms based on fusion similarity and class center is: First and foremost, the row (or column) of each R_d_r is located in the similarity matrix S, and each maximum $sim(d_i, R\_d_r)$ in the unlabeled part of these rows (or column) is found out. K unlabeled documents has the same label with R_d_r respectively after loop K times. Last but not least, we reconstruct the SCF set, recalculate class centers of each class as well as update similarity matrix S. Finally, this process is carried out iterative until all unlabeled document have been labeled. The pseudo code of algorithm as follows:

```
Program.2: SCFSC(D_L, D_U, S_r):
Input: D_L={d_1, d_2,..., d_L}, D_U={d_{L+1}, d_{L+2}, ...,d_{L+U}};
Output: cluster matrix Ω;
1 initialization: Cen=Φ;
2 Ω=C, Ω={Ω_1,...,Ω_r,...,Ω_K}
3 while(D_U≠Φ);
   3.1 Extracting strong category feature set:T=ESCF(D_L)
   3.2 for (r=1; r≤K; r++)
      3.2.1. Computing V_d(r);
      3.2.2. Cen=Cen∪{V_d(r)}
      //end for
   3.3 for each R_dr∈Cen, r=1...K,  do
      3.3.1. Max=0, index=0;
      3.3.2. for d_i∈D_U, i=1...U, do
         3.3.2.1. calculate sim(d_i,V_d(r));
         3.3.2.2. if sim(d_i,V_d(r))>Max,
               then Max=sim(d_i,V_d(r)),index = i;
            //end for
      3.3.3. Ω_r∪{d_{index} }, D_L=D_L∪{d_{index}}
      3.3.4. D_U=D_U/{ d_{index} }; //delete d_i from D_U;
      //end for
   3.4. update S;
   // end while
4 return Ω={Ω_1,...,Ω_r,...,Ω_K}
```

5 Experimental Results and Analysis

5.1 Experimental Data

The experimental dataset is a collection of 8442 piece of news posted from July to August 2016. We get 8 classes that contain 1000(env), 1184(com), 1014(tran), 1020 (edu), 1125(eco), 1049(Mil), 1004(med) and 1046(art) pieces of news, respectively. Eventually, only use the title of each piece of news for classification. We segment titles and remove the stop words by carrying out python program. At last, we keep two thirds pieces of news for each class as training samples and leave the remaining one thirds as test samples.in addition, 10% is labeled data, 90% is unlabeled data.

5.2 Results and Analysis

Three experiments were elaborately designed to verify the proposed semi-supervised short text classification algorithm:

(1) The effects of the parameters on the classification performance.
(2) By changing the threshold g to test the influence on the classification results and the number of strong category features
(3) Comparison of different short text classification algorithms.

In addition, three general metrics: Precise (P), Recall (R) and F_1 [13] are taken to measure the performance.

5.2.1   The Effect of Parameter α

Result is as shown in Tables 1 and 2, firstly, regardless of the value of threshold g, the SCSFC algorithm achieves the highest F_1 when setting parameter $\alpha = 0.35$. By observing the classification results, we can find that, on the one hand, the classification effect is poor in each class when the value of α is less than 0.35. The main reason is that the cosine similarity is too small in the process of similarity calculation, ignoring the important role of original features played in classification results; on the other hand, classification performance also arise a decline tendency when the value of α is greater than 0.35. Because the larger α is, the smaller proportion similarity based on SCF is, the weaker indicative of SCF set lead to the smaller impact on classification results.

Table 1. $g = 0.551$, the influence of α on F_1

Class	F_1					
	$\alpha = 0.3$	$\alpha = 0.35$	$\alpha = 0.4$	$\alpha = 0.5$	$\alpha = 0.6$	$\alpha = 0.7$
env	62.43	**71.91**	69.17	65.63	65.91	63.87
com	81.14	**83.57**	80.79	80.89	80.87	81.27
tran	54.32	**62.18**	60.57	58.94	58.01	56.36
edu	54.72	**62.37**	61.94	58.71	58.07	56.18
eco	58.43	**64.39**	62.82	61.79	61.13	60.51
mil	55.54	**62.71**	61.63	58.69	56.68	58.79
med	61.23	**67.53**	66.19	64.17	62.49	64.43
art	63.07	**70.28**	68.97	68.85	66.93	64.98

Table 2. $g = 0.555$, the influence of α on F_1

Class	F_1					
	$\alpha = 0.3$	$\alpha = 0.35$	$\alpha = 0.4$	$\alpha = 0.5$	$\alpha = 0.6$	$\alpha = 0.7$
env	58.71	**68.72**	65.19	66.67	61.37	59.09
com	77.92	**80.47**	79.77	79.88	77.14	75.89
tran	51.91	**61.88**	58.17	60.60	56.59	54.15
edu	52.63	**61.01**	58.04	59.63	55.64	54.59
eco	56.98	**63.97**	60.12	62.85	60.18	58.51
mil	53.88	**61.83**	58.06	61.24	58.06	56.37
med	57.29	**66.99**	62.50	66.14	62.11	60.68
art	62.18	**68.28**	66.82	67.14	63.68	61.60

5.2.2   The Influence of Threshold *G*

In this part, we set $\alpha = 0.35$ according to the result in the above section. It can be seen from Fig. 2 to Fig. 3 that the Precise and Recall of classification gradually increase with the increase of *g* when threshold $g \in [0.545, 0.551]$. However, the value of Precise and Recall start to reduce gradually when $g > 0.551$. The reason is that the smaller *g* is, the more the number of *SCF* are, so, indicative function of strong category features become to weaken. The tendency of the number of the *SCF*s is shown in Fig. 3. Conversely, the number of the *SCF*s is getting smaller when *g* is larger, which makes the algorithm less effective. Therefore, when $g = 0.551$, the mean number of *SCF* set reach to 35, the performance of the algorithm is the best.

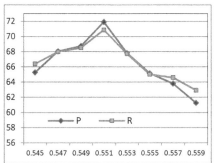
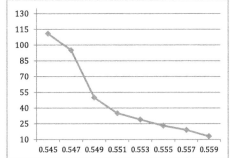

Fig. 2. Precise, recall of classification   **Fig. 3.** The number of strong category features

5.2.3   Comparison of Different Classification Algorithms

We have compared the performance of our *SCFSC* algorithm with two other classification methods: *SCSL* algorithm proposed in [6], *KNN* algorithm. It's well-known that *KNN* is a popular class-center based classification method, it achieves its best scores on smaller number of features. *SCSL* is a state-of-the-art short text classification approach.

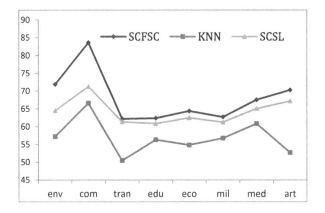

Fig. 4. Comparison of different classification algorithms

Figure 4 gives the graphical representation of classification methods with their correlation result. The F_1 score of *SCFSC* algorithm is larger than the other two algorithms on 8 classes, Especially, F_1 score of our method achieves 0.843 on the "com" class, and our method is far ahead of the other methods. F_1 score are computed for $k \in \{15,20,25\}$, as is seen, the performance of *KNN* in short text classification is worst though its performance is very desirable on traditional classification. *SCSL* algorithm is relatively stable but the performance is not good. So from this figure we can say that our method perform well among all the methods.

6 Conclusion

In this paper, we propose a semi-supervised short text classification algorithm based on fusion similarity and class center. First, an effective fusion similarity measure between short texts is designed combining cosine theorem and *SCF* based similarity. Then, the real class center of each class is determined by calculating the virtual center, and the purpose of classifying all unlabeled data correctly is achieved. The experimental results verify the effectiveness of the algorithm. We plan to select reasonable parameters and design a reasonable way to measure the similarity between short texts to improve the classification performance; we also plan to make a research on how to improve the quality of features of *SCF* set in the future work.

Acknowledgments. This work was supported in part by the Natural Science Foundation for Young Scientists of Gansu Province (No. 1606RJYA269), Project of Gansu Province Department of Education (No. 2015A-008), Youth Teacher Scientific Capability Promoting Project of NWNU (No. NWNU-LKQN-14-5, No. NWNU-LKQN-16-20).

References

1. He, H., Chen, B., Xu, W.: Short text feature extraction and clustering for web topic mining. In: Proceedings of IEEE 3rd International Conference on Semantics Knowledge and Grid (SKG 2007), pp. 382–385 (2007)
2. Wang, Z., Cheng, J., Wang, H., et al.: Short text understanding: a survey. J. Comput. Res. Dev. (2016)
3. Yang, Y., Liu, X.: A re-examination of text categorization methods. In: Proceedings of the 22nd Annual International ACM SIGIR Conference on Research and Development in Information Retrieval, pp. 42–49. ACM (1999)
4. Duan, L.G., Di, P., Li, A.P.: A new Naive Bayes text classification algorithm. Telkomnika Indonesian J. Electr. Eng. **12**(2), 51–58 (2014)
5. Joachims, T.: Learning to Classify Text using Support Vector Machines: Methods, Theory and Algorithms. Kluwer Academic Publishers, Berlin (2002)
6. Fan, Y., Liu, H.: Research on Chinese short text classification based on Wikipedia. New Technol. Libr. Inf. Serv. **3**, 47–52 (2012)
7. Zhang, Q., Liu, H.: An algorithm of short text classification based on semi-supervised learning. New Technol. Libr. Inf. Serv. **2**, 30–35 (2013)

8. Sriram, B., Fuhry, D., Demir, E., et al.: Short text classification in Twitter to improve information filtering. In: Proceedings of the 33rd International ACM SIGIR Conference on Research and Development in Information Retrieval, pp. 841–842. ACM (2010)
9. Yao, Q., Song, Z.: Research on text categorization based on LDA. Comput. Eng. Appl. **47** (13), 150–153 (2011)
10. Cai, Y., Zhu, Q., Sun, P., et al.: Semi-supervised short text categorization based on attribute selection. J. Comput. Appl. **4**, 049 (2010)
11. Ma, H., Zhou, R., Liu, F., Lu, X.: Effectively classifying short texts via improved lexical category and semantic features. In: Huang, D.-S., Bevilacqua, V., Premaratne, P. (eds.) ICIC 2016. LNCS, vol. 9771, pp. 163–174. Springer, Cham (2016). doi:10.1007/978-3-319-42291-6_16
12. Wen, H., Xiao, N.: A semi-supervised text clustering based on strong classification features affinity propagation. Pattern Recogn. Artif. Intell. **27**(7), 646–654 (2013)
13. Peat, H.J., Willet, P.: The limitations of term co-occurrence data for query expansion in document retrieval systems. J. Am. Soc. Inf. Sci. **42**(5), 378–383 (1991)

Unsupervised Learning

Clustering of High Dimensional Handwritten Data by an Improved Hypergraph Partition Method

Tian Wang[1], Yonggang Lu[1(✉)], and Yuxuan Han[2]

[1] School of Information Science and Engineering, Lanzhou University,
Lanzhou 730000, Gansu, China
ylu@lzu.edu.cn
[2] School of Computer and Communication, Lanzhou University of Technology,
Lanzhou 730050, Gansu, China

Abstract. High dimensional data clustering is a difficult task due to the curse of dimensionality. Traditional clustering methods usually fail to produce meaningful results for high dimensional data. Hypergraph partition is believed to be a promising method for dealing with this challenge. In this work, a new high dimensional clustering method called Merging Dense SubGraphs (MDSG) is proposed. A graph G is first constructed from the data by defining an adjacency relationship between the data points using Shared k Nearest Neighbors (SNN). Then a hypergraph is created from the graph G by defining the hyperedges to be all the maximal cliques in the graph. After the hypergraph is produced, an improved hypergraph partitioning method is used to produce the final clustering results. The proposed MDSG method is evaluated on several real high dimensional handwritten datasets, and the experimental results show that the proposed method is superior to the traditional clustering method and other hypergraph partition methods for high dimensional handwritten data clustering.

Keywords: Clustering · High dimensional data · Hypergraph · Hypergraph partition

1 Introduction

Data clustering is a discovery process that can group data into clusters such that the intra-cluster correlation is maximized and the inter-cluster correlation is minimized. It is one of the most popular techniques used in pattern recognition, machine learning and data mining [1]. Clustering in high dimensional space is a recurrent problem in many fields such as image analysis [2], medicine [3], bioinformatics [4] and the analysis of text documents [5]. However, traditional clustering methods such as the model-based methods and the distance-based methods often perform poorly in high dimensional space [6]. This is due to several challenges in high dimensional data clustering. Firstly, it is very difficult to define an appropriate similarity measure in the high dimensional space. Secondly, due to the inherent sparseness of the high dimensional space, it is not easy to determine the cluster centers using distance-based clustering method. Thirdly, the presence of irrelevant attributes may restrict the clustering performance. Finally, a

© Springer International Publishing AG 2017
D.-S. Huang et al. (Eds.): ICIC 2017, Part III, LNAI 10363, pp. 323–334, 2017.
DOI: 10.1007/978-3-319-63315-2_28

large number of attributes may cause the over-parameterized problem for the model-based clustering methods [7].

A number of approaches have been proposed on high dimensional clustering. Dimensionality reduction techniques have been used to overcome the problems caused by high dimensionality [8, 9]. Subspace clustering methods have been used to find clusters in different subspaces [10, 11]. With the recent improvements, the model-based clustering methods can now overcome the drawbacks of over-parameterized problem and thus can also be used to cluster the high dimensional data [12].

Hypergraph partition is believed to be a promising high dimensional clustering method. A hypergraph is a generalization of a graph in the sense that each hyperedge can connect more than two vertices, which can be used to represent relationships among subsets of a dataset. The main idea of the hypergraph-partition-based clustering is to map the relationships among the original data in high dimensional space into a hypergraph whose weights of hyperedges represent the strength of the affinity among the subsets of the data. Then the high dimensional clustering problem is transformed into a large-scale hypergraph partition problem.

There are two main problems in clustering using hypergraph partition: hypergraph construction and hypergraph partition. Many approaches have been proposed for hypergraph construction. For example, Association Rule Hypergraph Partition (ARHR) constructs hypergraphs whose hyperedges are defined as frequent item sets found by the association rule algorithm [13]. Sun et al. [14] carries out hypergraph construction by sequentially introducing new vertices into existing hyperedges using clique expansion. Huang et al. [15] proposes a probabilistic hypergraph model that assigns a vertex to a hyperedge in a probabilistic way according to the similarity between the vertex and the centroid of the hyperedge. Wang et al. [16] employs a learning-based method to build a hypergraph with a sparse representation strategy. In the work of Hu et al. [6], a hypergraph is constructed by defining clique hyperedges based on Shared Nearest Neighbors and Share Reverse Nearest Neighbors.

The majority of hypergraph partition methods divide a hypergraph into a pre-specified number of parts by optimizing some objective functions. For example, Fiduccia-Mattheyses algorithm [17] is a classic heuristic method which partitions a hypergraph into two parts under certain area ratio, such that the cut is minimized. The k-way Maximum Sum of Densities method (k-MSD) [18] divides a hypergraph into k parts, such that the sum of the densities of all k parts is maximized. The hMETIS method [19] is a multi-level hypergraph partition method, which simplifies the original hypergraphs followed by partitioning, then refines the partitions to minimize the weighted hypergraph cut. A number of methods [20–22] have been proposed based on the similar strategy of the hMETIS method. However, these heuristic methods usually produce different partitioning results under different objective functions and different configurations. Another recently proposed excellent method for hypergraph partition is called Dense Subgraph Partition (DSP) [23]. DSP can automatically, precisely and efficiently partition a hypergraph into dense subgraphs. The only drawback is that DSP cannot partition a hypergraph into a specified number of subgraphs. In this paper, a high dimensional clustering method called Merging Dense SubGraphs (MDSG) is proposed, which can be used to produce a specified number of clusters from the dense subgraphs. Experimental results show the superiority of the proposed method to the traditional clustering method and other hypergraph partition methods.

The outline of the remainder of the paper is organized as follows: In Sect. 2, the proposed method is introduced. In Sect. 3, the experimental results are reported. Finally, the conclusions are drawn in Sect. 4.

2   Method

The proposed MDSG method discovers the clusters in the data set by three steps. During the first step, data preprocessing is applied on the original handwritten image data. In the second step, a hypergraph is constructed and the DSP method is applied to partition the hypergraph into a large number of dense subgraphs (i.e. sub-clusters). In the third step, a new strategy is used to merge the dense subgraphs into the specified number of clusters. Figure 1 provides an illustration of the overall approach of the proposed method.

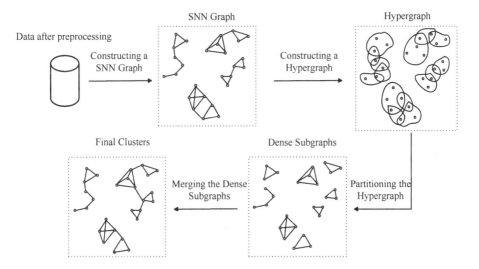

Fig. 1. The illustration of the proposed MDSG clustering method

2.1   Preprocessing

The preprocessing methods in the experiments include a Gaussian filtering method and two feature extraction methods for the input image data.

Gaussian Filtering. Gaussian filter is a linear smoothing filter for blurring images and removing noises. Mathematically, Gaussian filtering is the process of weighted averaging of the whole image. The specific operation of the Gaussian filter is to scan each pixel in the image with a template (or a mask), using the weighted average of the pixel values in the neighborhood determined by the template to replace the center pixel value of the template. There are two types of Gaussian templates used in the experiments depending on the size of the images: the first type is a 3 × 3 template with σ = 1.4, and the second type is a 5 × 5 template with σ = 1.4.

Feature Extraction. Two different feature extraction methods may be applied to the image data. The first method is projection histograms [24, 25] which is specially designed for handwritten characters. Since the projection on only one axis is not enough for recognizing the differences between characters because of the diversity in writing styles, in our method, projection histograms on four axis: x-axis, y-axis, y = x, y = −x are combined as the extracted features, as proposed by Tuba and Bacanin [25].

Another feature extraction method used in the experiments is t-Distributed Stochastic Neighbor Embedding (t-SNE). Stochastic Neighbor Embedding (SNE) approximates the probability distribution in the high dimensional space with the corresponding probability distribution in a lower dimensional space defined by neighboring points [26]. As a variation of SNE, t-SNE uses a Student-t distribution rather than a Gaussian distribution to compute the similarity between two points in the low dimensional space [27]. t-SNE is capable of capturing the local structure features of the high dimensional data, while also revealing global structure features such as the presence of clusters at several scales. In our method, the maximum iteration of t-SNE is set to 3000.

2.2   Hypergraph Construction and Partition

In the proposed method, a share nearest neighbors (SNN) graph is first constructed from the dataset. Then a hypergraph is constructed by defining the maximal cliques of the SNN graph as hyperedges. Finally, a hypergraph partition method called DSP is used to partition the hypergraph.

Hypergraph Construction. Since the distance-based similarity measures usually cannot represent well the similarity between different objects in the high dimensional space, the proposed method first constructs a SNN graph from the input data, then the maximal cliques of the SNN graph are used to construct the hypergraph.

First, the k nearest neighbors (kNN) of each data point are found using the Euclidean distance measure, where k is a parameter in our method. Then the similarity between two data points is defined in terms of how many k nearest neighbors the two points share. For any two objects, d1 and d2, from a dataset D, the number of share nearest neighbors (SNN) between d1 and d2 is defined as:

$$SNN(d_1, d_2) = |kNN(d_1) \cap kNN(d_2)| \tag{1}$$

If the number of SNN is larger than a threshold parameter Sc, an edge is created between the two objects in the SNN graph. The SNN graph can better represent the similarities among the data points, because the neighboring information is also considered in evaluating the similarities between the data points [6].

After getting the SNN graph, the next step is to find the maximal cliques of the SNN graph. A clique is defined as a complete subgraph of a graph, and a maximal clique is a clique that cannot be enlarged by introducing one more adjacent vertex. Finding maximal cliques is a NP hard problem. In our method, the improved Bron-Kerbosch method [28] is adopted to find the maximal cliques.

After getting the maximal cliques, all maximal cliques are defined as the hyperedges of the hypergraph. The weight of a hyperedge E is defined as:

$$W(E) = \frac{n|E|}{k+1} \left| \bigcap_{d_i \in E} kNN(d_i) \right| \tag{2}$$

where n is the number of vertices, and k is the same parameter used for defining kNN.

Hypergraph Partition. The DSP method is used to partition the hypergraph in the proposed method. DSP can automatically, precisely and efficiently partition a hypergraph into dense subgraphs based on the concepts of core subgraph (CS) and condition core subgraph (CCS) [23]. CS is a densest subgraph with maximal number of vertices. For a hypergraph $G = (V, E)$, a subset $U \subset V$, the subgraph induced by U is denoted by $G_U = (U, E_U)$, where $E_U = \{e|e \in E, e \subseteq U\}$. The density of a hypergraph G is defined to be $\rho(G) = \frac{w(G)}{|V|}$ where $w(G)$ is the total weight of G, and $|V|$ is the cardinality of V. The conditional density of a subgraph G_U conditioned on the subgraph G_S is defined to be $\rho(G_U|G_S) = \frac{w(G_U|G_S)}{|U|}$, where $w(G_U|G_S) = w(G_{U \cup S}) - w(G_S)$. CCS($G|$ G_S) is a subgraph G with maximal number of vertices whose conditional density reaches maximal conditioned on a subgraph G_S. DSP includes two layers of partitions. First, the hypergraph is sequentially partitioned into a sequence of conditional core subgraphs. Then, each conditional core subgraph is partitioned into pseudo-disjoint dense subgraphs by the operation of disjoint partition.

2.3   Producing the Specified Number of Clusters

A dense subgraph produced by DSP usually has strong connections among its vertices, so it can be viewed as a candidate cluster. However, the DSP method cannot partition a hypergraph into a specified number of dense subgraphs. So the relationship between the number of subgraphs produced by DSP and the parameter S_c for constructing SNN graph is studied. With the parameter $k = 32$, different S_c values ($\frac{k}{16}, \frac{k}{8}, \frac{k}{4}, \frac{k}{2}, \frac{3k}{4}, \frac{7k}{8}, \frac{15k}{16}$) are used to construct the SNN graph followed by partitioning using DSP. For the Semeion dataset containing 1,593 handwritten digits [29], the result is shown in Fig. 2. It can be seen from Fig. 2 that when S_c is small relative to k, the number of dense subgraphs produced by DSP is also small. When S_c is small, SNN graph becomes a dense graph, the size of the dense subgraphs produced by DSP will be large, so the number of the dense subgraphs will be small. It can also be seen from Fig. 2 that the number of dense

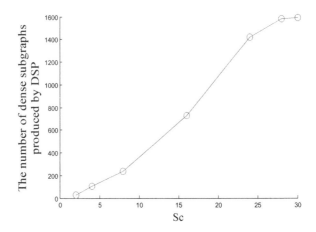

Fig. 2. The relationship between the number of subgraphs produced by DSP and the parameter Sc for the Semeion dataset.

subgraphs becomes similar to the number of input data points when S_c approaches $k = 32$. This is because many hyperedges of the hypergraph will only have a single vertex when S_c is too large relative to k.

Based on the observation, a new strategy is proposed to produce a specified number, Nc, clusters. First the parameter Sc is automatically adjusted to produce more than 2Nc dense subgraphs using DSP. In the merging phase, starting from the initial condition in which every dense subgraph is considered as a different cluster, the two most similar clusters are merged at each iteration until the specified number of clusters is produced. The flowchart of the proposed MDSG clustering method is shown in Fig. 3.

In the merging phase, the Reverse k Nearest Neighbors (RkNN) [6] is used as the similarity metric between the clusters. The similarity between cluster C1 and cluster C2 is defined as:

$$S(C_1, C_2) = \frac{\sum\limits_{V_1 \in C_1} \sum\limits_{V_2 \in C_2} \left| RkNN(V_1) \bigcap RkNN(V_2) \right|}{|C_1| \times |C_2|} \qquad (3)$$

where the Reverse k Nearest Neighbors (RkNN) is defined as follows: Given a dataset C, a query point $q \in C$, the parameter k, and a distance metric $dist(x, y)$, the set of revese k nearest neighbors of q, denoted as RkNN(q), is defined as:

$$RkNN(q) = \{p | p \in C, dist(p, q) \leq dist(p, p_k)\} \qquad (4)$$

where $p_k \in R$ is the k-th nearest neighbor of p.

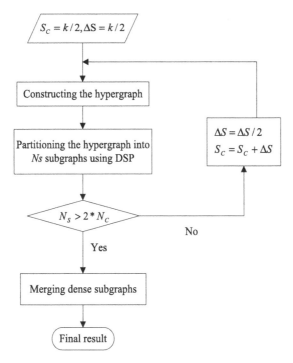

Fig. 3. The flowchart of the proposed MDSG clustering method

The pseudo-code for merging the dense subgraphs is shown in Algorithm 1.

Algorithm 1 MergingDenseSubgraphs

Input: C – The set of dense subgraph

   N_c – the number of clusters

Output: C – N_c clusters

Computing RkNN matrix using (4)

Computing the similarity matrix S for each pair of clusters in C using (3)

Repeat

$$(C_i, C_j) = \underset{C_m, C_n \in C}{argmax}[S(C_m, C_n)]$$

$$C_{new} \leftarrow C_i \cup C_j$$

$$C \leftarrow (C - \{C_i\} - \{C_j\}) \cup \{C_{new}\}$$

S is updated using (3) to reflect the correlation between the new cluster C_{new} and the other clusters.

Until $|C| = N_c$

Return C

---

3 Experimental Results

In the experiments, the parameter k for defining kNN is set to 15. Other values of k, such as 10, 20, 25, etc., are found to produce similar results. The clustering results are evaluated using the Fowlkes–Mallows index (FM-index) [30] whose range is between 0 and 1. Usually a larger FM-index indicates a better clustering result.

3.1 The Datasets

Five handwritten character datasets are used in our experiments. The MINST [31] dataset comprises of 60,000 grayscale image of handwritten digits and every image has $28 \times 28 = 784$ pixels. The USPS [32] dataset contains 11,000 grayscale images of handwritten digits with the resolution of 16×16. 1,000 images are randomly selected as the samples for both MINST and USPS in the experiments. The Semeion dataset [29] includes 1,593 handwritten digits from around 80 persons, which are stretched in a 16×16 rectangular box. The Binaryalphadigs dataset is publicly available at S. Roweis webpage [33]. The dataset contains binary images of the size 20×16, which contains the letters from A to Z and digits from 0 to 9. The Tilburg CHaracters dataset (TICH) [34] contains more than 40,000 images of handwritten characters of the size 56×56. The dataset comprise 35 classes: 25 upper-case character classes and 10 digit classes (the character 'X' is not included as a class in the dataset). 4,000 images of the TICH dataset are randomly chosen as the test samples.

3.2 Comparing Different Combinations of Preprocessing Methods and Clustering Methods

Five methods are used to do data preprocessing in the experiments, which are original data (Original), projection histograms (PH), Gaussian filtering+Projection Histograms (GF+PH), t-SNE, and Gaussian filtering+t-SNE (GF+t-SNE). Four clustering methods are used for high dimensional clustering, which are the proposed MDSG method, traditional k-mediods method [35], and two hypergraph partition methods: hMETIS and DSP+k-mediods [36]. Because the k-mediods method initializes cluster centers randomly, both k-mediods and DSP+k-mediods are run 100 times and the mean value of the 100 FM-indices is used to evaluate the result. The t-SNE method is used to reduce the dimensionalities of the USPS dataset, the Semeion dataset, the Binaryalphadigs dataset, the MINST dataset, and the TICH dataset to 50, 50, 50, 70, and 100 respectively. The results of different combinations of the methods on the five datasets are shown in Table 1.

In the experiments, the 5 different data preprocessing methods are all applied on the 5 different datasets. So, there are totally 25 cases studied. The results of the proposed MDSG method are better than the other three clustering methods for 24 out of the 25 cases. The only exception is the case of the USPS dataset preprocessed by t-SNE. For the MDSG method, the results of using the Gaussian filtering+t-SNE (GF+t-SNE) preprocessing method are better than the results of using the other 4 preprocessing

Table 1. The FM-indices produced by different combinations of preprocessing methods and clustering methods.

Dataset	Clustering method	Preprocessing				
		Original	PH[a]	GF + PH[b]	t-SNE	GF+t-SNE[c]
MINST	k-mediods	0.2893	0.2658	0.2672	0.5397	0.5408
	DSP+k-mediods	0.4661	0.4169	0.4247	0.5800	0.5807
	hMETIS	0.3149	0.2402	0.2503	0.5910	0.5151
	MDSG	**0.5660**	**0.4792**	**0.4260**	**0.6703**	**0.6580**
USPS	k-mediods	0.2580	0.2773	0.2894	0.4983	0.5725
	DSP+k-mediods	0.4154	0.3922	0.4940	**0.5443**	0.5627
	hMETIS	0.2909	0.3070	0.3594	0.3929	0.5594
	MDSG	**0.4675**	**0.5627**	**0.5448**	0.5300	**0.7017**
Semeion	k-mediods	0.3176	0.3572	0.3938	0.583	0.6771
	DSP+k-mediods	0.5153	0.3946	0.4498	0.4666	0.5029
	hMETIS	0.2478	0.2846	0.3702	0.4889	0.5944
	MDSG	**0.6069**	**0.6480**	**0.7242**	**0.6884**	**0.7925**
Binaryalphadigs	k-mediods	0.2103	0.2094	0.2110	0.3775	0.4062
	DSP+k-mediods	0.2018	0.2156	0.2573	0.3905	0.3809
	hMETIS	0.2075	0.1786	0.1897	0.3759	0.3999
	MDSG	**0.3889**	**0.3253**	**0.3124**	**0.4195**	**0.4452**
TICH	k-mediods	0.1255	0.1682	0.1956	0.3504	0.4223
	DSP+k-mediods	0.1664	0.2016	0.2139	0.3482	0.4635
	hMETIS	0.0656	0.0774	0.1008	0.2585	0.2681
	MDSG	**0.3916**	**0.4441**	**0.4459**	**0.3701**	**0.5158**

[a]projection histograms
[b]Gaussian filtering + projection histograms
[c]Gaussian filtering + t-SNE

methods, except for the MINST dataset. It shows that the combination of the Gaussian filtering+t-SNE (GF+t-SNE) preprocessing method and the proposed MDSG method usually produces the best result.

The other three clustering methods produce comparative results on Binaryalphadigs dataset, while DSP+k-mediods is better than k-mediods and hMETIS on MINST dataset, Semeion dataset and USPS dataset. For most cases, hMETIS method performs poorly compared to other three clustering methods. It can also be seen from Table 1 that the applying of Gaussian filtering usually improves the clustering results.

3.3 Evaluation of Proposed Method on Different Dimensionalities

To evaluate the effectiveness of the proposed method on high dimensional data, the MINST dataset is preprocessed by t-SNE to produce the embedded datasets at the dimensionalities of 2, 10, 70, 160, 250, 340, 430, 520, 610, 700, and 784 (original dimensionality). Then the proposed MDSG method is compared with the k-mediods method on the clustering of the embedded datasets. The results are shown in Fig. 4.

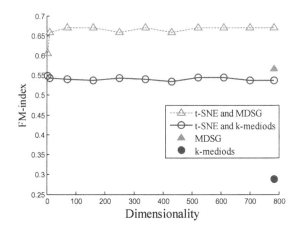

Fig. 4. The FM-indices produced by MDSG and k-mediods on the original MINST dataset (represented as filled markers) and the embedded MINST dataset (represented as unfilled markers) preprocessed by t-SNE under different dimensionalities.

It can be seen from Fig. 4 that MDSG can produce better results than k-mediods in all dimensionalities. The unfilled markers in Fig. 4 indicate the results on the embedded data using t-SNE. For the MDSG method, the results for high dimensionalities are better than the results when the dimensionality is 2, while the results of the k-mediods method are all similar for different dimensionalities. The filled markers in Fig. 4 indicate the results without using t-SNE. It can be seen that the result of MDSG is better than the result of k-mediods on both the original MINST dataset and the embedded data using t-SNE under different dimensionalities. These results show the effectiveness of the proposed MDSG method for clustering the high dimensional data. It can also be seen that the use of t-SNE can improve the clustering results regardless of the target dimensionalities.

4   Conclusions

In this work, a new high dimensional clustering method called Merging Dense Sub-Graphs (MDSG) is proposed. The method first constructs a hypergraph by finding all the maximal cliques of the SNN graph produced from the data after preprocessing. Then an improved hypergraph partition method is used to partition the hypergraph into the specified number of clusters. The experiments show that the proposed MDSG method can usually cluster the high dimensional handwritten data more effectively than the other traditional methods. In the future work, we will improve the proposed method and apply it on other types of high dimensional data.

Acknowledgement. This work is supported by the National Science Foundation of China (Grants No. 61272213) and the Fundamental Research Funds for the Central Universities (Grants No. lzujbky-2016-k07).

References

1. Hamerly, G., Elkan, C.: Alternatives to the k-means algorithm that find better clusterings. In: Proceedings of the Eleventh International Conference on Information and Knowledge Management, pp. 600–607. ACM (2002)
2. Matas, J., Kittler, J.: Spatial and feature space clustering: applications in image analysis. In: Hlaváč, V., Šára, R. (eds.) CAIP 1995. LNCS, vol. 970, pp. 162–173. Springer, Heidelberg (1995). doi:10.1007/3-540-60268-2_293
3. Natali, A., Toschi, E., Baldeweg, S., et al.: Clustering of insulin resistance with vascular dysfunction and low-grade inflammation in type 2 diabetes. Diabetes **55**(4), 1133–1140 (2006)
4. Ben-Dor, A., Shamir, R., Yakhini, Z.: Clustering gene expression patterns. J. Comput. Biol. **6**(3–4), 281–297 (1999)
5. Steinbach, M., Karypis, G., Kumar, V.: A comparison of document clustering techniques. In: KDD Workshop on Text Mining, vol. 400, no. (1), pp. 525–526 (2000)
6. Hu, T., Liu, C., Tang, Y., et al.: High-dimensional clustering: a clique-based hypergraph partitioning framework. Knowl. Inf. Syst. **39**(1), 61–88 (2014)
7. Bouveyron, C., Brunet-Saumard, C.: Model-based clustering of high-dimensional data: a review. Comput. Stat. Data Anal. **71**, 52–78 (2014)
8. Roweis, S.T., Saul, L.K.: Nonlinear dimensionality reduction by locally linear embedding. Science **290**(5500), pp. 2323–2326 (2000)
9. Zhu, X., Huang, Z., Yang, Y., et al.: Self-taught dimensionality reduction on the high-dimensional small-sized data. Pattern Recogn. **46**(1), 215–229 (2013)
10. Song, Q., Ni, J., Wang, G.: A fast clustering-based feature subset selection algorithm for high-dimensional data. IEEE Trans. Knowl. Data Eng. **25**(1), 1–14 (2013)
11. Soltanolkotabi, M., Elhamifar, E., Candes, E.J.: Robust subspace clustering. Ann. Stat. **42**(2), 669–699 (2014)
12. Bouveyron, C.: Model-based clustering of high-dimensional data in astrophysics. EAS Publ. Ser. **77**, 91–119 (2016)
13. Han, E.H., Karypis, G., Kumar, V., et al.: Hypergraph based clustering in high-dimensional data sets: a summary of results. IEEE Data Eng. Bull. **21**(1), 15–22 (1998)
14. Sun, L., Ji, S., Ye, J.: Hypergraph spectral learning for multi-label classification. In: Proceedings of the 14th ACM SIGKDD International Conference on Knowledge Discovery and Data Mining, pp. 668–676. ACM (2008)
15. Huang, Y., Liu, Q., Zhang, S., et al.: Image retrieval via probabilistic hypergraph ranking. In: 2010 IEEE Conference on Computer Vision and Pattern Recognition (CVPR), pp. 3376–3383. IEEE (2010)
16. Wang, M., Liu, X., Wu, X.: Visual classification by $\ell 1$-hypergraph modeling. IEEE Trans. Knowl. Data Eng. **27**(9), 2564–2574 (2015)
17. Fiduccia, C.M., Mattheyses, R.M.: A linear-time heuristic for improving network partitions. In: Papers on Twenty-Five Years of Electronic Design Automation, pp. 241–247. ACM (1988)

18. Huang, D.J.H., Kahng, A.B.: When clusters meet partitions: new density-based methods for circuit decomposition. In: Proceedings of the 1995 European Conference on Design and Test. IEEE Computer Society (1995)

19. Karypis, G., Aggarwal, R., Kumar, V., et al.: Multilevel hypergraph partitioning: applications in VLSI domain. IEEE Trans. Very Large Scale Integr. (VLSI) Syst. **7**(1), 69–79 (1999)

20. Cai, W., Young, E.F.Y.: A fast hypergraph bipartitioning algorithm. In: 2014 IEEE Computer Society Annual Symposium on VLSI (ISVLSI), pp. 607–612. IEEE (2014)

21. Lotfifar, F., Johnson, M.: A Serial Multilevel Hypergraph Partitioning Algorithm. arXiv preprint arXiv:1601.01336 (2016)

22. Henne, V., Meyerhenke, H., Sanders, P., et al.: n-Level Hypergraph Partitioning. arXiv preprint arXiv:1505.00693 (2015)

23. Liu, H., Latecki, L.J., Yan, S.: Dense subgraph partition of positive hypergraphs. IEEE Trans. Pattern Anal. Mach. Intell. **37**(3), 541–554 (2015)

24. Jagannathan, J., Sherajdheen, A., Deepak, R.M.V., et al.: License plate character segmentation using horizontal and vertical projection with dynamic thresholding. In: 2013 International Conference on Emerging Trends in Computing, Communication and Nanotechnology (ICE-CCN), pp. 700–705. IEEE (2013)

25. Tuba, E., Bacanin, N.: An algorithm for handwritten digit recognition using projection histograms and SVM classifier. In: 2015 23rd Telecommunications Forum Telfor (TELFOR), pp. 464–467. IEEE (2015)

26. Hinton, G., Roweis, S.: Stochastic neighbor embedding. In: NIPS, vol. 15, pp. 833–840 (2002)

27. Maaten, L., Hinton, G.: Visualizing data using t-SNE. J. Mach. Learn. Res. **9**(Nov), 2579–2605 (2008)

28. Eppstein, D., Löffle, M., Strash, D.: Listing all maximal cliques in sparse graphs in near-optimal time. In: Cheong, O., Chwa, K.Y., Park, K. (eds.) ISAAC 2010, vol. 6506, pp. 403–414. Springer, Heidelberg (2010). doi:10.1007/978-3-642-17517-6_36

29. The Semeion dataset. https://archive.ics.uci.edu/ml/datasets/Semeion+Handwritten+digit

30. Fowlkes, E.B., Mallows, C.L.: A method for comparing two hierarchical clusterings. J. Am. Stat. Assoc. **78**(383), 553–569 (1983)

31. The MNIST dataset. http://yann.lecun.com/exdb/mnist/index.html

32. The USPS dataset. http://www.cs.nyu.edu/~roweis/data/html

33. The Binaryalphadigs dataset. http://www.cs.toronto.edu/~roweis/data/binaryalphadigs.mat

34. Van der Maaten, L.: A new benchmark dataset for handwritten character recognition. Tilburg University, pp. 2–5 (2009)

35. Kaufman, L., Rousseeuw, P.: Clustering by Means of Medoids. North-Holland, Amsterdam (1987)

36. Sun, X., Tian, S., Lu, Y.: High dimensional data clustering by partitioning the hypergraphs using dense subgraph partition. In: Ninth International Symposium on Multispectral Image Processing and Pattern Recognition (MIPPR2015). International Society for Optics and Photonics (2015)

K-normal: An Improved K-means for Dealing with Clusters of Different Sizes

Yonggang Lu$^{(\boxtimes)}$, Jiangang Qiao, and Xiaochun Wang

School of Information Science and Engineering, Lanzhou University,
Lanzhou 730000, Gansu, China
ylu@lzu.edu.cn

Abstract. K-means is the most well-known and widely used classical clustering method, benefited from its efficiency and ease of implementation. But k-means has three main drawbacks: the selection of its initial cluster centers can greatly affect its final results, the number of clusters has to be predefined, and it can only find clusters of similar sizes. A lot of work has been done on improving the selection of the initial cluster centers and on determining the number of clusters. However, very little work has been done on improving k-means to deal with clusters of different sizes. In this paper, we have proposed a new clustering method, called k-normal, whose main idea is to learn cluster sizes during the same process of learning cluster centers. The proposed k-normal method can identify clusters of different sizes while keeping the efficiency of k-means. Although the Expectation Maximization (EM) method based on Gaussian mixture models can also identify the clusters of different sizes, it has a much higher computational complexity than both k-normal and k-means. Experiments on a synthetic dataset and seven real datasets show that, k-normal can outperform k-means on all the datasets. If compared with the EM method, k-normal still produces better results on six out of the eight datasets while enjoys a much higher efficiency.

Keywords: Clustering · K-means · Machine Learning

1 Introduction

Clustering is to divide data into groups that are useful and meaningful. Cluster analysis has played an important role in various fields: data mining, artificial intelligence, biology and statistics [1]. It has been shown that there is no clustering method which can produce desirable results under any conditions [2]. Actually, instead of seeking one universal method, a great number of clustering approaches have been developed to serve different purposes in different applications.

K-means is the most well-known clustering method. It is proposed a long time ago [3] and is still the most widely-used clustering method [4], benefited from its two main advantages: ease of implementation and efficiency [1, 5, 6]. It can deal with large datasets, and can usually produce acceptable results. The clustering in k-means is done by selecting initial cluster centers followed by iteratively optimizing the position of the cluster centers during a learning process [4, 7]. Because clusters are assumed to have

© Springer International Publishing AG 2017
D.-S. Huang et al. (Eds.): ICIC 2017, Part III, LNAI 10363, pp. 335–344, 2017.
DOI: 10.1007/978-3-319-63315-2_29

similar sizes, a data point is assigned to a cluster whose center is the nearest from the data point. A fuzzy version of k-means, called fuzzy c-means, has also been proposed [8], which assigns an object to a class with some degree of membership. It has been applied to image processing [9] and video processing [10].

However, the results of k-means are very sensitive to the selection of its initial cluster centers. Starting from different initial conditions, k-means usually produces different clustering results of different qualities. Therefore, many initialization methods have been proposed to improve the selection of the initial cluster centers. These methods include the random method used in the traditional k-means method, the k-means++ method using a probability model for selecting the initial cluster centers proposed by Arthur and Vassilvitskii [11], and the method using the maximum combined distances proposed by Erisoglu *et al.* [12].

The second issue of the k-means method is that the number of clusters needs to be given beforehand. Domain knowledge usually needs to be involved in determining a proper number of clusters [13]. However, automatic methods have also been proposed for selecting the number of clusters based on certain criterion. X-means can find the number of clusters by optimizing a criterion such as Akaike Information Criterion or Bayesian Information Criterion [14]. SHAKM selects the number of clusters within a range using a criterion called Cluster Evaluation Criterion [15].

Another issue of k-means which has been studied very little is that it can only identify the clusters of similar sizes, while most of the real datasets often contain clusters of various sizes. As far as we know, there are still no reports on how to improve k-means to deal with clusters of different sizes. In this paper, we have proposed a new clustering method, called k-normal, which can iteratively optimize both the position of the cluster centers and the sizes of the clusters at the same time, while still sharing the efficiency and the simplicity of the k-means method.

Although the Expectation Maximization (EM) method based on Gaussian mixture models [16] can also identify clusters of different sizes and shapes, it has a high computational complexity. The EM method needs to compute every element of the covariance matrix for each cluster at each iteration, which makes it a much slower method than both k-means and k-normal.

The rest of the paper is organized as follows: the statistical models of the k-means method and the proposed k-normal method are described in Sect. 2; experimental results are presented in Sect. 3; finally, the discussion and conclusion are given in Sect. 4.

2 The Development of the K-normal Method

In the following, a statistic model of the k-means clustering method is given before describing the proposed k-normal method.

2.1 A Statistic Model of K-means

If assuming that all clusters obey normal distribution, given a dataset $X = \{x_1, x_2, \ldots, x_N\}$ in a d-dimensional space and a set of k clusters $C_{SET} = \{C_1, C_2, \ldots, C_k\}$, the rule of clustering can be formulated as: assign x_i to C_j according to $P(x_i, C_j) = \max\limits_{C_V \in C_{SET}} P(x_i, C_V)$, where $P(x_i, C_j)$ is the probability that the point x_i belongs to the cluster C_j and can be expressed by the normal probability density function as:

$$P(x_i, C_j) = \frac{1}{\sqrt{(2\pi)^d \left| \sum_j \right|}} exp(-\frac{1}{2}(x_i - \mu_j)^T {\sum_j}^{-1} (x_i - \mu_j)) \tag{1}$$

where $\left| \sum_j \right|$ is the determinant of $\sum_j$, and $\sum_j$ and μ_j are the covariance matrix and the mean vector of the cluster C_j respectively.

In k-means, three additional assumptions are made:

(a) Different features are independent from each other, so the covariance matrix $\sum_j$ is a diagonal matrix;
(b) The shape of a cluster is a hyper-sphere, so the standard deviations of all the features in a cluster C_j are the same, i.e., $\sigma_{j1} = \sigma_{j2} = \ldots = \sigma_{jd} = \sigma_j$;
(c) The sizes of different clusters are the same, so their standard deviations are equal, i.e., $\sigma_1 = \sigma_2 = \ldots = \sigma_k = \sigma$.

Base on the three assumptions, the covariance matrices of different clusters are the same, which can be expressed as $\sum_j = \sigma_j^2 I = \sigma^2 I$, where I is the identity matrix. So, the probability of a point x_i belonging to a cluster C_j is given by:

$$\begin{aligned} P(x_i, C_j) &= \frac{1}{\sqrt{(2\pi)^d \left| \sum_j \right|}} exp(-\frac{1}{2}(x_i - \mu_j)^T {\sum_j}^{-1} (x_i - \mu_j)) \\ &= \frac{1}{(2\pi)^{d/2} \sigma_j^d} exp(-\frac{1}{2}(x_i - \mu_j)^T \frac{1}{\sigma_j^2} I(x_i - \mu_j)) \\ &= \frac{1}{(2\pi)^{d/2} \sigma^d} exp(-\frac{1}{2\sigma^2} \left\| x_i - \mu_j \right\|^2) \end{aligned} \tag{2}$$

It can be seen from (2) that the probability only depends on the distance from the point x_i to the center of the cluster C_j, which is $\left\| x_i - \mu_j \right\|$. So, a point is always assigned to the cluster whose center is the nearest from the point in k-means.

2.2   The K-normal Model for Clusters of Different Sizes

In most actual cases, the sizes of the clusters are different from each other. To identify clusters of different sizes, and to avoid the drawbacks of k-means, the statistic model described in Subsect. 2.1 is improved by using only two out of the three assumptions: the assumptions (a) and (b) are adopted while the assumption (c) is discarded.

Using the assumption (a), for each cluster C_j, all the features are assumed to be statistically independent, so the covariance matrix $\sum_j$ is a diagonal matrix, which can be described as:

$$\sum_j = \begin{bmatrix} \sigma_{j1}^2 & & & \\ & \sigma_{j2}^2 & & \\ & & \ddots & \\ & & & \sigma_{jd}^2 \end{bmatrix} \tag{3}$$

Using the assumption (b), the standard deviations of all the features in the same cluster are assumed to be equal, so (3) can be simplified as:

$$\sum_j = \begin{bmatrix} \sigma_j^2 & & & \\ & \sigma_j^2 & & \\ & & \ddots & \\ & & & \sigma_j^2 \end{bmatrix} = \sigma_j^2 I \tag{4}$$

where I is the identity matrix. It follows from (4) that $\left| \sum_j \right| = \sigma_j^{2d}$ and $\sum_j^{-1} = \sigma_j^{-2} I$. So, the probability given by (1) becomes:

$$p(x_i, C_j) = \frac{1}{(2\pi)^{d/2} \sigma_j^d} exp(-\frac{1}{2\sigma_j^2} \|x - \mu_j\|^2) \tag{5}$$

Because only the relative value of the probability matters during the clustering process, the constant $2\pi^{d/2}$ can be ignored, and (5) becomes:

$$p(x_i, C_j) \propto \frac{1}{\sigma_i^d} exp(-\frac{1}{2\sigma_j^2} \|x - \mu_j\|^2) \tag{6}$$

where μ_j is the center of the cluster C_j, and σ_j is the standard deviation of the cluster C_j, which is given by:

$$\sigma_j = \sqrt{\frac{1}{|C_j| - 1} \sum_{x \in C_j} \|x - \mu_j\|^2} \tag{7}$$

2.3   The Procedure of the K-normal Method

Based on the improved model described in Subsect. 2.2, the procedure of the k-normal method is designed as follows:

1. Initialization: choose k data points as the initial cluster centers: $\mu_1, \mu_2, \ldots, \mu_k$, using certain initialization method; and set the standard deviations of all the clusters to one: $\sigma_j = 1$ for $j \in \{1, 2, \ldots, k\}$;

2. For each data point $i \in \{1, 2, \ldots, N\}$, assign the data point x_i to the cluster C_m such that the probability computed using (6) is the maximum, *i.e.*, $C_m = \underset{C_j \in C_{SET}}{\arg\max} P(x_i, C_j)$;

3. For each cluster C_j, $j \in \{1, 2, \ldots, k\}$, compute the new center μ_j using $\mu_j = \frac{1}{|C_j|} \sum_{x \in C_j} x$;

4. For each cluster $C_j, j \in \{1, 2, \ldots, k\}$, compute the new standard deviation σ_j using (7);

5. Repeat Steps 2, 3 and 4 until the clustering result becomes stable or the maximum iteration is reached.

2.4   Illustration of the K-normal Method

To show the benefit of the proposed method, it is compared with k-means on a 2D synthetic dataset. The differences in the clustering process are shown in Fig. 1. The three plots on the first row from left to right show the initial stage, the intermediate stage and the final stage of the k-means clustering process respectively, and the three plots on the second row show these of k-normal. It can be seen that the proposed k-normal method can learn the cluster sizes during the clustering process and produces better clustering results than these of k-means at the final stage.

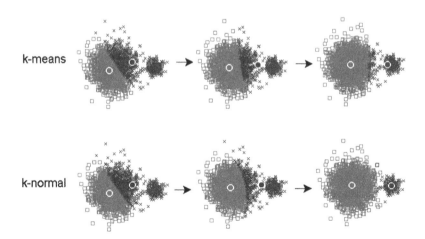

Fig. 1. The clustering process of k-normal *vs.* that of k-means.

2.5   Improving the Convergence of the K-normal Method

It is found in the experiments that if (7) is used to compute the new standard deviations of clusters as described in Subsect. 2.3, the value of the computed standard deviation may oscillate, which results in slow convergence rate in most cases. For example, when applied to the Wine dataset from UCI Machine Learning Repository [17], k-normal cannot converge within 27 h if (7) is used in our experiment.

To speed up the convergence of the proposed algorithm, after the first iteration, the weighted sum of the standard deviations computed at the previous iteration $t - 1$ and the one computed by (7) at the current iteration t are used as the new standard deviation in the step 4 described in Subsect. 2.3:

$$\sigma_j^{(t)} = w \times \sigma_j + (1 - w) \times \sigma_j^{(t-1)} \tag{8}$$

where σ_j is the standard deviation computed by (7) at the current iteration, and $\sigma_j^{(t-1)}$ is the standard deviation computed by (8) at the previous iteration for the same cluster C_j, and w is the weight which is set to 0.5 in our experiments.

3   Experimental Results

To evaluate the proposed k-normal method, it is compared with the k-means method and the EM method [16]. All the codes are implemented in MATLAB and the experiments are run on a desktop computer with an Intel 3.40 GHz CPU and 16 GB of RAM.

3.1   Criteria for Comparing Performance

The Fowlkes-Mallows index (FM-index) is usually used to determine the similarity between two clustering results [18]. In our experiments, the measure is selected to evaluate the consistency between the clustering results and the benchmark.

Given two different clustering results R1 and R2, the Fowlkes-Mallows index can be computed by: $FM = \sqrt{\frac{TP}{TP+FP} * \frac{TP}{TP+FN}}$, where TP is the number of true positives which are the pairs of points that are in the same cluster in both R1 and R2; FP is the number of false positives which are the pairs of points that are in the same cluster in R1 but not in R2; FN is the number of false negatives which are the pairs of points that are in the same cluster in R2 but not in R1; TN is the number of true negatives which are the pairs of points that are in different clusters in both R1 and R2. If the two clustering results match exactly, both FP and FN will be zero, and the Fowlkes-Mallows index will take the maximum value 1; if the two clustering results are completely different, TP will be zero and the Fowlkes-Mallows index will take the minimum value 0. So, a larger Fowlkes-Mallows index indicates a better match between the clustering result and the benchmark.

3.2 Datasets

The clustering methods are applied on a synthetic dataset and seven real datasets. For the synthetic dataset, it is composed of 4 spherical clusters of different sizes, with 2400 data points in 2-dimensional space. The seven real datasets are Wine (dimensionality: 13, data size: 178), Page Blocks Classification (dimensionality: 10, data size: 5473), Pima Indians Diabetes (dimensionality: 8, data size: 768), MAGIC Gamma Telescope (dimensionality: 10, data size: 19020), Blood Transfusion Service Center (dimensionality: 4, data size: 748), Haberman's Survial (dimensionality: 3, data size: 306), and Glass Identification (dimensionality: 10, data size: 214) downloaded from the UCI Machine Learning Repository [17].

3.3 Performance Comparison

The proposed k-normal method is compared with the k-means method and the EM method [16]. The initial cluster centers are choosing at random for all the three clustering methods. The only parameter for the k-means method, the k-normal method and the EM method is the number of clusters. In the experiments, the actual number of clusters is used for all the three methods. Because they are all the randomized methods, each of the methods is run 100 times, and both the maximum and the average FM-indices of the clustering results are recorded in Table 1, while their average running time in milliseconds are recorded in Table 2. For the running time of the three clustering methods on all the datasets, the average running time versus the size of the dataset is also plotted in Fig. 2.

Table 1. The FM-indices of the clustering results produced using the three different methods.

Datasets	k-means	EM		k-normal		
	Max	Mean	Max	Mean	Max	Mean
Wine	0.5994	0.5868	**0.8784**	**0.6491**	0.6839	0.6344
Pima Indians Diabetes	0.6308	0.6308	0.7093	0.5383	**0.7123**	**0.7118**
Page Blocks Classification	0.7385	0.7385	0.6606	0.5674	**0.9005**	**0.7979**
MAGIC Gamma Telescope	0.6309	0.6309	0.5704	0.5704	**0.7376**	**0.7374**
Blood Transfusion Service Center	0.7377	0.7114	0.7535	0.6158	**0.7864**	**0.7406**
Haberman's Survial	0.7329	0.5624	0.6043	0.6043	**0.7652**	**0.6131**
Glass Identification	0.5105	0.4603	0.5432	0.4529	**0.5488**	**0.4954**
The Synthetic Dataset	0.8879	0.8156	**0.9624**	**0.9430**	0.9446	0.8406

From the Table 1, it is obvious that the proposed k-normal method usually produces better FM-indices than both the k-means method and the EM method. Compared to the k-means method, the k-normal method produces better FM-indices for all the datasets. Compared to the EM method, the k-normal method produces better FM-indices for 6 out of the 8 datasets, where the only exceptions are the Wine and the synthetic dataset.

It is noted that k-normal produces the best results for the six datasets including the Page Blocks Classification, Pima Indians Diabetes, MAGIC Gamma Telescope

Table 2. The average running time in *milliseconds* of the three clustering methods.

Datasets	k-means	EM	k-normal
Wine	0.565	19.151	3.426
Pima Indians Diabetes	1.625	14.033	2.106
Page Blocks Classification	19.806	607.338	30.844
MAGIC Gamma Telescope	32.088	264.295	29.933
Blood Transfusion Service Center	0.900	23.843	1.729
Haberman's Survial	0.566	12.019	1.747
Glass Identification	0.859	21.403	2.984
The Synthetic Dataset	3.149	176.004	10.555

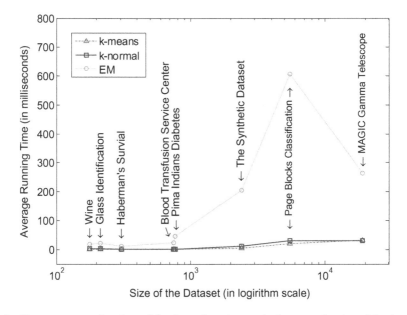

Fig. 2. The average running time of the three clustering methods versus the size of the datasets.

datasets, Blood Transfusion Service Center, Haberman's Survial and Glass Identification. For the Page Blocks Classification dataset, the average FM-indices produced by k-normal is 8% higher than that produced by k-means, while the maximum FM-indices produced by k-normal is more than 20% and 36% higher than these produced by k-means and EM. The average FM-indices produced by k-normal are 12.8% and 16.9% higher than these produced by k-means for the Pima Indians Diabetes dataset and the MAGIC Gamma Telescope dataset respectively. For the Blood Transfusion Service Center dataset, the average FM-indices produced by k-normal are 4.1% and 20.3% higher than these produced by k-means and EM respectively. For the Haberman's Survial dataset, the average FM-indices produced by k-normal is 9% higher than that produced by k-means. For the Glass Identification dataset, the average FM-indices produced by k-normal are 7.6% and 9.4% higher than these produced by k-means and EM respectively.

For the Wine dataset and the synthetic dataset, the EM method produces higher FM-indices than these produced by k-normal and k-means, while k-normal still produces higher FM-indices than k-means for the two datasets. The Wine dataset contains only 178 data points, which is the smallest dataset used in the experiments. The synthetic dataset is a two dimensional datasets, while the dimensionalities of the seven real datasets are ranged from 3 to 13. It can be seen that k-normal only fails to produce the best results for a dataset with the smallest data size and a dataset with the lowest dimensionality.

From Table 2 and Fig. 2, it can be seen that the average running time of k-normal is comparable to that of k-means, while the EM method usually takes much longer running time than both k-normal and k-means for all the datasets. It can be seen from Fig. 2 that, although the running time of k-normal is usually longer than that of k-means, the difference in the running time becomes smaller and even negligible with the increasing of the data size. For the Wine dataset with 178 data points, k-means is several times faster than k-normal. But for the MAGIC Gamma Telescope dataset with 19020 data points, k-means has a similar running time as that of k-normal. It can also be seen from Fig. 2 that, with the increasing of the data size, the running time of the EM method increases very rapidly, while the running time of k-normal and k-means only increases mildly.

Usually the running time increases with the increasing of the data sizes, except for the MAGIC Gamma Telescope dataset at the end of the curves. It is noticed that, although the size of the MAGIC Gamma Telescope dataset is the largest, it contains only 2 clusters, while the second largest dataset (Page Blocks Classification dataset) contains 5 clusters, which may be why the running time of the MAGIC Gamma Telescope dataset is shorter than that of the Page Blocks Classification dataset when the k-normal method or the EM method is used.

4   Discussion and Conclusion

Compared to the k-means method which only learns the cluster centers, the proposed k-normal method learns both the cluster centers and the cluster sizes during the clustering process. As a result, the k-normal method can be used to identify clusters of different sizes. Experimental results show that the k-normal method can consistently produce better results than the k-means method, while still enjoying the efficiency of the k-means method. Compared to another state-of-the-art method, the EM method, the k-normal method can still produce better results on most of the datasets. Experiments also show that the computational complexity of the k-normal method is much lower than that of the EM method. Although the EM method can also identify clusters of different sizes and different shapes, its performance may be deteriorated by its complicated models on large and high dimensional datasets, because it needs to compute the entire covariance matrix at each iteration. Different from the EM method, the k-normal method assumes that all the clusters are in globular shapes, which may seem to be a drawback, but it turns out that the simplification can bring the improvements on both the time complexity and the clustering quality.

In the further work, we will further improve the convergence of the k-normal method to improve its efficiency.

Acknowledgment. This work is supported by the National Natural Science Foundation of China (Grants No. 61272213) and the Fundamental research Funds for the Central Universities (Grants No. lzujbky-2016-k07).

References

1. Omran, M.G., Engelbrecht, A.P., Salman, A.: An overview of clustering methods. Intell. Data Anal. **11**, 583–605 (2007)
2. Kleinberg, J.: An impossibility theorem for clustering. In: Advances in Neural Information Processing Systems (NIPS) 15, Vancouver, British Columbia, Canada, 9–14 December, pp. 463–470 (2002)
3. Steinhaus, H.: Sur la division des corps matriels en parties. Bull. Acad. Polon. Sci. (in French) **4**, 801–804 (1957)
4. Jain, A.K.: Data clustering: 50 years beyond k-means. Pattern Recogn. Lett. **31**, 651–666 (2010)
5. Arthur, D., Vassilvitskii, S.: How slow is the k-means method. In: Proceedings of the Twenty-Second Annual Symposium on Computational Geometry, pp. 144–153 (2006)
6. Har-Peled, S., Sadri, B.: How fast is the k-means method. In: Proceedings of the Sixteenth Annual ACM-SIAM Symposium on Discrete Algorithms, Society for Industrial and Applied Mathematics, USA (2005)
7. MacKay, D.: An example inference task: clustering. In: Information Theory, Inference and Learning Algorithms, Cambridge University Press (2003)
8. Bezdek, J.: A convergence theorem for the fuzzy isodata clustering algorithms. Pattern Anal. Mach. Intell. **2**, 1–8 (1980)
9. Schaefer, G., Zhou, H.: Fuzzy clustering for colour reduction in images. Telecommun. Syst. **40**(1-2), 17–25 (2009)
10. Zhou, H., Sadka, A.H., Swash, M.R., Azizi, J., Sadiq, U.A.: Feature extraction and clustering for dynamic video summarisation. Neurocomputing **73**(10–12), 1718–1729 (2010)
11. Arthur, D., Vassilvitskii, S.: k-means++: the advantages of careful seeding. In: Proceedings of the Eighteenth Annual ACM-SIAM Symposium on Discrete Algorithms, pp. 1027–1035 (2007)
12. Erisoglu, M., Calis, N., Sakallioglu, S.: A new algorithm for initial cluster centers in k-means algorithm. Pattern Recogn. Lett. **32**, 1701–1705 (2011)
13. Hung, M.C., Wu, J., Chang, J.H., Yang, D.L.: An efficient k-Means clustering algorithm using simple partitioning. J. Inf. Sci. Eng. **21**, 1157–1177 (2005)
14. Pelleg, D., Moore, A.W.: X-means: extending k-means with efficient estimation of the number of clusters. In: Proceedings of the Seventeenth International Conference on Machine Learning, pp. 727–734 (2000)
15. Liao, K.Y., Liu, G.Z., Xiao, L., Liu, C.T.: A sample-based hierarchical adaptive k-means clustering method for large-scale video retrieval. Knowl.-Based Syst. **49**, 123–133 (2013)
16. Redner, R.A., Walker, H.F.: Mixture densities, maximum likelihood and the EM algorithm. SIAM Rev. **26**, 195–239 (1984)
17. UCI Machine Learning Repository. http://archive.ics.uci.edu/ml (2017)
18. Fowlkes, E.B., Mallows, C.L.: A method for comparing two hierarchical clusterings. J. Am. Stat. Assoc. **78**, 553–569 (1983)

Kernel Methods and Supporting Vector Machines

Kernel Based Non-Negative Matrix Factorization Method with General Kernel Functions

Wen-Sheng Chen[1,2], Liping Deng[1], Binbin Pan[1,2(✉)],
and Yang Zhao[3]

[1] College of Mathematics and Statistics, Shenzhen University, Shenzhen
518160, People's Republic of China
{chenws,pbb}@szu.edu.cn
[2] Shenzhen Key Laboratory of Media Security, Shenzhen University, Shenzhen
518160, People's Republic of China
[3] College of Information Engineering, Shenzhen University, Shenzhen 518160,
People's Republic of China

Abstract. Kernel based Non-Negative Matrix Factorizations (KNMFs) are one of the most important methods for non-negative nonlinear feature extractions and have achieved good performance in pattern classifications. However, most existing KNMF algorithms are merely valid for one special kernel function. Also, they model the pre-images inaccurately. In this paper, we utilize kernel matrix learning strategy to develop a Universal KNMF (UKNMF) algorithm, which is able to use all Mercer kernel functions. The proposed method avoids the pre-image learning simultaneously. We first establish three objective functions and then derive three update formula to determine three matrices, namely one feature matrix and two kernel matrices. The iterative rules are theoretically proven to be convergence by means of auxiliary function technique. Our UKNMF approaches with polynomial kernel and RBF kernel (UKNMF-Poly and UKNMF-RBF) are applied to face recognition respectively. The face databases, including ORL and Yale face databases, are selected for evaluations. Compared with some state of the art kernel based algorithms, experimental results show the effectiveness and superior performance of the proposed methods.

Keywords: Non-negative matrix factorization (NMF) · Non-negative feature extraction · Kernel method · Face recognition

1 Introduction

It is known that the feature extraction is a crucial step in face recognition. Three typical feature extraction approaches are principal component analysis (PCA) [1], linear discriminant analysis (LDA) [2] and locality preservation projections (LPP) [3]. These algorithms are based on different criterion and the features are able to be acquired by solving their corresponding eigen-systems. For these methods, we see that there are no restrictions on the sign of both basis-matrices and features. Unlike aforementioned

© Springer International Publishing AG 2017
D.-S. Huang et al. (Eds.): ICIC 2017, Part III, LNAI 10363, pp. 347–359, 2017.
DOI: 10.1007/978-3-319-63315-2_30

methods, non-negative matrix factorization (NMF) attempts to uncover the non-negative features from the non-negative data which widely exist in the real world, such as facial image data, hyper-spectral data [4], and gene expression data [5] etc. In detail, NMF approximately decomposes the non-negative data matrix X into two non-negative matrices W and H, namely, $X \approx WH, W, H \geq 0$, where W and H are called the basis matrix and the feature matrix respectively. Based on the measurements of Euclidean distance and asymmetric divergence, Lee and Seung [6] proposed two kinds of multiplicative update rules for NMF which are shown to be convergence. NMF, which has the ability of the parts-based feature representation, is actually a linear feature extraction method. Nevertheless, facial image data are non-linearly distributed in the pattern space because of pose and illumination variations. Therefore, the performance of NMF will be degraded when dealing with the nonlinear problem in face recognition. Kernel method is generally utilized to tackle the nonlinear problem. It is first to map the original data into a high dimensional Reproducing Kernel Hilbert Space (RKHS) via a nonlinear mapping such that the mapped data are linearly separated, and then performs the linear methods in the RKHS. Along this line, many linear algorithms have been extended to their kernel counterparts, such as kernel PCA (KPCA) [7], kernel LDA (KLDA) [9] kernel LPP (KLPP) [10] and so on. Empirical results demonstrate that kernel based methods outperform their linear versions. While for kernel based NMF approaches, they implement linear NMF in RKHS and find the decomposition as with $W, H \geq 0$, where W here is called the pre-image matrix and H is the feature matrix. Based on the loss function with Frobenius norm, a large number of KNMFs have been developed in recent years. The representative KNMF is the polynomial kernel non-negative matrix factorization (PNMF) [11]. However, most existing KNMF methods, involving PNMF, are only effective for one special kernel function. Their update rules fail to use the other general kernels. On the other hand, they need to model the pre-image matrix. It is difficult to obtained the pre-image accurately.

To overcome the universal kernel problem and pre-image learning problem, this paper propose a novel framework on universal KNMF (UKNMF) method. Meanwhile, our UKNMF can avoid the pre-image problem directly. We first establish three loss functions and then derive three update rules to determine three matrices, namely one feature matrix and two kernel matrices. The iterative rules are theoretically shown to be convergence by means of auxiliary function technique. Two kernels, namely polynomial kernel and RBF kernel, are exploited for our UKNMF approaches. They are called UKNMF-Poly and UKNMF-RBF respectively, and successfully applied to face recognition. Our KNMF approaches are tested on ORL and Yale face databases, respectively. Some state of the art kernel based algorithms are used for comparisons. Experimental results show that our UKNMF methods surpass the compared kernel-based methods.

The rest of this paper is organized as follows. In Sect. 2, we briefly introduce linear NMF and PNMF algorithms. The proposed UKNMF approach and the convergence analysis are given in Sect. 3. The experimental results on face recognition are reported in Sect. 4. The final conclusions are drawn in Sect. 5.

2   Related Works

This section will introduce linear NMF and KNMF with polynomial kernel function. Details are as follows.

2.1   Non-negative Matrix Factorization (NMF)

Let $X \in R_+^{m \times n}$ be a non-negative matrix containing n column training samples with dimension m. NMF aims to factorize X into two non-negative matrices $W \in R_+^{m \times r}$, $H \in R_+^{r \times n}$, namely $X \approx WH, W, H \geq 0$. Parameter r represents the number of features. Matrices W and H are called basis matrix and feature matrix respectively. To this end, we need to solve the following optimization problem:

$$\min_{W,H} \frac{1}{2} \|X - WH\|_F^2, \ W, H \geq 0 \tag{1}$$

The problem (1) can be converted to two convex optimization problems by fixing one of matrices W, H, The following update rules are finally acquired using gradient descent method [6]:

$$\begin{aligned} H^{(t+1)} &= H^{(t)} \otimes W^{(t)T} X \emptyset W^{(t)T} W^{(t)} H^{(t)}, \\ W^{(t+1)} &= W^{(t)} \otimes X H^{(t+1)T} \emptyset W^{(t)} H^{(t)} H^{(t+1)T}, \\ W^{(t+1)} &= W^{(t+1)} \emptyset S \end{aligned} \tag{2}$$

Where $\otimes$ and $\emptyset$ denotes element-wise multiplication and division respectively. The third equation in (2) is to normalize the basis images such that the sum of each column in W equals to one. The elements of S are given by $s_{ij} = \sum_{i=1}^r W_{ij}, j = 1, \cdots, r$. The update rules (2) have shown to be convergence to a local optimization solution.

2.2   Polynomial Kernel Non-negative Matrix Factorization

Kernel NMF attempts to represent the mapped training data $\phi(X)$ using the mapped pre-image in RKHS under the non-negative constraint. We denote $\phi(X)$ and $\phi(W)$ by $\phi(X) = [\phi(x_1), \phi(x_2), \cdots, \phi(x_n)]$, and $\phi(W) = [\phi(w_1), \phi(w_2), \cdots, \phi(w_r)]$ respectively. The feature matrix H is rewritten as $H = [h_1, h_2, \cdots, h_n]$. Then, $\phi(X)$ has the following representation under the mapped pre-images.

$$\phi(x_j) \approx \sum_j \phi(w_j) h_{ji}, \ i = 1, 2, \cdots, n,$$

where feature $h_{ji} \geq 0$ and pre-image $w_j \geq 0$. These representations can be reformulated in the matrix factorization form shown below.

$$\phi(X) \approx \phi(W)H \tag{3}$$

To obtain the decomposition (5), it usually employs the loss function $F_\phi(W,H)$ as $F_\phi(W,H) = \frac{1}{2}\|\phi(X) - \phi(W)H\|_F^2$. Hence, KNMF needs to solve the following constrained optimization problem:

$$\min_{W,H} F_\phi(W,H), \; subject\; to\; W, H \geq 0 \tag{4}$$

PNMF [8] approach resolves the problem (4) by adopting gradient descent method in polynomial feature space and derives out the following iterative formulae.

$$H^{(t+1)} = H^{(t)} \otimes K_{XW}^{(t)} \emptyset K_{WW}^{(t)} H^{(t)},$$
$$W^{(t+1)} = W^{(t)} \otimes X K_{WW}^{'(t)} \emptyset W^{(t)} \Omega K_{WW}^{'(t)}, \tag{5}$$
$$W^{(t+1)} = W^{(t+1)} \emptyset S$$

where $S = (S_{ij})_{m \times r}$ with $s_{ij} = \sum_{i=1}^{r} W_{ij}, j = 1, \cdots, r$, Ω is a diagonal matrix whose diagonal elements are given by $\Omega_{aa} = \sum_{j=1}^{n} H_{aj}$, $\left[K_{XW}^{'(t)}\right]_{ij} = d\left\langle x_i, w_j^{(t)} \right\rangle^{d-1}$ and $\left[K_{WW}^{'(t)}\right]_{ij} = d\left\langle w_i^{(t)}, w_j^{(t)} \right\rangle^{d-1}$. It can be seen that PNMF is not suitable for all the other kernel functions except polynomial kernel. In addition, for fixed H, the loss function $F_\phi(W,H)$ is not quadratic with respect to W. So, the pre-images learning of PKNMF is inaccurate.

3 Proposed Universal KNMF

In this section, we will propose a novel universal KNMF (UKNMF) approach which has two advantages over other kernel based NMF methods. Firstly, all kernel functions can be applied to our UKNMF directly. Secondly, we can avoid inaccurate pre-image learning. Different from the existing KNMF approaches, the proposed UKNMF is based on three loss functions and just needs to learn three non-negative matrices, namely one feature matrix H and two kernel matrices (K_{WX} and K_{WW}). The obtained update rules are listed below:

$$H^{(t+1)} = H^{(t)} \otimes K_{WX}^{(t)} \emptyset \left(K_{WW}^{(t)} H^{(t)}\right), \tag{6}$$

$$K_{WX}^{(t+1)} = K_{WX}^{(t)} \otimes \left(H^{(t+1)} K_{XX}\right) \emptyset \left(H^{(t+1)} H^{(t+1)T} K_{WX}^{(t)}\right), \tag{7}$$

$$K_{WW}^{(t+1)} = K_{WW}^{(t)} \otimes \left(K_{WX}^{(t+1)} H^{(t+1)T} + H^{(t+1)} K_{WX}^{(t+1)T}\right) \emptyset$$
$$\left(H^{(t+1)} H^{(t+1)T} K_{WW}^{(t)} + K_{WW}^{(t)} H^{(t+1)} H^{(t+1)T}\right), \tag{8}$$

where $K_{XX} = \phi^T(X)\phi(X)$, $K_{WX} = \phi^T(W)\phi(X)$, $K_{WW} = \phi^T(W)\phi(W)$. Notations $\otimes$ and $\oslash$ denote element-wise multiplicative and division respectively.

In detail, three objective functions are established in this paper as follows:

$$F_1(H) = \frac{1}{2}\|\phi(X) - \phi(W)H\|_F^2, \tag{9}$$

$$F_2(K_{WX}) = \frac{1}{2}\|K_{XX} - H^T K_{WX}\|_F^2, \tag{10}$$

$$F_3(K_{WW}) = \|K_{WX} - H^T K_{WW}\|_F^2, \tag{11}$$

In essence, these loss functions are able to estimate the degree of approximation between $\phi(X)$ and $\phi(W)$. To determine three unknown matrices H, K_{WX} and K_{WW}, we need to solve the following three optimization sub-problems, where two of H, K_{WX} and K_{WW} are fixed and the rest one will be resolved.

$$\min_H \quad F_1(H), \ s.t. \ H \geq 0, \tag{12}$$

$$\min_{K_{WX}} F_2(K_{WX}), \ s.t. \ K_{WX} \geq 0, \tag{13}$$

$$\min_{K_{WW}} F_3(K_{WW}), \ s.t. \ K_{WW} \geq 0, \tag{14}$$

For UKNMF-Poly and UKNMF-RBF, the main calculation complexity is that of K_{WW} which requires $O(mn^2 d)$ and $O((m+1)n^2)$ respectively because of $m \gg r$.

3.1   Updating Feature Matrix H

When the matrices K_{WX} and K_{WW} are fixed, the sub-problem (12) becomes a convex optimization problem. We update H using gradient descent method as follows:

$$H^{(t+1)} = H^{(t)} - \rho(H^{(t)}) \otimes \nabla F_1(H^{(t)}), \tag{15}$$

where $\rho(H^{(t)})$ is a step-size matrix and $\nabla F_1(H^{(t)})$ is the gradient of F_1 with respective to $H^{(t)}$. It can be obtained by direct calculation that

$$\nabla F(H^{(t)}) = -K_{WX} + H^{(t)T} K_{WW}. \tag{16}$$

In order to ensure the non-negativity of $H^{(t+1)}$, we properly choose that

$$\rho(H^{(t)}) = H^{(t)} \oslash (H^{(t)T} K_{WW} \tag{17}$$

Substituting (16) and (17) into (15), it yields the multiplicative update formula (6) directly.

3.2   Updating Kernel Matrix K_{WX}

For sub-problem (13) with both H and K_{WW} fixed, the kernel matrix K_{WX} is updated via gradient descent method, namely

$$K_{WX}^{(t+1)} = K_{WX}^{(t)} - \rho(K_{WX}^{(t)}) \otimes \nabla F_2(K_{WX}^{(t)}), \tag{18}$$

where $\rho(K_{WX}^{(t)})$ is a step-size matrix, $\nabla F_2(K_{WX}^{(t)})$ is the gradient of F_2 at point $K_{WX}^{(t)}$ and

$$\nabla F_2(K_{WX}^{(t)}) = -HK_{WW} + HH^T K_{WX}^{(t)}. \tag{19}$$

If setting the step-size to

$$\rho(K_{WX}^{(t)}) = K_{WX}^{(t)} \emptyset (K_{WW}^{(t)} HH^T + HH^T K_{WW}^{(t)}) \tag{20}$$

and substituting (19) and (20) into (18), we can drive the update rule of K_{WX} shown in (7).

3.3   Updating Kernel Matrix K_{WW}

Similarly, for fixed matrices H and K_{WX}, the sub-problem (14) can be resolved using gradient descent method as well. Thereby, we have

$$K_{WW}^{(t+1)} = K_{WW}^{(t)} - \rho(K_{WW}^{(t)}) \otimes \nabla F_3(K_{WW}^{(t)}), \tag{21}$$

where $\rho(K_{WX}^{(t)})$ is a step-size matrix, and the gradient $\nabla F_3(K_{WW}^{(t)})$ is calculated as

$$\nabla F_3(K_{WW}^{(t)}) = -K_{WX}H^T - HK_{WX}^T + K_{WW}^{(t)} HH^T + HH^T K_{WW}^{(t)}. \tag{22}$$

Let

$$\rho(K_{WW}^{(t)}) = K_{WW}^{(t)} \emptyset (K_{WW}^{(t)} HH^T + HH^T K_{WW}^{(t)}) \tag{23}$$

and substitute (22) and (23) into (21), the iterative formula (8) can be acquired.

3.4   The Convergence of Universal KNMF

In this section, we will discuss the the convergence of our UKNMF algorithm. For the three loss functions (9–11), we have the following theorems.

Theorem 1. *For fixed kernel matrices K_{WX} and K_{WW}, the loss function F_1 (9) is non-increasing under the update rule*

$$H^{(t+1)} = H^{(t)} \otimes K_{WX}^{(t)} / \left(K_{WW}^{(t)} H^{(t)} \right),$$

Theorem 2. *For fixed feature matrix H and kernel matrix K_{WW}, the loss function F_2 (10) is non-increasing under the update rule*

$$K_{WX}^{(t+1)} = K_{WX}^{(t)} \otimes \left(H^{(t+1)} K_{XX} \right) \emptyset \left(H^{(t+1)} H^{(t+1)T} K_{WX}^{(t)} \right)$$

Theorem 3. *For fixed feature matrix H and kernel matrix K_{WX}, the loss function $F_3(11)$ is non-increasing under update rule*

$$K_{WW}^{(t+1)} = K_{WW}^{(t)} \otimes \left(K_{WX}^{(t+1)} H^{(t+1)T} + H^{(t+1)} K_{WX}^{(t+1)T} \right) \emptyset$$
$$\left(H^{(t+1)} H^{(t+1)T} K_{WW}^{(t)} + K_{WW}^{(t)} H^{(t+1)} H^{(t+1)T} \right),$$

These theorems can be proved by means of auxiliary function technique. The definition and property of the auxiliary function are shown below.

Definition 1. *$G(h, h^{(t)})$ is called an auxiliary function of $F(h)$, if the conditions*

$$G(h, h^{(t)}) \geq F(h), \ G(h^{(t)}, h^{(t)}) \geq F(h^{(t)})$$

are satisfied.

Lemma 1. *If $G(h, h^{(t)})$ is an auxiliary function of $F(h)$, then $F(h)$ is non-increasing under the update rule:*

$$h^{(t+1)} = \arg \min G(h, h^{(t)})$$

In the sequel, we focus on the proof of theorem 3, the other two theorems can be proved similarly.

Theorem 4. *If L is a diagonal matrix with its entry*

$$L_{ab}(K_{Ww_j}^{(t)}) = \delta_{ab} \left[(K_{WW}^{(t)} H H^T)_{aj} + (H H^T K_{Ww_j}^{(t)})_a \right] \emptyset (K_{Ww_j}^{(t)})_a$$

Then

$$G(K_{Ww_j}, K_{Ww_j}^{(t)}) = F(K_{Ww_j}^{(t)}) + \left(K_{Ww_j} - K_{Ww_j}^{(t)} \right)^T \nabla F(K_{Ww_j}^{(t)}) + \frac{1}{2} \left(K_{Ww_j} - K_{Ww_j}^{(t)} \right)^T L(K_{Ww_j}^{(t)}) \left(K_{Ww_j} - K_{Ww_j}^{(t)} \right) \tag{24}$$

is an auxiliary function for

$$F(K_{Ww_j}) = \sum_i ((K_{Ww_j})_i - \sum_a H_{ai}(K_{Ww_j})_a) \tag{25}$$

Where $K_{Ww_j} = \left[k(w_1, w_j), k(w_2, w_j), \cdots, k(w_r, w_j) \right]^T$ *and* $K_{Ww_j} = \left[k(w_1, x_j), k(w_2, x_j), \cdots, . \ k(w_r, x_j) \right]^T$ *are the jth columns of* K_{WW} *and* K_{WX} *respectively.*

Due to the limitation of space, the proof is omitted here. By Theorem 4, the iterative formula (8) can be obtained via auxiliary function (25). To this end, we compute the gradient of G with respect to K_{Ww_j} and set it to zero, namely $\nabla G(K_{Ww_j}) = 0$. Subsequently, we derive that

$$K_{Ww_j}^{(t+1)} = K_{Ww_j}^{(t)} - L(K_{Ww_j}^{(t)})_a \nabla F(K_{Ww_j}^{(t)}) \tag{26}$$

It can be seen from Lemma 1 that $F(K_{Ww_j})$ is non-increasing under the update rule (26). We record the *ath* component of (26) as:

$$(K_{Ww_j}^{(t+1)})_a = (K_{Ww_j}^{(t)})_a - L(K_{Ww_j}^{(t)})_a^{-1} \nabla F(K_{Ww_j}^{(t)})_a \tag{27}$$

Where

$$L(K_{Ww_j}^{(t)})_{aa}^{-1} = (K_{Ww_j})_a / [(HH^T K_{Ww_j})_a + (K_{WW}HH^T)_{aj}] \tag{28}$$

And

$$\nabla F_{aj}(K_{Ww_j}^{(t)}) = -(K_{WX})_{aj} - (HK_{WX}^T)_{aj} + (HH^T K_{Ww_j})_a + (K_{WW}HH^T K_{WW})_{aj} \tag{29}$$

Substituting (28) and (29) into (27), we have

$$(K_{Ww_j}^{(t+1)})_a = (K_{Ww_j}^{(t)})_a (K_{WX}H^T + HK_{WX}^T)_{aj} / [(HH^T K_{Ww_j})_a + (K_{WW}HH^T)_{aj}]. \tag{30}$$

Rewriting (30) in matrix form, it gives the update rule (8). Figure 1 shows the convergence speed of our UKNMF algorithm with the addition of iteration number. X is a random grayscale image matrix with size 112×92, Y-axis means the total cost

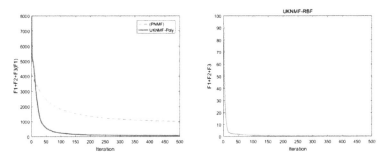

Fig. 1. r = 50, Iteration = 500, the Convergence of UKNMF-Poly and PNMF (left) with d = 2, UKNMF-RBF (right) with t = 10.

of three cost functions (9–11) for our UKNMF and the loss of cost function (9) for PNMF. Because of two distinct kernel functions are adopted, UKNMF-Poly has a higher loss at the end of 500 iterations than UKNMF-RBF. But they both have equal convergence rate. On the contrary, PNMF (the dotted line in left figure) is inferior to our method with regard to loss and convergence rate.

3.5   Feature Extraction and Classification

This section will discuss how to extract the feature of query sample using the proposed UKNMF method. Assume y is a query data, its feature h_y can be modeled according to the following procedure.

$$
\begin{aligned}
\phi(y) &= \phi(W)h_y \Rightarrow \phi^T(X)\phi(y) = \phi^T(X)\phi(W)h_y \Rightarrow \\
K_{Xy} &= K_{WX}^T h_y \Rightarrow h_y = (K_{WX}^T)^+ K_{Xy},
\end{aligned}
\tag{31}
$$

where $(K_{WX}^T)^+$ is the pseudo inverse of matrix K_{WX}^T. Based on above analysis, our UKNMF algorithm, including feature extraction and classification, is designed below.

Feature Extraction Stage

Step 1: Given feature number r, maximum number of iterations I_{max} and error bound ε; Initialize matrices $H^{(0)}$ and $W^{(0)}$.

Step 2: For a given kernel function $k(x,y)$, compute three kernel matrices K_{XX}, $K_{WX}^{(0)}$ and $K_{WW}^{(0)}$.

Step 3: Update the matrices H, K_{WX}, K_{WW}, using UKNMF update rules (6-8).

Step 4: Let the total loss function F be $F = F_1(H) + F_2(K_{WX}) + F_3(K_{WW})$. If $F < \varepsilon$ or the iterative number attains I_{max}, then stop the iteration and output the feature matrix H and two kernel K_{WW} and K_{WX} matrices. Otherwise, go to *step 3*.

Classification Stage

Step 5: For a query sample y, calculate its feature h_y using formula (31).

Step 6: If $l = \min\arg_j \|h_y - m_j\|$, then y is assigned to class l, where $\|\Box\|$ denotes Euclidean norm, and m_j represents the mean feature vector of the jth class.

---

4   Experimental Results

In this section, we will evaluate the performance of UKNMF for face recognition on two publicly available databases, namely ORL and Yale databases. In our experiments, NMF [6], PNMF [8], KPCA [9] are chosen for comparisons. For KPCA, we select those eigenvectors whose eigenvalues are larger than 10^{-3} as projection direction and adopt RBF kernel function $k(x,y) = \exp(-\|x-y\|^2/t)$ with $t = 10^3$. PNMF employs the polynomial kernel function $k(x, y) = \langle x, y \rangle^d$ with $d = 1.5$. Our UKNMF adopts both polynomial and RBF kernel functions with parameters $d = 1.5$, $t = 10^7$ respectively. The number of feature for NMF, PNMF and UKNMF are the same and set to 450 and the maximum number of iterations is 300. The experiments in each database are repeated 10 times, the average accuracy are recorded for comparisons.

4.1   Comparisons on ORL Database

ORL database consists of 40 people, each individual has 10 images which are taken in a uniform illumination environment, and have different facial expressions such as open or closed eyes and smiling or not. Ten facial images of one person from ORL data set are shown in Fig. 2.

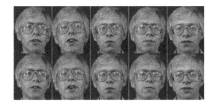

Fig. 2. Images of one person from ORL database

Table 1. Mean accuracy (%) versus training sample (TN) on ORL database

TN	2	3	4	5	6	7	8	9
NMF	76.132	82.68	86.75	89.35	90.63	92.58	92.63	94.00
KPCA	**80.97**	87.39	89.83	92.10	92.94	94.92	95.25	96.00
PNMF	79.47	85.36	89.04	90.09	92.38	95.00	95.86	94.75
UKNMF-Poly	80.75	**87.64**	91.42	93.60	94.81	96.68	97.13	96.75
UKNMF-RBF	80.64	87.40	**91.96**	**94.15**	**95.44**	**96.83**	**97.63**	**97.00**

We randomly select TN (TN $= 2, 3, \cdots, 9$) images from each person for training and the rest (10-TN) images are used for testing. The mean accuracy of all methods are tabulated in Table 1 and plotted in Fig. 2 (left) respectively. It can be seen from the Table 1 that the average recognition accuracy of our UKNMF-Poly method and UKNMF-RBF method increase from 80.75% and 80.64% with TN = 2 to 96.75% and 97.00% with TN = 9 respectively. Meanwhile, the mean accuracy of NMF, KPCA and PNMF raise from 76.13%, 80.97% and 79.47% with TN = 2 to 94.00%, 96.00% and 94.75% with TN = 9 respectively. Our UKNMF-Poly and UKNMF-RBF surpass KPCA, NMF and PNMF in all cases except KPCA with TN = 2. While for the comparisons between UKNMF-Poly and UKNMF-RBF, we find that the recognition rates of UKNMF-RBF outperform UKNMF-Poly except TN = 2 and 3.

4.2   Experiments on Yale Database

Yale database includes 165 images of 15 persons. Each individual contains 11 images with variations in lighting condition (left-light, center-light, right-light), facial expression (normal, happy, sad, sleepy, surprised, and wink), and with/without glasses. Figure 3 shows images of one person from Yale face database.

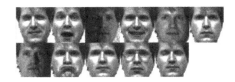

Fig. 3. Images of one person from Yale database

Similar to ORL database, we randomly choose TN (TN $= 2, 3, \cdots, 9$) images of one person as training samples, while the rest (11-TN) images are for testing. Each algorithm is run ten times under the same conditions. The average accuracy are recorded in Table 2 and plotted in Fig. 2 (right) respectively. It can be clearly seen from the Table 2 that the average accuracy of our UKNMF-Poly and UKNMF-RBF increase from 87.50% and 90.00% with TN = 3 to 97.33% and 96.00% with TN = 10, respectively, while the mean recognition rates of NMF, KPCA and PNMF raise from

Table 2. Mean accuracy (%) versus training sample (TN) on Yale database

TN	3	4	5	6	7	8	9	10
NMF	79.42	84.28	84.33	86.80	89.17	90.89	95.00	92.00
KPCA	75.42	76.10	77.89	78.27	79.17	78.67	84.33	86.00
PNMF	86.58	9.067	89.89	92.00	92.67	93.78	97.33	96.00
UKNMF-Poly	87.50	92.00	92.56	94.40	**96.17**	**97.56**	97.67	**97.33**
UKNMF-RBF	**90.00**	**94.95**	**93.78**	**95.20**	95.50	96.00	**98.67**	96.00

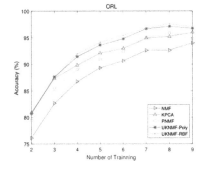
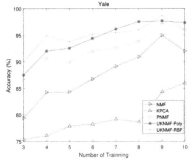

Fig. 4. Recognition rates on ORL (left) and Yale (right)

79.42%, 75.42% and 86.58% with TN = 3 to 92.00%, 86.00% and 96.00% with TN = 10 respectively. The results indicate that our UKNMF-Poly and UKNMF-RBF exceed KPCA, NMF and PNMF in all cases. We also see that UKNMF-RBF is superior to UKNMF-Poly except in the case of TN = 7, 8, 10. Above all, our UKNMF based methods achieve the best performance (Fig. 4).

5 Conclusion

This paper proposed a novel universal kernel NMF method which overcomes the drawbacks of existing KNMF methods. Our UKNMF method can not only use the general kernel functions, but also avoid the pre-image learning problem. The update rules of UKNMF are derived based on three loss functions and gradient descent method. It theoretically shows the convergence of UKNMF algorithm via auxiliary function technique. The proposed UKNMFs with polynomial kernel and RBF kernel are successfully applied to face recognition. Experimental results show the good performances of our UKNMF with different kernels.

Acknowledgments. This work was partially supported by the Natural Science Foundation of Guangdong Province under Grant 2015A030313544, and the Postgraduate Innovation Development Fund Project of Shenzhen University (PIDFP-ZR2017030), and the National Natural Science Foundation of China under Grant 61272252 and Grant 61602308. We would like to thank Olivetti Research Laboratory and Yale University for the contribution of the ORL database and Yale face database respectively.

References

1. Turk, M., Pentland, A.: Eigenfaces for recognition. J. Cogn. Neurosci. **3**, 71–86 (1991)
2. Belhumeur, P.N., Hespanha, J.P., Kriegman, D.J.: Eigenfaces vs. fisherfaces: recognition using class specific linear projection. IEEE Trans. Pattern Anal. Mach. Intell. **19**, 771–720 (1997)
3. He, X.F., Niyogi, P.: Locality preserving projections. Doctoral Dissertation, vol. 16, pp. 186–197 (2005)
4. Tong, L., Zhou, J., Qian, Y., Bai, X., Gao, Y.: Nonnegative-matrix-factorization- based hyperspectral unmixing with partially known endmembers. IEEE Trans. Geosci. Remote Sens. **54**(11), 6531–6544 (2016)
5. Li, Y., Ngom, A.: A new kernel non-negative matrix factorization and its application in microarray data analysis. In: IEEE CIS Society, Piscataway, pp. 371–378. IEEE Press (2012)
6. Lee, D.D., Seung, H.S.: Learning the parts of the objects by nonnegative matrix factorization. Nature **401**, 788–791 (1999)
7. Schölkopf, B., Smola, A., Müller, K.R.: Nonlinear component analysis as a kernel eigenvalue problem. Neural Comput. **10**, 1299–1319 (1998)
8. Ebied, R.M.: Feature extraction using PCA and Kernel-PCA for face recognition. In: 8th International Conference on Informatics and Systems, pp. 72–77. IEEE Press, New York (2012)

9. Salih, CK.: Non-linear kernel fisher discrimination analysis with application. IMP- ACT: Int. J. Res. Appl. Nat. Soc. Sci. (IMPACT:IJRANSS) **3**(8), 57–70 (2015)
10. Zhao, X., Zhang, S.: Facial expression recognition using kernel locality preserving projections. In: Jin, D., Lin, S. (eds.) Advances in Electronic Engineering, Communication and Management Vol.2. LNEE, vol. 140, pp. 713–718. Springer, Heidelberg (2012). doi:10. 1007/978-3-642-27296-7_107
11. Buciu, I., Nikolaidis, N., Pitas, I.: Nonnegative matrix factorization in polynomial feature space. IEEE Trans. Neural Netw. **19**, 711–720 (1997)

β-Barrel Transmembrane Protein Predicting Using Support Vector Machine

Cheng Chen, Hongjie Wu$^{(\boxtimes)}$, and Kaihui Bian

School of Electronic and Information Engineering,
Suzhou University of Science and Technology, Suzhou 215009, China
Hongjie.wu@qq.com

Abstract. Membrane protein is a kind of protein with unique transmembrane structure, which is the material basis for cells to perform various functions. It is an important biological signal molecule to assume the information transmission between the cell and the external environment. It is a precursor step to predict the classification of β-barrel transmembrane protein according to the protein sequence information for 3D structure modeling and function analysis. We firstly use the method of compromising features consist of the position information in sequence and the physiochemical properties of amino acid residues. Then a model by support vector machine algorithm (SVM) is built to predict the β-barrel transmembrane protein. The experimental results presented that transmembrane protein structure prediction based on SVM can provide valid enhancement to transmembrane protein 3D structure prediction and function analysis.

Keywords: Protein prediction · Feature extraction · Cross-test

1 Introduction

Membrane protein (MP) is a polypeptide chain embedded in the biomembrane, which has important biological functions in cells. It forms a variety of nerve signaling molecules, hormones and receptors. It is a transmembrane channel for various ion and a target of many drugs molecular [1]. Most membrane proteins can be divided into peripheral membrane proteins and intrinsic membrane proteins. The transmembrane structure of the inner membrane protein is mainly composed of α-helix and β-sheet. β-barrel structure can only form inter-chain hydrogen bonds between two adjacent β-sheets, so each β-sheet must have two adjacent β-sheets to form a β-barrel structure with the barrel axis is perpendicular to the membrane plane. The β-barrel structure of the membrane protein is found in Gram-negative and other bacteria, no membrane protein is found with both α-helix and β-barrel until now. At present, β-barrel trans-membrane protein is found only in the gram-negative bacteria and mitochondria and envelope membrane of chloroplast, the amount of data is relatively small. Therefore, the problem of how to use a small number of known β-barrel transmembrane protein to predict whether it belongs to β-barrel transmembrane protein according to the protein sequence information is a important pilot step of protein structure prediction and functional analysis of β-barrel transmembrane, but also a challenging issue in the field of protein prediction.

© Springer International Publishing AG 2017
D.-S. Huang et al. (Eds.): ICIC 2017, Part III, LNAI 10363, pp. 360–368, 2017.
DOI: 10.1007/978-3-319-63315-2_31

There has been a lot of progress in the prediction of β-barrel transmembrane protein. Hsieh et al. [2] proposed a knowledge-based potential method to predict the transmembrane regions of β-barrel membrane proteins. They have generated an equivalent potential for TM β-barrel proteins, and this potential can predict orientation within the membrane and identify functional residues involved in intermolecular interactions. Savojardo [3] developed two top-performing methods based on machine-learning approaches to predict β-barrel transmembrane protein. They tackled both the detection of TMBBs in sets of proteins and the prediction of their topology. Chen et al. [4] predicted the topology of β-barrel transmembrane proteins by using RBF networks. Xian and Xian [5] proposed a novel algorithm for predicting β-barrel outer membrane proteins by using ACO-based hyper-parameter selection for LS-SVMs. Savojardo [6] proposed a machine learning method to predict. The proposed approach is based on grammatical modeling and probabilistic discriminative models for sequence data labeling. The method was evaluated using a newly generated dataset of 38 TMBB proteins obtained from high-resolution data in the PDB.

Since the genus of the transmembrane protein is closely related to the physio-chemical properties of the individual amino acids in the sequence, and this correlation is not a simple one-to-one correspondence between the position and the attribute values [7]. Therefore, this paper presents a support vector machine prediction method based on the correlation of amino acid physicochemical index. In this study, 208 β-barrel transmembrane protein sequences were extracted by using the method of fuzzing amino acid composition and amino acid physicochemical properties, then predicted β-barrel transmembrane proteins by using support vector machine based on Gaussian radial basis. The classification accuracy and correlation coefficient reached 88.36% and 0.7723 respectively, which were higher than the existing mainstream method.

2   Data and Method

2.1   Support Vector Machine

Support vector machine (SVM) is a method based on statistical learning theory with a complete theoretical basis and a strict theoretical system, which can solve the problem of limited sample [8, 9]. SVM is built on the principle of structural risk minimization. One of its core ideas is introducing the kernel function technique, which solves the problem of "dimension disaster" calculated in high dimension feature space. As the kernel function determines the ability of nonlinear processing of SVM, the choice of kernel function and its parameters occupies a pivotal position in SVM and it is one of the hotspots of SVM research [10]. However, different kernel functions exhibit different characteristics, and choosing different kernel functions can lead to different performance of SVM. In this paper, Gaussian radial basis function and several different parameter values were attempted when we selected parameter values [11], when C is 30, Gamma g is 0.08, J is 0.1, we obtained the best prediction result which can be seen in Sect. 3. We used the SVM_Light [12] tool.

Assuming that there are multiple protein sequences, corresponding to the target values, SVM study amino acid sequence X. Here, the target values are set to +1 and −1.

+1 represents a transmembrane β-barrel protein (positive sample), and −1 represents a non-transmembrane β-barrel protein (negative sample).

2.2   Data Sets and Features

The data sets used in the training set and test set in this paper were collected by Gromiha and Suwa [13], then the data sets were obtained by removing redundant sequences with sequence similarity greater than 40% by Park et al. [14] with CD-HIT (http://bioinformatics.org/cd-hit/). The data sets consisted of TMB (208), GP (673, 155 of which were all α, 156 were all β, 183 α + β and 179 α/β, and the globular protein sequence was used for negative samples of the data sets), The detailed 3D structure information of the protein in the non-redundant data set can be obtained from the PDB database.

Amino acid residue index is the index to reflect the different physicochemical properties of amino acids, we select five most representative index from amino acid residue index database of Kawashima et al. [15]. They are hydrophobicity [16], hydrophilicity [17], flexibility [18], free energy transfer [19] and polarity [20], as shown in Table 1.

Table 1. Five normalized features of amino acids

Index value		Feature name									
		Hydrophobicity		Hydrophilicity		Flexibility		Free energy transfer		Polarity	
A	L	0.25	0.27	0.20	0.11	0.256	0.256	0.22	0.06	0.78	0.56
R	K	0.11	0.16	0.44	0.44	0.267	0.263	0.22	0.21	0.95	1.00
N	M	0.19	0.25	0.24	0.14	0.263	0.249	0.24	0.13	1.02	0.62
D	F	0.18	0.27	0.44	0.06	0.266	0.253	0.22	0.07	1.12	0.59
C	P	0.23	0.23	0.16	0.23	0.255	0.266	0.20	0.16	0.61	0.78
Q	S	0.18	0.21	0.24	0.25	0.265	0.266	0.24	0.20	0.95	0.86
E	T	0.19	0.22	0.44	0.20	0.265	0.261	0.21	0.20	1.07	0.82
G	W	0.24	0.26	0.23	0.00	0.268	0.252	0.23	0.09	0.84	0.60
H	Y	0.20	0.23	0.20	0.07	0.253	0.260	0.22	0.08	0.94	0.65
I	V	0.28	0.27	0.11	0.13	0.263	0.258	0.08	0.13	0.59	0.63

2.3   Feature Analysis from Sequence

In the aspect of selecting features vectors, we synthesized the correlation coefficient between the amino acid component and the physicochemical properties of the amino acid. The feature extraction algorithm based on amino acid component (AAC) was first proposed by Nakashima and Nishikawa [21]. The physicochemical properties of amino acids are important factors affecting the transmembrane protein type. Due to the differences in amino acid properties, the amino acid sequences of the various combinations have different structures and have their own specific physiological functions. Specific process is shown as Fig. 1.

Firstly, the protein sequence is expressed as: $\mathbf{X} = [f_1, f_2, \ldots, f_i, \ldots, f_{20}]^T$, where f_i is the frequency of the i-th amino acid appears in the protein sequence. $f_i = n_i/L$, where n_i is the number of times of the i-th amino acid appears in the sequence, L is the total

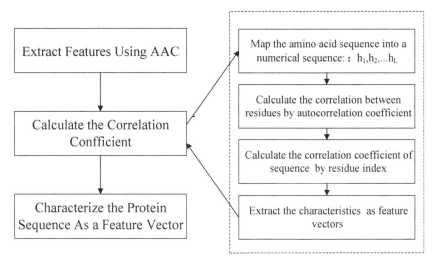

Fig. 1. Flow chart of feature extraction

number of amino acid residues in the protein sequence. The order of the elements in the feature vector X is in alphabetical order according to 20 natural amino acids.

Secondly, calculate the correlation coefficient of amino acid residue index. In the first step, map the β-barrel transmembrane protein amino acid sequence into the corresponding numerical sequence: $h_1, h_2, \ldots, h_{i,\ldots}, h_L$, where h represents the residue index corresponding to the i-th residue in the sequence, L is the length of transmembrane protein sequence.

In the second step, calculate the order correlation between the amino acid residues in the β-barrel transmembrane protein sequence by using the autocorrelation function:

$$r_n = \frac{1}{L-n} \sum_{i=1}^{L-n} h_i h_{i+n}, n = 1, 2, \ldots, m \tag{1}$$

Where m is the order of the correlation coefficient, $m < L$. When $m = 1$, it represents the first order correlation coefficient, which reflects the correlation between all adjacent two amino acid residues in the sequence; when $m = 2$, it represents the second order correlation coefficient, which reflects the correlation of all adjacent three residues, the other order and so on.

In the third step, calculate the correlation coefficient of the corresponding sequence by using the selected five amino acid residue indices. Extract the feature S_k of the transmembrane protein as a $5 \times m$ dimension eigenvector, characterized as:

$$AR_k = \left[r_{11}^k, r_{12}^k, \ldots, r_{1m}^k, \ldots, r_{w1}^k, r_{w2}^k, \ldots, r_{wm}^k \right]^T, w = 5 \tag{2}$$

Finally, summarize the correlation coefficient between the amino acid component and the physicochemical properties of the amino acid, a β-barrel transmembrane protein sequence is characterized as a $20 + 5 \times m$ dimension eigenvector, where the first 20 dimension is the amino acid component, and the rear $5 \times m$ dimension is the correlation coefficient obtained by the calculation of five amino acid residues index, the eigenvector is expressed as:

$$X = \left[c_1^k, c_2^k, \ldots, c_{20}^k, r_{11}^k, r_{12}^k, \ldots, r_{1m}^k, \ldots, r_{w1}^k, r_{w2}^k \ldots, r_{wm}^k\right]^T, w = 5 \quad (3)$$

Take into account the length of the selected β-barrel transmembrane protein sequence, select m as 30 Formula (3).

1. **Input:** β-barrel transmembrane protein sequence $\mathbf{X} = [\boldsymbol{f_1}, \boldsymbol{f_2}, \cdots, \boldsymbol{f_i}, \cdots, \boldsymbol{f_{20}}]^T$
2. **for each** f_i **in** $\mathbf{X} = [\boldsymbol{f_1}, \boldsymbol{f_2}, \cdots, \boldsymbol{f_i}, \cdots, \boldsymbol{f_{20}}]^T$ **do**
3.   Map f_i to h_i
4. **end for**
5. **for** n **in** m **do**
6. $Sum = 0$
7.   **for** i **in** L-n **do**
8. $Sum = Sum + h_i h_{i+n}$
9.   **end for**
10. $r_n = Sum/(L - n)$
11. $AR_k = [r_{11}^k, r_{12}^k, \ldots, r_{1m}^k, \ldots, r_{w1}^k, r_{w2}^k, \ldots, r_{wm}^k]^T$
12. **output:** $X = [c_1^k, c_2^k, \ldots, c_{20}^k, r_{11}^k, r_{12}^k, \ldots, r_{1m}^k, \ldots, r_{w1}^k, r_{w2}^k \ldots, r_{wm}^k]^T$

Fig. 2. The algorithm of feature extraction where f_i is the frequency of the i-th amino acid appears in the protein sequence and m is the order of the correlation coefficient, h represents the residue index corresponding to the i-th residue in the sequence, L is the length of sequence.

3   Experiments and Discussion

3.1   Evaluation Metric

Methods for testing performance of classification model usually include self-test and cross-test. Cross-test verify the generalization of the model, where the Jackknife test is considered the most reasonable cross-test method [22, 23], it eliminating the "memory effect" of the self-test, whereas the lack of Jackknife's test is too time-consuming. In the classification study, the k-sheet cross-validation of the test method is often used. First, divide the sample data randomly into approximately equal k subsets, take a subset as a test set successively, and the remaining k−1 subsets are used as training sets, alternate k times repeatedly and average the respective precision. In fact, when k is equal to the total number of samples, the k-sheet cross-test is the Jackknife test. We used [24] TP (True Positive), TN (True Negative), FP (False Positive), FN (False Negative), Classification Accuracy (ACC), Correlation Coefficient (MCC), Positive Sample Sensitivity, Positive Sample Specificity, Negative Sample Sensitivity, Negative Sample Specificity. FN is the number of transmembrane β-barrel protein classified correctly, TN is the number of non-transmembrane β-barrel protein classified correctly, FP is the number of non-transmembrane β-barrel protein but classified incorrectly, FN is the number of transmembrane β-barrel protein but classified incorrectly, ACC and MCC are calculated as:

$$ACC = \frac{TP + TN}{P + N} \tag{4}$$

$$MCC = \frac{TP \times TN - FP \times FN}{\sqrt{(TP + FP)(TP + FN)(TN + FP)(TN + FN)}} \tag{5}$$

P is the number of all transmembrane β-barrel proteins, N is the number of all non-transmembrane β-barrel proteins.

3.2   Experiments

Use the feature extraction method of compromising position weight amino acid composition and amino acid residue index, change the order of autocorrelation function, and the data set is further classified and predicted. The order m of the autocorrelation function is selected as 15, 20, 25, 30, 35, 40, 50 and 60 respectively, and the prediction results of 10-sheets cross-test are shown in the Table 2.

Table 2. The classification prediction results of different order (10-sheets cross-test)

Evaluating indicator	m = 15	m = 20	m = 25	m = 30	m = 35	m = 40	m = 50	m = 60
TP	381	381	381	380	383	380	382	382
TN	401	401	401	398	393	388	378	372
FP	13	13	13	16	21	26	36	42
FN	33	33	33	34	31	34	32	32
Specificity(+)	96.72%	97%	96.67%	95.96%	94.82%	93.59%	91.34%	90.0%
Sensitivity(+)	92.01%	92.07%	91.94%	91.69%	92.51%	91.89%	92.26%	92.3%
Specificity(-)	92.57%	92.62%	92.59%	92.38%	93.00%	92.27%	92.42%	92.4%
Sensitivity(-)	96.75%	96.75%	96.75%	96.00%	94.81%	93.56%	91.06%	89.6%
Accuracy	94.38%	94.41%	94.35%	93.85%	93.66%	92.73%	91.6%	91.0%
MCC	0.8902	0.8907	0.8898	0.8801	0.8757	0.8566	0.8354	0.8226

In the respect of prediction results from each group, different order within a certain range dose not has great influence on the prediction results of β-barrel protein prediction, that means selecting an order of the autocorrelation function within a certain range makes a very small difference in classification accuracy. For example, when the order is 15 and 60, the classification accuracy is 94.38% and 91.01%, both of them are excellent. On the other hand, it indicates that the selection of different order in a certain range is not the main factor to improve the performance of the classification model. Of course, the selection of different order makes the classification accuracy fluctuant. In the respect of results, when the order is from 15 to 60, the classification accuracy increases first and then decreases, when the order number is 20, the classification accuracy and correlation coefficient of MCC reach 94.41% and 0.8907 respectively.

The results in Table 3 show that classification accuracy and MCC correlation coefficient of the classification algorithm based on the combination of the characteristics are better than the results obtained by only comprising amino acid compositions as characters, the results are also superior to the classification method of combining amino acid and dipeptide. It is because the method takes into the frequency information of amino acids in the sequence, but also considers the function of physicochemical properties of amino acids which uses more sequence information, and the choice of physicochemical properties of amino acid also affects the prediction results. The results showed that it is effective to use the method of compromising amino acid composition and physicochemical properties.

Table 3. Comparison of cross test results for FSVM and SVM with different feature inputs on TMB data sets

Classification algorithm	ACC	MCC
SVM(one-vs-one) [24]	83.2	0.664
FSVM(cf) [24]	86.1	0.673
FSVM(aa) [24]	80.3	0.626
FSVM(aa + dispep) [24]	84.6	0.648
SVM(AAC + Hydi + Hydo + Flex + EnTr + Pol)	88.36	0.7723

4   Conclusions

Extract the feature of the transmembrane protein sequence (amino acid component and amino acid residue index) by using the method of comprising several characters and combine it with the classification algorithm of support vector machine, this method builds a novel classification model, which could describe the characteristics of the information contained in the transmembrane protein sequence more effectively, compared with the existing transmembrane protein classification and prediction method, this method is more adaptive and generalized, and gets better accuracy on classification and prediction. In addition, the method based on SVM and other machine learning methods are affected by the size of training set to varying degrees [25]. Moreover, the number of some classes of transmembrane protein sequences in today's protein sequence databases is relatively small, so the success rate of the model build above is lacking, but as the number of transmembrane protein sequences newly discovered increases, the data set will be gradually improved. Therefore, the success rate of β-barrel transmembrane protein prediction will be greatly improved in the future.

Acknowledgements. This paper is supported by grants no. SNG201610 under Science and Technology Development Project of Suzhou and grants no. SZS201609 under Virtual Reality Key Laboratory of Intelligent Interaction and Application Technology of Suzhou.

References

1. Wu, H.J., Li, H.O., Jiang, M., Chen, C., Lv, Q., Wu, C.: Identify high-quality protein structural models by enhanced-means. Biomed. Res. Int. **2017**, 9 (2017). Article ID 7294519
2. Hsieh, D., Davis, A., Nanda, V.: A knowledge-based potential highlights unique features of membrane α-helical and β-barrel protein insertion and folding. Protein Sci. **21**(1), 50–62 (2012)
3. Savojardo, C.: BETAWARE: a machine-learning tool to detect and predict transmembrane beta-barrel proteins in prokaryotes. Bioinformatics **29**(4), 504–505 (2013)
4. Chen, S.-A., Ou, Y.-Y., Gromiha, M.M.: Topology prediction of α-helical and β-barrel transmembrane proteins using RBF networks. In: Huang, D.-S., Zhao, Z., Bevilacqua, V., Figueroa, J.C. (eds.) ICIC 2010. LNCS, vol. 6215, pp. 642–649. Springer, Heidelberg (2010). doi:10.1007/978-3-642-14922-1_80
5. Xian, G., Xian, B.Z.: A novel algorithm for predicting β-barrel outer membrane proteins using ACO-based hyper-parameter selection for LS-SVMs. Atlantis Press (2012)
6. Savojardo, C.: Machine-learning methods for structure prediction of β-barrel membrane proteins (2013)
7. Wu, H.J., Wang, K., Lu, L.Y., Lv, Q., Jiang, M.: A deep conditional random field approach to transmembrane topology prediction and application to GPCR three-dimensional structure modeling. IEEE/ACM Trans. Comput. Biol. Bioinf. (2016). http://doi.ieeecomputersociety.org/10.1109/TCBB.2016.2602872
8. Ding, S., Zhang, N., Zhang, X.: Twin support vector machine: theory, algorithm and applications. Neural Comput. Appl. **27**, 1–12 (2016)
9. Chatterjee, P., Basu, S., Kundu, M., et al.: PPI_SVM: prediction of protein-protein interactions using machine learning, domain-domain affinities and frequency tables. Cell. Mol. Biol. Lett. **16**(2), 264–278 (2011)
10. Ding, S.F., Qi, B.J., Tan, H.Y.: An overview on theory and algorithm of support vector machines. J. Univ. Electron. Sci. Technol. China **40**(1), 2–10 (2011)
11. Ren, Y., Liu, H., Xue, C.: Classification study of skin sensitizers based on support vector machine and linear discriminant analysis. Anal. Chim. Acta **572**(2), 272–282 (2006)
12. Joachims, T.: Text categorization with support vector machines: learning with many relevant features. In: Nédellec, C., Rouveirol, C. (eds.) ECML 1998. LNCS, vol. 1398, pp. 137–142. Springer, Heidelberg (1998). doi:10.1007/BFb0026683
13. Gromiha, M., Suwa, M.: A simple statistical method for discriminating outer membrane proteins with better accuracy. Bioinformatics **21**, 961–968 (2005)
14. Park, K.J., Gromiha, M.M., Horton, P.: Discrimination of outer membrane proteins using support vector machines. Bioinformatics **21**(23), 4223–4229 (2005)
15. Kawashima, S., Ogata, H., Kanehisa, M.: AAindex: amino acid index database. Nucleic Acids Res. **27**(1), 368–369 (1999)
16. Eisenberg, D., Weiss, R.M., Terwilliger, T.C.: The hydrophobic moment detects periodicity in protein hydrophobicity. Proc. Natl. Acad. Sci. U.S.A. **81**(1), 140–144 (1984)
17. Hopp, T.P., Woods, K.R.: Prediction of protein antigenic determinants from amino acid sequecces. Proc. Natl. Acad. Sci. U.S.A. **78**(6), 3824–3828 (1981)
18. Bhaskaran, R., Ponnuswamy, P.K.: Positional flexibilities of amino acid residues in globular proteins. Int. J. Pept. Protein Res. **32**(4), 241–255 (1988)
19. Bull, H.B., Breese, K.: Surface tension of amino acid solutions: a hydrophobicity scale of the amino acid residues. Arch. Biochem. Biophys. **161**(2), 665–670 (1974)
20. Grantham, R.: Amino acid difference formula to help explain protein evolution. Science **185** (4154), 862–864 (1974)

21. Nakashima, H., Nishikawa, K.: Discrimination of intracellular and extracellular proteins using amino acid composition and residue-pair frequencies. J. Mol. Biol. **238**(1), 54–61 (1994)

22. Cheng, X.: Prediction of protein folding rates from hybrid primary sequences and its protein structure attributes. In: Eighth International Conference on Natural Computation, pp. 266–269. IEEE (2012)

23. Liang, R.P., Huang, S.Y., Shi, S.P.: A novel algorithm combining support vector machine with the discrete wavelet transform for the prediction of protein subcellular localization. Comput. Biol. Med. **42**(2), 180–187 (2012)

24. Bernhard, E.B., Isabelle, M.G., Vladimir, N.V.: A Training Algorithm for Optimal Margin Classifiers [EB/OL]

25. Baldi, P., Brunak, S., Chauvin, Y.: Assessing the accuracy of prediction algorithms for classification: an overview. Bioinformatics **16**(5), 412–424 (2000)

Knowledge Discovery and Data Mining

Innovative Research of Financial Risk Based on Financial Soliton Theory and Big Data Ideation

Yu-Shan Xue and Xian-Jun Yin[✉]

School of Mathematics and Statistics, Central University of Finance and
Economics, Beijing 100081, China
yinxj@cufe.edu.cn

Abstract. Nonlinear problems encountered in theoretical study on the financial risks are important objects of nonlinear science. Traditional financial risk management theory of nonlinear problems unresolved can be studied by using the method of nonlinear science, financial soliton theory and big data ideation. The theory can not only analyze the evolution of financial markets, the formation of risk transfer mechanism, but also it can more profoundly grasp the behaviors of financial markets, obtain new financial risk prediction, control concepts and methods.

Keywords: Financial risk management · Financial complex systems · Financial soliton · Financial market · Nonlinear equation

1 Introduction

Finance is a high-risk industry, scholars at home and abroad have been looking for scientific and effective financial risk control approaches, prediction methods and control measures [1–3]. Especially since the 90s of the 20th century, series of great financial risk events and financial crises have highlighted the complexity of financial risk, and the importance and urgency of financial risk prediction and control methods [4–6]. So far, the studies of financial market primarily base on the linear analysis of classic capital market theories.

Traditional financial risk management theories and methods do not take into account the interactions of internal subjects in the financial system, and between internal subjects and external factors. These theories and methods also ignore the internal nonlinear interaction mechanisms of financial market as a complex system and the resulting internal instability [7, 8]. However, the modern financial market is a real-time changing complex dynamical system with big data, and has many interaction factors and typical nonlinear dynamic properties [1–3, 9]. The core issue of finance is the control of financial risks [7]. The analysis of financial market evolution aims to better understand the mechanism and principle of financial risk generation, and lay the foundation for the scientific risk prevention [8]. In the framework of nonlinear dynamics, we can hopefully get a more scientific management concept, and then control the market behaviors more accurately [3–5].

© Springer International Publishing AG 2017
D.-S. Huang et al. (Eds.): ICIC 2017, Part III, LNAI 10363, pp. 371–376, 2017.
DOI: 10.1007/978-3-319-63315-2_32

The risk sources are mainly due to the uncertainty about the future [3–5]. Mathematically speaking, uncertainty is the probability of occurrence for one or more events. From the perspective of risk management, it is difficult for us to know the probability and their obedience distribution of future event occurrence. Therefore, we should study them from multiple perspectives including subjective judgments [6–8].

Financial risks refer to the exposure that results from the uncertainty about the future in financial activities. Generally speaking, financial risks can be divided into the following types: market risk, credit risk, liquidity risk and operation risk [10–12]. Market risk is inherent in the traditional financial system, and mainly caused by fluctuations in interest rates, exchange rates, stock prices and commodity prices. Credit risk involves almost all financial transactions, especially with the rapid development of the internet finance, credit risk of the network becomes increasingly highlighted. Liquidity risk is mainly due to the poor fluidity of financial participant assets which result in cash difficulties. Operation risk is generated by imperfect trading system and management failures. At present, the research and management on operation risk is becoming more and more important [13–15].

Financial market is the core of a country economic operation, so every country and government and individual devote themselves to maintain the stability of financial markets, and effectively avoid and manage the financial risk [3–6]. The consequences of individual or financial institution risks are often beyond the impact on their own. The direct consequence of financial risk may cause direct or potential economic losses for the economic entity. For example, if a bank has serious credit risk, consumers thereby worry about the deposit safety that will lead to the decrease of capital sources and deposits. Besides, financial institution risks arising from concrete deal activities not only may pose a great threat to the survival of the financial institutions themselves, but also bring crisis to the stable operation and healthy development of the entire financial system. Once the systemic risk of financial system emerges, the system will malfunction, which will inevitably lead to economic misorder of the whole society, and even cause serious economic crisis [13–15].

2 The Data Characteristics of Modern Financial Markets

Financial industry is one of the industries that have a huge amount of data in the world. China's financial industry have entered into the era of big data, and showed a rapid development trend. Big data is a set including large samples and high dimensional variables, and is the contemporary "telescope" and "microscope" [10, 13–15].

At present, it reaches a consensus all over the world that the financial industry data is an important asset. With the combination between the carrier of financial business and electronic commerce, especially with the rapid development of internet finance in recent years, analyzing 15\% of the original structured data can not satisfy the development demand. Thus, it is urgent for us by using the data strategy to break data borders to ensure more healthy and comprehensive market development trend [3–5]. Data can be regarded as the life of finance, the big data will play an important role in strengthening financial risk control, business innovation, fine management, and so on.

At first, in terms of risk control and management, we can increase risk auditability and management efforts based on big data. Secondly, big data support financial service innovation, which can achieve the goal of principle of "take the customer as the center". For example, we can improve customer conversion rates through the analysis of customer consumption behaviors, and develop different products to meet different customer demands, and implement diversified competition [5–8]. The foreign and domestic studies on big data simply considered the problems such as the data capitalization, credit digitization, and so on, they did not further discuss the risk environment of data block and data flow formed by the big data of financial markets.

3 The Nonlinear Analysis of Financial Markets

Modern financial market is not a simple and orderly system, but one complex and real-time changing dynamic system who has quantity of data. It also has many factors that are mutual affected, and presents typical nonlinear dynamics properties [11, 12]. As a complex system, the internal structure of financial market possesses the nonlinear interactions and the resulting internal instability, the risk intensity and frequency in financial markets are far beyond our imagination. Therefore, it will contribute to distinguishing the traditional theories and methods of financial risk management that realize the nonlinear features of financial markets. Thus, we can study the financial markets and control financial risks from the point of view of nonlinear theories [12–14].

The financial market, as a whole complex and developing evolution system, its internal submarkets such as bond market, stock market, and so on, influence and interact with each other. Financial markets also have many relations with the real economy, the real economy participants can make appropriate adjustments according to the financial market development, and in turn financial markets can be affected by the entity economy development. Financial markets have kinds of members include numerous individuals or organizations, individual behaviors do not exist in isolation, but can affect other economic participants, so if we just study the individual behaviors, the evolution process of the whole financial system will not be identified [5–7].

Financial market is essentially a nonlinear system. In nonlinear systems, nonlinear superposition principle is no longer applicable, so nonlinear systems can lead to diversity and complexity. The relationship between internal and external elements in nonlinear system is usually nonlinear, so is the relationship between the inside elements. Therefore, the financial markets in real life have the features of greater complexity, uncertainty and unbalance [9, 12].

4 The Brief Introduction of Soliton Theory

Nonlinear science is another scientific revolution after the quantum mechanics and relativity. Soliton, chaos and fractals constitute the three frontier theories of nonlinear science [9]. "Soliton" concept was firstly put forward in the 1950s, it is a kind of wave of nonlinear evolution equations (the nonlinear partial differential equations containing time independent variables, differential difference equations and differential integral equations) distributing within the scope of the limited space, and has stable structure and elastic collision characteristics. Because soliton has the features of energy maintenance and nondestructive transmission, in half a century, solition theory is rapidly developed and widely used in basic subjects and high-tech fields such as nonlinear optics, electromagnetism, fluid mechanics, condensed matter physics, biological physics, and so on [9, 12]. In recent years, its research production has also been started to be applied to financial fields, and as the theoretical basis of the studies of market evolution characteristics, certain research results has been obtained [9].

In soliton theory, soliton solution is a kind of special solution of nonlinear evolution equations. Finding analytical soliton solutions of nonlinear evolution equations not only can help to further understand the equation essential properties and algebraic structure, but also reasonably explain the relative nature phenomena in terms of applications [9]. However, the linear superposition theorem fails in solving nonlinear evolution equations, so far, there is not a common and effective method to construct the general solution of the nonlinear evolution equations [9]. Therefore, it is important and difficult to find the soliton solution of nonlinear evolution equations in nonlinear studies. In physics, through the study of nonlinear evolution equations, we can better explain nonlinear phenomena and object motion rules in various fields of natural science. In financial field, the discovery of financial soliton illustrates that there is a new form of matter and energy in financial transaction markets (futures, stocks). By using this theory, traditional research methods such as linear and perfectly rational balance paradigm and statistics, and so on, can be converted to the level of the non-linear, bounded rational non-equilibrium paradigm and complex systems of financial soliton [12].

5   The Financial Structure of Soliton Theory

The studies of financial soliton theory at home and abroad, mainly focused on predicting financial market price fluctuations and constructing investment models. However, they did not further systematically discuss the methods and measures of financial risk control and management based on the theory [4–6]. It is still a blank in the research fields both at home and abroad that using the financial soliton theory and big data logical ideation, combined with the basic methods of risk control and management, to study financial risk forecast and oriented control measures.

In the thinking of big data, nonlinear dynamics and financial soliton theory, with the real-time analysis of huge amounts of data, using the analysis of evolution, risk formation and transfer of financial markets with soliton, we can explore the generation mechanism of financial risk data, and gradually build a new theoretical system supporting risk management of financial complex system, which can grasp multitudinous risk behaviors in financial markets more deeply and scientifically, and better implement financial risk forecast and effective control.

Firstly, we proceed from actual conditions, combined with the basic methods of risk control, and four types of different sources of financial risks (market risk, credit risk, liquidity risk and operational risk). Secondly, we make the best of financial soliton theory and big data logical ideation, systematically construct and demonstrate the new ways and methods of financial risk forecast and control, and originally propose "financial risk soliton prediction methods" and "financial risk soliton oriented control methods". Finally, we build a "financial risk soliton forecast system", and a experimental "financial risk soliton prediction platform" on which financial high frequency data online flows. The platform is similar to the solitary wave phenomenon which is generated when a ship drives in the sea. The platform can in time broadcast the main information indexes of financial markets just as the weather forecast, and achieve the goal of effective control on the financial market risks. By using financial soliton, we can continuously track and make a analysis of financial data, predict and confirm the financial market risk indicators, and dig nonlinear laws of financial markets. These work can save time for financial risk controllers and help them to make decisions as early as possible.

The method explores the problem of financial risk forecast and control. It has the following three aspects of value and significance:

(1) With the help of nonlinear science, and based on big data ideation, we can solve the nonlinear problems that the traditional financial risk theories have not solved. It is significant for improving the level of financial risk control that conceive a new theory and method of risk prediction and control, which not only adapt to off-line traditional finance, but also the emerging online internet financial industry.

(2) It can give a new perspective and approach of financial risk prediction and control for financial authorities and financial participants. Furthermore, we can access to the information of risk occurrence in advance, gain time for risk control, and achieve the goal of safeguarding the regular orders of national financial markets.

(3) It focuses on maximumly combining with lots of information, big data ideation, modern information technologies, and nonlinear science. Thus, it can plug up loopholes in financial risk management, create a new height and mode of financial risk prediction and control methods in financial industries and academia.

6   Conclusions

The nonlinear problems encountered in financial risk theoretical studies are the important objects of nonlinear science studies. We can use nonlinear scientific methods, financial soliton theory, and big data ideation to analyze and study financial market evolution, risk formation and transfer mechanisms. Through the construction of "online real-time prediction platform of financial risk soliton", we can discover some nonlinear changes in financial markets, see the running status of financial markets at any time, which can ensure that the generation and evolution process of relative financial risks can be continuously followed and controlled, dig implied orderly structure in the disordered state of financial markets, and find various possible approaches of financial risks evolution process. Through optimizing financial market behaviors with the control of parameter change in financial markets, we can grasp financial market behaviors more deeply, obtain a new idea and method of financial risk prediction and control, can finally control the financial risk level within the scope that markets can bear and control.

Acknowledgments. We express our thanks to the referees and members of our discussion group for their valuable comments. This work is supported by National Natural Science Foundation of China (Grant Nos. 11447233, 11472315), Social Science Foundation of Beijing (Grant No. 15JGC184), Young Doctor Development Foundation of "121 Talent Project of CUFE" (Grant No. QBJ1420), Education Teaching Reform Fund 2016 of CUFE.

References

1. Matveev, V.B., Salle, M.A.: Scaling in stock market data: stable laws and beyond. In: Dubrulle, B., Graner, F., Sornette, D. (eds.) Scale Invariance and Beyond, pp. 15–18. Springer, Berlin (1997)
2. Dennis, K., Antonio, M.: Adding and subtracting Black-Scholes: a new approach to approximating derivative prices in continuous-time models. J. Financ. Econ. **102**(2), 390–415 (2011)
3. Liu, H.L., Wang, H.: Financial Risk Management, pp. 35–48. China Financial & Economic Publishing House, Beijing (2009)
4. Song, Q.H., Li, Z.H.: Financial Risk Management, pp. 25–38. China Financial & Economic Publishing House, Beijing (2003)
5. He, H.Q.: Financial risk and its prevention in the perspective of globalization. Econ. Soc. Dev. **7**, 34–36 (2009)
6. Robert, A.M.: Mondale Economics Corpus: Exchange Rate and the Optimal Currency Area, pp. 23–37. China Financial & Economic Publishing House, Beijing (2003)
7. Fu, S.C.: Financial information risk prevention countermeasures study. China Saf. Sci. J. **6**, 85–88 (2005)
8. Li, Y., Hu, B.: Global Financial Regulatory Reform in the Context of the Financial Crisis, pp. 2–7. Social Sciences Academic Press, Beijing (2010)
9. Huang, J.N., Xu, J.Z., Xiong, Y.T.: Solitions Concepts, Principles and Applications, pp. 10–27. Higher Education Press, Beijing (2004)
10. Glosten, R., Jagannathan, R., Runkle, D.: On the relation between the expected value and the volatility of nominal excess return on stocks. J. Finan. **46**, 1779–1801 (1992)
11. Guo, J.L.: Financial Evolution and Development in the Perspective of Complex System Model, pp. 13–26. China Financial & Economic Publishing House, Beijing (2007)
12. Ma, J.L., Ma, F.T.: Thinking through the numerical predicting of prices fluctuation in finance markets. Manage. Sci. China **19**(1), 78–84 (2006)
13. Chen, S.L.: Financial risk monitoring and early warning research. Econ. Sci. **3**, 28–36 (1997)
14. Li, G.J.: The scientific values of big data studies. Commun. China Comput. Fed. **8**(9), 8–15 (2012)
15. Liu, Y., Luo, M.X.: Internet financial model and risk regulatory thinking. China Mark. **43**, 29–36 (2013)

An Effective Sampling Strategy for Ensemble Learning with Imbalanced Data

Chen Zhang[1,2] and Xiaolong Zhang[1,2(✉)]

[1] School of Computer Science and Technology, Wuhan University of Science and Technology, Wuhan 430065, China
`251815407@qq.com, xiaolong.zhang@wust.edu.cn`
[2] Intelligent Information Processing and Real-Time Industrial Systems Hubei Province Key Laboratory, Wuhan 430065, China

Abstract. Classification of imbalanced datasets is one of the challenges in machine learning and data mining domains. The traditional classifiers still need to handle with minority instances. In this paper, we propose an effective method which applies sampling method based on ensemble learning. It uses Adaboost-SVM based on spectral clustering to boost the performance. This method also uses over-sampling and under-sampling methods based on the misclassified instances got by ensemble learning. Compared with the preview algorithms, the experiment results show that the proposed method is effective in dealing with imbalanced data in binary classification.

Keywords: Sampling · Ensemble learning · Spectral clustering · Misclassified samples

1 Introduction

Imbalanced classification has got more and more attention in recent years. It can be defined as the class imbalance problem when there are many more instances of some classes than others. In some practical applications, the minority class only takes up 1% of the majority classes [1]. Imbalanced classification problem becomes one of the challenges in data mining because traditional classifiers tend to classify all the instances as the majority class. In real life, it's quite usual to encounter the class imbalance problem such as medical diagnosis, fraud detection and telecom equipment fault detection. Usually, the minority class represents greater interest for us than the majority class. Thus, it is necessary to deal with imbalanced classification and improve the recognition ratio of instances of the minority class.

In recent years, some works have been done based on the imbalanced problem of binary classification [2]. Most traditional classification algorithms such as decision trees, SVM and KNN tend to maximize the overall classification accuracy. As a result, many minority class instances are ignored. In general, the classification methods for dealing with class imbalance problems can be mainly divided into three categories: data level approaches [3], cost-sensitive learning [4] and ensemble learning [5]. The data

© Springer International Publishing AG 2017
D.-S. Huang et al. (Eds.): ICIC 2017, Part III, LNAI 10363, pp. 377–388, 2017.
DOI: 10.1007/978-3-319-63315-2_33

level approaches focus on sampling techniques to solve the class imbalance problem. It's a common way to balance the class distribution by using over-sampling or under-sampling techniques. Original under-sampling refers to the process of removing instances of the majority class randomly, therefore under-sampling may delete some useful instances which may have great influence on the classification. Over-sampling especially randomly copying minority instances may lead to overfitting. To overcome the drawback of original sampling methods, some improved methods have been proposed. The SMOTE method selects the minority instance's nearest minority neighbors and generates synthetic minority samples along the lines between the minority instances and its nearest minority neighbors [6]. Based on SMOTE, Borderline-SMOTE [7] and Safelevel-SMOTE [8] were proposed. The combination of over-sampling and under-sampling might be more effective to solve the class imbalance problem. As for the cost-sensitive learning, it assigns different cost to each class rather than gives the equal weights to instances of any class. In this category, the cost-sensitive versions of Adaboost are well known [9]. Last, ensemble learning provides solutions to imbalanced classification problem. One strategy is based on combining ensemble learning with data level approaches such as SMOTEBagging [10], SMO-TEBoost [11] and RUSBoost [12]. The algorithm Adaboost-SVM-OBMS combines ensemble learning with over-sampling based on misclassified samples [13]. PCBoost applies random-sampling to synthesize minority class and uses Perturbation Correction for the synthetic minority samples [14]. The ClustFirstClass algorithm, which uses cluster method to divide the majority class into some subsets, randomly selects data from each subset to balance training data for each base classifier and then combines the results of these base classifiers with specific ensemble rules [15].

In this paper, we propose a new method named MSSC-Adaboost which is based on data level and boosting procedure. MSSC stands for Misclassified Sampling and Synthetic Correction. Before the Boosting, using spectral clustering to divide the majority class into several clusters. In each iteration of Adaboost, select a certain quantity of instances from each cluster due to the weights distribution and form the majority subsets. Then use SVM to train the subset of training data consisting of each majority subset and the whole minority class instances. Combine the results of these individual SVM and obtain the result of the strong classifier. Get the misclassified samples based on the result of the strong classifier, remove the qualified misclassified majority instances and apply SMOTE for those misclassified minority samples, this step is called misclassified sampling. The MSSC-Adaboost also uses a synthetic correction strategy to delete the misclassified synthetic samples.

The rest part of this paper is organized as follows: Sect. 2 introduces the related work. In Sect. 3, the proposed method is presented in details. As for Sect. 4, it provides the experimental setting and the experiment results of four comparative algorithms. Section 5 draws the conclusion.

2 Related Work

In this study, we focus on dealing with the imbalanced classification based on data level of imbalanced datasets. This part is a sort summary of comparative methods that have been proposed.

- SMOTE (Synthetic Minority Over-sampling Technique): This technique doesn't simply add the copies of existing instances, it randomly creates new samples in the line segments defined by the minority instance and its nearest minority neighbors. Borderline-Smote is an improved algorithm based on SMOTE, it only over-samples the minority instances which are near the borderline by applying SMOTE. Safelevel-SMOTE carefully samples minority instances along the same line with different weight degree.
- Ensemble learning: The diversity is the key point for ensembles. Sampling is an effective way to inject diversity into an ensemble [16]. SMOTEBagging is the method that applies SMOTE to synthesize minority samples in each iteration of Bagging. In Adaboost, how to select instances to train is based on the weight distribution. SMOTEBoost and RUSBoost are the algorithms that combine Boosting with SMOTE or random under-sampling in each iteration. The algorithm Adaboost-SVM-OBMS uses over-sampling for both minority and majority misclassified instances got by each iteration of Adaboost-SVM. ClustFirstClass is the method that applies cluster method for the majority class instances and randomly selects a quantity of instances from each cluster to form some majority subsets before training. It trains the different combinations of each subset and the minority class instances and combines the classification results of these base classifiers.

2.1 Adaboost-SVM Algorithm

Ensemble learning makes use of multiple weak classifiers to get a strong classifier. In this way, the adaptability of general classification algorithm is improved.

In 1990, Schapire [17] used the constructive method to prove the equivalence of weak classifier and strong classifier, the original Boosting algorithm is proposed. Later, Freund and Schapire [18] proposed Adaboost algorithm, it avoids the difficulty of getting lower bound of weak classifier's accuracy in the practical problems. In order to make the classification interface linearly separable, SVM maps the nonlinear classification interface to the high-dimensional feature space by applying kernel function method. Adaboost-SVM applies re-weighting technique to update the weights of training instances. The proposed Adaboost-SVM demonstrates better generalization performance than SVM imbalanced classification problems [19].

2.2   Performance Evaluation in Imbalanced Domains

The evaluation criterion is important for imbalanced data classification. How to evaluate a classifier for imbalanced classification is of great importance. Focusing on the binary classification problems, whether the instances are correctly or incorrectly predicted can be recorded in the confusion matrix (Table 1).

Table 1. Confusion matrix for binary classification

Class	Predicted positive	Predicted negative
Actual positive	TP (True positive)	FN (False negative)
Actual negative	FP (False positive)	TN (True negative)

According to the confusion matrix, several different measures can be deduced to evaluate the effect of the classifier for the imbalanced classification:

- $TP_{rate} = TP/(TP + FN)$
- $TN_{rate} = TN/(FP + TN)$
- $FP_{rate} = FP/(FP + TN)$
- $FN_{rate} = FN/(TP + FN)$
- G-mean (Geometry-mean) = $\sqrt{TP_{rate} * TN_{rate}}$
- AUC (Area Under ROC Curve) = $(1 + TP_{rate} - FP_{rate})/2$
- F-measure = $\dfrac{(1+\beta^2) \times TP_{rate} \times \frac{TP}{TP+FP}}{\beta^2 \times TP_{rate} + \frac{TP}{TP+FP}}$

ROC curve is an effective visual tool to judge the validity of classification. ROC curve reflects well on the classification of both classes at the same time, it provides the visual of the trade-off between TP_{rate} and FP_{rate} [20]. Since no specific values are given, it's not convenient to directly compare the results of different classifiers. By calculating the area under ROC curve, we can get the number AUC which can provide a scalar measure of the performance of a classifier [21].

3   Proposed Method for Imbalanced Classification

In this section, the method MSSC-Adaboost for the classification of imbalanced data is presented. Most of the misclassified samples are the boundary samples, applying under-sampling for the qualified misclassified samples of majority class and SMOTE for the misclassified samples of minority class can help to distinguish the boundary samples.

MSSC-Adaboost can be mainly divided into three stages: Firstly, normalize the continuous attributes of original datasets, set the outer loop iteration times and the iteration termination condition. Secondly, we take Adaboost of SVM based on spectral clustering algorithm to boost the performance. Lastly, based on the misclassified samples got by the second step, filter the misclassified negative class samples and delete them, using SMOTE to synthesize new positive samples between the misclassified positive samples and their nearest positive samples. The misclassified synthetic samples will be deleted timely during the next iteration. Repeat the second and third step until the loop condition is reached. The positive samples refer to the minority and the negative samples refer to the majority.

3.1   Boosting-SVM Based on Spectral Clustering

The traditional under-sampling methods do not take the effectiveness of training subsets into account, although the size of each class is balanced, but the impact on the SVM hyperplane may be not good or even worse. The traditional clustering algorithms such as k-means and EM algorithm are based on a spherical sample space, so the algorithm will easily fall into local optimum due to that sample space is not convex. Spectral clustering can be used in kernel space clustering, so the algorithm can be well combined with SVM algorithm. In order to maintain the space structure of positive class samples and make the SVM classification interface biased towards the majority samples, this method uses spectral clustering to divide the majority class into several clusters. During each iteration of Adaboost, select samples from each cluster to form some subsets due to the weight distribution, and train the different combinations of each majority subset and minority class by SVM. In this way, the selected training subsets own spatial representation and certain differences, and in each iteration of Adaboost, the majority samples with higher weights in each cluster are more likely to be chosen.

The advantage of spectral clustering algorithm is that it can avoid local optimization and converge the whole optimization. It is based on the spectral graph theory of graph theory [22].

Discompose the similarity matrix of samples, use clustering algorithm for the feature vector of the similarity matrix. Suppose that $S = \{s_1, s_2, \cdots, s_n\}$ is the dataset, n is the number of dataset, d is the samples dimensions. The steps of this spectral clustering can be described as following:

- Construct a similarity matrix $A = (A_{ij})_{n \times n}$.
- Construct the matrix $L = D^{-1/2} A D^{1/2}$, $D = (d_{ij})_{n \times n}$ is the diagonal matrix, the diagonal elements $d_{ii} = \sum A_{ij}$, the rest elements are zero.

- Calculate the eigenvectors and eigenvalues of L, the matrix $X = (X_1, X_2, \cdots, X_3) \in R^{n \times n}$ is consist of the eigenvectors based on descending order of corresponding eigenvalues.
- Normalize the row vectors of X and then get matrix Y, view each row of Y as one sample of R^d. The simple is divided into j cluster when Y_i is divided into j class.

3.2   MSSC-Adaboost Method

Ensemble learning has better superior generalization performance than one single weak learner by combining multiple individual learners. The misclassified samples got by integrated classification algorithm own the stronger stability and representation. The misclassified samples are mainly boundary samples. Use KNN algorithm to filter the noises or dangerous samples from misclassified negative samples and then remove them. Synthesize K samples among the misclassified positive samples and their k-nearest neighbors of correctly classified positive samples by applying SMOTE. It helps to recognize the positive samples around the boundary.

Algorithm 1. MSSC(Misclassified Sampling and Synthetic Correction).

Input: S_{train} is the training set, $S\{(x_i, y_i)\}(i = 1, 2, \cdots, n, y_i \in \{-1, +1\})$ is the misclassified training set

Output: New dataset

Steps:

for $i = 1, 2, \cdots, n$

1. If y_i=-1, use KNN to get it's k_1 nearest neighbors in S_{train}, m is the number of positive samples in it's k_1 nearest neighbors.
2. If $m \geq k_1/2$, remove it
3. If y_i=+1, use SMOTE to generate k_2 synthetic minority samples
4. Apply synthetic correction to timely delete misclassified synthetic samples got by step 3

End for

Algorithm 1 uses under-sampling and over-sampling based on the misclassified data. For each sample of S, if $y_i = -1$, S_i is the positive or minority class instance, if $y_i = +1$, S_i is the negative or majority class instance. The misclassified negative instances can be viewed as the noises or the danger points when m is more than the half of k_1. SMOTE is specially used for misclassified positive samples to synthesize new positive samples.

Algorithm 2. MSSC-Adaboost(spectral clustering Adaboost combined with MSSC)

Input: The training set S of examples $\{(x_1, y_1), (x_2, y_2), \cdots (x_n, y_n)\}, x_i \in X \subseteq R^n$, $y_i \in \{-1, +1\}$

Output: The last result of the classification

Steps:

1. *Miterate*(outer iteration), k(number of clusters), T(iteration of Adaboost), MIR(max ratio)
2. Initial the outer iteration: *Iterate*=0.
3. $S_{Neg} = \{(x_i, y_i)\}$, $S_{Neg} \subseteq S$, $i = 1, 2, \cdots, N$ and $y_i = -1$. $S_{Pos} = \{(x_i, y_i)\}$, $S_{Pos} \subseteq S$, $i = 1, 2, \cdots, M$ and $y_i = 1$.
4. Use spectral clustering for S_{Neg} to get the clusters: $A_1, A_2, \cdots, A_K (K \leq M)$.
   (a) Iteration of Adaboost: t=0. $D(i)$=1/($N+M$) for i=1,2, $\cdots$, $N + M$.
   (b) For t=1,2$\cdots$,T:
       (i) Combine the minority instances with instances selected from each cluster to form S_t:

$$M_{num(j)} = (C_{size(A_j)}/N) * M \tag{1}$$

   (ii) Apply SVM to train S_t and get the base classifier h_t.
   (iii) Calculate the error rate of the classifier h_t:

$$\varepsilon_t = \sum D_{t(i)}(h_t(x_i) \neq y_i) \tag{2}$$

   (iv) If $\varepsilon_t \geq 0.5$, T=t. Otherwise

$$\alpha_t = 0.5 * \ln((1 - \varepsilon_t)/\varepsilon_t) \tag{3}$$

   (v) Update $D(t)$:

$$D_{t+1}(i) = \frac{D_t(i)}{\sum_{i=1}^{N+M} D_t(i)} \quad (h_t(x_i) = y_i) \tag{4}$$

$$D_{t+1}(i) = \frac{D_t(i)}{\sum_{i=1}^{N+M} D_t(i)} * e^{\alpha_t} \quad (h_t(x_i) = y_i) \tag{5}$$

   (c) End for
   (d) The result of the strong classifier:

$$H_{iterate}(x_i) = sign(\sum_{t=1}^{T} h_t(x_i)\alpha_t) \tag{6}$$

5. Use Algorithm1 to get the new training set. If $M/N \leq MIR$ and *Iterate* $\leq$ *Miterate* go back to 3. Otherwise *final*=*Iterate*.
6. The last strong classifier: H_{final}.

In Algorithm 2, *MIR* is the max ratio between the minority and the majority. Before the step of Adaboost, spectral clustering is used to divide the majority into several clusters. In each iteration of Adaboost, the weights of instances are updated. The instances with higher weights in each cluster are firstly considered to be selected. The number of samples selected from each cluster is $M_{num(j)}$.

4 Empirical Study

4.1 Datasets

In this paper, we have employed 8 imbalanced datasets from UCI machine learning databases. Among these 8 datasets, some of them are originally binary and others are multi-class datasets. Transform the multi-classification problem into binary classification. These 8 datasets have different number of samples, attributes and class imbalance ratio. The negative samples are the majority and positive samples are the minority.

Table 2 introduces the information of each dataset in details. It includes the total number of classes (NC), the number of samples (Ins), the number of majority samples (Neg), the number of minority samples (Pos), the number of total attributes (Attr) and the imbalance ratio (IR).

Table 2. Statistic information of the datasets

Dataset	NC	Ins	Pos	Neg	Attr	IR
haberman	2	306	81	225	4	2.778
ionosphere	2	351	126	225	35	1.786
vowel2	11	528	48	480	11	10
wdbc	2	568	212	357	32	1.684
wpdc	2	198	47	151	35	3.213
yeast7	10	1484	244	1240	9	5.082
letter4	17	20000	805	19195	17	23.845
flag4	6	194	52	141	28	2.711

4.2 Experimental Setup

Divide each dataset into training set and test set based on the ratio 7:3. Set the experimental parameters: the number of Adaboost iteration is 10. The number of clusters is the number of the minority class samples. *Miterate, T, MIR, k_1* and k_2 are the parameters proposed in Algorithms 1 and 2. *MIR* = 0.75, *Miterate* = 5, *T* = 10, k_1 = 7 and k_2 = 3. Thus synthesize 3 new positive samples based on each misclassified positive samples.

4.3 Experiment Results and Analysis

In order to confirm the algorithm's effectiveness for imbalanced datasets, choose three kinds of methods ascomparisons. Adaboost-SVM-OBMS is the algorithm that combines Adaboost-SVM and over-sampling based on misclassified samples. ClusterFirstClass is the method that combines ensemble learning and cluster-based under-sampling. The following table shows G-means, AUC and F-measure based on different methods. The two figures followed by Table 3 show the comparison of line chart based on G-means and F-measure (Figs. 1 and 2).

Table 3. Different evaluation values of four methods on 8 datasets

Dataset	Algorithm	G-means	AUC	F-measure
haberman	SMOTE	0.521	0.625	0.412
	Adaboost-SVM-OBMS	0.666	0.676	0.528
	ClustFirstClass	0.603	0.655	0.488
	MSSC-Adaboost	**0.682**	**0.708**	**0.578**
ionosphere	SMOTE	0.827	0.836	0.801
	Adaboost-SVM-OBMS	0.829	0.835	0.8
	ClustFirstClass	0.821	0.831	0.795
	MSSC-Adaboost	**0.842**	**0.844**	**0.810**
vowel2	SMOTE	0.885	0.886	0.498
	Adaboost-SVM-OBMS	0.859	0.861	**0.619**
	ClustFirstClass	0.869	0.870	0.435
	MSSC-Adaboost	**0.897**	**0.897**	0.526
wdbc	SMOTE	0.973	0.973	0.968
	Adaboost-SVM-OBMS	0.966	0.966	0.958
	ClustFirstClass	0.951	0.951	0.938
	MSSC-Adaboost	**0.978**	**0.978**	**0.972**
wpdc	SMOTE	0.713	0.718	0.568
	Adaboost-SVM-OBMS	0.680	0.7	0.552
	ClustFirstClass	0.722	0.722	0.564
	MSSC-Adaboost	**0.755**	**0.756**	**0.611**
yeast7	SMOTE	0.753	0.760	**0.586**
	Adaboost-SVM-OBMS	0.776	0.776	0.569
	ClustFirstClass	**0.781**	**0.781**	0.583
	MSSC-Adaboost	0.762	0.763	0.534
Letter4	SMOTE	0.881	0.881	0.407
	Adaboost-SVM-OBMS	0.913	0.914	0.389
	ClustFirstClass	0.911	0.914	0.324
	MSSC-Adaboost	**0.928**	**0.929**	**0.438**
flag4	SMOTE	0.776	0.787	0.608
	Adaboost-SVM-OBMS	**0.838**	**0.842**	**0.684**
	ClustFirstClass	0.774	0.787	0.605
	MSSC-Adaboost	0.817	0.833	0.651

The results show that over haberman, ionosphere, vowel, wdbc, wpdc and letter4, MSSC-Adaboost has better performance compared to other methods regarding G-means and AUC as the evaluation criterions. As for the other two datasets, MSSC-Adaboost also gets considerable G-means and AUC values compared to the best results. In terms of F-measure, MSSC-Adaboost achieves higher values based on five datasets. We can conclude that our proposed method has considerably better performance compared to the other three methods in most cases.

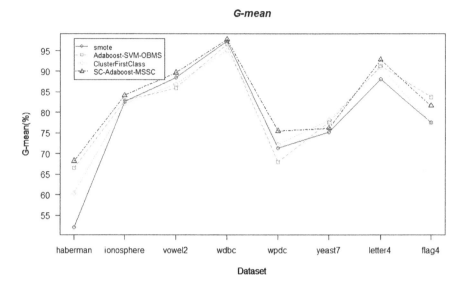

Fig. 1. G-means for datasets of different methods

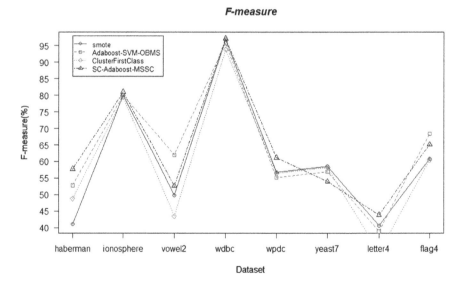

Fig. 2. F-measure for datasets of different methods

5 Conclusion

In this paper, we propose a combined method for classification learning from imbalanced data. We divide the majority class into several clusters by applying spectral clustering, delete qualified misclassified majority samples and synthesize new samples.

Misclassified synthetic samples may have bad influence on the result of classification, so it is necessary to correct and delete them. In addition, the algorithm proposed in this paper greatly reduce the number of synthetic samples compared to SMOTE. Compared with Adaboost-SVM-OBMS, it selects majority samples with space characteristic as the training subsets and timely deletes misclassified synthetic samples. For the most datasets chosen in this paper, our proposed method achieves higher performance compared to other methods.

Acknowledgement. This work was supported in part by National Natural Science Foundation of China (61273225, 61273303, 61373109), the Program for Outstanding Young Science and Technology Innovation Teams in Higher Education Institutions of Hubei Province (No. T201202), and the Program of Wuhan Subject Chief Scientist (201150530152), as well as National "Twelfth Five-Year" Plan for Science & Technology Support (2012BAC22B01).

References

1. Chawla, N.V., Japkowicz, N., Kotcz, A.: Editorial: special issue on learning from imbalanced data sets. J. Acm SIGKDD Explor. Newslett. **6**, 1–6 (2004)
2. Gao, J.W., Liang, J.Y.: Research and advancement of classification method of imbalanced data sets. J. Comput. Sci. **35**, 10–13 (2008)
3. Tahir, M.A., Kittler, J., Yan, F.: Inverse random under sampling for class imbalance problem and its application to multi-label classification. J. Pattern Recogn. **45**, 3738–3750 (2012)
4. Chawla, N.V., Cieslak, D.A., Hall, L.O.: Automatically countering imbalance and its empirical relationship to cost. J. Data Mining Knowl. Discov. **17**, 225–252 (2008)
5. Sun, Z., Song, Q., Zhu, X.: A novel ensemble method for classifying imbalanced data. J. Pattern Recogn. **48**, 1623–1637 (2015)
6. Chawla, N.V., Bowyer, K.W., Hall, L.O.: Smote: synthetic minority over-sampling technique. J. Artif. Intell. Res. **16**, 321–357 (2002)
7. Han, H., Wang, W.Y., Mao, B.H.: Borderline-smote: a new over-sampling method in imbalanced data sets learning. J. Lect. Notes Comput. Sci. **3644**, 878–887 (2005)
8. Bunkhumpornpat, C., Sinapiromsaran, K., Lursinsap, C.: Safe-level-smote: safe-level-synthetic minority over-sampling technique for handling the class imbalanced problem. Adv. Knowl. Discov. Data Mining **5476**, 475–482 (2009)
9. Fan, W., Stolfo, S.J, Zhang, J.: AdaCost: misclassification cost-sensitive boosting. In: Sixteenth International Conference on Machine Learning, pp. 97–105 . Morgan Kaufmann Publishers Inc. (1999)
10. Lertampaiporn, S., Thammarongtham, C., Nukoolkit, C.: Heterogeneous ensemble approach with discriminative features and modified-SMOTEbagging for pre-miRNA classification. J. Nucleic Acids Res. **41**, e21 (2013)
11. Chawla, N.V., Lazarevic, A., Hall, L.O.: Smoteboost: improving prediction of the minority class in boosting. J. Lect. Notes Comput. Sci. **2838**, 107–119 (2003)
12. Seiffert, C., Khoshgoftaar, T.M., Hulse, J.V.: Rusboost: a hybrid approach to alleviating class imbalance. J IEEE Trans. Syst. Man Cybern. **40**, 185–197 (2010)
13. Wang, C., Hongye, S.U., Yu, Q.U.: Imbalanced data sets classification method based on over-sampling technique. J. Comput. Eng. Appl. **47**, 139–143 (2011)
14. Li, X.F., Li, J., Dong, Y.F.: A new learning algorithm for imbalanced data—pcboost. J. Chinese J. Comput. **2**, 202–209 (2012)

15. Sobhani, P., Viktor, H., Matwin, S.: Learning from imbalanced data using ensemble methods and cluster-based undersampling. In: Appice, A., Ceci, M., Loglisci, C., Manco, G., Masciari, E., Ras, Z.W. (eds.) NFMCP 2014. LNCS, vol. 8983, pp. 69–83. Springer, Cham (2015). doi:10.1007/978-3-319-17876-9_5

16. Sun, Z., Song, Q., Zhu, X.: Using coding-based ensemble learning to improve software defect prediction. J. IEEE Trans. Syst. Man Cybern. Part C **42**, 1806–1817 (2012)

17. Schapire, R.E.: The strength of weak learnability. J. Mach. Learn. **5**, 197–227 (1990)

18. Freund, Y., Schapire, R.E.: A decision-theoretic generalization of on-line learning and an application to boosting. J. Comput. Syst. Sci. **55**, 119–139 (1999)

19. Li, X., Wang, L., Sung, E.: Adaboost with SVM-based component classifiers. J. Eng. Appl. Artif. Intell. **21**, 785–795 (2008)

20. Bradley, A.P.: The use of the area under the ROC curve in the evaluation of machine learning algorithms. J. Pattern Recogn. **30**, 1145–1159 (1997)

21. Huang, J., Ling, C.X.: Using AUC and accuracy in evaluating learning algorithms. J. IEEE Trans. Knowl. Data Eng. **17**, 299–310 (2005)

22. Luxburg, U.V., Belkin, M., Bousquet, O.: Consistency of spectral clustering. J. Ann. Stat. **36**, 555–586 (2008)

Adaptive Kendall's τ Correlation in Bipartite Network for Recommendation

Xihan Shan[1] and Junlong Zhao[2(✉)]

[1] Beihang University, Beijing, China
[2] Beijing Normal University, Beijing, China
zhaojunlong928@126.com

Abstract. The commonly used algorithms in recommender system tend to recommend popular items. The recently proposed algorithm, denoted as G-CosRA, shows good performance in handling this problem, with two parameters to control the popularity of items and activeness of users. In this paper, we refine this algorithm and propose a new recommendation algorithm based on adaptive Kendall's τ correlation, where only one tuning parameter is involved. The proposal has better performance in accuracy, popularity and diversity, compared with G-CosRA and other existing algorithms. A parameter-free version, named weighted Kendall, is also proposed for better efficiency in computing.

Keywords: Recommendation · Kendall's τ correlation · Diversity · Accuracy

1 Introduction

With the rapid development of e-commerce, more and more people shop online and express their opinions upon different commodities on the Internet [1]. How to provide users with authentic information effectively has become an important issue for both consumers and sellers in modern society [2]. Recommender systems could help users quickly seek out their preferable items from massive redundant information based on their historical ratings and choices, and also help sellers improve their sales.

Collaborative filtering (CF) is one of the most popular approaches in the recommender systems. The Neighborhood-based CF methods focus on the similarity between items (ICF), or alternatively, between users (UCF) [3]. Taking UCF as an example, for user u_i, CF works by searching users who have similar taste for items as u_i and creating a ranked list of suggestions based on what they like. The key point of CF algorithms is how to define the similarities between users or objects [4]. There are many similarity measurements [3, 5, 6], among which Pearson and Cosine similarities are two most widely used indices [7, 8]. However, these two measurements tend to recommend popular items.

Methods based on network have been used in recommendation. Motivated by mass diffusion (MD) and heat conduction (HC) in physical dynamics, Zhou et al. [9] proposed a weighting method, treating the problem as resource-allocation process(RA) in user-item bipartite network. Later, Zhou et al. [10] proposed a recommendation algorithm with degree-dependent initial configuration. Zhou et al. [11] proposed a

© Springer International Publishing AG 2017
D.-S. Huang et al. (Eds.): ICIC 2017, Part III, LNAI 10363, pp. 389–399, 2017.
DOI: 10.1007/978-3-319-63315-2_34

hybrid method that combines both MD and HC. These methods have better accuracy than traditional CF [10], but still tend to distribute more resource to nodes of large-degree in the network, and consequently result in lower diversity.

To overcome the limitations of RA based algorithms, Liu et al. [12] proposed MCF index, where a tuning parameter is used to control the effect of popular items. Moreover, Chen et al. [13] refined RA and proposed CosRA algorithm. As an extension of CosRA, Chen et al. [13] proposed a more general algorithm, denoted as G-CosRA, with two tuning parameters controlling the popularity of items and the activeness of users respectively. Our simulations in Sect. 3 show that G-CosRA is more accurate than CosRA when the network is highly sparse and the sample size is relatively small. However, choosing two tuning parameters in G-CosRA can be computationally intensive.

In this paper, we show that G-CosRA can be viewed as a weighted inner product. Note that recommendation actually concerns about the order of items. A new algorithm, called adaptive Kendall (A-Kendall), is proposed for personalized recommendation. A-Kendall method involves only one tuning parameter, and consequently is more efficient in computation. Moreover, based on our experiment, A-Kendall has overall better performance than that of G-CosRA and other benchmark methods in terms of accuracy, popularity and diversity. Similar to CosRA, we define a parameter-free version of A-Kendall, named weighted Kendall (W-Kendall), by fixing the value of the parameter at -0.5, which also shows outstanding performance. The effectiveness of the proposal is demonstrated on several real data sets.

The main contents of this paper are arranged as follows. In Sect. 2, we review the related works briefly and propose our approaches, named A-Kendall and W-Kendall. In Sect. 3, we compare our methods with other methods on several real data sets. And a short discussion is presented in Sect. 4.

2 Recommendation with Kendall's τ Correlation in Bipartite Network

2.1 Brief Review of Similarity Indices Based on Bipartite Network

Zhou et al. [9] viewed the recommendation as a problem of allocating the resource on the user-item bipartite network. Specifically, for $1 \leq i \leq m$ and $1 \leq l \leq n$, define $a_{il} = 1$ if user u_i selects the item o_l, otherwise $a_{il} = 0$. Then an adjacent matrix $A = (a_{il}) \in R^{m \times n}$ of the network can be constructed. Liu et al. [14] proposed the following measurement of similarity

$$S_{ij}^{SA-CF} = \sum_{l=1}^{n} \frac{a_{il}}{k_{u_j}} \frac{a_{jl}}{k_{o_l}},$$

where k_{u_i} is degree of user u_i, and k_{o_l} is the degree of o_l. Moreover, Liu et al. [12] proposed a modified version of the above measurement

$$S_{ij}^{MCF} = \frac{1}{k_{u_j}} \sum_{l=1}^{n} \frac{a_{il}(k_{u_j}k_{o_l})^{\lambda} a_{jl}(k_{u_i}k_{o_l})^{\lambda}}{k_{o_l}},$$

where a tuning parameter is used to control the degree of item and user. Chen et al. [13] then proposed G-CosRA

$$S_{ij}^{G-CosRA} = \frac{1}{(k_{u_i}k_{u_j})^{-\lambda_1}} \sum_{l=1}^{n} \frac{a_{il}a_{jl}}{k_{o_l}^{-\lambda_2}}.$$

CosRA is a special case of G-CosRA with parameter $\lambda_1 = \lambda_2 = -0.5$. CosRA is parameter-free, but has poor popularity and diversity compared with G-CosRA, which is extremely important for a recommendation system.

Denote $D_u = \text{diag}(d_{u_1}, \ldots, d_{u_n}) \in R^{n \times n}$ and $D_o = \text{diag}(d_{o_1}, \ldots, d_{o_m}) \in R^{m \times m}$, where $d_{u_i} = k_{u_i}^{\lambda_1}$ and $d_{o_l} = k_{o_l}^{\lambda_2}$. Define $\tilde{A} = D_u A D_o$. With the above definition, it is easy to see that

$$S_{ij}^{G-CosRA} = \langle \tilde{A}_{i\cdot}, \tilde{A}_{j\cdot} \rangle,$$

where $\tilde{A}_{i\cdot}$ is the i-th row of $\tilde{A}$. That is, $S_{ij}^{G-CosRA}$ is the inner product of i-th and j-th rows of $\tilde{A}$.

2.2 Recommendation with Adaptive Kendall's τ Correlation

Note that the order of recommended items is the main concern of recommendation. And the inner product used in G-CosRA is not a good measurement on similarity of the order. It is well known that Kendall's τ correlation uses only the information of order. In this subsection, we extend the idea of [12], and propose the following adaptive Kendall's τ correlation, denoted as A-Kendall. Recall the definition $\tilde{A} = D_u A D_o$ and denote $\tilde{A} = (\tilde{a}_{il})$. A-Kendall similarity between users u_i and u_j is defined as follows.

$$\tau_{ij} = \frac{2}{n(n-1)} \sum_{1 \le k < l \le n} sign(\tilde{a}_{ik} - \tilde{a}_{il}) \, sign(\tilde{a}_{jk} - \tilde{a}_{jl}).$$

By the definition of $\tilde{a}_{il}$, it is easy to see that

$$sign(\tilde{a}_{ik} - \tilde{a}_{il}) \, sign(\tilde{a}_{jk} - \tilde{a}_{jl}) = sign(a_{ik}d_{o_k} - a_{il}d_{o_l}) \, sign(a_{jk}d_{o_k} - a_{jl}d_{o_l}). \quad (2.1)$$

The right side of (2.1) depends only on the weights d_{o_l} of the items, and is free from the weights of users. Recalling $d_{o_l} = k_{o_l}^{\lambda_2}$, one can see that τ_{ij} involves only one parameter λ_2 which controls the impact of popular items, and consequently is computationally more efficient than G-CosRA. Moreover, it is well known that Kendall's correlation is more robust, insensitive to noise in the data.

To further improve the computational efficiency, we define a parameter-free similarity index, called weighted Kendall's τ correlation (W-Kendall), by fixing $\lambda_2 = -0.5$ in A-Kendall, which is the counterpart of CosRA. Real data analysis confirms the good performance of both A-Kendall and W-Kendall.

Given the similarity between users, if u_i has not yet collected o_l (i.e. $a_{il} = 0$), the predicted score, denoted by v_{il}, is given as

$$v_{il} = \frac{\sum_{j=1}^{m} \tau_{ij} a_{jl}}{\sum_{j=1}^{m} \tau_{ij}}. \tag{2.2}$$

We compute the scores of uncollected items for each user. Larger the score is more likely the item is being recommended. The process of recommendation is shown in Fig. 1.

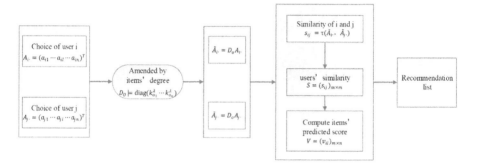

Fig. 1. The procedure of A-Kendall method

In summary, our proposal has the following advantages. First, A-Kendall and W-Kendall consider the order of the item, which is more coincident with the goal of recommendation. Second, A-Kendall, involving only one tuning parameter, is computationally more efficient compared with G-CosRA. Finally, Kendall's τ correlation is robust to noise.

3   Numerical Result

In this section, we compare experimentally our approaches with some benchmark methods mentioned in Sect. 2.1 including CF, MCF, as well as CosRA and G-CosRA which have been tested to have better performance than other widely used algorithms [13]. Replacing the τ_{ij} in (2.2) with the similarities defined in these benchmark methods, one can compute the predicted scores for unselected items and recommend the items of higher score. Three criteria are considered: accuracy, popularity and diversity. In addition, the computation cost of G-CosRA and A-Kendall are compared. Details on these criteria are presented in Sect. 3.2.

3.1 Description of Data Set

Three commonly studied rating data sets, Movielens [15], Epinion [16] and Filmtrust [17], are used to test the algorithm.

Movielens. Movielens-1M is a data set including 700 users who rated at least twice among 9000 different movies, and the ratings of the movies are scaled from 0.5 (min) to 5.0 (max) with step 0.5. We first apply a coarse-grain method [10] to construct the adjacent matrix. Specifically, there is an edge between user and movie only if the corresponding rating is larger than 3. Then we select a relatively smaller group of ratings from Jan. 1995 to Dec. 1998 with 161 users, 1224 movies and 10207 edges as our experimental data.

Epinion. Epinion was collected from the Epinions.com Web site, consisting of 49,290 users who rated a total of 139,738 different items. The ratings are also scaled from 0.5 to 5. After applying coarse-grain, we select 95 users who have rated at least 2 items by simple random sampling for experimenting.

Filmtrust. Filmtrust is a small dataset crawled from the entire FilmTrust website with 35497 item ratings scaling from 0.5 to 5. We apply coarse-grain on it and select 164 users by simple random sampling with their ratings as experimental data. Several topological properties of the real data are presented in Table 1.

Table 1. Basic statistics of three real rating data sets.

Data	User	Item	Edges	Sparsity
Movielens	161	1224	10207	0.0518
Eponion	84	584	1364	0.0278
Filmtrust	164	530	2537	0.0291

In the following experiments, a 5-folder cross-validation strategy is used to evaluate the performance of recommendation. Specifically, all the ratings of users would be randomly divided into five groups $\{G_i, 1 \leq i \leq 5\}$ of equal size. For $i = 1, \cdots, 5$, by taking data in G_i as test set, and the data in the remaining four groups as training set, we evaluate the performance on test set G_i in terms of given criterion, and then compare the average values over G_i's.

3.2 Algorithmic Performance Metrics

We use the following criteria to compare different methods [10, 13].

Accuracy. For each user u_i, $1 \leq i \leq m$, the algorithm will generate a recommendation list of length L_i, consisting of items that have not been chosen by u_i and the uncollected items are sorted by their scores. For an arbitrary dyad $u_i - o_l$ in test set, the ranking score of o_l is defined as $r_{il} = R_{il}/L_i$ where R_{il} is the ordinal number of o_l in recommendation list of u_i. The mean value of ranking score, averaged over entries in the test set, is defined as average ranking score $\langle r \rangle$. Since all items in test set are actually chosen by users, a more accurate algorithm ought to have smaller $\langle r \rangle$.

Popularity. To test whether an algorithm is more likely to recommend popular items, we consider average degree $\langle d \rangle$ which is defined as the average degree of all the first L items of users' recommendation list. The smaller $\langle d \rangle$ is, more likely that minority or unpopular items are to be recommended.

Diversity. Given the recommendation lists of two users u_i and u_j, Hamming distance is defined as $H_{ij} = 1 - Q_{ij}/L$, where Q_{ij} is the number of overlapped items among the first L elements in the lists of u_i and u_j. The average Hamming distance over all users, $\langle H \rangle$, measures the recommendation diversity of the algorithm. $H = 0$ means lists of all users are same and $H = 1$ means lists of all users are totally different when the algorithm has the best diversity. Therefore, a large H is preferred. Without loss of generality, we take $L = 50$ in this paper.

Computation Efficiency. Computing time is considered as an important measurement. We report the computing times for G-CosRA and A-Kendall, denoted as T.

3.3   Analysis of Result

Numerical results on three data sets are shown in Table 2 and The parameters of A-Kendall, G-CosRA and MCF are set as the ones corresponding to the lowest ranking scores. It can be seen that A-Kendall and W-Kendall have better performance than other four methods in most cases. A-Kendall has the best accuracy and good diversity on all three datasets.

Compared with CosRA, accuracy $\langle r \rangle$ of A-Kendall is improved by 11.84%, 29.47% and 46.79% respectively for the three data sets; and Popularity and diversity are also obviously improved, especially on Movielens, by 4.67% and 10.05%.

The parameter-free method W-Kendall has performance slightly inferior to A-Kendall, but remarkably better than CosRA. Even if G-CosRA could lead to better results than CosRA by properly adjusting parameters, it is still much inferior to A-Kendall and W-Kendall on the three metrics. MCF and CF are two weakest among all six methods notwithstanding that CF is slightly better in terms of popularity when applying to Epinion. Actually it is hard to solve the accuracy-diversity dilemma in recommender systems and although A-Kendall is not the best on all data sets over three metrics, it is still good and quite close to the best one. Based on these analysis, it can be concluded that Kendall's τ based methods have the overall best accuracy, popularity and diversity.

Remind that G-CosRA involves parameters (λ_1, λ_2), while A-Kendall involves only λ_2. For λ_1 and λ_2 respectively, we choose the optimal one in the candidate set of size n. Therefore, there are n possible values for parameter in A-Kendall, and n^2 possible values for parameters in G-CosRA. Figure 2 reports the relationship between running time T and n in MovieLens data. Obviously, the running time T is a quadratic function of n for G-CosRA, but is a linear function for A-Kendall. When $n > 19$, computing time of A-Kendall is apparently less than G-CosRA.

Table 2. Comparison of different methods. Smaller values of $\langle r \rangle$ and $\langle d \rangle$, and larger values of $\langle H \rangle$ are preferred. The best values are emphasized by bold. The parameters of A-Kendall, G-CosRA and MCF are set as the ones corresponding to the lowest ranking scores. And the percentages represent the improvement of A-Kendall compared over other methods, where '+' indicates A-Kendall is comparatively better and vice versa.

	$\langle r \rangle$		$\langle d \rangle$		$\langle H \rangle$	
MovieLens						
A-Kendall	**0.1109**	-	51.8510	-	**0.6341**	-
W-Kendall	0.1184	+6.33%	**50.0607**	−3.58%	0.6255	+1.37%
G-CosRA	0.1244	+10.85%	52.0088	+0.30%	0.6316	+0.40%
CosRA	0.1258	+11.84%	54.3883	+4.67%	0.5762	+10.05%
MCF	0.1352	+17.97%	52.4176	+1.08%	0.5930	+6.93%
CF	0.1417	+21.74%	55.3080	+6.50%	0.5062	+25.27%
Epinion						
A-Kendall	**0.2702**	-	4.9685	-	0.8912	-
W-Kendall	0.2725	+0.84%	5.0959	+2.50%	**0.8959**	−0.52%
G-CosRA	0.3792	+28.74%	5.5929	+11.16%	0.8909	+0.03%
CosRA	0.3831	+29.47%	5.5532	+10.53%	0.8800	+1.27%
MCF	0.3825	+29.36%	5.8343	+14.84%	0.8843	+0.78%
CF	0.3902	+30.75%	**4.4449**	−11.78%	0.8827	+0.96%
Filmtrust						
A-Kendall	**0.0721**	-	**17.3541**	-	**0.4419**	-
W-Kendall	0.0829	+13.03%	18.6502	+6.95%	0.3996	+10.59%
G-CosRA	0.1350	+46.59%	18.5195	+6.29%	0.4191	+5.44%
CosRA	0.1355	+46.79%	18.4592	+5.99%	0.4152	+6.43%
MCF	0.1355	+46.79%	18.9358	+8.35%	0.3875	+14.04%
CF	0.1367	+47.26%	18.3635	+5.50%	0.3946	+11.99%

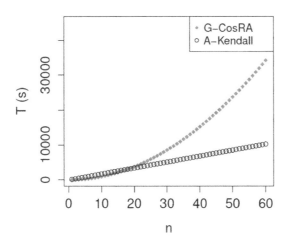

Fig. 2. Running time T (in seconds) for A-Kendall and G-CosRA in MovieLens data.

Denoting the parameter λ_2 as λ, plots (a), (b) and (c) in Fig. 3 report the accuracy $\langle r \rangle$ of A-Kendall as a function of λ on test set. In plot (a) of Fig. 3, the red horizontal line denotes $\langle r \rangle = 0.1244$, corresponding to the optimal accuracy of G-CosRA (see also Fig. 3, plot (d)), and the green one denotes $\langle r \rangle = 0.1258$, corresponding to the accuracy of CosRA. When $\lambda_2 \in [-40, 0]$, the accuracy of A-Kendall is completely below the dotted lines, indicating that A-Kendall outperforms CosRA and G-CosRA in large part of its parameter space. By the plots (b)–(e) and (c)–(f) of Fig. 3, one can see that A-Kendall has better accuracy in whole parameter space from -100 to 50.

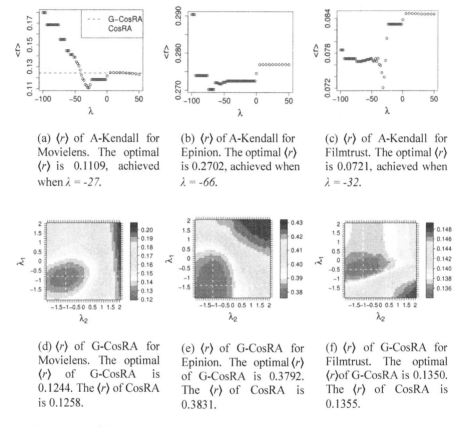

(a) $\langle r \rangle$ of A-Kendall for Movielens. The optimal $\langle r \rangle$ is 0.1109, achieved when $\lambda = -27$.

(b) $\langle r \rangle$ of A-Kendall for Epinion. The optimal $\langle r \rangle$ is 0.2702, achieved when $\lambda = -66$.

(c) $\langle r \rangle$ of A-Kendall for Filmtrust. The optimal $\langle r \rangle$ is 0.0721, achieved when $\lambda = -32$.

(d) $\langle r \rangle$ of G-CosRA for Movielens. The optimal $\langle r \rangle$ of G-CosRA is 0.1244. The $\langle r \rangle$ of CosRA is 0.1258.

(e) $\langle r \rangle$ of G-CosRA for Epinion. The optimal $\langle r \rangle$ of G-CosRA is 0.3792. The $\langle r \rangle$ of CosRA is 0.3831.

(f) $\langle r \rangle$ of G-CosRA for Filmtrust. The optimal $\langle r \rangle$ of G-CosRA is 0.1350. The $\langle r \rangle$ of CosRA is 0.1355.

Fig. 3. Accuracy $\langle r \rangle$ of A-Kendall and G-CosRA/CosRA on test set of three data sets. The parameter of A-Kendall, denoted as λ, varies from -100 to 50. Each of the parameters λ_1 and λ_2 in G-CosRA varies from -2 to 2. The point of crossing of vertical and horizontal dash lines corresponds to the optimal parameter in terms of $\langle r \rangle$.

Figure 4 presents the results of $\langle d \rangle$ and $\langle H \rangle$ with different λ on A-Kendall. As we expected, parameter λ controls the influence of popular items. Taking the result of Movielens as an example, as λ increases from -100 to -27, the popularity decreases significantly, and the diversity $\langle H \rangle$ is improved by 61.2%, from 0.3934 to 0.6341.

Moreover, Figs. 3 and 4 show that A-Kendall is insensitive to λ, providing the same recommendation list for users when λ in a certain interval, such as $\lambda \in (-46, 0)$ on Epinion in plot (b) of Fig. 3.

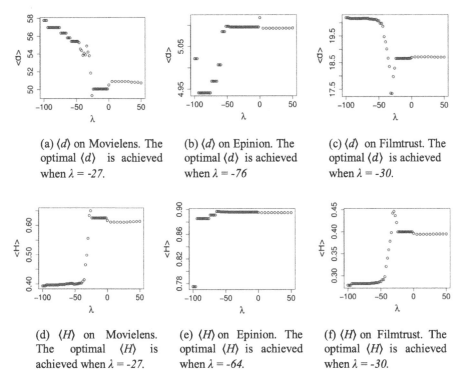

(a) $\langle d \rangle$ on Movielens. The optimal $\langle d \rangle$ is achieved when $\lambda = -27$.

(b) $\langle d \rangle$ on Epinion. The optimal $\langle d \rangle$ is achieved when $\lambda = -76$

(c) $\langle d \rangle$ on Filmtrust. The optimal $\langle d \rangle$ is achieved when $\lambda = -30$.

(d) $\langle H \rangle$ on Movielens. The optimal $\langle H \rangle$ is achieved when $\lambda = -27$.

(e) $\langle H \rangle$ on Epinion. The optimal $\langle H \rangle$ is achieved when $\lambda = -64$.

(f) $\langle H \rangle$ on Filmtrust. The optimal $\langle H \rangle$ is achieved when $\lambda = -30$.

Fig. 4. Popularity $\langle d \rangle$ and diversity $\langle H \rangle$ on the test sets in three data sets. λ are varying from -100 to 50. The length of recommendation list is set as $L = 50$.

Moreover, we notice that for Movielens and Filmtrust, the optimal values of all three metrics in A-Kendall could be reached almost simultaneously at the same λ (i.e. -27 for Movielens; -30 for Filmtrust), and for sparser Epinion, A-Kendall also has good performance over three metrics when $\lambda = -64$. On the other hand, the optimal values of the parameters for three metrics are significantly different in G-CosRA, which means that accuracy, popularity and diversity cannot be controlled simultaneously. This brings difficulties for G-CosRA in applications. From the above analysis, we can conclude that A-Kendall experimentally has better performance on accuracy, popularity, diversity, as well as lower computing complexity compared with G-CosRA. And W-Kendall, which is computationally more efficient than A-Kendall, also has good performance in terms of the above metrics. People could choose either one based on their demand.

4 Conclusion

In this paper, we propose a new recommendation algorithm, named A-Kendall, based on Kendall correlation. This algorithm considers the order of the items, which is more compatible to the concerns of recommendation. In addition, A-Kendall involves only one tuning parameter. A parameter free version W-Kendall is proposed for computational efficiency. Empirical results confirm the effectiveness of the algorithm. Also, Kendall-based methods well overcome the accuracy-diversity dilemma of previous algorithm and is very simple but effective in recommendation.

By using suitable similarity index, the performance of the algorithm could be effectively enhanced. Nevertheless, our simulation still needs further improvement and more experiments on different kind of data will be conducted. Moreover, Kendall's τ correlation works well on measuring the similarity of the orderings of the data and more application on problems about order is expected to be explored. We believe the current work can enlighten readers in their future work.

References

1. Moradi, P., Ahmadian, S.: A Reliability-Based Recommendation Method to Improve Trust-aware Recommender Systems. Pergamon Press, Oxford (2015)
2. Kohrs, A., Merialdo, B.: Improving collaborative filtering for new users by smart object selection (2001)
3. Sarwar, B., Karypis, G., Konstan, J.: Item-based collaborative filtering recommendation algorithms. In: International Conference on World Wide Web, pp. 285–295 (2001)
4. Goldberg, D., Nichols, D., Oki, B.M.: Using collaborative filtering to weave an information tapestry. Commun. ACM **35**(12), 61–70 (1992)
5. Sun, H., Peng, Y., Chen, J.: A new similarity measure based on adjusted euclidean distance for memory-based collaborative filtering. J. Softw. **6**(6), 993–1000 (2011)
6. Schafer, J.B., Frankowski, D., Herlocker, J., Sen, S.: Collaborative filtering recommender systems. In: Brusilovsky, P., Kobsa, A., Nejdl, W. (eds.) The Adaptive Web. LNCS, vol. 4321, pp. 291–324. Springer, Heidelberg (2007). doi:10.1007/978-3-540-72079-9_9
7. Symeonidis, P., Nanopoulos, A., Papadopoulos, A.N., Manolopoulos, Y.: Collaborative filtering: fallacies and insights in measuring similarity. In: Berendt, B. (ed.) (2006)
8. Lü, L., Medo, M., Chi, H.Y., Zhang, Y.C., Zhang, Z.K., Zhou, T.: Recommender systems. Phys. Rep. **519**(1), 1–49 (2012)
9. Zhou, T., Ren, J., Medo, M., Zhang, Y.C.: Bipartite network projection and personal recommendation. Phys. Rev. E: Stat., Nonlin. Soft Matter Phys. **76**(2), 046115 (2007)
10. Zhou, T., Jiang, L.L., Su, R.Q., Zhang, Y.C.: Effect of initial configuration on network-based recommendation. Physics **81**(5), 58004 (2008)
11. Zhou, T., Kuscsik, Z., Liu, J.G., Medo, M., Wakeling, J.R., Zhang, Y.C.: Solving the apparent diversity-accuracy dilemma of recommender systems. Proc. Natl. Acad. Sci. **107** (10), 4511–4515 (2010)
12. Liu, J., Zhou, T., Wang, B.H., Zhang, Y.C., Guo, Q.: Degree correlation of bipartite network on personalized recommendation. Int. J. Mod. Phys. C **21**(1), 137–147 (2011)
13. Chen, L.J., Zhang, Z.K., Liu, J.H., Gao, J., Zhou, T.: A vertex similarity index for better personalized recommendation. Phys. A **466**, 607–615 (2015)

14. Liu, J.G., Wang, B.H., Guo, Q.: Improved collaborative filtering algorithm via information transformation. Int. J. Mod. Phys. C **20**(2), 285–293 (2009)
15. GroupLens: Movielens. https://grouplens.org/datasets/movielens/
16. Massa, P., Souren, K., Salvetti, M., Tomasoni, D.: Trustlet, open research on trust metrics. Scalable Comput. Pract. Exp. **9**(4), 31–43 (2008)
17. Guo, G., Zhang, J., Yorke-Smith, N.: A novel bayesian similarity measure for recommender systems. In: International Joint Conference on Artificial Intelligence, pp. 2619–2625. AAAI Press (2013)

Using Ontology and Cluster Ensembles for Geospatial Clustering Analysis

Xin Wang[1,2(✉)] and Wei Gu[1]

[1] Department of Geomatics Engineering, University of Calgary,
Calgary, AB, Canada
xcwang@ucalgary.ca
[2] School of Information and Technology, Northwest University, Xi'an, China

Abstract. Geospatial clustering is an important topic in spatial analysis and knowledge discovery research. However, most existing clustering methods clusters geospatial data at data level without considering domain knowledge and users' goals during the clustering process. In this paper, we propose an ontology-based geospatial cluster ensemble approach to produce better clustering results with the consideration of domain knowledge and users' goals. The approach includes two components: an ontology-based expert system and a cluster ensemble method. The ontology-based expert system is to represent geospatial and clustering domain knowledge and to identify the appropriate clustering components (e.g., geospatial datasets, attributes of the datasets and clustering methods) based on a specific application requirement. The cluster ensemble is to combine a diverse set of clustering results which is produced by recommended clustering components into an optimal clustering result. A real case study has been conducted to demonstrate the efficiency and practicality of the approach.

Keywords: Spatial analysis · Ontology · Cluster ensemble · Facility location problem

1 Introduction

Geospatial clustering is an important topic in spatial analysis and knowledge discovery research. It aims to place similar objects into the same group (called a cluster) based on their connectivity, density and reachability in geographical space while placing dissimilar objects in different groups [1, 2]. It can be used to find natural clusters (e.g., extracting the type of land use from the satellite imagery), to identify hot spots (e.g., epidemics, crime, traffic accidents), and to partition an area based on utility (e.g., market area assignment by minimizing the distance to customers).

Domain knowledge and users' goals play important roles during geospatial clustering [3–5]. The background knowledge concerning the domain described by the geospatial data is called domain knowledge. In geospatial clustering analysis, a user seeks to discover knowledge from geospatial data based on a particular goal by applying clustering methods. However, most existing clustering processes and clustering methods focus solely on the data itself without considering domain knowledge.

© Springer International Publishing AG 2017
D.-S. Huang et al. (Eds.): ICIC 2017, Part III, LNAI 10363, pp. 400–410, 2017.
DOI: 10.1007/978-3-319-63315-2_35

Thus, clustering occurs at the data level instead of the knowledge level, which prevents the user from precisely understanding the clustering results and achieving his or her goal.

A few options for handling the problem seem apparent. One option is to develop new geospatial clustering methods exactly tailored for users' applications. The customized methods should consider which attributes of the geospatial data are needed and what kind of the domain knowledge has to be included. Some customized clustering methods called constrained-based clustering methods, have been proposed [8, 9, 11]. Since they only consider limited knowledge concerning the domain and the user's goals, they are typically difficult to reuse. In particular, they usually have very restricted means of incorporating domain-related information from non-geospatial attributes.

The second option can be defined by considering the overall nature of the clustering process and building a knowledge-based system to support the integration of knowledge in the geospatial clustering process. The clustering process consists of all steps required to accomplish a clustering task given by a user. It starts from data pre-processing (including data cleaning, data integration, data selection, and data transformation), then applies clustering methods on the datasets, and finally presents the clustering results to the user [12]. Applying a clustering method is only one step of the overall process. Thus, the second option is to incorporate domain knowledge and users' goals into the clustering process, which allows an informed choice to be made from choosing the available datasets, suitable attributes of the datasets and clustering methods. However, this option can only get the best clustering results by using the existing most suitable clustering method and are not helpful when none of the existing clustering method can provide good clustering results for a specific application.

The third option is to apply cluster ensembles [13] to geospatial clustering analysis. By applying available clustering methods to different attributes of datasets, cluster ensembles can obtain a large set of clustering results and finally combine them into a single consolidated clustering result. The result contains all information in the ensemble. However, it is time consuming to get a diverse set of clustering results. Previous research also shows that it is not always the best to include all available clustering results in the ensemble [14]. Thus, there is an emerging interest on reducing the number of clustering results in the ensemble.

In this paper, we propose a novel approach to geospatial clustering analysis which combines an ontology-based expert system with a cluster ensemble method. Specifically, we first build an ontology to represent geospatial and clustering domain knowledge and then develop an expert system to help identify appropriate geospatial datasets, attributes of the datasets and clustering methods for a specific application. Next, all the datasets, the attributes of the datasets and the clustering methods recommended by the ontology-based expert system are used to produce a diverse set of clustering results. Finally, with the help of the domain knowledge in the expert system, a subset of clustering results are selected from all the clustering results and are combined into a single clustering result. The approach can perform better than existing clustering methods because of the following reasons:

First, instead of developing a new clustering method for every specific application, the approach considers knowledge reuse. The domain knowledge learned from previous applications is formalized into the ontology-based expert system and would be reused for the similar applications in the future.

Second, with the help of the domain knowledge, the approach can identify appropriate components, including appropriate datasets, attributes of datasets, and clustering methods, according to users' goals.

Third, the approach combines the clustering results produced by the appropriate clustering components into one best clustering result using the cluster ensemble method. It provides more comprehensive clustering results when none of the clustering results produced by the appropriate clustering components is good enough.

The rest of the paper is organized as follows. The ontology-based geospatial clustering ensemble approach is proposed in the Sect. 2. A case study of the approach to do facility location analysis in South Carolina, USA is presented in Sect. 3. Section 4 summarizes the paper and discusses the future work.

2 An Ontology-Based Geospatial Cluster Ensemble Approach

In this section, we present an ontology-based cluster ensemble approach for geospatial clustering analysis. The approach includes two components: the ontology-based geospatial clustering system and a clustering ensemble method. In the following, we start with the general workflow of the approach and then we will introduce each component in detail.

2.1 Work Flow of the Approach

As shown in Fig. 1, users' clustering goals are first sent to the ontology-based geospatial clustering system, *GEO_CLUST*. The *GEO_CLUST* identifies appropriate geospatial datasets, attributes of the datasets and clustering methods according to the goals and the domain knowledge. Then, a diverse set of clustering results are produced by applying different clustering methods to different datasets and attributes of datasets multiple times.

For a better understanding of the approach, we treat each clustering result of a geospatial clustering application as a solution to that application and plotted them into a solution space. Figure 2 shows the changes in a solution space when applying the ontology-based geospatial cluster ensemble approach under different statuses. At status 1, only the solutions within the circle are left for the following analysis. Second, a set of clustering results are sent the cluster ensemble method. In the first step of the method, a subset of the clustering results is selected with the criteria of high quality and diversity. The domain knowledge may be extracted from the *GEO_CLUST* to measure the quality. Thus at status 2, the solution space is further reduced, as shown the smaller rectangle in Fig. 2. Finally, the second step of the cluster ensemble method combines the selected clustering results into one optimal combined clustering result (as shown the black point in Fig. 2). According to the Eq. 6, the combination may need the domain knowledge which could be extracted from the *GEO_CLUST*.

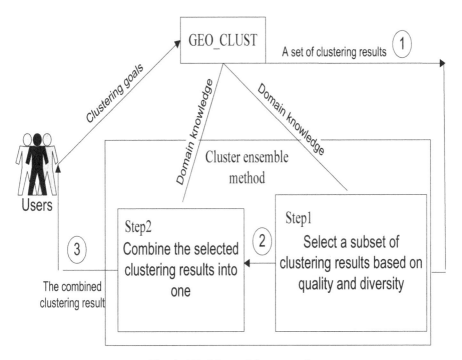

Fig. 1. Workflow of the approach

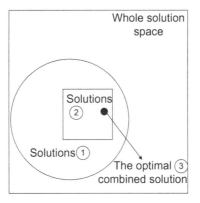

Fig. 2. Changes in solution space

2.2   An Ontology-Based Geospatial Clustering System

In this section, we present an ontology-based expert system, named GEO_CLUST, for performing geospatial clustering. The goal of the system is to make better use of geospatial and clustering knowledge to select proper methods and datasets to achieve clustering results that better meet users' requirements. The system consists of the *GeoCO* ontology for geospatial clustering and the Ontology Reasoner reasoning mechanism. The *GeoCO* ontology is used to represent geospatial and clustering domain knowledge. The Ontology Reasoner uses classification and decomposition techniques to specify users' tasks.

An *ontology* is a formal explicit specification of a shared conceptualization [15]. It provides domain knowledge relevant to the conceptualization and axioms for reasoning with it. For geospatial clustering, an appropriate ontology must include a rich set of geospatial and clustering concepts. Therefore, it can provide a knowledge source that supplements domain experts. Since the ontology in the geospatial domain is complex and varies according to the application [18], we build *GeoCO* at a high generic level such that it can be extended and materialized for specific applications. The *GeoCO* geospatial clustering ontology is a domain-specific ontology. The ontology includes two types of the knowledge: geospatial knowledge and clustering knowledge. Domain ontology can vary between applications. The classes included in *GeoCO* are neither complete nor universal for all applications. The purpose of introducing the top-level ontology is to let readers understand how the system works with an ontology. We will discuss each component in the top-level ontology of *GeoCO* in detail.

The *GeoCO* geospatial clustering ontology has been represented in using Protégé-OWL [24] and the detail information about it can be found in [25].

The structure of the GEO_CLUST system for ontology-based clustering is shown in Fig. 3. It includes five components: the Geospatial Clustering Ontology GeoCo, the Ontology Reasoner, the Clustering Methods, the Data, and the Graphical User Interface (GUI). The *geospatial clustering ontology* component is used when identifying the clustering problem and the relevant data. Within this component, the *task model* specifies the data and methods that may potentially be suitable for meeting the user's goals, and *GeoCO* includes all classes, instances, and axioms in a geospatial clustering domain. Through classification and decomposition conducted in the Ontology Reasoner, proper clustering data and methods can be indentified from the ontology.

The system works as follows: With the system, the user first gives his or her goals for clustering through the GUI. To be able to find proper data and clustering methods in ontology, the goal needs formalizing as a task instance. A *task instance* describes the specific problem to be solved. An example of a user's goal is to identify the best locations for five hospitals in South Carolina. A task instance "determine best locations of five hospitals in South Carolina" is created. The task instance is refined in the task model by using Ontology Reasoner and the refined elementary sub-tasks are used to search the domain ontology. For the example above, one of elementary tasks of "determine the best locations of five hospitals in South Carolina" is "to find population data in South Carolina," which can be implemented through database queries. The results of these queries identify the proper clustering methods and the appropriate data sets. Based on these results, clustering is conducted. The clustering results can be used for statistical analysis or they can be interpreted using the task ontology and the domain ontology.

In the system, the Ontology Reasoner is used to reason about knowledge represented in the ontology. The input of the reasoner is the user's goals, and the output is a set of appropriate geospatial clustering methods and datasets. The reasoner performs the following steps. First, it builds a task instance [26] to associate the reasoning with the geospatial clustering ontology and the user's requirements. For the above example,

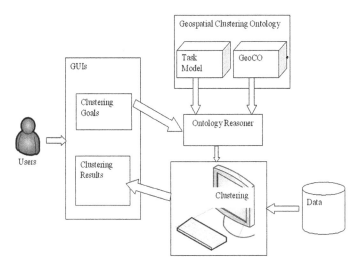

Fig. 3. The GEO_CLUST system for ontology-based clustering

the task instance "determine best locations of five hospitals in South Carolina" is created based on a *Partitioning-Clustering* task in the task model, because the final results of clustering are to form five population clusters assigned to individual hospitals. Second, each available geospatial clustering method described in the ontology is considered either as an elementary task (which is accomplished by a simple primitive function) or as a complex task (which is accomplished via a task decomposition method represented in some problem solving strategy). The task "determine best locations of five hospitals in South Carolina" is a complex task because we cannot solve the task by simply calling an existing primitive function. For the task, we first need to find the proper data, such as South Carolina population data, and then identify the proper partitioning clustering method based on the characteristics of the data and the user's specification of "best locations". So the task needs to be decomposed. Finally each complex task is recursively decomposed into elementary sub-tasks [27]. The detailed description about the Ontological Reasoner component in GEO_CLUST is in [25], including task model, inference engine, and classification algorithm.

2.3   A Cluster Ensemble Method

Given a set of clustering results, the cluster ensemble method used in the approach aims to select a subset of the clustering results and then combines them into a new clustering result which is better than the best result in the given result set. We extend cluster ensemble selection method in [14] by cooperating with the domain knowledge. The method includes two steps.

Step one: select a subset of clustering results with high *quality* and *diversity*, where quality measures the accuracy of clustering results in the subset and diversity measures the difference among the clustering results. According to the research did by Fern and

Lin [14], selecting only a subset of clustering results based quality and diversity could improve the accuracy of the final ensemble result as well as reducing the execution time. Particularly, in this paper, given a set of clustering results CA (i.e., $CA = \{C_1, C_2, \ldots, C_r\}$) and a subset $C \subset CA$, the way to measure the quality of C is separated into two conditions:

(1) Without the domain knowledge. The quality of $C_i(C_i \in C)$ is defined as the similarity between it and the other clustering results in CA (as shown in Eq. 1) and the quality of C is defined as the sum of the quality of all the clustering results in it (as shown in Eq. 2).

$$Quality(C_i) = \sum_{k=1}^{r} NMI(C_i, C_k) \tag{1}$$

$$Quality(C) = \sum_{C_i \in C} Quality(C_i) \tag{2}$$

where $NMI(C_i, C_k)$ is the normalized mutual information[1] between clustering C_i and C_k. We adopt the Eq. (3) in [13] to estimate the NMI value between two clustering results. According to [13], if two clusterings are completely independent partition, their NMI value is 0, vice versa. Thus, the larger the value, the higher is the quality.

(2) With the domain knowledge. The quality of $C_i(C_i \in C)$ is measured by the external objective function according to the domain knowledge and the quality of C is defined as the sum of the quality of all the clustering results in it (as shown in Eq. 3).

$$Quality(C) = \sum_{C_i \in C} EF(C_i), \tag{3}$$

where $EF(C_i)$ is the external objective function value of C_i. For instance, when applying geospatial clustering analysis for the facility location planning [7], all the demand nodes in the target region are clustered into different groups and the demand nodes in each group are served by one facility. According to the domain knowledge in facility location planning, the external objective function value of a clustering result is the total travelling distance from the demand nodes to their assigned facilities.

The diversity of C is defined as the sum of all pairwise similarities in the set (as shown the Eq. 4). The lower the value, the higher is the diversity. We measure the diversity as follows because it has been proved to impact the cluster ensemble performance [14].

---

[1] Mutual information is a symmetric measure to quality the statistical information shared between two distributions.

$$Diversity(C) = \sum_{i \neq j, C_i, C_j \in C} NMI(C_i, C_j) \tag{4}$$

Since high quality and diversity are two objectives to be achieved during the clustering result selection, we treat it as a bi-objective optimization problem [6] and solve it by using a bi-objective function. Specifically, given a set of clustering results, a subset of clustering results C is selected from the whole set that minimizes the value of $BOF(C)$ in the following.

$$BOF(C) = \alpha Quality(C) + (1 - \alpha)Diversity(C) \tag{5}$$

In the Eq. (5), α is defined as a co-efficient for balancing the quality objective and diversity objective, which is within [0, 1]. The parameter α is a constant to control the weight of each objective.

Selecting a subset of solutions to minimize the value of $BOF(C)$ is a NP-hard problem. In the research, we adopt a greedy procedure to perform the selection. It begins with a C that only contains the single solution of the lowest linear normalized cost value and then incrementally adds one solution at a time into C to minimize the value of $BOF(C)$. The procedure stops when the size of C reaches the predefined number.

Step two: combine the selected clustering results into one optimal combined clustering result. The optimal clustering result C^{opt} should reach the objective function (Eq. 6), which asks for the optimal result has maximal mutual information with other selected clustering results and achieves higher external objective function value if domain knowledge is applied. We adopt the greedy optimization approach in [13] to solve the Eq. 6.

$$C^{opt} = \arg \max_{\hat{C}} \sum_{C_i \in C} (\beta EF(\hat{C}) + (1 - \beta)NMI(\hat{C}, C_i)) \tag{6}$$

3   Case Study

In this section, we apply the approach to a real case study, finding the best locations for breast cancer screening clinics in Alberta (AB), Canada.

A population-based program to increase the number of Alberta women screened regularly for breast cancer was implemented in 1990 and today the Alberta Breast Cancer Screening Program (ABCSP) recommends Alberta women between the ages of 50 and 69 have a screening mammogram at least once every two years [10]. A key challenge is to determine the best locations for screening clinics to minimize the average demand–weighted distance from demand nodes to their assigned clinics. In this case, we perform the approach to separate the whole province into 26^2 screening

---

[2] The number of current cancer care facilities in Alberta is 26.

clusters and for each cluster find a suitable location for building a clinic to serve the people within it.

In the approach, the first step is to find appropriate datasets, attributes of datasets, and clustering methods by sending the clustering goal to the ontology-based geospatial clustering system. The user interface and parameter setting of the ontology-based system can be found in [25]. Here we only list the results as below:

Dataset: Census dataset in AB
Attributes of datasets: women from 50 to 69 in DA level
Similarity Measurement: Euclidean distance
Clustering method: Capability K-MEANS

The dataset is the 2006 Canadian census data in AB. Estimates of the screening population (Alberta women aged 50 to 69 years) were derived from census data at the Dissemination Area (DA) level [16]. There are 327830 women within the target age in Alberta. A total of 5180 DAs were used in the research. Their values range from 0 to 920. The system adopts Euclidean distance to measure the similarity among the

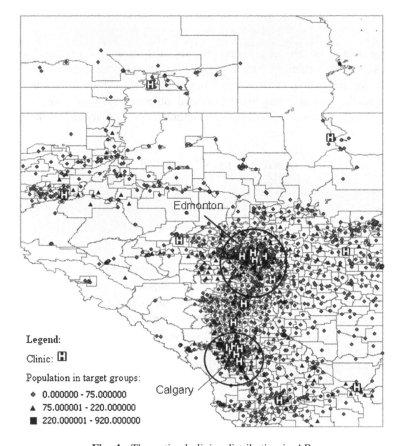

Fig. 4. The optimal clinics distribution in AB

locations. The recommended clustering method is Capability K-MEANS [17]. In order to meet the overall demands in the province, the capacity of each clinic is set to 15,000. We generate 30 clustering results by applying Capability K-MEANS with different initializations.

The second step is to select a subset of clustering results based on quality and diversity. The Eq. 3 in 3.1 is chosen to measure the quality. The external objective function value is the total travelling distance from the DAs to their assigned clinics. The diversity is measured by the Eq. 4 in 3.1. α in the Eq. 5 is set to 0.5. The number of the selected solutions is set to 5.

In the third step, the selected 5 results are combined into one optimal result. Since the case has a clear clustering goal, minimizing the average demand–weighted distance from demand nodes to their assigned clinics, the way to choose the optimal result is only based on the goal and thus β in the Eq. 6 is set to 1. Figure 4 shows the clinic locations in the optimal result. We compare the optimal result with the one produced by the Interchange algorithm [28], the most popular algorithm used in facility location planning. The average demand–weighted distance are 24.12 km for the optimal result and 26.64 for the one produced by the Interchange algorithm. In this case, the approach achieves better results than by simply picking a standard facility location solution approach.

4   Conclusion and Future Work

In this paper, we present an ontology-based cluster ensemble approach to produce good clustering results for geospatial applications. The approach includes two components: an ontology-based expert system for formalizing the domain knowledge in the geospatial clustering and a cluster ensemble method for combing good clustering results into an optimal result. To our best knowledge, it is the first research work to combine geospatial ontology and cluster ensembles for geospatial clustering analysis. The real case study did in Alberta, Canada shows that it is practical that combine the ontology and the cluster ensembles together for geospatial analysis. In the future, the approach is to be applied on other application scenarios.

References

1. Ng, R., Han, J.: Efficient and effective clustering method for spatial data mining. In: Proceedings of 20th International Conference on Very Large Data Bases (1994)
2. Shekhar, S., Chawla, S.: Spatial Databases: A Tour. Prentice Hall, Upper Saddle River (2003)
3. Graco, W., Semenova, T., Dubossarsky, E.: Toward knowledge-driven data mining. In: International Workshop on Domain Driven Data Mining at 13th ACM SIGKDD (2007)
4. Tung, A.K.H., Han, J., Lakshmanan, L.V.S., Ng, R.T.: Constraint-based clustering in large databases. In: Proceedings of International Conference on Database Theory (2001)
5. Wang, X., Hamilton, H.J.: Towards an ontology-based spatial clustering framework. In: Proceedings of 18th Canadian Artificial Intelligence Conference (2005)

6. Mitropoulos, P., Mitropoulos, I., Giannikos, I., Sissouras, A.: A biobjective model for the locational planning of hospitals and health centers. Health Care Manag. Sci. **9**, 171–179 (2006)

7. Liao, K., Guo, D.: A clustering-based approach to the capacitated facility location problem. Trans. GIS **12**, 323–339 (2008)

8. Han, J., Lakshmanan, L.V.S., Ng, R.T.: Constraint-based multidimensional data mining. Computer **32**, 46–50 (1999)

9. Wang, X., Rostoker, C., Hamilton, H.J.: Density-based spatial clustering in the presence of obstacles and facilitators. In: Proceedings of 8th European Conference on Principles and Practice of Knowledge Discovery in Databases (2004)

10. Alberta Breast Cancer Screening Program website. http://www.cancerboard.ab.ca/abcsp/program.html

11. Breaux, T.D., Reed, J.W.: Using ontology in hierarchical information clustering. In Proceedings of 38th Annual Hawaii International Conference on System Sciences (2005)

12. Han, J., Kamber, M.: Data Mining: Concepts and Techniques, 2nd edn. Morgan Kaufmann, Burlington (2006)

13. Strehl, A., Ghosh, J.: Cluster ensembles – a knowledge reuse framework for combining multiple partitions. Mach. Learn. Res. **3**, 583–617 (2002)

14. Fern, X.Z., Lin, W.: Cluster ensemble selection. J. Stat. Anal. Data Min. **1**, 128–141 (2008)

15. Gruber, T.R.: A translation approach to portable ontologies. Knowl. Acquis. **5**, 199–220 (1993)

16. Data quality index for census geographies. http://www12.statcan.ca.ezproxy.lib.ucalgary.ca/census-recensement/2006/ref/notes/DQ-QD_geo-eng.cfm

17. Ng, M.K.: A note on constrained k-means algorithms. Pattern Recogn. **33**, 515–519 (2000)

18. Fonseca, F., Egenhofer, M., Agouris, P., Câmara, G.: Using ontologies for integrated geographic information systems. Trans. GIS **6**, 231–257 (2002)

19. Maedche, A., Zacharias, V.: Clustering ontology-based metadata in the semantic web. In: Proceedings of 6th European Conference on Principles of Data Mining and Knowledge Discovery (2002)

20. Worboys, M.F.: Metrics and topologies for geographic space. In: Advances in Geographic Information Systems Research II: International Symposium on Spatial Data Handling (1996)

21. Egenhofer, M.J., Clementini, E., di Felice, P.: Topological relations between regions with holes. Int. J. Geogr. Inf. Sci. **8**, 129–142 (1994)

22. Papadias, D., Egenhofer, M.: Hierarchical spatial reasoning about direction relations. GeoInformatica **1**, 251–273 (1997)

23. Egenhofer, M.J., Franzosa, R.D.: Point-set topological spatial relations. Int. J. Geogr. Inf. Sci. **5**, 161–174 (1991)

24. Protégé web site. http://protege.stanford.edu/index.html

25. Wang, X., Gu, W., Ziébelin, D., Hamilton, H.: An ontology-based framework for geospatial clustering. Int. J. Geogr. Inf. Sci. **24**(11), 1601–1630 (2010)

26. Crubézy, M., Musen, M.: Ontologies in support of problem solving. In: Staab, S., Studer, R. (eds.) Handbook on Ontologies. International Handbooks on Information Systems, pp. 321–341. Springer, Heidelberg (2004). doi:10.1007/978-3-540-24750-0_16

27. Parmentier, T., Ziebelin, D.: Distributed problem solving environment dedicated to DNA sequence annotation. In: Proceedings of 11th European Workshop on Knowledge Acquisition, Modeling and Management (1999)

28. Teitz, M.B., Bart, P.: Heuristic methods for estimating the generalized vertex median of a weighted graph. Oper. Res. **16**, 955–961 (1968)

Natural Language Processing and Computational Linguistics

Mining Implicit Intention Using Attention-Based RNN Encoder-Decoder Model

Chenxing Li[(⊠)], Yajun Du, and SiDa Wang

School of Computer and Software Engineering,
Xihua University, Chengdu 610039, China
272463637@qq.com

Abstract. Nowadays, people are increasingly inclined to use social tools to express their intentions explicitly and implicitly. Most of the work is dedicated to solving the explicit intention detection, ignoring the implicit intention detection, as the former is relatively easy to solve with the classification method. In this work, we use the Attention-Based Encoder-Decoder model which is specified for the sequence-to-sequence task for user implicit intention detection. Our key idea is to leverage the model to "translate" the implicit intention into the corresponding explicit intent by using the parallel corpora built on the social data. Specifically, our model has domain adaptability since the way people express implicit intentions for different domain is variable, while the way to express explicit intentions is mostly in the same form, such as "I want to do sth". In order to demonstrate the effectiveness of our method, we conduct experiments in four domains. The results show that our method offers a powerful "translation" for the implicit intentions and consequently identifies them.

Keywords: Implicit intent detection · Recurrent neural networks · Attention · Encoder-Decoder model

1 Introduction

Social platforms, such as Twitter, Facebook, etc., provide a platform for users to express their thoughts, wishes, and needs. With the popularity of these platforms, more and more users settle in these platforms and generate massive amount of their status messages. Making use of and understanding these content has been an ongoing endeavor over the past years including tasks such as sentiment analysis [1] and opinion mining [2], in which intention analysis has received more and more attention in recent years. By analyzing intention in user status, product manufacturers or service providers can provide relevant items or targeted ads to users. Meanwhile, such recommendation can also benefit users since they can find products or services they need without searching for them.

Usually, users express their intentions in an explicit or implicit way [3] and researchers find that the latter approach is more common. For example, the tweet "I want to buy a jacket" explicitly indicates the intention that the user want to buy a jacket. In the other tweet "I do not have a suitable jacket to wear.", it also expresses the intention to buy a coat. The difference between the two tweets is that the latter requires

© Springer International Publishing AG 2017
D.-S. Huang et al. (Eds.): ICIC 2017, Part III, LNAI 10363, pp. 413–424, 2017.
DOI: 10.1007/978-3-319-63315-2_36

inference to obtain the intention, that is, it expresses the intention implicitly. However, most of the existing work focus on detection of the explicit intention [3, 4, 13], and only few work aim at the identification of implicit intention [5, 14].

In this paper, our task is to recognize the implicit intention in the tweets. Since it is easy to identify tweets with explicit intention by formulating it as a classification problem. Therefore, we can firstly "translate" tweets with implicit intention to corresponding tweets with the explicit intention using the Encoder-Decoder models, and then classify the "translated" tweets with classifier.

The Encoder-Decoder models have been successfully applied in many sequence-to-sequence learning problems such as machine translation [6, 8, 16] and speech recognition [7, 15]. The main idea of the model is to encode input sequence into a dense vector, and then use this vector to generate corresponding output sequence. Specially, Bahdanau et al. [8] introduce attention mechanism which releases the constraint that the input sequence has to be encoded into a fixed-length context vector. It only encodes the parts of the input sentence that surround a particular word accurately. This mechanism is particular useful for the long sequence, and we adapt the Attention-Based Encoder-Decoder model in this paper.

We build parallel corpora containing both implicit and the corresponding explicit intention tweets to train the Encoder-Decoder models where tweets with implicit intention are input sequences and tweets with the corresponding explicit intention are output sequences. Once we obtain the explicit expression of the tweets, the next step is to classify whether it has particular intention using classifier.

The remainder of the paper is organized as follows. We review related work in Sect. 2. In Sect. 3, we introduce two Encoder-Decoder models, the RNN Encoder-Decoder model and the Attention-Based RNN Encoder-Decoder model. In Sect. 4, we describe how to mine implicit intentions using Attention-Based RNN Encoder-Decoder model. Section 5 discusses the experiment setup and results. Section 6 concludes the work.

2 Related Work

Recently, intention analysis has attracted researchers' much attention, which can be categorized into two main research domains: one is mining the intention of queries and the other is identifying the intention(especially the consumption intention) in social networks.

For the query intention, Broder [9] classified queries into three classes: navigational, informational, and transactional. Dai et al. [10] were the first to identify search queries containing online commercial intention by training a binary classifier on a large amount of labeled queries enriched with search result snippets. Similarly, Ashkan and Clarke [11] exploited the click-through behavior data to enrich queries while Guo and Agichtein [12] used users' mouse movement behaviors or scrolling behaviors data. However, query is different from tweets which is too short to identify the intention. Such as the query "Paris" does not explicitly express user intents. Conversely, tweets often contain sentences that explicitly express the user intent, such as "I want to travel to Paris." .

For the intention analysis on social media, Hollerit et al. [3] firstly proposed the commercial intention at the tweet-level as a first step to bring together buyers and sellers. They found that there were two ways to express the intention, explicitly or implicitly, and the latter was more common. However, most of the research work is aimed at the detection of the explicit intention. Chen et al. [13] aimed at identifying explicit intents expressed in posts of forums. Wang et al. [4] proposed a semi-supervised learning approach to classify tweets with explicit intention into six categories. Only a few work is about the implicit intention. Ding et al. [5] developed a Consumption Intention Mining Model (CIMM) based on convolutional neural network (CNN) to identify whether the user has a implicit consumption intention. Different with Ding et al., we firstly "translate" the implicit intention to the corresponding explicit intention and then classify it. Park et al. [14] inferred the implicit intent in social media text by leveraging the parallel corpora to simply find the corresponding explicit intention. They followed the information retrieval approach to match texts with its similar implicit intention texts in parallel corpora. If the expression of implicit intention did not exist in the corpora, then expression of the corresponding explicit intention can not be found either. While our Encoder-Decoder model can learn a semantically meaningful representation of linguistic phrases.

3   Encoder-Decoder Models

In this paper, we mainly introduce two Encoder-Decoder models, one is the conventional RNN Encoder-Decoder model and the other is Attention-Based Encoder-Decoder model which both have made great achievements in the field of machine translation.

3.1   RNN Encoder-Decoder Model

RNN Encoder-Decoder model was first proposed by Cho et al. [6] to learn to encode a sequence $(x_1, \ldots, x_{T_x})$ into a fixed-length vector representation and to decode a given fixed-length vector representation back into another sequence $(y_1, \ldots, y_{T_y})$ where the length of the input sequence T_x and the length of the output sequence length T_y may differ.

The encoder is an RNN that reads an input sequence into a vector C. The decoder is another RNN which is trained to generate the output sequence y, and the probability of y is as follows:

$$P(y) = \prod_{t=1}^{T_y} P(y_t | y_1, \ldots, y_{t-1}, C) \tag{1}$$

The probability of the predicted words y_t given the context $y_1 \sim y_t$ is:

$$P(y_t|y_1,\ldots,y_{t-1}) = g(s_t, y_{t-1}, C) \tag{2}$$

Where g is a nonlinear function and s_t is the hidden state of the RNN at time t. It should be noted that encoder and decoder can be different combinations of RNN, BiRNN or even CNN.

3.2   Attention-Based RNN Encoder-Decoder Model

Bahdanau et al. [8] proposed the Attention-Based RNN Encoder-Decoder model which freed the conventional model from the constraint that the model had to encode the input sequence into a fixed-length vector.

The encoder that Bahdanau et al. [8] used was a bidirectional RNN. For the decoder, similarly, the conditional probability of is as follows:

$$P(y_i|y_1,\ldots,y_{i-1},X) = g(s_i, y_{i-1}, c_i) \tag{3}$$

Where s_i represents the hidden state of RNN at time i and it is computed as:

$$s_i = f(s_{i-1}, y_{i-1}, c_i) \tag{4}$$

It should be noted that the conditional probability in Eq. (3) is related to the content vector c_i. While in Eq. (2), there is only one content vector C. In fact, c_i is the weighted sum of the hidden vector sequence $(h_1,\ldots,h_{T_x})$ of the encoder, which can be presented as:

$$c_i = \sum_{j=1}^{T_x} \alpha_{ij} h_j \tag{5}$$

Where α_{ij} is the weight of h_j indicating the attention of the j-th input when generating the i-th output. The higher the value of α_{ij} is, the more attention is allocated to the i-th output on the j-th input. α_{ij} is determined by the output hidden state s_{i-1} and the respective hidden states in the input, and it is represented as:

$$\alpha_{ij} = \frac{\exp(e_{ij})}{\sum_{k=1}^{T_x} \exp(e_{ik})} \tag{6}$$

$$e_{ij} = a(s_{i-1}, h_j) \tag{7}$$

Where a is the alignment model which is parametrized as a feed-forward neural network and jointly trained with all the other components of the proposed system.

Compared to the conventional Encoder-Decoder model, the encoder of the Attention model needs to encode the input into a sequence of vectors and at the process of decoding, each step selects a subset of the vector sequences for further processing. In this way, each time the output is generated, it is possible to make full use of the information carried by the input sequence, and this method has achieved good results in the translation task.

4   Mining Implicit Intention Using Attention-Based RNN Encoder-Decoder Model

According to Hollerit et al. [3], there are two categories of intention, explicit intention and implicit intention. Consider the following two real tweets:

- Example 1: I want to buy a coat.
- Example 2: When the season changes, I always feel no clothes to wear.

Example 1 explicitly states the user's consumption intention of buying a coat. Example 2 also reflects the user's wish to buy clothes to a certain extent although it does not express the intention directly. Compared to the way that user express their wishes explicitly, it has been observed that users are more inclined to express their intention implicitly.

However, most of the work is now devoted to solving the identification of explicit intention by treating it as the classification problem. With the SVM classifier or simple rule-based methods, more than 80% of the explicit intention could be identified. While for implicit intention, to the best of our knowledge, there is even no professional definition. Since the implicit intention often needs to make inference to determine whether there is some specific intention. Some research work define the implicit intention manually. Ding et al. [5] annotated tweets containing implicit consumption intention by asking two annotators to label tweets. Then they trained a CNN classifier with the annotated corpora to classify whether a tweet has the consumption. However, they do not solve the problem essentially, that is, how to define a tweet with implicit intention. In this paper, we solve the problem with the Parallel Corpora build from social media and the Attention-Based RNN Encoder-Decoder model. It also should be noted that in this paper the implicit intentions could be a various of intentions, not just the consumption intention. Specifically, we choose four intention domains, which include buying clothes, eating food, traveling and fitting.

Parallel corpora contains texts with implicit intent and the corresponding texts with explicit intent. For example, tweets such as "Whenever the season changed, there is always no suitable clothes to wear, I want to buy clothes.", "I am too hungry to work, it is time to have something eat.", "May Day holiday come, I want to travel.", "If you do not lose weight in March, then you will be sad in April. So it is time to design fitness plan." and so on. The first half parts of the sentences do not reveal the intention while the second parts of the sentences explicitly express the intention. Using these tweets, we can build parallel corpora and therefore find the explicit intentions of the tweets whose intentions are expressed implicitly. However, if a tweet with implicit intention does not exist in the corpora, then we can not find the corresponding expression of explicit intention. This is where Attention-Based RNN Encoder-Decoder model come. The model can "translate" text with implicit intent into text with explicit intent. For the "translated" text, we can then use classifier to classify it. Thus we solve the problem of how to define the tweet with implicit intention. To the best of our knowledge, mining user's implicit intent by translating it to the corresponding texts with explicit intent has not been addressed in previous work.

Firstly, we build the parallel corpora based on the Sina Weibo, then the corpora is used to train the Attention-Based RNN Encoder-Decoder model. Once the model is established, it can be used to translate the tweet with implicit intention to the corresponding texts with explicit intention in the parallel corpora. Finally, a classifier which is trained with the tweets with explicit intention will be applied to classify the "translated" tweets. See Fig. 1 for a graphical depiction of how the proposed model "translates" the tweets with implicit intention to the corresponding tweets with explicit intention.

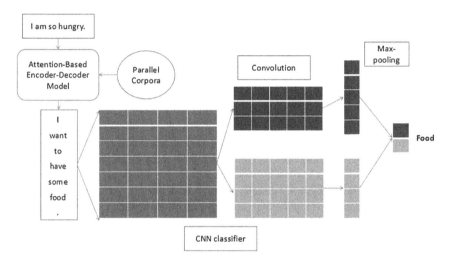

Fig. 1. The architecture of our model

5   Experiments

5.1   Dataset and Models

Each domain has 50000 positive instances and 50000 negative instances. These instances are crawled from Sina Weibo (the most popular microblogging service in China) using the Scripy-based python crawler framework. After a usual tokenization, we use a list of 300,000 most frequent words to train our models. Any word not included in the list is mapped to a special token ([UNK]). We then split the data into train set and test set according to the ratio of 10: 1.

In the experiment, we train the basic RNN Encoder-Decoder model and the Attention-Based RNN Encoder-Decoder model. The encoder and decoder of the two models are both recurrent neural networks (RNN) and each has 1000 hidden units. Specially, the encoder of the two model is bidirectional RNN. The basic RNN unit is the gated hidden unit [6] which is similar to a long short-term memory (LSTM) unit. The gated hidden unit shares the LSTM's ability to better model and learn long-term dependencies. We use a minibatch stochastic gradient descent (SGD) algorithm toge-ther with Adam to train each model and the minibatch size is 128. In order to minimize the waste of the computation, we sort the sentences according to the length and split them into 4 buckets which are (6,6), (10,10), (50,15), (105,25). Once the model is trained, we use a beam search [19, 20] to find a translation that approximately maxi-mizes the conditional probability. In both cases, we use a multilayer network with a single maxout [17] hidden layer to compute the conditional probability of each target word [18].

5.2   Compared Methods

In the experiment, we compare the following three methods:

- CIMM (consumption Intention Mining Model) [5]: It exploits the convolution neural network (CNN) to classify whether the tweet has consumption intention. We take it as the baseline method.
- RNN Encoder-Decoder+CIMM: The sentences are firstly translated by the RNN Encoder-Decoder Model and then the "translated" sentences are classified by the CIMM classifier. The parameters used for RNN Encoder-Decoder model are the same as the Attention-Based RNN Encoder-Decoder model.
- Attention-Based RNN Encoder-Decoder+CIMM: The sentences are firstly trans-lated by the Attention-Based RNN Encoder-Decoder Model and then the "trans-lated" sentences are classified by the CIMM classifier.

5.3   Results and Analysis

We made ten cross validation on the dataset, the accuracy of the baseline systems as well as the proposed model on implicit intention identification was shown in Fig. 2. We can draw the following results from Fig. 2:

(1) Encoder-Decoder+CIMM models achieve better performance than the baseline method. This is likely due to the reason that CIMM uses sentences with explicit intention as training sets, and it can not capture the semantics of sentences with implicit intention. Naturally, it does not perform well in these sentences. After "translation" by Encoder-Decoder models, the accuracy of the classifier increases by 4.2%, which shows that Encoder-Decoder models accurately capture the sentences semantics and translate them into another explicit way.

(2) Encoder-Decoder model with attention mechanism performs better than the tra-ditional model. That is because the basic Encoder-Decoder model has the limi-tation that the encoder compresses the entire sequence of information into a fixed-length semantic vector. There are two drawbacks about this model, one is that the semantic vector can not fully represent the entire sequence of information, another is that the information in front of the sentence will be covered by the information behind the sentence. The longer the input sequence is, the more serious the phenomenon is. This makes it impossible to obtain enough

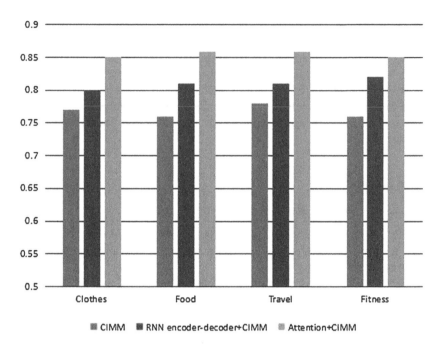

Fig. 2. The accuracy of the three methods in clothes, food, travel, and fitness

information for the input sequence at the beginning of the decoding, and the accuracy of the decoding will naturally decline. Consequently, its performance is worse than the model with attention mechanism.

To fully understand the limitation of the basic Encoder-Decoder approach, we compare the performance of the two Encoder-Decoder models on different lengths of sentences. The lengths of sentences determined by buckets are (6,6), (10,10), (50,15), and (105,25). The first number in the tuple represents the length of the input sequence, and the second number represents the length of the output sequence. For example, the input sequence "I have no clothes to wear." and the corresponding output sequence "I want to buy clothes." belong to the bucket (6,6). If the length of sequence is less than the number of the bucket it belongs to, then it will be padded with special <PAD> sign.

In Fig. 3, we use "RNN" to represent the basic RNN Encoder-Decoder model and "Attention" for Attention-Based Encoder-Decoder model. We can see that as the length of the sentence increases, the accuracy of the two models decreases. In particular, for the "RNN" model, its accuracy rate drops sharply when the length of the sentence is greater than 50. On the contrary, for the "Attention" model, although its accuracy rate drops after the length of the sentence exceeds 50, the accuracy rate has not been greatly affected, even with sentences of length 105.

It can be clearly seen from the Fig. 3 that the attention model is much better than the conventional model, especially for the long sentences. This is likely due to the fact that the attention model breaks the constraint that it has to encode a long sentence into a fixed-length vector, and it encodes the parts of the input sentence that surround a

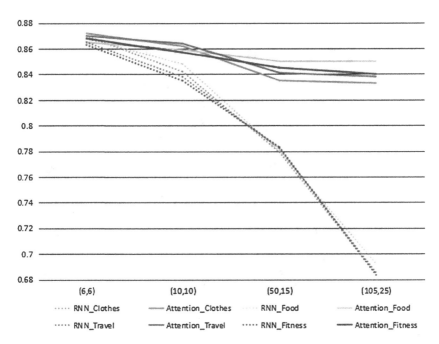

Fig. 3. Different lengthes of training data

particular word. There is a concrete example from the test set: "Less than 72 h from the starting and there are so many things that have to picked up, I would like to take eyelashes, nails, cut hair... and I have no clothes to wear.". The "RNN" model translated this sentence into: "I want to pick eyelashes, do nails and cut hair.". On the other hand, the "Attention" translated the sentence into "I want to pick eyelashes, do nails, cut my hair and buy clothes.", preserving all the intentions of the input sentence. As with the example, the "RNN" model began omitting the intention of the source sentence after generating approximately 33 words. After that point, the model can not translate the intention. While the "Attention" model is able to translate all the intentions of the long sentence correctly. These qualitative observations confirm our hypotheses that the "Attention" model enables far more reliable translation of long sentences than the standard "RNN" model.

The "Attention" model provides an intuitive way to inspect the soft-alignment between the words in a generated translation and those in a source sentence. Figure 4 shows the soft-alignment by visualizing the annotation weights α_{ij} from Eq. (6), where each row indicates the weights associated with the annotations. The weights indicate the importance of the input word to the output word. As shown in the Fig. 4, the source sentence is "Whenever the season changed, I have no clothes to wear." and the generated translation is "I want to buy clothes.". The word "season"and "changed" in the source sentence have a large effect on the generation of word "want". This is in line with our expectations, because there are a lot of data shows that users want to buy clothes mainly because there is no clothes to wear season. Compared with the word

Fig. 4. A sample alignment found by "Attention" model

"have" and "no", the latter has larger weight for the generating of "want". This is also consistent with the common sense and the dataset that "no" is more easily to trigger the intention word, that is, "want". At the same time, we see strong weights along the diagonal of each matrix.

6   Conclusions

In this paper, we make the first attempt to identify user implicit intention from social media using the Attention-Based Encoder-Decoder model. We firstly translate the texts with implicit intention into the corresponding expression with explicit intention and then the classify the explicit intention using the CNN-based classifier. Through experimental results, we show that our method is better than the conventional method that classifies the texts with implicit intention directly. Specifically, we find that the Encoder-Decoder model with attention mechanism performs better than traditional RNN Encoder-Decoder model since it focuses on the part of the input sentence when generating the target word.

Acknowledgement. This work is supported by the National Nature Science Foundation of China (No. 61271413, 61472329, 61532009). Innovation Fund of Postgraduate, Xihua University.

References

1. Agarwal, A., Xie, B., Vovsha, I., Rambow, O., Passonneau, R.: Sentiment analysis of Twitter data. In: Workshop on Languages in Social Media, pp. 30–38. Association for Computational Linguistics (2011)
2. Maynard, D., Funk, A.: Automatic detection of political opinions in tweets. In: Proceedings of the 1st Workshop on Making Sense of Microposts at ESWC 11 (2011)
3. Hollerit, B., Kröll, M., Strohmaier, M.: Towards linking buyers and sellers: detecting commercial intent on Twitter. In: the 22nd International Conference on World Wide Web Companion, pp. 629–632. International World Wide Web Conferences Steering Committee, Rio de Janeiro (2013)
4. Wang, J., Cong, G., Zhao, X.W., Li, X.: Mining user intents in Twitter: a semi-supervised approach to inferring intent categories for tweets. In: Twenty-Ninth AAAI Conference on Artificial Intelligence, Texas (2015)
5. Ding, X., Liu, T., Duan, J., Nie, J.-Y.: Mining user consumption intention from social media using domain adaptive convolutional neural network. In: AAAI (2015)
6. Cho, K., van Merrienboer, B., Gulcehre, C., Bougares, F., Schwenk, H., Bengio, Y.: Learning phrase representations using RNN Encoder-Decoder for statistical machine translation. In: Proceedings of the Empiricial Methods in Natural Language Processing (2014)
7. Chan, W., Jaitly, N., Le, Q.V., Vinyals, O.: Listen, attend and spell. arXiv preprint arXiv: 1508.01211 (2015)
8. Bahdanau, D., Cho, K., Bengio, Y.: Neural machine translation by jointly learning to align and translate. arXiv preprint arXiv:1409.0473 (2014)
9. Broder, A.: A taxonomy of web search. In: ACM SIGIR Forum, vol. 36, pp. 3–10. ACM (2002)
10. Dai, H.K., Zhao, L., Nie, Z., Wen, J.R., Wang, L., Li, Y.: Detecting online commercial intention (OCI). In: The 15th International Conference on World Wide Web, pp. 829–837. ACM, Edinburgh (2006)
11. Ashkan, A., Clarke, C.L.: Impact of query intent and search context on clickthrough behavior in sponsored search. J. KIS 34(2), 425–452 (2013)
12. Guo, Q., Agichtein, E.: Ready to buy or just browsing?: detecting web searcher goals from interaction data. In: Proceedings of the 33rd International ACM SIGIR Conference on Research and Development in Information Retrieval (2010)
13. Chen, Z., Liu, B., Hsu, M., Castellanos, M., Ghosh, R.: Identifying intention posts in discussion forums. In: HLT-NAACL, Atlanta, pp. 1041–1050 (2013)
14. Park, D.H., et al.: Mobile App Retrieval for Social Media Users via Inference of Implicit Intent in Social Media Text. In: the 25th ACM International Conference on Information and Knowledge. (2016)
15. Graves, A., Jaitly, N., Mohamed, A.-R.: Hybrid speech recognition with deep bidirectional LSTM. In: IEEE Workshop on Automatic Speech Recognition and Understanding (ASRU) (2013)
16. Sutskever, I., Vinyals, O., Le, Q.: Sequence to sequence learning with neural networks. In: Advances in Neural Information Processing Systems (2014)
17. Goodfellow, I., Warde-Farley, D., Mirza, M., Courville, A., Bengio, Y.: Maxout networks. In: Proceedings of the 30th International Conference on Machine Learning, pp. 1319–1327 (2013)

18. Pascanu, R., Gulcehre, C., Cho, K., Bengio, Y.: How to construct deep recurrent neural networks. In: Proceedings of the Second International Conference on Learning Representations (2014)
19. Graves, A.: Sequence transduction with recurrent neural networks. In: Proceedings of the 29th International Conference on Machine Learning (2012)
20. Boulanger-Lewandowski, N., Bengio, Y., Vincent, P.: Audio chord recognition with recurrent neural networks. In: ISMIR (2013)

POS-Tagging Enhanced Korean Text Summarization

Wuying Liu[1] and Lin Wang[2(✉)]

[1] Laboratory of Language Engineering and Computing,
Guangdong University of Foreign Studies,
Guangzhou 510420, Guangdong, China
wyliu@gdufs.edu.cn
[2] Center for Translation Studies, Guangdong University of Foreign Studies,
Guangzhou 510420, Guangdong, China
wanglin@nudt.edu.cn

Abstract. Information explosion causes a serious scarcity of people's time and a severe divergence of people's attention. This paper addresses the issue of automatic summarization for Korean texts and presents a novel keyword-extraction-based Korean text summarization (KKTS) algorithm. We investigate the enhancement of POS-tagging to the KKTS algorithm according to three kinds of text feature: noun words, predicate words, and all words. The experimental results show that our POS-tagging enhanced KKTS algorithm according to noun words can achieve the best performance in the Korean summarization task.

Keywords: Korean text summarization · POS-tagging · Keyword extraction · Noun words · ROUGE

1 Introduction

With the rapid development of computing science and language technology, mankind has entered the era of big data. A large number of ubiquitous smart devices make the unprecedented improvement of human production capacity of language data. For the present instance of Korean language, the official language in North Korea (조선말) and South Korea (한국어), there are more than 45 millions Internet users in the total 70 millions Korean speakers. And the number of websites displayed by Korean texts is as high as about 7 millions. With the explosive growth of Korean texts, the above numbers are constantly being broken and refreshed.

Nowadays, huge information brings people an overload of data, which greatly reduces the scarcity of information and extremely improves that of people's time. The value of information is not just the number of words, but more the concentration of key information. Automatic text summarization, the process of reducing a text document with a computer program to create a summary that retains the most important points of the original document, can help people fish out the informative contents from an information ocean in the limited time. In order to deal with the information explosion, text summarization has been widely investigated since the early days of natural

© Springer International Publishing AG 2017
D.-S. Huang et al. (Eds.): ICIC 2017, Part III, LNAI 10363, pp. 425–435, 2017.
DOI: 10.1007/978-3-319-63315-2_37

language processing, and many effective text summarization algorithms have been proposed [1] till now.

2   Related Works

Back in the past, the concept of automatic text summarization has been firstly proposed [2] in 1958. The early text summarization algorithms mainly used the information about documental structures (title, subtitle, the first few sentences of a paragraph, the conclusion sentence, etc.) to identify the essence of a text document. At the end of 1993, the seminar about *Summarizing Text for Intelligent Communication* was held in Germany. And in 1995, the *Journal of Information Processing and Management* published a special issue on Text Summarization [3]. Both of the two opened a new era of the text summarization investigation.

Subsequently, various kinds of supervised text summarization algorithms have been proposed [4], which can learn features of sentences that make them good candidates for inclusion in the summary from a collection of documents and human-generated summaries. Almost at the same time, many semi-supervised [5] and unsupervised [6] text summarization algorithms were booming too. Some unsupervised summarization algorithms are based on finding a centroid sentence.

And then later, people paid more attention to the investigation of multi-document summarization [7], and pursued concise and comprehensive both. A multi-document summary is a text that covers the content of more than one text document, and is usually used only when the input documents are thematically relevant [8].

Recently, the focus of the text summarization investigation has changed from single language to multiple languages [9]. More and more text summarization algorithms are proposed for non-general languages [10]. Among the investigations, the study on Korean text summarization (KTS) is relatively rare [11]. In this paper, we try to study the scientific problem of KTS.

3   POS-Tagging Enhanced KTS

3.1   POS-Tagging Enhanced Framework

According to real-world application requirements, we proposes a novel POS-tagging enhanced framework for efficient text summarization. Figure 1 shows the detailed framework, which includes seven processing modules around the POS-tagging module. When a Korean text document arrives, the module of Paragraph Partitioning will be triggered firstly. The Paragraph Partitioning module will segment the whole document into some paragraphs according to its natural documental structure. Secondly, the Sentence Splitting module will scan the sentence stop punctuation to segment each paragraph into a list of sentences. After the above two preprocessing steps, the POS-tagging module segments each sentence into a list of words, and labels a POS tag for each Korean word, which will output all words with POS tags of the given document. Parallelly, the Nounword Extracting module and Predicateword Extracting module receives the all words with POS tags, and distills the noun words and the predicate words according to POS tags. Subsequently, the Keyword Extracting module respectively ranks three groups of words by their frequencies [12] and truncates the top *n* words according to a preset shared *TopN* value. Finally, the Summarizing module selects informative sentences according to the keywords and produces a Korean text summary, while the MMR Summarizing module adds a MMR (maximal marginal relevance) function.

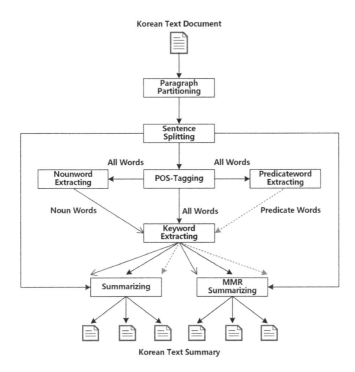

Fig. 1. POS-tagging enhanced framework

3.2   Algorithm

The only real difference between keyword extraction and our text summarization is the text granularity—word or sentence [13]. The keyword extraction is a word-level text summarization, while our text summarization is a sentence-level key "word" extraction. So, within the above POS-tagging enhanced framework, we design a novel keyword-extraction-based Korean text summarization (KKTS) algorithm, which is just the algorithmic bridge connecting keyword extraction and text summarization.

Figure 2 shows the detailed pseudocode, in which there is a preset list of arguments (*Ratio*, *TopN*). The *Ratio* argument indicates a summarization rate, whose value is the final summary length divided by the total length of document. And the *TopN* argument has been explained in the Sect. 3.1. In our KKTS algorithm, there are mainly eight procedures (*ParagraphPartitioning*, *SentenceSplitting*, *POSTagging*, *NounwordExtracting*,

```
1.// Keyword-Extraction-based Korean Text Summarization
  (KKTS) Algorithm
2.PRESET(Ratio, TopN)
3.INPUT(D) // D: Korean Text Document
4.OUTPUT(S[], Smmr[]) // S[], Smmr[]: Korean Text Summaries
5.String[]: AllWords;
6.String[]: Paras←ParagraphPartitioning(D);
7.For (Each String: Para In Paras) Loop
8.    String[]: Sens←SentenceSplitting(Para);
9.    For (Each String: Sen In Sens) Loop
10.        String[]: Words←POSTagging(Sen);
11.        AllWords.add(Words);
12.    End For
13.End For
14.String[]: NounWords←NounwordExtracting(AllWords);
15.String[]:
   PredicateWords←PredicatewordExtracting(AllWords);
16.String[][]:    Keywords[0]←KeywordExtracting(NounWords,
   TopN);
17.String[][]:    Keywords[1]←KeywordExtracting(AllWords,
   TopN);
18.String[][]:
   Keywords[2]←KeywordExtracting(PredicateWords, TopN);
19.For (Each Integer: i In {0, 1, 2}) Loop
20.    Float: F←S[i].length/D.length;
21.    While (F<Ratio) Loop
22.        String           KeySen←Summarizing.keysen(Sens,
           Keywords[i]);
23.        S[i].add(KeySen);
24.        F←S[i].length/D.length;
25.    End While
26.    F←Smmr[i].length/D.length;
27.    While (F<Ratio) Loop
28.        String           KeySen←MMRSummarizing.keysen(Sens,
           Keywords[i]);
29.        Smmr[i].add(KeySen);
30.        Sens←MMRSummarizing.mmr(Sens, KeySen);
31.        F←Smmr[i].length/D.length;
32.    End While
33.End For
34.Return S[], Smmr[].
```

Fig. 2. Keyword-extraction-based Korean text summarization algorithm

PredicatewordExtracting, *KeywordExtracting*, *Summarizing* and *MMRSummarizing*). The *NounwordExtracting* procedure can extract 16-class nouns, while the *PredicatewordExtracting* procedure can extract 5-class predicates. Table 1 shows the detailed POS classes of the 16-class nouns and the 5-class predicates [14].

The major time cost of our KKTS algorithm is the time in the loop (line 7 of Fig. 2) and the loops (line 21, 27 of Fig. 2). The time cost of the first loop is only proportional to the total length of text document. And that of the second loops is related to the Ratio value, which is normally a small floating number. There is no time-consuming operations. Above time complexity is acceptable in the practical KTS application.

Table 1. Noun and predicate tags

Noun	Tag	Noun	Tag	Noun	Tag	Noun	Tag
동작명사	ncpa	고유명사_성	nqpa	단위성 의존명사	nbu	지시대명사	npd
상태명사	ncps	고유명사_이름	nqpb	비단위성 의존명사	nbn	양수사	nnc
비서술성 명사	ncn	고유명사_성+이름	nqpc	비단위성 의존명사_하다 붙는 것	nbs	서수사	nno
비서술성 직위 명사	ncr	고유명사_기타_일반	nqq	인칭대명사	npp	사임 추가	nq
Predicate	**Tag**	**Predicate**	**Tag**	**Predicate**	**Tag**	**Predicate**	**Tag**
지시 동사	pvd	일반 동사	pvg	지시형용사	pad	성상형용사	paa
보조용언	px						

4 Experiment

4.1 Evaluation and Corpus

The classical evaluation measure of ROUGE (recall-oriented understudy for gisting evaluation) and associated evaluation methodology are applied. Here, ROUGEk is a k-gram recall between an automatic summary and a set of manual summaries [15]. At the end of the experiment, we report the ROUGE1, ROUGE2, ROUGE3, and ROUGE4 to evaluate the result of experimental runs.

During the preparation of experimental corpus, we firstly crawl the eleven Korean journals on the Internet and obtain 8298 articles in the pdf format. Secondly, we extract the Abstract, BodyText of each article to form a text document with two text fields of <ManualSummary> and <BodyText> . Finally, we generate a KorSummBank (V1.1) corpus including 8298 text documents. Table 2 shows the detailed source journals and document numbers in the KorSummBank (V1.1) corpus.

Table 2. Source journals and document numbers

Source Journals	Number
Journal of the Korean Biblia Society for Library and Information Science	262
Journal of Korea Spatial Information System Society	331
Journal of Korea Game Society	333
KIPS Transactions on Software and Data Engineering	344
Convergence Security Journal	412
Journal of the Korean Data & Information Science Society	478
Journal of the Korean Society of Civil Engineers	628
Journal of the Institute of Electronics and Information Engineers	632
Journal of the Korean Society of Hazard Mitigation	921
Journal of Digital Convergence	1187
The Journal of the Korea Contents Association	2770
Total Number of Documents	8298

4.2 Implementation and Runs

On one hand, we implement the complete KKTS algorithm and run it under different *TopN* value from 10 to 100 in the experiment. Finally, there will be total following six results—"Noun", "All", "Predicate", "Noun + MMR", "All + MMR", and "Predicate + MMR". On the other hand, we also implement a pure structure-based Korean text summarizer as a baseline.

Table 3. Experimental result of "Noun"

TopN	ROUGE1	ROUGE2	ROUGE3	ROUGE4
10	0.1997	0.0772	0.0422	0.0283
20	0.2087	0.0829	0.0461	0.0312
30	0.2124	0.0853	**0.0478**	0.0326
40	0.2139	0.0855	**0.0478**	**0.0328**
50	**0.2153**	**0.0857**	**0.0478**	**0.0328**
60	0.2146	0.0849	0.0475	**0.0328**
70	0.2140	0.0838	0.0467	0.0321
80	0.2142	0.0836	0.0466	0.0321
90	0.2134	0.0827	0.0459	0.0315
100	0.2120	0.0816	0.0451	0.0310

4.3   Results and Discussion

Table 3 shows the experimental result of "Noun". We find that the performance of the KKTS algorithm improves gradually with the increase of the *TopN* value from 10 to 50. While the performance decreases gradually when the *TopN* value is greater than 50. When the *TopN* value is equal to 50, the four measures (ROUGE1 = 0.2153, ROUGE2 = 0.0857, ROUGE3 = 0.0478, and ROUGE4 = 0.0328) can achieve the best performance.

Table 4 shows the experimental result of "Noun + MMR". Comparing the result in Table 3, we find that there is a significant improvement of each corresponding ROUGE1 value. For instance, when the TopN value is equal to 50, the measure (ROUGE1 = 0.2278) rise 0.0125. However, the other ROUGE values are slightly

Table 4. Experimental result of "Noun + MMR"

TopN	ROUGE1	ROUGE2	ROUGE3	ROUGE4
10	0.2100	0.0748	0.0383	0.0249
20	0.2209	0.0805	0.0420	0.0276
30	0.2255	0.0820	0.0428	0.0283
40	0.2274	**0.0831**	**0.0437**	**0.0292**
50	**0.2278**	0.0821	0.0429	0.0286
60	0.2275	0.0815	0.0425	0.0284
70	0.2275	0.0810	0.0424	0.0283
80	0.2268	0.0803	0.0419	0.0279
90	0.2250	0.0794	0.0415	0.0278
100	0.2240	0.0790	0.0416	0.0281

Table 5. Experimental result of "All"

TopN	ROUGE1	ROUGE2	ROUGE3	ROUGE4
10	0.1539	0.0598	0.0332	0.0227
20	0.1712	0.0692	0.0391	0.0268
30	0.1798	0.0734	0.0413	0.0282
40	0.1861	0.0768	0.0437	0.0301
50	0.1897	0.0784	0.0447	0.0306
60	0.1920	0.0795	0.0454	0.0311
70	0.1941	0.0805	0.0459	0.0316
80	0.1961	0.0812	0.0462	0.0317
90	0.1980	0.0823	0.0469	0.0323
100	**0.1991**	**0.0827**	**0.0471**	**0.0324**

Table 6. Experimental result of "All + MMR"

TopN	ROUGE1	ROUGE2	ROUGE3	ROUGE4
10	0.1677	0.0622	0.0324	0.0212
20	0.1892	0.0728	0.0386	0.0255
30	0.2000	0.0773	0.0412	0.0272
40	0.2089	0.0814	0.0437	0.0289
50	0.2141	0.0833	0.0447	0.0297
60	0.2178	0.0845	0.0452	0.0300
70	0.2208	0.0853	0.0456	0.0303
80	0.2237	0.0869	**0.0470**	**0.0315**
90	0.2255	0.0871	0.0468	0.0312
100	**0.2265**	**0.0875**	**0.0470**	0.0314

decreased. The experimental results indicate that the MMR strategy is effective to enhance ROUGE1 performance.

Table 5 shows the experimental result of "All". We find that the performance improves gradually with the increase of the TopN value. When the TopN is equal to

Table 7. Experimental result of "Predicate"

TopN	ROUGE1	ROUGE2	ROUGE3	ROUGE4
10	0.1682	0.0523	0.0256	0.0167
20	0.1791	0.0584	0.0302	0.0204
30	0.1852	0.0623	0.0331	0.0229
40	0.1871	0.0638	0.0347	0.0246
50	**0.1872**	0.0644	0.0360	0.0259
60	0.1859	**0.0648**	**0.0369**	**0.0271**
70	0.1830	0.0633	0.0365	0.0270
80	0.1793	0.0618	0.0361	0.0270
90	0.1760	0.0605	0.0357	0.0270
100	0.1721	0.0588	0.0349	0.0267

100, the four measures (ROUGE1 = 0.1991, ROUGE2 = 0.0827, ROUGE3 = 0.0471, and ROUGE4 = 0.0324) attain its top performance. But obviously, the performance of "Noun" or "Noun + MMR" is better than that of "All".

Table 6 shows the experimental result of "All + MMR". Comparing the result in Table 5, we find that the MMR strategy is also effective to enhance ROUGE1 and ROUGE2 performance, while ROUGE3 and ROUGE4 values are slightly decreased. The performance of "All + MMR" is not as good as that of "Noun + MMR".

Table 7 shows the experimental result of "Predicate". We find that the performance of "Predicate" is not as good as that of "Noun" and "All", which indicates that the predicate words can not catch the semantic key points.

Table 8. Experimental result of "Predicate + MMR"

TopN	ROUGE1	ROUGE2	ROUGE3	ROUGE4
10	0.1755	**0.0531**	0.0254	0.0164
20	**0.1777**	0.0530	**0.0261**	**0.0172**
30	0.1757	0.0518	0.0256	0.0171
40	0.1732	0.0508	0.0254	0.0170
50	0.1698	0.0495	0.0251	0.0171
60	0.1668	0.0483	0.0246	0.0168
70	0.1645	0.0475	0.0244	0.0169
80	0.1616	0.0462	0.0237	0.0164
90	0.1593	0.0452	0.0233	0.0162
100	0.1565	0.0438	0.0225	0.0156

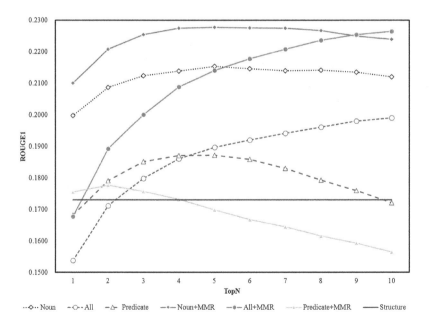

Fig. 3. Experimental results of ROUGE1

Table 8 shows the experimental result of "Predicate + MMR". Comparing the result in Table 7, we find that the MMR strategy can not enhance ROUGEk performance of "Predicate".

Table 9. Example

\<Manual Summary\>스마트폰 게임시장은 스마트폰이 시장에 나타나고 얼마 되지 않아 엄청난 속도로 발전하며 세대교체가 일어나고 있다. 일반적으로 스마트폰 게임은 수명이 짧지만 Supercell 사의 클래시 오브 클랜은 2 년 넘는 시간동안 꾸준한 인기를 얻고 있다. 그렇기 때문에 클래시 오브 클랜이 가지는 가장 특징적인 게임 요소인 수성과 약탈을 주제로 잡아 재미요인을 분석하였다. 수성과 약탈 요소를 분석하기 위해서 클래시 오브 클랜과 다른 형태의 수성과 약탈요소를 가진 게임인 도둑의 왕을 사례로 하였다. 두 가지의 사례를 라프 코스터의 재미이론에 따라 준비, 공간감, 정연한 핵심구조, 일련의 도전, 대결을 위해 필요한 능력, 능력을 사용하기 위한 기술로 분석하여 수성과 약탈 요소의 재미 요인을 알아 보았다.

\<Noun+MMR Summary\>그렇기 때문에 클래시 오브 클랜이 가지는 가장 특징적인 게임 요소인 수성과 약탈을 주제로 잡아 재미요인을 분석하였다. 수성과 약탈 요소를 분석하기 위해서 클래시 오브 클랜과 다른 형태의 수성과 약탈요소를 가진 게임인 도둑의 왕을 사례로 하였다. 두 가지의 사례를 라프 코스터의 재미이론에 따라 준비, 공간감, 정연한 핵심구조, 일련의 도전, 대결을 위해 필요한 능력, 능력을 사용하기 위한 기술로 분석하여 수성과 약탈 요소의 재미 요인을 알아 보았다. 주제어: 스마트폰 융복합 게임, 클래시 오브 클랜, 수성과 약탈, 라프 코스터, 재미이론, 융복합스마트폰 융복합 게임의 재미요인 분석 -수성과 약탈을 중심으로 만약 플레이어가 공격에 실패할 경우엔 사용되었던 병사들이 사라지기 때문에 다시 자원을 사용하여 병사들을 훈련시켜야 한다.

\<All+MMR Summary\>수성과 약탈 요소를 분석하기 위해서 클래시 오브 클랜과 다른 형태의 수성과 약탈요소를 가진 게임인 도둑의 왕을 사례로 하였다. 주제어: 스마트폰 융복합 게임, 클래시 오브 클랜, 수성과 약탈, 라프 코스터, 재미이론, 융복합스마트폰 융복합 게임의 재미요인 분석 -수성과 약탈을 중심으로 그렇기 때문에 본 논문에서는 클래시 오브 클랜이 가지고 있는 게임의 특징 요소인 수성과 약탈에 집중하여 수성과 약탈 요소를 가진 다른 스마트폰 네트워크 게임 중 클래시 오브 클랜과 유사하지 않은 게임인 도둑의 왕을 추가적인 사례로 들어 수성과 약탈 요소가 가지는 재미 요인을 기존에 검증되어진 라프 코스터의 재미이론에 입각해 분석해본다. 스마트폰 네트워크 게임의 종류 중 하나는 소셜 네트워크 게임이 있다. 플레이어는 자신의 자원을 지키기 위해 마을을 안전하게 꾸미는 것 뿐만 아니라 다른 플레이어의 자원을 약탈하기 위해 각 특성을 가지고 있는 병사들을 훈련시켜 다른 플레이어의 마을을 공격한다.

\<Predicate+MMR Summary\>수성과 약탈 요소를 분석하기 위해서 클래시 오브 클랜과 다른 형태의 수성과 약탈요소를 가진 게임인 도둑의 왕을 사례로 하였다. 하지만 2014년 4 분기 SNS 와 게임이 결합한 형태의 SNG 가 아닌 Supercell 사의 스마트폰 네트워크 게임인 클래시 오브 클랜의 대대적인 광고와 함께 최고 매출 순위의 상위권을 위치하게 된다. 스마트폰 게임은 휴대 전화나 스마트폰, PDA, 포터블 미디어 플레이어 등의 휴대용 기기를 통해 즐길 수 있는 게임의 일종이다. 능력을 사용하기 위한 기술은 플레이어가 게임에 승리하기 위해 이용할 수 있는 모든 것이며 잘못된 선택은 플레이어를 게임에서 패배하게 만든다. 만약 플레이어가 공격에 실패할 경우엔 사용되었던 병사들이 사라지기 때문에 다시 자원을 사용하여 병사들을 훈련시켜야 한다.

Figure 3 shows the experimental results of ROUGE1. Among the results of our KKTS algorithm, the overall performance of "Noun + MMR" is optimal.

All the experimental results indicate that the generalization from keyword to summary is an effective KTS method. In order to further explain the advantage of our KKTS algorithm, we show an example in Table 9, where the "Noun + MMR" summary generated by our algorithm almost coincides exactly with the manual summary.

5 Conclusion

In this paper, we suggests a POS-tagging enhanced framework and a KKTS algorithm to solve the scientific problem of KTS. The experimental results in the KorSummBank (V1.1) corpus have proved the effectiveness of our framework and algorithm. Especially, the MMR function is a powerful strategy to promote the performance of our "Noun" and "All" summarization.

However, our extraction method by selecting a key subset of existing sentences to shorten original text document will also cause some disfluencies between sentences in the summary. Therefore, further research will concern abstraction method by Korean semantic representation and generation techniques to create the summary close to the one generated by human beings. We also expect to transfer our above framework and algorithm to Korean multi-document summarization.

Acknowledgements. The research is supported by the Key Project of State Language Commission of China (No. ZDI135-26) and the Featured Innovation Project of Guangdong Province (No. 2015KTSCX035).

References

1. Saggion, H., Poibeau, T.: Automatic text summarization: past, present and future. In: Poibeau, T., Saggion, H., Piskorski, J., Yangarber, R. (eds.) Multi-source, Multilingual Information Extraction and Summarization, pp. 3–21. Springer, Heidelberg (2013). doi:10. 1007/978-3-642-28569-1_1
2. Luhn, H.P.: The automatic creation of literature abstracts. IBM J. Res. Dev. 2(2), 159–165 (1958)
3. Jones, K.S., Brigitte, E.-N.: Introduction: automatic summarizing. Inf. Process. Manag. 31 (5), 625–630 (1995)
4. Yu, L., Ren, F.: A study on cross-language text summarization using supervised methods. In: Proceedings of the International Conference on Natural Language Processing and Knowledge Engineering (2009)
5. Amini, M-R., Gallinari, P.: The use of unlabeled data to improve supervised learning for text summarization. In: SIGIR Forum, pp. 105–112 (2002)
6. Nomoto, T., Matsumoto, Y.: An experimental comparison of supervised and unsupervised approaches to text summarization. In: Proceedings of the IEEE International Conference on Data Mining, pp. 630–632 (2001)
7. Lin, C.Y., Hovy, E.: From single to multi-document summarization: a prototype system and its evaluation. In: Proceedings of the ACL, pp. 457–464 (2002)

8. Yih, W., Goodman, J., Vanderwende, L., Suzuki, H.: Multi-document summarization by maximizing informative content-words. In: Proceedings of the International Joint Conference on Artificial Intelligence, pp. 1776–1782 (2007)

9. Riadh, B., Ahmed, G.: A supervised approach to arabic text summarization using adaboost. Adv. Intell. Syst. Comput. **353**, 227–236 (2015)

10. Vishal, G., Narvinder, K.: A novel hybrid text summarization system for punjabi text. Cogn. Comput. **8**(2), 261–277 (2016)

11. Kim, J.-H., Kim, J.-H., Hwang, D.: Korean text summarization using an aggregate similarity. In: Proceedings of the International Workshop on Information Retrieval with Asian Languages, pp. 111–118 (2000)

12. Nenkova, A., Vanderwende, L.: The impact of frequency on summarization. Technical report, MSR-TR-2005-101 (2005)

13. Jayashree, R., Srikanta Murthy, K., Sunny, K.: Keyword extraction based summarization of categorized Kannada text documents. Int. J. Soft Comput. **2**(4), 81–93 (2011)

14. Park, S., Choi, D.H., Kim, E.-K., Choi, K.-S.: A plug-in component-based Korean morphological analyzer. In: Proceedings of HCLT 2010, pp. 197–201 (2010)

15. Lin, C.-Y.: ROUGE: a package for automatic evaluation of summaries. In: Proceedings of the Workshop on Text Summarization (2004)

A Supervised Term Weighting
Scheme for Multi-class Text Categorization

Yiwei Gu and Xiaodong Gu$^{(\boxtimes)}$

Department of Electronic Engineering, Fudan University,
Shanghai 200433, China
xdgu@fudan.edu.cn

Abstract. Most supervised term weighting (STW) schemes can only be applied to binary text classification tasks such as sentiment analysis (SA) rather than text classification with more than two categories. In this paper, we proposed a new supervised term weighting scheme for multi-class text categorization. The so-called inverse term entropy (*ite*) measures the distribution of different terms across all the categories according to the definition of entropy in information theory. We present experimental results obtained on the *20NewsGroup* dataset with a popular classifier learning method, support vector machine (SVM). Our weighting scheme *ite* achieved the best result in classification accuracy compared with other existing methods. And *ite* has the most stable performance with the reduction of training samples as well. Furthermore, our method has a built-in property to prevent over-weighting in STW. Over-weighting is a newly proposed concept especially with supervised term weightings in our earlier work and re-introduced here. Caused by the improper singular terms and too large ratios between term weights, over-weighting could deprive the performance of text classification tasks.

Keywords: Text categorization · Supervised term weighting · Machine learning

1 Introduction

Text categorization (TC) is the activity to automatically assign a textural document to a set of pre-defined categories, by means of traditional machine learning (ML) techniques usually. According to Debole and Sebastiani [3], the construction of a text categorization may be seen as consisting of essentially two phases:

1. A phase of document indexing or document representation. This typically consists in:
   (a) a phase of term selection, i.e. a common used representation is bag-of-term model. In such representations, a textual document is represented as a vector contains of all the terms that occur in the documents. Here terms can be words, phrases, or other more complicated units identifying the contents of a document.; and
   (b) a phase of term weighting, in which, for every term t_k selected in phase (1a) and for every document d_j, a weight $0 \leq \omega_{kj} \leq 1$ is computed which represents how much term t_k contributes to the discriminative semantics of document d_j

© Springer International Publishing AG 2017
D.-S. Huang et al. (Eds.): ICIC 2017, Part III, LNAI 10363, pp. 436–447, 2017.
DOI: 10.1007/978-3-319-63315-2_38

2. A phase of classifier induction, i.e. the creation of a classifier by learning from the internal representations of the training documents.

This paper mainly focuses on the term weighting schemes, in phase (1b) above.

Term weighting schemes usually can be divided into two different categories, the unsupervised one and the supervised one. An unsupervised term weighting scheme does not take category information into account when computing the weights. A commonly used unsupervised scheme in TC is the inverse document frequency (*idf*). In contrast, the supervised one embraces the category information of training documents into the term weights. Many studies in the literature have proven that supervised term weighting is an efficient method in most of the binary text categorization tasks [1, 3, 14, 15]. With the extra information of documents categories, STW can improve the performance of text classifier in most cases. While most of these supervised term weighting schemes cannot be applied to multi-class text categorization directly. To fill in this gap, we proposed a new supervised term weighting scheme, inverse term entropy (*ite*), which is effectual for classification tasks with arbitrary number of text categories. Inverse term entropy measures the distribution of different terms across all categories under the guide of information theory [13]. In the theory, Shannon entropy is a measure of unpredictability of information content. In other words, if the state of a term is unpredictable, it will have a large entropy. When it comes to term weighting, an unpredictable term usually is assigned with lower weight as we cannot tell the label of a documents according to such terms. Usually, an imbalanced distributed term carries more information about which category the document with such term belongs to. While the even distributed one have limited contribution to the discriminative semantics of document.

Over-weighting, a newly proposed concept in our earlier work [15] and re-introduced here, would occur due to the improper handling of singular terms and unreasonably too large ratios between term weights. To reduce over-weighting, we employed add-one smoothing and bias item in our proposed inverse term entropy. Add-one smoothing is a commonly used technique to handle the singular terms which feature high imbalanced distribution across categories but could be noisy or useless terms. Bias item shrinks the ratios between term weights.

We present experimental results obtained on several datasets constructed from the *20NewsGroup* and two sentimental corpuses with support vector machines (SVM) under different experimental setups. The results show that our term weighting scheme, inverse term entropy, outperformed other existing methods with multi-class text categorization tasks. And the superiority becomes more apparent as the topics are more similar across different categories. Also, our method has a more stable performance when the size of training set changes. And the results become better when over-weighting problem is prevented.

The paper is organized as follows. Section 2 reviews existing term weighting schemes. Section 3 describes the methodology of the new proposed term weighting scheme, inverse term entropy. Section 4 introduces the datasets we used and experimental settings. Section 5 shows the experimental results and followed by the conclusion in Sect. 6.

2 Reviews of Term Weighting Schemes

Vector Space Model (VSM) is a commonly used method to represent documents as vector of terms in text categorization. And term weighting is usually tackled by means of methods borrowed from Information Retrieval (IR), aiming to evaluate the relative importance of different terms.

A typical weighting scheme contains three different components: local weight, global weights and normalization factor [12]

$$\omega_{ij} = l_{ij} \times g_i \times n_j \qquad (1)$$

here ω_{ij} is the weight of the i_{th} term in the j_{th} document, l_{ij} is the local weight of the i_{th} term in the j_{th} document, g_i is the global weight of the i_{th} item and n_j is the normalization factor of the j_{th} document. A common method is cosine normalization. Suppose that t_{ij} represents weight of ith term in the j_{th} document, then the cosine normalization factor is defined as $1/ \sum_i t_{ij}^2$.

2.1 Local Weighting Scheme

Local weight only considers the appearance of a term within the document. Here we introduce three widely used local weighting scheme used in this paper: *term frequency* (*tf*), *term presence* (*tp*) and *augmented term frequency* (*atf*) as what presented in Table 1. Raw term frequency, the most popular local weighting scheme, uses occurrence time in document as the weight of a term. More frequently appeared term in one document is assigned with higher local weight. While term presence ignores the occurrence time of the term in *tf*. This simplest binary representation is proved to be efficient in some cases that the number of times a term appears is not such important, e.g. sentiment analysis. Augmented term frequency is designed to prevent a bias towards longer documents by dividing the raw term frequency by the maximum *tf* of any term in the document [2].

Table 1. Local term weighting schemes.

Local weight scheme	Denoted by	Mathematical form
Term frequency	*tf*	*tf*
Term presence	*tp*	$\begin{cases} 1, & \text{if } tf > 0 \\ 0, & \text{otherwise} \end{cases}$
Augmented term frequency	*atf*	$k + (1-k)\frac{tf}{\max_t(tf)}$

2.2 Global Weighting Scheme

While local weight only considers the *tf* or *tp* within one document, global weight depends on the whole set of training documents, including the category information. Hence global weight methods can be either unsupervised or supervised. Table 2 presents three commonly used existing unsupervised global term weighting schemes.

Table 2. Unsupervised global term weighting schemes.

Global weight scheme	Denoted by	Mathematical form
Inverse document frequency	idf	$\log_2 \frac{N}{|\{d \in D: t \in d\}|}$
Probabilistic inverse document frequency	$pidf$	$\log_2 \frac{N - n_t}{n_t}$
BM 25 inverse document frequency	$bidf$	$\log_2 \frac{\bar{n}_t + 0.5}{n_t + 0.5}$

idf [5] stands for inverse document frequency which is a measure of how much information the word or term provides. And $pidf$ [16] and $bidf$ [6] are two modifications of the original idf.

Borrowed from IR, most global weighting methods don't involve category information. Based on and modified from these global weight schemes, several supervised methods have been proposed by many researchers. Dobel and Sebastiani [3] replaced idf with ig, gr, and chi in global term weighting. Martineau and Finin [9] presented a new supervised scheme Delta idf for sentiment analysis. $didf$ is the difference of a term's idf values in the positive and negative categories. Paltoglou and Thelwall [10], presented a smoothed version of $didf$, delta smoothed idf ($dsidf$). They also explored other more sophisticated methods originated from IR such as delta BM25 idf ($dbidf$). Table 3 shows the definitions and descriptions of these supervised method mentioned above.

Table 3. Supervised global term weighting schemes. a(c) is the number of training documents in the positive(negative) category containing term t_i. N^+/N^- is the number of training documents in the positive/negative category.

Global weight scheme	Denoted by	Mathematical form
Chi-square	chi	$\frac{[P(t_k,c_i)P(\bar{t}_k,\bar{c}_i) - P(t_k,\bar{c}_i)P(\bar{t}_i,c_i)]^2}{P(t_k)P(\bar{t}_k)P(c_i)P(\bar{c}_i)}$
Information gain	ig	$\sum\limits_{c \in \{c_i, \bar{c}_i\}} \sum\limits_{t \in \{t_k, \bar{t}_i\}} P(t, c) \log_2 \frac{P(t,c)}{P(t)P(c)}$
Delta idf	$didf$	$\log_2 \frac{N^- a}{N^+ c}$
Delta smoothed idf	$dsidf$	$\log_2 \frac{N^- a + 0.5}{N^+ c + 0.5}$
Delta BM25 idf	$dbidf$	$\log_2 \frac{(N^- - c + 0.5)(a + 0.5)}{(N^+ - a + 0.5)(c + 0.5)}$

All these supervised weighting methods share the same principle, that is, the best discriminators are the terms that are distributed most differently in the sets of positive and negative training examples. The more the terms are differently distributed across categories, the weight is higher. Besides, there is another common point of these methods that they are all limited when it comes to multi-class problems, since they all treat the training documents as two subsets of positives and negatives. A popular method to apply these supervised term weighting schemes to multi-class classification tasks is the one-vs-rest model. In such model, one multi-class problem is divided into N binary problems (N is the total number of different categories). For each given category, a training set of documents is built with a subset of positive samples (documents belonged to the given category) and a subset of negative samples (documents not belonged to the given category). And N classifiers are trained independently. Under this setup, the global term weight is different and independent across categories, thus for every category the term weights have to be calculated severally. Though these supervised weighting methods outperformed the unsupervised ones in some cases, the computation complexity increases along with the number of categories. Following the principle that most differently distributed terms are weighted higher, we proposed a new supervised term weighting scheme named *inverse term entropy* (*ite*) which can be used in text classification tasks with arbitrary number of categories, which is going to be introduced in next section.

3   The Method

In this section, we introduce our new supervised term weighting scheme which is suitable for text classification tasks with arbitrary number of categories. Here is the simplified mathematical form of the weighting scheme, *inverse term entropy* (*ite*), of term t_k

$$g_k = b_0 + (1 - b_0)\tilde{h}(t_k) \tag{2}$$

where $b_0 \in [0, 1]$ is the bias term, whose value controls the trade-off between over-weighting and under-weighting. And, $\tilde{h}(t_k)$, which is going to be detailed in this section, is the measure of how much term t_k contributes to the task of text categorization.

3.1   Inverse Term Entropy

According to the information theory, entropy is a measure of unpredictability of information content. Named after Boltzmann's H-theorem, Shannon defined the entropy H of a discrete random variable X with possible values $\{x_1, \ldots, x_n\}$ and probability mass function $P(X)$ as

$$H(X) = E[I(X)] = E[-\ln(P(X))] \tag{3}$$

here E is the expected value operator, and I is the information content if X. $I(X)$ is itself a random variable. The entropy can be more explicitly written as

$$H(X) = \sum_{i=1}^{n} P(x_i)I(x_i) = -\sum_{i=1}^{n} P(x_i) \log_b P(x_i) \tag{4}$$

where b is the base of the logarithm used. Common values of b are 2, Eular's number e, and 10. And $P(x_i)$ is the probability that the random variable X equals to x_i. In the case of $P(x_i) = 0$ for some i, the value of the corresponding summand $0\log_b(0)$ is taken to be 0, which is consistent with the limit

$$\lim_{p \to 0+} p \log(p) = 0 \tag{5}$$

In such representation, random variable X with more imbalanced probabilistic distribution over all possible values x_i has a higher entropy $H(X)$. And it can be confirmed using the Jensen inequality that

$$H(X) = E\left[\log_b\left(\frac{1}{p(X)}\right)\right] \leq \log_b\left(E\left[\frac{1}{p(X)}\right]\right) = \log_b(n) \tag{6}$$

This maximal entropy of $\log_b(n)$ is effectively attained by a source alphabet having a uniform probability distribution. In the case of multi-class text categorization, we assume that there are totally $|C|$ different categories of the whole training documents. For one certain document d containing term t_k, let $p(t_k, c_i)$ be the probability of a document d belonging to category c_i. Then the probability $p(t_k, c_i)$ can be estimated as

$$p(t_k, c_i) \approx \frac{n_i}{\sum\limits_{i \in [1, |C|]} n_i} \tag{7}$$

where n_i is the number of documents belonging to category c_i which contain term t_k. And the denominator in the right side of equality (7) indicates the total number of documents containing term t_k in the whole training documents corpus. In order to deal with imbalanced distributed corpus, we normalized the probability in equality (7) as

$$p(t_k, c_i) \approx \frac{n_i/N_i}{\sum\limits_{i \in [1, |C|]} n_i/N_i} \tag{8}$$

where N_i is the total number of documents belonging to category c_i. Thus, we define our term entropy for text categorization tasks in the form of equality (4) as follows

$$h(t_k) = - \sum\limits_{i \in [1, |C|]} p(t_k, c_i) \log_2 p(t_k, c_i) \tag{9}$$

Specifically, we choose 2 as the base of the logarithm used. $h(t_k)$ is a measure of the uncertainty of a document contains term t_k belonging to a certain category. Same as equality (6), the maximum of term entropy, $h(t_k)$, is $\log_2(|C|)$ and can be effectively attained when term t_k is evenly distributed across all the categories. An effective global weighting scheme should always assign terms that are most unevenly distributed across categories with the highest weights, that is, a term with lager term entropy should be assigned with lower weight. Under this assumption, we construct the original formula of our weighting scheme, *invers term entropy*, as follows:

$$\tilde{h}(t_k) = \log_2(|C|) - h(t_k). \tag{10}$$

3.2 Reduction of Over-Weighting

Experimental results in Sect. 5 has affirmed that our original formula of *ite* in equality (10) outperformed most existing global weighting schemes, while the results could become better with the reduction of over-weighting. Over-weighting (and under-weighting) is a newly proposed concept in our earlier work [15]. In practice we identify that over-weighting is caused by the improper handling of singular terms and too large ratios between term weights. Here we define singular terms as terms with high imbalanced distributions. Singular terms with high frequency could be very discriminative, should be assigned with high term weights. While singular terms with low frequency could be noisy or useless words, especially when such terms only exist in one of the categories. These terms should have low weights but usually are also assigned with high weights. Specifically, let's take a binary classification problem as an example. Suppose the training corpus is balanced with 1000 documents both in positive and negative categories, term t_1 occurs in 100 positive documents while 0 in the negatives, and term t_2 occurs in 1 positive documents while 0 in the negatives. According to the original *ite* scheme, the global weight of t_1 and t_2 both should be 1. Obviously the weight of term t_2 is unreasonable and violates common sense. It is distinctly a noisy term and should have low weights. We named this improper weighting problem over-weighting. Over-weighting problem occurs even more frequently in most existing supervised global weighting schemes. To reduce over-weighting in term weighting schemes, add-one smoothing is one of the effective techniques. Using add-one smoothing, t_2's weight in the example changes into a more reasonable value of 0.09 while the weight of t_1 maintains at the same level with a value of 0.92.

To deal with the too large ratios between term weights, we introduced bias term in our original term weighting scheme. With a bias term shrinking the ratios between term weights, the weighting scheme is transformed into equality (2). Controlling the trade-off between over-weighting and under-weighting, b_0's value should neither be too large, nor too small. The optimal value of b_0 can be obtained via cross-validation, a model selection technique widely used in machine learning.

With the deduction and the over-weighting reducing techniques introduced above, the finall mathematical form of our supervised global weighting scheme, *inverse term entropy* (*ite*), is shown as follows:

$$g_k = b_0 + (1 - b_0) \left[\log_2(|C|) + \sum_{i \in [1, |C|]} \frac{(n_i + 1)/N_i}{\sum_{i \in [1, |C|]} (n_i + 1)/N_i} \log_2 \frac{(n_i + 1)/N_i}{\sum_{i \in [1, |C|]} (n_i + 1)/N_i} \right]. \quad (11)$$

4 Dataset and Experimental Setup

4.1 Dataset

We conduct our experiments on two multi-class topical classification datasets constructed from the *20NewsGroup* corpus [7]. The *20NewsGroup* dataset is a collection of approximately 20,000 newsgroup documents, partitioned (nearly) evenly across 20 different newsgroups, each corresponding to a topic.

The first experimental dataset (NewsGroups-Unr) contains 5 different highly unrelated topics chosen from the original newsgroups. They are: sci.crypt, alt.atheism, comp.graphics, misc.forsale and rec.autos.

Another experimental dataset (NewsGroups-Rel) contains newsgroups with 5 closely related topics under the subject of *"computer"*. They are: comp.os.ms-windows. misc, comp.sys.ibm.pc.hardware, comp.sys.mac.hardware, comp.graphics and comp. windows.x.

In experiments of multi-class text categorization, our scheme is compared with existing unsupervised global weighting methods (*idf, pidf, bidf*) that are directly suitable for the tasks.

Besides, to authenticate the ability of handling text classification tasks with arbitrary number of categories and better performance, we also compared our scheme with existing supervised schemes on two binary sentiment classification datasets. They are:

MR-2k: 2000 full-length movie reviews that has become the de facto benchmark for sentiment analysis [11]. The reviews are balanced across sentiment polarities.

IMDB: A large movie review dataset containing 50k reviews (25k training, 25k test), collected from Internet Movie Database [8]. The reviews are also balanced across sentiment polarities. Detailed statistics of these datasets are shown in Tables 4 and 5.

Table 4. Dataset statistics of NewsGroups datasets. L: average length of document.

Dataset	L	Vocab size	Topics	Train	Test
NewsGroups-Unr	281	13822	alt.atheism	480	319
			comp.graphics	584	389
			misc.forsale	585	390
			rec.autos	594	396
			sci.crypt	595	396
NewsGroups-Rel	328	15505	comp.graphics	584	389
			comp.os.ms-windows.misc	591	394
			comp.sys.ibm.pc.hardware	590	392
			comp.sys.mac.hardware	578	385
			comp.windows.x	593	395

4.2 Experimental Setup

For pre-processing of the corpuses, we lowercase all of the words but no stemming nor lemmatization is performed. Unigram representation is used to construct the vocabulary in most cases. And only the tokens that occurred at least 4 times in the whole training set are considered as the valid terms of the vocab. We use the implementation software LIBLINEAR as the solution and adopt the L2-regularized L2-loss linear SVM as the classifier [4]. Tuning of the bias b_0 is done by testing the models on a portion of training data with possible values of b_0.

Table 5. Dataset statistics. (CV means that there is no separate test set for this dataset and thus a 10-fold cross-validation was used to calculate accuracies).

Dataset	Length	Pos	Neg	Test
MR-2k	762	1000	1000	CV
IMDB	271	12.5k	12.5k	25k

For newsgroups datasets, three local term weighting scheme (tf, tp, atf) are used. While for sentiment analysis on MR-2k and IMDB datasets, only tp is adopted. Classification accuracy is used as the evaluation metric of the performance.

5 Experimental Results

5.1 *ite* Outperforms Existing Unsupervised Method in Multi-class Categorization

Table 6 shows the comparisons of results on multi-class text categorization tasks. Our method outperforms the traditional unsupervised term weighting schemes on all benchmarks for both the datasets with highly unrelated topics and closely related topics. Summarization detailed as follows.

Table 6 shows that, our method achieves both the highest classification accuracy on these two datasets with different local term weighting schemes. And even without the over-weighting reduction techniques, the original form of our method (*ite-ori*) still has the best results in most cases. In NewsGroups-Unrelated datasets, since the topics across categories are highly unrelated, all the method in the table achieve high accuracies, while our method has the highest. Specifically, our method combined with local weight scheme tf (*tf-ite*) outperforms the most widely used scheme tf-idf by 0.9524 (95.8201 vs. 94.8677). In NewsGroups-Related datasets, the accuracies of different methods all suffer from a substantial decline because of the closely related topics. It is harder for the classifier to distinguish one topic from others. Our method is still the best one. Specifically, an even larger difference of 1.8926 could be found between our method combined with local weight scheme tf (*tf-ite*) and tf-idf (83.7852 vs. 81.8926).

Table 6. Classification accuracy of multi-class text categorizations. *no* indicates that no global scheme is used. *itr-ori* is the original version of invers term entropy without over-weighting reduction in equality (10)

NewsGroups-Unrelated				NewsGroups-Related			
Local weighting schmem				Local weighting schmem			
	tf	*tp*	*atf*		*tf*	*tp*	*atf*
no	94.0212	94.0741	94.1270	*no*	79.1304	79.5908	80.1023
idf	94.8677	95.2910	95.4497	*idf*	81.8926	83.0691	83.5294
pidf	94.4444	94.3915	94.7090	*pidf*	82.3018	82.8133	83.5294
bidf	94.3915	94.8677	95.1852	*bidf*	82.0460	82.5575	83.1202
ite-ori	94.9206	95.5556	95.3968	*ite-ori*	81.8926	83.4271	84.0921
ite	**95.8201**	**95.9259**	**95.8730**	*ite*	**83.7852**	**83.9386**	**84.7059**

5.2   *ite* is a More Stable Scheme with Different Size of Training Sets

To further illustrate the robustness of our method, we reduced the size of training set on datasets NewsGroups-Unrelated and NewsGroups-Related and compared *ite* with *no*, *idf*, *pidf* and *bidf*. Local weighting scheme *atf* is used in this experiment. Figure 1 shows the result.

As is shown in Fig. 1, the accuracies of all the existing weighting methods suffer from a substantial decline. But our methods maintain in a steady level of high classification accuracy. Overall, our methods (*ite-ori* and *ite*) perform much better than others when the training set have a small amount of examples. That is, our method can achieve better results with smaller size of training datasets.

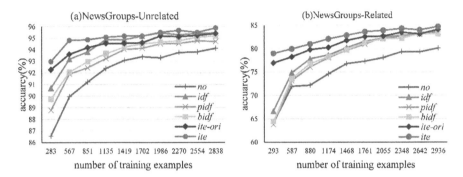

Fig. 1. Results of *ite* (*ite-ori*) against existing schemes with different size of training datasets. All the training examples are randomly and evenly selected from the original training sets.

5.3   *ite* Outperforms Existing Supervised Method in Binary Text Categorization

Furthermore, in order to illustrate that our method is also a better scheme among the supervised ones. As most of the existing supervised term weighting schemes cannot be applied to multi-class tasks directly, we also compared our weighting method with four commonly used methods on two sentiment classification datasets (MR-2k and IMDB). Table 7 shows the results.

Table 7 shows that our method *ite* outperforms all the commonly used supervised weighting schemes mentioned in the table. In general, our method has the ability to

Table 7. Results of our methods against existing supervised term weighting schemes. (1) indicates that only unigram is used, (2) indicates both unigrams and bigrams are used.

Dataset	*ig*	*chi*	*dsidf*	*dbidf*	*ite*
MR-2k (1)	86.90	86.60	88.25	88.40	**89.30**
MR-2k (2)	88.00	87.55	89.75	89.95	**90.30**
IMDB (1)	86.73	86.65	89.26	89.36	**89.68**
IMDB (2)	88.04	87.92	91.76	91.75	**91.86**

handle text classification tasks with arbitrary number of categories and has the better performance than existing weighting schemes on different datasets.

In addition, all the experimental results above have illustrated that with the using of over-weighting reduction techniques, the accuracy of text categorization tasks can be further improved.

6 Conclusions

In this paper, we proposed a new supervised term weighting scheme named *inverse term entropy* (*ite*) for multi-class text categorizations. Based on Shannon entropy, inverse term entropy measures the distribution of a term across different categories. For all we know, our method is the first ever supervised term weighting scheme that can handle text classification tasks with arbitrary number of categories directly. Various experiments that have been taken illustrated that our weighting scheme achieves the highest accuracy on different datasets, including binary and multi-class corpuses. And the superiority of our method becomes more apparent as the topics are closely related across different categories. Additionally, our method is a more robust scheme than other existing ones. With a reduction of training examples, the performance of our term weighting scheme maintains in a steady high level. While, other schemes perform terribly when the training dataset has a small size.

Furthermore, our inverse term entropy has built-in techniques of add-one smoothing and bias term to prevent the scheme from over-weighting problem. Over-weighting is a newly proposed concept in our earlier work [15]. Caused by improper weight of singular terms and too large ratios between weights of different terms. Experimental results have shown that the accuracy of text categorization tasks can be further improved with the over-weighting reduction techniques within the proposed weighting schemes.

Acknowledgments. This work was supported in part by National Natural Science Foundation of China under grant 61371148.

References

1. Bata, I., Hauskrecht, M.: Boosting KNN text classification accuracy by using supervised term weighting schemes. In: Proceedings of the 18th ACM Conference on Information and Knowledge Management, pp. 2041–2044. ACM, November 2009
2. Croft, W.B.: Experiments with representation in a document-retrieval system. Inf. Technol.-Res. Dev. Appl. **2**(1), 1–21 (1983)
3. Debole, F., Sebastiani, F.: Supervised term weighting for automated text categorization. In: Sirmakessis, S. (ed.) Text Mining and Its Applications. Springer, Heidelberg, pp. 81–97 (2004)
4. Fan, R.E., Chang, K.W., Hsieh, C.J., Wang, X.R., Lin, C.J.: LIBLINEAR: a library for large linear classification. J. Mach. Learn. Res. **9**(1), 1871–1874 (2008)

5. Jones, K.S.: A statistical interpretation of term specificity and its application in retrieval. J. Doc. **28**(1), 11–21 (1972)
6. Jones, K.S., Walker, S., Robertson, S.E.: A probabilistic model of information retrieval: development and comparative experiments. Inf. Process. Manag. **36**(6), 779–808 (2000)
7. Lang, K.: Newsweeder: learning to filter netnews. In: Proceedings of the 12th International Conference on Machine Learning, pp. 331–339, July 1995
8. Maas, A.L., Daly, R.E., Pham, P.T., Huang, D., Ng, A.Y., Potts, C.: Learning word vectors for sentiment analysis. In: Proceedings of the 49th Annual Meeting of the Association for Computational Linguistics: Human Language Technologies, vol. 1, pp. 142–150. Association for Computational Linguistics, June 2011
9. Martineau, J., Finin, T.: Delta TFIDF: an improved feature space for sentiment analysis. ICWSM **9**, 106 (2009)
10. Paltoglou, G., Thelwall, M.: A study of information retrieval weighting schemes for sentiment analysis. In: Proceedings of the 48th Annual Meeting of the Association for Computational Linguistics, pp. 1386–1395. Association for Computational Linguistics, July 2010
11. Pang, B., Lee, L.: A sentimental education: sentiment analysis using subjectivity summarization based on minimum cuts. In: Proceedings of the 42nd Annual Meeting on Association for Computational Linguistics, p. 271. Association for Computational Linguistics, July 2004
12. Salton, G., Buckley, C.: Term-weighting approaches in automatic text retrieval. Inf. Process. Manag. **24**(5), 513–523 (1988)
13. Shannon, C.E.: A mathematical theory of communication. Bell Syst. Tech. J. **27**(3), 379–423 (1948)
14. Soucy, P., Mineau, G.W.: Beyond TFIDF weighting for text categorization in the vector space model. IJCAI **5**, 1130–1135 (2005)
15. Wu, H., Gu, X., Gu, Y.: Balancing between over-weighting and under-weighting in supervised term weighting. Inf. Process. Manag. **53**(2), 547–557 (2017). doi:10.1016/j.ipm.2016.10.003
16. Wu, H., Salton, G.: A comparison of search term weighting: term relevance vs. inverse document frequency. In: ACM SIGIR Forum, vol. 16, no. 1, pp. 30–39. ACM, May 1981

Recognizing Text Entailment via Bidirectional LSTM Model with Inner-Attention

Chengjie Sun$^{(\boxtimes)}$, Yang Liu, Chang'e Jia, Bingquan Liu,
and Lei Lin$^{(\boxtimes)}$

Harbin Institute of Technology, Harbin 150001, China
{cjsun,cejia,liubq,linl}@insun.hit.edu.cn,
14s103044@hit.edu.cn

Abstract. In this paper, we propose a sentence encoding-based model for recognizing text entailment (RTE). In our approach, the encoding process of sentences consists of two stages. Firstly, average pooling is used over word-level bidirectional LSTM (biLSTM) to generate a first-stage sentence representation. Secondly, attention mechanism is employed to replace average pooling on the same sentence for better representations. Instead of using target sentence to attend words in source sentence, we utilize the sentence's first-stage representation to attend words appeared in itself, which is called "Inner-Attention" in our paper. Experiments conducted on Stanford Natural Language Inference (SNLI) Corpus has proved the effectiveness of "Inner-Attention" mechanism. With less number of parameters, our model outperformed the existing best sentence encoding-based approach by a large margin.

Keywords: RTE · biLSTM · Inner-Attention · SNLI

1 Introduction

Given a pair of sentences, the goal of recognizing text entailment (RTE) is to determine whether the hypothesis can reasonably be inferred from the premises. There were three types of relation in RTE, Entailment (inferred to be true), Contradiction (inferred to be false) and Neutral (two irrelevant sentences). A few examples were given in Table 1.

Table 1. Examples of three types of label in RTE, where P stands for Premises and H stands for Hypothesis.

P	The boy is running through a grassy area	
H	The boy is in his room	C
	A boy is running outside	E
	The boy is in a park	N

Traditional methods for RTE has been the dominion of classifiers employing hand engineered features, which heavily relied on natural language processing pipelines and external resources [1, 2]. Formal reasoning methods [3] were also explored by many

© Springer International Publishing AG 2017
D.-S. Huang et al. (Eds.): ICIC 2017, Part III, LNAI 10363, pp. 448–457, 2017.
DOI: 10.1007/978-3-319-63315-2_39

researchers, but not been widely used because of their complexity and domain limitations.

Recently published Stanford Natural Language Inference (SNLI[1]) corpus makes it possible to use deep learning methods to solve RTE problems. So far proposed deep learning approaches for RTE can be roughly categorized into two groups: sentence encoding-based models and matching encoding-based models. As the name implies, the encoding of sentence is the core of former methods, while the latter methods directly model the relation between two sentences and didn't generate sentence representations at all.

We focused our efforts on sentence encoding-based model in this paper. Existing methods of this kind including: LSTMs-based model, GRUs-based model, TBCNN-based model and SPINN-based model. Single directional LSTMs [4] and GRUs [5] suffer a weakness of not utilizing the contextual information from the future tokens and Convolutional Neural Networks (CNN) didn't make full use of information contained in word order. Bidirectional LSTM utilizes both the previous and future context by processing the sequence on two directions which helps to address the drawbacks mentioned above [6]. A recent work by [7] improved the performance by applying a neural attention model that didn't yield sentence embedding. [8] proposed a Hierarchical Attention model on the task of document classification, which also used sentence encoding and the target representation in their attention mechanism is randomly initialized.

However, all those previous work via attention concentrated on the interrelation between two sentences, while ignoring the weight proportion of tokens in a sentence. In fact, two distinct words in the same sentence would attract people's attention in various degrees. Traditional sentence representations based on LSTM aggregate the information and reduce the sentence vector space by average pooling or max pooling. Both mean and max pooling suffered from their shortcomings [9]: in average pooling operation, each time step output of the LSTM is considered equally, which may weaken or overlook some strong activation values; the max pooling gets rid of that deficiency and used more widely in CNN, while it may cause strong over-fitting on the training set [10], which may lead to be poor generalize beyond the test data. [10] proposed some stochastic pooling to mitigate over-fitting. Recently, the generalization of max pooling employs k-max value instead extracted in the original order [11], which could extract more secondary information from sentence. Beyond that, [11] also proposed the dynamic k-max pooling, and the sentence size and the level within the convolution hierarchy decide the value of k.

In this paper, we proposed a unified deep learning framework for recognizing textual entailment which does not require any feature engineering, or external resources. The basic model is based on building biLSTM models on top of both premises and hypothesis. The mean pooling encoder can roughly form a intuition about what this sentence is talking about. Based on this representation, we extended the model by utilize an Inner-Attention mechanism on both sides. This mechanism helps generate more accurate and focused sentence representations for classification. In

---

[1] http://nlp.stanford.edu/projects/snli/.

addition, we introduce a simple effective input strategy that get rid of same words in hypothesis and premise, which further boosts our performance. Without parameter tuning, we improve the state-of-the-art performance of sentence encoding-based model by nearly 2%.

Overall, the main contributions of this work are two-fold: (1) We are the first to introduce idea to teach sentence to attend by itself, and propose an Inner-Attention Model based on bidirectional LSTM architecture which outperformed the state-of-the-art RTE results by a large margin; (2) We come up with a new input strategy which helps boost the performance of RTE.

2 Our Approach

In our work, we treated RTE task as a supervised three-way classification problem. The overall architecture of our model is shown in Fig. 1. We follow the idea of Siamese Network [12] to design this model, that the two identical sentence encoders share the same set of weights during training, and the two sentence representations then are combined together to generated a "matching vector" for classification. As we can see from the figure, the model mainly consists of three parts: (A) The sentence input module; (B) The sentence encoding module; (C) The sentence matching module. We will explain the last two parts in detail in the following subsection. And the sentence input module will be introduced in Sect. 3.3.

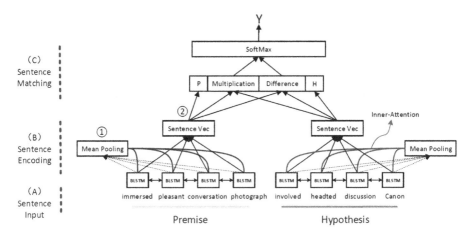

Fig. 1. Architecture of bidirectional LSTM model with Inner-Attention

2.1 Sentence Encoding Module

Sentence encoding module is the fundamental part of this architecture. To generate better sentence embedding, we employed a two-step strategy to encode sentences. Firstly, average pooling layer was built on top of word-level biLSTMs to produce sentence vector. This simple encoder combined with sentence matching module formed the basic architecture of our model. With much less parameters, this basic model alone can outperform state-of-the-art method by a small margin (refer to Table 4). Secondly, attention mechanism was employed to replace mean pooling for better representation, instead of using target representation to attend words in source sentence, we used the representation generated in previous stage to attend words appeared in itself. More attention was given to important words.

The idea of "Inner-attention" was inspired by the observation that when human read one sentence, people usually can roughly form an intuition about which part of the sentence is more important according past experience. And we implemented this idea using attention mechanism in our model. The inner-attention mechanism is formalized as follows:

$$M = \tanh(W^h Y + W^h R_{ave} \otimes e_L) \tag{1}$$

$$\alpha = softmax(w^T M) \tag{2}$$

$$R_{att} = Y\alpha^T \tag{3}$$

where Y is a matrix consisting of output vectors of biLSTM, R_{ave} is the output of mean pooling layer, the outer product $W^h R_{ave} \otimes e_L$ repeats the linearly transformed R_{ave} as many times as there are words in the sentence. α denotes the attention vector and R_{att} is the *attention-weighted* sentence representation.

2.2 Sentence Matching Module

After generating the sentence embedding, three matching methods are applied to extract relations between premise and hypothesis:

Concatenation of the Two Representations. The concatenation of the two representations is one of the most widely used matching strategies in sentence matching module. It just concatenates the two sentences directly and usually feed the con-vector into DNN to encode and decode. The concatenation equation is formalized as follows.

$$SentCat = [S_1; S_2] \tag{4}$$

Where S_1 and S_2 denote the premise and hypothesis sentence vector respectively.

Element-Wise Product. Element-wise product is a common method of vector interaction in deep learning area. It is effective for many tasks and formalized as follows.

$$SentMul = [S_1 \circ S_2] \tag{5}$$

Element-Wise Difference. Element-wise difference is to get the difference within each element pair from two sentences, and always uses absolute value of the difference, which can be formalized as follows.

$$SentDiff = [S_1 - S_2] \qquad (6)$$

Euclidean Distance. Euclidean Distance is the most commonly chosen type of distance. It is the geometric distance in the multidimensional space. We apply it to compute the distance of two vector S_1 and S_2, and the formula showed in Eq. 7. The distance value regards as an element of the matching vector.

$$Euclidean\_Dis = \sqrt{\sum_{i=1}^{n} (S_1^i - S_2^i)^2} \qquad (7)$$

Cosine Similarity. Cosine similarity [13], or cosine distance computes similarity value between two vectors S_1 and S_2 of an inner product space which use the cosine of the angle. The distance value regards as an element of the matching vector. Cosine Similarity uses commonly in the tasks of Semantic similarity of sentences and Q&A sentences matching. The formula is showed in Eq. 8.

$$Cosine\_Dis = \frac{\sum_{i=1}^{n} S_1^i S_2^i}{\sqrt{\sum_{i=1}^{n} (S_1^i)^2} \sqrt{\sum_{i=1}^{n} (S_2^i)^2}} \qquad (8)$$

Finally, we conduct our experiment with adding the above five matching strategies to the model, the results of experiments are showed in Table 2. Table 2 clearly shows that concatenation of the two representations, Element-wise product and Element-wise difference can obtain good results. While, the combination with two similarity computation methods cannot improve the accuracy. We believe that this might be due to the reason the combination strategies contribute to the similarity of the sentences-pair. However, in our task, the non-similarity is beneficial to the judgment between contraction and neutral, so the similarity would obstruct the ability of classification of contraction and neutral.

Table 2. The combination of sentence matching strategies (The SentCat, SentMul and SentDiff denote concatenation of the two representations, element-wise product and element-wise difference respectively, ' + ' denote the vectors concatenation)

Match structures	Acc (train)	Acc (test)	Numbers (parameters)
Concat	84.36%	70.85%	1683203
Concat + Mul	85.26%	82.98%	1833903
Concat + Diff	85.41%	81.13%	1833903
Concat + Mul + Diff	86.38%	**83.25%**	1964403
All five operations	86.63%	82.74%	1966019

This matching architecture was first used by [14]. The concatenation is the most widely used sentence matching method, and the element-wise product and element-wise difference have been used in many sentence matching models. The effectiveness of the two methods, element-wise product and difference, have been proved [14].

As illustrated in Fig. 1, the three matching layers are further concatenated to form a matching vector. Finally, a SoftMax layer was used over the generated matching vector for classification.

3   Experiments

3.1   DataSet

To evaluate the performance of our model, we conducted our experiments on Stanford Natural Language Inference (SNLI) corpus [15]. SNLI contains 570K pairs, which is two orders of magnitude larger than all other resources of its type. Each sentence in the dataset is written by human. The target labels comprise three classes: Entailment, Contradiction, and Neutral. And we applied the standard train/validation/test split, containing 550k, 10k, and 10k samples, respectively.

3.2   Parameter Setting

The training objective of our model is cross-entropy loss, and we used RMSProp [16] for optimization. The batch size is 128. A dropout [17] layer was applied in the output of the network with the dropout rate set to 0.25. In our model, we used pretrained 300D GloVe 840B vectors [18] to initialize the word embedding. Out-of-vocabulary words in the training set are randomly initialized by sampling values uniformly from (–0.05, 0.05). All of these embedding are not updated during training. We didn't tune representations of words for two reasons: (1) Reduce the number of parameters need to be trained. (2) Ensure their representations stay close to unseen similar words for which we have GloVe embeddings. And this model is implemented using open-source framework Keras[2].

3.3   The Input Strategy

In this part, we investigated four strategies to modify input on our basic model which help us increase performance. The four strategies are:

- Inverting Premises [19]
- Doubling Premises [20]
- Doubling Hypothesis
- Differentiating Inputs (Removing same words appeared in premises and hypothesis)

---

[2] http://keras.io/.

Experimental results were illustrated in Table 3. As we can see, doubling hypothesis and differentiating inputs both improved our model's performance. While hypothesises are usually shorter than premises, doubling hypothesis may absorb this difference and emphasize the meaning twice. Differentiating input strategy forces the model to focus on different part of the two sentences, which helps the model capture contradiction information need in classification of neutral and contradiction examples. The original input sentences in Fig. 1 are:

Table 3. Comparison of different input strategies

Input strategy	Train Acc.	Test Acc.	# of iteration
Original sequences	86.38%	83.24%	20
Inverting premises	86.07%	82.60%	20
Doubling premises	87.73%	82.83%	22
Doubling hypothesis	87.73%	83.66%	29
Differentiating inputs	85.46%	83.72%	17

- **Premise:** Two man in polo shirts and tan pants immersed in a pleasant conversation about photograph.
- **Hypothesis:** Two man in polo shirts and tan pants involved in a heated discussion about Canon.
- **Label:** Contradiction.

While most words in this pair are same or close in semantic, it is hard to distinguish the difference between them, which resulted in classifying this example to Neutral or Entailment. Through differentiating input strategy, this kind of problem can be handled well.

3.4   Comparison Methods

In this part, we compared our model against the following strong baseline approaches:

- **LSTM enc:** 100D LSTM encoders + MLP [15].
- **GRU enc:** 1024D GRU encoders + skip-thoughts + **cat**, - [21].
- **TBCNN enc:** 300D Tree-based CNN encoders + **cat**, ○, - [14].
- **SPINN enc:** 300D SPINN-NP encoders + **cat**, ○, - [22].
- **Static-Attention:** 100D LSTM + static attention [7].
- **WbW-Attention:** 100D LSTM + word-by-word attention [7].

The **cat** refers to concatenation, - and ○ denoted element-wise difference and product, respectively. The results are illustrated in Table 4.

Table 4. Performance comparison of different models on SNLI.

Model	Params	Test Acc.
Sentence encoding-based models		
LSTM enc	3.0 M	80.6%
GRU enc	15 M	81.4%
TBCNN enc	3.5 M	82.1%
SPINN enc	3.7 M	83.2%
Basic model + Inner-attention + Differentiating input	**2.0 M**	83.3%
	2.8 M	84.2%
	2.8 M	**85.0%**
Matching encoding-based models		
Static-attention	242 K	82.4%
WbW-attention	252 K	83.5%

3.5   Results and Qualitative Analysis

Although the classification of RTE example is not solely relying on representations obtained from attention, it is still instructive to analysis Inner-Attention mechanism as we witnessed a large performance increase after employing it. We hand-picked several examples from the dataset to visualize them. In order to make the weights more discriminative, we didn't use uniform color atlas cross sentences. In each example, the lightest color and the darkest color denoted the smallest attention weight and the biggest value within the sentence, respectively. Visualizations of Inner-Attention on these examples are depicted in Fig. 2.

We observed that more attention was given to Nouns, Verbs and Adjectives. This conform to our experience that these words are more semantic richer than function words. And we also found that the mechanism tend to assign smaller weights to **man** and **women** although they are all Nouns. This may because they appeared very frequently in SNLI corpus. Approximately 30% of sentences contain them as shown in Table 5. While mean pooling regarded each word as equal importance, the attention mechanism helps re-weight words according to their importance. And more focused and accurate sentence representations were generated based on produced attention vectors.

Fig. 2. Inner-Attention visualizations

Table 5. Frequency of common words in SINL.

Word	*"man"*	*"men"*	*"woman"*	*"women"*
Frequency	28.8%	4.3%	15.7%	7.2%

4 Conclusion and Future Work

In this paper, we proposed a bidirectional LSTM-based model with Inner-Attention for RTE task. We come up with an idea to utilize attention mechanism for sentence encoding. Experimental results demonstrated the effectiveness of our model. In addition, the simple effective differentiating input strategy introduced by us further boosted our results. And this model can be easily adapted to other sentence-matching models. we will employ this architecture on other sentence-matching tasks in future.

Acknowledgment. This work is sponsored by the National High Technology Research and Development Program of China (2015AA015405) and National Natural Science Foundation of China (61572151 and 61602131).

References

1. Dagan, I., Roth, D., Sammons, M., Zanzotto, F.M.: Recognizing textual entailment: models and applications. Synth. Lect. Hum. Lang. Technol. **6**, 1–220 (2013)
2. Dagan, I., Dolan, B., Magnini, B., Roth, D.: Recognizing textual entailment: rational, evaluation and approaches -erratum. Nat. Lang. Eng. **16**, 105 (2010)
3. Bos, J., Markert, K.: Recognising textual entailment with logical inference. In: Proceedings of the conference on Human Language Technology and Empirical Methods in Natural Language Processing, pp. 628–635 (2005)
4. Hochreiter, S., Schmidhuber, J.: Long short-term memory. Neural Comput. **9**, 1735–1780 (1997)
5. Chung, J., Gulcehre, C., Cho, K., Bengio, Y.: Empirical evaluation of gated recurrent neural networks on sequence modeling. In: NIPS Deep Learning Workshop (2014)
6. Tan, M., Xiang, B., Zhou, B.: LSTM-based deep learning models for non-factoid answer selection. arXiv Prepr. arXiv:1511.04108 (2015)
7. Rocktäschel, T., Grefenstette, E., Hermann, K.M., Kočiský, T., Blunsom, P.: Reasoning about entailment with neural attention. In: International Conference on Learning Representations (2016)
8. Yang, Z., Yang, D., Dyer, C., He, X., Smola, A., Hovy, E.: Hierarchical attention networks for document classification. In: Proceedings of the 2016 Conference of the North American Chapter of the Association for Computational Linguistics: Human Language Technologies, pp. 1480–1489 (2016)
9. Severyn, A., Moschitti, A.: Learning to rank short text pairs with convolutional deep neural networks. In: Proceedings of the 38th International ACM SIGIR Conference, pp. 373–382 (2015)
10. Zeiler, M.D., Fergus, R.: Stochastic pooling for regularization of deep convolutional neural networks. In: International Conference on Representation Learning, pp. 1–9 (2013)

11. Kalchbrenner, N., Grefenstette, E., Blunsom, P.: A convolutional neural network for modelling sentences. In: Proceedings of the 52nd Annual Meeting of the Association for Computational Linguistics, pp. 655–665 (2014)
12. Chopra, S., Hadsell, R., LeCun, Y.: Learning a similarity metric discriminatively, with application to face verification. In: Proceedings of IEEE Conference on Computer Vision and Pattern Recognition, pp. 349–356 (2005)
13. Wan, S., Lan, Y., Guo, J., Xu, J., Pang, L., Cheng, X.: A deep architecture for semantic matching with multiple positional sentence representations. In: AAAI, pp. 2835–2841 (2016)
14. Mou, L., Rui, M., Li, G., Xu, Y., Zhang, L., Yan, R., Jin, Z.: Natural language inference by tree-based convolution and heuristic matching. In: Proceedings of the 54th Annual Meeting of the Association for Computational Linguistics, pp. 130–136 (2016)
15. Bowman, S.R., Angeli, G., Potts, C., Manning, C.D.: A large annotated corpus for learning natural language inference. In: Proceedings of the 2015 Conference on Empirical Methods in Natural Language Processing (EMNLP), pp. 632–642. Association for Computational Linguistics (2015)
16. Tieleman, T., Hinton, G.: Lecture 6.5-rmsprop. COURSERA Neural Netw. Mach. Learn. **4**, 26–31 (2012)
17. Srivastava, N., Hinton, G., Krizhevsky, A., Sutskever, I., Salakhutdinov, R.: Dropout: A simple way to prevent neural networks from overfitting. J. Mach. Learn. Res. **15**, 1929–1958 (2014)
18. Pennington, J., Socher, R., Manning, C.D.: Glove: global vectors for word representation. In: EMNLP, pp. 1532–1543 (2014)
19. Sutskever, I., Vinyals, O., Le, Q.V.: Sequence to sequence learning with neural networks. In: Advances in Neural Information Processing Systems, pp. 3104–3112 (2014)
20. Zaremba, W., Sutskever, I.: Tom: learning to execute. arXiv Prepr. arXiv:1410.4615 (2014)
21. Vendrov, I., Kiros, R., Fidler, S., Urtasun, R.: Order-embeddings of images and language. arXiv Prepr. arXiv:1511.06361 (2015)
22. Bowman, S.R., Gauthier, J., Rastogi, A., Gupta, R., Manning, C.D., Potts, C.: A fast unified model for parsing and sentence understanding. In: Proceedings of the 54th Annual Meeting of the Association for Computational Linguistics (2016)

Advances of Soft Computing: Algorithms and Its Applications - Rozaida Ghazali

Design of Smart Garden System Using Particle Filter for Monitoring and Controlling the Plant Cultivation

Yang-Weon Lee[(✉)]

Honam University, 417, Eodeung-daero, Gwangsan-gu, Gwangju 62399,
Republic of Korea
ywlee@honam.ac.kr

Abstract. This paper designs and implements a smart garden system that can be used inside the house. To do this, we used particle filter and environmental sensors, which are open hardware controllers, and designed to control and observe automatic water supply, lighting, and growth monitoring with three wireless systems (Bluetooth, Ethernet, Wi-Fi). This system has been developed to make it possible to use it in an indoor space such as an apartment, rather than a large-scale cultivation system such as a conventional plant factory which has already been widely used. The developed system collects environmental data by using soil sensor, illuminance sensor, humidity sensor and temperature sensor as well as control through smartphone app, analyzes the collected data, and controls water pump, LED lamp, air ventilation fan and so on. As a wireless remote control method, we implemented Bluetooth, Ethernet and Wi-Fi. Finally, it is designed for users to enable remote control and monitoring when the user is not in the house.

Keywords: Particle filter · Smart garden system · Wireless sensor network

1 Introduction

It is well known fact that environmental monitoring is remote and widely distributed. The traditional approach to analyzing soil parameters is doing an on-the-spot evaluation, which is always a very inconvenient procedure and require additional labor. In this paper, we develop the automatic control of droughts using estimation and prediction of particle filters.

As shown in Figs. 1 and 2, we designed a wireless drought monitoring system based on wireless sensor network. The proposed system can be applied as smart garden at indoor environmental.

© Springer International Publishing AG 2017
D.-S. Huang et al. (Eds.): ICIC 2017, Part III, LNAI 10363, pp. 461–466, 2017.
DOI: 10.1007/978-3-319-63315-2_40

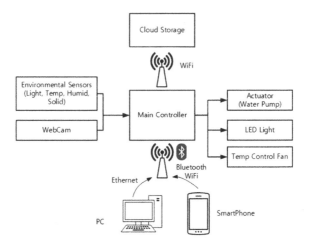

Fig. 1. System overall functional diagram

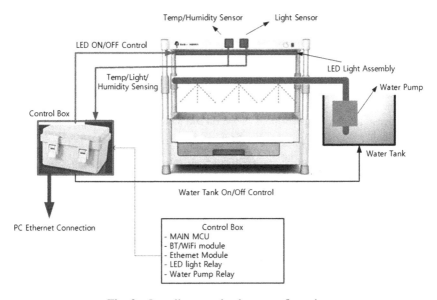

Fig. 2. Overall system hardware configuration

2 Modeling of Particle Filter for Smart Garden System

The general particle filter approach to track trajectory of flying objects, also known as the condensation algorithm [1] and Monte Carlo localization [2, 3], uses a large number of particles to explore the state space. Each particle represents a hypothesized target location in state space. Initially the particles are uniformly randomly distributed across the state space, and each subsequent frame the algorithm cycles through the steps illustrated in Fig. 1:

1. To initialize, draw N particles from an *a priori* distribution $p(x_0)$. Set all weights to be normalized and equal.

$$x_0^i \sim p(x_0), \quad i = 1 : N \tag{1}$$

$$m_0^i = \frac{1}{N}, \quad i = 1 : N \tag{2}$$

2. Propagate the particle set according to a process model of target dynamics.

$$x_k^i = f(x_{k-1}^i) + w_{k-1}^i, \quad i = 1 : N \tag{3}$$

Here $f(x_{k-1}^i)$ is the dynamics model and w_{k-1}^i is the noise.

3. Update probability density function (PDF) based on the received measurement z_k and the likelihood function of receiving that measurement given the current particle value.

$$m_k^i = m_{k-1}^i p(z_k | x_k^i), \quad i = 1 : N \tag{4}$$

4. Normalize the weights.

$$m_k^i = \frac{m_k^i}{\sum_{i=1}^{N} m_k^i} \tag{5}$$

5. Compute the effective sample size. The bounds on the effective sample size are given by $1 \leq N_{eff} \leq N$.

$$N_{eff} = \frac{1}{\sum_{i=1}^{N} (m_k^i)^2} \tag{6}$$

6. Compare N_{eff} to the resampling threshold. Resample if necessary.
$N_{eff} < N_{thr} \rightarrow$ Resample and GO TO STEP 2

$$N_{eff} \geq N_{thr} \rightarrow \text{GO TO STEP 2} \tag{7}$$

Here, N_{eff} is used to determine if the particles are unevenly distributed. Also if a few particles have large weighting and the other have small weighting values, resulting in N_{eff} will be close to 1. In contrast, if all the particles have equal weighting values, N_{eff} will be equal to N (Fig. 3).

At any time step, the value of the estimate is as follows:

$$\bar{x}_k = \sum_{i=1}^{N} (x_k^i - \bar{x}_k)^2 m_k^i \tag{8}$$

The covariance can be computes using

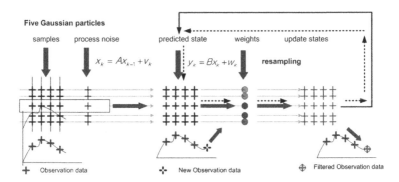

Fig. 3. Particle filter calculation process

$$P_k = \sum_{i=1}^{N} \left(x_k^i - \bar{x}_k \right)^2 m_k^i \qquad (9)$$

This results in particles congregating in regions of high probability and dispersing from other regions, thus the particle density indicates the most likely target states. See [3] for a comprehensive discussion of this method. The key strengths of the particle filter approach to localization and tracking are its scalability (computational requirement varies linearly with the number of particles), and its ability to deal with multiple hypotheses (and thus more readily recover from tracking errors). However, the particle filter was applied here for several additional reasons:

– It provides an efficient means of searching for a target in a multi-dimensional state space.
– Reduces the search problem to a verification problem, i.e. is a given hypothesis face-like according to the sensor information?
– Allows fusion of cues running at different frequencies.

3   Implementation of Smart Garden System

Five drought monitoring wireless units were deployed around a smart vase. These units continuously monitor the soil and send data t the coordinator, which is inside the control box.

Once all the system hardware is connected, the serial port is opened on the particle filter program. The filter initially record and displays the number of sensors in the network (Figs. 4 and 5).

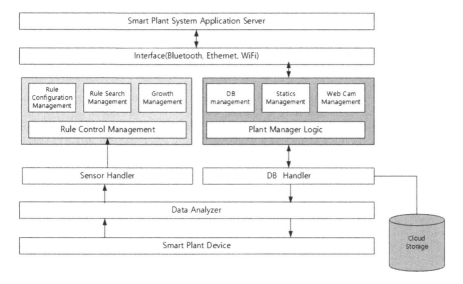

Fig. 4. System operation s/w structure

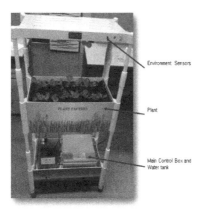

Fig. 5. Outlook of smart veranda system hardware

4 Conclusion and Future Work

Smart veranda system has been implemented using Arduino and sensor which are basic hardware of the Internet based on particle filter tracking algorithm. The results of this study are expected to be commercialized for start-up who wish to start business or other related companies. In this paper, we develops that a wireless sensor network based drought estimation and prediction using particle filters. It can monitor four different soil/atmospheric parameters.

Acknowledgement. This work is resulted from financially supported by KCA.

References

1. Fortmann, T.E., Bar-Shalom, Y., Scheffe, M.: Sonar tracking of multiple targets using joint probabilistic data association. IEEE J. Oceanic Eng. **8**(3), 173–184 (1983)
2. Gordon, N., Salmond, D., Smith, A.: Novel approach to nonlinear/non-Gaussian Bayesian state estimation. In: IEE Proceedings F, Radar and Signal Processing, pp. 107–113 (1993)
3. Isard, M., Blake, A.: Condensation-conditional density propagation for visual tracking. Int. J. Comput. Vis. 5–28 (1998)

Hybrid Global Crossover Bees Algorithm for Solving Boolean Function Classification Task

Habib Shah[1(✉)], Nasser Tairan[1], Wali Khan Mashwani[2],
Abdulrahman Ahmad Al-Sewari[3], Muhammad Asif Jan[2],
and Gran Badshah[4]

[1] College of Computer Science, Department of Computer Science,
King Khalid University, Abha, Kingdom of Saudi Arabia
habibshah.uthm@gmail.com, nasser@kku.edu.sa
[2] Department of Mathematics, Kohat University of Science and Technology,
Kohat, Khyber Pukhtunkhwa, Pakistan
mashwanigr8@gmail.com, majan@kust.edu.pk
[3] Faculty of Computer Systems and Software Engineering (FSKKP),
Universiti Malaysia Pahang, Malaysia, Malaysia
alsewari@gmail.com
[4] Department of Computer Science, Abdul Wali Khan University,
Timergara Campus, Mardan, Khyber Pukhtunkhwa, Pakistan
granl6178@gmail.com

Abstract. Using typical algorithms for training multilayer perceptron (MLP) creates some difficulties like slow convergence speed and local minima trapping in the solution space. Bio-inspired learning algorithms are famous for solving linear and nonlinear combinatorial problems. Artificial Bee Colony (ABC) algorithm is one among the famous bio-inspired algorithms. However, due to slow exploration process, it has been focused by researchers for further enhancement in optimization area. Therefore, this paper proposed a new hybrid swarm based learning algorithm called Global Crossover Artificial Bee Colony (GCABC) algorithm for training MLP for solving boolean classification problems. The simulation results of proposed GCABC algorithm compared with standard bio-inspired algorithms such as ABC, and Global Artificial Bee Colony (GABC) show that the proposed algorithm is achievable and efficient results in benchmark boolean function classification, with fast convergence speed.

Keywords: Hybrid Global Crossover Honey Bees algorithm · Artificial Bee Colony · Multilayer perceptron · Crossover operator

1 Introduction

Artificial Neural Network (ANN) is an in information-processing paradigm that is inspired by the way biological nervous systems, such as the brain, process in information [1]. It is composed of a huge number of highly interconnected processing neurons working in unity to solve specific problems. ANNs simulate like a human

© Springer International Publishing AG 2017
D.-S. Huang et al. (Eds.): ICIC 2017, Part III, LNAI 10363, pp. 467–478, 2017.
DOI: 10.1007/978-3-319-63315-2_41

brain for solving a given tasks. Feed Forward Neural Network (FFNN) consists of a large number of simple processing units called perceptron organized in multiple hidden layers, also called Multilayer Perceptron (MLP) is the famous model for solving different science, engineering and optimized mathematical problems like time series prediction, clustering, traveling salesman, clustering, character recognition and nonlinear classification task [2]. It is being successfully applied to different optimization and mathematical problems such as object recognition, medicine, electronic nose, security, and cloud classification, image processing, function optimization, signal processing and numeric function [3]. An ANN is configured for a particular task such as classification and prediction through different supervised, unsupervised learning algorithms. Learning in biological systems adds adjustments to the synaptic connections that exist between the neurons [6].

There are several learning techniques used in favorable performance for training MLP such as evolutionary algorithms (EA), genetic algorithm (GA), Particle Swarm Optimization (PSO), ant colony optimization, Cuckoo Search (CS), backpropagation (BP) and Bat algorithm and so on [4–6]. Bio-inspired learning algorithms are much efficient and famous due fast convergence, high efficiency and robustness. The main objective of the abovementioned training process is to find the set of optimal weight values that will cause the output from the MLP to match the actual target values as closely as possible. These learning techniques are used for initial weights, parameters, activation function, and selection of a proper network structure as well.

In early BP algorithm used to train MLP for updating the network weights for minimizing output error. It has a high success rate in solving some optimization problems, but it still has some drawbacks for solving some complex problems, especially when setting parameter values like initial values of connection weights, value for learning rate, and momentum, local minimum trapping and slow convergence [7, 8].

In order to overcome the drawbacks of standard BP and improve the performance of ANN models, many global optimization techniques developed such as ABC, Ant Bee Colony (HABC), Local population-based, cooperative based [6, 9], Ant Bee Colony algorithms, Bat algorithm, CS algorithm can provide the best possible solutions to optimization problems [8–12]. These bio-inspired training algorithms are more efficient than typical learning techniques. ABC is among famous bio-inspired algorithms which used to solve mathematics, science and engineering problems. However due to the poor process of exploration and exploitation, here the hybrid version called Global Crossover Artificial Bee Colony (GCABC) algorithm has been proposed for boolean function classification. The details of typical and proposed bio-inspired algorithm are given in Sects. 2 and 3 respectively. The simulation results and conclusion are added in Sects. 4 and 5.

2 Bio-Inspired-Artificial Bee Colony Algorithm

The techniques which used the success nature characteristics of social insects like bees, ants, birds, fishes for solving complex problems are called bio-inspired or nature inspired algorithms. In the most generic term, the main source of inspiration is nature. Among bio-inspired algorithms, special classes of algorithms have been developed by drawing inspiration from social insect's behaviors. Therefore, almost all new algorithms can be referred to as nature-inspired or swarm intelligence based [10]. Artificial Bee Colony (ABC) developed by Karaboga is one of the most famous, robust and attractive algorithms among bio-inspired algorithms. Initially, ABC algorithm used for training feed forward neural network for solving XOR classification problems. After getting hive efficiency in XOR classification, it has been applied to many optimization problems in the field of science and engineering era [11]. It's using the searching, dancing, walking, sharing, and selecting habits of honey bees.

Three types of bees are employed, onlooker and scout which solve the given problems by their nature and combined strategy [12]. The first two bees are used to process exploitation for a given problem towards the best solution. The onlooker bee uses for exploration process. Bees share information with each other through dancing strategy. Flowchart of ABC algorithm is shown in Fig. 1.

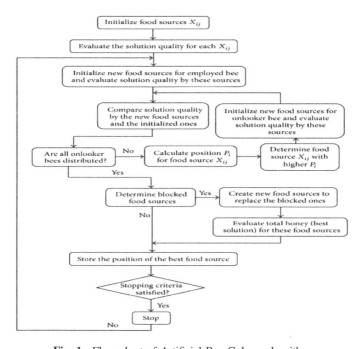

Fig. 1. Flow chart of Artificial Bee Colony algorithm

The employed bees share the nectar information of the best food sources with the onlooker bees. Then each employed bee returns to the food source and then selects new food source based on the neighbor information. The onlooker bee (exploration process) uses the information retrieved from the employed bees at the dance area to select a good food source. Mathematically, the exploitation processes are performed by employed and Onlooker's bees by Eq. 1, the selection condition performed by scout bees represented in Eq. (2) as:

$$X_{ij} = x_{ij} + \theta_{ij}\left(x_{ij} - x_{kj}\right) \tag{1}$$

$$x_{ij}^{rand} = x_{ij}^{min} + rand(0, 1)\left(x_{ij}^{max} - x_{ij}^{min}\right) \tag{2}$$

The employed and onlookers used same Eq. (1) for exploitation process which cannot guaranty for finding optimal solution. The researchers improved typical ABC algorithm by different strategies such as global and guided bees.

2.1   Improved Bio Inspired-Global Artificial Bee Colony Search Algorithm

The ABC algorithm has a big achievement history, especially in the field of computer science and engineering. The performance ratio obtained by ABC algorithm was much better than typical techniques [13]. However, the researchers have been motivated to improve the performance of standard ABC algorithm through the true characteristics of bio-insects. Therefore, Global Artificial Bee Colony Search (GABCS) algorithm is a population-based metaheuristic algorithm that simulates the foraging behavior of artificial global honey bee swarms [14]. The implementation of GABCS technique is very easy and effective for combinatorial problems. There are three groups of bees in the GABCS are global employed bees, global onlooker bees and global scout bees. Usually, in bee swarm, the experienced foragers can use previous knowledge of position and nectar quantity of food source to regulate their group directions in the search space. The GABCS approach has merged their best finding in standard ABC algorithm by the following two procedures.

Phase 1: Global employed and onlooker bees section

$$v_{ij} = x_{ij} + \varphi_{ij}\left(x_{ij} - x_{kj}\right) + c_1 rand(0, 1)\left(x_j^{best} - x_{ij}\right) + c_2 rand(0, 1)\left(y_j^{best} - x_{ij}\right) \tag{3}$$

Phase 2: Global scout section as

$$x_{ij}^{rand} = x_{ij}^{min} + rand(0, 1)\left(x_{ij}^{max} - x_{ij}^{min}\right) \tag{4}$$

If rand (0, 1) $\leq$ 0.5, then

$$x_{ij}^{mutation} = x_{ij} + rand(0, 1)\left(1 - \frac{iter}{iter_{max}}\right)^b + \left(x_j^{best} - x_{ij}\right) \tag{5}$$

else

$$x_{ij}^{mutation} = x_{ij} + rand(0, 1)\left(1 - \frac{iter}{iter_{max}}\right)^b + \left(y_j^{best} - x_{ij}\right) \tag{6}$$

Typical ABC algorithm exploration and exploitation steps (Eqs. 1 and 2,) have been modified by the GABCS algorithm through gbest strategy with Eqs. 3, 4 and 5 respectively.

2.2   The Genetic Algorithm-Crossover Operator

The Genetic Algorithm (GA) was invented by John Holland in the 1960s, a professor of Michigan University to solve specific problems. The performance of GA depends on various operators, such as crossover operator, mutation, and selection. GA searches a result equal to or close to the answer through the new generation of a given problem. Basic strategy used in GA to create the best solutions/offspring is to crossover the parent genes. The crossover is the well-known operator of GA used to producing a new generation from parents, which offer a route whereby different good properties of parents can be recombined to provide a child that is better than either parent [15]. In other words, the best part of the one solution of the population placed into the mating pool. The crossover operator then used to change their parts mutually of the two parents. The crossover operator combines parts of good solution to create a new optimal solution [16].

The genetic algorithm crossover operator have classified in one point crossover, K-point crossover, shuffle crossover, discrete crossover and flat crossover, single-point crossover, two-point crossover [17], cut and splice, Uniform crossover and half uniform crossover, crossover biases, order crossover, average crossover, cycle crossover and so on [18]. The single point and two point crossover are the most popular operators widely used for solving problems in various applications. These crossover operators have been used successfully in solving many mathematical, science and engineering problems in different techniques such as particle swarm optimization [19], hybrid artificial bee colony [20] and so on. From the literature review, it can be seen that crossover operators have improved the performance of the given algorithm. Some examples of crossover operators are:

Parent	1	: 5 4 7 6 5 2 6 1 3	Parent 1	: 1 1 1 0 1 0 0 1 0
Parent	2	: 5 3 3 2 3 9 7 6 5	Parent 2	: 1 0 0 0 1 0 1 1 0
Average Offspring		: 5 3 5 4 4 5 6 3 4	**Discrete** Offspring :	1 1 0 0 1 0 1 1 0

3   Hybrid Global Crossover Artificial Bee Colony Algorithm

The crossover operator has been used successfully in typical computational algorithms especially in evolutionary algorithms, GA and of course in latest bio-inspired algorithms. By using the crossover operator the performance has been increased rapidly for complex optimization problems. Besides that, the efficiency of the standard ABC algorithm can be improved through the exploration and exploitation balance amount. The employed bees and onlookers bees are responsible for performing exploring and exploiting process efficiently. However they used same random way of searching and selecting (Eq. 3) which can't provide the best solution, so sometimes they took long convergence time with low accuracy as well. Furthermore, for increasing exploitation, Guo Peng in 2011 suggested GABCS algorithms based on global best approach with time, and the bee's aim is to discover the places of global food sources [21]. Inappropriately, GABCS algorithm sometimes cannot provide the balance quantity of exploitation due to the same approach between employed and onlookers' bees as in

Eq. (3). Therefore, to obtain the advantages of crossover operator and gbest searching strategy in global optimization algorithm, the new hybrid algorithm is proposed called Global Crossover Artificial Bee Colony (GCABC) algorithm.

The main aim of GCABC algorithm is to speed up the exploration and exploitation process of typical ABC algorithm for solving complex problems. The crossover operator is introduced in the global onlooker bee phase to improve solution based on their fitness and balance exploitation process. The flowchart of GCABC algorithm is shown in Fig. 2. The detailed pseudo-code of GCABC algorithm is shown as follows.

1: Initialise the population of solutions x_i where i=1...N
2: Evaluate the population.
3: Cycle=1
4: Repeat from step 2 to step 14.
5: Produce new solutions (food source positions) v_{ij} in the neighbourhood of x_{ij} for the global employed bees using the eq (3).
6: Apply the Greedy Selection process between processes.
7: Calculate the probability values p_i for the solutions x_i by means of their fitness values by using formula

$$p_i = \frac{fit_i}{\sum_{k=1}^{n} fit_k} \tag{7}$$

The calculation of fitness values of solution is given as

$$fit_i = \begin{cases} \dfrac{1}{1 + fit_i}, & \text{for} \quad f_i \geq 0 \cdot \\ 1 + \text{abs}\left(fit_i\right), & \text{for} \quad f_i < 0 \end{cases} \tag{8}$$

Normalise P_i values into [0,1].
8: Apply crossover operation
  {New off-springs generated from parents as a result of crossover. Replace the worst parent with the best new Offspring if it is better}
9: Produce the new solutions (new positions) x_i for the global onlookers from the solutions x_i, selected depending on P_i, and evaluate them by eq (3)
10: Apply the Greedy Selection process for the onlookers between x_i and v_i
11: Determine the abandoned solution (source), if exists, replace it with a new randomly produced solution x_i for the scout using the eq (5, 6 and 7)
12: Memorise the best food source position (solution) achieved so far
13: cycle=cycle+1
14: until cycle= Maximum Cycle Number (MCN)

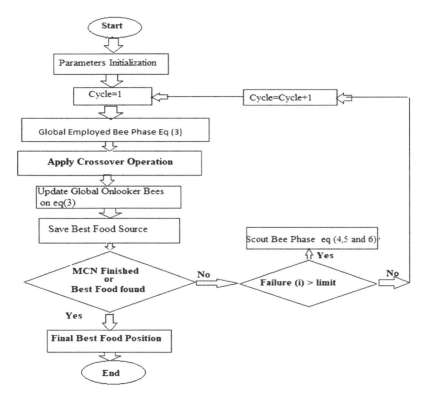

Fig. 2. Flow chart of proposed Global Crossover Artificial Bee Colony algorithm

4   Experimental Setting and Results

The proposed GCABC algorithm trained and tested on three dataset for Boolean dataset for classification task. There are many types of Boolean operations; however, commonly used are XOR, 3-Bit Parity and Encoder/Decoder operators.

Definition. The exclusive-or function, or parity function, is the Boolean function $\oplus$: $B2{\to}B$ defined as $\oplus (x_1, x_2) = x_1\overline{x_2} \vee \overline{x_1}x_2$ for all x1, x2 $\in$ **B**, which usually can be write x1 $\oplus$ x2 instead of $\oplus$ (x1, x2). XOR is a basic dataset that is widely used to train and test ANN, which has historically been considered as good test of a network model and learning algorithm as well [22]. The seemingly simple XOR-function cannot be computed by a one-layer network of sigmoid neurons because of its non-linearity and thereby caused a great deal of disillusions in ANN research in the 70s [23]. The XOR classification is often used as a first benchmark for ANN supervised learning algorithm due to the small size of its data set and nonlinear function. Secondly, 3-Bit Parity Problem has been addressed because they are nonlinearly separable like XOR, and hence cannot be solved by single layer perceptron [24]. In other words, if the number of binary inputs is odd, the output is 1, otherwise it is 0. The 3-bit parity problem can be written as Parity $\leq$ ((p xor q) xor r).

Thirdly, 4-Bit Encoder/Decoder Problem is the third classification task is 4-Bit Encoder/Decoder which is well-known in computer science. The network is presented with 4 distinct input patterns, each having only one bit turned on. The 4-Bit Encoder/Decoder is quite close to the real-world pattern classification task, where small changes in the input pattern cause small changes in the output pattern [11]. Here, Feedforward neural network (FFNN) trained and tested through proposed GCABC and typical ABC and GABC algorithms for boolean classification with the following configuration (Table 1).

Table 1. Setting of FFNN topologies for boolean function classification task

Boolean function	No of inputs	Hidden nodes	CS	MCN
XOR	2	2–7	20	1000
3 Bit Parity	3	3–7	20	1000
Encoder/Decoder	4	3–9	20	1000

Mean of Mean Square Error (MSE), success rate, S.D of MSE and accuracy measured by calculating the performance proposed and typical learning algorithms for Boolean function classification, using Matlab 2010a. During the experimentation, 5 trials performed for training. The sigmoid function used as activation function for network production. The values of C1 and C2 were selected 1.5 and -1.5, respectively. The average results have been mentioned in the following tables and figures.

From Table 2, the GCABC-FFNN algorithm outperformed in training phase than GABC-FFNN and ABC-FFNN algorithms for XOR, 3 Bit Parity and Encoder/Decoder classification problems. Using the proposed GCABC algorithm the mean MSE for XOR, 3 Bit Parity and Encoder/ Decoder with seven and nine hidden nodes reached to 1.01E−16, 1.28E−10 and 2.26E−13 respectively, which is much better than typical algorithms like ABC, and GABC.

Table 2. Average MSE training for boolean function classification task

Boolean function	FFNN structure	ABC-FFNN	GABC-FFNN	GCABC-FFNN
XOR	2-2-1	3.09E−04	6.73E−06	**9.28E−13**
	2-3-1	4.00E−04	9.99E−05	**9.21E−12**
	2-5-1	5.01E−04	7.89E−05	**7.12E−08**
	2-7-1	1.70E−04	9.87E−07	**1.01E−16**
3 Bit Parity	3-3-1	1.59E−04	9.82E−05	**9.28E−12**
	3-4-1	2.76E−04	9.87E−05	**1.21E−13**
	3-6-1	3.59E−04	6.23E−05	**2.71E−12**
	3-7-1	1.70E−05	1.98E−05	**1.28E−10**
Encoder/Decoder	4-3-4	3.00E−04	2.98E−04	**1.21E−10**
	4-6-4	5.92E−04	1.99E−05	**3.49E−10**
	4-8-4	1.52E−04	2.98E−05	**1.01E−13**
	4-9-4	6.98E−04	6.32E−06	**2.26E−13**

In term of MSE out of samples for boolean function classification task, the proposed GCABC-FFNN reached to the minimum error (8.76E−16, 1.09E−11, and 9.12E −13) in testing phase through optimal weight values. Furthermore, the GABC-FFNN algorithm also success for optimal classification of Boolean function classification task as shown in Table 3. The SD-MSE of typical ABC, GABC and proposed GCABC are given in Table 4, where the ABC-FFNN got 7.61E−04, 2.39E−06 and 3.63E−04, GABC-FFNN 9.89E−06, 1.98E−04 and 6.92E−05, and proposed GCABC-FFNN 1.21E−09, 1.98E−08 and 9.11E−11 respectively.

Table 3. Average MSE out of samples for boolean function classficiation task

Boolean function	ABC-FFNN	GABC-FFNN	GCABC-FFNN
XOR	8.63E−05	6.28E−07	9.12E−10
	8.63E−05	9.96E−06	9.22E−08
	5.96E−05	7.01E−06	7.11E−09
	1.75E−05	9.11E−08	8.76E−16
3 Bit Parity	1.85E−05	9.81E−06	9.10E−13
	2.11E−05	9.04E−06	1.92E−14
	3.73E−05	6.13E−06	2.90E−12
	1.81E−06	1.87E−06	1.09E−11
Encoder/Decoder	9.64E−05	2.09E−05	6.42E−11
	6.00E−04	1.85E−05	3.92E−11
	1.59E−04	9.71E−06	8.09E−13
	6.02E−04	6.97E−06	9.12E−13

Figure 3 obtained by learning algorithms during the training phase for XOR classification task. The errors for the training data sets are the smallest using GCABC then GABC and ABC for XOR classification. The GCABC converged fastly than others. For 3 Bit Parity problem as given in Fig. 4, again the GCABC-FFNN

Table 4. Average S.D-MSE out of samples for boolean function classficiation task

Boolean Function	ABC-FFNN	GABC-FFNN	GCABC-FFNN
XOR	9.86E−04	6.12E−04	**9.55E−07**
	9.88E−04	9.19E−05	**9.12E−06**
	6.72E−04	7.10E−05	**7.78E−07**
	7.61E−04	9.89E−06	**1.21E−09**
3 Bit Parity	9.89E−04	9.00E−04	**1.60E−09**
	8.95E−04	9.04E−04	**1.09E−10**
	3.73E−05	6.19E−04	**2.10E−08**
	2.39E−06	1.98E−04	**1.98E−08**
Encoder/Decoder	9.19E−05	2.74E−04	**6.09E−09**
	6.11E−04	1.31E−04	**3.10E−08**
	1.93E−04	9.71E−05	**8.10E−11**
	3.63E−04	6.92E−05	**9.11E−11**

converged earlier than above mentioned typical algorithms. The starting MSE of proposed GCABC-FFNN has a smaller value in comparison to others learning techniques. Using ABC for XOR, 3 Bit Parity and 4 bit, the MSE is not stable and decreased slowly than GABC-FFNN and GCABC-FFNN. Overall, the proposed GCABC algorithm converged fastly than ABC algorithm for all type of boolean function classification problems (Fig. 5).

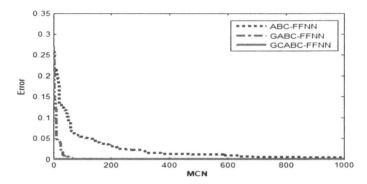

Fig. 3. Error curve of ABC, GABC and GCABC algorithms for XOR classification

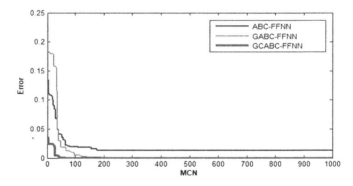

Fig. 4. Error curve of ABC, GABC and GCABC algorithms for 3 Bit Parity problem

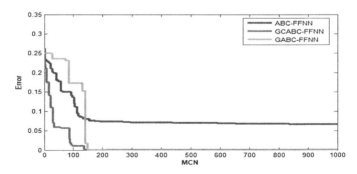

Fig. 5. Error curve of ABC, GABC and GCABC algorithms for Encoder/Decoder

5 Conclusion

This research work explores two typical bio-inspired learning algorithms and one proposed are ABC, GABC and GCABC algorithms for training FFNN to get the best solutions for the Boolean classification task. The GCABC used two strategies global best from GABC and crossover operator from GA algorithm. The GCABC achieved the enough amounts of exploration and exploitation with balance quality and quantity for solving boolean function classification task. The GCABC algorithm for training FFNN on Boolean function classification task outperformed than typical ABC and improved version GABC algorithms in term of error and convergence speed. The proposed approach can be further use for time series forecasting, clustering and numerical function optimization.

References

1. Rosenblatt, F.: The perception: a probabilistic model for information storage and organization in the brain. In: James, A.A., Edward, R. (eds.) Neurocomputing: Foundations of Research, pp. 89–114. MIT Press, Cambridge (1988)
2. Du, K.L.: Clustering: a neural network approach. Neural Netw. **23**(1), 89–107 (2010)
3. Karaboga, D.: An idea based on honey bee swarm for numerical optimization, pp. 1–10. Technical report TR06, Erciyes University, Engineering Faculty, Computer Engineering DepartmenTR06 (2005)
4. Suganthan, P.N.: Differential evolution algorithm: recent advances. In: Dediu, A.-H., Martín-Vide, C., Truthe, B. (eds.) TPNC 2012. LNCS, vol. 7505, pp. 30–46. Springer, Heidelberg (2012). doi:10.1007/978-3-642-33860-1_4
5. Kennedy, J., Eberhart, R.: Particle swarm optimization. In: Proceeding of IEEE International Conference on Neural Network 4, Australia, pp. 1942–1948 (1995)
6. Tairan, N.: Cooperative guided local search. Ph.D., University of Essex, UK, 573069 (2012)
7. Otair, M.A., Salameh, W.A.: Speeding up back-propagation neural networks. In: Proceeding of the 2005 Informing Science and IT Education Joint Conference, Flagstaff, Arizona, USA, pp. 167–173 (2005)
8. Nawi, N.M., Ransing, M.R., Ransing, R.S.: An improved learning algorithm based on the Broyden-Fletcher-Goldfarb-Shanno (BFGS) method for back propagation neural networks. In: Sixth International Conference on Intelligent Systems Design and Applications, ISDA 2006, vol. 1, pp. 152–157 (2006)
9. Alarifi, A.S.N., Alarifi, N.S.N., Al-Humidan, S.: Earthquakes magnitude predication using artificial neural network in northern Red Sea area. J. King Saud Univ. - Sci. **24**(4), 301–313 (2012)
10. Cai, X., Gao, X.-Z., Xue, Y.: Improved bat algorithm with optimal forage strategy and random disturbance strategy. Int. J. Bio-Inspired Comput. **8**(4), 205–214 (2016)
11. Karaboga, D., Akay, B., Ozturk, C.: Artificial bee colony (ABC) optimization algorithm for training feed-forward neural networks. In: Torra, V., Narukawa, Y., Yoshida, Y. (eds.) MDAI 2007. LNCS, vol. 4617, pp. 318–329. Springer, Heidelberg (2007). doi:10.1007/978-3-540-73729-2_30

12. Bullinaria, J.A., AlYahya, K.: Artificial bee colony training of neural networks. In: Terrazas, G., Otero, F.E.B., Masegosa, A.D. (eds.) Nature Inspired Cooperative Strategies for Optimization (NICSO 2013): Learning, Optimization and Interdisciplinary Applications, pp. 191–201. Springer, Cham (2014)

13. Olariu, S., Zomaya, A.Y.: Handbook of Bioinspired Algorithms and Applications. Chapman and Hall/CRC, Boca Raton (2005)

14. Guo, P., Cheng, W., Liang, J.: Global artificial bee colony search algorithm for numerical function optimization. In: 2011 Seventh International Conference on Natural Computation (ICNC), vol. 3, pp. 1280–1283 (2011)

15. Chambers, L.D.: Practical Handbook of Genetic Algorithms: New Frontiers, p. 448. CRC Press Inc., Boca Raton (1995)

16. Karaboga, D., Gorkemli, B., Ozturk, C., Karaboga, N.: A comprehensive survey: artificial bee colony (ABC) algorithm and applications. Artif. Intell. Rev. **42**(1), 21–57 (2014)

17. Haupt, R.L., Haupt, S.E.: Introduction to optimization. In: Practical Genetic Algorithms, pp. 1–25. Wiley (2004)

18. Jalali Varnamkhasti, M., Lee, L.S., Abu Bakar, M.R., Leong, W.J.: A genetic algorithm with fuzzy crossover operator and probability. Adv. Oper. Res. **2012**, 16 (2012). Article no. 956498

19. Zhang, T., Hu, T., Guo, X., Chen, Z., Zheng, Y.: Solving high dimensional bilevel multiobjective programming problem using a hybrid particle swarm optimization algorithm with crossover operator. Knowl.-Based Syst. **53**, 13–19 (2013)

20. Ma, L., Hu, K., Zhu, Y., Chen, H.: A hybrid artificial bee colony optimizer by combining with life-cycle, Powell's search and crossover. Appl. Math. Comput. **252**, 133–154 (2015)

21. Zhu, G., Kwong, S.: Gbest-guided artificial bee colony algorithm for numerical function optimization. Appl. Math. Comput. **217**(7), 3166–3173 (2010)

22. Toh, K.-A., Lu, J., Yau, W.-Y.: Global feedforward neural network learning for classification and regression. In: Figueiredo, M., Zerubia, J., Jain, Anil K. (eds.) EMMCVPR 2001. LNCS, vol. 2134, pp. 407–422. Springer, Heidelberg (2001). doi:10.1007/3-540-44745-8_27

23. Aijun, L., Yun-Hui, L., Si-Wei, L.: On the solution of the XOR problem using the decision tree-based neural network. In: 2003 International Conference on Machine Learning and Cybernetics, vol. 2, pp. 1048–1052 (2003)

24. Minsky, M., Papert, S.A.: Perceptrons: An Introduction to Computational Geometry. MIT Press, Cambridge (1969)

Advances in Swarm Intelligence Algorithm

A Hybrid Approach Based on CS and GA for Cluster Analysis

Xiaofeng Li and Hongqing Zheng[(✉)]

College of Information Engineering, Guangxi University of Foreign Languages,
Nanning 530222, China
zhq7972@sina.com

Abstract. After analyzing the disadvantages of the classical K-means clustering problem, an improved cuckoo search algorithm (ICS) is applied to cluster
analysis, and this paper proposes a novel hybrid clustering algorithm based on
Genetic algorithm (GA). The hybrid algorithm includes two modules. At the
initial stage, the cuckoo search algorithm (CS) is executed, the clusters' result
are used to the crossover and mutation of genetic algorithm for local search.
Comparision of the performance of the proposed approach with the cluster
method based on CS and GA algorithm are experimented. The experimental
result show the proposed meth has not only higher accuracy bust also higher
level of stability. And the faster convergence speed can also be validated by
statistical results.

Keywords: Clustering problem · Cuckoo search algorithm · Genetic algorithm

1 Introduce

The K-means algorithm is one of the most commonly used clustering techniques,
which uses the data reassignment method to repeatedly optimize clustering [1]. The
main goal of clustering is to generate compact groups of objects or data that share
similar patterns within the same cluster, and isolate these groups from those which
contain elements with different characteristics. Although the K-means algorithm has
features such as simplicity and high convergence speed, it is totally dependent on the
initial centroids which are randomly selected in the first phase of the algorithm. Due to
this random selection, the algorithm may converge to locally optimal solutions [2]. For
overcoming this problem, many scholars began to solve the clustering problem using
meta-heuristic algorithms. Niknam et al. have proposed an efficient hybrid evolutionary
algorithm based on combining ACO and SA for clustering problem [3]. Then Shelokar
and Kao solved the clustering problem using the ACO algorithm [4, 5]. Kennedy and
Eberhart have proposed a particle swarm optimization (PSO) algorithm which simulates the movement of organisms in bird flock or fish school in 1995 [6]. The algorithm
also has been adopted to solve this problem by Omran and Merwe [7, 8]. Kao et al.
have presented a hybrid approach according to combination of the k-means algorithm,
Nelder–Mead simplex search and PSO for clustering analysis [9]. Kevin et al. have
used an evolutionary-based rough clustering algorithm for the clustering problem [10].

© Springer International Publishing AG 2017
D.-S. Huang et al. (Eds.): ICIC 2017, Part III, LNAI 10363, pp. 481–490, 2017.
DOI: 10.1007/978-3-319-63315-2_42

In this paper, we propose a novel clustering approach called ICS to provide more accurate clustering results compared to the K-means algorithm and CS versions. The proposed clustering method combines the crossover and mutation of genetic algorithm for local search. The experiment proves that the effect of the ICS is well.

This article is organized as follows: Sect. 2 describes the mathematical model of clustering. Section 3 discusses an improved search algorithm cuckoo combination genetic operators. The validity of the algorithm is verified in Sect. 4. Finally, we end this paper with some conclusions and future work in Sect. 5.

2   The Mathematical Model of Clustering

2.1   Mathematical Definition of Data Clustering

In order to explain the definition, we supposed that there exists a data set $D = \{d_1, d_2, \cdots d_n\}$. Each individual $d_i (i = 1, 2, \cdots n)$ has many features. If the dimension is m, each individual can be shown as $d_i = (l1, l2, \cdots, lm)$. Data clustering is a process which can classify the given data set D into a certain numbers of clusters $G_1, G_2, \cdots G_k$ (assume K clusters) based on the similarity of individuals. And $G_1, G_2, \cdots G_k$ should satisfy the following formulas [11]:

$$G_i \neq \emptyset. \quad i = 1, 2, \cdots k.$$

$$G_i \cap G_j = \emptyset. \quad i, j = 1, 2, \cdots k. \ i \neq j.$$

$$\bigcup_{i=1}^{k} G_i = \{d_1, d_2, \cdots d_n\}$$

2.2   The Principle of Data Clustering

In the clustering process, if the given data set D should be divided into K clusters $(G_1, G_2, \cdots G_k)$, and each cluster must have one center $C_j (j = 1, 2, \cdots, k)$.

The main idea of clustering is to define K centers, one for each cluster. These centers should be placed in a crafty way, because different location will causes different result. Therefore, the better choice is to place them as far away from each other as possible. In this paper, we will use Euclidean metric as a distance metric. The expression is given as follows:

$$d(d_i, c_j) = \sqrt{\sum_{k=1}^{m} (d_{ik} - c_{jk})^2} \tag{1}$$

where $d_i (i = 1, 2, \cdots, n)$ is an individual in the given dataset D, m is the number of individual features, $c_j (j = 1, 2, \cdots, k)$ is the center of jth subset. Because individual has m features, c_j can be presented by $(c_{j1}, c_{j2}, \cdots, c_{jm})$. In order to confirm which subset di belongs to, the distances between di and c_j should be calculated via (1).

2.3 Evaluation of Data Clustering

In order to optimize the coordinates of centers of K subsets, it is easily to find that the dimension of solution should be $k * m$. The individual in the population can be described as $s = (c_1, c_2, \cdots, c_k)$. A great classification should minimize the sum of distances value. The proposed algorithm aims at minimizing the objective function, which can be expressed as following:

$$f(D, C) = \sum_{i=1}^{n} \min\{\|d_i - c_k\| \| k = 1, 2, \cdots, k\} \tag{2}$$

where $D = (d_1, d_2, \cdots, d_n)$ is the given data set, $C = (c_1, c_2, \cdots, c_k)$ is the centers of subsets $(G_1, G_2, \cdots, G_k)$.

3 Clustering Combined with GA Algorithm

3.1 Improved CS Algorithm (ICS)

Cuckoo Search algorithm is by the scholars of the University of Cambridge, UK, YANG Xin - she and DEB Suash, to simulate the behavior that Cuckoos search nest to lay eggs. Then, they found a new Search algorithm, namely the Cuckoo Search algorithm [12]. Because of this algorithm is simple, efficient and easy to implement, it is widely used in various fields. The algorithm simulates the natural process that Cuckoo search nest to lay eggs, where the parameters of handling problem are coded to form the bird's nest. Population consists of lots of bird's nests. The individuals of population choose bird's nest by Levy flight of Cuckoo and determine to abandon the bird's nest to update population according to a certain probability. To iterate until it gets the final optimization results.

The updating formula of path and position that Cuckoo searches nest is as follows:

$$x_i^{(t+1)} = x_i^{(t)} + \partial \oplus L(\lambda), \ i = 1, 2, \cdots, n \tag{3}$$

In the formula, $x_i^{(t)}$ represents the position of the i-th one in the t-th generation of nests, $\oplus$ represents dot product, ∂ represents Step Length, $L(\lambda)$ represents path that Levy randomly searches. $L \sim u = t^{-\tau}, 1 < \tau \leq 3$.

∂ in basic cuckoo search algorithm is a constant, it is not conducive to rapid convergence and accuracies of the algorithm. So do the following adjustments:

$$\partial = \partial_{max} - \frac{\partial_{max} - \partial_{min}}{itermax} \times t \tag{4}$$

In the formula, ∂_{max} represents the maximum step size, ∂_{min} represents the minimum step size, $itermax$ represents the maximum number of iterations, t represents the current number of iterations.

3.2   CS Algorithm Combined with Genetic Algorithm Operators

Genetic algorithm is one of the earliest intelligent algorithms, which simulates the survival of the fittest in nature. With the aid of heredity and variation operation, it can solve the problem and get an approximate optimal solution. Scholars study it successfully. Its application fields are also quite extensive. Intelligent algorithms are easy to fall into local optimum prevalent drawbacks commonly. In order to make the cuckoo search algorithm out of local optimal solution, to speed up the search efficiency, it combines the crossover and mutation operator of genetic algorithm.

3.3   Algorithm Steps

Step 1: Initialize parameters and randomly initialize solutions.

Step 2: Using the formula (2) to undertake an evaluation of all the nests to obtain *fmin* and *best*.

Step 3: With the formulas (3) and (4), generating a new nest and obtaining *fmin*1 and *best*1 by evaluated fitness value. If *fmin*1 < *fmin*, replacing the original optimal value and optimal solution with *fmin*1 and *fbest*1, otherwise being unchanged.

Step 4: Using the ideas of genetic algorithms to transform the nest.

Step 5: Obtain *fmin*2 and *best*2 by evaluating fitness value again. If *fmin*2 < *fmin*, replacing the original optimal value and the optimal solution with *fbest*2 and *fmin*2, otherwise unchanged.

Step 6: Determine whether the cycle ends, if it is, outputting waypoint information, otherwise, skipping to step 3.

4   The Simulation Examples and Analysis

In order to verify the performance of the proposed algorithm, using standard cuckoo search algorithm and improved cuckoo search algorithm to perform six experiments respectively, using MATLAB R2010a to write Codes, and all instances run in Celeron processor (R) dual-core CPU T3100, 1.90 GHZ, 2 G memory for PC. Parameter is set to: population size is 50, the total number of iterations is 200, the rate of abandoning colonies $pa = 0.25$, the ratio of crossing $pc = 0.45$, $\partial_{max} = 0.1$, $\partial_{min} = 0.01$. Each experiment independently operates 20 times.

4.1   Numerical Compared

4.1.1   Artificial Data Set One

Artificial data set one from literature [13]. The comparison of algorithms for artificial data set one is listed in Table 1. And the convergence curves of algorithms are shown in Fig. 1. In Fig. 2, it is the result of anova test for artificial data set one. From Table 1, the all value of ICS are much better than those of other algorithms such as CS and k-Means. Two convergence curves can be seen in Fig. 1, which is the comparison for artificial data set one. We can discover that the curve of ICS is much smoother than CS. And its convergence speed is faster. Figure 2 is the anova test for data set art1. We can find that the stability of ICS is well.

4.1.2   Artificial Data Set Two

Artificial data set two from literature [13]. The comparison of algorithms for artificial data set one is listed in Table 2. And the convergence curves of algorithms are shown

Table 1. Results obtained by the algorithms for 20 different runs on artificial dataset one.

Algorithm	Best	Worst	Mean	Std.
ICS	1718.3149	1719.2733	1718.7016	0.3096
CS	1793.3761	2136.7171	1917.1253	87.3800
k-means	1747.3859	2507.9091	1991.9351	342.2974

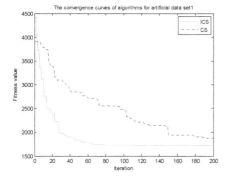

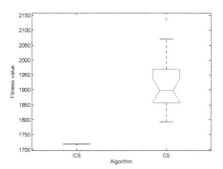

Fig. 1. The convergence curves of algorithm for art1

Fig. 2. The anova test of algorithms for art1

in Fig. 3. In Fig. 4, it is the result of anova test for artificial data set two. From Table 2, we can discover that the best value, worst value, mean value and standard deviation of ICS are much better than those of other algorithms. Two convergence curves can be

Table 2. Results obtained by the algorithms for 20 different runs on artificial dataset two.

Algorithm	Best	Worst	Mean	Std.
ICS	*513.9034*	*513.9658*	*513.9066*	*0.0139*
CS	514.0863	516.4402	514.8671	0.6379
k-means	525.5957	907.1413	694.4421	191.4831

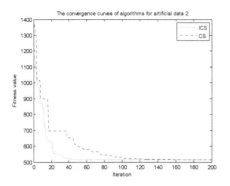

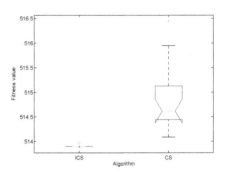

Fig. 3. The convergence curves of algorithm for art2

Fig. 4. The anova test of algorithms for art2

seen in Fig. 3, the curve of ICS is smooth than others can be discovered easily. Figure 4 show that the ICS has a high degree of stability.

4.1.3   Data Set Iris

The data set was used by Fisher in his initiation of the linear-discriminant-function technique [13]. The comparison of algorithms for artificial data set one is listed in Table 3. And the convergence curves of algorithms are shown in Fig. 5. In Fig. 6, it is the result of anona test for Iris. The below table and figures all show that the performance of ICS.

4.1.4   Data Set Seeds

Table 3. Results obtained by the algorithms for 20 different runs on Iris.

Algorithm	Best	Worst	Mean	Std.
ICS	*96.6575*	*96.8548*	*96.6919*	*0.0450*
CS	97.2717	99.3541	98.5180	0.4787
k-means	97.1901	121.3554	100.8866	8.7805

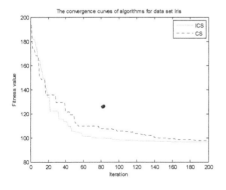

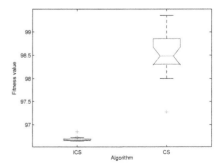

Fig. 5. The convergence curves of algorithm for Iris

Fig. 6. The anova test of algorithms for Iris

Seeds data from the literature [13], the comparison of algorithms for artificial data set one is listed in Table 4. And the convergence curves of algorithms are shown in Fig. 7. In Fig. 8, it is the result of anona test for Iris. The below table and figures all show that the performance of ICS.

Table 4. Results obtained by the algorithms for 20 different runs on seeds

Algorithm	Best	Worst	Mean	Std.
ICS	*311.8807*	*313.5267*	*312.5886*	*0.4908*
CS	319.3362	329.9818	324.7536	3.2572
k-means	313.1428	313.7343	313.4799	0.2687

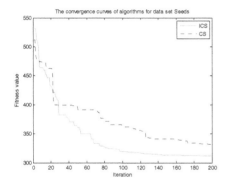

Fig. 7. The convergence curves of algorithm for seeds

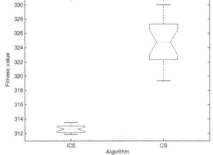

Fig. 8. The anova test of algorithms for seeds

4.1.5 Data Set Heart

The data set is a heart disease database similar to a database already present in the repository but in a slightly different form [13]. The comparison of algorithms for artificial data set one is listed in Table 5. And the convergence curves of algorithms are shown in Fig. 9. In Fig. 10, it is the result of anona test for Iris. We can discover that ICS has a faster convergence speed and high level of stability easily.

Table 5. Results obtained by the algorithms for 20 different runs on heart

Algorithm	Best	Worst	Mean	Std.
ICS	*10623.4060*	*10625.4460*	*10623.8259*	*0.4618*
CS	10635.5420	10657.8780	10649.1222	5.6361
k-means	10682.0809	10700.8385	10691.7056	8.2080

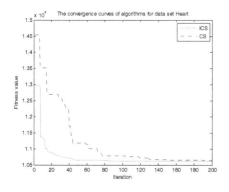

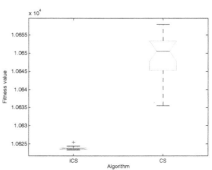

Fig. 9. The convergence curves of algorithm for heart

Fig. 10. The anova test of algorithms for heart

4.1.6 Data Set Haberman's Survival

The data set contains cases from a study that was conducted between 1958 and 1970. The comparison of algorithms for artificial data set one is listed in Table 6. And the convergence curves of algorithms are shown in Fig. 11. In Fig. 12, it is the result of anona test for Iris. It shows clearly that ICS has a faster convergence speed and high level of stability easily.

Table 6. Results obtained by the algorithms for 20 different runs on survival.

Algorithm	Best	Worst	Mean	Std.
ICS	*2566.9889*	*2566.9889*	*2566.9889*	*9.3312e−013*
CS	2567.0062	2567.2512	2567.0975	0.0640
k-means	2625.1076	3193.5941	2655.1274	126.7500

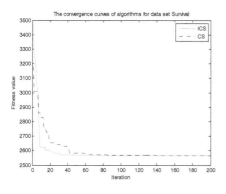

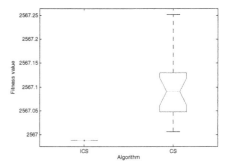

Fig. 11. The convergence curves of algorithm for survival

Fig. 12. The anova test of algorithms for survival

And navigation route is smooth, and above shows the feasibility of the algorithm. However, the advantages of the improved algorithm is not obvious, in order to illustrate the efficiency of the improved algorithm preferably, at this time, increasing the number of control points and threat point, continue to experiment two and three.

4.2   The Clustering Process of ICS

In this subsection, artificial data set one is selected to show the clustering process of proposed ICS. In order to show the whole process of clustering, the iterations are 0,10,20,50, which are shown in Figs. 13, 14, 15, 16, and 17. Figure 13 shows the distribution of artificial data set one, Fig. 14 is the clustering result when iteration is 10, From the Fig. 14 can be seen that Data are divided into 5 categories, but it has Cross phenomenon of classification. Figure 16 show that these five classes are separated successfully by proposed ICS.

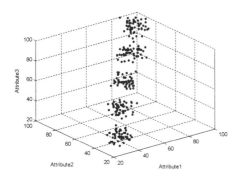

Fig. 13. The distribution of artificial data set1

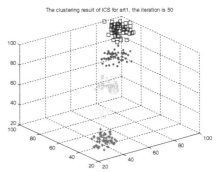

Fig. 14. The clustering result of ICS for art1, the iteration is 10

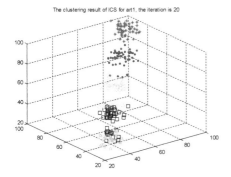

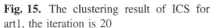

Fig. 15. The clustering result of ICS for art1, the iteration is 20

Fig. 16. The clustering result of ICS for art1, the iteration is 50

5 Conclusion

Clustering analysis is an important method in data mining, which is widely used in various fields. So far, researchers have proposed many methods for clustering. The proposed clustering method is a combination of cuckoo search algorithm, genetic algorithm that presents more accurate clustering results on the real datasets, when compared to other algorithms, can clearly see from Tables 1, 2, 3, 4, 5, 6 and Figs. 1, 2, 3, 4, 5, 6, 7, 8, 9, 10, 11 and 12. In addition, Figs. 14, 15 and 16 show the classification obtained by ICS is much faster.

Acknowledgments. This work is supported by the Project of Guangxi High School Science Foundation under Grant no. KY2015YB539, Project of Guangxi Education Science Foundation under Grant no. 2013C118.

References

1. Lloyd, S.P.: Least squares quantization in PCM. IEEE Trans. Inf. Theory **28**(2), 129–136 (1982)
2. Celebi, M.E., Kingravi, H., Vela, P.A.: A comparative study of efficient initialization methods for the K-means clustering algorithm. Expert Syst. Appl. **40**(1), 200–210 (2013)
3. Niknam, T., Olamaie, J., Amiri, B.: A hybrid evolutionary algorithm based on ACO and SA for cluster analysis. J. Appl. Sci. **8**(15), 2695–2702 (2008)
4. Shelokar, P.S., Jayaraman, V.K., Kulkarni, B.D.: An ant colony approach for clustering. Anal. Chim. Acta **509**(2), 187–195 (2004)
5. Kao, Y., Cheng, K.: An ACO-Based Clustering Algorithm, pp. 340–347. Springer, Heidelberg (2006)
6. Kennedy, J., Eberhart, R.C.: Particle swarm optimization. In: Proceedings of the IEEE International Conference on Neural Networks, pp. 1942–1948. IEEE Service Center, Piscataway (1995)
7. Omran, M., Engelbrecht, A.P., Salman, A.: Particle swarm optimization method for image clustering. Int. J. Pattern Recogni. Artif. Intell. **19**(3), 297–321 (2005)
8. Merwe, V.D., Engelbrecht, A.P.: Data clustering using particle swarm optimization. In: Proceedings of IEEE Congress on Evolutionary Computation, CEC03, pp. 215–220 (2003)
9. Kao, Y.T., Zahara, E., Kao, I.W.: A hybridized approach to data clustering. Expert Syst. Appl. **34**(3), 1754–1762 (2008)
10. Voges, K.E., Pope, N.K.L.: Rough clustering using an evolutionary algorithm. In: IEEE 2012 45th Hawaii International Conference on System Science, HICSS, pp. 1138–1145 (2012). Zheng, R., Feng, Z., Lu, M.: Application of particle genetic algorithm to plan planning of unmanned aerial vehicle. Comput. Simul. **28**(6), 88–91 (2011)
11. Wang, R., Zhou, Y., Qiao, S.: Flower pollination algorithm with bee pollinator for cluster analysis. Inf. Process. Lett. **116**, 1–14 (2016)
12. Yang, X.S., Deb, S.: Cuckoo search via Levy flights. In: proceedings of World Congress on Nature & Biologically Inspired Computing, pp. 210–214. IEEE Publications, India (2009)
13. Blake, C.L., Merz, C.J.: UCI repository of machine learning databases. http://archive.ics.uci.edu/ml/datasets.html

Solving 0–1 Knapsack Problems by Binary Dragonfly Algorithm

Mohamed Abdel-Basset[1], Qifang Luo[2,3(✉)], Fahui Miao[2],
and Yongquan Zhou[2,3]

[1] Head of Department of Operations Research,
Faculty of Computers and Informatics, Zagazig University, Zagazig, Egypt
[2] College of Information Science and Engineering,
Guangxi University for Nationalities, Nanning 530006, China
l.qf@163.com, yongquanzhou@126.com
[3] Guangxi High School Key Laboratory of Complex System
and Computational Intelligence, Nanning 530006, China

Abstract. The 0–1 knapsack problem (0–1KP) is a well-known combinatorial optimization problem. It is an NP-hard problem which plays significant roles in many real life applications. Dragonfly algorithm (DA) a novel swarm intelligence optimization algorithm, inspired by the nature of static and dynamic swarming behaviors of dragonflies. DA has demonstrated excellent performance in solving multimodal continuous problems and engineering optimization problems. This paper proposes a binary version of dragonfly algorithm (BDA) to solve 0–1 knapsack problem. Experimental results have proven the superior performance of BDA compared with other algorithms in literature.

Keywords: Dragonfly algorithm · Meta-heuristics · Combinatorial optimization · 0–1 knapsack problem

1 Introduction

The 0–1 knapsack problem (0–1KP) is known to be a combinatorial optimization problem. The knapsack problem has a variety of practical applications such as cutting stock problems, portfolio optimization, scheduling problems and cryptography [1]. The knapsack appears as a sub problem in many complex mathematical models of real world problems. Given a set of n items, each item has a weight w_i and a profit p_i, and the maximum weight capacity W, where x_i represents the number of copies of items i to be included in the knapsack. The target is to maximize the profit of the items in the knapsack so that overall weights are less than or equal to the knapsack capacity. It is an NP-Hard problem and hence it does not have a polynomial time algorithm unless P = NP [1–3]. The problem may be mathematically modeled as follows:

© Springer International Publishing AG 2017
D.-S. Huang et al. (Eds.): ICIC 2017, Part III, LNAI 10363, pp. 491–502, 2017.
DOI: 10.1007/978-3-319-63315-2_43

$$Maximize \sum_{i=1}^{n} p_i x_i \tag{1}$$

$$Subject\ to \sum_{i=1}^{n} w_i x_i \leq W, x_i \in \{0, 1\} \tag{2}$$

where x_i takes values either 1 or 0 which represents the selection or rejection of the *ith* item. There are several methods that are applied to solve the knapsack which is categorized into exact algorithms and metaheuristic algorithms. Exact algorithms contain dynamic programming, branch and bound and so on, which can give the exact solutions. The worst case occurs when the problem size is increased as the computation time increases exponentially. The metaheuristic algorithms can give approximate solutions, but in reasonable times compared to exact algorithms [2, 3].

Recently, A novel swarm intelligence optimization algorithm, called dragonfly algorithm (DA), has been developed by Mirjalili [32]. The main inspiration of the DA algorithm originates from the static and dynamic swarming behaviors of dragonflies in nature. Two essential phases of optimization, exploration and exploitation, are designed by modeling the social interaction of dragonflies in navigating, searching for foods, and avoiding enemies when swarming dynamically or statistically.

2   Literature Review

In recent years, many metaheuristic algorithms (specially swarm-based) have been employed to solve 0–1KPs. Zhou et al. [7] proposed an improved version of Monkey Algorithm that includes two modified processes the somersault and cooperation in order to increase exploration and convergence rate respectively. In addition, the greedy algorithm was used in order to increase intensification. In [16], the complex-valued encoding method introduced to the bat algorithm for solving both small-scale and large-scale 0–1KPs. Meanwhile in [5] both techniques: The complex-valued encoding and the greedy algorithm are considered to enhance the exploration and the exploitation of wind driven optimization. The experimental result shows that the proposed algorithm is suitable standard, small-scale, and large-scale 0–1KPs. Gherboudj et al. [23] combined monarch butterfly optimization algorithm with chaos maps and Gaussian mutation process for solving three different types of large-scale 0–1KPs. In [6], a new hybrid algorithm called greedy degree and expectation efficiency was considered for solving 0–1KPs. The authors hybridized greedy strategy degree model with static dynamic expectation efficiency strategy that supported with parallel computing for accelerating the algorithm. Although the efficiency and stability of the proposed algorithm for solving 0–1KPs, it still needs more enhancement as the parallel computing causes higher space complexity and greedy degree model makes the optimal solution is difficult to be obtained. Kulkarni and Shabir [4] firstly used Cohort

Intelligence Algorithm for solving 0–1KP. The proposed algorithm integrated with a generic accepting random behavior approach for exploration enhancement. Pavithr [10] integrated the Quantum-inspired evolutionary algorithm principles with Social Evolution algorithm. Zou et al. [24] proposed a new Harmony Search algorithm (HS) variant with position updating and genetic mutation operations. The computational results show that the proposed algorithm has better performance for solving large-scale KPs. Also, Layeb [22] hybridized the exploitation capabilities of the quantum inspired computing with HS. Xiang et al. [28] considered a novel global-best HS with a two-phase repair function to increase the feasibility rate. Kong et al. [13] introduced a simplified binary version of HS with a modified improvisation process and adaptive harmony memory considering rate. Also, a two-stage greedy procedure is employed to guarantee the feasibility of the solutions. Fang et al. [20] combined clonal selection algorithm with Ant colony optimization (ACO) for more exploration. Changdar et al. [29] proposed an improved ACO algorithm for solving fuzzy 0–1KPs where profits and weights are trapezoidal fuzzy numbers. Du and Zu [15] considered a self-adaption variant of ACO that based on the greedy strategy and normal distribution. The experimental result shows that the proposed algorithm is more efficient that the traditional ACO for solving 0–1KPs. Bansal and Deep [30] improved the exploration of binary Particle Swarm Optimization (BPSO) by using a new probability function that maintains the swarm diversity. Nguyen et al. [9] presented a binary version of PSO with two new constraints handling operators: penalty factor and repair function which depends on both the greedy strategy and random selection. For solving multidimensional 0–1KPs. Lin et al. [11] proposed a binary version of PSO based on surrogate information with proportional acceleration coefficients. The proposed algorithm incorporate cognition learning factor and social learning factor for identifying particles' movement. Razavi and Sajedi [17] used two new cognitive and social components to guide the search procedure of Gravitational Search Algorithm (GSA). In [12], Truong et al. hybridized Chemical Reaction Optimization Algorithm (CRO) with greedy strategy and random selection. Also, CRO was hybridized by Yan et al. [14] but with tabu search for improving exploitation. The computational results show that the proposed algorithm has better performance than the basic CRO for solving 0–1KPs. Ma and Wan [31] improved Genetic Algorithm (GA) with adaptive adjustment of crossover and mutation operators. In addition, it was combined with the greedy algorithm to repair the infeasible solutions. Gupta [21] introduced a fast variant of GA by adding a new function check the population after each mutation process and replace all infeasible solutions by new random ones. In [8], Lim et al. explored a subclass of Genetic Algorithm (GA) called monogamous pairs Genetic Algorithm (MopGA) for solving 0–1 knapsack problem. The computational results show that the MopGA is more efficient, especially for high-dimensionality 0–1KPs. Baykasoğlu and Ozsoydan [19] introduced an adaptive firefly algorithm with a priority ranking technique for solving static and dynamic multidimensional KPs. Gong et al. [25] applied Artificial glowworm swarm optimization (AGSO) algorithm for solving 0-1KP. The experimental results show that AGSO is more suitable for solving small instances of 0-1KP. Layeb [26] hybridized Cuckoo Search (CS) with quantum computing principles. Gherboudj et al. [23] introduced a binary version of CS based on sigmoid transformation. Feng et al.

[18] used position updating and genetic mutation operations with a greedy strategy to introduce an improved binary version of CS algorithm.

3   Dragonfly Algorithm (DA)

The main inspiration of DA originates from static and dynamic swarming behaviors. These two swarming behaviors are very similar to the two main phases of optimization using meta-heuristics: exploration and exploitation. Dragonflies create sub swarms and fly over different areas in a static swarm, which is the main objective of the exploration phase. In the static swarm, however, dragonflies fly in bigger swarms and along one direction, which is favorable in the exploitation phase. A conceptual model of the dynamic and static swarms are shown in Fig. 1.

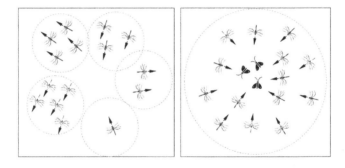

Fig. 1.   Static *vs.* dynamic

For simulating the swarming behaviour of dragonflies, the three primitive principles of swarming in insects proposed by Reynold as well as two other new concepts as shown in Fig. 2. These five concepts allowed me to simulate the behaviour of dragonflies in both dynamic and static swarms. The basic steps of DA can be summarized as the pseudo-code shown in Algorithm 1.

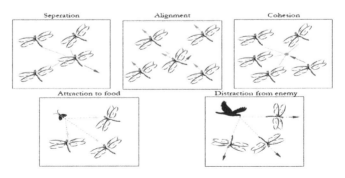

Fig. 2.   The five primitive principles of swarming

Dragonfly algorithm
Initialize the dragonflies population X_i (i = 1, 2, ..., n)

Initialize step vectors ΔX_i (i = 1, 2, ..., n)
while *the end condition is not satisfied*
Calculate the objective values of all dragonflies
Update the food source and enemy
Update w, s, a, c, f, and e
Calculate S, A, C, F, and E by using the following Eqs.

$$S_i = -\sum_{j=1}^{n} X - X_j$$

$$A_i = \frac{\sum_{j=1}^{N} V_i}{N}$$

$$C_i = \frac{\sum_{j=1}^{N} X_j}{N} - X$$

$$F_i = X^+ - X$$
$$E_i = X + X$$

Update neighbouring radius
if *a dragonfly has at least one neighbouring dragonfly*
Update velocity vector using the following Eq.
$\Delta X_{t+1} = (sS_i + aA_i + cC_i + fF_i + eE_i) + w\Delta X_t$
Update position vector by
$X_{t+1} = X_t + \Delta X_{t+1}$
else
Update position vector by using the following Eq.
$X_{t+1} = X_i + Levy(d) \times X_t$
end if
Check and correct the new positions based on the boundaries of variables
end while

Algorithm 1 Pseudo code of the Dragonfly algorithm

4   A Binary Dragonfly Algorithm for Solving 0–1 KP

The DA algorithm is only capable to solve continuous optimization problems, Mirjalili [32] introduced the binary version of this algorithm called Binary DA (BDA) for solving combinatorial problems. So that, he employed a v-shaped transfer function on the standard DA. Transfer functions receive velocity (step) values as inputs and return a number in [0,1], which defines the probability of changing positions. The output of such functions is directly proportional to the value of the velocity vector. Therefore, a large value for the velocity of a search agent makes it very likely to update its position.

This method simulates abrupt changes in particles with large velocity values similarly to continuous optimization. The basic steps of BDA can be summarized as the pseudo-code shown in Algorithm 2.

Binary Dragonfly Algorithm

Initialize the dragonflies population Xi (i = 1, 2, ..., n)
Initialize step vectors ΔXi (i = 1, 2, ..., n)
while *the end condition is not satisfied*
Calculate the objective values of all dragonflies
Update the food source and enemy
Update w, s, a, c, f, and e
Calculate S, A, C, F, and E by usingthe following Eqs.

$$S_i = -\sum_{j=1}^{n} X - X_j$$

$$A_i = \frac{\sum_{j=1}^{N} V_i}{N}$$

$$C_i = \frac{\sum_{j=1}^{N} X_j}{N} - X$$

$$F_i = X^+ - X$$
$$E_i = X + X$$

Update step vectors by this Eq.
$$\Delta X_{t+1} = (sS_i + aA_i + cC_i + fF_i + eE_i) + w\Delta X_t$$
Calculate the probabilities by this Eq.

$$T(\Delta x) = \left| \frac{\Delta x}{\sqrt{\Delta x^2 + 1}} \right|$$

Update position vectors by this Eq.
$$X_{t+1} = \begin{cases} \neg X_t & r < T(\Delta x_{t+1}) \\ X_t & r \geq T(\Delta x_{t+1}) \end{cases}$$

end while

Algorithm 2 Pseudo code of the BDA

---

5   Computational Results

All the experiments were achieved on a Windows 7 Ultimate 64-bit operating system; processor Intel Core to Duo CPU2.20 GHZ; 4 GB of RAM and codes were implemented in MATLAB. The parameters of the proposed algorithm set to be, 30 search agents over 1000 iterations. All of the test cases were evaluated by 50 independent experiments.

5.1 Small-Scale Knapsack Problems

To measure the efficiency of BDA, several experiments were achieved. The standard test cases are presented in Table 2 [24]. We have matched our results by those given by [4, 5, 22, 24, 26] the results are given in Table 1. In this experiment, BDA is better than the results by [4, 24] and has the same results in the other algorithms [5, 22, 26]. Table 3 shows the optimal solutions of small-scale KB obtained by our algorithm.

Table 1. Results for small instances (NA means not available)

Instance	Optimal value	Algorithm	Best	Mean	Worst	SD	Median
TP#1	295	BDA	295	295	295	0	295
		CWDO [5]	295	295	295	0	NA
		IC [4]	295	267.46	260	0	NA
		HS [24]	295	295	295	0	295
		IHS [24]	295	294.78	288	1.06	295
		NGHS [24]	295	295	295	0	295
TP#2	1024	BDA	1024	1024	1024	0	1024
		CWDO [5]	1024	1024	1024	0	NA
		IC [4]	1024	1020.5	1009	0	NA
		HS	1024	1024	1024	0	1024
		IHS	1024	1024	1024	0	1024
		NGHS	1024	1024	1024	0	1024
TP#3	35	BDA	35	35	35	0	35
		CWDO [5]	35	35	35	0	35
		IC [4]	35	34.55	28	0	NA
		HS	35	35	35	0	35
		IHS	35	34.58	27	1.68	35
		NGHS	35	35	35	0	35
TP#4	23	BDA	23	23	23	0	23
		CWDO [5]	23	23	23	0	23
		IC [4]	23	22.06	16	0.64	NA
		HS	23	23	23	0	23
		IHS	23	23	23	0	23
		NGHS	23	23	23	0	23
TP#5	481.1	BDA	481.1	481.07	481.1	0	481.07
		CWDO [5]	481.1	481.07	481.1	2.89E-13	NA
		IC [4]	481.1	449.98	412.6	10.7	NA
		HS	481.1	481.07	481.1	0	481.07
		IHS	481.1	478.48	437.9	10.4	481.07
		NGHS	481.1	481.07	481.1	0	481.07
TP#6	52	BDA	52	52	52	0	52
		CWDO [5]	52	52	52	0	NA

(continued)

Table 1. (*continued*)

Instance	Optimal value	Algorithm	Best	Mean	Worst	SD	Median
		IC [4]	51	50.73	49	0.66	NA
		HS	50	50	50	0	50
		IHS	50	49.2	44	1.85	50
		NGHS	50	50	50	0	50
TP#7	107	BDA	107	107	107	0	107
		CWDO [5]	107	107	107	0	NA
		IC [4]	105	86.6	79	2.99	NA
		HS	107	106.8	105	0.61	107
		IHS	107	103.98	93	4.48	105
		NGHS	107	107	107	0	107
TP#8	9767	BDA	9767	9767	9767	0	9767
		CWDO [5]	9767	9767	9767	0	NA
		IC [4]	9759	9753.3	9710	11.5	NA
		HS	9767	9767	9767	0	9767
		IHS	9767	9767	9767	0	9767
		NGHS	9767	9767	9767	0	9767
TP#9	130	BDA	130	130	130	0	130
		CWDO [5]	130	130	130	0	NA
		IC [4]	130	124.6	106	2.89	NA
		HS	130	130	130	0	130
		IHS	130	130	130	0	130
		NGHS	130	130	130	0	130
TP#10	1025	BDA	1025	1025	1025	0	1025
		CWDO [5]	1025	1025	1025	0	NA
		IC [4]	1025	997.7	892	18.6	NA
		HS	1025	1025	1025	0	1025
		IHS	1025	1025	1025	0	1025
		NGHS	1025	1025	1025	0	1025

Table 2. Optimal solution of small-scale knapsack problems obtained by BDA

Instance	Optimal Sol.	Optimal solution vector	Time	SR (%)
TP#1	295	0111000111	0.012	100
TP#2	1024	111111111111110101011	0.003	100
TP#3	35	1101	0.002	100
TP#4	23	0101	0.021	100
TP#5	481.07	001010110111011	0.003	100
TP#6	52	0010111111	0.0024	100
TP#7	107	1001000	0.004	100
TP#8	9767	111111100100001000000	0.005	.100
TP#9	130	11110	0.003	100
TP#1	1025	11111111101111010111	0.005	100

5.2   The Multidimensional Knapsack Problems

In this section, some multidimensional knapsack instances are used to evaluate the performance of BDA. Instances are taken from the well-known benchmark library [22, 26] OR-Library (mknap1).

Table 3. Experimental results of *mknap1*

Instance	No. of items	Best solution	BDA	QICSA [26]	PSO-P/PSO-R [27]	QIHSA [22]
TP#1	6	3800	3800	3800	3800	3800
TP#2	10	8706.1	8706.1	8706.1	8706.1	8706.1
TP#3	15	4015	4015	4015	4015	4015
TP#4	20	6120	6120	6120	6120	6120
TP#5	28	12400	12400	12400	12400	12400
TP#6	39	10618	10618	10618	10618	10618
TP#7	50	16537	16537	16537	16491/16537	16537

As shown in Table 3, BDA is able to find the best solution for all the instances of mknap1.

5.3   The Hard Multidimensional Knapsack Problems

A hard multidimensional knapsack instances, taken from [22, 26], which are considered to be rather difficult for optimization approaches. We have used 5 tests of the benchmarks maknapcb1, which have 5 constraints and 100 items, and we have used 5 tests of the benchmarks maknapcb4, which have 10 constraints and 100 items [27] (see Table 4).

In Table 4, we have compared the results of BDA against the results that obtained by [23, 26, 27]. The results of this test are summarized in Table 4. As we can see, the results of the proposed algorithm are clearly better than other algorithms.

Table 4. Experimental results of *mknapcb1* and *mknapcb4*

Instance	Best known	BDA	BCS [23]	QICSA [26]	PSO-P [27]	PSO-R [27]
5.100.00	24381	24381	23510	23416	22525	24381
5.100.01	24274	24274	22938	22880	22244	24258
5.100.02	23551	23551	22518	22525	21822	23551
5.100.03	23534	23534	22677	22727	22057	23527
5.100.04	23991	23991	23232	22854	22167	23966
10.100.00	23064	23064	21841	21796	20895	23057
10.100.01	22801	22801	21708	21348	20663	22781
10.100.02	22131	22131	20945	20691	20058	22131
10.100.03	22772	22772	21395	21377	20908	22772
10.100.04	22571	22571	21453	21251	20488	22751

5.4   A Random Generator Test Problems

In this section, we have used knapsack tests generated using a random generator. Seven experiment instances with large scales are given in Table 5 to testify the efficiency of the BDA. The values of weight and profit are generated randomly.

Table 5. Results for random experiments

Instance	Size	Algorithm	Best	Mean	Worst	SD	Median
TP#1	200	BDA	4375	4375	4375	0	4375
		Exact	4375	4375	4375	0	4375
TP#2	500	BDA	16847	16847	16847	0	16847
		Exact	16847	16847	16847	0	16847
TP#3	800	BDA	12462	12462	12462	0	12462
		Exact	12462	12462	12462	0	12462
TP#4	1000	BDA	21513	21513	21513	0	21513
		Exact	21513	21513	21513	0	21513
TP#5	1500	BDA	30405	3043.43	30403	0.73	30405
		Exact	30405	30405	30405	0	30405
TP#6	2000	BDA	47264	74260.3	47259	1.65	47264
		Exact	47264	47264	47264	0	47264
TP#7	2500	BDA	65317	65298.24	65272	10.34	65317
		Exact	65317	65317	65317	0	65317

We have compared the results of BDA against the dynamic programming method (exact method) as shown in Table 5. The results show that the proposed algorithm is efficiency.

6   Conclusions and Future Works

In this paper, the performance of the BDA has been extensively investigated by using a large number of experimental studies. The experimental results show that the BDA has demonstrated strong convergence and stability for 0–1 knapsack problems. The proposed algorithm thus provides a new method for 0–1 knapsack problems, and it may find the required optima in cases when the problem to be solved is too complicated and complex. Future work includes multidimensional 0–1 knapsack problem, quadratic knapsack problem (QKP), and different combinatorial optimization problems such as, travelling salesman problem, quadratic assignment problem and flow shop scheduling problems can be solved using this algorithm.

Acknowledgments. This work is supported by National Science Foundation of China under Grants Nos. 61563008; 61463007. Project of Guangxi Science Foundation under Grant No. 2016GXNSFAA380264.

References

1. Hu, T.C., Kahng, A.B.: The knapsack problem. In: Hu, T.C., Kahng, Andrew B. (eds.) Linear and Integer Programming Made Easy, pp. 87–101. Springer, Cham (2016). doi:10.1007/978-3-319-24001-5_8
2. Martello, S., Pisinger, D., Toth, P.: New trends in exact algorithms for the 0–1 knapsack problem. Eur. J. Oper. Res. **123**(2), 325–332 (2000)
3. Plateau, G., Nagih, A.: 0–1 knapsack problems. In: Paradigms of Combinatorial Optimization, 2nd edn., pp. 215–242 (2010)
4. Kulkarni, A.J., Krishnasamy, G., Abraham, A.: Solution to 0–1 knapsack problem using cohort intelligence algorithm. In: Kulkarni, A.J., Krishnasamy, G., Abraham, A. (eds.) Cohort Intelligence: A Socio-inspired Optimization Method. ISRL, vol. 114, pp. 55–74. Springer, Cham (2017). doi:10.1007/978-3-319-44254-9_5
5. Zhou, Y., Bao, Z., Luo, Q., Zhang, S.: A complex-valued encoding wind driven optimization for the 0–1 knapsack problem. Appl. Intell. 1–19 (2016)
6. Lv, J., Wang, X., Huang, M., Cheng, H., Li, F.: Solving 0-1 knapsack problem by greedy degree and expectation efficiency. Appl. Soft Comput. **41**, 94–103 (2016)
7. Zhou, Y., Chen, X., Zhou, G.: An improved monkey algorithm for a 0-1 knapsack problem. Appl. Soft Comput. **38**, 817–830 (2016)
8. Lim, T.Y., Al-Betar, M.A., Khader, A.T.: Taming the 0/1 knapsack problem with monogamous pairs genetic algorithm. Expert Syst. Appl. **54**, 241–250 (2016)
9. Nguyen, P.H., Wang, D., Truong, T.K.: A new hybrid particle swarm optimization and greedy for 0-1 knapsack problem. Indones. J. Electr. Eng. Comput. Sci. **1**(3), 411–418 (2016)
10. Pavithr, R.S.: Quantum inspired social evolution (QSE) algorithm for 0–1 knapsack problem. In: Swarm and Evolutionary Computation (2016, in press)
11. Lin, C.J., Chern, M.S., Chih, M.: A binary particle swarm optimization based on the surrogate information with proportional acceleration coefficients for the 0-1 multidimensional knapsack problem. J. Indu. Prod. Eng. **33**(2), 77–102 (2016)
12. Truong, T.K., Li, K., Xu, Y., Ouyang, A., Nguyen, T.T.: Solving 0-1 knapsack problem by artificial chemical reaction optimization algorithm with a greedy strategy. J. Intell. Fuzzy Syst. **28**(5), 2179–2186 (2015)
13. Kong, X., Gao, L., Ouyang, H., Li, S.: A simplified binary harmony search algorithm for large scale 0–1 knapsack problems. Expert Syst. Appl. **42**(12), 5337–5355 (2015)
14. Yan, C., Gao, S., Luo, H., Hu, Z.: A hybrid algorithm based on tabu search and chemical reaction optimization for 0-1 knapsack problem. In: Tan, Y., Shi, Y., Buarque, F., Gelbukh, A., Das, S., Engelbrecht, A. (eds.) ICSI 2015. LNCS, vol. 9141, pp. 229–237. Springer, Cham (2015). doi:10.1007/978-3-319-20472-7_25
15. Du, D.-P., Zu, Y.-R.: Greedy strategy based self-adaption ant colony algorithm for 0/1 knapsack problem. In: Park, J.J., Pan, Y., Chao, H.-C., Yi, G. (eds.) Ubiquitous Computing Application and Wireless Sensor. LNEE, vol. 331, pp. 663–670. Springer, Dordrecht (2015). doi:10.1007/978-94-017-9618-7_70
16. Zhou, Y., Li, L., Ma, M.: A complex-valued encoding bat algorithm for solving 0–1 knapsack problem. Neural Process. Lett. **44**(2), 407–430 (2016)
17. Razavi, S.F., Sajedi, H.: Cognitive discrete gravitational search algorithm for solving 0-1 knapsack problem. J. Intell. Fuzzy Syst. **29**(5), 2247–2258 (2015)
18. Feng, Y., Jia, K., He, Y.: An improved hybrid encoding cuckoo search algorithm for 0-1 knapsack problems. Comput. Intell. Neurosci. **2014**, 1 (2014)

19. Cheng, K., Ma, L.: Artificial glowworm swarm optimization algorithm for 0-1 knapsack problem. Appl. Res. Comput. **4**, 009 (2013)
20. Fang, Z., Yu-Lei, M., Jun-Peng, Z.: Solving 0-1 knapsack problem based on immune clonal algorithm and ant colony algorithm. In: Yang, G. (ed.) Proceedings of the 2012 International Conference on Communication, Electronics and Automation Engineering. AISC, vol. 181, pp. 1047–1053. Springer, Heidelberg (2013). doi:10.1007/978-3-642-31698-2_148
21. Gupta, M.: A fast and efficient genetic algorithm to solve 0-1 Knapsack problem. Int. J. Digit. Appl. Contemp. Res **1**(6), 1–5 (2013)
22. Layeb, A.: A hybrid quantum inspired harmony search algorithm for 0–1 optimization problems. J. Comput. Appl. Math. **253**, 14–25 (2013)
23. Gherboudj, A., Layeb, A., Chikhi, S.: Solving 0-1 knapsack problems by a discrete binary version of cuckoo search algorithm. Int. J. Bio-Inspired Comput. **4**(4), 229–236 (2012)
24. Zou, D., Gao, L., Li, S., Wu, J.: Solving 0–1 knapsack problem by a novel global harmony search algorithm. Appl. Soft Comput. **11**(2), 1556–1564 (2011)
25. Gong, Q.Q., Zhou, Y.Q., Yang, Y.: Artificial glowworm swarm optimization algorithm for solving 0-1 knapsack problem. In: Advanced Materials Research, vol. 143, pp. 166–171. Trans Tech Publications (2011)
26. Layeb, A.: A novel quantum inspired cuckoo search for knapsack prob lems. Int. J. Bio-Inspired Comput. **3**(5), 297–305 (2011)
27. Kong, M., Tian, P.: Apply the particle swarm optimization to the multidimensional knapsack problem. In: Rutkowski, L., Tadeusiewicz, R., Zadeh, L.A., Żurada, J.M. (eds.) ICAISC 2006. LNCS, vol. 4029, pp. 1140–1149. Springer, Heidelberg (2006). doi:10.1007/11785231_119
28. Xiang, W.L., An, M.Q., Li, Y.Z., He, R.C., Zhang, J.F.: A novel discrete global-best harmony search algorithm for solving 0-1 knapsack problems. Discret. Dyn. Nat. Soc. **2014**, 19 p. (2014). Article no. 637412. http://dx.doi.org/10.1155/2014/637412
29. Changdar, C., Mahapatra, G.S., Pal, R.K.: An ant colony optimization approach for binary knapsack problem under fuzziness. Appl. Math. Comput. **223**, 243–253 (2013)
30. Bansal, J.C., Deep, K.: A modified binary particle swarm optimization for knapsack problems. Appl. Math. Comput. **218**(22), 11042–11061 (2012)
31. Ma, Y., Wan, J.: Improved hybrid adaptive genetic algorithm for solving knapsack problem. In: 2011 2nd International Conference on Intelligent Control and Information Processing (2011)
32. Mirjalili, S.: Dragonfly algorithm: a new meta-heuristic optimization technique for solving single-objective, discrete, and multi-objective problems. Neural Comput. Appl. **27**(4), 1053–1073 (2016)

Moth Swarm Algorithm for Clustering Analysis

Xiao Yang[1], Qifang Luo[1,2(✉)], Jinzhong Zhang[1], Xiaopeng Wu[1],
and Yongquan Zhou[1,2]

[1] College of Information Science and Engineering,
Guangxi University for Nationalities, Nanning 530006, China
l.qf@163.com
[2] Guangxi High School Key Laboratory of Complex System
and Computational Intelligence, Nanning 530006, China

Abstract. Moth Swarm Algorithm (MSA) is a new swarm intelligent algorithm, it is inspired by the moth looking for food, phototaxis and celestial navigation in the dark environment, proposed a moth search algorithm. Because the algorithm has good convergence speed and high convergence precision, it is applied in many fields. Cluster analysis, as an effective tool in data mining, has attracted widespread attention and has been developed rapidly and has been successfully applied in recent years. Among the many clustering algorithms, the K-means clustering algorithm is easy to implement, so it is widely used. However, the K-means algorithm also has the disadvantages of large computational complexity and clustering effect depending on the selection of the initial clustering center, which seriously affects the clustering effect, and the algorithm is easy to fall into the local optimum. To solve these problems, The MSA is applied to cluster analysis, the results show that the MSA not only achieves superior accuracy, but also exhibits a higher level of stability.

Keywords: Cluster analysis · Moth swarm algorithm · K-means algorithm · Swarm intelligent algorithm

1 Introduction

Metaheuristic algorithm [1] is an improvement of heuristic algorithm, which is a combination of random algorithm and local search algorithm. Heuristic algorithm is a technique that can find the best solution within an acceptable computational cost, but does not necessarily guarantee the feasibility and optimality of the solution, and even in most cases it cannot explain the solution And the approximate degree of the optimal solution. Heuristic algorithms include genetic algorithm (GA) [2], Firefly algorithm (FA) [3], particle swarm optimization (PSO) [4], grey wolf optimizer (GWO) [5] and flower pollination algorithm (FPA) [6].

Moth swarm algorithm (MSA) [12] is also a heuristic algorithm proposed to solve complex problems. Moth Swarm Algorithm is proposed by Al-Attar Ali Mohamed in 2016, it is inspired by the moth looking for food and phototaxis at night. The moth is a generic term for insects that are close to the butterfly, both of which belong to the

© Springer International Publishing AG 2017
D.-S. Huang et al. (Eds.): ICIC 2017, Part III, LNAI 10363, pp. 503–514, 2017.
DOI: 10.1007/978-3-319-63315-2_44

Lepidoptera. Moths and more activities in the night, like to gather in the light, so the proverb 'moth fire from the burning body' argument. Clustering [7] is a process of sorting a number of data vectors based on the similarity between data. Clustering algorithms [8, 9] have been widely used in various fields, such as data analysis, data mining [10], image segmentation and mathematical programming. The k-means clustering method is one of the most commonly used partitional methods. The combination of MSA algorithm and k-means clustering algorithm eliminates the need of calculating the clustering center in the k-means algorithm, reduces the computational overhead, improves the algorithm as an optimization tool to optimize the clustering model, solve the computational complexity of the k-means algorithm and clustering effect depended on the initial clustering center selection and so on.

2   The Background and Concept of Basic MSA Algorithms

In the moth search algorithm, the position of the light source represents the feasible solution of the optimization problem, the luminous intensity of the light source represents the solution fitness value. In addition, the moths are considered divided into three parts, which are pathfinders, prospectors, onlookers, respectively. They will be described as follows:

At the start of the flight, for the d-dimensional space and the number of n populations, the initial number of moths $(n \times d)$ are randomly generated as follows:

$$x_{ij} = rand[0, 1] \cdot \left(x_j^{\max} - x_j^{\min}\right) + x_j^{\min} \tag{1}$$

where x_{ij} is the position of the i-th moth at j-dimensional space, $x_j^{\max}$ an $x_j^{\min}$ are the upper and lower bound, respectively. In some cases, the moths may fall into local optimum. The pathfinders play the role and fly for a long distance with adaptive-crossover and lévy-mutation, which is described as follows.

Cross points are selected in order to improve diversity. At t interation, the normalized dispersal degree σ_j^t in the j-th dimension is described as follows:

$$\sigma_j^t = \frac{\sqrt{\frac{1}{n_p} \sum_1^{n_p} \left(x_{ij}^t - \bar{x}_j^t\right)^2}}{\bar{x}_j^t}$$

where, n_p is the number of pathfinders. Then the variation coefficient μ^t, which is a calculation for relative dispersion, be formulated as follows:

$$\mu^t = \frac{1}{d} \sum_{j=1}^{d} \sigma_j^t$$

The crossover points groups c_p is described as:

$$j \in c_p \text{ if } \sigma_j^t \leq \mu^t$$

As the algorithm progresses, the crossover points is dynamically changed. Lévy flight is a random process with α-stable distribution that moves a longer distance in the different size of step. Mantegna's algorithm [11] is taked to simulate the α-stable distribution by producing random samples L_i that have the same behaviour of the Lévy-flights, as follows:

$$L_i \sim step \oplus Levy(\alpha) \sim 0.01 \frac{\mu}{|y^{1/\alpha}|}$$

where, *step* is size of the scale of the interests of the problem, $\oplus$ is the entrywise multiplications, $\mu = N(0, \sigma_u^2)$ and $y = N(0, \sigma_y^2)$ are two normal stochastic distributions with $\sigma_u = \left[\frac{\Gamma(1+\alpha) \sin \pi\alpha/2}{\Gamma(1+\alpha)/2\alpha 2^{(\alpha-1)/2}} \right]^{1/\alpha}$ and $\sigma_y = 1$.

For $n_c \in n_p$, this process is mirrored to Differential Evolution model. By disturbing the selected host vector $\vec{x}_p = \left[x_{p1}, x_{p2}, \ldots x_{pn_c} \right]$, the algorithm forms the sub-trial vector $\vec{v}_p = \left[v_{p1}, v_{p2}, \ldots, v_{pn_c} \right]$ with correlatived vectors (e.g., $\vec{x}_{r^1} = \left[x_{r^1 1}, x_{r^1 2}, \ldots x_{r^1 n_c} \right]$). The sub-trial vector is described as follow:

$$\vec{v_p^t} = \vec{x_{r^1}^t} + L_{p^1}^t \cdot \left(\vec{x_{r^2}^t} - \vec{x_{r^3}^t} \right) + L_{p^2}^t \cdot \left(\vec{x_{r^4}^t} - \vec{x_{r^5}^t} \right)$$

$$\forall r^1 \neq r^2 \neq r^3 \neq r^4 \neq r^5 \neq p \in \{1, 2, \ldots n_p\}$$

where L_{p2} and L_{p2} are created by Lévy α-stable distribution using the 5th expression.

One of the path finder (host vector) renovates its position by amalgamated the mutated variables of the sub-trail vector with the matching variables of host vector. The completed solution v_{pj} be described as:

$$v_{pj}^t = \begin{cases} v_{pj}^t & if \quad j \in c_p \\ x_{pj}^t & if \quad j \notin c_p \end{cases}$$

After the above process is completed, calculate the sub-trail fitness value solution compared to the host solution. The next generation of solutions, which may be defined by the minimization problem as follows:

$$\vec{x_p^{t+1}} = \begin{cases} \vec{x_p^t} & if \quad f(\vec{v_p^t}) \geq f(\vec{x_p^t}) \\ \vec{v_p^t} & if \quad f(\vec{v_p^t}) < f(\vec{x_p^t}) \end{cases}$$

The luminous intensity fit_p is proportional to the estimated probability value p_e, as follows:

$$p_e = \frac{fit_p}{\sum_{p=1}^{n_p} fit_p}$$

$$fit_p = \begin{cases} \frac{1}{1+f_p} & \text{for} \quad f_p \geq 0 \\ 1 + |f_p| & \text{for} \quad f_p < 0 \end{cases}$$

As a result of celestial navigation, the moths fly in logarithmic helix. When the path finders moth completes the search, the prospectors n_e make a deep search around the light source in the form of a logarithmic spiral. And the position of each moth x_i is updated with the following formula:

$$n_e = round\left((n - n_c) \times \left(1 - \frac{t}{T}\right)\right)$$

$$x_p^{t+1} = \left|x_i^t - x_p^t\right| \cdot e^\theta \cdot \cos 2\pi\theta + x_p^t \tag{2}$$

$$\forall p \in \left\{1, 2, \ldots n_p\right\}, i \in \left\{n_p + 1, n_p + 2, \ldots, n_f\right\}$$

where, $\theta \in [r, 1]$ is used to define the spiral shape as a randsom number and $r = -1 - t/T$. The expression is same as in Moth-flame Optimization (MFO) [12] algorithm.

In the MSA algorithm, moths with low fitness values are considered as onlookers ($n_u = n - n_e - n_c$). In this phase, the algorithm is designed to power the onlookers to probe more effectively around the hot spots of the prospector. Divided into the following two cases.

In first part, the number is equals to $n_o = round(n_u/2)$. The new onlookers fly based on Gaussian distributions using Eq. (3). This part of moth in this subblock x_i^{t+1} fly with progression steps of Gaussian walks using Eq. (4), which be expressed as follows:

Gaussian distribution, $q \sim N(\mu, \sigma_G^2)$ density is:

$$f(q) = \frac{1}{\sqrt{2\pi}\sigma_G} \exp\left(-\frac{(q - u)^2}{2\sigma_G^2}\right) - \infty < q < \infty \quad q \sim N(\mu,) \tag{3}$$

$$x_i^{t+1} = x_i^t + \varepsilon_1 + \lfloor \varepsilon_2 \times gbest^t - \varepsilon_3 \times x_i^t \rfloor \quad \forall i \in \{1, 2, \ldots n_o\} \tag{4}$$

$$\varepsilon_1 \sim round(size(d)) \oplus N(best_g^t, \frac{\log t}{t}(x_i^t - best_g^t))$$

where, ε_1 is a random sample of the size of the group from the Gaussian distribution, $best_g$ is the global best solution, ε_2 and ε_3 are random numbers from [0, 1].

The number is equal to the first part $n_m = n_u - n_o$. In many heuristic algorithms, there is a memory that conveys information to the next generation. But, the reality is that the moths fall into the fire because they do not have an evolutionary memory. Associative learning and short-term memory (1–3 s) [13, 14] have a strong impact on the behavior of moths. Associate learning plays an important role in the exchange of information among moths [15, 16]. This part of the moth is designed to drift to the moonlight according to the association learning and short-term memory to imitate the moth's actual behavior. The immediate memory is initialized from $x_i^t - x_i^{min}$ to $x_i^{max} - x_i^t$ based on the continuous uniform Gaussian distribution. The update formula is as follows:

$$x_i^{t+1} = x_i^t + 0.001 \cdot G\lfloor x_i^t - x_i^{min}, x_i^{max} - x_i^t \rfloor + (1 - g/G) \cdot r_1(best_p^t - x_i^t)$$
$$+ 2g/G \cdot r_2 \cdot (best_g^t - x_i^t) \tag{5}$$

where, $i \in \{1, 2, \ldots, n_m\}$, $2g/G$ and $1 - g/G$ is the social factor and the cognitive factor respectively, r_1 and r_2 are random number from [0, 1]. Be same as the first part, $best_p$ is a light source randomly chosen from the new pathfinders group based on the probability value of its corresponding solution. Each iteration is completed and the type of each moth will be redefined.

3   The Typical Clustering Algorithm K-Means

Classification issues can be seen everywhere in everyday life. The purpose of classification is to classify data sets into fixed numbers. For a clear understanding of the concept, we assume that there is a data set $D = \{d_i, i = 1, 2, \ldots, n\}$, where d_i is k-dimensional vector. The process of clustering is to divide D into several subclasses $G = \{G_1, G_2, \ldots, G_m\}$ according to similarity. The G satisfies the following formulas:

(1) $G_i \neq \emptyset, i = 1, 2, \ldots, m$
(2) $G_i \cap G_j = \emptyset, i, j = 1, 2, \ldots, m, i \neq j$
(3) $D = \bigcup\limits_{i=1}^{m} G_i$

3.1   A Typical Clustering Algorithm K-Means

The clustering problem in R^n class space can be described as n points $(x_1, x_2, \ldots, x_n)$ in a given R^n, and they are transformed into K classes according to the similarity between them. k-mean clustering algorithm is as follows:

Step 1: Given the number of classes K
Step 2: K points $(c_1, c_2, \ldots, c_k)$ are randomly selected from $(x_1, x_2, \ldots, x_n)$ as the center of the K clusters
Step 3: The set of $(x_1, x_2, \ldots, x_n)$ is divided by $(c_1, c_2, \ldots, c_k)$ as the center, the division of the principle is:

If $Distance(x_i, c_j) < Distance(x_i, c_m), j, m = 1, 2, \ldots, k$ and $j \neq m$,
Then x_i is divided into the classification set G_j

Step 4: Calculate the new center point $(c_i^*, c_2^*, \ldots, c_k^*)$ based on the points in the category collection $(G_1, G_2, \ldots, G_k)$

$$c_i^* = \frac{1}{|G_i|} \sum_{x_j \in G_i} x_j, i = 1, 2, \ldots, k$$

where $|G_i|$ is the number of points G_i

Step 5: If $c_i^* = c_i, i = 1, 2, \ldots, k$, the calculation ends. The current center point is the result of the clustering, otherwise it returns (3)

3.2   The MSA Algorithm on Clustering

In the MSA algorithm, the position of each moth is equivalent to m clustering centers, so each moth is represented as a $k * m$ dimension vector. For each moth i, its position is expressed as $X_i = \{x_{i,1}, x_{i,2}, \ldots, x_{i,k*m}\}$. The main steps are shown in Algorithm 1.

Algorithm 1. MSA for clustering analysis
Step 1: Initialize the moth group(x_i)
Step 2: Calculate the fitness using Eq(6)
Step 3: Each moth represents all center points of clusters
    For each pathfinders moth
      Using Eq(1) update x_i ,Calculate the fitness using Eq(6)

      Sort according to fitness values, choose the better fitness values
      For each prospector moth
      Update the position using Eq(2) ,Calculate the new fitness.
      Created the new light sources.
      For each onlooker ,
    If ($i \in n_0$),Generate Gaussian walks steps Update the positionx_i with Gaussian
      walk using Eq(4).
    Else
    Update the position$x_i, i \in n_m$ using Eq(5) .
    End If.
      Calculate the fitness.
    End For.
    Redefined the type of each moth.
    Get clustered results.

4   Data Clustering Experimental Results and Discussion

To verify the expression of the MSA algorithm in the cluster analysis, we use different algorithms (for example K-means, ABC, FPA, grey wolf optimizer(GWO), CS) to test some data sets selected from the UCI machine learning repository [17]. The comparison of the experimental results reflects the potential of the moth algorithm in the cluster analysis.

Artificial data set 2 [18, 19] ($N = 600, d = 2, and\, k = 4$): The data set has 600 data items, each data item has two attributes, data set is divided into four categories. The attribute values of each data item in the class are subject to a 2-dimensional random distribution:

Balance Scale data set ($N = 625, d = 4, k = 3$) [21]: Generate this data set to simulate the results of a mental experiment. The data set is divided into leftward, rightward, and centered in terms of equilibrium. Each data item in the data set has four attributes (left-weight, left distance, right weight, right distance).

Contraceptive Method Choice data set (CMC) [21] ($N = 1473, d = 10, k = 3$): The contraceptive method selection (CMC) data set comes from a sample of the national contraceptive prevalence survey conducted in 1987 in Indonesia.

Statlog (Heart) data set ($N = 270, d = 13, k = 2$) [21]: The data set is taken from the heart disease database, but with different forms of expression. There are 270 data items in the data set, each data item contains 13 attributes, the data set of the data items are divided into two categories.

4.1   Experimental Setup

The entire algorithm was programmed in MATLAB R2012a, numerical experiment was set up on AMD Athlon™Π*4 640 processor and 4 GB memory.

4.2   Parameters Setting

By comparing the moth swarm algorithm, the typical clustering algorithm K-means, and other mainstream cluster intelligence algorithms (ABC, FPA, GWO, CS). It is reflected in the performance of clustering problems. The given parameters of the algorithm are as follows:

ABC setting: the number of the colony is 50, the maximum number of iterations is 200, the parameters of limit is 10.

FPA setting: the population number is 50, the number of iterations is 200, the probability of p is 0.8.

GWO setting: $\overrightarrow{\alpha}$ is linear decreased from 2 to 0, the population and iterations is the same as FPA.

CS setting: the discovery probability is 0.25, the number of cuckoo is 50, the run algebra is 200.

These algorithms run independently for 20 times as shown in Table 1:

Table 1. Comparison of intracluster distances for the six clustering algorithms

Data set	Criteria	K-means	ABC	FPA	GWO	CS	KM-MSA
Art2	Best	514.6623	514.7803	569.621	514.054	513.9036	513.9030
	Worst	910.2391	524.5923	682.7043	519.6674	515.8	862.5598
	Mean	643.9603	517.0688	628.6757	516.0698	514.0997	531.8124
	Std	180.953	2.184732	32.70766	1.591735	0.482156	77.8783
Balance Scale	Best	1423.851	1425.441	1431.868	1423.84	1424.51	1423.822
	Worst	1433.84	1428.707	1439.367	1425.742	1427.721	1433.452
	Mean	1426.725	1427.045	1434.883	1424.049	1426.542	1426.258
	Std	3.169121	0.835866	1.911942	0.558875	0.76072	1.967701
CMC	Best	5703.344	5821.235	6254.709	5753.163	5705.238	5700.776
	Worst	5706.091	6077.472	6750.034	6075.156	5752.669	5733.758
	Mean	5704.3	5929.711	6426.311	5865.125	5724.3	5712.237
	Std	1.003552	84.58559	107.4847	75.84728	12.80705	8.766082
Heart	Best	10681.06	10655.65	10813.11	10632.39	10623.34	10622.26
	Worst	10700.84	11288.53	11297.73	10885.44	10624.56	10631.78
	Mean	10691.48	10807.02	11084.19	10673.72	10623.78	10628
	Std	8.992255	153.0101	116.391	54.50476	0.370144	2.177681

From Table 1, the optimal value found by the MSA algorithm is smaller than the other algorithms in the data set. In these algorithms, the value found by the FPA algorithm is the largest, indicating that the expression of the FPA is not good for several other algorithms for the clustering problem.

In Fig. 1, the convergence rate of the algorithm is faster than that of other algorithms before the 10 generation, the algorithm achieves a stable point before generation 50. In Fig. 2, the MSA algorithm is faster than other algorithms in the rate of convergence. The algorithm achieves a stable point over a few iterations rather than other algorithms. In Figs. 3 and 4, the CS algorithm converges faster than other algorithms, but there is no comparable MSA algorithm.

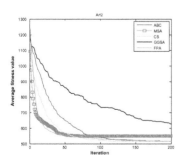

Fig. 1. The evolution curve of Art2 data set

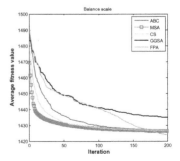

Fig. 2. The evolution curve of Balance Scale data set

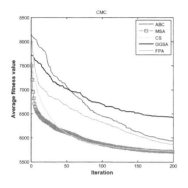
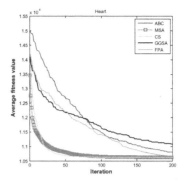

Fig. 3. The evolution curve of CMC data set   **Fig. 4.** The evolution curve of Heart data set

Figures 5, 6, 7, 8 and 9 is the anova test of several Swarm intelligent algorithms on clustering data set over 20 independent runs. In Fig. 5, the anova test value of the FPA algorithm is outstanding, the anova test value of the MSA and CS algorithm is stable in this five algorithm. In Fig. 6, the anova test value of the GWO algorithm is stable. In

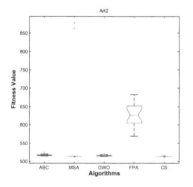
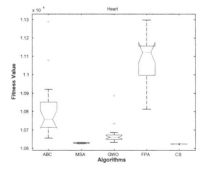

Fig. 5. The anova test of Art2 data set   **Fig. 6.** The anova test of Balance Scale data set

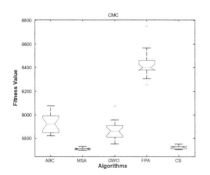

Fig. 7. The anova test of CMC data set   **Fig. 8.** The anova test of Heart data set

Figs. 7 and 8, the MSA algorithm is relatively stable compared to several other algorithms. In Fig. 9, the MSA algorithm is more stable than ABC algorithm and FPA algorithm.

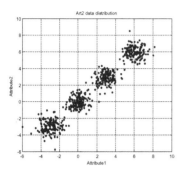

Fig. 9. The Art2 data distribution

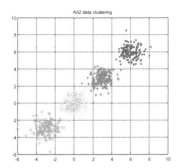

Fig. 10. The results for Art2 (MSA, iter = 20)

In this paper, the Table 2 shows the p value, indicating that the result of MSA is significant compared with ABC, GWO, FPA and CS. In order to verify the experimental results, with a pair of Wilkesson's rank-sum test [22], it is a nonparametric statistical test at 0.05 significance level to determine the differents between MSA and the other five algorithms in statistical methods.

Table 2. The Wilcoxon rank-sum test results (p ≥ 0.05) have been underlined

MSA VS	ABC	GWO	FPA	CS	K-means
Art2	1.6E−05	1.6E−05	1.6E−05	0.228694	3.36E−06
CMC	0.15557	4.54E−07	6.8E−08	7.9E−08	6E−07
Heart	6.8E−08	6.8E−08	6.8E−08	1.06E−07	6.28E−08
Balance Scale	4.17E−05	0.000161	1.25E−05	0.989209	6.24E−08

From Table 2, we conclude that the performance of MSA is better than other algorithms for Iris data set. The p-value of MSA versus GWO, FPA, K-means is more than 0.05 for Art1 data set, there is no statistical difference between them. For Art2 datasets, MSA is superior to ABC, GWO, FPA, K-means except CS. For the Haberman survival data set, compared the MSA and the other five algorithms, we obtain that performance of the MSA is better in the two groups of the five comparison algorithms. So, we summary that the MSA algorithm obtains better results than the other five algorithms for these data sets.

4.3   Clustering Results

We apply the MSA algorithm to the data set Art2, CMC, Heart, and the results are shown in the following figure:

In Figs. 9, 10, 11, 12 and 13, the experimental results of the MSA algorithm in clustering analysis is clearly described. For different experimental data, MSA algorithm can be very clear to classify them, so that the data classification results more clearly presented.

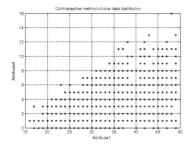

Fig. 11. The CMC data distribution

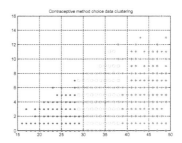

Fig. 12. The results for CMC (MSA, iter = 20)

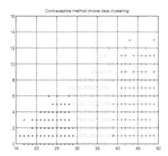

Fig. 13. The results for CMC (GWO, iter = 20)

5   Discussion and Conclusion

This paper, we use MSA, GWO, FPA, CS, ABC and traditional clustering method k-means comparison, which reflects the broad application prospect of swarm intelligence algorithm in clustering analysis. These algorithms and the K-means method are applied to the artificial data set and several real data sets, which proves the performance of the MSA algorithm in the clustering problem. The experimental results show that the performance of MSA algorithm in clustering analysis outperform other algorithms. The MSA algorithm has a high efficiency in solving complex optimization problems.

Acknowledgments. This work is supported by National Science Foundation of China under Grant Nos. 61563008, and 61463007, and Project of Guangxi Natural Science Foundation under Grant No. 2016GXNSFAA380264.

References

1. Yang, X.S.: Nature-Inspired Metaheuristic Algorithms. Luniver Press, Bristol (2010)
2. Holland, J.H.: Genetic algorithms. Sci. Am. **267**(1), 66–72 (1992)
3. Yang, X.S.: Firefly algorithm, stochastic test functions and design optimization. Int. J. Bio-Inspired Comput. **2**(2), 78–84 (2010)

4. Kennedy, J., Eberhart, R.: Particle swarm optimization. In: Proceedings of the IEEE International Conference on Neural Networks, Perth, Australia, vol. IV, pp. 1942–1948 (1995)
5. Mirjalili, S., Mirjalili, S.M., Lewis, A.: Grey wolf optimizer. Adv. Eng. Softw. **69**, 46–61 (2014)
6. Yang, X.-S.: Flower Pollination Algorithm for Global Optimization. In: Durand-Lose, J., Jonoska, N. (eds.) UCNC 2012. LNCS, vol. 7445, pp. 240–249. Springer, Heidelberg (2012). doi:10.1007/978-3-642-32894-7_27
7. Mohamed, A.A.A., Mohamed, Y.S., El-Gaafary, A.A.M., et al.: Optimal power flow using moth swarm algorithm. Electr. Power Syst. Res. **142**, 190–206 (2017)
8. Mohamed, A.A.A., El-Gaafary, A.A., Mohamed, Y.S., Hemeida, A.M.: Multi-objective states of matter search algorithm for TCSC-based smart controller design. Electr. Power Syst. Res. **140**, 874–885 (2016)
9. Hartigan, J.A.: Clustering Algorithms. Wiley, Hoboken (1975)
10. Jain, A.K., Dubes, R.C.: Algorithms for Clustering Data. Prentice-Hall, Inc., Upper Saddle River (1988)
11. Fayyad, U.M., Piatetsky-Shapiro, G., Smyth, P., et al.: Advances in Knowledge Discovery and Data Mining. MIT Press, Cambridge (1996)
12. Mantegna, R.N.: Fast, accurate algorithm for numerical simulation of Levy stable stochastic processes. Phys. Rev. E: Stat. Nonlinear Soft Matter Phys. **49**(5), 4677–4683 (1994)
13. Mirjalili, S.: Moth-flame optimization algorithm: a novel nature-inspired heuristic paradigm. Knowl. Based Syst. **89**, 228–249 (2015)
14. Cunningham, J.P., Moore, C.J., Zalucki, M.P., West, S.A.: Learning, odour preference and flower foraging in moths. J. Exp. Biol. **207**(1), 87–94 (2004)
15. Menzel, R., Greggers, U., Hammer, M.: Functional organisation of appetitive learning and memory in a generalist pollinator, the honey bee. In: Lewis, A.C. (ed.) Insect Learning: Ecological and Evolutionary Perspectives, pp. 79–125. Chapman and Hall, London (1993). doi:10.1007/978-1-4615-2814-2_4
16. Fan, R.J., Anderson, P., Hansson, B.: Behavioural analysis of olfactory conditioning in the moth Spodoptera littoralis (Boisd.) (Lepidoptera: Noctuidae). J. Exp. Biol. **200**(23), 2969–2976 (1997)
17. Skiri, H.T., Stranden, M., Sandoz, J.C., Menzel, R., Mustaparta, H.: Associative learning of plant odorants activating the same or different receptor neurones in the moth Heliothis virescens. J. Exp. Biol. **208**(4), 787–796 (2005)
18. Cattell, R.B.: The description of personality: basic traits resolved into clusters. J. Abnorm. Soc. Psychol. **38**(4), 476 (1943)
19. Han, J.W., Kamber, M., Pei, J.: Data Mining: Concepts and Techniques. Morgan Kaufmann Publishers, Burlington (2011)
20. Blake, C., Merz, C.J.: UCI Repository of Machine Learning Databases (1998). http://www.mendeley.com/research/uci-repository-of-machine-learning-databases/
21. Kao, Y.T., Zahara, E., Kao, I.W.: A hybridized approach to data clustering. Expert Syst. Appl. **34**(3), 1754–1762 (2008)
22. Niknam, T., Amiri, B.: An efficient hybrid approach based on PSO, ACO and k-means for cluster analysis. Appl. Soft Comput. **10**(1), 183–197 (2010)

BPSO Optimizing for Least Squares Twin Parametric Insensitive Support Vector Regression

Xiuxi Wei[1] and Huajuan Huang[2(✉)]

[1] Information Engineering Department, Guangxi International Business
Vocational College, Nanning 530007, China
[2] College of Information Science and Engineering,
Guangxi University for Nationalities, Nanning 530006, China
hhj-025@163.com

Abstract. The recently proposed twin parametric insensitive support vector regression, denoted by TPISVR, which solves two dual quadratic programming problems (QPPs). However, TPISVR has at least four regularization parameters that need regulating. In this paper, we increase the efficiency of TPISVR from two aspects. Fist, we propose a novel least squares twin parametric insensitive support vector regression, called LSTPISVR for short. Compared with the traditional solution method, LSTPISVR can improve the training speed without loss of generalization. Second, a discrete binary particle swarm optimization (BPSO) algorithm is introduced to do the parameter selection. Computational results on several synthetic as well as benchmark datasets confirm the great improvements on the training process of our LSTPISVR.

Keywords: Support vector regression · Twin support vector regression · Least squares · Twin parametric insensitive support vector regression · BPSO

1 Introduction

In 2010, Peng [1] introduced a new nonparallel plane regression, termed the twin support vector regression (TSVR). TSVR aims at generating two nonparallel functions such that each function determines the ε-insensitive down- or up- bounds of the unknown regressor. TSVR only needs to solve a pair of smaller QPPs, instead of solving the large one in SVR. Furthermore, the number of constraints of each QPP in STVR is only half of the classical SVR, which makes TSVR work faster than SVR.

We often encounter the heteroscedastic noise, which is dependent on the regional position and not consistent. In order to solve this problem, in 2010, Shao [2] proposed the parametric-insensitive v-support vector regression (par-v-SVR) introducing a parameter insensitive loss function. Compared with SVR, par-v-SVR is more suitable for solving the heteroscedastic noise. However, the training speed of par-v-SVR is not satisfactory. In 2012, Peng [3] proposed twin parametric insensitive support vector regression (TPISVR). The TPISVR determines the regression function through two nonparallel parametric insensitive up- and down-bound functions. Compared with

© Springer International Publishing AG 2017
D.-S. Huang et al. (Eds.): ICIC 2017, Part III, LNAI 10363, pp. 515–521, 2017.
DOI: 10.1007/978-3-319-63315-2_45

TSVR and SVR, TPISVR is suitable for many cases and owns better generalization performance. Similar to TSVR, TPISVR also solves the QPPs in the dual space. However, this solving method will be affected by time and memory constraints when dealing with the large datasets, which would make the learning speed low.

In order to improve the learning speed of TPISVR, in this paper we will enhance TPISVR to least squares TPISVR (LSTPISVR) by introducing the least squares method and then solve LSTPISVR directly in the primal space instead of the dual space. Therefore, our LSTPISVR can accurately solve large datasets without any external optimizers. Like TPISVR, the LSTPISVR has four parameters that need to be adjusted. In this paper, a discrete binary particle swarm optimization (BPSO) algorithm is introduced to do the parameter selection. The experimental results on artificial and benchmark datasets show that LSTPISVR surpasses TSVR and par-v-SVR in speed without loss of generalization.

2 Least Squares Twin Parametric Insensitive Support Vector Regression (LSTPISVR)

2.1 Linear LSTPISVR

In this section, we elaborate the formulation of the LSTPISVR which we name as least squares twin parametric insensitive support vector regression (LSTPISVR). To obtain our LSTPISVR formulations, we solve the following optimization problems

$$\min \frac{1}{2}(\|w_1\|^2 + b_1^2) - v_1 e^T (Aw_1 + b_1 e) + \frac{c_1}{2} \xi^T \xi, \tag{1}$$
$$s.t. \, Y = Aw_1 + b_1 e - \xi,$$

and

$$\min \frac{1}{2}(\|w_2\|^2 + b_2^2) + v_2 e^T (Aw_2 + b_2 e) + \frac{c_2}{2} \eta^T \eta, \tag{2}$$
$$s.t. \, Y = Aw_2 + b_2 e + \eta,$$

where v_1, c_1, v_2 and c_2 are positive parameters, and ξ, η are the slack variables.

To solve (1), on substituting the equality constraints into the objective function

$$L = \frac{1}{2}(\|w_1\|^2 + b_1^2) - v_1 e^T (Aw_1 + b_1 e) + \frac{c_1}{2} \|Aw_1 + b_1 e - Y\|^2 \tag{3}$$

Setting the gradient of (3) with respect to w_1 and b_1 to zero, gives

$$\frac{\partial L}{\partial w_1} = w_1 - v_1 A^T e + c_1 A^T (Aw_1 + b_1 e - Y) = 0 \tag{4}$$

$$\frac{\partial L}{\partial b_1} = b_1 - v_1 e^T e + c_1 e^T (A w_1 + b_1 e - Y) = 0 \tag{5}$$

Arranging (4) and (5) in matrix form and solving for w_1 and b_1 gives

$$c_1 \begin{bmatrix} A^T A + \frac{1}{c_1} I & A^T e \\ e^T A & e^T e + \frac{1}{c_1} \end{bmatrix} \begin{bmatrix} w_1 \\ b_1 \end{bmatrix} - v_1 \begin{bmatrix} A^T e \\ e^T e \end{bmatrix} - c_1 \begin{bmatrix} A^T Y \\ e^T Y \end{bmatrix} = 0 \tag{6}$$

Defining $H = [A \quad e]$, the solution becomes

$$\begin{bmatrix} w_1 \\ b_1 \end{bmatrix} = (c_1 H^T H + I)^{-1} (c_1 H^T Y + v_1 H^T e) \tag{7}$$

In an exactly similar way the solution of QPP (24) can be shown to be as

$$\begin{bmatrix} w_2 \\ b_2 \end{bmatrix} = (c_2 H^T H + I)^{-1} (c_2 H^T Y - v_2 H^T e) \tag{8}$$

After optimizing (1) and (2), we construct the estimated regressor as follows:

$$f(x) = \frac{1}{2} (f_1(x) + f_2(x)) \tag{9}$$

2.2   Nonlinear LSTPISVR

For the nonlinear case, according to a similar idea, we consider the following functions with kernel.

$$f_1(x) = K(x^T, A^T) w_1 + b_1, f_2(x) = K(x^T, A^T) w_2 + b_2 \tag{10}$$

where $K(x^T, A^T)$ is an appropriately chosen kernel.

Similar to the linear cases, the above two functions are obtained through the following two QPPs:

$$\min \frac{1}{2} (\|w_1\|^2 + b_1^2) - v_1 e^T (K(A, A^T) w_1 + b_1 e) + \frac{c_1}{2} \xi^T \xi,$$
$$s.t.\ Y = K(A, A^T) w_1 + b_1 e - \xi, \tag{11}$$

$$\min \frac{1}{2} (\|w_2\|^2 + b_2^2) + v_2 e^T (K(A, A^T) w_2 + b_2 e) + \frac{c_2}{2} \eta^T \eta,$$
$$s.t.\ Y = K(A, A^T) w_2 + b_2 e + \eta, \tag{12}$$

and the parameters are as described in above subsection for linear case.

Similar to linear cases, the solutions of the problems (11) and (12) are

$$\begin{bmatrix} w_1 \\ b_1 \end{bmatrix} = (c_1 R^T R + I)^{-1} (c_1 R^T Y + v_1 R^T e), \tag{13}$$

and

$$\begin{bmatrix} w_2 \\ b_2 \end{bmatrix} = (c_2 R^T R + I)^{-1} (c_2 R^T Y - v_2 R^T e), \tag{14}$$

respectively, where $R = [K(A, A^T) \, e]$.

2.3   Parameter Selection for LSTPISVR

LSTPISVR are some parameters that should be selected. They are the penalty parameters c_1, c_2, v_1, v_2 and an extra kernel parameter σ for nonlinear case. In our implementation, the discrete binary particle swarm optimization (BPSO) [4] algorithm is introduced to the parameter selection.

At each time step t, by using the individual best position, $p_i(t)$, and global best position, $p_g(t)$, a new velocity for particle i is updated by

$$v_i(t+1) = v_i(t) + c_1 \phi_1 (p_i(t) - x_i(t)) + c_2 \phi_2 (p_g(t) - x_i(t)) \tag{15}$$

where c_1 and c_2 are positive constants, ϕ_1 and ϕ_2 are uniformly distributed random numbers in [0,1]. Based on the updated velocities, each particle changes its position according to the following:

$$x_i(t+1) = x_i(t) + v_i(t+1) \tag{16}$$

Kennedy [5] proposed a binary PSO (BPSO) algorithm in 1997. In BPSO, each particle is encoded as a binary vector.

According to the above procedures, we conduct our BPSO as the parameters selection method for LSTPISVR.

2.3.1   The Design of Particle
As we know, there are some parameters need to be predetermined. They are the penalty parameters c_1, c_2, v_1, v_2 and an extra kernel parameter σ for nonlinear case. Figure 1 shows the representation of a particle which is a bit string.

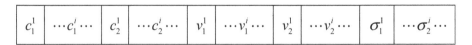

Fig. 1. The structure of a particle

For (c_1, c_2, v_1, v_2) part, there are two components (the first one and the other ones). The first one denotes this parameter to be bigger than 1 or not (0 for bigger, 1 for not), while the others is a binary coding system. The length of this part relies on the accuracy requirement. The σ part is related to the nonlinear model.

2.3.2   The Design of Fitness Function

In this section, we design the fitness function of BPSO as follows.

$$fitness = \frac{1}{(w_a \times LSTPISVR\_RMSE)^2}, \tag{17}$$

where w_a represents the weight of regression accuracy, $LSTPISVR\_RMSE$ represents the regression accuracy.

3   Experimental Results

To check the validity of our LSTPISVR, we compare it with the TPISVR, par-v-SVR and TSVR on several datasets. In our implementation, we use grid search approach to obtain the optimal parameters for TPISVR, par-v-SVR and TSVR, and the BPSO parameter selection for our LSTPISVR from the set $\{2^{-8}, \ldots, 2^7\}$. In this paper, for the nonlinear case, we only consider the Gauss kernel function. For BPSO, we set the number of the first generation offsprings to 250, the upper bond of iterations to 200, the study factors of BPSO $c_1, c_2 = 2.0$ and $w_a = 0.8$, respectively.

3.1   Artificial Data Set

To demonstrate the regression performance of our algorithm in the noise environment, we apply LSTPISVR, TPISVR, par-v-SVR and TSVR on a synthetic dataset with noise.

Sinc function [3] is usually used to test the performance of the regression methods. The formula of sinc function is

$$y = \sin c(x_i) = \frac{\sin x_i}{x_i} + e_i, \ x_i \sim U[-3\pi, 3\pi].$$

To effectively check the performance of LSTPISVR, training data points are perturbed by two kinds of noises with heteroscedastic error structures, which are

Type A: $e_i = (-\frac{|x_i|}{8\pi} + 0.5)\varepsilon_i, \quad \varepsilon_i \sim U[-0.5, 0.5],$

Type B: $e_i = (-\frac{|x_i|}{8\pi} + 0.5)\varepsilon_i, \quad \varepsilon_i \sim N(0, 0.25^2),$

where $U[a, b]$ represents the uniformly random variable in $[a, b]$ and $N(c, d^2)$ represents the Gaussian random variable with c means and variance d^2.

Table 1 lists the average test performance of these methods on the two types of noise with 10 independent runs. Figures 2 and 3 show the one-run results of LSTPISVR, TPISVR, par-v-SVR and TSVR with the two types of noises.

Table 1. Results of four algorithms on Sinc function with two types of noises on linear case

Dataset	-	LSTPISVR	TPISVR	TSVR	Par-v-SVR
Type A	RMSE	0.094 ± 0.001	0.094 ± 0.004	0.101 ± 0.002	0.094 ± 0.002
	Time (s)	0.0684	0.0824	0.0917	0.3611
Type B	RMSE	0.085 ± 0.002	0.084 ± 0.002	0.087 ± 0.002	0.086 ± 0.001
	Time (s)	0.0586	0.0782	0.0816	0.3711

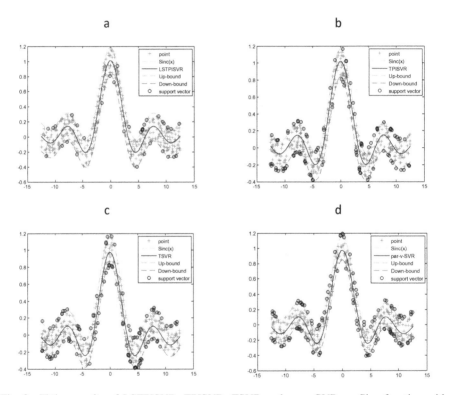

Fig. 2. Fitting results of LSTPISVR, TPISVR, TSVR and par-v-SVR on Sinc function with noise Type A (a–d)

From Table 1, we can see that the LSTPISVR obtains the better prediction performance than other three methods. As for the learning CPU time of these algorithms, it can be found that the LSTPISVR is much faster than other methods. In short, LSTPISVR obtains the better simulation performance than other methods with much less learning time. From Fig. 2, the results show that the LSTPISVR obtains the good regressors for the two types of noises. Figure 2 also shows the support vectors (SVs) marked by extra circles of these algorithms. It can be found that our LSTPISVR obtains less numbers of SVs than other methods, which indicates the LSTPISVR is also a parse regression method.

4 Conclusions

In this paper, we have proposed a novel least squares twin parametric insensitive support vector regression (LSTPISVR). The contributions are as follow. First, our LSTPISVR optimizes a pair of QPPs in the primal space instead of the dual space, which can obviously improve the regression precision of LSTPISVR without loss of the generalization. Second, a binary PSO-based parameters selection for our LSTPISVR was further suggested. Computational results on several benchmark data-sets and one synthetic dataset confirm the great improvements of our algorithm in the training procedure.

Acknowledgments. This work is supported by the National Natural Science Foundation under Grant no. 61662005, the Science and Technology Research Project of Guangxi University under Grant no. KY2015YB076, the Talent Research Projects of Guangxi University for Nationalities under Grant no. 2014MDQD018 and the open fund of Key Laboratory of Guangxi High Schools for Complex System & Computational Intelligence (No. 15CI01D).

References

1. Peng, X.J.: TSVR: an efficient twin support vector machine for regression. Neural Netw. **23**, 365–372 (2010)
2. Shao, P.Y.: New support vector algorithms with parametric insensitive/margin model. Neural Netw. **23**(1), 60–73 (2010)
3. Peng, X.J.: Efficient twin parametric insensitive support vector regression model. Neuro-computing **79**, 26–38 (2012)
4. Wong, T.C., Ngan, S.C.: A comparison of hybrid genetic algorithm and hybrid particle swarm optimization to minimize make-span for assembly job shop. Appl. Soft Comput. **13**(3), 1391–1399 (2013)
5. Jin, W., Zhang, J.Q., Zhang, X.A.: Face recognition method based on support vector machine and particle swarm optimization. Expert Syst. Appl. **38**(4), 4390–4393 (2011)

Computational Intelligence and Security for Image Applications in Social Network

A Multi-modal SPM Model for Image Classification

Peng Zheng, Zhong-Qiu Zhao[⊠], and Jun Gao

College of Computer and Information, Hefei University of Technology,
Hefei, China
z.zhao@hfut.edu.cn

Abstract. The BoF (bag-of-features) model is one of the most famous models applied to many fields in computer vision and has achieved impressive results. However, the SIFT/HOG visual words have a limit discriminative power which is partly due to the fact that it only describes the local gradient distribution. In the meanwhile, there is still redundancy and hidden information existed in the formed histogram. Considering these respects, we propose a multi-modal SPM model which fuses global features to complement traditional local ones and conducts dimensionality reduction in local spaces for mining possible feature dependencies. Experimental results show the efficiency of the proposed method in comparison with the existing counterparts.

Keywords: Multi-modal · Dimensionality reduction · SPM

1 Introduction

In recent years, many models have been proposed to different research fields in computer vision [19–21]. Among them, the BoF model is one of the most famous, which has been widely applied to object recognition [23], scene categorization [24] or object retrieval [3] etc. It quantizes the unordered appearance descriptors extracted from local patches into discrete "visual words" by the vector quantization (VQ) scheme and takes advantages of the formed histogram representation to accomplish the semantic image classification tasks.

As the BoF method does not consider the spatial order of local descriptors, it makes the representation obtained less discriminative and descriptive. To solve this problem, Lazebnik et al. [10] proposed an extension of the traditional BoF model called spatial pyramid matching (SPM) which obtained promising results on several benchmarks like Caltech-101 [11] and Caltech-256 [8] and also became the major component of the state-of-the-art systems [1].

However, there are still two problems existed in the models especially for SPM based ones. One is the limit discriminative power of local descriptors such as SIFT/HOG which can only describe the local gradient distribution. For lack of certainty of global distribution, it sometimes results in degraded performance. So it's admirable to provide complementary information by feature fusion with low correlated ones. The other one is high dimensionality caused by multiple scales. Although this operation

© Springer International Publishing AG 2017
D.-S. Huang et al. (Eds.): ICIC 2017, Part III, LNAI 10363, pp. 525–535, 2017.
DOI: 10.1007/978-3-319-63315-2_46

usually obtains higher recognition accuracy, it will bring high burden for memory cost and efficient calculation and make extensions to other applications such as image retrieval difficult. And directly applying dimensionality reduction will usually destroy the local structures. So it is better to do dimensionality reduction in local spaces. By combining these concepts, we modified the SPM models and proposed a multi-modal version. It makes use of the SPM model to fuse global features to complement traditional local ones and conducts dimensionality reduction in local spaces to mine possible feature dependencies. Then experiments on several datasets show the effectiveness of the proposed method.

In summary, the contributions of this paper are as follows:

(1) We propose a multi-modal feature fusion method which embeds global features into the SPM model to provide complementary information for traditional SIFT based one.
(2) We propose a local dimensionality reduction scheme which can preserve local structures of the concatenated feature vector and dig out valuable hidden information.
(3) By making use of different mid-level representation, local dimensionality reduction scheme and other fusion schemes, we propose the multi-modal SPM model which inherits the advantages of them and can achieve enhanced recognition performance.

The reminder of the paper is organized as follows. Section 2 reviews the related works. Section 3 describes the traditional SPM and ScSPM models. Section 4 presents details of the proposed model. Then, Sect. 5 shows the experimental results and makes some discussions. Finally, conclusions and future works are provided in Sect. 6.

2 Related Work

Although there have been some researches which extend the traditional BoF model, such as generative methods [6, 16] for modeling the co-occurrence of the codewords, discriminative codebook learning in [5, 9] which replaces standard unsupervised K-means clustering and the spatial pyramid matching kernel (SPM) [10] which represents the spatial layout of the objects in the image is particularly successful.

Subsequently, some researchers extended the SPM model. Yang et al. [23] proposed the ScSPM model by replacing the K-means vector quantization with sparse coding and adapting max pooling scheme in each segment. Extended versions were then achieved based on the ScSPM model. Several kinds of constraints were added to the optimization objective function of sparse coding to obtain different kinds of promotion of the ScSPM model, such as the Laplacian term in [7], the attribute-to-word co-occurrence in [2] and the empirical Maximum Mean Discrepancy in [12].

Different from those focusing on the promotion of dictionary, some researchers cared about the features used. Oliva and Torralba argued that global features were of importance for recognizing different scenes [15]. Wu and Rehg [22] used the Census transform to get local features and combined SPM model to form partial PACT.

However, they abandoned the value information brought by global features. Wang et al. [18] used partial pyramid matching kernel as the likelihood function to generate a set of probability maps with different cues and achieved considerable performance on several challenging public video sequences. Zheng et al. [27] proposed a coupled Multi-Index (c-MI) framework to perform feature fusion at indexing level which got a remarkable result in image retrieval task. However, no previous work has been reported to embed global feature into the SPM model to accomplish image classification task. In the meanwhile, to reduce the dimensionality of SPM based features, most of researchers just made use of dimensionality reduction algorithms directly. For example, Zhao et al. [26] reduced the feature vectors in ScSPM by PCA and utilized them in co-training method to accomplish image annotation task. Seldom of them discussed about whether this scheme was reasonable or not. In contrast, we take these problems into consideration in our work and propose corresponding algorithms to solve them.

3   The BoF Models

In this section, we review two different BoF models: the traditional SPM model by Lazebnik et al. [10] and the extended ScSPM model by Yang et al. [23]. The traditional SPM model usually has four stages: feature extraction-which extracts local descriptors such as SIFT [13], HOG [4] from overlapping and densely-sampling patches; vector quantization- which finds the nearest word from the codebook for each patch; spatial pooling-which divides the images into parts according to different scales and the codes are used to form the histograms; classification-which feeds the features into multi-class SVMs to do training and classification.

The differences between the traditional SPM and the ScSPM model are mainly in the 2nd and 3rd steps. The ScSPM model replaces the VQ operation with sparse coding which is more robust and accurate for less quantization loss,

$$\min_{\mathbf{U},\mathbf{V}} \sum_{m=1}^{M} \|\mathbf{x}_m - \mathbf{u}_m \mathbf{V}\|^2 + \lambda|\mathbf{u}_m|$$
$$subject\,to\|\mathbf{v}_k\| \leq 1, \forall k = 1, 2, \cdots, K \tag{1}$$

where the codebook $\mathbf{V}$ is an overcomplete basis set and the L2-norm constraint on $\mathbf{v}_k$ is applied to avoid trivial solutions in which $\mathbf{v}_k$ is k-th column of $\mathbf{V}$. Also the max pooling is adopted to get the maximum response in each region.

4   Proposed Approach

To make full use of different characteristics of global features and the SPM method, we proposed a multi-modal SPM model. This model can be divided into three major processes—namely multi-modal feature fusion, dimensionality reduction and fusion which is shown in Fig. 1. The details of the proposed model are as follows.

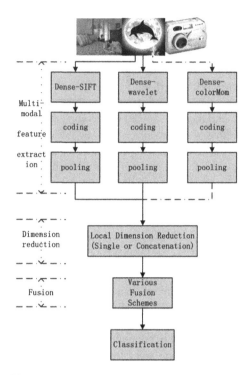

Fig. 1. The process of multi-modal SPM model.

4.1   Multi-modal Feature Extraction and Direct Fusion

As mentioned above, the SIFT descriptors can only describe the local gradient distribution. Therefore, if we could provide complementary information by fusing global
features such as color, wavelet, the performance is expected to be better. However, it's
useless to directly enlarge the obtained long feature vector after spatial pooling by
adding a short global feature.

Refer to Fig. 2, firstly as the way SIFT descriptors are extracted, we consider a local
patch with an area proportional to the scale around each sampled keypoint. Then, color
moments [17] or wavelet features [14] for each pixel in this area can be calculated.

The part for dense-colorMom is optional as the dotted line in Fig. 1 since not all
images have discriminative color information. Subsequently, following the traditional
SPM model—feature coding and spatial pooling, we will get a final feature vector for
each global feature similar to those SIFT based ones. Finally the concatenated feature
can be fed to multi-class SVMs directly to obtain the results.

4.2   Local Dimensionality Reduction

As we know, the concatenated feature is quite long especially after the multi-modal
feature fusion—doubling even tripling the dimension of single feature based one,
which will bring much burden for both memory and computational cost. However,

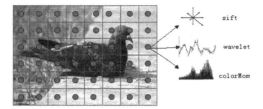

Fig. 2. Multi-modal feature extraction

directly conducting linear dimensionality reduction methods, e.g. Principal Component Analysis (PCA), on the features will be likely to damage the local structure and do harm to classification procedure afterwards. Accordingly, we proposed a kind of local dimensionality reduction scheme as Fig. 3 exhibits.

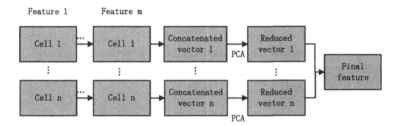

Fig. 3. The process of local dimensionality reduction scheme (here PCA is taken as the example).

In the figure, n represents the number of cells and m stands for the number of different kinds of features. Firstly, we need to divide the histograms into cells according to regions. By regarding different sub-histograms from various features formed in the same local cell as a whole, dimensionality reduction methods such as PCA can be applied to reduce the dimension.

Then the divided and reduced vectors will be concatenated again to form the final feature for classification. If we reduce a cell's dimension to k, then the dimension of the resulting feature will be n * k. And if m has the value of one, then this scheme reduces to apply local PCA on single feature vector—called single feature local PCA. By doing this, we can pick out some useful information hidden in the local fusion structure on one hand and improve the efficiency of the algorithm on the other hand with smaller data size.

4.3 Different Fusion Schemes

To make full use of the formed different mid-level features, we provide other two fusion schemes besides the mentioned above. All of them are summed up as follows:

(1) Concatenated feature vectors with global PCA (global);
(2) Concatenated feature vectors with local PCA (local);
(3) Considering the representation of each feature channel as the input of a classifier to form the posterior probability of each class and using the weight probabilities as the decision rule (posterior).

$$p(y = l|x) = \sum_{f=1}^{m} w_f p_f(y = l|x) \tag{2}$$

w_f is the weight for each feature channel.

(4) Taking mid-level feature to form the similarity matrix of different channels in MTJSRC (multi-task joint sparse representation and classification) model and using the information of all the features to make decision (MTJSRC) [25].

$$\min_{W} \frac{1}{2} \sum_{k=1}^{K} \left\| y^k - \sum_{j=}^{J} X_j^k w_j^k \right\|_2^2 + \lambda \sum_{j=1}^{J} \|w_j\|_2^2 \tag{3}$$

J and K are the number of classes and modalities; y is the testing sample, X is the training feature matrix and w_j is the weight for different classes. And the decision rule for the model is as follows:

$$j^* = \arg\min_{j} \sum_{k=1}^{K} \theta^k \left\| y^k - X_j^k \hat{w}_j^k \right\|_2^2 \tag{4}$$

θ^k is the weight that measures the confidence of different tasks in final decision.

5 Experiments and Results

In this section, the proposed method is compared with the competing models on two publicly available datasets which will be described briefly in the following part.

5.1 Experimental Settings and Datasets

15 Scenes Categorization consists of totally 4485 images from 15 scene categories, with the number of images for each category ranging from 200 to 400. The 15 categories vary from living room and kitchen to street and industrial. The Caltech-101 dataset contains 102 classes (including 101 classes of objects such as animals, vehicles, flowers and one class of background) with high shape variability. The number of images per category varies from 31 to 800. Imageclef 2012 Plant Dataset consists of 126 tree species from the French Mediterranean area and can be divided into three different kinds of pictures: scans, scan-like photos and free natural photos. Here the training set of this dataset is considered in our experimental study, which contains 8422 images (4870 scans, 1819 scan-like photos, 1733 natural photos).

All the images are resized to about 300 × 300. The features utilized in this paper are 128-dimensional SIFT, 81-dimensional Color Moment and 81-dimensional wavelet. And they are all extracted from 16 × 16 pixel patches densely on a grid with a step size of 8 pixels. The codebook is formed by using K-means and the size of codebook was 1024. We take multi-class linear SVMs as the classifiers. And all the three features are utilized for Caltech 101 while Color Moment is deleted from 15 Scenes Dataset and Imageclef 2012 plant Dataset for lack of color information and low discriminative power. So m in Fig. 3 is set to 3, 2 and 2 respectively and n has the value of 21 as we utilized a three-level SPM structure with linear SVM as the classifier (LSPM). Following the previous researches, for Caltech 101 and 15 Scenes Dataset, we take 30 and 100 images per class for training respectively with the rest for testing. For Imageclef 2012 Plant Dataset, as there are some categories with too few images, we only choose those categories whose size is larger than 30 for experiments. Finally, there are 7877 images falling into 81 different classes. For each class, half is randomly selected for training and the rest is kept for testing.

The experiments for each dataset were all repeated for 10 times with different randomly selected training and testing images. The mean of the recognition rates (%) across different categories were taken as the evaluation measure. All the parameters are validated with cross-validations.

5.2 Necessity for Multi-modal Feature Fusion

Column 1 in Table 1 shows the effects of multi-modal feature fusion on the datasets. Although the recognition rates for wavelet and colorMom are not so promising, from the table we can see that the feature fusion operation achieves enhanced performance. And the improvements are more apparent for Scenes 15 Dataset which obtains an increment of 3% and 1.7% for SPM and ScSPM respectively than SIFT only based ones. The irregular case for Imageclef 2012 Plant Dataset may be caused by improper choice of global features. More benefits on SPM model may be caused by the loss of quantization by VQ. The improvement of the performance of multi-modal feature fusion (direct fusion) scheme against those of single features is caused by the additional discriminant information brought by the cooperation and complementary between different feature channels.

5.3 Necessity for Local Dimensionality Reduction

Table 1 also exhibits the effects of different dimensionality reduction schemes on the datasets. The dimensions are set as the same for fair comparison. From it, we can reach the conclusion that mostly adopting dimension reduction in local levels will bring better results and dimension reduction is necessary especially the local one as global one sometimes will reduce the accuracy. Also recognition accuracy is further improved by combining multi-modal feature fusion and local dimensionality reduction. At last, we observe that the two operations have different effects on different datasets–15 Scenes Dataset benefits more from multi-modal feature fusion while Caltech 101

Table 1. Results for multi-modal fusion and dimension reduction. Bold numbers indicate the best for each feature.

Feature	Models					
	LSPM			ScSPM		
	Original	Global	Local	Original	Global	Local
colorMom	36.9 ± 0.8	**37.5 ± 2.0**	37.4 ± 1.1	51.6 ± 1.6	48.0 ± 1.6	**52.3 ± 1.3**
Wavelet	35.8 ± 0.5	36.1 ± 1.4	**37.1 ± 0.8**	53.8 ± 1.0	51.4 ± 0.8	**56.6 ± 1.4**
SIFT	48.2 ± 1.0	46.8 ± 1.0	**49.5 ± 1.3**	70.1 ± 1.1	70.9 ± 1.4	**72.5 ± 0.7**
Concatenation	49.7 ± 0.8	48.5 ± 1.0	**52.0 ± 0.9**	71.6 ± 1.1	63.8 ± 1.0	**73.5 ± 1.4**
(a) Caltech 101						
wavelet	65.9 ± 0.3	66.6 ± 0.9	67.8 ± 0.6	73.6 ± 0.3	73.8 ± 0.6	74.9 ± 0.5
SIFT	74.2 ± 0.4	74.4 ± 0.4	75.6 ± 0.5	80.3 ± 0.6	79.8 ± 0.4	81.5 ± 0.7
concatenation	77.2 ± 1.0	77.2 ± 0.3	78.2 ± 0.5	82.0 ± 0.5	82.0 ± 0.4	83.4 ± 0.5
(b) 15 Scenes						
Wavelet	46.0 ± 0.9	45.4 ± 0.8	**47.4 ± 1.3**	64.6 ± 0.7	70.4 ± 1.3	**71.5 ± 0.6**
SIFT	**70.5 ± 0.6**	69.1 ± 1.0	69.5 ± 0.7	84.3 ± 0.3	83.1 ± 0.5	**85.5 ± 0.2**
Concatenation	**72.5 ± 0.7**	71.1 ± 1.0	71.7 ± 0.6	83.2 ± 0.4	85.2 ± 0.7	**86.8 ± 0.4**
(c) Imageclef 2012 Plant						

Dataset and Imageclef 2012 Plant Dataset benefit more from local dimension reduction strategy. These experimental results demonstrate the effectiveness of the proposed operations and show that preserving local structure is of significance for improving recognizing performance.

Here we give a detailed explanation on the irregular case of Imageclef 2012 Plant Dataset. The multi-modal feature fusion operation has an opposite effect on the two models—positive on the SPM model and negative on the ScSPM model, due to the different characteristics of the models and adopted features. In the SPM model, the dimensionality reduction operation has a negative effect on the main feature (SIFT) which reduces the recognition accuracy of the whole multi-modal SPM model. However, the dimensionality reduction operation has a positive effect on both features in the ScSPM model, which in turn improves the degraded results caused by improper multi-modal feature fusion. These results demonstrate that the performance of the model is affected more by the dominant feature and the two components are both necessary for improving the performance.

5.4   Effectiveness of Multi-modal SPM Model

Table 2 provides the results for multi-modal SPM model with different fusion schemes introduced in Sect. 4.3. The results for 'global' can refer to Table 1. It can be observed that by combining the two operations in this paper with other fusion schemes will boost recognition rates again.

Table 2. Results for different fusion schemes. Bold numbers indicate the best performance. They are all based on local dimension reduction pipeline.

Models	Schemes		
	Concatenation	Posterior	MTJSRC
LSPM	52.0 ± 0.9	54.0 ± 1.3	**55.4 ± 0.9**
ScSPM	73.5 ± 1.4	74.0 ± 0.8	**75.3 ± 0.3**
(a) Caltech 101			
LSPM	**78.2 ± 0.5**	77.8 ± 0.6	76.6 ± 1.2
ScSPM	83.4 ± 0.5	82.8 ± 0.6	**84.2 ± 0.4**
(b) 15 Scenes			
LSPM	71.7 ± 0.6	71.0 ± 0.8	**79.3 ± 0.8**
ScSPM	86.8 ± 0.4	86.8 ± 0.4	**88.6 ± 0.5**
(c) Imageclef 2012 Plant			

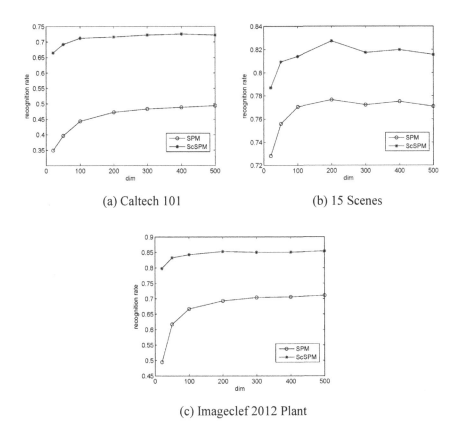

(a) Caltech 101   (b) 15 Scenes

(c) Imageclef 2012 Plant

Fig. 4. Effects of dimension. (Dim) stands for the dimension of a concentrated cell and the dimension of final feature is 21*dim.

5.5 Effects of Dimension

In Fig. 3, it analyzes the effect of dimension for multi-modal SPM model. From the figure, we can find that the trend of recognition rate is similar on all the datasets. It increases rapidly at first and becomes stable at about 200 for each cell which proves the robustness of the algorithm (Fig. 4).

6 Conclusion

In this paper, we proposed a multi-modal SPM model which fused various features to make full use of different characteristics of them. To reduce the dimension of combined feature and dig out hidden information in the local structure, we proposed a local dimensionality reduction scheme. The experiments on several public datasets demonstrate the effectiveness of the proposed algorithm.

However, there are still some open issues in the future. One is the selection of more appropriate feature as the feature combination doesn't always behave positively. The other is to design algorithms to capture feature dependencies by mining meaningful combinations of visual features more accurately.

References

1. Bosch, A., Zisserman, A., Muoz, X.: Scene classification using a hybrid generative/discriminative approach. IEEE Trans. Pattern Anal. Mach. Intell. **30**(4), 712–727 (2008)
2. Cao, L., Ji, R., Gao, Y., Yang, Y., Tian, Q.: Weakly supervised sparse coding with geometric consistency pooling. In: IEEE Conference on Computer Vision and Pattern Recognition (CVPR), 2012, pp. 3578–3585. IEEE (2012)
3. Chum, O., Philbin, J., Sivic, J., Isard, M., Zisserman, A.: Total recall: automaticquery expansion with a generative feature model for object retrieval. In: ICCV, pp. 1–8 (2007)
4. Dalal, N., Triggs, B.: Histograms of oriented gradients for human detection. In: Computer Vision and Pattern Recognition. vol. 1, pp. 886–893. IEEE (2005)
5. Elad, M., Aharon, M.: Image denoising via sparse and redundant representations over learned dictionaries. IEEE Trans. Image Proc. **15**(12), 3736–3745 (2006)
6. Fei-Fei, L., Perona, P.: A bayesian hierarchical model for learning natural scene categories. In: CVPR, vol. 2, pp. 524–531. IEEE (2005)
7. Gao, S., Tsang, I.W., Chia, L.T., Zhao, P.: Local features are not lonely–laplacian sparse coding for image classification. In: CVPR, pp. 3555–3561. IEEE (2010)
8. Griffin, G., Holub, A., Perona, P.: Caltech-256 object category dataset (2007)
9. Jurie, F., Triggs, B.: Creating efficient codebooks for visual recognition. In: ICCV, vol. 1, pp. 604–610. IEEE (2005)
10. Lazebnik, S., Schmid, C., Ponce, J.: Beyond bags of features: spatial pyramid matching for recognizing natural scene categories. In: CVPR, pp. 2169–2178 (2006)
11. Li, F.F., Fergus, R., Perona, P.: Learning generative visual models from few training examples: an incremental bayesian approach tested on 101 object categories. Comput. Vis. Image Underst. **106**(1), 59–70 (2007)

12. Long, M., Ding, G., Wang, J., Sun, J., Guo, Y., Yu, P.S.: Transfer sparse coding for robust image representation. In: CVPR, pp. 407–414. IEEE (2013)
13. Lowe, D.G.: Distinctive image features from scale-invariant keypoints. Int. J. Comput. Vis. **60**(2), 91–110 (2004)
14. Manjunath, B., Ma, W.: Texture features for browsing and retrieval of image data. IEEE Trans. Pattern Anal. Mach. Intell. **18**(8), 837–842 (1996)
15. Oliva, A., Torralba, A.: Building the gist of a scene: the role of global image features in recognition. Prog. Brain Res. **155**, 23–36 (2006)
16. Quelhas, P., Monay, F., Odobez, J.M., Gatica-Perez, D., Tuytelaars, T., Van Gool, L.: Modeling scenes with local descriptors and latent aspects. In: ICCV, vol. 1, pp. 883–890. IEEE (2005)
17. Stricker, M., Orengo, M.: Similarity of color images. In: SPIE Conference on Storage and Retrieval for Image and Video Databases, vol. 2420, pp. 381–392, San Jose, USA (1995)
18. Wang, D., Lu, H., Chen, Y.W.: Object tracking by multi-cues spatial pyramid matching. In: ICIP, pp. 3957–3960. IEEE (2010)
19. Wang, M., Gao, Y., Lu, K., Rui, Y.: View-based discriminative probabilistic modeling for 3d object retrieval and recognition. IEEE Trans. Image Proc. **22**(4), 1395–1407 (2013)
20. Wang, M., Li, W., Liu, D., Ni, B., Shen, J., Yan, S.: Facilitating image search with a scalable and compact semantic mapping. IEEE Trans. Cybern. **45**(8), 1561–1574 (2015)
21. Wang, M., Liu, X., Wu, X.: Visual classification by l1-hypergraph modeling. IEEE Trans. Knowl. Data Eng. **27**(9), 2564–2574 (2015)
22. Wu, J.X., Rehg, J.M.: Where am i: place instance and category recognition using spatial pact. In: CVPR, pp. 1–8. IEEE (2008)
23. Yang, J., Yu, K., Gong, Y., Huang, T.: Linear spatial pyramid matching using sparse coding for image classification. In: CVPR, pp. 1794–1801. IEEE (2009)
24. Yin, H., Cao, Y., Sun, H.: Combining pyramid representation and adaboost for urban scene classification using high-resolution synthetic aperture radar images. Radar Sonar Navig. IET **5**(1), 58–64 (2011)
25. Yuan, X.T., Liu, X., Yan, S.: Visual classification with multitask joint sparse representation. IEEE Trans. Image Proc. **21**(10), 4349–4360 (2012)
26. Zhao, Z.Q., Glotin, H., Xie, Z., Gao, J., Wu, X.D.: Cooperative sparse representation in two opposite directions for semi-supervised image annotation. IEEE Trans. Image Proc. **21**(9), 4218–4231 (2012)
27. Zheng, L., Wang, S., Liu, Z., Tian, Q.: Packing and padding: coupled multi-index for accurate image retrieval. In: CVPR (2014)

Coverless Information Hiding Based on Robust Image Hashing

Shuli Zheng[1], Liang Wang[1], Baohong Ling[2], and Donghui Hu[1(✉)]

[1] School of Computer and Information,
Hefei University of Technology, Hefei 230009, China
hudh@hfut.edu.cn
[2] Department of Information Engineering,
Anhui Broadcasting, Movie and Television College, Hefei 230022, China

Abstract. Traditional image steganography modifies the content of the image more or less, it is hard to resist the detection of image steganalysis tools. New kind of steganography methods, coverless steganography methods, attract research attention recently due to its virtue of do not modify the content of the stego image at all. In this paper, we propose a new coverless steganography method based on robust image hashing. Firstly, we design an effective and stable image hash by using the orientation information of the SIFT feature points. Then the local image database is created and the corresponding hash values of these images in the database are computed. Secondly, the secret message is divided into segments with the same length as the hash sequences. And a series of images are chosen from the image database by matching the secret information segments and the hash sequences of all the images. Finally, these images are transmitted as the carriers of the secret information. When the receiver receives these images, the secret information is extracted by using the shared hash method. Due to the characteristics that SIFT features can resist common image attacks in a certain extent, the secret information corresponding to the hash has strong robustness. To improve the retrieval and matching efficiency of the hashing system, an inverted index of quadtree structure is designed. Compared with the traditional image steganography, this method does not modify the content of the image itself, therefore, can effectively resist steganalysis tools. Furthermore, we compare the proposed method with state-of-art coverless steganography method which also based on image hash, and experimental results show that our method has higher capacity, robustness and security than the method proposed in [15].

Keywords: Information hiding · Coverless · Image hash · SIFT feature

1 Introduction

With the rapid development of computer network technology, the research on information hiding is developing rapidly, and has become an important research area in the field of information security. As the important research direction of information hiding, steganography constantly promotes the development of information hiding technology. In the way of traditional steganography, the secret information is embedded into the

© Springer International Publishing AG 2017
D.-S. Huang et al. (Eds.): ICIC 2017, Part III, LNAI 10363, pp. 536–547, 2017.
DOI: 10.1007/978-3-319-63315-2_47

carriers which is used as the cover of secret information by an invisible way, so as to achieve the purpose of secret communication. The carriers include common digital images, audios, videos and so on [1]. Mostly, traditional steganography directly modifies the carrier to embed secret data. Owing to the partial distortion of the carrier, the third party can detect the existence of hidden secret information by finding the statistic evidence introduced by the embedding method.

Among all the carriers of information hiding, digital images are the most widely used. In traditional image steganography, pixel values are modified to achieve the embedding of secret information. According to the different ways of hiding, the common steganography methods can be classified into two categories: hiding methods in spatial domain and transform domain. Steganography in the spatial domain, such as the method proposed in [2] which replace the LSB (least significant bits) of the image with secret data. In order to improve the capacity and invisibility of hiding, an adaptive LSB hiding method is proposed in [3] by PVD (pixel-value difference); The transform domain method is to modify the host image data to change some statistical features to achieve data hiding [4–11], such as the hidden method in DCT (discrete cosine transform) domain [5–7], DWT (discrete wavelet transform) domain [8, 9], and DFT (discrete Fourier transform) domain [10, 11].

These methods modify the carrier images according to certain rules to embed the secret information, it is inevitable to leave some traces of modification on the carrier. Hence, we are facing such a problem: these steganography methods cannot resist the detection of existing steganalysis tools. For example, the method proposed by [16] can effectively detect the carrier image which is embedded using LSB replacement. To solve this problem, researchers have proposed the concept of coverless information hiding. And in recent years, some coverless information hiding has appeared [12–15]. These methods do not modify the carrier image, so that the secret information cannot be detected by steganalysis tools, which greatly enhances the security of information hiding. The methods can be classified into two types: one is generating digital images driven by secret data and some function relationship [12, 13], such as hiding secret data by generating texture images; the other is to establish the mapping relationship between the original image and the secret information to express the secret information [14, 15]. The secret information is represented by establishing the correspondence between the image and the Chinese vocabulary [14]. Recently, literature [15] proposed a method which directly mapping image to secret information by image hashing. Although these new methods can effectively resist the existed steganalysis tool, but the capacities of these methods are small. In this paper, we propose a new method of information hiding from the point of view of both enhancing capacity and security, which constructing a robust image hash with SIFT algorithm to hide the secret data. The hash sequence of each image is not unique, in other words, each image can be represented by a great deal of 18 bits' binary hash sequences. This method doubled the capacity of steganography in [15], and can also effectively resist common steganalysis tools.

The rest of this paper is organized as follows. Section 2 describes the proposed method, Sect. 3 shows performance analysis and experiments, and conclusions are presented in Sect. 4.

2   The Proposed Image Steganography Without Embedding

The flowchart of the proposed coverless image steganography algorithm is illustrated as Fig. 1. We propose a hash algorithm based on SIFT feature which can extract 18-bits binary sequences from each image as the robust hash value. Rather than directly embed the secret data into images, we establish the relationship between the images and secret data by hash map. Before the secret communication, it is necessary to set up a local image database, which should contain images with the same hash values as all of 18-bits binary sequences. Owing to 18-bits binary hash of each image, we need 2^{18} images with different hashes to represent 18-bits different binary information. It is not practical to create such a huge image database. To solve this problem, we adjust the originally hash scheme in which an image can only correspond to one kind of hash sequences, we make each image represented by a number of different hash sequences, which reduce the volume of image database significantly, moreover, increase the security of secret data. We randomly downloaded 5000 images from the Internet and normalized these images to the size of 512*512. Then we calculate the hash values of all the images and store them into the quadtree index structure for fast retrieval. The secret information is divided into 18-bits segments to match the quadtree. According to the information stored in the leaf node of the quadtree, the corresponding image is selected as a carrier of the current 18-bits secret data.

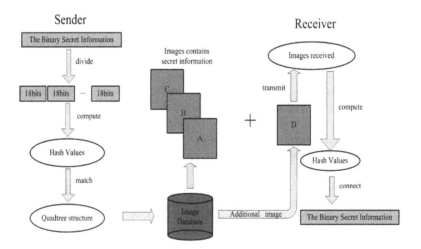

Fig. 1. The flowchart of the proposed steganographic algorithm

Considering the length of information is possible not multiples of 18, the last image perhaps contains less than 18-bits binary information. In order to find the general applicability of the algorithm, the information containing the binary secret data length and the direct correspondence between the hash sequences and the image is hidden in another image picked randomly from the image database with LSB algorithm. Thus, the images containing secret information are actually a series of images without being embedded and one containing additional information. The receiver accepts these images in turn and get some necessary information from the additional image firstly, then extract secret information from the other images using the shared hash algorithm.

2.1   Robust Image Hashing Method

In the process of transmission, images may encounter various content changes, such as luminance change, contrast enhancement, JPEG compression, rescaling and noise adding etc. In case of these random tampering in transmission, image hash used for information hiding must be robust against various attacks as much as possible. We proposed a robust image hash based on image SIFT features algorithm. As shown in the Fig. 2, we get the 512*512 image by image normalization, then divide the image into 3*3 segments.

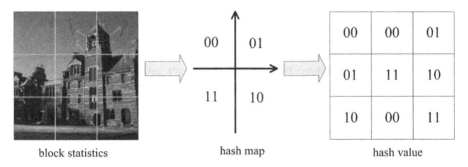

block statistics   hash map   hash value

Fig. 2.   The robust image hash scheme (Color figure online)

It is well known that the SIFT features are very stable, which is usually detected at the edges and corners of the image and some areas where brightness changes significantly. Therefore, points computed by SIFT algorithm will not disappear with these various content changes, such as luminance change, contrast enhancement, JPEG compression, rescaling, noise adding and so on. We make use of SIFT's stability to form the robust image hash. The local extreme point of each layer of the image difference pyramid is calculated on each image block. Usually the number of such stable points will be very considerable, there may be dozens which is beyond our demand. By adjusting the threshold, we give up some redundant points and make a few stable points preserved. For each stable point extracted on image blocks, the appropriate size of a window is selected around the point (circular area). The gradient directions of all the sampling points in the window are accumulated to form a histogram. Then we compare all the values of this histogram to get the max one, if the max one locates between 0 and 90°, the hash value of this image block is set to 00; otherwise, if the max one locates between 90 and 180°, the hash value is set to 01, and so on. Especially, supposing there is no SIFT extreme point on the block, the value of this block is also set to 00. To better show the image hash process, we draw the SIFT feature points (the green arrow) in Fig. 2 calculated by the algorithm. In SIFT algorithm process, the low contrast feature points are eliminated by adjusting the threshold. We can see in Fig. 2 that there are appropriate number of feature points left in each block by adjusting the threshold value from the default to 0.2. The interval values of the statistical histogram of all the feature points in each block are accumulated. The max value in this histogram is illustrated in Fig. 2 by red arrow.

We count the orientation information of selected obvious extreme points of each block and connect the hash values of nine segments in certain order to form final image hash. It is worth noting that rather than computing the completed SIFT descriptor, we only need to count the gradient of the extreme points in the whole calculation process. Furthermore, only a few representative feature points are extracted, the cost of calculation is greatly reduced.

2.2   Inverted Index of Quadtree Structure

The secret information is divided into 18-bits segments which need to match images. Consequently, there are millions different hash values to meet the demand of 18 bits meta data. There is an equivalent amount of local hash records that must be matched accurately. But it is time consuming to compare the hash sequences with so massive records. In order to speed up the matching of secret information and images, as illustrated in Fig. 3, the quadtree index structure is created. The height of this quadtree is 10, each node has four child nodes whose values are 00, 01, 10, 11, corresponding to four different values of image blocks. There is a list in each leaf node respectively to store the information about images, the hash sequences of which in some certain order of blocks is the same as current secret information segment. The order here is set to

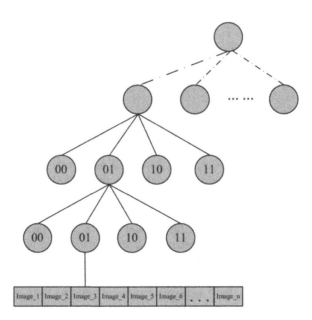

Fig. 3. The inverted index of quadtree structure

0–255, which is selected from all arrangements of the 9 image blocks. In the best case, each image can represent 256 different 18-bits binary sequences, so that the size of image database will be reduced to less than 2000. As the matter of fact, each leaf node stores only one image, the whole secret communication can be carried out normally. Of course, to keep secret information imperceptible, there are at least two images information stored in each leaf node.

2.3   Process of Information Hiding and Extraction

Process of Hiding

Step 1:   As result of that one image represents 18-bits binary information, the secret information must be divided into n secret binary segments. Supposing the length of secret data is l, n and l satisfy the following relationship:

$$n = \begin{cases} l/18 & if\ l\%18 = 0 \\ l/18 + 1 & if\ l\%18! = 0 \end{cases} \tag{1}$$

Noting that the length of last segment divided by secret data may be less than 18, if so, several zeros is added to the end of this segment to ensure the length of 18.

Step 2:   Matching inverted index quadtree structure with every two bits of these secret data segments computed by Step 1. Each segment will correspond to a leaf node. And each image stored in the leaf node can be used as the carrier of this confidential data segment. We know from Sect. 2.2 that there are at least two images stored in each leaf node. If there are the same segments appear in communication, we choose the different images stored in the same leaf nodes in turn. For example, there is 10b secret data which contains some same 18-bits segments. We need 5 images ((10*8)/18 + 1 = 5) as carriers. The last image actually contains only 8 bits (80−(4*18) = 8) data following 10 auxiliary zeros. To avoid duplication of stego images, different images stored in the same leaf node will be selected in turn as stego images of these same segments.

Step 3:   An image is selected randomly from the image database, then the information that the block orders of these images matched from Step 2 and the length of secret information l is hidden into this image with LSB replacement steganography. It is worth noting that the block orders of these stego image are also stored in the leaf node of the quadtree index structure. Finally, n images containing secret information and an additional image are sent to the receiver.

Process of Extraction
The receiver sequentially receives all images, and extracts the information that the order of image blocks and the length of secret data from the additional image. To

ensure the integrity of secret information, the data length s contained in the last image is computed as follows:

$$s = \begin{cases} l \, \% 18 & \textit{if } l \, \% 18! = 0 \\ 18 & \textit{if } l \, \% 18 = 0 \end{cases} \tag{2}$$

The secret information hidden in each image is actually connected by hash value of each image block in certain order. There are n integers between 0–255 extracted from the additional image as the orders of these stego images. These orders are selected from all arrangements of the 9 image blocks and critical to the extraction of secret data. Then the hash sequences of the stego images are calculated by shared hash algorithm according to these orders. Finally, the hash sequences of these images are connected to acquire the secret data.

3   Performance Analysis and Experiments

In this paper, the experiment and analysis include the following four parts: steganographic capacity, robustness, security and resist to detection. All the experiments run in VS2013 environment with part of the functions provided by opencv3.0. The image database is created by 5000 images downloaded randomly from the Internet.

3.1   Capacity of Steganography

Due to the direct mapping from secret information to image hash sequences, the steganographic capacity is limited by the length of the image hash. This phenomenon is inevitable that the capacity of the method proposed by [15] is obviously less than that of traditional image steganography. In [15] each image transmits 8 bits of binary information, by contrast, our method enlarges the capacity by one time (Table 1).

Table 1. The number of images needed when the same data is hidden

	lb	10b	100b	Ikb
Method proposed by [15]	1	10	100	1024
Our method	2	6	46	457

We randomly generate some binary bit streams as secret information to be transmitted, experiments are carried out by the method in [15] and the method proposed by this paper, the number of images needed are observed. Experiments show that the steganographic capacity of our method is two times that of the method in [15].

3.2   Robustness to Common Attacks

In the transmission process, it is inevitable to encounter all kinds of content damage, such as image noise, JPEG compression, rescaling, luminance change, contrast enhancement, and so on. The information extracted from the image must be able to resist these factors. That is to say, the hash algorithm is robust to these attacks. The SIFT feature is invariant to rotation, scaling, and brightness variations, and is a very stable local feature. We effectively exploit the robustness of this feature. And instead of directly adopting the SIFT algorithm, only the histogram statistics of the image gradient are adopted to form the final hash sequences. This approach further increases the robustness of the algorithm. 100 images were randomly selected from the image database as the experimental objects. To give a measure of robustness, we define success rate of secret data extraction $s = 1 - \sum_{i=1}^{n} \{d(M_i, M_i')\}/nl$, where $d(x,y)$ denotes the Hamming distance between vectors x and y, M_i and M_i' indicate respectively the embedded information and extracted information from image i. l is the embedding capacity of each image, n is sample size and it is set to 100 in experiments. We compared the performance of secret information extraction under common image attacks for LSB-replacement, JSteg, our method and method proposed by [15].

From Fig. 4, we can find that with the improvement of the JPEG compression quality, the success rate of our method is obviously higher than that of the method proposed by [15] and it reaches 1 when the compression quality is 90%. Performance of JSteg approaches to that of method in [15]. LSB-replacement is sensitive to JPEG compression. In Fig. 5, according to the different noise ratio, a fixed number of pixels are randomly selected from image and their value is modified as 0 or 255. We can see that with the increase of the noise ratio, the success rate of extraction using our method decrease more slowly than that by using the method in [15]. And LSB-replacement performs better than three other methods.

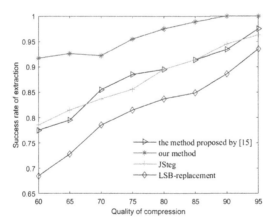

Fig. 4. Success rate of extraction of secret data when stego images under varying JPEG compression quality

We add varying degrees of Gauss noise to images by standard Gauss function in which the parameter u is set to 0 and σ is set from 0.5 to 4. Figure 6 shows that the success rate of extraction of our method stays nearly above 90%. But the success rate of

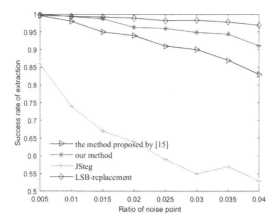

Fig. 5. Success rate of extraction of secret data when stego images under varying ratio of noise point.

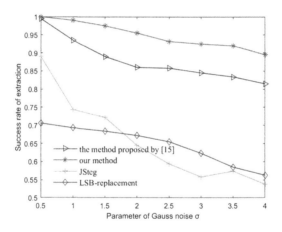

Fig. 6. Success rate of extraction of secret data when stego images under varying parameter of Gauss noise.

extraction by using method in [15] descends sharply with enlargement of σ. JSteg and LSB-replacement show a high sensitivity to Gauss noise. Our method and the method proposed by [15] both can achieve very high success rate (100%) of extraction when the images are modified by rescaling changes, which is shown in Fig. 7. Similarly, both two methods have same performance when they deal with brightness and contrast changes.

Stego images are subject to various types of attacks and signal processing in transmission. To accurately reflect the circumstances of images conveyed in the real world, we give the following experiments on different combinations of some distortions in Tables 2 and 3, where σ is a parameter in the generation function of Gauss noise, ρ is the ratio of random noise and q is JPEG compression quality. From results of

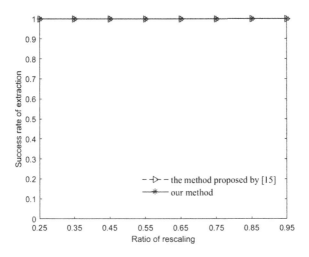

Fig. 7. Success rate of extraction of secret data when stego images under varying ratio of scaling

Table 2. Success rate of extraction under varying degrees of Gauss noise and random noise

	$\sigma = 0.5$ $p = 0.01$	$\sigma = 1$ $p = 0.015$	$\sigma = 1.5$ $p = 0.02$	$\sigma = 2$ $p = 0.025$	$\sigma = 2.5$ $p = 0.03$	$\sigma = 3$ $p = 0.035$	$\sigma = 3.5$ $p = 0.04$	$\sigma = 4$ $p = 0.005$
Method in [15]	0.976	0.915	0.873	0.854	0.838	0.832	0.827	0.796
Our method	0.988	0.974	0.970	0.943	0.933	0.926	0.915	0.883
Jsteg	0.843	0.729	0.681	0.643	0.586	0.579	0.530	0.487
LSB-replacement	0.703	0.680	0.564	0.535	0.549	0.474	0.536	0.529

Table 3. Success rate of extraction under varying degrees of Gauss noise and JPEG compression

	$\sigma = 0.5$ $q = 0.95$	$\sigma = 1$ $q = 0.90$	$\sigma = 1.5$ $q = 0.85$	$\sigma = 2$ $q = 0.80$	$\sigma = 2.5$ $q = 0.75$	$\sigma = 3$ $q = 0.70$	$\sigma = 3.5$ $q = 0.65$	$\sigma = 4$ $q = 0.60$
Method in [15]	0.932	0.897	0.846	0.828	0.824	0.817	0.794	0.783
Our method	0.966	0.953	0.945	0.943	0.939	0.910	0.891	0.875
Jsteg	0.817	0.694	0.630	0.619	0.605	0.569	0.514	0.496
LSB-replacement	0.698	0.626	0.613	0.551	0.514	0.509	0.487	0.493

experiments in tables we draw that our method is more robust than three other method despite images encounter two types of distortion. Overall, experimental results show that our method has some competitive advantages when it deals with JPEG compression, Gauss noise and random noise.

3.3   Security of Communication

In this paper, the steganographic method is proposed to ensure the security of the stego images. Compared with traditional information encryption and transmission method, which modify the carrier data so that the secret information has certain perceptibility, our method does not modify the carrier image except the additional image. And the same hash sequences are not mapping to the same image when choosing an image containing a secret message. In this way, the disadvantage that the same images may appear in transmission is avoided. Moreover, the order of image blocks is not fixed so that each image can correspond to multiple hash sequences. Even if the same images appear in transmission, the same images represent different secret data because of their different blocks order. In general, this algorithm has high security, so that it is difficult for malicious attackers to decipher the secret information.

3.4   Resist to Detection

According to the mapping relationship between the secret information and robust image hash, the secret information can be converted into the corresponding image sequences. Which means that except for the additional image that containing some additional information, the carriers are the original images without any modification. It is obvious that the method proposed by this paper can resist the detection of common steganalysis tools [16] based on statistical analysis.

4   Conclusions

In this paper, we proposed a new coverless steonagaphy method based on robust image hashing. The mapping relationship between the secret information segments and the stego image is created without modifying the original images. Because the carrier image is not modified, the third party will not easily perceive anomalies about stego images. Furthermore, this method can resist detection of all the existing steganalysis tools. And the hash method applies the orientation information of the SIFT feature points which made the robustness of image hash enhance effectively. Experimental results show that the hash can resist common image attacks. And the inverted index of quadtree structure is created to store hash sequences of all the images in image database, which speeds up the matching between secret information segments and images. Because of the 18-bits image hash sequences, there are at least 2^{18} images in the image database. Each image can form many different hash sequences by connecting the image blocks in different orders which reduces the size of the image database greatly. In general, the new steganography proposed by this paper has some advantages, such as security and robustness. We doubled the capacity of steganography compared with the method in [15], but the capacity is still much less than that of traditional steganography. It is necessary to set up a larger image database with the increasing of capacity. In future work, we will focus on how to further improve the capacity of steganography while ensuring the normal volume of the image database.

Acknowledgments. This work was supported in part by the Natural Science Research Project of Colleges and Universities in Anhui Province (NO. KJ2017A734), and in part by the Natural Science Foundation of Anhui Province (NO. 1508085MF115).

References

1. Fridrich, J.: Steganography in Digital Media: Principles. Algorithms and Applications. Cambridge University Press, Cambridge (2009)
2. Wu, H.C., Wu, N., Tsai, C.S., Hwang, M.S.: Image steganographic scheme based on pixel-value differencing and LSB replacement methods. Proc.-Vis. Image Sig. Process. **152**, 611–615 (2005)
3. Yang, C.H., Weng, C.Y., Wang, S.J.: Adaptive data hiding in edge areas of images with spatial LSB domain systems. IEEE Trans. Inf. Forensics Secur. **3**(3), 488–497 (2008)
4. Li, Z., Chen, X., Pan, X., Zheng, X.: Lossless data hiding scheme based on adjacent pixel difference. In: Proceedings of the International Conference on Computer Engineering and Technology, pp. 588–592 (2009)
5. Penvny, T., Fridrich, J.: Merging Markov and DCT features for multi-class JPEG steganalysis. In: Electronic Imaging, Security, Steganography, and Watermarking of Multimedia Contents IX, Proceedings of SPIE, vol. 6505, pp. 3–4 (2007)
6. Ruanaidh, J.J.K.O., Dowling, W.J., Boland, F.M.: Phase watermarking of digital images. In: International Conference on Image Processing, pp. 239–242 (1996)
7. Fridrich, J., Goljan, M.: Practical steganalysis of digital images-state of the art. In: Security and Watermarking of Multimedia Contents IV, Proceedings of SPIE, vol. 4675, pp. 1–13 (2002)
8. Cox, I.J., Kilian, J., Leighton, F.T.: Secure spread spectrum watermarking for multimedia. IEEE Trans. Image Process. **6**(12), 1673–1687 (2010)
9. Chen, W.Y.: Color image steganography scheme using set partitioning in hierarchical treescoding, digital Fourier transform and adaptive phase modulation. Appl. Math. Comput. **185**(1), 432–448 (2007)
10. Hsieh, M.S., Tseng, D.C., Huang, Y.H.: Hiding digital watermarks using multiresolution wavelet transform. IEEE Trans. Industr. Electron. **48**(5), 875–882 (2001)
11. Lin, W.H., Horng, S.J., Kao, T.W.: An efficient watermarking method based on significant difference of wavelet coefficient quantization. IEEE Trans. Multimedia **10**(5), 746–757 (2008)
12. Wu, K.C., Wang, C.M.: Steganography using reversible texture synthesis. IEEE Trans. Image Proc. **24**(1), 130–139 (2015)
13. Xu, J., Mao, X., Jin, X.: Hidden message in a deformation-based texture. Vis. Comput. **31**, 1653–1669 (2015)
14. Chen, X., Sun, H., Tobe, Y., Zhou, Z., Sun, X.: Coverless information hiding method based on the chinese mathematical expression. In: Huang, Z., Sun, X., Luo, J., Wang, J. (eds.) ICCCS 2015. LNCS, vol. 9483, pp. 133–143. Springer, Cham (2015). doi:10.1007/978-3-319-27051-7_12
15. Zhou, Z., Sun, H., Harit, R., Chen, X., Sun, X.: Coverless image steganography without embedding. In: Huang, Z., Sun, X., Luo, J., Wang, J. (eds.) ICCCS 2015. LNCS, vol. 9483, pp. 123–132. Springer, Cham (2015). doi:10.1007/978-3-319-27051-7_11
16. Xia, Z.H., Wang, X.H., Sun, X.M., Wang, B.W.: Steganalysis of least significant bit matching using multi-order differences. Secur. Commun. Netw. **7**(8), 1283–1291 (2014)

Image Firewall for Filtering Privacy or Sensitive Image Content Based on Joint Sparse Representation

Zhan Wang, Ning Ling, Donghui Hu$^{(\boxtimes)}$, Xiaoxia Hu, Tao Zhang, and Zhong-qiu Zhao

School of Computer Science and Information, Hefei University of Technology, Hefei 230009, China
hudh@hfut.edu.cn

Abstract. As the commonest part of social networks, sharing an image in social networks transmits not only can provide more information, but also more intuitive than any text. However, images also can leak out information more easily than text, so the audit of image content is particularly essential. The disclosure of a tiny image, which involves sensitive information about individual, society even the state, may trigger a series of serious problems. In this paper, we design an image firewall to detect sensitive image content through joint sparse representation on features. We take LBP, SIFT and Wavelet features into consideration, trying to find an effective combination among these features. We also find some features, which have the same accuracy but less time cost. In addition, we consider the spatial relation of the detected objects, especially the distance between the persons appeared in an image. Experimental results show the effectiveness of the proposed methods.

Keywords: Semantic-based rules · Joint sparse representation · Image classification · Application level firewall

1 Introduction

In recent years, Online Social Networks (OSNs) such as Facebook, Twitter, Weibo, and Wechat are becoming more and more influential and important in people's daily life. Although these OSNs can shorten our distance to the whole world, inappropriate social behaviors can lead to unnecessary troubles. A simple image can show not only one's daily life, but also some extra information, such as life track, habit. Some social network users may accidentally publish sensitive images that may disclose personal information, leave business secrets or even endanger the safety of a country. After the instance of Prism, people pay more and more attention to the issues of protection the privacy of published information. Though protecting the privacy of digital images has become a very important research topic, the related researches in this area are still limited.

Over the last decades, researchers tried to design different kinds of features to describe an image, like LBP and SIFT etc. Combining different features together to describe an image usually causes a high dimension of images, which may affect the

© Springer International Publishing AG 2017
D.-S. Huang et al. (Eds.): ICIC 2017, Part III, LNAI 10363, pp. 548–559, 2017.
DOI: 10.1007/978-3-319-63315-2_48

efficiency of the image process. Therefore, some works to reduce the dimension have been proposed, such as the PCA [1] method and the sparse representation method. Sparse representation was first used in face recognition in 2009 [6]. It means that sparse coding has a stronger representation power to describe an image. As an improvement of sparse representation, multi-task joint sparse representation classifier (MTJSRC) [2] combines different kinds of features, making it more accurate on classification. In this paper, we propose an image firewall based on joint sparse representation. Compared with previous work, our proposed model has the following two attributes: first, joint sparse representation is used for private image registration and privacy rule matching, which can provide more accurate and robust privacy prediction; second, the spatial relationships, like the distance and position of the detected objects, are considered in the new proposed method.

The remainder of the paper is organized as follows. In Sect. 2, we discuss related work including sparse representation, multitask joint sparse representation, and digital image privacy decision recommendation. In Sect. 3, we describe our firewall model and its implementation in detail. Section 4 presents evaluation results of subtasks including face recognition, object categorization, distance and position detection. Finally, in Sect. 5, we draw some conclusions and discuss directions in the future work.

2 Related Work

2.1 Sparse Representation

Sparse representation is an interesting research area over the past 20 years. Many efficient algorithms [4] about sparse representation were presented, and some excellent theoretical frameworks appeared [5]. It has been demonstrated in [6] that the sparse representation model is discriminative and particularly useful for building robust multi-class visual classification systems. Let us consider a K class classification problem, where a matrix $X \in \mathbb{R}^{d \times n}$ represent n columns of training images feature vectors of dimension d, and $X_k \in \mathbb{R}^{d \times n_k}$ ($\sum_{k=1}^{K} n_k = n$) denotes K classes of image set. Given a testing image with feature $y \in \mathbb{R}^d$, the sparse linear representation model seeks to solve the following optimization problem:

$$\widehat{w} = \arg\min_{w} \|w\|_0, s.t. \|y - Xw\| \leq \in \tag{1}$$

where $\|\cdot\|_0$ symbolize $\ell_0 - norm$ which counts the number of non-zero entries in a vector and $\in$ is the noise level parameter. Unfortunately, $\ell_0 - norm$ is a NP-hard problem. Results [5, 6] show that under the assumption that the solution w is sparse enough, the sparse representation can be recovered by solving the $\ell_1 - norm$ minimization:

$$\widehat{w} = \arg\min_{w} \|w\|_1, s.t. \|y - Xw\| \leq \in \tag{2}$$

where the $\ell_1 - norm$ is a convex optimization of $\ell_0 - norm$, and it can be efficiently optimized through several well toolboxes, such as NESTA [7]. Given the optimal solution $\widehat{w}$, the class label of y is decided on the following equation of minimum reconstruction error:

$$\widehat{k} = \arg\min_{k \in \{1,...,K\}} \|y - X_k \widehat{w}_k\| \tag{3}$$

where $\widehat{w}_k$ is the components of $\widehat{w}$ restricted on class k. The model (2) together with the decision rule (3) is known as sparse representation-based classification (SRC) in the study of face recognition [6].

2.2   Multitask Joint Sparse Representation and Classification

Suppose we have a training set with P classes and each sample has Q different modalities of features (e.g., color, shape and texture). For each modality index $q = 1,...,Q$, $X^q \in \mathbb{R}^{d_q \times n}$ denotes the training feature matrix. Let $X_p^q \in \mathbb{R}^{d_q \times n_p}$ be the n_pth columns of X^q associated with the qth class. In addition, we suppose that a testing sample is given by $y = \{y^{qr} \in \mathbb{R}^{d_q}, q = 1, \ldots, Q, r = 1, \ldots, R\}$ as an ensemble of R different instances (e.g., multiple views of a human face), each of which is represented by the same Q modalities as training images. Let us consider the following $Q \times R$ linear representation models:

$$y^{qr} = X^q W^{qr} + \varepsilon^{qr}, \quad q = 1, \ldots, Q, \quad r = 1, \ldots, R \tag{4}$$

where $W^{qr} \in \mathbb{R}^n$ is the coefficient vector for y^{qr} and $\varepsilon^{qr} \in \mathbb{R}^{d_q}$ is the residual term, which is i.i.d. Gaussian noise. Then $\{W^{qr}\}$ can be estimated by fitting the following least squared regression (LSR) model:

$$\min_W \left\{ f(W) := \frac{1}{2} \sum_{q=1}^Q \sum_{r=1}^R \|y^{qr} - X^q W^{qr}\|^2 \right\} \tag{5}$$

where $W \in \mathbb{R}^{n \times QR}$ is the matrix stacked by $Q \times R$ columns of coefficient vectors $\{W^{qr}\}$. To avoid singularity of linear systems, an additional regularization term $\lambda \|W\|_F^2$ is typically imposed in (5). From the viewpoint of multi-task learning, the problem of (5) is a multi-task regression model with $Q \times R$ independent LSR.

Given the optimal coefficient matrix $\widehat{W}$, one can approximate feature y^{qr} as $\widehat{y}^{qr} = X^q \widehat{W}_p^{qr}$. The decision is made in favor of the class with the lowest reconstruction error accumulated over all the $Q \times R$ tasks:

$$\widehat{p} = \arg\min_{p \in \{1,...,p\}} \sum_{q=1}^Q \sum_{r=1}^R \left\| y^{qr} - X_p^q \widehat{W}_p^{qr} \right\|^2 \tag{6}$$

In the work of Yuan et al. [2], they call the model (5) together with the decision rule (6) as multi-task joint sparse representation and classification (MTJSRC). Particularly, when $L = K = 1$, the MTJSRC reduces to a regularization form of SRC [6].

2.3 Digital Image Privacy Decision Recommendation for Sharing Images in OSNs

It is a threat to privacy when sharing the digital image blindly in the social networks. However, due to the complexity of privacy semantic understanding and the uncertain of privacy decision, there are no effective models for privacy decision when sharing image in social networks. On account of that condition, Hu et al. [8] concluded some constructive conclusions, by investigating the purposes, attitudes, preferences, modes and recommendations of OSNs users when sharing images among OSNs. They found that based on either the privacy levels or the attribute tags of images shared in social networks, there is a partial order, which can be used as guidance in the design of the privacy decision or access control methods. Hu et al. [3] designed an automatic recommendation system for social network privacy decisions based on robust perceptual hashing and semantic privacy rules, in order to compute the privacy level of an image after the user registered a small number of private images. However, in this method, the features are quantified by a hashing function, which may make some details of the features lost, and finally affects the accuracy of the classification.

3 Our Model

Overview. We build a firewall that could detect sensitive object defined previously. It could be deployed on personal computer or gateway, what the administrators need to do is just setting rules in the firewall, and the pictures are decided to pass or not pass according to the rules. Different privacy level are returned as different rules are matched, which can provide user as an alert or privacy decision reference information. The firewall includes three parts, object detection module, distance and position detection module, and the matching rules module. The detail of the three parts are described as follows.

3.1 Object Detection Module

To illustrate how the module work, we will elaborate on these three aspects:

• Feature extraction

To describe the person and ordinary object features appeared in an image, we use the LBP and SIFT feature. To get more rich features, we also introduce other common features, such as Wavelet, HOG, and gist. All of the features are finally combined in our model based on sparse representation.

- Feature process

In our previous explanation, we need a matrix $X \in \mathbb{R}^{d \times n}$, which represent n columns of training image feature vectors of dimension d. However, how to ensure that all the images have the same dimension. For the global features (Wavelet, LBP etc.), we can resize the images to make them the same dimension. SIFT is not suitable, because SIFT feature is a local feature, and it depends on the complexity of edges and corners. In extreme case, SIFT would lose its effectiveness, and the number of feature point is zero. Different image has a different shape and texture, how to transform them to a certain dimension is a problem. We need some pretreatment before sparse representation. Sparse Pyramid Matching (SPM) [9] is a method to solve this problem. The basic idea is mapping extracted feature points into a fixed dimension based on global geometric correspondence. Support X is a matrix of $N \times D$ dimension, $X = [x_1, \ldots, x_N] \in \mathbb{R}^{D \times N}$, given a code book $B = [b_1, \ldots, b_N] \in \mathbb{R}^{D \times M}$ which contain M entries, different coding schemes convert each descriptor into a M-dimensional code to generate the final image representation. Like VQ in [9], scSPM in [10], here, we use their improvement program [11], Locality-constrained Linear Coding (LLC). It uses the following criteria:

$$\min_{C} \sum_{i=1}^{N} \|x_i - Bc_i\|^2 + \|\lambda d_i \odot c_i\|^2 \tag{7}$$

where $\odot$ denotes the element-wise multiplication, and $d_i \in \mathbb{R}^M$ is the locality adaptor that gives different freedom for each basis vector proportional to its similarity to the input descriptor x_i. Specifically,

$$d_i = \exp\left(\frac{dist(x_i, B)}{\sigma}\right) \tag{8}$$

where $dist(x_i, B) = [dist(x_i, b_1), \ldots, dist(x_i, b_M)]^T$, and $dist(x_i, b_j)$ is the Euclidean distance between x_i and b_j. σ is used for adjusting the weight decay speed for the locality adaptor. Usually we further normalize d_i to be between $(0,1]$ by subtracting $\max(dist(x_i, B))$ from $dist(x_i, B)$. The constraint $\mathbf{1}^T c_i = 1$ follows the shift-invariant requirements of the LLC code. Note that the LLC code in Eq. (7) is not sparse in the sense of $\ell_0 - norm$, but is sparse in the sense that the solution only has few significant values. In practice, we simply threshold those small coefficients to be zero. In our model, for the cause of a memory shortage, we choose a 512-dimensional codebook under the circumstance that it is over-complete in our situation, while Wang et al. [11] choose 2048 in their experiments.

- Image registration

In order to demonstrate the procedure of image registration more clearly, we draw a flow chart in Fig. 1. From the chart, we can see that our images were transformed into

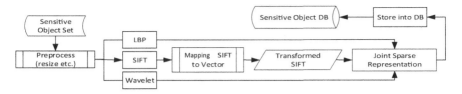

Fig. 1. The procedure of image registration. It includes preprocess, extraction, post-process and registration.

lots of numbers via Joint Sparse Representation. Moreover, this transformation is a one-way function, or we say it is hard to recovery an image only according to these features. Furthermore, the procedure of SPM is a coding procedure, which increased entropy of the data. So do not worry about the image safety, we only store features.

3.2   Distance and Position Detection Module

In this section, we will introduce the distance and position detection module. We find it essential in considering whether an image is sensitive. For example, when two faces are quite close, we think it may be a couple kissing each other. Therefore, we build a module to measure distance and position parameters. Details for every step are presented below.

- Step 1. Get all the position of feature description point. In our model, we choose some features that show position, like SIFT.
- Step 2. Calculate Centroid through these positions. We regard the region where there are dense feature points as the calculating region, and calculate the average of these point position.
- Step 3. Detect the distance and position information. We concentrate on not only the existence of an object, but also the relationships of the object with others.

Before detection, we should find a way to measure the real distance through an image. As we know, the pupil distance is about six centimeters ($D_e = 6$ cm), according to this, we can detect the distance between two persons. As for object, we think it is not important in distance, but position is. Since the image is 2-dimension space, it is hard to calculate the distance when two objects stand a front-behind position. Therefore, we compare the size of two faces, when the size of two faces is very different, we regard them as a front-behind position, and we will not calculate the distance.

Suppose we extracted N feature points of object A and M feature points of object B in the dense feature region, and we use Cartesian Coordinates $\left(x_i^A, y_i^A\right)$ and $\left(x_j^B, y_j^B\right)$ to describe the position of feature points. And we conclude that the center of A:

$\left(\bar{x}^A = \frac{1}{N}\sum_{i=1}^{N} x_i^A, \bar{y}^A = \frac{1}{N}\sum_{i=1}^{N} y_i^A\right)$, B: $\left(\bar{x}^B = \frac{1}{M}\sum_{j=1}^{M} x_j^B, \bar{y}^B = \frac{1}{M}\sum_{j=1}^{M} y_j^B\right)$. Therefore, the Euclidean distance between A and B is:

$$d = \sqrt{\left(\bar{x}^A - \bar{x}^B\right)^2 + \left(\bar{y}^A - \bar{y}^B\right)^2} \tag{9}$$

And we can infer the real distance between A and B as:

$$D = D_e \frac{d}{d_e} \tag{10}$$

where d_e is the pixel distance obtained from eye detection algorithm supported by OpenCV[1]. Similar to that, we can describe the position. Suppose $d_x = \bar{x}^A - \bar{x}^B$ and $d_y = \bar{y}^A - \bar{y}^B$ are the distance in x direction and y direction. We decide the position by comparing the d_x and d_y as:

$$position = \begin{cases} left - right, \ if \max\{d_x, d_y\} = d_x \\ up - down, \ if \max\{d_x, d_y\} = d_y \end{cases}$$

In summary, we get three possible position relationship: front-behind (F-B), left-right (L-R), and up-down (U-D).

3.3 Matching Rules Module

A user-defined rule in our image firewall system consists 7 parts, including operating symbol (action), operating object (object 1 and object 2), relationship symbol (logic) and some auxiliary fields (position, distance and level), which is shown in Fig. 2. It is similar to a traditional firewall rule, people can deploy the firewall easily. The relationship symbols are used to describe the logic condition of the objects under which an action (deny or accept) a suggested to given and a privacy level is recommended. In this paper, we simply define two relationship symbols, and relationship that is more complex can be achieved by the combination of these two symbols. Besides, three kinds of position relationships are defined in position detection module above. For the distance relationship, according to Edward T. Hall's theory[2], an intimate distance is 15–45 cm, thus, we set the threshold to 15 cm in the experimental Section. We think if the distance of two persons appeared in an image is less than 15 cm, than the image is an intimate and privacy image, otherwise it is a normal image. Users can set a privacy level from 1 to 7, which has different alert meanings taking effect along with the action of Deny. The bigger the level, the stricter the action.

---

[1] http://docs.opencv.org/2.4/modules/objdetect/doc/cascade_classification.html.

[2] https://en.wikipedia.org/wiki/Edward_T._Hall.

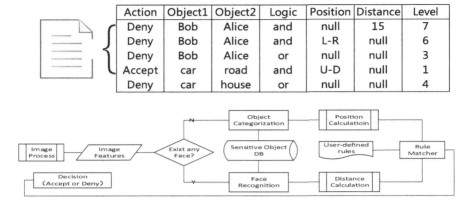

Action	Object1	Object2	Logic	Position	Distance	Level
Deny	Bob	Alice	and	null	15	7
Deny	Bob	Alice	and	L-R	null	6
Deny	Bob	Alice	or	null	null	3
Accept	car	road	and	U-D	null	1
Deny	car	house	or	null	null	4

Fig. 2. Top: How to deploy the rules in our firewall? Here are some demo rules. For example, the first rule was set to deny the image that contains Bob and Alice when their distance is less than 15 cm. Bottom: Our method detects and determines whether an image should be passed.

4   Experiments

We evaluate the performance of the proposed method on several benchmark datasets including Caltech101[3] dataset and Caltech256[4] dataset. In addition, we collect some images, which contain only two persons and at least one eyepair detected by harrcascade classifier[5] from Visual Genome[6].

4.1   Experimental Settings

To test the performance on face recognition, we take out two classes of face image named as faces_easy (aligned) and faces (unaligned) in dataset Caltech101, and classify them manually into 15 persons. We divided 10 images for training set, the other 10 images for the test set.

For the purpose of testing model performance on higher sample capacity, we choose 15 categories of the objects which is common in life (personal computer, private car etc.), and randomly select 20 images as training set, the other 30 as the test image set.

We use the accuracy of classification and average time-cost of single-image classification as two indicators to evaluate our model. We use an open experimental approach to explore different image features with different accuracy and efficiency, randomly combine different feature extraction method for a better joint.

---

[3] http://www.vision.caltech.edu/Image_Datasets/Caltech101/.

[4] http://www.vision.caltech.edu/Image_Datasets/Caltech256/.

[5] http://docs.opencv.org/2.4/modules/objdetect/doc/cascade_classification.html.

[6] https://cs.stanford.edu/people/rak248/VG_100K_2/images.zip.

As for distance and position detection, we first carried out human-computer interaction research. 176 images containing two persons and at least one eyepair were labeled by respondents by considering: Whether the persons on the image are intimate? What is their position? After that, we got these labeled images for our experiments. The distance contains three classes: intimate, which means the distance is less than 15; normal, which means the distance is 15 or more; and unknown, which means the distance is hard to calculate, including different face scales, and a face in image recognized by mistake.

4.2   Results and Analysis

In this section, we will present our experimental results on face recognition, object categorization, distance and position detection using our model.

First, the results on face recognition with different joint sparse representation of features are shown in Table 1.

Table 1. Performance on face recognition

Image set & features	Faces_easy (300[a])		Faces (300)	
	Time (s)	Accuracy (%)	Time (s)	Accuracy (%)
Original LBP	4.35	41.3	6.429	2
LBP uniform pattern	1.84	41.3	5.982	4.3
SIFT[b]	161.55 + 4.812	100	323.1 + 6.829	96.3
Wavelet	4.75	93.3	6.505	28.7
Wavelet + LBP uniform pattern	4.788	76.7	9.427	28
SIFT + LBP uniform pattern	4.842	74	9.01	93.7
SIFT + wavelet	5.107	99.3	9.6	52.7
SIFT + LBP uniform pattern + wavelet	7.297	91.3	13.88	51.7

[a]Totally 15 persons, each has 10 training images and 10 test images.
[b]Only SIFT needs a pretreatment. A combination of SIFT and other features requires the same SIFT pretreatment. Tables 1 and 2 are the same. See more details in Sect. 3.1 feature process.

From the table we could see that SIFT feature work best among these features regardless of unaligned or aligned. Wavelet feature has a good performance when the image is aligned, while has a poor performance when the image is unaligned. LBP feature, whether original LBP or its uniform pattern, have a general performance in aligned image classification, not to mention it in unaligned. Therefore, we conclude that the original LBP and its uniform pattern are not suitable in our face recognition model. In the following experiment, we regard them as the same feature, and LBP uniform pattern represents a series of LBP pattern. We also noticed that a combination of two features might not achieve better performance than single of original feature. It means that some combination would be out of expected effect, so our question is how to choose a feature combination. The experimental results show that SIFT is an excellent feature in sparse coding face recognition.

Second, the results of object categorization in a larger sample capacity are presented in Table 2.

Table 2. Performance on object categorization

Image set & features	Caltech256 (450[a])	
	Time (s)	Accuracy (%)
Original LBP	10.502	42
SIFT	484.65 + 11.2	81.6
Wavelet	10.423	42.4
Wavelet + LBP uniform pattern	18.259	53.3
SIFT + LBP uniform pattern	18.173	65.3
SIFT + wavelet	16.946	60
SIFT + LBP uniform pattern + wavelet	21.91	66

[a]20 for training, 30 for test, 450 in total.

Apparently, feature combination works better than a single feature in object categorization. SIFT feature is the most effective one again in this experiment, although its accuracy is reduced. To test whether the accuracy would go on decrease with the sample capacity grow up when using the SIFT feature, and to test the performance of other sparse coding base features in large and complex scale circumstance, we continue to improve sample capacity. In this time, HOG and gist features are introduced to rich the combinations of the features. The results are shown in Table 3.

Table 3. Change features and improve sample capacity

Image set & features		SIFT + LBP + wavelet + HOG + gist	HOG	gist	SIFT[a]	HOG + gist
Caltech101	Time (s)	295.69	280.61	283.72	286.15	284.28
	Accuracy (%)	52.45	33.53	48.33	56.67	48.33
Caltech256	Time (s)	2051	2049	2036	2046	2069
	Accuracy (%)	20.47	11.75	17.04	26.77	18.33

[a]The pretreatment time is too long, not comparable with classification time, so we did not record this.

In our previous experiments, we found that SIFT works best, so we choose it as representative to compete with HOG and gist. From the table, we conclude that SIFT works still very well, but its accuracy drops sharply. To balance the factors of accuracy and time cost, we can use HOG + gist instead.

At last, the experimental results on distance and position detection are shown in Table 4.

Table 4. The results of distance and position detection on faces

			Human being				
			Intimate		Normal		Unknown
			L-R	U-D	L-R	U-D	F-B
Our model	Intimate	L-R	16	0	1	0	5
		U-D	17	8	0	0	7
	Normal	L-R	3	0	24	6	9
		U-D	1	0	25	10	10
	Unknown	F-B	3	1	1	1	28

Our model works well on distance detection, which means that it is an effective way to measure the distance between two faces. As shown in Table 4, the different evaluation results between human being perception and our model are shown. For example, the first column show the number of intimate images that human being evaluated altogether is 49. Our model accesses 41 images as intimate in these images. We assume that the human being is 100% correct with no recognition bias, hence we can calculate that the accuracy of our model on detecting intimate relationship is 83.7%, and the accuracy of our model on detecting normal relationship is 95.6%.

In fact, there are over 600 images, which have at least two faces in the Visual Genome dataset, while only 176 of them satisfy our requirement with at least an eyepair. Since our model is based on pupil distance, and eyepair is essential. What's more, the result of the position detection is somewhat different from human perception. People tend to regard an up-down position as a left-right position in some special situations. We think there are two possible reasons: one may be the recognition bias in human, the other is that our experimental images is not appropriate for position detection since all the respondents are told that these images are photos contained at least two persons, which gives a hint of left-right position.

From the experiments above, we can see that joint sparse representation depends on not only dictionary, but also the combination of features. To find a suitable combination among those diverse features is difficult, especially with a large scale. The accuracy drops sharply even we choose a good feature SIFT.

5   Conclusions and Future Work

In this paper, we design a privacy relationship detection model based on joint sparse representation. For the implement of firewall, we design a series of formatted rules, which appropriately describe privacy relationship. In addition, we propose an algorithm on distance and position detection, which is significant in our relationship detection model. The model achieves remarkable results in human spatial relationship detection. In our future work, we will try to find a better combination of features, at the same time, to design more effective privacy relationship detection model based on deep learning. In this paper, our module on distance detection is limited by the existence of eyepair, and thereby in the future, we will find a better way to detect the distance of objects or persons. Besides, we will also consider more complex space relationships such as 'on', 'under', and so on.

Acknowledgements. This work was supported by the National Natural Science Foundation of China (Nos. 61375047 & 61672203), National Innovation and Entrepreneurship Training Program project funding (No. 201510359033) in Hefei University of Technology, 2015. Our server is supported by Network Center of Hefei University of Technology.

References

1. Smith, L.I.: A tutorial on principal components analysis. Inf. Fusion **51**(3), 52 (2002)
2. Yuan, X.T., Liu, X., Yan, S.: Visual classification with multitask joint sparse representation. IEEE Trans. Image Process. Publ. IEEE Sig. Process. Soc. **21**(10), 4349–4360 (2012)
3. Hu, D., et al.: A framework of privacy decision recommendation for image sharing in online social networks. In: IEEE First International Conference on Data Science in Cyberspace IEEE Computer Society, pp. 243–251 (2016)
4. Beck, A., Teboulle, M.: A fast iterative shrinkage-thresholding algorithm for linear inverse problems. SIAM J. Imaging Sci. **2**(1), 183–202 (2009)
5. Candès, E.J., Romberg, J.K., Tao, T.: Stable signal recovery from incomplete and inaccurate measurements. Commun. Pure Appl. Math. **59**(8), 1207–1223 (2005)
6. Wright, J., et al.: Robust face recognition via sparse representation. IEEE Trans. Pattern Anal. Mach. Intell. **31**(2), 210 (2009)
7. Becker, S., Bobin, J., Candès, E.J.: NESTA: a fast and accurate first-order method for sparse recovery. SIAM J. Imaging Sci. **4**(1), 1–39 (2009)
8. Hu, X., et al.: How people share digital images in social networks: a questionnaire-based study of privacy decisions and access control. Multimedia Tools Appl. 1–23 (2017). doi:10. 1007/s11042-017-4402-x
9. Lazebnik, S., Schmid, C., Ponce, J.: Beyond bags of features: spatial pyramid matching for recognizing natural scene categories. In: 2006 IEEE Computer Society Conference on IEEE Computer Vision and Pattern Recognition, pp. 2169–2178 (2006)
10. Yang, J., et al.: Linear spatial pyramid matching using sparse coding for image classification, pp. 1794–1801 (2009)
11. Wang, J., et al.: Locality-constrained linear coding for image classification. In: Computer Vision and Pattern Recognition IEEE, pp. 3360–3367 (2010)

Biomedical Image Analysis

Hierarchical Skull Registration Method with a Bounded Rotation Angle

Xiaoning Liu$^{(\boxtimes)}$, Lipin Zhu, Xiongle Liu, Yanning Lu, and Xiaodong Wang

College of Information Science and Technology, Northwest University, Xi'an, China xnliu@nwu.edu.cn, 690285098@qq.com

Abstract. To reconstruct the appearance of an unknown skull based on knowledge, the most similar skull should be retrieved from the skull Database. The process is called skull registration. A hierarchical skull registration method with bounded rotation angle is proposed in this paper. The surface of the skull is divided into concave or convex regions by K-means. The optimal 3D transform is searched for each potential pair of matched feature regions approximately to align the skulls roughly. And then the novel ICP (Iterative Closest Point) algorithm with a bounded rotation angle (BRA-ICP) is applied for fine registration. To show the generality, the proposed algorithm is applied not only for skull registration but also for the public data registration. Experiments show that the proposed algorithm can achieve better registration accuracy and higher iterative convergence speed in the fine registration stage and the entire process is completed without human intervention.

Keywords: Bounded rotation angle · Skull registration · Feature regions · Integral invariant · Principal component analysis

1 Introduction

The craniofacial reconstruction has important application in criminal investigation and ancient human archaeology [1–3]. The method based on knowledge is one important way to rebuild the face for a skull. For an unknown skull, named U, which to be reconstructed, we need to search a most similar skull, named S, in the skull database. The process of searching skull S which is similar to the skull U in the Database is 3D skull registration.

3D skull registration is a kind of point cloud rigid registration [4]. The methods for rigid registration can be divided into coarse registration and fine registration. Normal and curvature [5] are commonly used in coarse registration to describe the local features, which are invariant to rotation and translation, but sensitive to noise. Integral invariant [6] are a geometric invariant to describe the intrinsic properties of surfaces. It is noise insensitivity and can accurately describe the local characteristics of the surface.

On the stage of fine registration, ICP (Iterative Closest Point) proposed by Besl and McKay [7] is widely used. It requires the inclusion relation between the two point sets and the good initial position. To improve ICP, Xie et al. [8] presented FICP (Fractional

© Springer International Publishing AG 2017
D.-S. Huang et al. (Eds.): ICIC 2017, Part III, LNAI 10363, pp. 563–573, 2017.
DOI: 10.1007/978-3-319-63315-2_49

ICP) algorithm. It effectively removed outliers and improved the robustness of ICP algorithm. Choi et al. [9] proposed an improved k-d tree traversal method to speed up the searching process of the nearest point, and the convergence speed of the algorithm is improved effectively. Bae and Lichti [10] presented an improved ICP algorithm based on the feature points of the bound of point cloud to improve the efficiency and accuracy. So far, the ICP algorithm is still the main method for fine registration [11, 12].

ICP registration based on bounded rotation angle is proposed by Du et al. [13]. It can effectively improve the matching accuracy, robustness and convergence speed, and can also avoid the interference of noise and outer points to a certain extent. In this paper, a hierarchical skull registration method with bounded rotation angle is proposed. The main steps are as follows. (1) The concavity and convexity of all the points on skull are calculated according to the integral invariant. In addition, the points are divided into three categories, namely concave vertex, convex vertex and planar vertex. (2) The vertices are clustered to form the convex or concave regions according to the connectivity and the K-means. (3) Principal component analysis is used to calculate the similarity of feature regions, and then determine the optimal matching of feature regions. (4) Improved ICP with bounded rotation angle (BRA-ICP) is used for the fine registration.

2 Pre-processing

2.1 Simplification

The data used in this paper were collected by Northwest University in Xianyang Hospital located in western China. All the data came from the living volunteers that mostly came from Han ethnic group in northern China. We wrote program to automatically extract the outline of the skull and reconstruct the 3D skull model. The result is as shown in Fig. 1(a). It includes too many points. By combining the point cloud curvature and normal compaction algorithm, we deleted almost 50% points off. Figure 1(b) shows the skull with 229984 points after simplification. The shape has not been significantly changed when the number of points was deleted by half.

     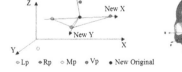

(a)459970 points   (b)229984 points   (a) Coordinate of FP   (b) Normalized skull in FP

Fig. 1. Comparison of simplification   **Fig. 2.** Normalized skull in coordinate of FP

2.2 Normalization

It is necessary to convert all skulls to a same size but not changing their shapes. In this paper, the Frankfurt coordinate system is used to unify the size. Frankfurt plane (FP) is determined by Lp, Rp, Mp. As shown in formula (1).

$$P_F = M_P L_P \times M_P R_P \qquad (1)$$

where Lp, Rp, Mp represents the midpoint of the left ear hole, the midpoint of the light ear hole and midpoint of infraorbital margin of left eye respectively. In addition, Vp is eyebrow point. The Frankfurt coordinate system is established as Fig. 2(a). The skull model is shown in it as Fig. 2(b).

After unified in the Frankfurt coordinate system, scale normalization is also needed. Set the distance between the Lp - Rp of all skull models to one. The skull scale transformation is as follows:

$$\begin{bmatrix} x' \\ y' \\ z' \end{bmatrix} = \frac{1}{L_p - R_p} \begin{bmatrix} x \\ y \\ z \end{bmatrix} \qquad (2)$$

where (x, y, z) is the coordinates of the point before scaling, (x', y', z') is the new coordinates of the points after scaling.

3 Coarse Registration of Skull

Because the surface of the skull has different concave and convex areas, it will contain more feature points. According to the characteristics of the skull surface, the coarse registration of skull is applied firstly.

3.1 Determine the Concave and Convex for Skull Vertex

In this paper, the volume integral invariant is used to determine the concavity and convexity of each point. As is shown in Fig. 3, S is the surface of the three dimensional solid model D. Take the p as the center of any point on the S and the sphere with r as the radius of the sphere neighborhood Br(p) of the point p. The volume integral invariant formula of the point is as follow:

$$V_r(p) = \int_{B_r(p)} I_D(x) dx \qquad (3)$$

Geometric meaning of integral invariant: The volume of Br(p) in the outer part of a curved surface is Vr(p). As shown in Fig. 3, the value of Vr(p) is related to the degree of concave and convex surface and in proportional to the size of r. In addition, Vr(p) is not affected by the noise points inside the Br(p). Therefore, Vr(p) reflects the concave

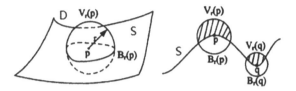

(a)Volume integral invariant (b) Two dimensional

Fig. 3. Three dimensional representation of representation

and convex degree of the point p neighborhood surface, and can be used to calculate the different radius.

Here $\varphi_r(p)$ is used to describe the concavity and convexity of the vertex p in the area of r. It is defined as (4).

$$\varphi_r(p) = \frac{3}{4\pi\,r^3}\,V_r(p) - \frac{1}{2} \tag{4}$$

where, if $\varphi_r(p) > 0$, $V_r(p) > \frac{2}{3}\pi\,r^3$, p is a convex vertex. If $\varphi_r(p) < 0$, $V_r(p) < \frac{2}{3}\pi\,r^3$, p is a concave vertex. If $\varphi_r(p) = 0$, $V_r(p) = \frac{2}{3}\pi\,r^3$, p is a planar vertex.

We need to set the threshold ε. If $\varphi_r(p) > \varepsilon$, p is a convex vertex. If $\varphi_r(p) < -\varepsilon$, p is a concave vertex. If $\varphi_r(p) \in [-\varepsilon, \varepsilon]$, p is a planar vertex.

3.2   Extract Feature Regions

A characteristic region is a connected concave or convex or planar local area. According to the calculation of volume integral invariant in the Sect. 3.1, the K-Means clustering method is used to cluster the points into convex and concave regions. The algorithm for extracting feature regions is as follows,

1. Calculate the $V_r(p)$ and $\phi_r(p)$ of each vertex of the skull surface.
2. Classify the vertex of the skull surface according to $\phi_r(p)$. If the convex vertices satisfying $\phi_r(p) > \varepsilon$, it is recorded in the collection ϕ_a. If the concave vertices satisfying $\phi_r(p) < -\varepsilon$, it is recorded in the collection ϕ_b.
3. Cluster. According to the connectivity of vertices, the vertices of the set ϕ_a and ϕ_b are clustered by K-Means. The skull surface is divided into a plurality of concave or convex feature regions. If the number of vertices in a feature region is less than a given threshold, the region can be removed.

3.3   Search the Similar Feature Regions and Roughly Registration

The local region sets of the skull to be restored and the similar skull are R_u and R_s. If two regions from R_u and R_s meet the following conditions, they are considered to be similar regions.

1. The same type, that is, the same convex or concave area, regardless of the situation of the plane area.
2. Have similar principal components.
3. Nearly same area size.

In this paper, we use principal component analysis (PCA) to compare the similarity of feature regions. Coarse registration steps are as follows,

1. Calculate principal component information of vertex set of feature region.

Set feature area R contains n vertices, which denoted as $P_i \in R(i = 1, 2, \ldots, n)$. The coordinates of the vertices are recorded as (x_i, y_i, z_i), and the centroid and the covariance matrix $M \in 3 \times 3$ of R are

$$\bar{p} = \frac{1}{n} \sum_{i=1}^{n} p_i \tag{5}$$

$$M = \sum_{i=1}^{n} [(p_i - \bar{p}) \bullet (p_i - \bar{p})^{\mathrm{T}}] \tag{6}$$

The eigenvalue of M is $\lambda_0^R \geq \lambda_1^R \geq \lambda_2^R$, which can be defined as the principal component of regional R. The corresponding eigenvector is n_0^R, n_1^R, n_2^R, which is the main direction of the region R.

2. Compare the similarity of feature regions.

The definition of dimension feature $\phi(R)$ and the anisotropic characteristics $A(R)$ are

$$\phi(R) = (\lambda_0^R + \lambda_1^R + \lambda_2^R)^{1/2} \tag{7}$$

$$A(R) = \left| \frac{\lambda_1^R}{\lambda_0^R} \right|^{1/2} \tag{8}$$

There are two local regions R_1 and R_2, which belong to the collection of R_u and R_s, respectively, defined as $\phi(R_1, R_2)$ and $A(R_1, R_2)$

$$\phi(R_1, R_2) = \left| \frac{S(R_1) - S(R_2)}{S(R_1) + S(R_2)} \right| \tag{9}$$

$$A(R_1, R_2) = \left| \frac{A(R_1) - A(R_2)}{A(R_1) + A(R_2)} \right| \tag{10}$$

If $\varphi(R_1, R_2) < \varepsilon_\varphi$ and $A(R_1, R_2) < \varepsilon_A$. It is considered that R_1 and R_2 has similar principal components.

The area of R_1 and R_2 is set to be $N(R_1)$ and $N(R_1)$, respectively, and the definition of $N(R_1, R_2)$ is as follows:

$$N(R_1, R_2) = \frac{|N(R_1) - N(R_2)|}{|N(R_1) + N(R_2)|} \tag{11}$$

If $N(R_1, R_2) < \varepsilon_n$, it is considered that the area size of R_1 and R_2 is similar. Where $\varepsilon_\varphi, \varepsilon_A, \varepsilon_n$ is threshold have been given.

3. Remove the wrong match.

After calculating the similar pairs, there are often some wrong matches. The clustering algorithm based on distance principal direction constraint is used to eliminate the false matching pairs. There are limited regions, so we use exhaustive method to calculate the 3D transformation of each possible match, and the surface of the two skulls is roughly aligned.

4   Fine Registration of Skull

In order to improve the robustness and reduce the diversity of rotation, the rotation angle θ is added in the iteration process. The process consists of two steps. Firstly, the boundary of the rotation angle (the upper and lower bounds) is estimated by the best matching criterion, and the registration model is established. Then the improved ICP algorithm is used to match the point cloud data.

4.1   Bounded Rotation Angle

The 3D point cloud registration based on bounded rotation angle is described as (12):

$$\begin{cases} \min\limits_{P,q,c(i)_{i=1}^{i=Ns}} \sum\limits_{i=1}^{Nu} ||(Pu_i + q) - s_{c(i)}||^2 \\ P^{\mathrm{T}}P = I_n, \det(R) = 1 \\ \theta_x \in [\theta_{xb} - \Delta\theta_x, \theta_{xb} + \Delta\theta_x] \\ \theta_y \in [\theta_{yb} - \Delta\theta_y, \theta_{yb} + \Delta\theta_y] \\ \theta_z \in [\theta_{zb} - \Delta\theta_z, \theta_{zb} + \Delta\theta_z] \end{cases} \tag{12}$$

where, $P = P_x P_y P_z$.

$$P_x = \begin{bmatrix} 1 & 0 & 0 \\ 0 & \cos\theta_x & -\sin\theta_x \\ 0 & \sin\theta_x & \cos\theta_x \end{bmatrix} \quad P_y = \begin{bmatrix} \cos\theta_y & 0 & \sin\theta_y \\ 0 & 1 & 0 \\ -\sin\theta_y & 0 & \cos\theta_y \end{bmatrix} \quad P_z = \begin{bmatrix} \cos\theta_z & -\sin\theta_z & 0 \\ \sin\theta_z & \cos\theta_z & 0 \\ 0 & 0 & 1 \end{bmatrix}$$

where, $\theta_x, \theta_y, \theta_z$ are rotation angles, $\theta_{xb}, \theta_{yb}, \theta_{zb}$ are mean values of rotation angles, $\Delta\theta_x, \Delta\theta_y, \Delta\theta_z$ are deviation of rotation angles.

4.2   ICP Registration Algorithm with Bounded Rotation Angle (BRA-ICP)

The $U = \{u_i\}_{i=1}^{Nu}$, $S = \{s_i\}_{i=1}^{Ns}$ are the two skulls to be registered. The registration between U and S is to compute an optimal transform F so that U can be better matched with S. The process of seeking F is a least squares problem, as is shown in formula (13).

$$\min_F \sum_{i=1}^{Nu} ||F(u_i) - s_{c(i)}||^2 \tag{13}$$

where $F(u_i)$ is the U point cloud transform, $s_{c(i)}$ is the corresponding point in S. In fine registration, F can be expressed as $F(u_i) = Pu_i + q$, where $P \in n \times n$ is a rotation matrix, $q \in 1 \times n$ is a translation vector. According to the literature [7], this paper proposes the following improvements:

1. Estimate the boundary of the three rotation angles $\theta_x, \theta_y, \theta_z$, as is shown in formula (12).
2. Set the initial transform P_0 and q_0, $U_0 = P_0 U + q_0$, $k = 0$.
3. Repeat, k = k+1,and establish the correspondence $\{c_k(i)\}_{i=1}^{i=Ns}$.
4. Calculate the optimal rotation angle $\theta_{xk}, \theta_{yk}, \theta_{zk}$, optimal rotation matrix $P_{opt} = P_{xk}P_{yk}P_{zk}$ and translation vector $q_k = \frac{1}{Nu}(-P_k \sum_{i=1}^{Nu} u_i + \sum_{i=1}^{Nu} s_{ck(i)})$.
5. Apply transformation according to the $U_k = P_k M + q_k$ and calculate the root mean square error of mismatch. $D_k = \sqrt{\frac{1}{Nu} \sum_{i=1}^{Nu} ||u_i - s_{ck(i)}||^2}, u_i \in U_k$.
6. Judge termination condition. Until $|D_k - D_{k-1}| < T_{\min}$ or reach the maximum iterative number, where $T_{\min}$ is a given threshold.

5   Experimental Results and Comparative Analysis

In order to prove the effectiveness and efficiency of the proposed algorithm, experiments were done compared with the classical ICP algorithm of literature [7] and the improved ICP algorithm of literature [12].

5.1   Experiment 1

In this section, we collected 308 skulls as skull database. For an unknown skull U, one or several similar skulls are retrieved in the skull database. Figure 4 are the two skull to be registered. Figure 5 is the result after the coarse registration. And then three algorithms are used for fine registration. Figures 6, 7 and 8 show the fine registration. Table 1 shows the detailed information of skull registration after 20 experiments of each algorithm. Among them, run time, Iterations times and Matching points are the average number of successful tests taken. The $\Delta\theta_x, \Delta\theta_y, \Delta\theta_z$ is set $8°$.

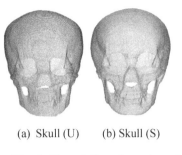

(a) Skull (U)   (b) Skull (S)

Fig. 4. Two skulls to be registered

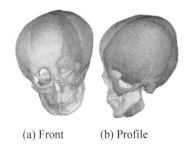

(a) Front   (b) Profile

Fig. 5. PCA coarse registration

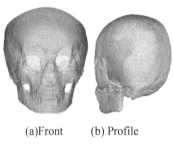

(a)Front   (b) Profile

Fig. 6. Literature [7] ICP registration

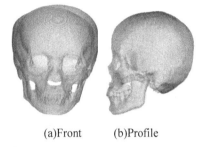

(a)Front   (b)Profile

Fig. 7. Literature [12] ICP registration

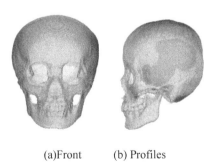

(a)Front   (b) Profiles

Fig. 8. The improved ICP registration in this paper

5.2   Experiment 2

In this part, two dragons from Stanford 3D graphics library are registered. Figure 9 is the fine registration. Table 2 shows the detailed information after 15 times iterations.

Table 1. Algorithm operation analysis table

Skull registration	ICP of [7]	ICP of [12]	BRA-ICP
Skull U	163288	163288	163288
Skull S	211523	211523	211523
Registration times	20	20	20
Failure times	3	1	0
Run time (s)	6.23	3.02	2.87
Iterations times	56	39	31
Matching points	43126	57661	68173

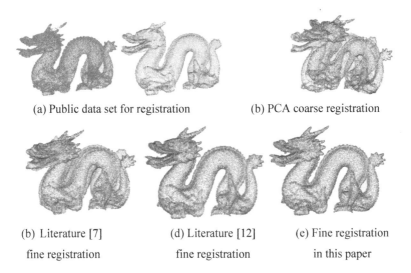

(a) Public data set for registration   (b) PCA coarse registration

(b) Literature [7]   (d) Literature [12]   (e) Fine registration
fine registration   fine registration   in this paper

Fig. 9. Dragons' registration

Table 2. Algorithm operation analysis table

Common data set registration	ICP of [7]	ICP of [12]	BRA-ICP
Model 1 points	20167	20167	20167
Model 2 points	19220	19220	19220
Registration times	9	10	10
Failure times	1	0	0
Run time(s)	4.17	2.83	2.04
Iterations times	43	26	19
Matching points	4697	7015	8423

From the two group's experiments, comparing algorithm in this paper with the classical ICP algorithm and the registration effect of the literature [12], the times of iterations are reduced and the efficiency of the point cloud registration are further improved by adding the bounded rotation angle. The success rate of registration is better than literature [12]. The proposed algorithm is a quite good point cloud registration algorithm with high precision and fast speed.

6   Conclusion

On the coarse registration stage, by calculating the volume integral invariant point and clustering feature region, we can avoid marking the feature points manually. On the fine registration stage, by adding a bounded rotation angle, the accuracy and speed of registration are both improved greatly. Experiments show that the proposed method has strong robustness and anti-noise performance.

The experimental results also show that different sets of data need to be set up with different iteration factors. Though the presented algorithm is effective for skull registration, there are still some problems in the accuracy of the automatic matching of the fragments of Terracotta Army. Further research is needed on the automatically setting the iteration factor by feedback.

Acknowledgments. The authors would like to thank the reviewers and Visualization Institute of Northwest University for providing skull data. This work is supported by the National Natural Science Foundation of China (61305032, 61572400).

References

1. Shui, W.Y., Zhou, M.Q., Deng, Q.Q., et al.: Densely calculated facial soft tissue thickness for craniofacial reconstruction in Chinese adults. Forensic Sci. Int. **266**, 573.e1–573.e12 (2016)
2. Eloy, J.A., Marchiano, E., Vázquez, A., et al.: Management of skull base defects after surgical resection of sinonasal and ventral skull base malignancies. Otolaryngol. Clin. North Am. **50**(2), 397–417 (2017)
3. Lin, S.Z., Wang, D.J., Zhong, J.R., et al.: Approach to reassembling virtual small bronze fragments using the curvature feature. J. Xidian Univ. **43**(6), 141–146 (2016)
4. Yan, M.H.: The Research on Feature Points Extraction From Skull Based on Statical Methods. Northwest University, Xi'an (2011)
5. Izumiya, S., Nabarro, A.C., de Jesus Sacramento, A.: Pseudo-spherical normal darboux images of curves on a timelike surface in three dimensional lorentz-minkowski space. J. Geom. Phys. **97**, 105–118 (2015)
6. Cao, Z.C., Ma, F.L., Fu, Y.L., et al.: A scale invariant interest point detector in gabor based energy space. Acta Automatica Sin. **40**(10), 2356–2363 (2014)
7. Besl, P.J., McKay, N.D.: A method for registration of 3-D shapes. IEEE Trans. Pattern Anal. Mach. Intell. **14**(2), 239–256 (1992)
8. Xie, Z.X., Xu, S., Li, X.Y.: A high-accuracy method for fine registration of overlapping point clouds. Image Vis. Comput. **28**(4), 563–570 (2010)

9. Choi, W.S., Kim, Y.S., Oh, S.Y., et al.: Fast iterative closest point framework for 3D LIDAR data in intelligent vehicle. In: Proceedings of 2012 IEEE Intelligent Vehicles Symposium, Alcala de Henares, Madrid, Spain, pp. 1029–1034 (2012)
10. Bae, K.H., Lichti, D.D.: A method for automated registration of unorganised point clouds. ISPRS J. Photogramm. Remote Sens. **63**(1), 36–54 (2008)
11. Ge, Y.Q., Wang, B.Y., Nie, J.H., et al.: A point cloud registration method combining enhanced particle swarm optimization and iterative closest point method. In: Proceedings of 2016 Chinese Control and Decision Conference. IEEE, Yinchuan (2016)
12. Li, W.M., Song, P.F.: A modified ICP algorithm based on dynamic adjustment factor for registration of point cloud and CAD model. Pattern Recogn. Lett. **65**, 88–94 (2015)
13. Du, S., Zhang, C., Wu, Z., et al.: Robust isotropic scaling ICP algorithm with bidirectional distance and bounded rotation angle. Neuro Computing. **215**, 160–168 (2016)

Sex Determination of Incomplete Skull of Han Ethnic in China

Xiaoning Liu$^{(\boxtimes)}$, Xiongle Liu, Lipin Zhu, Qianna Zhao,
and Guohua Geng

College of Information Science and Technology,
Northwest University, Xi'an, China
xnliu@nwu.edu.cn, liuxiongle@stumail.nwu.edu.cn

Abstract. Sex determination is the first step in criminal investigation. Due to its easiness for protection, skull is considered to be the second important skeleton for sex determination. However, not all criminal cases can provide complete skull. In this paper, we present a sex determination model for incomplete skull. First, the skull is divided into seven partitions, the feature points are marked and the unmeasurable features are quantized. Then, the optimal feature subset of each partition is selected by using forward stepwise regression method based on maximum likelihood estimation. Seven partition sex determination decision models were set up and tested by using leave-one-out test. Finally, the final sex determination for incomplete female and male skull were constructed. Experiments show that any 3 partitions are enough to determine the sex of a skull with a high accuracy.

Keywords: Sex determination of skull · Incomplete skull · Skull partition · Logistic regression · Forward stepwise regression

1 Introduction

Pelvis and skull are the well-known first and second most reliable indicator of sex determination in forensic anthropology. It can narrow the scope of investigation by 50%. However, not all criminal cases can provide complete skeleton. Due to its difficult to destroy, skull plays important role in sex determination.

Though experienced experts can identify the sex of skull in high accuracy rate, there still has a certain degree of subjectivity. To reduce the subjectivity and dependence of experts, researchers began to use computer analysis to determine the skull's sex.

Robinson and Bidmos [1] collected 230 South African human skull samples and extracted 12 measured characteristics. Then he established five discriminant function equations and got 72.0–95.5 accuracy for different equations. Li et al. [2] reported his accuracy 89.2% (male) and 90% (female) for adult skull in Southwest China. Yoshinori et al. [3] measured ten indicators of modern Japanese skull samples and established nine discriminant equations. The classification accuracy is between 80% and 90%. Ramsthaler et al. [4] and Guyomarc'h et al. [5] extracted skull feature by using Fordisc software and established discriminant function for skull of Thailand, France and Germany. The identification accuracy was reported between 52.5% and 77.8% for

© Springer International Publishing AG 2017
D.-S. Huang et al. (Eds.): ICIC 2017, Part III, LNAI 10363, pp. 574–585, 2017.
DOI: 10.1007/978-3-319-63315-2_50

Thailand and France skull, 86% for German race. Franklin et al. [6] used OsiriX to mark the feature points of 31 skulls for Australian 3D skull and reached 90% identification rate. Shui et al. [7] extracted 14 skull features of Xi'an and used Step Fisher Discriminant to get a discrimination rates 87.5% for male and 86.67% for female. Liu [8] collected 142 adult skulls' X-ray film of North China and used Photoshop to measure the 13 characteristic indicators and got 95% accuracy of the sex determination. Ramamurthy et al. [9] collected seventy adult skull of South Indian. Twenty-six craniometrics parameters were analyzed by using the SPSS discriminant function. The analysis of stepwise, multivariate, and univariate discriminant function gave an accuracy of 77.1%, 85.7%, and 72.9% respectively. Li et al. [10] combines the advantages of metrical and morphological methods. With a group of Chinese skulls choosing 92 males and 58 females to establish the discriminant model, the correct rate is 95.7% and 91.4% for females and males.

Some researchers [11–15] focus on gender determination based on CT (Computed Tomography) of the maxillary sinus. Though the data is from CT images, the measurement was done on the maxillary bone.

All the above methods are designed for complete skull. However, the investigator sometimes will encounter the incomplete skull. The above algorithms will not work. In this paper, we aimed to solve the sex determination of incomplete skull of Han ethnic in China. We collected 120 3D skulls of Han by CT scan, who are living in all over the province of China. We used logistic regression method to establish the sex decision model for each skull partition (totally 7 partitions), and then integrated the seven partition decision models to construct an incomplete skull sex determination model. The experiments show that three partitions are enough to effectively determine the sex of an incomplete skull.

2   Skull Partition and Feature Extraction

2.1   Skull Partition

Skull is divided into seven sections according to the unique characteristics. They are Orbital Bone, Maxillary Bone, mandibular Bone, Frontal bone, Parietal Bone, Cheek Bone and Occipital Bone respectively, as shown in Fig. 1(a).

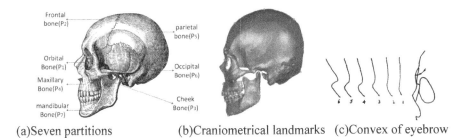

(a)Seven partitions   (b)Craniometrical landmarks   (c)Convex of eyebrow

Fig. 1. Skull information

Craniometrical Landmarks

In this section, according to the research of forensic science and human morphology, 50 (12 points on the midline and 19 points on each symmetrical part) feature points are marked by hand for sex identification. As is shown in Fig. 1(b) and Table 1.

Table 1. Feature points of skull

Partition	Feature point/abbreviation							
Orbital Bone	ectoconchiohec	ek	*nasion	n	daltryon	d
	supraorbital	os	orbitale	or	eyebrow arch	ea
Frontal bone	*bregma	b	coronale	co	fmntotemporale	ft
	frontozygomatic temporal	fmt	fronlomalare orbitale	fmo		
Cheek bone	zygion	zy	zygonmaxillareh	zm		
Maxillary bone	*nasospinale	ns	alar base	al	*pposthion	pr
	endomalare	enm	*alveolon	alv	maxillofrontale	mf
	ectomalare	ecm	*ovrale	ol	*staphylm	sta
Parietal bone	*lambdoid	l	*baihui	b		
Occipital bone	*inion	i				
Mandibular bone	mandibular notch	mn	*infradentale	id	*gnathion	gn
	condylion laterale	cdl	gonion	go	coronion	cr

The "*" represent the feature points of the skull's centerline.

Feature Extraction

Euclidean Distance Feature

The Euclidean distance is the shortest distance between two points. According to the unit's proportion between the real CT scan and the real head, we got the linear distance.

Angle Feature

According to the 3D coordinates of the three feature points, we can get the vector of two straight lines, and then calculate the angle between the two lines.

Curve Distance Feature

In this paper, we transformed the 3D surface curve into 2D curve to calculate the curve distance of the 3D skull model.

For the two feature points A and B on the skull, there are multiple curves between A and B. It is quite difficult to determine the shortest one. The method used in this paper is to manually mark a point C in the direction of A to B, so that a plane can be determined by A, B and C, as the auxiliary plane P. The auxiliary plane P and the surface of the skull model can be used to determine a curve, which defines the shortedt length of the curve between A and B.

Unmeasurable Features

For some unmeasurable features, we fit a curve or surface by LS (Least Square) and quantize them as the parameters. Figure 1(c) is a convex show of eyebrow. We use six numbers to describe the eyebrow's convex. The eyebrow is seen as a part of the circle.

By using the LS fitting, we can get the radius of the circle, which can be used to describe the convex.

In this paper, 43 kinds of characteristic indexes of skull are extracted, as is shown in Table 2.

Table 2. Features extracted from skull

Partition	No/feature/remarks					
Orbital bone	X1	The first orbital width of right eye	d-ek	X2	The second orbital width of right eye 2	mf-ek
	X3	Two eyes wide	ek-ek	X4	Right eye high	os-or
	X5	Distance between two inner eyes	d-d			
Frontal bone	X6	Minimum frontal breadth	ft-ft	X7	The maximum width of the frontal bone	co-co
	X8	Above section width	fmt-fmt	X9	Orbital width	fmo-fmo
	X10	Convexity of eyebrow	ea	X11	Frontal string	n-b
Cheek bone	X12	Zygomaxillare wide	zm-zm	X13	Zygion wide	zy-zy
Maxillary bone	X14	Nasal height	n-ns	X15	Nasal width	al-al
	X16	Superior alveolar arch width	ecm-ecm	X17	The maxillary alveolar arch length	pr-alv
	X18	Palatal length	ol-sta	X19	Palatal width	enm-enm
	X20	Palatal height	ecm-pr-ecm			
Parietal bone	X21	Parietal chord	b-l			
Occipital bone	X22	The angle of the occipital torus	o			
Mandibular bone	X23	Bigonial breadth	go-go	X24	Mandibular joint height	id-gn
	X25	Right mandibular support height	cdl-go	X26	Mandibular condylar width	cdl-cdl
	X27	Coracoid width	cr-cr	X28	Mandibular notch width	cdl-cr
	X29	The mandibular notch depth	cdl-mn	X30	Height of mandibular ramus	cdl-go
	X31	Mandibular body thickness	go-go	X32	Mandibular angle	mdb
Whole	X33	Maximum skull length	g-op	X34	Skull base line length	n-ba
	X35	Maximum skull width	eu-eu	X36	The first cranial height	b-ba
	X37	Facial width	zy-zy	X38	Upper facial height	n-pr
	X39	The second cranial height	v-gn	X40	The nose forehead height	g-ns
	X41	Distance from occipital to right mastoid point	op-ms	X42	Maximum cranial circumference	mcc
	X43	Middle facial width	zm-zm			

3 Sex Determination Models for Incomplete Skull

For a given skull training data set $T = \{(x_1, y_1), (x_2, y_2), \cdots, (x_N, y_N)\}$, $x_i \in R^n$, $y_i \in \{0, 1\}$ which R^n is the skull and y_i is a binary variable indicating the skull is female or male. When the Logistic regression model is learned, the maximum likelihood estimation method can be used to estimate the parameters of the model.

We define the probability of male skull as $P(y_i = 1|x_i) = \pi(x_i)$, and female skull as $P(y_i = 0|x_i) = 1 - \pi(x_i)$. Then the Likelihood function is as (1):

$$\prod_{i=1}^{N} [\pi(x_i)]^{y_i} [1 - \pi(x_i)]^{1-y_i} \tag{1}$$

Log likelihood function is as (2):

$$\begin{aligned}
L(\alpha, \beta) &= \sum_{i=1}^{N} [y_i \log \pi(x_i) + (1 - y_i) \log(1 - \pi(x_i))] \\
&= \sum_{i=1}^{N} [y_i \log \frac{\pi(x_i)}{1 - \pi(x_i)} + \log(1 - \pi(x_i))] \\
&= \sum_{i=1}^{N} [y_i(\alpha + \beta x_i) - \log(1 + e^{\alpha + \beta x_i})]
\end{aligned} \tag{2}$$

If we can get the maximum value of L, we can get the α and β. The problem becomes an optimization problem. In this paper, we use Newton method to solve the problem. Assuming the estimated values of α and β are $\hat{\alpha}$ and $\hat{\beta}$, the Logistic regression model is as (3):

$$\begin{aligned}
P(y_i = 1|x_i) &= \frac{e^{\hat{\alpha} + \hat{\beta} x_i}}{1 + e^{\hat{\alpha} + \hat{\beta} x_i}} \\
P(y_i = 0|x_i) &= \frac{1}{1 + e^{\hat{\alpha} + \hat{\beta} x_i}}
\end{aligned} \tag{3}$$

where $P(y_i = 1|x_i)$ is the probability of x_i is male skull, $P(y_i = 0|x_i)$ the probability of x_i is female skull.

We use Logistic classification method to construct the gender identification model, denoted by $h(R_i)$, where $R_i (i = 1, \ldots 7)$ represents partition i. a_{im} and a_{if} are defined as the accuracy of whether a skull is determined to be male or female by partition i. Combining with the sex determination model of each partition and the accuracy, the sex determination model for incomplete skull can be created.

Establishment of Sex Determination Model for Incomplete Skull
The Fig. 2 shows the establishment process of the sex determination model of incomplete skull based on partition of skull. The incomplete skull sex determination model based on partition skull can be established by the following four steps:

Extracting the Characteristics of Skull
According to Table 2, extracting the features of the skull.

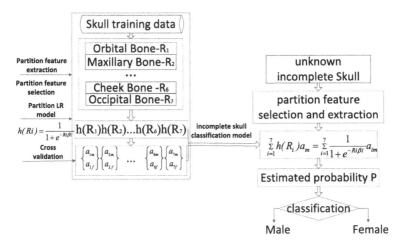

Fig. 2. The flow of the sex determination model of incomplete skull based on partition of skull

Constructing a Gender Decision Model for Each Partition

In view of the superiority of the Logistic regression method in the identification of the gender for intact skull (The features are less and the identification accuracy is high). Therefore, the Logistic regression method is also used to construct the sex determination model for every partition of the skull. The decision equations of each partition can be expressed as (4):

$$h(R_i) = \frac{1}{1 + e^{-R_i \beta_i^T}} \tag{4}$$

where β is the regression coefficient and the sex determination model of formula (5) can be rewritten as:

$$P(y_m|x) = \sum_{i=1}^{7} h(R_i) a_{im}$$
$$P(y_f|x) = \sum_{i=1}^{7} (1 - h(R_i)) a_{if} \tag{5}$$

First of all, the forward stepwise regression method based on the maximum likelihood estimation is used to select the optimal feature subset of each partition. The results are shown in Table 3.

Then, according to the selected optimal feature subset and logistic regression method, using 125 skulls specimens (68 males and 57 females) to construct a gender identification model for each region, as is shown in Table 4.

Verify the Accuracy of the Sex Determination Model of the Skull for Every Partition

Based on the samples of skull, the method of leave-one-out cross validation is used to verify the accuracy of the discriminant model of each partition. The result is shown in Table 4 and the following conclusions can be drawn from Table 4.

Table 3. The best feature subset selection based on partition of the skull

Partition	The best feature subset index
Orbital bone	$X3$
Frontal bone	$X6, X10, X11$
Cheek bone	$X12, X13$
Maxillary bone	$X14, X15$
Parietal bone	$X21$
Occipital bone	$X22$
Mandibular bone	$X23, X25, X32$

Table 4. Partition based decision model and cross validation accuracy

Partition R_i	Decision model	Determination accuracy		Average
		a_{im}	a_{if}	
Orbital R_1	$\dfrac{\exp(0.371 \times X3 - 36.222)}{1 + \exp(0.371 \times X3 - 36.222)}$	67.8	72.0	69.9
Frontal R_2	$\frac{\exp(0.303 \times X6 + 0.177 \times X11 - 0.101 \times X10 - 42.818)}{1 + \exp(0.303 \times X6 + 0.177 \times X11 - 0.101 \times X10 - 42.818)}$	89.4	85.0	87.2
Cheek R_3	$\dfrac{\exp(0.533 \times X12 + 0.166 \times X13 - 36.342)}{1 + \exp(0.533 \times X12 + 0.166 \times X13 - 36.342)}$	81.6	86.2	83.9
Maxillary R_4	$\dfrac{\exp(0.252 \times X14 + 0.290 \times X15 - 21.254)}{1 + \exp(0.252 \times X14 + 0.290 \times X15 - 21.254)}$	65.6	72.0	68.8
Parietal R_5	$\dfrac{\exp(0.194 \times X21 - 22.316)}{1 + \exp(0.194 \times X21 - 22.316)}$	68.6	67.6	68.1
Occipital R_6	$\dfrac{\exp(-0.377 \times X22 + 65.895)}{1 + \exp(-0.377 \times X22 + 65.895)}$	82.2	78.6	80.4
Mandibular R_7	$\frac{\exp(0.243 \times X23 + 0.384 \times X25 - 0.171 \times X32 - 28.222)}{1 + \exp(0.243 \times X23 + 0.384 \times X25 - 0.171 \times X32 - 28.222)}$	87.4	89.4	88.4

- In the seven partition of the skull, the sex determination accuracy of the partition for Frontal bone and Mandibular Bone are highest, their average accuracy is 88.4% and 87.2%.
- The sex determination accuracy of partition for Cheek Bone and Occipital Bone are centered. Their average accuracy is 83.9% and 80.4%.
- The sex determination accuracy of partition for Orbital Bone and parietal bone are low and the average accuracy is 69.9% and 68.1%.

Constructing Sex Determination Model of Incomplete Skull

Taking the accuracy of the model in Table 4 for skull gender identification into formula (6), the model of sex determination can be produced as follows

$$P(y_m|x) = \sum_{i=1}^{7} h(R_i)a_{im} = h(R_1) \times 0.678 + h(R_2) \times 0.894 + h(R_3) \times 0.816$$

$$+ h(R_4) \times 0.656 + h(R_5) \times 0.686 + h(R_6) \times 0.822 + h(R_7) \times 0.874$$

$$P(y_f|x) = \sum_{i=1}^{7} (1 - h(R_i))a_{if} = (1 - h(R_1)) \times 0.720 + (1 - h(R_2)) \times 0.850$$

$$+ (1 - h(R_3)) \times 0.862 + (1 - h(R_4)) \times 0.720 + (1 - h(R_5)) \times 0.676$$

$$+ (1 - h(R_6)) \times 0.786 + (1 - h(R_7)) \times 0.894$$

$$(6)$$

Logistic was used to determine the sex of the skull. Its advantage lies in the independent variable can be continuous or discrete and there is no need to meet the normal distribution assumption, so it has a wider range of applicability than other models. In the learning of Logistic regression model, the maximum likelihood estimation method is used to estimate the parameters of the model. The algorithm is used to analyze two dependent variables, which is a nonlinear S type curve function, and the maximum likelihood estimation is used to ensure the fitting of each point. Therefore, the algorithm can be used to determine the sex of the incomplete skull well.

4 Experimental Results and Discussions

Feature Statistics
We had done statistics about the mean value, standard deviation and significant difference P index for male and female skull respectively. The result is as shown in Table 5. The following conclusions can be concluded:

- In mean value of the 43 cranial features, three features of females are greater than males, they are X10, X22 and X32.
- There was overlap in the indexes of male and female.
- Except X20 and X31, the P values of the other 41 indicators are less than 0.05. The results show that there are significant differences in the 41 indexes.

Experimental Results
Figure 3 shows the determination accuracy of male, female and average for each skull partition.

As can be seen from Fig. 3, Mandibular Bone and Cheek Bone are the two regions with the highest and lowest recognition rates. The average determination accuracy of the partition can be obtained from high to low order R7 > R2 > R3 > R6 > R1 > R4 > R5.

Assuming that there are only Frontal bone and Mandibular Bone left, according to the formula (6), the sex determination model of the incomplete skull is as (7):

Table 5. Statistical feature data for male and female of our data

Index	Male		Female		P	Index	Male		Female		P
	Mean value/standard deviation						Mean value/standard deviation				
X1	39.3	7.1	37.1	2.0	0.028	X2	41.2	2.4	39.9	1.9	0.002
X3	98.9	3.0	95.2	3.8	0.000	X4	38.1	2.1	37.2	2.0	0.042
X5	25.4	2.9	23.3	2.0	0.000	X6	92.9	4.4	89.6	4.2	0.000
X7	112.8	7.2	108.8	7.1	0.008	X8	108.1	5.1	103.8	4.6	0.000
X9	97.4	3.2	93.8	3.9	0.000	X10	31.1	11.9	115.7	100.5	0.000
X11	113.5	4.9	109.1	5.9	0.000	X12	36.1	3.6	31.6	3.4	0.000
X13	110.9	5.9	106.9	5.0	0.001	X14	55.1	3.4	52.5	3.3	0.000
X15	26.5	1.9	25.7	1.5	0.028	X16	64.6	4.0	62.4	4.4	0.007
X17	53.0	3.8	50.8	5.0	0.005	X18	47.0	4.5	45.2	5.0	0.027
X19	40.7	3.3	39.1	2.9	0.023	X20	10.9	2.4	10.8	2.3	0.401
X21	118.4	5.4	112.1	6.2	0.001	X22	167.0	18.1	177.8	3.0	0.000
X23	102.9	6.7	96.9	5.7	0.000	X24	26.4	2.9	25.1	3.6	0.010
X25	65.5	5.4	57.9	4.8	0.000	X26	128.2	5.7	120.5	5.0	0.000
X27	103.6	4.6	99.3	4.3	0.000	X28	35.4	2.9	33.2	3.6	0.003
X29	14.4	2.1	12.7	2.1	0.000	X30	45.7	3.9	42.6	3.0	0.000
X31	9.5	1.3	9.3	1.4	0.270	X32	111.2	6.1	117.6	6.0	0.000
X33	178.7	7.8	168.5	5.7	0.000	X34	138.5	5.5	133.1	5.1	0.000
X35	138.7	7.3	135.3	6.8	0.026	X36	157.2	4.6	151.9	5.0	0.000
X37	135.1	5.2	126.7	4.7	0.000	X38	71.9	3.9	69.2	4.4	0.000
X39	235.6	7.6	226.7	8.6	0.000	X40	65.9	4.1	63.5	3.5	0.001
X41	94.2	5.0	90.2	4.8	0.000	X42	522.7	7.8	502.9	6.7	0.000
X43	101.0	4.4	96.4	4.8	0.000						

Note: the distance measurement unit is mm; the range of the angle is [0, 180].

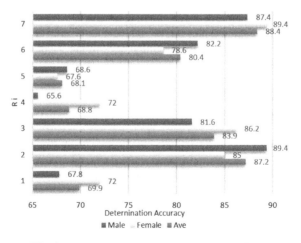

Fig. 3. Determination accuracy of every partition

$$P(y_m|x) = h(R_2) \times 0.894 + h(R_7) \times 0.874$$
$$P(y_f|x) = (1 - h(R_2)) \times 0.850 + (1 - h(R_7)) \times 0.894 \tag{7}$$

Table 6 shows the comparison of the accuracy by using LOOCV (Leave One Out Cross Validation) test.

It can be seen from Table 6, the joint decision accuracy of Frontal bone and Mandibular bone is higher than accuracy of single area for Frontal bone and Mandibular Bone.

Table 6. A comparison of the accuracy of single and combined zoning

Decision accuracy	Frontal bone	Mandibular bone	Frontal + Mandibular
Male	89.4	87.4	90.8
Female	85.0	89.4	90.2
Average (%)	87.2	88.4	90.5

In the practical skull sex determination, which partitions of the incomplete skull contains are uncertain. It can be concluded from Fig. 4(a) that the rank of the accuracy is R7 > R2 > R3 > R6 > R1 > R4 > R5. To verify the combination accuracy, we started sex determination from R7, followed by adding R2, R3, R6, R1, R4, R5. In Fig. 4(a), the recognition rate tends to increase slowly. R7, R2 and R3 are enough for gender determination.

In Fig. 4(b), the recognition started from R5, followed by adding R4, R1, R6, R3, R2, R7 in. We can see that the recognition rate also tends to increase slowly after R5, R4 and R1 partitions are added.

To a certain extent, it is only necessary to obtain a partial partition of the skull instead of complete skull to determine the sex of the skull. So the effectiveness of the method based on partition is confirmed. In other words, the increase of the partition does not significantly improve the accuracy of the skull. It can be seen in Fig. 4 that when the number of partitions is only about two, the average recognition rate can reach

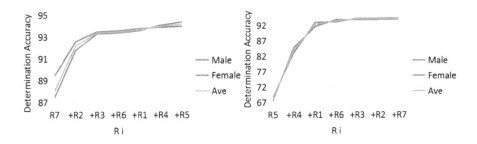

(a) adding partitions with the accuracy from high to low order

(b) adding partitions with the accuracy from low to high order

Fig. 4. Cumulative accuracy

up to 84.05% using the worst two partitions. When the number of arbitrary partitions is reach to 3, the accuracy will reach 92.2%. When the number of arbitrary partitions is known to reach 3, the recognition accuracy will reach 92.2%. When the number of partitions reaches 3 or more, the accuracy is very close to the intact skull. Therefore, although the skull is incomplete, this method can identify the sex well.

Compare the Effect with Other Methods
The method proposed in this paper is to identify the sex of the incomplete skull, and there are few references in the literature to identify the incomplete skull. In this paper, we choose the method in the literature 8 and literature 10 which has high accuracy as contrast. When the number of partitions selected is 2,3,4 and 5, the results are shown in Table 7:

Table 7. A comparison of the accuracy of single and combined zoning

Method/selected partition	$P_5 + P_4$	$P_5 + P_4 + P_1$	$P_5 + P_4$ $+ P_1 + P_6$	$P_5 + P_4 + P_1 + P_6 + P_3$	Whole
Literature 8	70.2	75.3	79.6	82.1	95
Literature 10	72.8	76.2	80.1	82.3	93.55
This paper	84.05	92.2	93.45	94	94.2

It can be concluded from the Table 6 that although the accuracy of the intact skull is very high in the method of literature 8 and literature 10, the accuracy is substantially decreased when the skull is incomplete. When the number of partition is two, their accuracy of is 70.2% and 72.8%. When the number of partition is up to five, the method of this paper is very close to that of the intact skull, however the accuracy of contrast method is only 82.1% and 82.3%.

5   Conclusions

Sex determination of skull is very important in criminal case and craniofacial reconstruction. In this paper, we calculated the measureable features and quantized the unmeasurable features, which are discriminative in sex determination. And a model of Logistic regression is constructed to predict the sex of incomplete skull. Experiments show that the presented model are useful for the sex identification of incomplete skull. Any three partitions are enough to determine the gender of a skull. The average accuracy rate can reach over 92.2%. However, we only quantize the eyebrow's convex, there are still some unmeasurable features need to be quantize, such as the forehead and Eye sockets.

Acknowledgements. This work was partially supported by the National Natural Science Foundation of China (No. 61305032, No. 61673319).

References

1. Robinson, M., Bidmos, M.: The skull and humerus in the determination of sex: reliability of discriminant function equations. Forensic Sci. Int. **186**(1), 86.e1–86.e5 (2009)
2. Li, M., Fan, Y.N., Yu, Y.M., et al.: Sex assessment of adult from southwest area of China by bones of facial cranium. Chin. J. Forensic Med. **27**(2), 132–134 (2012)
3. Yoshinori, O., Kazuhiko, I., Sachio, M., et al.: Discriminant functions for sex estimation of modern Japanese skulls. J. Forensic Leg. Med. **20**(4), 234–238 (2013)
4. Ramsthaler, F., Kreutz, K., Verhoff, M.A.: Accuracy of metric sex analysis of skeletal remains using Fordisc® based on a recent skull collection. Int. J. Legal Med. **121**(6), 477–482 (2007)
5. Guyomarc'h, P., Bruzek, J.: Accuracy and reliability in sex determination from skulls: a comparison of Fordisc® 3.0 and the discriminant function analysis. Forensic Sci. Int. **208**(1), 180.e1–180.e6 (2011)
6. Franklin, D., Cardini, A., Flavel, A., et al.: Estimation of sex form cranial measurements in a Western Australian population. Forensic Sci. Int. **229**(1–3), 158.e1–158.e8 (2013)
7. Shui, W., Yin, R.C., Zhou, M., et al.: Sex determination from digital skull model for the Han people in China. Chin. J. Forensic Med. **28**(6), 461–468 (2013)
8. Liu, Y.Y.: The sex determination of han nationality in north China by adult facial skull x-ray. Chin. J. Forensic Sci. **1**, 26–31 (2016)
9. Ramamoorthy, B., et al.: Assessment of craniometric traits in South Indian dry skulls for sex determination. J. Forensic Legal Med. **37**, 8–14 (2016)
10. Li, L., Wang, M.Y., Yun, T., et al.: Automatic sex determination of skulls based on a statistical shape model. Comput. Math. Methods Med. 251628:1–251628:6 (2013)
11. Mohammed, F.A., et al.: Sex identification in Egyptian population using multidetector computed tomography of the maxillary sinus. J. Forensic Legal Med. **19**(2), 65–69 (2012)
12. Teke, H.Y., Duran, S., Canturk, N., Canturk, G.: Determination of gender by measuring the size of the maxillary sinuses in computerized tomography scans. Surg. Radiol. Anat. **29**(1), 9–13 (2007)
13. Sharma, S.K., Jehan, M., Kumar, A.: Measurements of maxillary sinus volume and dimensions by computed tomography scan for gender determination. J. Anat. soc. India **63**(1), 36–42 (2014)
14. Prabhat, M., Rai, S., Kaur, M., Prabhat, K., Bhatnagar, P., Panjwani, S.: Computed tomography based forensic gender determination by measuring the size and volume of the maxillary sinuses. Forensic Dent. Sci. **8**(1), 40–46 (2016)
15. Kanthem, R.K., Guttikonda, V.R., Yeluri, S., Kumari, G.: Sex determination using maxillary sinus. Forensic Dent. Sci. **7**(2), 163–167 (2015)

Normalized Euclidean Super-Pixels for Medical Image Segmentation

Feihong Liu[1], Jun Feng[1(✉)], Wenhuo Su[2], Zhaohui Lv[1], Fang Xiao[1],
and Shi Qiu[3]

[1] School of Information and Technology, Northwest University, Xi'an, China
fengjun@nwu.edu.cn
[2] Center for Nonlinear Studies, Department of Mathematicals,
Northwest University, Xi'an, China
[3] Xi'an Institute of Optics and Precision Mechanics of CAS, Xi'an, China

Abstract. We propose a super-pixel segmentation algorithm based on normalized Euclidean distance for handling the uncertainty and complexity in medical image. Benefited from the statistic characteristics, compactness within super-pixels is described by normalized Euclidean distance. Our algorithm banishes the balance factor of the Simple Linear Iterative Clustering framework. In this way, our algorithm properly responses to the lesion tissues, such as tiny lung nodules, which have a little difference in luminance with their neighbors. The effectiveness of proposed algorithm is verified in The Cancer Imaging Archive (TCIA) database. Compared with Simple Linear Iterative Clustering (SLIC) and Linear Spectral Clustering (LSC), the experiment results show that, the proposed algorithm achieves competitive performance over super-pixel segmentation in the state of art.

Keywords: Medical image processing · Segmentation · Super-pixels · Local compactness

1 Introduction

Medical image segmentation is the first step of medical image processing including detection of interest region, lesions classification, benign or malignant analysis and so on. Compared with osseous organs, soft tissues are more complex and deformable in medical image series such as CT, MRI etc. Especially in small lesion tissues, their imaging appearance of shape or luminance is more uncertain. What is more, due to the interference and diffraction, it is difficult to distinguish the object from blurred pixels along with the edges.

In our previous work, an abdominal tissues segmentation method was proposed based on Simple Linear Iterative Clustering (SLIC) framework [1]. It achieved a better performance when specified a proper balance factor by manual. In the control of this preinstalled parameter, rules to guide the process of pixels clustering are set by the tradeoff between proximity and similarity. Proximity and similarity are two concepts borrowed from Gestalt psychology [2]. The balance factor would mislead the clustering results if discrepancies in shape and luminance are both small.

© Springer International Publishing AG 2017
D.-S. Huang et al. (Eds.): ICIC 2017, Part III, LNAI 10363, pp. 586–597, 2017.
DOI: 10.1007/978-3-319-63315-2_51

As mentioned above, the distance between pixels in feature space would be small because the uncertainty in shape or luminance which also makes the preinstalled parameter invalid. It would generate mistakes if a super-pixel contains a large number of pixels with close location which belong to both sides of an edge. They are easily marked a wrong uniform label. Mean values of these super-pixels are not the representative feature in fact.

Inspired by Bilateral filtering methods, cross-edge impact is diminished by the neighbor compactness of each pixel [3]. Statistical characteristics of neighbors generated from domain and range filters help to describe the compactness, which intuitively correspond to the similarity concept of Gestalt.

In this paper, we propose a super-pixel method equipped a measurement for compactness of super-pixel which does not need any preinstalled parameters by manual. After several mathematical induction, this measurement has the same form of normalized Euclidean distance. Hence, "*Normalized Euclidean Simple Linear Iterative Clustering*" (NE-SLIC) is named. We also evaluated NE-SLIC on TCIA [4] database. Experiment results show that the performance of NE-SLIC is the best.

The rest of this paper is organized as follows. Section 2 shows two major branches of super-pixel methods. Section 3 shows characteristics of medical images. Section 4 introduces the NE-SLIC method and shows the details. While in Sect. 5, we test our method in public dataset. Section 6 concludes this paper with a discussion.

2   Related Works

Similar pixels aggregate to a super-pixel, as a whole, covers small discrepancy with no interest. There are two major branches of super-pixel methods which are clustering-based and graph-based.

The graph-based methods follow the rules minimizing variance within class meanwhile maximizing variance between classes. They could be considered as top-down process. After transforming images to graph, vertices are partitioned into expected number of parts [5]. When Normalized Cut (N-cut) is proven as a special type of weighted kernel K-means method, graph-based methods could get the efficiency in time [6]. And a speed-up version of weighted kernel K-means is implemented to generate super-pixels which is sensitive to contents [7].

As for clustering-based methods, the way that pixels aggregate to a clump follows Gestalt group rules which is considered as bottom-up process.

In the combination 5-d feature space, SLIC measures the similarity and proximity by Euclidean distance. Pixel aggregates to tie together by iteratively comparing Euclidean distance with the clustering centers which would be updated when an iteration has finished [8]. Due to the aggregation procedure, SLIC lacks the measurement of compactness. In order to introduce compactness measurement into super-pixel, a Geodesic distances is introduced into super-pixel [9].

In addition, the vision could be described as the combination of bottom-up and top-down process [10]. It is useful to extend the neighbor scope to a super-pixels scope. Top-down method needs bottom-up way to speed up and the bottom-up method needs more restricted information. Additional information guarantees the compactness. In this paper, combining the global and local information, we construct a measurement for this combination and propose a NE-SLIC to banish the parameters specified by manual.

3 Characteristics of Soft-Tissues in Medical Images

Medical images have a larger encoding length than 8-bit of natural images. Larger encoding length contains more information as more grayscale. Nonetheless, tissues in medical image appear at regular position and in relative fixed luminance range which is a stronger priori knowledge.

Benefiting from a larger encoding length, medical images contain evidences to support doctors make a reliable judgment. But doctors can't process the information at the same time. It is common to observe images at several windows of different luminance range according to different tissues. There is a special range of luminance should be paid much intention, as is shown in Fig. 1.

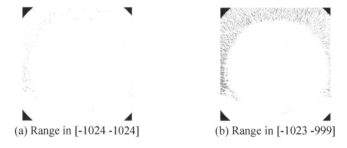

(a) Range in [-1024 -1024]   (b) Range in [-1023 -999]

Fig. 1. Displays on different gray scale range.

−1024 is the lowest intensity in ABD_LYMPH Dataset [11] which represents the black areas in Fig. 1(a). They are possibly the CT machine and some noisy points. Noisy points in Fig. 1(b) locate the air area within the machine which seem not same as the usual observation, as is shown in Fig. 2(a).

Figure 2(a) shows one abdominal slice of a patient's CT series. There is a blue rectangular at the left bottom which covers a part of right kidney. Figure 2(b) shows the details within the rectangular after gray stretching which is depicted in Fig. 2(c). The shape of the histogram appears to be a mixture of two Gaussian models. In addition, as interference and diffraction, edges are not a clear one-dot line, which is shown in Fig. 2(b).

The complexity of medical imaging is also depicted in Fig. 3 where is the histogram of the same slice. In order to highlight the complexity of its histogram, the vertical coordinate is transformed to logarithm.

From the whole perspective, big amplitude in Fig. 3 could easily cover the small amplitude when sampling in an average way. As pathological changes of tissues would trigger small changes of intensity, average value of super-pixel in the control of a manual threshold can't discriminate the unexpected lesion information appropriately.

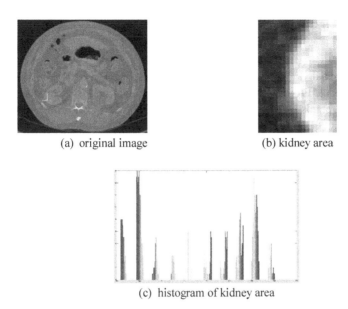

(a) original image   (b) kidney area

(c) histogram of kidney area

Fig. 2. An image details of interference & diffraction

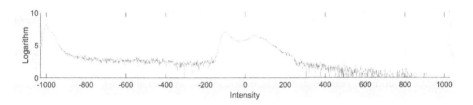

Fig. 3. Local complexity signed by histogram of an abdominal CT image

4 Normalized Euclidean SLIC for Medical Image Segmentation

In order to discriminate the unexpected lesion, we propose a super-pixels method which is more sensitive to intensity. Statistic characteristics are used to response to local compactness.

4.1 Formulization on Discrepancy Between Pixels

Inspired by the edge response function in Bilateral filtering, local compactness is measured by neighboring pixels which restrain the impact crossing an edges. The expansion of neighbor scope is much larger than 4-neighbors, as is shown in Eq. (1).

$$\text{BF} = \frac{1}{W_p} \sum_{q \in \mathcal{N}(p)} G_{\sigma_s}(q - p) G_{\sigma_r}(I_q - I_p) I_p \tag{1}$$

In the expanded neighboring area, the variance of these pixels are used to form a Gaussian model. From the insight of multidimensional Gaussian model, the discrepancy of two pixels is measured by the exponential part which is the Mahalanobis distance. If all the non-diagonal elements of covariance matrix are zero, it is also called normalized Euclidean distance.

Utilizing the negative logarithm to extract the exponential part of Eq. (1), the result is rewritten as Eq. (2),

$$D = (x - \mu)^T \Sigma^{-1}(x - \mu) \tag{2}$$

where x is the 3-d vector of each pixel and μ corresponds to its mean value, Σ is covariance matrix defined by formula (3)

$$\Sigma = \begin{bmatrix} \sigma_{I_k} & 0 & 0 \\ 0 & \sigma_{x_k} & 0 \\ 0 & 0 & \sigma_{y_k} \end{bmatrix} \tag{3}$$

where σ_{I_k} is the intensity variance of the super-pixel with label k and σ_{x_k}, σ_{y_k} corresponds to the variance of coordinate position.

The distance measurement mentioned above has a problem. When the group size of super-pixels is small enough, the intensity variance turns to zero which leads the dividend in Eq. (2) turns to zero too. It is impossible to find the inverse of covariance matrix. On the other hand, if σ_{I_k} equals to zero, all the pixels are same which is the ideal condition to compactness. But the distance can't be measured by Eq. (2).

4.2 A Solution to Positive Definite Problem of Covariance Matrix

It is able to measure the distance between pixels and its center when the covariance matrix is positive definite in Eq. (2). But if the determinant of covariance matrix equal to zero, the non-positive definite problem of covariance matrix occurs. The matrix in formula (3) need to be modulated properly.

It is common to plus a positive number in each diagonal position and which influence should be eliminated by add a factor in the quadratic term. Adding a unit matrix I and introducing an assuming factor θ, Eq. (2) is rewritten in Eq. (4).

$$\tilde{D} = (x - \mu + \theta)^T \tilde{\Sigma}^{-1}(x - \mu + \theta) \tag{4}$$

Unfold Eq. (4), as is shown in Eq. (5),

$$\tilde{D} = \begin{bmatrix} x_{I_{ij}} - \mu_{I_k} + \theta_{I_k} \\ i - \mu_{x_k} + \theta_i \\ j - \mu_{y_k} + \theta_j \end{bmatrix}^T \begin{bmatrix} \frac{1}{\sigma_{I_k}+1} & 0 & 0 \\ 0 & \frac{1}{\sigma_{x_k}+1} & 0 \\ 0 & 0 & \frac{1}{\sigma_{y_k}+1} \end{bmatrix} \begin{bmatrix} x_{I_{ij}} - \mu_{I_k} + \theta_{I_k} \\ i - \mu_{x_k} + \theta_i \\ j - \mu_{y_k} + \theta_j \end{bmatrix} \tag{5}$$

where the θ in each item has an identical form. $\tilde{D}$ would equate with Euclidean distance when the σ approaches to zero. Each θ is designed in the same form of Eq. (6).

$$\theta = (x - \mu)\left(-(\sigma^2 + 1) + \left(\frac{1}{\sqrt{\sigma^2 + 1}}\right)\right) \tag{6}$$

If σ^2 turns to zero, θ turns to zero too and the Σ degenerates to a unit matrix. If the Σ is a unit matrix normalized Euclidean distance turns to Euclidean distance. And if σ^2 turns to ∞, $\tilde{D}$ is equivalents to the infinitesimal with high order of $o(\sigma^2)$ than D. $\tilde{D}$ is the high local compactness distance measurement.

4.3 NE-SLIC Algorithm for Medical Image Segmentation

With the same framework of SLIC, equipped with high local compactness distance, our algorithm is shown in Table 1.

The distance measurement is determined by θ which is defined by Eq. (4). When applying Eq. (4) to generate a super-pixel, pixels within boundary of super-pixel would be more compact.

5 Experiments and Discussion

Including SLIC and LSC, experiments are all applied to original DICOM (Digital Imaging and Communications in Medicine) images from TCIA [4]. The testing data are from both ABD_LYMPH (Abdominal Lymph Nodes) [11] and LIDC-IDRI (The Lung Image Database Consortium image collection) [12].

5.1 Medical Image Super-Pixels

As for SLIC and LSC, the factor specified by manual is 20 and 0.065 respectively which makes a tradeoff between proximity and similarity. As for NE-SLIC, it does not need this tradeoff. The results for ABD_LYMPH are shown in Fig. 4.

As is shown in Fig. 4, all the algorithms respond more precisely to small difference of intensity when K increases. But when the K increases, clustering result is affected more seriously due to interference and diffraction. Because of the measurement of normalized Euclidean distance, as a regularization cue, local compactness improves the clustering results. NE-SLIC is more sensitive to this imperceptible information which is shown by the partial enlarged details in Fig. 4.

Table 1. Algorithm of local compactness SLIC [8]

Algorithm: NE-SLIC superpixel

Input: An image I with N pixels.
The desired number of super-pixels K.
The convergence threshold T, E.
Output : K super-pixels keeping edges appropriately.
Algorithm:
  1: Initialize cluster centers $C_k = [l_k, x_k, y_k]^T$ at regular
     grids with a step $S = \sqrt{\frac{N}{K}}$.
  2: Initialize parameters of each cluster, $[\mu_{l_k} = l_k, \mu_{x_k} =$
     $x_k, \mu_{y_k} = y_k]^T$, $[\sigma_{l_k} = 1, \sigma_{x_k} = 1, \sigma_{y_k} = 1]^T$.
  3: Move each seed to the lowest gradient position in a
     3 * 3 window.
  4: Set label $l(p) = -1$ and distance $d(p) = \infty$ at each posi-
     tion p.
  5: Initialize the residual error $\varepsilon = \infty$ and $iter = 0$.
  6: **while** $\varepsilon > E$ and $iter \leq T$
  7:   **for** each cluster center C_k
  8:     **for** each pixel in a $2S \times 2S$ region around C_k
  9:       Compute $D(C_k, p)$ using equation (4)
 10:       **if** D < d(p)
 11:         $d(p) = D(C_k, p)$;
 12:         $l(p) = k$;
 13:       **end**
 14:     **end**
 15:   **end**
 16:   compute new cluster center C_k.
 17:   compute residual error ε.
 18:   $iter = iter + 1$;
 19: **end while**

As for high uncertainty characteristic of small lesion tissues, lung nodules on LIDC-IDRI dataset are tested. All the parameters are same in our experiment. Experiments results are shown in Fig. 5.

Red circles in Fig. 5(a) and (e) are the boundaries of lung nodules which are marked by radiologists. The circles roughly sign the position of lung nodules. Figure 5 (b\c\d) show the partial enlarged details of a tiny nodule and Fig. 5(f\g\h) show the segmentation results of a big nodule. However, NE-SLIC are more close to its real contour.

SLIC   LSC   NE-SLIC

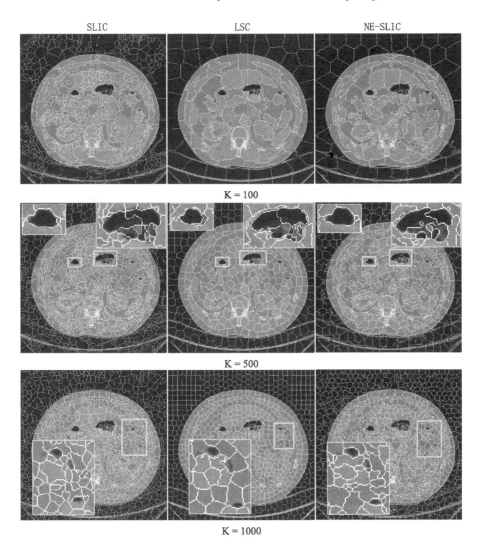

K = 100

K = 500

K = 1000

Fig. 4. Experiment results on an abdomen DICOM image

5.2 Metric on Local Compactness

As labeled data are always not exact in medical images, a metric are proposed to evaluate the performance of these clustering results which based on joint edges classification accuracy, as is shown in Fig. 6.

There are two super-pixels labeled as A and B respectively. Pixels falling into these joint edge are either similar to A or to B which are easily clustered to an inappropriate label. Small discrepancy of intensity between two compact super-pixels is easily

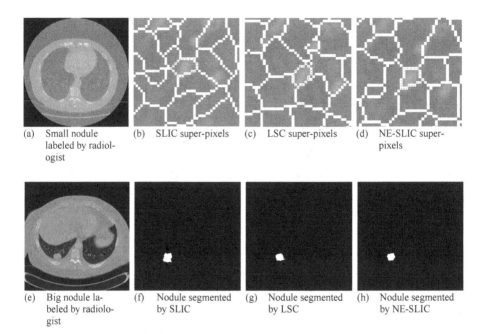

(a) Small nodule labeled by radiologist	(b) SLIC super-pixels	(c) LSC super-pixels	(d) NE-SLIC super-pixels

(e) Big nodule labeled by radiologist	(f) Nodule segmented by SLIC	(g) Nodule segmented by LSC	(h) Nodule segmented by NE-SLIC

Fig. 5. Super-pixels in lung nodule (k = 2000) (Color figure online)

Fig. 6. Joint edge of two adjacent super-pixels

influenced by position. In fact, intensity information takes the leading role in local scope which is used as reference to evaluate the local compactness.

Local compactness of super-pixel is modeled by its intensity histogram distribution. Figure 7(a) and (b) show histograms of the super-pixels A and B in Fig. 6 respectively. The affiliation of pixels falling into the joint edge are signed by GMM (Gaussian mixture model), as is shown in Fig. 7(c). The parameters of Gaussian distributions are acquired from the statistical characteristics of super-pixel.

GMM signs labels of the pixels in joint edge only by its intensity distribution which is defined by formula (7)

$$\mathcal{L}_{ij} = arg \max_{k} P(I_{ij}|\mu_k, \sigma_k^2) \qquad (7)$$

where $P(I_{ij}|\mu_k, \sigma_k^2)$ is the probability computed by these Gaussian model with mean μ_k and variance σ_k^2. $\mathcal{L}_{ij}$ is the most probable label. The classification accuracy of the pixels in joint edge are defined by formula (8).

$$T = \sum \delta(\mathcal{L}_{ij}, L_{ij}) \qquad (8)$$

where δ is the Kronecker delta. This accuracy number interprets the compactness in the boundary of super-pixels. And the L_{ij} indicate the label generate by super-pixel methods. The K are set from 100 to 1000 with an interval of 100. A list of classification accuracy number are acquired from SLIC, LSC and our method which are visualized in Fig. 7.

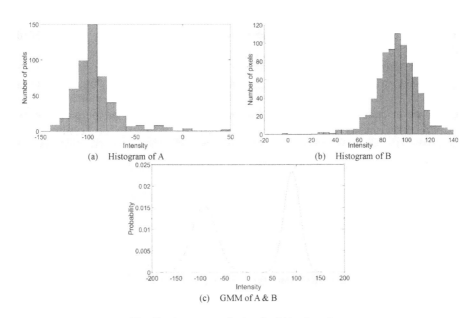

(a)   Histogram of A

(b)   Histogram of B

(c)   GMM of A & B

Fig. 7. Accuracy calculated within the edges

According to Fig. 8, when the clustering number are increase to 700, NE-SLIC exceeds the others in classification accuracy.

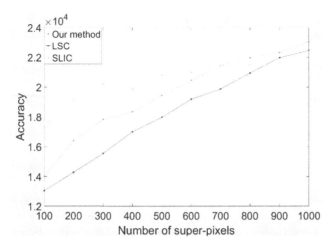

Fig. 8. The classification performance between different super-pixels methods.

6 Conclusion

This paper proposed a local compactness super-pixel algorithm which is called NE-SLIC. The algorithm banishes the balance factor to adjust the ratio between similarity and proximity which are inspired by Bilateral filtering by extending its 4-neighbor to a larger area with more pixels. The effectiveness of this algorithm is tested in TCIA database and the experiment results show that our algorithm has the best appearance in compactness. According the visualization and the metric proposed by this paper, experiments result shows that our algorithm is the best in local compactness.

Acknowledgement. This work was supported by National Natural Science Foundation of China (No. 61372046).

References

1. Li, P., Feng, J., Bu, Q., Liu, F., Wang, H.: Multi-object segmentation for abdominal CT image based on visual patch classification. In: Zha, H., Chen, X., Wang, L., Miao, Q. (eds.) CCCV 2015. CCIS, vol. 547, pp. 130–138. Springer, Heidelberg (2015). doi:10.1007/978-3-662-48570-5_13
2. Wagemans, J., Elder, J.H., Kubovy, M., Palmer, S.E., Peterson, M.A., Singh, M., von der Heydt, R.: A century of Gestalt psychology in Visual Perception: I. Perceptual Grouping and Figure-Ground Organization. Psychol. Bull. **138**(6), 1172–1217 (2012)
3. Tomasi, C., Manduchi, R.: Bilateral filtering for gray and color images. In: International Conference on Computer Vision, pp. 839–846 (1998)
4. The Cancer Imaging Archive. http://www.cancerimagingarchive.net
5. Shi, J., Malik, J.: Normalized cuts and image segmentation. IEEE Trans. Pattern Anal. Mach. Intell. **22**(8), 888–905 (2000)

6. Dhillon, I.S., Guan, Y., Kulis, B.: Kernel k-means: spectral clustering and normalized cuts. In: Proceedings of the tenth ACM SIGKDD, pp. 551–556 (2004)
7. Li, Z., Chen, J.: Superpixel segmentation using linear spectral clustering. In: Proceedings of the IEEE Conference on Computer Vision and Pattern Recognition, pp. 1356–1363 (2015)
8. Achanta, R., Shaji, A., Smith, K., Lucchi, A., Fua, P., Süsstrunk, S.: SLIC superpixels compared to state-of-the-art superpixel methods. IEEE Trans. Pattern Anal. Mach. Intell. **34** (11), 2274–2282 (2012)
9. Liu, Y.J., Yu, C.C., Yu, M.J., He, Y.: Manifold slic: a fast method to compute content-sensitive superpixels. In: Proceedings of the IEEE Conference on Computer Vision and Pattern Recognition, pp. 651–659 (2016)
10. Yuille, A., Kersten, D.: Vision as Bayesian inference: analysis by synthesis? Trends cogn. Sci. **10**(7), 301–308 (2006)
11. Roth, H., Lu, L., Seff, A., Cherry, K.M., Hoffman, J., Wang, S., Summers, R.M.: A new 2.5 D representation for lymph node detection in CT. The Cancer Imaging Archive (2015). http://doi.org/10.7937/K9/TCIA.2015.AQIIDCNM
12. Armato III, S.G., McLennan, G., Bidaut, L., McNitt-Gray, M.F., Meyer, C.R., Reeves, A.P., Clarke, L.P.: Data from LIDC-IDRI. The Cancer Imaging Archive (2015). http://doi.org/10. 7937/K9/TCIA.2015.LO9QL9SX

A Computer Aided Ophthalmic Diagnosis System Based on Tomographic Features

Vitoantonio Bevilacqua[1(✉)], Sergio Simeone[1], Antonio Brunetti[1],
Claudio Loconsole[1], Gianpaolo Francesco Trotta[2],
Salvatore Tramacere[3], Antonio Argentieri[3], Francesco Ragni[3],
Giuseppe Criscenti[3], Andrea Fornaro[3], Rosalina Mastronardi[3],
Serena Cassetta[3], and Giuseppe D'Ippolito[3]

[1] Department of Electrical and Information Engineering (DEI),
Polytechnic University of Bari, Via Orabona 4, 70126 Bari, Italy
vitoantonio.bevilacqua@poliba.it
[2] Department of Mechanics, Mathematics and Management (DMMM),
Polytechnic University of Bari, Viale Japigia 182, 70126 Bari, Italy
[3] Ligi Tecnologie Medicali, S.p.A. Via Luigi Corsi 50, 74121 Taranto, Italy

Abstract. Keratoconus is a bilateral progressive corneal disease characterized by thinning and apical protrusion; its early diagnosis is fundamental since it allows one to treat this rare disease by cross-linking approach, thus preventing a major corneal deformation and avoiding more invasive and risky surgical therapies, such as cornea transplant. Ophthalmology improvements have allowed a more rapid, precise and painless acquisition of corneal biometric parameters which are useful to evaluate alterations and abnormalities of eye's outer structure. This paper presents a study about Keratoconus diagnosis based on a machine learning approach using corneal physical and morphological parameters obtained through Precisio™ tomographic examination. Artificial Neural Networks (ANNs) have been used for classification; in particular, a mono-objective Genetic Algorithm has been used to obtain the best topology for the neural classifiers for different input datasets obtained from features ranking. High levels of accuracy (higher than 90%) have been reached for all types of classification; in particular, binary classification has showed the best discrimination capability for Keratoconus identification.

Keywords: Keratoconus · Rare disease · Corneal Tomography · Artificial Neural Networks · Genetic algorithm · Decision support systems

1 Introduction

Keratoconus is a progressive non-inflammatory rare disease characterized by anterior protrusion of the cornea (ectasia) which induces high order aberration and causes significant distortion of vision. It usually hits in puberty and its advancement stops in the third or fourth decade [1].

© Springer International Publishing AG 2017
D.-S. Huang et al. (Eds.): ICIC 2017, Part III, LNAI 10363, pp. 598–609, 2017.
DOI: 10.1007/978-3-319-63315-2_52

Keratoconus is one of the leading causes of corneal transplant, which is an invasive and complex surgical therapy enabling to restore patient's functional vision. However, corneal transplant is risky and may cause both intra- and post-operative complications. An alternative method to treat this disease is the cross-linking [2, 3]; it is an alternative semi-invasive and effective treatment to halt or even to aid the regression of the progressive and irregular changes in the corneal shape of Keratoconus. This technique uses riboflavin (a photosensitizing agent) and ultraviolet-A light to strengthen chemical bonds in the cornea. Cross-linking permits also a long-term stabilization and its benefits have been documented in literature [4]. Therefore, Keratoconus early diagnosis is essential as it allows the cross-linking treatment, thus preventing a major corneal protrusion.

In order to diagnose Keratoconus, there are several systems based on imaging. Photokeratoscopy and videokeratography, also known as Corneal Topography, are non-invasive medical imaging techniques which strongly contribute to Keratoconus clinical diagnosis. In details, they map the surface curvature of the cornea detecting topographic changes and extracting quantitative biometric corneal parameters for qualitative interpretation of the disease. However, these techniques, which are based on Placido-disk system [5], feature some disadvantages that substantially alter the measurement (e.g., they assume cornea to be spherical and are limited to the anterior corneal surface losing accuracy both in peripheral corneal zones and in the tear film [6]). It must be pointed out that the Placido-disc systems do not allow the detection of the corneal morphology, which is related to the elevation profile: possible elevation maps provided by the Placido Disc devices are generated by an arbitrary integration process in the radial direction, based on the hypothesis that the center of acquisition (corresponding to the center of the Placido Disc, where no information is available) is the point of higher corneal elevation. This recursive process is also subject to a cumulative reading error which propagates in the radial direction (Fig. 1).

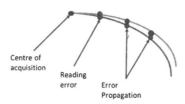

Fig. 1. Error propagation for corneal topographers.

Despite the previously mentioned problems, good values of accuracy in the discrimination of different subtypes of Keratoconus were reached in using a combination of corneal topographic and optical features [7].

The principle of triangulation can be employed to determine the morphology of the corneal surface, by means of a specific profile illuminated by a source projecting from a known angle and which is detected by a camera placed at a specific distance. Therefore, the recognition of the whole cornea elevation profile requires several multiple scans to cover the whole area to detect. Systems based on a such approach are called Corneal

Tomographers. Recently, Corneal Tomography contributed to a great improvement in ophthalmology overcoming topography limits. This kind of imaging technique, based on the Scheimpflug Camera system, can extract features from both anterior and posterior corneal surfaces with a high level of accuracy; at the same time, it is not affected by the artifacts produced by the tear film reflection. In general, a tomographic system is composed of: (i) a monochromatic blue light source, which not induces pupillary miosis; (ii) a fixed central camera; (iii) a camera rotating around the optical axis of the eye. The light beam illuminates only a thin portion of the eye and, since the cornea is not completely transparent, the light diffuses (scattering principle) through the entire anterior chamber up to the iris, thus highlighting the different corneal tissues. The rotating camera (oriented according to the Scheimpflug's principle to focus the entire image plane) resumes the radial sections of the illuminated eye's anterior chamber performing a rotation of 360° around the optical axis of the eye.

However, unlike the parameters extracted by using the Corneal Topography, the analysis and interpretation of the biometric parameters coming from the Corneal Tomography examination may be difficult. In fact, at certain stages, Keratoconus disease may be wrongly diagnosed as Irregular Astigmatism.

Since the identification of Keratoconus disease is fundamental for a correct treatment, making it less invasive and less risky, in this paper a novel approach for the identification of Keratoconus from physical and morphological indexes of the cornea obtained through an innovative tomographic examination is proposed.

In literature, there are several supervised approaches that process ophthalmic features for different purposes: biometry, early diagnosis and detection [8–10]. Moreover, there are several approaches that use a combination of features from both topographic and tomographic examinations; although authors generally got good results in terms of accuracy, the patients generally undergo two different examinations, thus increasing the timing for an accurate diagnosis [11, 12].

In this paper, an Artificial Neural Network (ANN) is used to detect Keratoconus through different features extracted from the tomographic examination: in the first stage of the approach, a binary classifier was used to discriminate between Keratoconus and the group composed by Irregular Astigmatisms and No-Irregular Disorders; then, a multiclass classifier was used to classify among the three classes of disease.

The following sections are organized as follows: details about the dataset are provided in Sect. 2; the ANN approach and the description of the used classifiers are presented in Sect. 3; the results are reported in Sect. 4, whereas the conclusions in Sect. 5.

2   Materials

2.1   The Acquisition System

The used acquisition system is the Precisio™ corneal tomographer (Fig. 2), which is a new generation ophthalmic device used to measure, in high definition, the elevation data ("morphology") of the anterior and posterior corneal surfaces from limbus-to-limbus, corneal pachymetry, epithelial and stromal thickness, corneal refractive power, anterior chamber depth, detailed corneal refractive data including

Fig. 2. Precisio™ tomograph used for corneal examination

corneal high order aberrations (HOA) and asphericity. The fundamental basis of Precisio™ tomographer is a triangulated Scheimpflug based measurements of a rotating ultra-thin blue slit that scans the anterior segment of the eye, by means of a synchronous stereo high resolution CMOS tandem cameras and an optic system with selective filters to exclude parasite light signals. This geometrical rule allows obtaining focused images of a whole corneal cross section produced by means of the slit light projected on the patient eye.

During the acquisition, the head rotates by 360° in 1.14 s and 60 images of corneal cross section are acquired by the MAIN camera (1 image every 6°). A second camera, placed exactly in front of the eye, is used for iris recognition, relevant points registration and eye tracking functions. The integrated eye tracking system registers and compensates the eye movements in the 6-spatial degree of freedom (X, Y, Z, θx, θy, θz). Using the images acquired by the MAIN camera, the elaboration unit performs an edge-detection algorithm to detect on each cross section, from limbus to limbus, the anterior and posterior corneal contours, the anterior iris contour and the lens contour. The results are processed together to produce a raw data set for each surface to be calculated considering at least 100,000 points. The FOCUS camera is synchronized to the MAIN camera and acquires 60 frames used by the eye tracker. This one exploits the spots produced on the cornea by the 8 IR led in front of the eye to track the eye on the vertical plane during the acquisition. To perform the acquisition, the patient, voice guided by Precisio™ control system, places his head on the motorized chinrest, and his eye is automatically aligned in the correct position. The patient stares at a blinking red alignment laser diode placed in front of his eye. The chinrest can move along the 3 axes X, Y and Z. A representative scheme of the acquisition procedure is shown in Fig. 3.

2.2   The Acquired Data

In a time period from 2015 to 2016, 774 eyes examinations were performed at IVIS Centre of Refractive Surgery in "San Camillo" Hospital (Trento, Italy). All the examinations were performed using the Precisio™ corneal tomograph (http://www.ivistechnologies.it/precisio.php), and biometric features were extracted from each

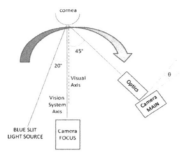

Fig. 3. Scheme of vision system during the acquisition procedure

examination. In detail, data were divided into 3 groups, not-equally distributed since Keratoconus is a rare disease:

- 96 cases of Keratoconus (K);
- 126 cases of Irregular Astigmatism (IA);
- 552 cases of No-Irregular Disorder (ND), i.e. eyes that have no ocular diseases, irregular astigmatism and which have not undergone refractive surgery.

Among all the parameters obtained from the tomographic examination, an initial set **S1** of 14 physical and morphological features was considered for subsequent classification; in particular, the selected features were:

- minimum pachymetry (corneal thickness) and its ordinate **[F1, F2]**, since Keratoconus mostly starts in the inferotemporal part of the cornea;
- minimum and maximum curvature in diopters for both anterior **[F3, F4]** and posterior **[F5, F6]** corneal surfaces, since Keratoconus causes ectasia, or dilation of these surfaces;
- amplitude and axis tilt of the aconic's cylindrical component (astigmatism) of both anterior **[F7, F8]** and posterior **[F9, F10]** corneal surfaces;
- aconic asphericity, an index measuring refractive power variation along each emimeridian of both posterior **[F11]** and anterior **[F12]** corneal surface;
- mean value and standard deviation of indexes measuring the corneal spherical aberration **[F13, F14]**.

Then, a subset **S2** of 9 corneal physical and morphological indexes was obtained reducing the dimensionality of the previous set **S1**. In particular, the subset **S2** was obtained using a features selection algorithm based on the "greedy hill climbing" attribute ranking [13] available in WEKA software [14]. The resulting dataset consisted of the following 9 parameters:

- minimum pachymetry (corneal thickness) and its ordinate **[F1, F2]**;
- minimum and maximum curvature in diopters of the posterior corneal surface **[F5, F6]**;
- amplitude of the aconic's cylindrical component (astigmatism) of anterior corneal surface **[F7, F8]**;
- aconic asphericity of the anterior corneal surface **[F11]**;

- mean value and standard deviation of indexes measuring the corneal spherical aberration **[F13, F14]**.

3 Methods

The main goal of this work is to detect and differentiate Keratoconus (K) from Irregular Astigmatism (IA) and No-Irregular Disorder (ND). For this reason, we preliminarily created two sets (S1 and S2 types) grouping IA and ND samples (IA_ND). After the datasets creation, a binary classifier was designed to discriminate between the two groups, considering K as Positive Class (P) and IA_ND as Negative Class (N). To optimize classification performance, an evolutionary approach based on Genetic Algorithm was exploited. The optimization of classification systems is an interesting topic in literature; if fact, different works use evolutionary algorithms for the optimization of classifiers (e.g., using mono- and multi-objective Genetic Algorithms [15, 16], or other kinds of evolutionary systems [17]), as well as other algorithms (e.g. multi-kernel learning based algorithm [18]).

In this work, the optimal topology of the Artificial Neural Network (ANN) classifier was found through a mono-objective genetic algorithm (GA) considering a 30-bits chromosome composed of 6 genes coding the number of neurons and the activation functions for each hidden layer. The maximum number of neurons for each hidden layer was 256, while the encoded activation functions were the log-sigmoid (*logsig*), the hyperpolic tangent sigmoid (*tansig*), the pure linear (*purelin*) and the symmetric saturating linear (*satlins*). The resilient backpropagation algorithm [19] was set for weights and bias update, whereas the activation function set in the output layer (consisting of a single neuron) was the hyperbolic tangent sigmoid (*tansig*). An initial population of 100 individuals randomly generated was considered, and the operators employed in the GA set up were: (i) crossover with 2 points, with a probability of 0.8; (ii) mutation, with a probability of 0.2; (iii) selection system: elitism; (iv) stop criteria set to maximum generations numbers (100), or 30 consecutive generations with fitness value unchanged.

As in [16], since Accuracy (Eq. 1) is an index of the discrimination capabilities of a neural classifier, each individual was evaluated using a fitness function (Eq. 2) based on the mean Accuracy on test set computed on a fixed number of iterations of training, validation and test of each ANN in the population of the current generation. In this work, the number of iterations was set to 100 and the Confusion Matrix model reported in Table 1 was used to evaluate the Accuracy.

Table 1. Confusion matrix

Predicted condition	True condition	
	Positive	*Negative*
Positive	TP	FP
Negative	FN	TN

$$Accuracy = \frac{TP + TN}{TP + TN + FP + FN} \tag{1}$$

$$Fitness\,function = \frac{\sum_{n=1}^{iterations} Accuracy_n}{iterations} \tag{2}$$

Following the evolutionary approach previously described, a second binary classifier was designed to classify between IA, considered as Positive Class (P), and ND, considered as Negative Class (N).

After the binary classification, the Genetic Algorithm was properly adapted to work with ANNs able to perform multi-class classification. In detail, the output layer of the ANN was modified to classify among the three classes (K, IA and ND) using the *softmax* function, while the cross-entropy was used as cost function for the minimization of the training error [20]. In all the reported cases, input data was normalized using z-score normalization [21].

4   Results

In this section, the results for all classifications are reported. In particular, four different classifications were performed evaluating

- *Case 1:* K vs IA_ND considering the dataset S1 composed of 14 features;
- *Case 2:* K vs IA_ND considering the dataset S2 composed of 9 features;
- *Case 3:* IA vs ND considering the dataset S2 composed of 9 features;
- *Case 4:* K vs IA vs ND considering the dataset S2 composed of 9 features.

The results are expressed in terms of mean values of Accuracy (Eq. 1), Sensitivity (Eq. 3), which represents the classifier capability to correctly discriminate the Positive samples, and Specificity (Eq. 4), which measures the negatives samples that are correctly classified as negatives.

$$Sensitivity = \frac{TP}{TP + FN} = \frac{TP}{P} \tag{3}$$

$$Specificity = \frac{TN}{TN + FP} = \frac{TN}{N} \tag{4}$$

For *Case 1*, the optimal ANN topology found by the Genetic Algorithm for the Keratoconus classification using the dataset with 14 features was composed of 2 layers, with 34 neurons for the hidden layer, and 1 neuron in the output layer. The activation functions found by the GA was *logsig* for the hidden layer, while the *tansig* function was set for the neuron in the output layer (Fig. 4).

For *Case 2*, the optimal ANN topology found by the Genetic Algorithm for the same task of the previous case, but considering the dataset with 9 features, was composed of 4 layers, with 213 neurons for the first hidden layer, 66 neurons for the second hidden layer, 2 neurons for the third hidden layer, and 1 neuron in the output

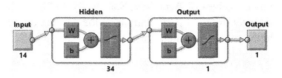

Fig. 4. A schematic representation of the best ANN topology found by the GA for case 1

layer. The activation functions found by the GA were *logsig* for all the hidden layers, while the *tansig* function was set for the neuron in the output layer (Fig. 5).

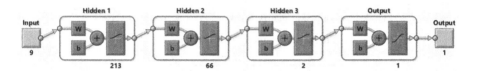

Fig. 5. A schematic representation of the best ANN topology found by the GA for case 2

For *Case 3*, the optimal ANN topology found by the Genetic Algorithm for the classification of Irregular Astigmatism and No Irregular Disorder was composed of 2 layers, with 13 neurons for the hidden layer, and 1 neuron in the output layer. The activation functions found by the GA was *logsig* for the hidden layer, while the *tansig* function was set for the neuron in the output layer (Fig. 6).

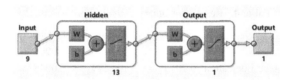

Fig. 6. A schematic representation of the best ANN topology found by the GA for case 3

For *Case 4*, the optimal ANN topology found by the Genetic Algorithm for the multi-class classification was composed of 4 layers, with 78 neurons for the first hidden layer, 95 neurons for the second hidden layer, 20 neurons for the third hidden layer, and 3 neurons in the output layer. The activation functions found by the GA were *purelin* for the first layer, *logsig* for the second and the third hidden layers, while the *softmax* function was set for the output layer (Fig. 7).

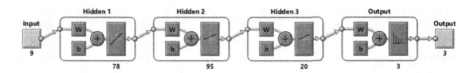

Fig. 7. A schematic representation of the best ANN topology found by the GA for case 4

In Table 2, the mean values of Accuracy, Sensitivity and Specificity, evaluated on 100 iterations considering different permutations of the dataset, are reported for case 1. In this case, the higher value reached for accuracy was 99.36%, whose confusion matrix is reported in Table 3.

Table 2. Results for binary classification of Case 1

	Accuracy (%)	Sensitivity	Specificity
Mean	98,01	0,8785	0,9949
Standard deviation	0,8329	0,0528	0,0058

Table 3. Confusion matrix for the highest accuracy in Case 1

Predicted condition	True condition	
	Positive	Negative
Positive	19	0
Negative	1	137

In Table 4, the mean values of Accuracy, Sensitivity and Specificity, evaluated on 100 iterations considering different permutations of the dataset, are reported for case 2. As for the previous one, even in this case the higher value reached for accuracy was 99.36%, whose confusion matrix is reported in Table 5.

Table 4. Results for binary classification of Case 2

	Accuracy (%)	Sensitivity	Specificity
Mean	98,09	0,8740	0,9965
Standard deviation	0,7995	0,0579	0,0045

Table 5. Confusion matrix for the highest accuracy in Case 2

Predicted condition	True condition	
	Positive	Negative
Positive	20	1
Negative	0	136

In Table 6, the mean value of Accuracy, Sensitivity and Specificity, evaluated on 100 iterations considering different permutations of the dataset, are reported for case 3. In this case, where the classes to discriminate were Irregular Astigmatism and No Irregular Disease, the higher value of accuracy was 97.08% whose confusion matrix is reported in Table 7.

About the multi-class approach, the confusion matrix for the best case is reported in Table 8, showing an accuracy of 96.18%. As for the previous cases, the mean values of

Table 6. Results for binary classification of Case 3

	Accuracy (%)	Sensitivity	Specificity
Mean	92,41	0,7358	0,9682
Standard deviation	1,9739	0,0991	0,0207

Table 7. Confusion matrix for the highest accuracy in Case 3

Predicted condition	True condition	
	Positive	Negative
Positive	19	0
Negative	1	137

Table 8. Confusion matrix for the highest accuracy in Case 4

Predicted condition	True condition		
	Keratoconus (K)	Irregular astigmatism (IA)	No irregular disorder (ND)
Keratoconus (K)	20	1	0
Irregular astigmatism (IA)	0	22	2
No irregular disorder (ND)	0	3	109

Table 9. Results for multi-class classification of Case 4

	Accuracy (%)	Sensitivity$_K$	Sensitivity$_{IA}$	Sensitivity$_{ND}$
Mean	90,15	0,9055	0,6842	0,9516
Standard deviation	1,9278	0,0531	0,0905	0,0217

Accuracy, Sensitivity for all the three classes, evaluated on 100 iterations considering different permutations of the dataset, are reported in Table 9.

5 Discussion and Conclusions

In this work, several ANN classifiers were obtained and evaluated using an evolutionary approach, based on mono-objective genetic algorithm to detect and classify Keratoconus, a rare disease affecting the eye.

As it could be seen from the result tables reported in the previous section, high levels of accuracy were reached by the classification process for all cases. In particular, these results pointed out that: (i) Keratoconus disease is discriminable from other diseases by using physical and morphological features obtainable from a tomographic examination, and (ii) the evolutionary approach is able to obtain the optimal ANN topology.

In detail, the binary classification approach for Keratoconus classification seems to work better than the multi-class approach; in fact, the mean accuracy on random permutations of the dataset is higher for both 9-features and 14-features binary classifiers (about 98%), while it is lower for the multi-class classifier (90.15%). Specifically, the binary classification considering only 9 features shows the better result in terms of accuracy, even though it is comparable to that obtained using the dataset with more features. This result is acceptable since, observing the results reported in Table 9, the lowest sensitivity is obtained for Irregular Astigmatism, which could be considered as a central pathological state between Keratoconus and No-Irregular Disorder.

Moreover, since the multi-class approach shows a lower accuracy with respect to the previously discussed cases, 2 binary cascading classifiers could be used to obtain the same multi-class classification: the first one should be used for the discrimination between Keratoconus (P) and other diseases (N), and in case of a negative result, a second classifier should be used for the classification between Irregular Astigmatism and No-Irregular Disorder, since this latter classifier shows about 92% of accuracy on 100 random permutations of the dataset, considering the set composed of 9 features.

As future work, new methods for classification, considering other types of classifiers, or deep learning approaches will be investigated to obtain an improvement of the performance in this classification process.

References

1. Rabinowitz, Y.S.: Keratoconus. Surv. Ophthalmol. **42**, 297–319 (1998)
2. Meek, K.M., Hayes, S.: Corneal cross-linking–a review. Ophthalmic Physiol. Opt. **33**, 78–93 (2013)
3. Vinciguerra, R., Romano, M.R., Camesasca, F.I., Azzolini, C., Trazza, S., Morenghi, E., Vinciguerra, P.: Corneal cross-linking as a treatment for keratoconus: four-year morphologic and clinical outcomes with respect to patient age. Ophthalmology **120**, 908–916 (2013)
4. Raiskup-Wolf, F., Hoyer, A., Spoerl, E., Pillunat, L.E.: Collagen crosslinking with riboflavin and ultraviolet - a light in keratoconus: long-term results. J. Cataract Refract. Surg. **34**, 796–801 (2008)
5. Rand, R.H., Howland, H.C., Applegate, R.A.: Mathematical model of a Placido disk keratometer and its implications for recovery of corneal topography. Optom. Vis. Sci. **74**, 926–930 (1997)
6. Guirao, A., Artal, P.: Corneal wave aberration from videokeratography: accuracy and limitations of the procedure. JOSA A **17**, 955–965 (2000)
7. Rabinowitz, Y.S., Li, X., Canedo, A.L.C., Ambrósio, R., Bykhovskaya, Y.: Optical coherence tomography combined with videokeratography to differentiate mild keratoconus subtypes. J. Refract. Surg. **30**, 80–86 (2014)
8. Carnimeo, L., Bevilacqua, V., Cariello, L., Mastronardi, G.: Retinal vessel extraction by a combined neural network–wavelet enhancement method. In: Huang, D.-S., Jo, K.-H., Lee, H.-H., Kang, H.-J., Bevilacqua, V. (eds.) ICIC 2009. LNCS, vol. 5755, pp. 1106–1116. Springer, Heidelberg (2009). doi:10.1007/978-3-642-04020-7_118
9. Bevilacqua, V., Carnimeo, L., Mastronardi, G., Santarcangelo, V., Scaramuzzi, R.: On the comparison of NN-based architectures for diabetic damage detection in retinal images. J. Circuits Syst. Comput. **18**, 1369–1380 (2009)

10. Bevilacqua, V., Cariello, L., Columbo, D., Daleno, D., Dellisanti Fabiano, M., Giannini, M., Mastronardi, G., Castellano, M.: Retinal fundus biometric analysis for personal identifications. Adv. Intell. Comput. Theor. Appl. Asp. Artif. Intell. 1229–1237 (2008)

11. Arbelaez, M.C., Versaci, F., Vestri, G., Barboni, P., Savini, G.: Use of a support vector machine for keratoconus and subclinical keratoconus detection by topographic and tomographic data. Ophthalmology **119**, 2231–2238 (2012)

12. Li, Y., Meisler, D.M., Tang, M., Lu, A.T., Thakrar, V., Reiser, B.J., Huang, D.: Keratoconus diagnosis with optical coherence tomography pachymetry mapping. Ophthalmology **115**, 2159–2166 (2008)

13. Freitag, D.: Greedy attribute selection. In: Machine Learning Proceedings 1994: Proceedings of 8th International Conference, p. 28. Morgan Kaufmann (2014)

14. Hall, M., Frank, E., Holmes, G., Pfahringer, B., Reutemann, P., Witten, I.H.: The WEKA data mining software: an update. ACM SIGKDD Explor. Newsl. **11**, 10–18 (2009)

15. Bevilacqua, V., Mastronardi, G., Menolascina, F., Pannarale, P., Pedone, A.: A novel multi-objective genetic algorithm approach to artificial neural network topology optimisation: the breast cancer classification problem. In: International Joint Conference on Neural Networks, 2006, IJCNN 2006, pp. 1958–1965. IEEE (2006)

16. Bevilacqua, V., Brunetti, A., Triggiani, M., Magaletti, D., Telegrafo, M., Moschetta, M.: An optimized feed-forward artificial neural network topology to support radiologists in breast lesions classification. In: Proceedings of 2016 on Genetic and Evolutionary Computation Conference Companion, pp. 1385–1392. ACM (2016)

17. Du, J.-X., Huang, D.-S., Wang, X.-F., Gu, X.: Shape recognition based on neural networks trained by differential evolution algorithm. Neurocomputing **70**, 896–903 (2007)

18. Wang, J., Wang, H., Zhou, Y., McDonald, N.: Multiple kernel multivariate performance learning using cutting plane algorithm. In: 2015 IEEE International Conference on Systems, Man, and Cybernetics (SMC), pp. 1870–1875. IEEE (2015)

19. Riedmiller, M., Braun, H.: A direct adaptive method for faster backpropagation learning: The RPROP algorithm. In: IEEE International Conference on Neural Networks, pp. 586–591. IEEE (1993)

20. De Boer, P.-T., Kroese, D.P., Mannor, S., Rubinstein, R.Y.: A tutorial on the cross-entropy method. Ann. Oper. Res. **134**, 19–67 (2005)

21. Zill, D., Wright, W.S., Cullen, M.R.: Advanced Engineering Mathematics. Jones & Bartlett Learning, Burlington (2011)

Information Security

A Reversible Data Hiding Method Based on HEVC Without Distortion Drift

Si Liu, Yunxia Liu$^{(\boxtimes)}$, Cong Feng, and Hongguo Zhao

College of Information Science and Technology, Zhengzhou Normal University,
Zhengzhou, China
liuyunxia0110@hust.edu.cn

Abstract. This paper presents a reversible data hiding algorithm without intra-frame distortion drift based HEVC. We embed the secret information into the multivariate array of the 4 × 4 luminance DST blocks to avert the distortion drift. With the inverse operation of multivariate array in decoder, the embedded video is perfectly reconstructed as the original encoded video. Moreover, the entire process of embedding and extracting is very easy to operate. The superiority of the presented algorithm is verified through experiments.

Keywords: HEVC · Reversible data hiding · Multivariate array · Intra-frame distortion drift

1 Introduction

Data hiding [1, 2] is an effective technique to embed secondary data into digital cover media for many applications such as law enforcement, medical systems and perceptual transparency, copyright protection, user identification, video indexing, and access control, etc. There exists a drawback in many data hiding methods that the cover media is permanently distorted since irreversible operations such as quantization and truncation, etc., which makes the application of data hiding prohibited in the fields of law enforcement, medical systems, perceptual transparency, etc. Then it is desired to reverse the marked media back to the original cover media without any distortions after the hidden data is extracted.

HEVC is the latest and most advanced standard for video compression with high compression efficiency [3, 4]. It is well adapted for network transmission. However, most research for reversible data hiding has focused on H.264 video [5, 6], only a handful of studies have begun to turn to HEVC [7, 8]. Therefore, research on HEVC reversible data hiding methods is very valuable.

Intra-frame distortion drift is a huge problem of data hiding in HEVC because this technique increases the correlation of the neighboring blocks. When one frame is changed by data hiding, the reconstructed pixels of related frames will be influenced, i.e., intra-frame distortion drift happens. The DCT (discrete cosine transform) coefficient of a block in an intra-frame technique is one of the most popular transform domain techniques and adopted in HEVC. However, the reversible data hiding scheme that embed the information into the DCT coefficients for hiding data is not correct for

© Springer International Publishing AG 2017
D.-S. Huang et al. (Eds.): ICIC 2017, Part III, LNAI 10363, pp. 613–624, 2017.
DOI: 10.1007/978-3-319-63315-2_53

HEVC video streams because of the intra-frame distortion drift. Thus, it is necessary to introduce a without distortion drift mechanism to HEVC when data is hidden into the DCT coefficients.

From the beginning of the H.264/AVC period, some researchers have proposed methods for without intra-frame distortion drift data hiding algorithm [9–13].The method proposed in [9] employed the paired-coefficients of a 4×4 DCT block for embedding data to compensate the intra-frame distortion, which is readable and covert. The method proposed in [10], which improve the algorithm [9], employed the paired-coefficients and the directions of intra-frame prediction to avert the distortion drift. However, these algorithms developed based on H.264/AVC cannot be applied to HEVC [14–17]. Especially for the 4×4 luminance blocks in HEVC, where the DST (discrete sine transform) is used instead of DCT. To solve the DST coefficient distortion drift problem in HEVC, [7, 8] proposed a group of triple-coefficients which had a similar effect to the paired-coefficients used in [10]. However, the implementations of these methods need for complex judgment on the conditions of adjacent blocks.

In this paper, we find out a multivariate array applicable to 4×4 luminance DST blocks for embedding data to compensate the intra-frame distortion, and the embedding process does not need to judge the condition of adjacent blocks. At the decoder, we can conduct the inverse operation of multivariate array to the embedded 4×4 luminance DST blocks to achieve reversible data hiding.

The rest of the paper is organized as follows. Section 2 describes the theoretical framework of the proposed algorithm. Section 3 describes the proposed algorithm. Experimental results are presented in Sect. 4 and conclusions are in Sect. 5.

2   Theoretical Framework

2.1   Intra-frame Prediction

A prediction block of HEVC intra prediction method is formed based on previously encoded adjacent blocks. The sixteen pixels in the 4×4 block are predicted by using the boundary pixels of the upper and left blocks which are previously obtained, which use a prediction formula corresponding to the selected optimal prediction mode, as shown in Fig. 1.

Each 4×4 block has 33 angular prediction modes (mode 2–34), as shown in Fig. 2.

2.2   Intra-frame Distortion Drift

The intra-frame distortion drift emerges because we embed bits into I frames. As illustrated in Fig. 3, we assume that current prediction block is $B_{i,j}$, then each sample of $B_{i,j}$ is the sum of the predicted value and the residual value. Since the predicted value is calculated by using the samples which are gray in Fig. 3. The embedding induced errors in blocks $B_{i-1,j-1}$, $B_{i,j-1}$, $B_{i-1,j}$, and $B_{i-1,j+1}$ would propagate to $B_{i,j}$ because of using intra-frame prediction. This visual distortion that accumulates from the upper left to the lower right is defined as intra-frame distortion drift.

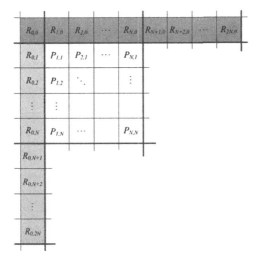

Fig. 1. Labeling of prediction samples

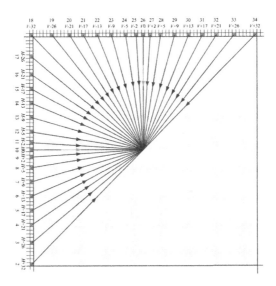

Fig. 2. 4×4 luminance block prediction modes

3   Description of Algorithm Process

3.1   Embedding

According to the intra angular prediction modes of these five adjacent blocks, it can be judged that if the current block is embedded, whether the embedding error will be transmitted to the adjacent blocks by the intra-frame prediction process.

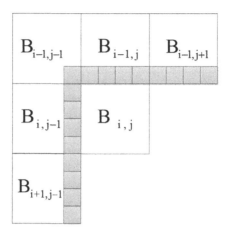

Fig. 3. The prediction block $B_{i,j}$ and the adjacent encoded blocks

In other words, when the intra prediction mode of the five adjacent blocks satisfies certain conditions, if the embedding error just changed the other pixels of the current block instead of the edge pixels used for intra-frame angular prediction reference, then the distortion drift can be avoided. In [7–13], the authors proposed 3–5 conditions to prevent the distortion drift specifically, which are complicated and prone to errors.

In this paper, we proposed a multivariate array to solve the distortion drift problem which does not need to judge the conditions of adjacent blocks. Moreover, with the inverse operation of multivariate array in decoder, the embedded video is perfectly reconstructed as the original encoded video.

The integer discrete sine transform (IST) which is developed from the DST is used in HEVC standard. The transform based on 4×4 blocks is shown in (1). $Y_{4 \times 4}$, which is the 'core' part of the transform, is the matrix of DST coefficients corresponding to the residual block $X_{4 \times 4}$.

$$Y = C_f X C_f^T / (128 * 128) \tag{1}$$

Where

$$C_f = \begin{pmatrix} 29 & 55 & 74 & 84 \\ 74 & 74 & 0 & -74 \\ 84 & -29 & -74 & 55 \\ 55 & -84 & 74 & -29 \end{pmatrix}$$

At the decoder, after the inverse IST step we can get the residual block R', as shown in (2).

$$R' = C_f^T Y C_f / (128 * 128) \tag{2}$$

In data hiding algorithm, the encoded information is embedded into the luminance DST coefficients as in (3):

$$Y' = Y + \Delta \tag{3}$$

Where Δ is the error matrix added to the DST coefficient matrix $Y_{4\times4}$ by data hiding, $\Delta = (a_{ij})_{4\times4}$.

Then we can get the residual block R'' after embedding Δ, as shown in (4).

$$R'' = (C_f^T Y C_f + C_f^T \Delta C_f)/(128 * 128) \tag{4}$$

The deviation of the pixel luminance value between the original block and the one after embedding is $E_{4\times4}$, where $E_{4\times4} = (e_{ij})_{4\times4}$, which can be calculated according to (5).

$$E = R'' - R' = C_f^T \Delta C_f/(128 * 128) \tag{5}$$

Using B ($B = (b_{ij})_{4\times4}$) to express $C_f^T \Delta C_f$ if the Δ could make both $b_{i3} = 0(i = 0, 1, 2, 3)$ and $b_{3j} = 0(j = 0, 1, 2, 3)$, the distortion drift can be completely prevented which does not need to judge the prediction modes of adjacent blocks. We proposed a multivariate array used as Δ can meet this requirement when embedded in.

The multivariate array can be defined as a nine-element combination $(C_1, C_2, C_3, \ldots, C_9)$, C_9 is used for bit embedding, and C_1 to C_8 are used for distortion compensation. In the actual embedding process, $C_9 = 1$ or 0. We can define them as follow, where $C_9 \in Z$, $C_1 = C_3 = C_5 = C_7 = C_9$, $-C_1 = C_2 = C_4 = C_6 = C_8$:

$$\begin{pmatrix} C1 & 0 & C2 & C3 \\ 0 & 0 & 0 & 0 \\ C4 & 0 & C5 & C6 \\ C7 & 0 & C8 & C9 \end{pmatrix}$$

Proposition 1. If

$$\Delta = \begin{pmatrix} a_{00} & 0 & -a_{00} & a_{00} \\ 0 & 0 & 0 & 0 \\ -a_{00} & 0 & a_{00} & -a_{00} \\ a_{00} & 0 & -a_{00} & a_{00} \end{pmatrix}, \tag{6}$$

Then $b_{i3} = 0(i = 0, 1, 2, 3)$
Where

$$a_{ij} \in Z(i = 0, 1, 2, 3; j = 0, 1).$$

Proof: Since $B = C_f^T \Delta C_f$
If $b_{i3} = 0(i = 0, 1, 2, 3)$ then

$$(84a_{00} - 74a_{01} + 55a_{02} - 29a_{03}) * \vec{h}_0 +$$
$$(84a_{10} - 74a_{11} + 55a_{12} - 29a_{13}) * \vec{h}_1 +$$
$$(84a_{20} - 74a_{21} + 55a_{22} - 29a_{23}) * \vec{h}_2 +$$
$$(84a_{30} - 74a_{31} + 55a_{32} - 29a_{33}) * \vec{h}_3 = 0$$

(7)

Where

$$
\vec{h}_0 = \begin{bmatrix} 29 \\ 55 \\ 74 \\ 84 \end{bmatrix}, \vec{h}_1 = \begin{bmatrix} 74 \\ 74 \\ 0 \\ -74 \end{bmatrix}, \vec{h}_2 = \begin{bmatrix} 84 \\ -29 \\ -74 \\ 55 \end{bmatrix}, \vec{h}_3 = \begin{bmatrix} 55 \\ -84 \\ 74 \\ -29 \end{bmatrix}
$$

Since $\vec{h}_0, \vec{h}_1, \vec{h}_2$ and $\vec{h}_3$ are linearly independence, then we can get that

$$
\begin{cases}
(84a_{00} - 74a_{01} + 55a_{02} - 29a_{03}) = 0 \\
(84a_{10} - 74a_{11} + 55a_{12} - 29a_{13}) = 0 \\
(84a_{20} - 74a_{21} + 55a_{22} - 29a_{23}) = 0 \\
(84a_{30} - 74a_{31} + 55a_{32} - 29a_{33}) = 0
\end{cases}
$$

(8)

Plug the multivariate array in (6) can make (7) and (8) equal to 0. So that

$$b_{i3} = 0 (i = 0, 1, 2, 3)$$

Proposition 2. If

$$
\Delta = \begin{pmatrix}
a_{00} & 0 & -a_{00} & a_{00} \\
0 & 0 & 0 & 0 \\
-a_{00} & 0 & a_{00} & -a_{00} \\
a_{00} & 0 & -a_{00} & a_{00}
\end{pmatrix},
$$

(9)

Then $b_{3j} = 0 (j = 0, 1, 2, 3)$
Where

$$a_{i,j} \in Z (j = 0, 1, 2, 3; i = 0, 1).$$

Proof: Since B = $C_f^T \Delta C_f$
If $b_{3j} = 0 (j = 0, 1, 2, 3)$, then

$$
\begin{aligned}
&(84a_{00} - 74a_{10} + 55a_{20} - 29a_{30}) * \vec{h}_0 + \\
&(84a_{01} - 74a_{11} + 55a_{21} - 29a_{31}) * \vec{h}_1 + \\
&(84a_{02} - 74a_{12} + 55a_{22} - 29a_{32}) * \vec{h}_2 + \\
&(84a_{03} - 74a_{13} + 55a_{23} - 29a_{33}) * \vec{h}_3 = 0
\end{aligned}
\tag{10}
$$

Where

$$
\vec{h}_0 = \begin{bmatrix} 29 \\ 55 \\ 74 \\ 84 \end{bmatrix}, \;
\vec{h}_1 = \begin{bmatrix} 74 \\ 74 \\ 0 \\ -74 \end{bmatrix}, \;
\vec{h}_2 = \begin{bmatrix} 84 \\ -29 \\ -74 \\ 55 \end{bmatrix}, \;
\vec{h}_3 = \begin{bmatrix} 55 \\ -84 \\ 74 \\ -29 \end{bmatrix}
$$

Since $\vec{h}_0, \vec{h}_1, \vec{h}_2$ and $\vec{h}_3$ are linearly independence, then

$$
\begin{cases}
(84a_{00} - 74a_{10} + 55a_{20} - 29a_{30}) = 0 \\
(84a_{01} - 74a_{11} + 55a_{21} - 29a_{31}) = 0 \\
(84a_{02} - 74a_{12} + 55a_{22} - 29a_{32}) = 0 \\
(84a_{03} - 74a_{13} + 55a_{23} - 29a_{33}) = 0
\end{cases}
\tag{11}
$$

Plug the multivariate array in (9) can make (10) and (11) equal to 0.
So that

$$
b_{3j} = 0 (j = 0, 1, 2, 3)
$$

Since the quantization process of DST transform made the number of zero values is greatly increased and the nonzero values are concentrated in the upper left corner of the quantized DST coefficients matrix, in most cases the value of the elements in the lower right corner of the matrix is equal to 0, for example, the a_9.

Assume $\begin{pmatrix} a1 & 0 & a2 & a3 \\ 0 & 0 & 0 & 0 \\ a4 & 0 & a5 & a6 \\ a7 & 0 & a8 & a9 \end{pmatrix}$ is the selected quantized DST coefficients block to be

embedded, where a_9 is used to hide information, a_1 to a_8 are used to compensate the intra-frame distortion.

(1) If $a_9 \neq 0$, then no embedding operations to this block, a_1 to a_9 are modified as follows:
If $a_9 > 0$, then $a_1 = a_1 + 1$, $a_2 = a_2 - 1$, $a_3 = a_3 + 1$, $a_4 = a_4 - 1$, $a_5 = a_5 + 1$, $a_6 = a_6 - 1$, $a_7 = a_7 + 1$, $a_8 = a_8 - 1$, $a_9 = a_9 + 1$. If $a_9 < 0$, then $a_1 = a_1 - 1$, $a_2 = a_2 + 1$, $a_3 = a_3 - 1$, $a_4 = a_4 + 1$, $a_5 = a_5 - 1$, $a_6 = a_6 + 1$, $a_7 = a_7 - 1$, $a_8 = a_8 + 1$, $a_9 = a_9 - 1$.
(2) If $a_9 = 0$ and the embedded bit is 1, a_1 to a_9 are modified as follows:
$a_1 = a_1 + 1$, $a_2 = a_2 - 1$, $a_3 = a_3 + 1$, $a_4 = a_4 - 1$, $a_5 = a_5 + 1$, $a_6 = a_6 - 1$, $a_7 = a_7 + 1$, $a_8 = a_8 - 1$, $a_9 = a_9 + 1$.

(3) If $a_9 = 0$ and the embedded bit is 0, a_1 to a_9 are modified as follows:

$a_1 = a_1, a_2 = a_2, a_3 = a_3, a_4 = a_4, a_5 = a_5, a_6 = a_6, a_7 = a_7, a_8 = a_8, a_9 = a_9.$

3.2 Data Extraction and Restoration

After entropy decoding of the HEVC, we choose the embeddable blocks of one frame and decode the embedded data. Then, we extract the hidden data M as follows:

$$M = \begin{cases} 1 & if\ \tilde{Y}_{33} = 1 \\ 0 & if\ \tilde{Y}_{33} = 0 \end{cases}$$

Assume $\begin{pmatrix} a1 & 0 & a2 & a3 \\ 0 & 0 & 0 & 0 \\ a4 & 0 & a5 & a6 \\ a7 & 0 & a8 & a9 \end{pmatrix}$ is the selected quantized DST coefficients block to be restored.

Conduct the inverse operation of the multivariate array to this block, as shown in (12) and (13).

If $a_9 > 0$, then

$$\begin{pmatrix} a1 & 0 & a2 & a3 \\ 0 & 0 & 0 & 0 \\ a4 & 0 & a5 & a6 \\ a7 & 0 & a8 & a9 \end{pmatrix} = \begin{pmatrix} a1 & 0 & a2 & a3 \\ 0 & 0 & 0 & 0 \\ a4 & 0 & a5 & a6 \\ a7 & 0 & a8 & a9 \end{pmatrix} + \begin{pmatrix} -1 & 0 & 1 & -1 \\ 0 & 0 & 0 & 0 \\ 1 & 0 & -1 & 1 \\ -1 & 0 & 1 & -1 \end{pmatrix} \quad (12)$$

If $a_9 < 0$, then

$$\begin{pmatrix} a1 & 0 & a2 & a3 \\ 0 & 0 & 0 & 0 \\ a4 & 0 & a5 & a6 \\ a7 & 0 & a8 & a9 \end{pmatrix} = \begin{pmatrix} a1 & 0 & a2 & a3 \\ 0 & 0 & 0 & 0 \\ a4 & 0 & a5 & a6 \\ a7 & 0 & a8 & a9 \end{pmatrix} - \begin{pmatrix} -1 & 0 & 1 & -1 \\ 0 & 0 & 0 & 0 \\ 1 & 0 & -1 & 1 \\ -1 & 0 & 1 & -1 \end{pmatrix} \quad (13)$$

Because the inverse matrix $\begin{pmatrix} -1 & 0 & 1 & -1 \\ 0 & 0 & 0 & 0 \\ 1 & 0 & -1 & 1 \\ -1 & 0 & 1 & -1 \end{pmatrix}$ also consistent with the form of multivariate array, so even this restore operation is not suit for every quantized DST coefficients block, the inverse operation would not produce intra-frame distortion drift in the decoding process. In fact, the simulation experiment shows that all the test videos are perfectly restored.

4 Case Study

The proposed method has been implemented in the HEVC reference software version HM16.0. In this paper we take "Coastguard" (176 * 144), "Container" (176 * 144) and "Claire" (176 * 144) as test video. The number of encoded I-frame is 20 and the values of QP are set to be 16, 24, 32 and 40. In all experiments, the extracted videos are completely restored to the original video, and the embedded information is extracted accurately. As shown from Figs. 4, 5, 6, 7, 8 and 9.

Fig. 4. The original decode frame of Coastguard (QP = 16)

Fig. 5. The restored frame of Coastguard (QP = 16)

Fig. 6. The original decode frame of Container (QP = 24)

Fig. 7. The restored frame of Container (QP = 24)

Fig. 8. The original decode frame of Claire (QP = 32)

Fig. 9. The restored frame of Claire (QP = 32)

5 Conclusion

This paper proposed to utilize a multivariate array to prevent the intra-frame distortion drift. With the inverse operation of multivariate array in decoder, the embedded video is perfectly reconstructed as the original encoded video. Experimental results demonstrate the feasibility and superiority of the proposed method.

Acknowledgment. This paper is sponsored by the National Natural Science Foundation of China (NSFC, Grant 61572447).

References

1. Shaikh, S., Sayyad, S.: Data hiding in encrypted HEVC/AVC video streams. Int. J. Adv. Res. Comput. Commun. Eng. **5**(8), 60–65 (2016)
2. Wang, J.-J., Wang, R.-D., Xu, D.-W., Li, W.: An information hiding algorithm for HEVC based on angle differences of intra prediction mode angle differences of intra prediction mode. J. Softw. **10**(2), 213–221 (2015)
3. Kim, I., Min, J., Lee, T., Han, W., Park, J.: Block partitioning structure in the HEVC standard. IEEE Trans. Circuits Syst. Video Technol. **22**(12), 1697–1706 (2012)
4. Tew, Y., Wong, K.S., et al.: HEVC video authentication using data embedding technique. In: ICIP 2015, pp. 1265–1269 (2015)
5. Zhao, J., Li, Z.T., Feng, B.: A novel two-dimensional histogram modification for reversible data embedding into stereo H.264 video. Multimedia Tools Appl. (2015)
6. Zhao, J., Li, Z.T.: Three-dimensional histogram shifting for reversible data hiding. Multimedia Syst. (2016)
7. Chang, P.-C., Chung, K.-L., Chen, J.-J., Lin, C.-H.: An error propagation free data hiding algorithm in HEVC intra-coded frames. In: Signal & Information Processing Association Summit & Conference, pp. 1–9 (2013)
8. Chang, P.-C., Chung, K.-L., Chen, J.-J., Lin, C.-H., et al.: A DCT/DST-based error propagation-free data hiding algorithm for HEVC intra-coded frames. J. Vis. Commun. Image Represent. **25**(2), 239–253 (2013)
9. Ma, X.J., Li, Z.T., Lv, J., Wang, W.D.: Data hiding in H.264/AVC streams with limited intra-frame distortion drift. In: Computer Network and Multimedia Technology, CNMT (2009)
10. Ma, X.J., Li, Z.T., Tu, H., Zhang, B.: A data hiding algorithm for H.264/AVC video streams without intra frame distortion drift. IEEE Trans. Circuits Syst. Video Technol. **20**(10), 1320–1330 (2010)
11. Liu, Y.X., Li, Z.T., Ma, X.J.: Reversible data hiding scheme based on H.264/AVC without distortion drift. J. Syst. Softw. **7**(5), 1059–1065 (2012)
12. Liu, Y.X., Li, Z.T., Ma, X.J., Liu, J.: A robust without intra-frame distortion drift data hiding algorithm based on H.264/AVC. Multimedia Tools Appl. (2013). (Springer)
13. Liu, Y.X., Li, Z.T., Ma, X.J., Liu, J.: A robust data hiding algorithm for H.264/AVC video streams. J. Syst. Softw. (2013). (Elsevier)
14. Liu, Y.X., Hu, M.S., Ma, X.J., Zhao, H.G.: A new robust data hiding method for H.264/AVC without intra-frame distortion drift. Neurocomputing **151**, 1076–1085 (2014)

15. Liu, Y.X., Ju, L.M., Hu, M.S., Ma, X.J., Zhao, H.G.: A robust reversible data hiding scheme for H.264 without distortion drift. Neurocomputing **151**, 1053–1062 (2014)
16. Liu, Y.X., Jia, S.M., Hu, M.S., et al.: A robust reversible data hiding scheme for H.264 based on secret sharing. In: ICIC 2014, pp. 553–559 (2014)
17. Liu, Y.X., Jia, S.M., Hu, M.S., Jia, Z.J., Chen, L., Zhao, H.G.: A reversible data hiding method for H.264 with Shamir's (t, n)-threshold secret sharing. Neurocomputing (2014)

A Protection Method of Wavelength Security Based on DWDM Optical Networks

Yuan Chen[1(✉)], Shuang Liang[2], and Zhen He[1]

[1] Department of Information Management, Henan Arts and Crafts School,
Zhengzhou, China
beatink@yeah.net
[2] Department of Software Application,
Zhengzhou Information Technology School, Zhengzhou, China

Abstract. A little link and wavelength failure of the optical network would result in a great traffic loss. Therefore a well-performed protection strategy is of great importance. We present a wavelength protection algorithm for the optical network. Compared with the previous works, this method achieves better protection effect for the wavelength than the algorithms proposed ever before, and this algorithm provides a higher wavelength utilization ratio and lower wavelength blocking probability.

Keywords: Optical network · Wavelength protection · Utilization ratio · Real-time protection

1 Introduction

Along with the developing of communications technologies, the request for wider networks bandwidth is rapidly presented. Dense Wavelength-Division Multiplexing (DWDM) is emerging as the dominant technology for next-generation optical networks [1], for it can greatly expand the inherent capacity of optical fibers [2]. Currently considerable interest is given to future developments of passive optical networks (PON) following the demands for increasing traffic, higher bandwidth and extended reach. Future developments expect longer lifetimes of implemented technologies and low upgrade costs for progressing technologies. Legacy Gigabit-capable passive optical networks (GPON) are known as most currently deployed high-speed and high-capacity optical access technologies. The first progression growth aspirant is a 10-Gigabit-capable passive optical network (XG-PON) based on minimal equipment investments. In GPON and XG-PON all signals are distributed through the same optical distribution network (ODN) from the optical line terminal (OLT) to every end user's optical network termination unit (ONU) connected on the same PON branch. The most promising candidate to progress is a hybrid PON based on the simultaneous usage of time (TDM) and wavelength (WDM) division multiplexed PON which has been still under standardization process. If we combine WDM and TDM PON technologies together, we may obtain TWDM-PON access network that offer an affordable progress. It has a good compatibility with XG-PON and moderate compatibility with

D.-S. Huang et al. (Eds.): ICIC 2017, Part III, LNAI 10363, pp. 625–633, 2017.
DOI: 10.1007/978-3-319-63315-2_54

GPON and can provide economically reasonable migration from current TDM-PONs to future TWDM.

The specific wavelength bands for present and future PON technologies are or should be allocated by International Telecommunication Union (ITU). However, it is necessary to protect present and future PON signals in the ONUs from interference. To guarantee this, some protective measures should be taken to avoid interferences between GPON, XG-PON and TWDM-PON signals due to common broadband ONU receivers. One of these measures is a precise scheme of the wavelength allocation together with the so-called guard bands. Due to the increasing bandwidth demands, the guard bands should be narrowed under increasing traffic. Therefore, the implementation of specific wavelength blocking filters is generally accepted and recommended by ITU standards. Extensive research is necessary to design appropriate and low-cost candidates for blocking filters. Thin-film interference filters (TFF) are suitable and low-cost, coexisting (i.e. ONU-independent), temperature-resistant and simple operation candidates.

In the DWDM system, each fiber has several (for example N) available wavelength channels. When a signal flow comes, it is firstly modulated into a wavelength channel by an appropriate tuned laser setup at the transmitter. And then, all modulated signals are combined by a multiplexer and transported through the fiber. When the combined signal arrives at the other end of the fiber, a de-multiplexer will demodulate it into N signals in different wavelengths and crossly connect them to the corresponding receivers. In the above-mentioned process, any malfunction of laser, multiplexer, or de-multiplexer would lead to the failure of one or several wavelength channels. Besides, in dense wavelength division multiplexing (DWDM) optical networks, a wavelength channel has a transmission rate of several gigabits or more per second [3]. A wavelength failure in fiber link is shown in Fig. 1. If a wavelength fails, a lot of business may be blocked. Therefore, wavelength protection plays an important role of DWDM optical networks.

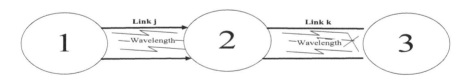

Fig. 1. Wavelength failure

There are several techniques that have been proposed in literature to realize survivable optical network. Various authors have investigated the link failure problem and presented the algorithms that is called complete path-shared protection (CPSP) and differentiated path-shared protection (DPSP). The basic idea of CPSP is to assign one working path and two link-disjoint backup paths to each connection request. In the worst case, if the working path traverses a failed link and the first backup path traverses another failed link, the second backup path also can be available to transmit the traffic. Furthermore the DPSP, which utilizes the idea of differentiated reliable protection, can dynamically establish the connection according to the requirements of users, and thus, it can save significant resources and reduce the blocking probability.

However, in actual networks the fiber links may undergo correlated failures because they may share some common physical resources, which results in a high blocking probability of CPSP and DPSP. Moreover, they do not utilize the wavelength resources in the failed link, so they have a low wavelength resource utilization ratio. And thus, in this paper we proposes an algorithm named wavelength utilization ratio based wavelength protection algorithm (WUR-WP) in single fiber link, which has a better wavelength resource utilization ratio and a lower wavelength blocking probability than the previous works. We analyze the various failure scenarios, the appropriate survivability schemes and their related implementation issues in addition to a discussion on the recent work in this area.

The conventional protection schemes include link-dedicated protection, shared protection, link-shared protection, path-shared protection, and segment-shared protection. The conventional protection schemes are shown in Fig. 2. From the figure we can see that link protection is the most mature strategy with the least restoration time [4], so our proposed algorithm choose the simplest and most easily achieved protection mechanism based on single fiber link. Another reason why we choose the link protection scheme is that it can improve the accuracy and flexibility of the protection and increase the transmission reliability of key businesses. Moreover this algorithm adopts the sharing resources scheme; obviously this scheme may increase the utilization ratio of link resources.

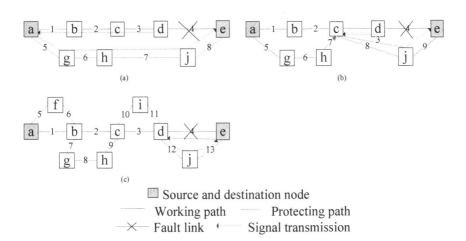

Source and destination node

—— Working path   ⋯⋯ Protecting path

—✕— Fault link   ⟨⋯⋯ Signal transmission

Fig. 2. Different protection schemes: (a) path protection; (b) sub-path protection; (c) link protection.

As the number of wavelength channels keeps increasing, the risk of wavelength fault becomes higher and higher. And the protection design for wavelength failures has been considered in DWDM optical networks [5]. However, the research of wavelength protection in the single-link is quite insufficient, so in this paper we proposes a algorithm named wavelength utilization ratio based wavelength protection algorithm (WUR-WP) in single fiber link, which has a better wavelength resource utilization ratio and a lower wavelength blocking probability than the previous works. The basic idea of this algorithm is stated as follows. When a wavelength failure emerges, we find out a protection wavelength from these free wavelengths firstly. If there are more than one such protection wavelengths, anyone of them can be used for replacing the failing one. If there is not any a free wavelength, search the wavelength with the minimum wavelength utilization ratio. After such refinement, if the system has a lower wavelength utilization rate than the failure one, then we should do a substitution; otherwise, the protection algorithm is failed [6].

The rest of the paper is organized as follows. In Sect. 2, we briefly expound the problem to be solved. In Sect. 3, we propose the construction of the protection algorithm and the simulation results together with the analysis of the algorithm is presented in Sect. 4. Finally, we conclude in Sect. 5.

2 Problem Statement

A network for a given survivable meshed DWDM optical network can be expressed as $G(V, E, W)$ [7], where V is the set of optical nodes, E is a set of bidirectional links in which each link is made up by a pair unidirectional fibers with opposite direction, and W is the set of available wavelengths per fiber link. The expressions $|N|$, $|E|$, and $|W|$ denote the numbers of all the nodes, links, and wavelengths in the network, respectively. We assume that connection requests arrive at the network dynamically, but just one request at a time. And each wavelength channel can bear only one connection requirement at a time either. We also suppose that each node has the capacity of optical-electronic-optical (OEO) wavelength conversion and wavelength-routing switching [8]. The Dijkstra's algorithm, a classical minimal-cost path algorithm, is applied for the computation of the routes.

Next, we introduce some notations and assumptions used in our algorithm [9]. In this paper a bidirectional fiber link between a node pair in G is denoted as l, $l \in E$, and $cost_l$ expresses its cost. The cost of a link is dynamically changed along with the state of the transmission business of the network, such as working wavelengths, free wavelengths, reserved wavelengths, paths' reliabilities. WW_l, FW_l, PW_l represents the number of working wavelengths, free wavelengths, protecting wavelengths on link l, respectively [10].

In DWDM optical networks, there are three kinds of main constraints related to wavelengths: wavelength continuity constraint (WCC), distinct wavelength assignment constraint (DWAC), and non-wavelength continuity constraint (NWCC) [11]. In WCC, the wavelength of the links between the nodes of the selected route must be the same. For DWAC, the assigned wavelength on the fibers of randomly chosen two light-paths must not be the same. And in NWCC, different wavelengths can be used on the links between the nodes of the selected route, but the nodes should have the capability of wavelength conversion. Wavelength conversion is the ability to convert the data which is presented by a certain wavelength to a different one. Eliminating the wavelength conversion ability of a switching node would significantly reduce the cost of the network [12]. However, it also may reduce network efficiency, because more wavelengths might be used in the transmission. For such a reason, we require that the nodes adjacent to the protected link have the capacity of optical wavelength conversion in this paper, and as a result, the NWCC strategy is adopted [13, 14].

Two wavelength assignment strategies are used in our wavelength assignment algorithm and wavelength protection algorithm. One is the Most-Used (MU) strategy, and another one is the Least-Used (LU) strategy. In the MU strategy, the wavelength that is used by most of the links in the network is firstly tried, and then the wavelength with the second-highest number of connections, this procedure is not ended until the available wavelength is found. This strategy attempts to provide maximum wavelength reuse in the network and leave the maximum wavelengths underutilized for the future connection requirement. On the contrary, the LU strategy selects the wavelength which is the least used on links in the network, and it attempts to spread the load evenly across all wavelengths. The LU strategy is the opposite of the MU strategy in that it attempts to select the least-used wavelength in the network [15].

In this paper, the Most-Used (MU) Strategy is used for searching for working wavelength, and the Least-Used (LU) Strategy is selected to search for protecting wavelength (PW). A protecting wavelength can be a free wavelength or a least-used protecting wavelength, and the most-used working (MWW) wavelength with high wavelength utilization ratio is protected by the free wavelength or least-used protecting wavelength with the smallest wavelength utilization ratio when the MWW fails.

3   Proposed Algorithm

Most of the previous works investigate the failure of single-link, which is dominant case in the DWDM optical networks. Hence this paper adopts the same protection setting. And then we will illustrate the proposed algorithm named Wavelength Utilization Ratio based Wavelength Protection algorithm (WUR-WP). In the proposed algorithm, the wavelength with high wavelength utilization ratio is preferentially protected. The wavelength utilization ratio of the i-th wavelength of a link can be expressed as

$$U_i = \frac{m}{n},$$

where m is the number of times that wavelength i is occupied in unit time period, and n is the number of times that the total business are transported in the fault fiber link in unit time period.

According to the case of transmission businesses, the endpoint of a link calculates the wavelength utilization ratio of each wavelength, and stores the results in an array $w[\ ]$. When a wavelength, for example wavelength p, fails, the algorithm searches for a new free wavelength. If there are some free wavelengths, it chooses the one with the smallest wavelength utilization ratio to replace the failed one. Else, the algorithm selects a busy wavelength with the minimum utilization ratio, say wavelength q, in the array $w[\ ]$: if $w[p] > w[q]$, replace the failed one by wavelength q, else, terminate the algorithm and it means that the protection is failed. The procedures of the WUR-WP algorithm are presented as follows.

Step 1. Fault detection: the fault is defined as the wavelength failure in the link k, and we fix the fault on wavelength i.

Step 2. Calculate all the wavelengths' utilization ratio in the failing optical link.

Step 3. Store all the results in an array $w[\ ]$ (the same calculation results, including free wavelengths' are reserved according to the calculation orders).

Step 4. Protect the fault working wavelength (FWW):

　　(i)　Search for available free wavelength firstly, if there is any, then use it to replace the FWW. Else, go to (ii).

　　(ii)　Extract the fault wavelength's utilization ratio $w[p]$ in the array $w[\ ]$, then seek the protecting wavelength with the smallest wavelength utilization ratio in $w[\ ]$ of descending order, assumed to be $w[q]$.

　　(iii)　Judge whether $w[p] > w[q]$? If the inequality holds, use the protecting wavelength (PW) with the smallest wavelength utilization ratio $w[q]$ to replace the FWW. Else, the protection algorithm fails.

This algorithm introduces the conception of wavelength utilization ratio, which is a Real-time protection strategy. It adopts the Seizing Protection Mechanism (that is, it could occupy the wavelength with small wavelength utilization ratio when the network is competing for resources) and the resources are shared in the whole network, which ensures the maximum protection of the working wavelength with high wavelength utilization ratio. Besides, it also improves the transmission reliability of the key wavelengths, which is suitable for the transmission of the significant businesses. The flow chart of our proposed algorithm is presented in Fig. 3.

4   Case Study

We assume that the connection requests arrival according to an independent Poisson process, and the holding time of each connection is a negative exponentially distributed with rate μ. Then, the network load is Erlang business with large capacity. We assume that $|W| = 16$ in each bidirectional fiber and each required bandwidth is a wavelength granularity. Additionally, all nodes are assumed to have wavelength conversion capacities (OEO). The test network is shown as Fig. 2(c), where each node pair is interconnected by a bidirectional fiber link. The Dijkstra's shortest path algorithm is used to find the path from source to destination. In addition, we require the nodes adjacent to the protected link have the capacity of optical wavelength conversion, so we adopts the NWCC strategy. Afterwards, the Most-Used (MU) Strategy is used to search for working wavelength (WW), and the Least-Used (LU) Strategy is used to search for protecting wavelength.

We compare WUR-WP with the previous Link Protection (LP) on the performances of Wavelength Utilization Ratio (WUR) and Wavelength Blocking Probability (WBP). The simulation results are shown in Fig. 4, where the horizontal level presents the network traffic density in Erlang and the vertical level presents the different standards (WUR and WBP) separately.

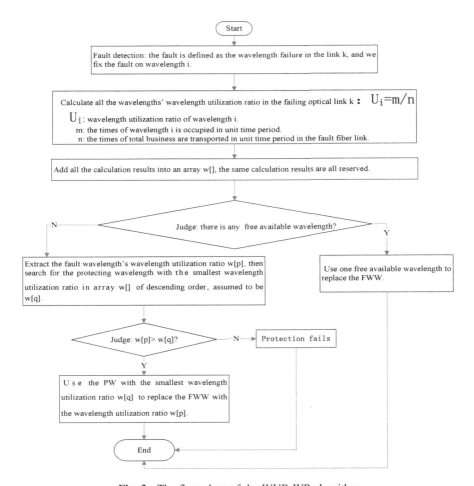

Fig. 3. The flow chart of the WUR-WP algorithm

From the first figure, we can see that WUR-WP has higher wavelength utilization ratio than LP. There are two reasons for this: (1) The WUR-WP algorithm protects against the wavelength fault, which can guarantee precision of protection. However, the LP algorithm protects against the link fault, when the working wavelength has failed, it searches for another normal link to transport business alternatively, but the WUR-WP algorithm searches for another available wavelength in the fault link, and more wavelength resources will be saved; (2) The WUR-WP algorithm adopts sharing resources protection mechanism, and thus more free wavelength resources and sharing protecting wavelengths can be used by the subsequent requests. So the wavelength utilization ratio will be higher than LP.

From the second figure, we find that the WUR-WP has lower wavelength blocking probability than LP. The reason for this is that, the WUR-WP adopts the real time protection strategy, which can improve the efficiency of recovery and reduce the wavelength blocking probability.

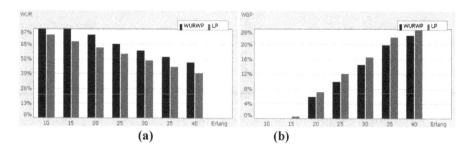

Fig. 4. Performance of WUR-WP and LP with different standards: (a) wavelength utilization ratio (WUR); (b) wavelength blocking probability (WBP).

We can also notice that the cost of the WUR-WP is higher than LP's. The reason is that all nodes are required to have wavelength conversion capacities in the WUR-WP, which the LP does not require, so LP can reduce the network costs significantly. In addition, the WUR-WP adopts the priority protection model, so that it can provides a maximum wavelength protection of high utilization ratio.

In summary, we can thus conclude that the proposed WUR-WP can obtain better performances in wavelength utilization ratio and blocking probability than the previous LP. However, the network cost of WUR-WP is higher than LP.

5 Conclusion

In this paper, we present a new algorithm to protect against wavelength failures in DWDM optical networks. This work adopts the priority protection model and the sharing resources protection mechanism (that is, the network can occupy the wavelength with low utilization ratio when the network is competing for resources), so it ensures a maximum wavelength protection of high utilization ratio, and also it improves transmission reliability about the important business and significant wavelength. Besides, it is applicable to the important business with large-capacity transmission. This algorithm provides a lower level of network wavelength congestion. At the same time, it is a real-time protection strategy, so it can improve the efficiency of recovery. The simulation results show that, compared to previous LP, WUR-WP can provide a higher wavelength utilization ratio and lower wavelength blocking probability.

References

1. Rani, S., Sharma, A.K., Singh, P.: Efficient restoration strategy for DWDM multifiber optical networks. In: Proceedings of International Conference on Challenges and Opportunities in IT Industry, PCTE, Baddowal, Ludhiana, Punjab, India (2005)
2. Krishnaswamy, R.M., Sivarajan, K.N.: Algorithms for routing and wavelength assignment based on solutions of LP-relaxations. IEEE Commun. Lett. **5**(10), 435–437 (2001)

3. He, R., Wen, H., Li, L., Wang, G.X.: Shared sub-path protection algorithm in traffic-grooming DWDM mesh networks. Photon. Netw. Commun. **8**, 239–249 (2004)

4. Ou, C., Zang, H.: Sub-path protection for scalability and fast recovery in optical WDM mesh networks. In: Proceedings of Optical Fiber Communication Conference, Anaheim, CA, pp. 495–497, March 2002

5. Zang, H., Jue, J.P., Mukherjee, B.: A review of routing and wavelength assignment approaches for wavelength-routed optical DWDM networks. Opt. Netw. Mag. **1**(1), 47–60 (2000)

6. Singh, P., Rani, S., Sharma, A.K.: Efficient wavelength assignment strategy for wavelength-division multiplexing optical networks. Opt. Eng. **46**(8), 085009 (2007)

7. Saad, M., Luo, Z.: On the routing and wavelength assignment in multifiber WDM networks. IEEE J. Sel. Areas Commun. **22**(9), 1708–1717 (2004)

8. Data, R., Mitra, B., Ghose, R., Sengupta, I.: An algorithm for optimal assignment of a wavelength in a tree topology and its application in WDM networks. IEEE J. Sel. Areas Commun. **22**(9), 1589–1600 (2004)

9. Guo, L., Li, L.M., Yu, H.F.: Heuristic of differentiated path-shared protection for dual failures in restorable optical networks. Opt. Eng. **46**(2), 025009 (2007)

10. Ho, P., Mouftah, H.: A novel survivable routing algorithm for shared segment protection in mesh WDM networks with partial wavelength conversion. IEEE J. Sel. Areas Commun. **22**, 1548–1560 (2004)

11. He, W., Somani, A.: Path-based protection for surviving double link failures in mesh-restorable optical networks. In: Proceeding of IEEE GLOBECOM 2003, pp. 2558–2563 (2003)

12. Saradhi, C.V., Murthy, C.S.R.: Routing differentiated reliable connections in WDM optical networks. Opt. Netw. Mag. **3**(3), 50–67 (2002)

13. Yu, H., Wen, H., Wang, S., Li, L.M.: Shared-path protection algorithm with differentiated reliability in meshed WDM networks. In: APOC 2003, Proceeding of SPIE, vol. 5282, pp. 682–687 (2003)

14. Ramamurthy, B., Mukherjee, B.: Wavelength conversion in WDM networking. IEEE J. Sel. Areas Commun. **16**(7), 1061–1073 (1998)

Modulation Technology of Humanized Voice in Computer Music

Zhiqi Zhao[✉]

Communication Institute, Zhengzhou Normal University, Zhengzhou, China
1019398616@qq.com

Abstract. Computer music humming tone modulation technology is mainly to enhance the computer music sound of human nature. Such as enhancing the authenticity of the sound and beauty or enhance the performance of the human and virtual sound field effects, etc. It mainly uses the existing technical means of hardware and software to change the basic attributes of sound, which mainly includes the sound envelope and the virtual sound field effect modulation. Through the modulation of the preset tone, you can improve the characteristics of human voice performance.

Keywords: Computer music · Sound · Modulation · Humanize

1 Introduction

Today, computer music has developed to a very high level, the high development of computer music technology and art has had a great impact on the creation, performance, dissemination and appreciation of traditional music. In music creation, tools and tools have changed, in the performance, the emergence of electronic musical instruments, virtual sound performance and the disappearance of the traditional performance practice is an important feature of computer music. The emergence of non-traditional sounds and new ideas, as well as the sound of material and sound effects of the composer and appreciation of the music have a broader understanding; And the development of media technology and make the spread of music more convenient and extensive.

In the twentieth century, the traditional music concept has undergone tremendous changes. At the same time, music has experienced the development of tape music, electronic music and computer music in science and technology. In this process, the expansion of the sound is the core of its development. In this century, computer music technology is basically mature, from the source of music practice music creation, it completely changed the people's traditional understanding of music. Technology development to the present, computer music technology production works have been everywhere, and many professionals and ordinary audiences have basically accepted it. But in the practice of art, people think that computer music has some lack of human problem. The author believes that the problem of humanistic lack of computer music is multifaceted, but one of the most important aspects is the lack of certain humanization of computer music.

© Springer International Publishing AG 2017
D.-S. Huang et al. (Eds.): ICIC 2017, Part III, LNAI 10363, pp. 634–641, 2017.
DOI: 10.1007/978-3-319-63315-2_55

We know that most of the traditional music creation is such that the composer on the piano to complete the pre-work, and then carry out the equipment or composer directly writes the band score. Because the creation is the first part of the practice of music art, limited to the specific conditions, the composer can only be based on the existing sound auditory experience creation, other artistic practice but also rely on the performer and conductor to perform. Unlike traditional music creation, in most cases, computer music creation and production can be done by a person, mainly because of the computer music production platform.

However, in the current technical conditions, the computer music has some problems of human nature, in which the voice of the human problem is particularly prominent. Such as the lack of authenticity and beauty of the preset tone or playing the law is too single lack of dynamic and random; sound is not like the real performance, and virtual sound field effect is not true and not perfect. In practice, we can use some technical means to try to solve the problem of humanized lack of computer music. Due to space reasons, I only explore the computer virtual tone of human modulation technology.

The sound of traditional acoustic instruments is relatively fixed, but because the player, playing techniques, playing style and venue is different, especially for different tone performance needs, it is from the different dynamic and randomness and the interaction of human and musical instruments to reflect its humane. Practice tells us that due to the special form of computer music and the particularity of playing (Such as pre-characterization of performance), it cannot give us all the necessary sound, especially it cannot provide personalized and random sound. However, we can use the existing computer music technology in the production of the early to get some of the human needs of the preset voice. (That is, the physical properties of the sound are changed.) In the production practice, Edirol HQ Qrchestral, East West EWQLSO Gold Edition, Real Guitar 2.0, Expansion BFD, Vienna Symphonic Library is the more commonly used computer audio and sound. These sources can be modulated by their own module, and the tone can be loaded into the sampler to be modulated by the sampler. Of course, theoretically, the humanized effects of computer music can be modulated before or after playing, but it is based on the specific situation. For example, It has been determined that the modulation of the tone of the playing method should be completed in the early stage, and the preset humane tone for the sound of the human performance has a direct effect. From another technical level, some sounds can be made on a soft sound source (A collection of virtual music for computer music), and some can also be on the MIDI keyboard, both can be directly applied to the preset computer virtual tone. The author through the practice proved that: Through the appropriate playing before the tone of the modulation, you can enhance the authenticity of the traditional music of the computer music, non-traditional sound beauty, sound performance and randomness of the virtual sound field effects and other human characteristics. The following discussion is the computer soft sound source of the two main tone modulation technologies.

2   The Envelope of the Sound

After selecting the sound we want to use, but the preset tone has not yet fully meet our performance requirements, we can use the necessary technical means to make some of the necessary modulation of the tone. The first thing that can be done is the envelope of the sound, where the envelope refers to a graphical representation of the sound from the volume to the disappearance of the volume (Early in the authenticity of the sound on the people pay great attention to the sound spectrum, and later found that the dynamic characteristics of the sound of human nature also plays an important role, so today's sound source basically have sound and randomness of the sound modulation envelope module). The envelope of the sound can generally be divided into five parts: attack, hold, decay, sustain, release. The adjustment of the sound can adjust the softness of the sound head (The volume reaches the maximum value). For example, we chose a violin bow sound, that is not in place, you can adjust the value of Attack, Attack the size of the value of the hardness of the adjustment at the same time, also affect the size of the sound volume. When a sound starts to reach the maximum volume, it will decay until it disappears. Hold is the maximum volume duration. Adjusting its value can shorten the maximum volume hold time (That is, when the sound reaches the maximum volume immediately decay.

Decay is the ratio of the sound from the maximum volume to the normal volume, keeping its numeric sound faster to reach the master volume from the maximum volume. Sustain mainly adjusts the size of the master volume. To adjust it to the value of the objective so that we feel a time to increase the performance of a single tone. Release adjusts the length of the lingering delay of the sound. The adjustment of the parameters of the sound envelope (Mainly including the rise time Attack, fading time Decay, duration Sustain, release time Release) can make the sound characteristics change. Figure 1 shows a complete envelope with a soft sound and no sound. It is characterized by rising time and fading time is relatively long. We can change the envelope parameters to modify the timbre. Figure 2 in the thin line envelope shows the

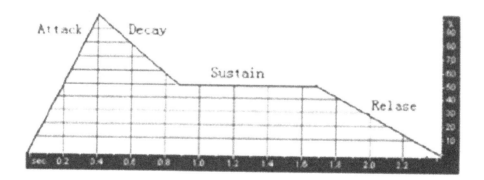

1 The envelope of the sound

Fig. 1. The envelope of the sound 1

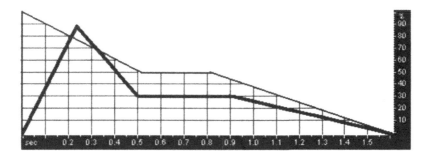

2 The envelope of the sound

Fig. 2. The envelope of the sound 2

sound of the piano state, and now to make this sound quality is hard, the sound of the sound head becomes soft, you can change the envelope parameters to achieve: We increase the rise of this sound envelope, which weakened the impact of the sound and weaken the sound of the sound head; then the duration and decline in time increases, which increases the tone of the tail, the overall tone feeling softened, as shown in Fig. 2 in the thick envelope. Through the different envelopes of the preset timbre, we can get personalized tones with rich playing methods.

The envelope of the sound in our production can be based on different musical performance of the corresponding tone of the envelope processing to enhance the authenticity of the sound. It can also enhance the dynamic and random playing and the interaction between human and musical instruments and other human characteristics. The envelope of the tone can also be made on a MIDI keyboard, such as the use of envelopes on the CME UF70 MIDI keyboard. The CME UF70 MIDI keyboard also envelops the tone through different modulations of the velocity curve and the trailing curve, but the envelope processing on the MIDI keyboard only works for the responsive tone.

3   The Modulation of Virtual Sound Field

Sound field is the sound of the environment, in real life, the sound always exists in a specific natural environment. After the early recording technology was developed, people saved a real sound scene by recording live sound reductions. Live sound though real, but there are many people do not need the noise to disturb people to listen.

With the sound of the aesthetic taste of the improvement and the development of audio technology, people began to use the recording studio recording and later added a variety of sound field, which is what we now call the virtual sound field technology. Modern virtual sound field technology can simulate the real sound field effect. Effects can be divided into MIDI effects and audio effects from use. Formally divided into hardware effects and software effects. In the work mode can be divided into frequency change effects, dynamic effects and time changes effect. Here, their principles and methods of use are not the focus of our discussion (Of course, do not mix the computer music production technology as the main body of our discussion). We only discuss what the effect of the addition can make us to a more real, aesthetic, with the real effect of the human nature of the sound. Through the practice of the author believes that in the early music production, the following two aspects of the modulation of the computer music sound virtual sound field of humanized effect has a direct and important role.

After the development of the history of the traditional band formed a relatively fixed position, such as the symphony orchestra instrumental position, 1. According to the different styles of music, in the process of creating music should be based on different styles of music to choose the appropriate tone. 2. According to the objective conditions of the sound under the different properties. Sound in the transition of time and the content of its objectivity has a very close contact, the sound characteristics are not the same, so the sound is not the same. So, when we choose the sound, we should be familiar with the concept of sound, so it is easy to understand the sound and other equipment and explore the characteristics of the sound. 3. Depending on the type of instrument. Each instrument is of a different type, its sound is very different, like keyboard, brass and other instruments. By grasping each instrument with different sound quality, it is easy to grasp the target sound. 4. According to the function of the classification of knowledge. Music level is very obvious, like the whisper layer, harmony layer and so on. The role of the sound in the music to play the role is not the same, may be solo type, it may be percussion type, the creator according to these different standards for voice classification, knowledge of the function is not the same, the sound is not the same. If the use of appropriate, the uniqueness is very obvious.

The string in front of the music after the treble in the left bass in the right. Since then, the band's position more or less refer to the way the symphony orchestra, like jazz, rock music, light music and the Chinese national band. This way is suitable for the performance of the sound effects of the venue, resulting in the sound effect gradually formed a people's music appreciation habits. Later, with the development of recording technology, people can adjust the voice of the sound through the technology, and sometimes even can be modulated by the traditional way can not do the effect and better sound. The phonological modulation of the computer music with the traditional way is more than it is, it can not take into account the actual reality of the sound field of the acoustic phase, but it must be made to the sound field uniformity (That is, all the sounds sound is roughly in a sound environment). Now we have the sound source of the sound, and some no sound of the preset, such as many sampling sound, and some preset the sound phase, and some of these sound can change some can not change, such as some drum group voice and Orchestral Sound Edirol HQ Orchestral. Because most of the time in practice, we will use different sound sources, So the modulation of the acoustic phase is very important, if the modulation is not appropriate in the hearing will cause unnatural and no human feelings. In fact, in practice, even with the same source, due to music style and other reasons the sound of the sound phase should also be random and dynamic.

We use different tones or different instruments of the sample, the specific position of the acoustic phase will be different, and if the music in the process of using different sources, It will make people feel the player on the stage of the seat changed. Therefore, the sound source or the sound on the sampler for the unity of sound and reasonable arrangements, which is conducive to the sound of the virtual sound field of authenticity and humanity. Computer technology in the technical superiority also allows us to do the traditional way cannot handle but very user-friendly sound effects (This sound and sound effect of the sound is not based on traditional auditory experience and habits, but based on psychological subjective feelings). In real life, if we look forward to the appearance of a sound and a sound is very prominent, in the hearing it will become our center sound or we will turn around with the sound source to listen to the direction. People in the hearing of this habit to become our modulation of the sound and the sound in the vocal phase with the characteristics of human nature. When a sound is not in the center of the sound, but it becomes the main tone, its sound should be moved to the center.

We can put them on the left and right sides when we have two instruments of the same sound, but when one becomes the main tone, we can put it in the center. Some sound types of sound is very sporty, we can carry out the operation of the wandering. As in Nuendo 3, we can take full advantage of the Write automation and Read automation functions in the audio track real-time read and write changes in the sound phase.

4   Reverberation of Modulation

Sound propagation in the natural environment will encounter obstacles, and it will produce reflection diffraction and scattering. So when there is a sound source sound, you will hear the sound to reach the point of listening to the direct sound, less reflection of the early reflection of the sound, and repeated reflection of the sound formation of the reverberation in this environment. And the role of computer music effects is to simulate this effect to achieve the real sound field effects and user-friendly sound effects. In the reverberation of the sound, for different styles of music on the tone of the treatment is not the same. The traditional orchestral music need to add some room effect, of course, this is related to the traditional music playing style and people's long-term appreciation of the habit. For the style of today's popular music, we can focus on adding the plate reverb effect, which is associated with pop music performances and appreciation of the venue, people's appreciation experience and the characteristics of the interaction between performers and listeners are closely related. With the development of science and technology, and the improvement of aesthetic needs and the emergence of aesthetic habits of audiovisual combination, the traditional sound field concept is no longer the only reference for sound field processing, such as creating some fantastic sounds, horror sounds, mysterious wide space tones and so on. Sometimes these sounds may be just a single tone, but because of the reverberation effect, it makes it very strong. At this time, the tone of the expression is far greater than the melody of the expression. When people hear the sound of this effect, they immediately think of the corresponding unreal scene. Today, audio-visual art is very developed, which has become another kind of people's auditory experience. We will use a different source in the actual creation of many times, and the sound of these voices has their own advantages and characteristics. In the creation we will choose the sound we need, but at the same time we will encounter such a problem, that is, when the combination of different tone reverb effect will be different, Some of the sound reverb effect is more prominent, such as East West EWQLSO Gold Edition, and some sounds are basically dry words such as the Vienna Symphonic Library, this situation will cause the same work in the sound of the sound field inconsistency and sound (Reverberation of the different will cause the tone in particular the depth of the different, the greater the reverberation sound farther, the smaller the reverberation sound closer). So we can first sound the necessary reverb and unified modulation, the unity of the reverberation can make different timbre has unified and real virtual sound field.

5   Conclusion

Computer music sound of different humanized modulation technology can be directly or indirectly to strengthen and improve the sound of human nature. Of course, there are different in the practical application of the focus, Tone of the envelope is not only the authenticity of virtual sound and beauty plays an important role in these two aspects, and it is very important to strengthen the human also to play, And virtual sound field effect of the authenticity of human nature not only strengthen the timbre modulation sound field, it also strengthens the authenticity of the composite sound (This article refers to the combination of vertical multi-tone sound). It is because the sound of human nature, melody and computer music works will be more humane.

Acknowledgment. This paper is sponsored by the topic of soft science of Henan Science and Technology Department in 2017 (Subject ID: 172400410136).

References

1. Ng, R., Han, J.: Efficient and effective clustering method for spatial data mining. In: Proceedings of 20th International Conference on Very Large Data Bases (1994)
2. Shekhar, S., Chawla, S.: Spatial Databases: A Tour. Prentice Hall, Upper Saddle River (2003)
3. Graco, W., Semenova, T., Dubossarsky, E.: Toward knowledge-driven data mining. In: International Workshop on Domain Driven Data Mining at 13th ACM SIGKDD (2007)
4. Tung, A.K.H., Han, J., Lakshmanan, L.V.S., Ng, R.T.: Constraint-based clustering in large databases. In: Proceedings of International Conference on Database Theory (2001)
5. Wang, X., Hamilton, H.J.: Towards an ontology-based spatial clustering framework. In: Proceedings of 18th Canadian Artificial Intelligence Conference (2005)
6. Mitropoulos, P., Mitropoulos, I., Giannikos, I., Sissouras, A.: A biobjective model for the locational planning of hospitals and health centers. Health Care Manag. Sci. **9**, 171–179 (2006)
7. Liao, K., Guo, D.: A clustering-based approach to the capacitated facility location problem. Trans. GIS **12**, 323–339 (2008)
8. Han, J., Lakshmanan, L.V.S., Ng, R.T.: Constraint-based multidimensional data mining. Computer **32**, 46–50 (1999)

9. Wang, X., Rostoker, C., Hamilton, H.J.: Density-based spatial clustering in the presence of obstacles and facilitators. In: Proceedings of 8th European Conference on Principles and Practice Of Knowledge Discovery in Databases (2004)

10. Alberta Breast Cancer Screening Program. http://www.cancerboard.ab.ca/abcsp/program. html

11. Breaux, T.D., Reed, J.W.: Using ontology in hierarchical information clustering. In: Proceedings of 38th Annual Hawaii International Conference on System Sciences (2005)

A Data Hiding Method for H.265 Without Intra-frame Distortion Drift

Yunxia Liu[1(✉)], Shuyang Liu[2], Hongguo Zhao[1], Si Liu[1],
and Cong Feng[1]

[1] College of Information Science and Technology,
Zhengzhou Normal University, Zhengzhou, China
liuyunxia0ll0@hust.edu.cn
[2] School of Mathematics and Statistics, Lanzhou University, Lanzhou, China

Abstract. This paper presents a readable H.265/HEVC data hiding algorithm. To avert intra-frame distortion drift, we first give the Condition of the directions of intra-frame prediction. Then we embed the message into the multi- coefficients of the 4×4 luminance DCT blocks of the selected frames which meet the Condition. The experimental results show that this data hiding algorithm can effectively avert intra-frame distortion drift, and get good visual quality.

Keywords: Data hiding · H.265/HEVC · Intra-frame distortion drift

1 Introduction

Data hiding is a technique that embeds message into cover media contents, which is used in many applications such as medical systems, law enforcement, copyright protection and access control, etc. And the technique is well researched for the previous compression standards, such as MPEG1/2 and H.264/AVC/SVC [1–3]. The existing video data hiding schemes can be selected part of the video coding structure, including intra prediction [4], motion vector [5], DCT (discrete cosine transform) coefficient [6], etc., to embed message.

H.265/HEVC (high efficiency video coding) is the latest video coding standard published by ITUTVCEG and ISO/IEC MPEG [8]. H.265's main achievement is its significant improvement in compression performance when compared to the previous state-of-the-art standard with at least 50% reduction in bitrate for producing video of similar perceptual quality [9], and it is well adapted for network transmission.

Data hiding for compressed media usually uses DCT domain because DCT is a used mechanism in compression standards such as JPEG, MPEG and H.264/AVC [12]. And the DCT coefficient of a block in an intra-frame technique is one of the most popular transform domain techniques and adopted in the compression standards [7]. A consensus in traditional algorithms for hiding data into images is "the less number of modification to the DCT coefficients, the less amount of distortion in the image" [19]. However, the scheme that embed the message into the DCT coefficients of I frames for hiding data is not correct for H.265 video streams because of the intra-frame distortion

© Springer International Publishing AG 2017
D.-S. Huang et al. (Eds.): ICIC 2017, Part III, LNAI 10363, pp. 642–650, 2017.
DOI: 10.1007/978-3-319-63315-2_56

drift. Thus, it is necessary to introduce a without distortion drift mechanism to H.265 when data hide into the DCT coefficients.

In recent research, some researchers have proposed methods for without Intra-frame distortion drift data hiding algorithm in H.264 [2, 3, 7, 13–18]. The method proposed in [13] employed the multi-coefficients of a 4 × 4 DCT block for embedding data to compensate the intra-frame distortion, which is readable and blind. The method proposed in [7], which improve the algorithm [13], employed the multi-coefficients and the directions of intra-frame prediction to avert the distortion drift. [2] embedded the data into 4 × 4 DCT block and proposed a reversible data hiding scheme without distortion drift in H.264. [13–18] presented the robust methods and can improve the robustness of [2, 13]. Although there are a few of H.264 series video data hiding algorithms, they cannot be applied directly to H.265 because H.265 uses different transformation and block sizes. Therefore, research on H.265 data hiding methods is very valuable.

The existing H.265 video data hiding schemes are studied by few scholars since H.265 is recently finalized [10, 11]. [10] proposed for H.265 video by using the coding block size feature in HEVC, the nonzero DCT coefficients are manipulated based on the transform block size in all slices and an data hiding technique is proposed to adaptively manipulate the prediction block size. These techniques have the potential to be further fine-tuned to handle. [11] modified the LSB of the selected QTCs and embedded one of the watermark bit (Mb) in each QTC. To the best of our knowledge, the data hiding method using DCT domain has not been investigated to be compatible to H.265standard. Consequently, further study and investigation are required.

In this paper, we present a data hiding algorithm that is readable and can be used in video watermarking, covert communication and error concealment. The main contributions of our work are as follows. To prevent the distortion drift, we give the Condition. And we embed the data into the coefficients of the 4 × 4 DCT blocks of the selected frames which meet the Condition to avert the distortion drift.

The rest of the paper is organized as follows. Section 2 describes the theoretical framework of the proposed algorithm. Section 3 describes the proposed algorithm. Experimental results are presented in Sect. 3.3 and conclusions are in Sect. 4.

2   Theoretical Framework

A. Intra-frame Prediction

A prediction block of H.265 intra prediction method is formed based on previously encoded adjacent blocks. The sixteen pixels in the 4 × 4 block are predicted by using the boundary pixels of the upper and left blocks which are previously obtained, which use a prediction formula corresponding to the selected optimal prediction mode, as shown in Fig. 1. Each 4 × 4 block has 33 angular prediction modes, as shown in Fig. 2.

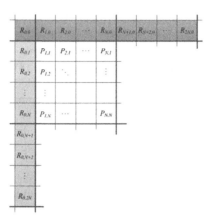

Fig. 1. Labeling of prediction samples

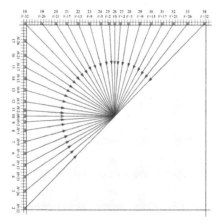

Fig. 2. 4×4 luminance block prediction modes

B. Intra-frame Distortion Drift

The intra-frame distortion drift emerges because we embed the message into the block of I frames. As illustrated in Fig. 3, we assume that current prediction block is $B_{i,j}$, then each reference sample of $B_{i,j}$ is the sum of the predicted value and the residual value. Since the predicted value is calculated by using the reference samples which are gray in Fig. 3. The embedding induced errors in blocks $B_{i-1,j-1}$, $B_{i,j-1}$, $B_{i-1,j}$, and $B_{i-1,j} + 1$ would propagate to $B_{i,j}$ because of using intra-frame prediction. This visual distortion that accumulates from the changes of reference blocks to the current block is defined as intra-frame distortion drift.

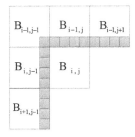

Fig. 3. The prediction block Bi,j and the adjacent encoded blocks

C. Intra-frame Distortion Drift Prevention

The algorithm only embeds data into 4 × 4 luminance blocks, because the human eyes are less sensitive to the brightness. For convenience, we give several definitions, the 4 × 4 block on the right of the current block is defined as right-block; the 4 × 4 block on the top of the right-block is defined as top- right–block; the right-block has the prediction unit prediction modes 26–34 and the top- right–block has the prediction unit prediction modes 10–34 and DC is defined as Condition as shown in Fig. 4. If a prediction block meets the Condition, the pixel values of its samples on the rightmost column will not be used in the intra-frame prediction of its right-block. That is, the embedding induced distortion will not propagate to its right-block and top- right–block.

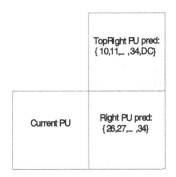

Fig. 4. The current block meets condition.

3 Proposed Scheme

The proposed method hides message in H.265 bit stream. Figure 5 depicts the method of our proposed algorithm.

3.1 Embedding

In order to avert the distortion drift, the current block must meet above-mentioned Condition.

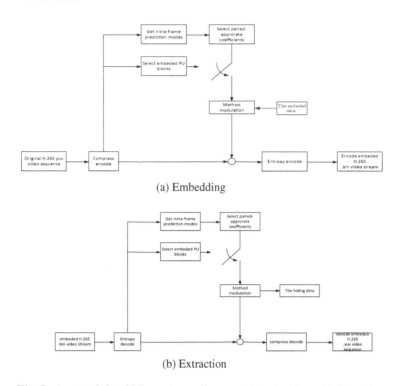

(a) Embedding

(b) Extraction

Fig. 5. Proposed data hiding scheme diagram. (a) Embedding. (b) Extraction.

The embedding operation is shown in Fig. 5(a). The original video is first before entropy encoded to get the intra-frame prediction modes of prediction unit and quantized DCT coefficients. And the appropriate DCT coefficients of 4×4 luminance DCT blocks which meet our proposed conditions are selected to embed the message. Then, the encoded message is embedded into appropriate coefficients based on modulo modulation. All the quantized DCT coefficients are entropy encoded to get the target embedded video.

To make the embedding procedure more clear, the block which have a coefficient $\tilde{Y}_{i,j}$ (i,j = 0,1,2,3)is exploited for data hiding as an example.

1. Encode embedded data.

   We first encode the embedded data by binary code before data hiding.

2. Select embeddable blocks.

   If the current block meets above-mentioned predictions (for example, right-mode is 15, top-right-mode is 30, as depicted in Fig. 6), the 4×4 luminance DCT block for embedding is selected.

   Modulation Method:

   If the embedded bit is 1, $\tilde{Y}_{i,j}$ is modified according to the formula (16);

   If the embedded bit is 0, $\tilde{Y}_{i,j}$ is not modified according to the formula (17).

(a)   (b)

Fig. 6. (a) The original frame of BasketballPass. (b) The embedded frame of BasketballPass.

$$Y_{i,j} = \begin{cases} Y_{i,j}+1, & if \ Y_{i,j}\%2=0, \ Y_{i+2,j}=Y_{i+2,j}-1, \ Y_{i+3,j}=Y_{i+3,j}+1 \\ Y_{i,j}, & if \ Y_{i,j}\%2=1 \end{cases} \qquad (16)$$

$$Y_{i,j} = \begin{cases} Y_{i,j}+1, & if \ Y_{i,j}\%2=1, \ Y_{i+2,j}=Y_{i+2,j}-1, \ Y_{i+3,j}=Y_{i+3,j}+1 \\ Y_{i,j}, & if \ Y_{i,j}\%2=0 \end{cases} \qquad (17)$$

3.2 Data Extraction

The extraction operation is shown in Fig. 6(b). After the entropy decoding of the H.265 video, we choose the embeddable blocks and extract the hidden data M′.

After selecting appropriate prediction modes of adjacent prediction unit blocks, we can get the hidden data M ($M = m1m2, \ldots, mi, \ mi \in \{0, 1\}$). Then we extract the hidden data M as follows:

If $|Y_{i,j}|\%2 = 1$, then the extracted bit is 1.
If $|Y_{i,j}|\%2 = 0$, then the extracted bit is 0.

3.3 Experimental Results

The proposed method has been implemented in the H.265/AVC reference software version HM16.0. The original H.265 encoder used a fixed quantization step parameter of 32 for all I-frames. The 6 video sequences BasketballPass, Keiba, BlowingBubbles, RaceHorses, BQMall, and KristenAndSara are used to be test samples. The standard video sequences are encoded into 20 frames at 30 frame/s and with an intra-period of 1 (group of picture I).Each sequence has twenty I frames.

Table 1 shows the embedding capacity performances of our algorithm using multi-coefficients (Y00,-Y20,Y30), (Y01,-Y21,Y31), (Y02,-Y22,Y32), (Y03,-Y23, Y33). Y00,Y01,Y02,Y03 are used for message embedding, and Y20,Y30, Y21,Y31, Y22,Y32,Y23,Y33 are used for distortion compensation. The video sequences which have different resolutions are used to be test samples. We use capacity proportion to describe the proportion of appropriate predictions prediction unit blocks weight in all of 4 × 4 prediction unit blocks. In Table 1, we show the comparison of embedding capacity with different resolution test videos.

Table 1. Embedding performances of our method

Video sequence	Resolution	Frame number	QP	Embeded capacity	Capcaity proportion
BasketballPass	416*240	20	32	1972	13.78%
Keiba	416*240	20	32	1024	6.41%
BlowingBubbles	416*240	20	32	5316	18.21%
RaceHorses	832*480	20	32	8104	12.25%
BQMall	832*480	20	32	15592	16.87%
KristenAndSara	1280*720	20	32	9456	16.96%

Table 2 shows the perceptual quality performances of our algorithm using peak signal-to-noise ratio (PSNR). The PSNR value "PSNR1" in this paper is calculated compared to the original video of the corresponding H.265 yuv files. The PSNR value "PSNR2" in this paper is calculated compared to the embedded video of the corresponding H.265 decoded yuv files. Embedding number means different numbers of the embedded data. Bit-rate shows the change caused by different number of embedding message.

Table 2. Visual quality of our method

Video sequence	Embedding number	PSNR1	PSNR2	Bit-rate
BasketballPass	96	35.49	35.46	1683.28
	960	35.49	34.54	1693.24
	1972	35.49	33.45	1703.8
Keiba	96	34.76	34.73	2033.28
	480	34.76	34.47	2039.42
	1024	34.76	34.22	2046.74
BlowingBubbles	96	40.67	40.39	2921.16
	2400	40.67	39.32	2947.78
	5316	40.67	38.129	2976.46
RaceHorses	96	34.62	34.58	9921.4
	1920	34.62	34.34	9968.32
	4052	34.62	33.35	10018.88
BQMall	96	41.91	41.87	9935.13
	2880	41.91	40.02	9988.52
	7796	41.91	39.12	10192.32
KristenAndSara	96	46.5	46.33	7596.57
	960	46.5	45.73	7660.03
	2364	46.5	44.31	7729.05

Table 1 provides different embedded capacity performance of six video sequences with our proposed algorithm. The average capacity proportion of the six sequences is more than 12% and the embedded capacity is more than 2000 bits excluding Keiba sequence which often varies smoothly.

Table 2 provides different visual quality performance of the six video sequences with our proposed algorithm. The average difference between PSNR1 and PSNR2 of the six sequences is less than 2 dB, the bit-rate increase by embedding message is less than 100 bits/sec excluding, and the bit-rate increase of BQMall and KristenAndSara sequences is more than 100 bits/sec because of a greater resolution.

Figures 6 and 7 give BasketballPass and Keiba as examples to test the visual quality of our method, which show the original frame of the test video sequences and the embedded frame of the test video sequences. It can be seen that the data hiding can obtain a good visual quality.

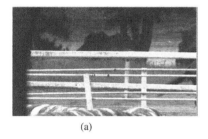 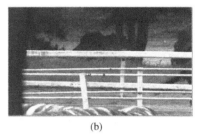

(a)   (b)

Fig. 7. (a) The original frame of Keiba (b) The embedded frame of Keiba.

4 Conclusion

In this paper, a data hiding method e without distortion drift based on H.265/AVC, which can avoid the distortion drift, is presented. The experiment results show that the method with prediction mode and multi- coefficients can get a little bit-rate increase, good visual quality and sufficient embedding capacity. In our future work, we will try our best to improve our method and give the performance comparison between our algorithm and other new methods.

Acknowledgment. This paper is sponsored by the National Natural Science Foundation of China (NSFC, Grant 61572447).

References

1. Wiegand, T., Sullivan, G.J., Bjontegaard, G., Luthra, A.: Overview of the H.264/AVC video coding standard. IEEE Trans. Circ. Syst. Video Technol. **13**(7), 560–576 (2003)
2. Liu, Y.X., Li, Z.T., Ma, X.J.: Reversible data hiding scheme based on H.264/AVC without distortion drift. J. Syst. Softw. **7**(5), 1059–1065 (2012)
3. Liu, Y.X., Li, Z.T., Ma, X.J., Liu, J.: A robust without intra-frame distortion drift data hiding algorithm based on H.264/AVC. Multimedia Tools Appl. **72**(1), 613–636 (2013). Springer
4. Kim, D.-W., Choi, Y.-G., Kim, H.-S., Yoo, J.-S., Choi, H.-J., Seo, Y.-H.: The problems in digital watermarking into intra-frames of H.264/AVC. Image Vis. Comput. **28**(8), 1220–1228 (2010)

5. Guo, Y., Pan, F.: Information hiding for H.264 in video stream switching application. In: Proceedings of IEEE International Conference on Information Theory and Information Security, pp. 419–421, December 2010

6. Lin, T.-J., Chung, K.-L., Chang, P.-C., Huang, Y.-H., Liao, H.-Y.M., Fang, C.-Y.: An improved DCT-based perturbation scheme for high capacity data hiding in H.264/AVC intra frames. J. Syst. Softw. **86**(3), 604–614 (2013)

7. Ma, X.J., Li, Z.T., Tu, H., Zhang, B.: Data hiding algorithm for H.264/AVC video streams without intra frame distortion drift. IEEE Trans. Circ. Syst. Video Technol. **20**(10), 1320–1330 (2010)

8. Sullivan, G.J., Ohm, J.-R., Han, W.-J., Wiegand, T.: Overview of the high efficiency video coding (HEVC) standard. IEEE Trans. Circ. Syst. Video Technol. **22**(12), 1649–1668 (2012)

9. Ohm, J.-R., Sullivan, G.J., Schwarz, H., Tan, T.K., Wiegand, T.: Comparison of the coding efficiency of video coding standards - including high efficiency video coding (HEVC). IEEE Trans. Circuits Syst. Video Technol. **22**(12), 1669–1684 (2012)

10. Tew, Y., Wong, K.: Information hiding in HEVC standard using adaptive coding block size decision. In: ICIP 2014, pp. 5502–5506 (2014)

11. Swati, S., Hayat, K., Shahid, Z.: A watermarking scheme for high efficiency video coding (HEVC). PLoS ONE **9**(8), e105613 (2014). doi:10.1371/journal.pone.0105613

12. Shanableh, T.: Data hiding in MPEG video files using multivariate regression and flexible macroblock ordering. IEEE Trans. Inf. Forensics Secur. **7**(2), 455–464 (2012)

13. Ma, X.J., Li, Z.T., Lv, J., Wang, W.D.: Data hiding in H.264/AVC streams with limited intra-frame distortion drift. In: Computer Network and Multimedia Technology, CNMT 2009 (2009)

14. Liu, Y.X., Li, Z.T., Ma, X.J., Liu, J.: A robust data hiding algorithm for H.264/AVC video streams. J. Syst. Softw. **86**(8), 2174–2183 (2013). Elsevier

15. Liu, Y.X., Hu, M.S., Ma, X.J., Zhao, H.G.: A new robust data hiding method for H.264/AVC without intra-frame distortion drift. Neurocomputing **151**, 1076–1085 (2014)

16. Liu, Y.X., Ju, L.M., Hu, M.S., Ma, X.J., Zhao, H.G.: A robust reversible data hiding scheme for H.264 without distortion drift. Neurocomputing **151**, 1053–1062 (2014)

17. Liu, Y., Jia, S., Hu, M., Jia, Z., Chen, L., Zhao, H.: A robust reversible data hiding scheme for H.264 based on secret sharing. In: Huang, D.-S., Bevilacqua, V., Premaratne, P. (eds.) ICIC 2014. LNCS, vol. 8588, pp. 553–559. Springer, Cham (2014). doi:10.1007/978-3-319-09333-8_61

18. Liu, Y.X., Jia, S.M., Hu, M.S., Jia, Z.J., Chen, L., Zhao, H.G.: A reversible data hiding method for H.264 with shamir's (t, n)-threshold secret sharing. Neurocomputing **188**, 63–70 (2014)

19. Noorkami, M., Mersereau, R.M.: A framework for robust watermarking of H.264-encoded video with controllable detection performance. IEEE Trans. Inf. Forensic Secur. **2**(1), 14–23 (2007)

Machine Learning

Study on Updating Algorithm of Attribute Coordinate Evaluation Model

Xiaolin Xu[1($\boxtimes$)], Guanglin Xu[2], and Jiali Feng[3]

[1] Shanghai Polytechnic University, Shanghai, China
xlxu2001@163.com
[2] Shanghai Lixin University of Commerce, Shanghai, China
glxu@outlook.com
[3] Shanghai Maritime University, Shanghai, China
jlfeng@shmtu.edu.cn

Abstract. Evaluation model based on attribute coordinate has made some achievements in both theoretical research and practical applications. However, if the new evaluation samples are added, the evaluation model needs to be reconstructed rather than the dynamic updating. Almost no progress has been made on how to dynamically update the evaluation model. Thus, this paper puts forward a dynamic updating algorithm based on barycentric coordinates and satisfaction function to effectively solve this problem. The experiment results show the reasonability and effectiveness of this algorithm.

Keywords: Comprehensive evaluation · Attribute coordinate evaluation · Barycentric coordinates · Global satisfaction

1 Introduction

The main difficulty of comprehensive evaluation is the weight determination of each attribute of evaluated objects by evaluators. The characteristic of evaluation method of attribute coordinate is that the evaluation way is very close to the normal thinking pattern of people and the corresponding preference curve can be accurately constructed according to the preference of the evaluator. Hence, this method can not only learn about the experience of experts but also give full play to the advantages of machine learning [1–5]. After more than ten years of research and application, the evaluation method of attribute coordinate has made a certain achievements in many fields [6–14]. But after the new evaluation samples are added, the evaluation model needs to be reconstructed rather than the dynamic update. There has been no progress on the study on how to dynamically update the evaluation model. This paper proposes a kind of update model to solve this problem. The remaining parts of this paper are organized as follows. Section 2 introduces the related work including the introduction to updating strategies and evaluation method of attribute coordinate. Section 3 elaborates the algorithm through the flow chart, which combines the evaluation method with strategy

The work was supported by the Key Disciplines of Computer Science and Technology of Shanghai Polytechnic University (No. XXKZD1604).

© Springer International Publishing AG 2017
D.-S. Huang et al. (Eds.): ICIC 2017, Part III, LNAI 10363, pp. 653–662, 2017.
DOI: 10.1007/978-3-319-63315-2_57

III. Section 4 carries out the experiments and validates the updating algorithm. Section 5 comes to the conclusion.

2 Related Work

2.1 Classification on Updating Strategy

(1) The updating of evaluation model

After the completion of evaluation model, if new evaluation sample data are collected, the evaluation model needs to be correspondingly updated. So the updating process refers to the process changing from the evaluation on old samples to that on both new and old samples. For the comprehensive evaluation, the updating of the evaluation model is an essential process. The purpose and effect of evaluation model update are that the original understanding is amended by the latest information to decrease the subjective or objective inaccuracy caused by the imperfect understanding. Moreover, the latest information could be various and contain other information in the comprehensive evaluation, such as the expert experiences and preferences.

The updating mode of evaluation model of attribute coordinate is shown in Fig. 1. The figure indicates that, in the evaluation depending on the weight obtained from samples, when new samples are added, they are likely to have a certain impact on old samples like changing the weight of samples, which further gives rise to the change of evaluation result. Old samples and new samples need to work together to generate new samples in light of a certain weight calculation method.

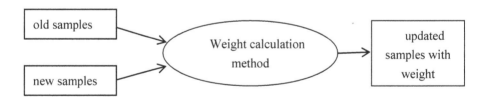

Fig. 1. Updating mode of samples with weight

(2) Updating Strategy

We divide the updating methods of evaluation model into three strategies: Strategy I, Strategy II and Strategy III.

Strategy I
Strategy I refers to the algorithm that after the new data are added to the original data, apply the original evaluation model to update the evaluation results. Its main characteristic is that both old sample data and comprehensive evaluation model are known. Although the updating method is easy, the result may be beyond accuracy, thus the updating is meaningless.

Strategy II

Strategy II refers to the algorithm that all the data join to calculate the weight. Its main characteristic is that the new and old data are thought of as the original data to reconstruct the new evaluation model. Strictly, this kind of updating is the standard to verify the effectiveness of other updating algorithms. However the drawback is it might take enormously long time.

Strategy III

Strategy III refers to the algorithm that the newly collected data are used to supplement and affect the original comprehensive evaluation model. Its core is how to impose the new weight on the latest information.

Obviously, from Strategy I to Strategy III, as the constraint conditions are more and more complex, the updating algorithm will be more and more difficult.

2.2   Introduction to Attribute Coordinate Evaluation

Comprehensive evaluation is to evaluate whether the evaluated objects with multi attributes are good or not. After attribute values with certain unified dimension are given to all the evaluated objects, the optimal evaluated object is the solution A = (10, ...,10) that each attribute value is full score (assuming the full score is 10). However, there does not exist the optimal solution in the comprehensive evaluation but generally exists the satisfactory solution. Hence, the satisfactory solution can be obtained instead of the optimal solution in the practical decisions. Thus the comprehensive evaluation could only require the most satisfactory solution meeting some weighted conditions. When evaluators evaluate multi-attribute objects, they often think that some attributes are important, namely, attributes can be given certain weights in the evaluation model. The importance of different attributes would change with the degree to which good evaluated objects or bad evaluated objects belong, namely, the weight of an attribute would dynamically change in the evaluation model.

(1) Solution of attribute barycentric coordinate

It is assumed that $S_T = \left\{ x_i = (x_{i1}, \ldots, x_{im}) \middle| \sum_{j=1}^{m} x_{ij} = T \right\}$ is a hyper plane with the total

score T. $x_i = (x_{i1}, \ldots, x_{im})$ are supposed to be independent attribute values. The intersection of S_T and X $(S_T \cap X)$ is a $(n-1)$-dimensional simplex (shown in Fig. 2, e.g. $S_{100} \cap X = \Delta ABC$, and $\Delta A'B'C'$ is a two-dimensional simplex that the total score hyperplane S_T of $10 < T < 10 \times n$ intersects X, $(S_T \cap X) = \Delta A'B'C'$).

Let $\{x_k, k = 1, \ldots, s\} \subseteq ST \cap X$ be the sample solution set and $Z = \{z_1, \cdots, z_n\}$ be the evaluator set. The

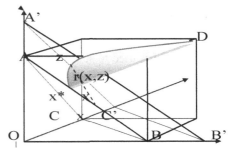

Fig. 2. Two-dimensional simplex from $ST \cap X = \Delta A'B'C'$

decision maker z_i selects t sets of satisfactory solutions $\{x_h, h = 1,...,t\}$, and each solution is marked as $w_h(x_h)$. Because the solution space X is a convex set with respect to score $w_h(x_h)$, the barycenter $b(\{x_h(z_i)\})$ of $\{x_h(z_i)\}$ can be solved as formula (1) by the weighted average method ($w_h(x_h)$ is taken as the weight).

$$b(\{x_h(z_i)\}) = \left(\frac{\sum\limits_{h=1}^{t} w_h^i x_{1h}}{\sum\limits_{h=1}^{t} w_{1h}^i}, \cdots, \frac{\sum\limits_{h=1}^{t} w_{mh}^i x_{mh}}{\sum\limits_{h=1}^{t} w_{mh}^i} \right) \qquad (1)$$

(2) Solution of attribute barycentric curve

Obviously, when we have large enough training sample set $\{x_k\}$ and more enough training times, the solution sets selected by decision makers z_i are enough, the barycenter $b(\{x_h(z_i)\})$ would be close to the local most satisfactory solution x*|T, namely, $\lim\limits_{h\to\infty} b(\{x_h(z)\}) \to x^*|T$. When the number of decision makers Z is more than 1, $\lim\limits_{h\to\infty} b'(\{x_h(Z)\})$ is the local most satisfactory solution of decision makers Z in $S_T \cap X$, namely, the barycenter of decision makers. The set of all local most satisfactory solutions $\{b'(\{x_h(Z)\})|T \in [10, 10 \times m]\}$ can be obtained after T covers the interval $[10, 10 \times m]$. The set would be a line, written as $L(b'(\{x_h(Z)\}))$, and this line is called as the local most satisfactory linear solution of decision makers Z. The line can be obtained by polynomial curve fitting method such as the following polynomial function (2):

$$G(T) = a_0 + a_1 T + a_2 T_2 \qquad (2)$$

In this situation, three local most satisfactory solutions of decision makers in S100 $\cap$ X, ST $\cap$ X and S10 $\times$ m $\cap$ X are taken as three interpolation points to be substituted into Lagrange interpolation formula (3) to calculate the local most satisfactory solution line.

$$g_i(T) = \frac{(T - x_1^*)(T - x_2^*)}{(x_0^* - x_1^*)(x_0^* - x_2^*)} a_{i0} + \frac{(T - x_0^*)(T - x_2^*)}{(x_1^* - x_0^*)(x_1^* - x_2^*)} a_{i1} + \frac{(T - x_0^*)(T - x_1^*)}{(x_2^* - x_0^*)(x_2^* - x_1^*)} a_{i2} \quad (3)$$

2.3   The Work of the Paper

After the calculation of barycentric point $b'(\{x_h\})$, if the new sample data are collected again, Strategy II is used to substitute both new and old sample data into the formula (3) for recalculation. The advantage of this method is that the calculation results are very accurate, but the disadvantages are that repeated marks on old samples are unavoidable. Besides, the calculation will not be feasible if the old data are lost. However, if Strategy III is used, only the marks on new samples are needed, hence this algorithm is simple and effective. In the paper, we integrate the Strategy III into the evaluation model of attribute coordinate to accomplish the updating of the model in the case of new sample data being added into the current model. The combined algorithm is described as follows.

3   Updating Algorithm of Evaluation Model of Attribute Coordinate

3.1   Algorithm Process

Figure 3 illustrates the process of the algorithm in the way of flow chart.

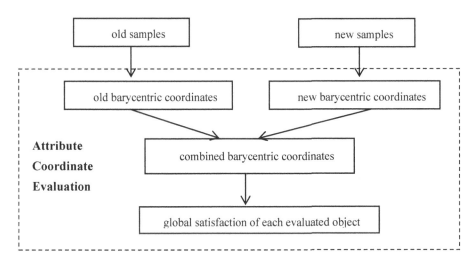

Fig. 3. The process of the algorithm

3.2   Updating of Attribute Barycentric Coordinates

After calculating old barycentric coordinates, if t new samples are added and they are marked as $w_t(\{x_h\})$, respectively, the formula (1) can be updated as formula (4). Where s is the number of old samples, $x_1, \ldots, x_m$ are the old barycentric coordinates.

$$b^u = \left[\frac{\sum\limits_{h=1}^{t} w_h^i x_{1h}}{\sum\limits_{h=1}^{t} w_{1h}^i} + sx_1, \cdots, \frac{\sum\limits_{h=1}^{t} w_{mh}^i x_{mh}}{\sum\limits_{h=1}^{t} w_{mh}^i} + sx_m \right] \tag{4}$$

The new barycentric coordinate figured out is $b^u(\{x_h(z_i)\})$, and the barycentric coordinate of old model is $b^u(\{x_h(z_i)\})$. Their norm can be solved according to the formula (5) and be calculated by Euclidean distance, as green segment shown in Fig. 4.

$$e = \|b^u(\{x_h(z_i)\}) - b'(\{x_h(z_i)\})\| \tag{5}$$

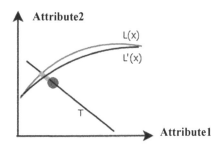

Fig. 4. The old and updated barycentric curve (Color figure online)

Figure 4 also shows the old psychological barycenter curve L(x) and the updated psychological barycentric curve L'(x). T represents a certain total score obtained from Attribute1 and Attribute2.

3.3 Satisfaction Calculation after Updating Barycentric Coordinates

When there exists the new psychological standard point $b^u(\{x_h(z_i)\})|T$ of Z in $S_T \cap X$, taking $b^u(\{x_h(z_i)\})|T$ as the standard, the global satisfaction function (6) is used for the satisfaction evaluation of all the solutions in $S_T \cap X$.

$$sat(x, Z) = \lambda(x, Z) * \exp\left(-\frac{\sum\limits_{j=1}^{m} w_j\left|x_{ij} - b_{ij}^u(\{x_h(Z)\})\right|}{\sum\limits_{j=1}^{m} \delta_j(b_{ij}^u(\{x_h(Z)\}) - \delta_j)}\right) \tag{6}$$

Where $sat(x, Z)$ is the satisfaction of evaluated object x_i from evaluator Z, whose value is expected to be between 0 and 1. x_{ij} is each attribute value. $\left|x_{ij} - b_{ij}^u(\{x_h(Z)\})\right|$ is to measure the difference between each attribute value and the corresponding barycentric value. w_j and δ_j are used as the factor which can be adjusted to make the satisfaction comparable value in the case where the original results are not desirable. It is found from the above learning that all the solutions in $L(b^u(\{x_h(Z)\}))$ are determined by taking the local most satisfactory solution as the standard in different simplexes $S_T \cap X$ as well as calculating the maximal satisfaction by formula (6). The correction coeffi-

cient $\lambda(x, Z)$ is described as $\lambda(x, Z) = \left(\dfrac{\sum\limits_{i=1}^{m} x_{ij}}{\sum\limits_{j=1}^{m} X_j}\right)^S$, Where $\sum\limits_{j=1}^{m} X_j$ is the sum of X_j with

each attribute value the full score, $\sum\limits_{ij=1}^{m} x_{ij}$ is the sum of each attribute value x_{ij} of solution

x_i, and $S = \left(\dfrac{\sum\limits_{j=1}^{m} Xj}{\sum\limits_{i=1}^{m} xij} \right)$. Formula (6) can reach consistent satisfaction $sat(x, Z)$ in the

entire solution space, so $\lambda(x,Z)$ is called the global consistent coefficient.

4 Experiments

The data set comes from the scores of college entrance examination of a certain city in 2015 and contains more than 4,000 objects to be evaluated. Moreover, each object contains four attributes: Chinese, mathematics, English and comprehensive discipline. This simulation adopts the first three attributes in order to get a more intuitive three-dimensional chart.

4.1 Experiments of Updated Barycentric Coordinates

According to evaluation model of attribute coordinates, firstly, a hyper plane with the total score of 370 is selected, on which there are 26 sample points. Then the 26 sample points are marked by five experts. It is assumed that 18 sample points are initial samples and the rest (8 sample points)are the supplementary samples. The above attribute coordinate updating algorithm is used to update the barycentric coordinates of attributes.

(a) Calculate the original barycentric coordinates. According to formula (1), the barycentric coordinates of 18 sample points are figured out, as black points shown in Fig. 5.
(b) Calculate the new barycentric coordinates by the updating method of Strategy II. According to formula (1), the barycentric coordinates of 26 sample points are figured out, as green points shown in Fig. 5.

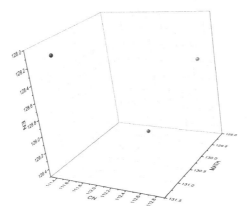

Fig. 5. Comparison among original and updated barycentric coordinates with Strategy II and III (Color figure online)

Table 1. Comparison of original and updated barycentric coordinates

Barycentric coordinates	Chinese	Maths	English
Original barycentric coordinates	111.47	131.11	128.12
Updated coordinates by Strategy II	112.74	129.24	128.51
Updated coordinates by Strategy III	111.99	129.50	129.41

(c) Calculate the new barycentric coordinates by the updating method of Strategy III. The barycentric coordinates integrated with the other 8 sample points are obtained according to formula (4), as red points shown in Fig. 5.

Table 1 shows the results of barycentric coordinates in terms of the above three cases.

As we had stated in Sect. 2, the updated barycentric coordinates by Strategy III provides more accurate result as the standard to verify the effectiveness of other updating algorithms. It can be seen from Table 1 that the updated barycentric coordinates by Strategy III cannot be completely equal to the updated barycentric coordinates by Strategy II but be close to it.

Figure 5 illustrates the position of the three barycentric coordinates in three dimentional space, each point including three attributes indicated with CN (Chinese), MATH (mathematics), and EN (English).

In Fig. 6, the black point is the barycentric coordinates which are figured out from the original samples. Then the satisfaction curve illustrates the degrees of satisfaction of samples according to formula (6), as the red line shown in Fig. 6. The barycentric coordinates of 8 sample points evaluated by experts are shown as the blue point. The new barycentric coordinates updated by formula (4) is shown as the red point. Finally, the updated satisfaction curve is figured out as the yellow line (solid line).

From Fig. 6, the satisfaction curve after adding sample is basically close to the original satisfaction curve (dotted line), which indicates that our updating algorithm is effective. Compared with the Strategy II which needs to calculate the barycentric coordinates for all the samples again, this algorithm greatly avoids the repeated evaluation on old samples and effectively reduces the time of evaluation.

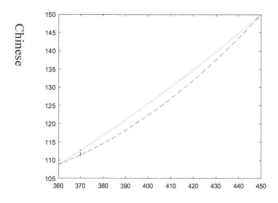

Fig. 6. Comparison of original and updated barycentric curves (Color figure online)

5   Conclusions

This paper studies how to update the existing evaluation model based on the evaluation of attribute coordinate after the new sample data are added and presents the updating algorithms of barycentric coordinates and satisfaction function. The simulation of comparison is carried out according to the real data. It can be clearly seen from the calculation results that the updating strategy of Strategy III is used to figure out the new barycentric coordinates and then to calculate the new satisfaction for all samples by calculating the barycentric coordinates of supplementary samples. This can not only realize the effective update of evaluation system but also greatly decrease the calculation work. In the further work, we would conduct the multi-expert evaluation on the increased samples, integrate the supplementary samples into the old evaluation model, and then observe how they affect the evaluation results.

References

1. Feng, J.L., Zhao, T.A.: KDD model based on conversion of quantity-quality features of attributes. J. Comput. Res. Dev. **37**(9), 1114–1119 (2000)
2. Feng, J.L., Feng, J.J.: Granularity transformation of qualitative criterion, orthogonality of qualitative mapping system, and pattern recognition. In: IEEE International Conference on Granular Computing, pp. 211–216 (2006)
3. Xu, G.L., Xu, X.L.: Study on evaluation model of attribute barycentric coordinates. Int. J. Grid Distrib. Comput. **9**(9), 115–128 (2016)
4. Wu, Q.F., Feng, J.L.: A kind of evaluation and decision model based on analysis and learning of attribute coordinate. J. Nanjing Univ. (Nat. Sci.) **39**(2), 182–188 (2006)
5. Xu, X.L., Feng, J.L.: A quantification method of qualitative indices based on inverse conversion degree functions. In: Enterprise Systems Conference, pp. 261–264 (2014)
6. Duan, X.Y., Yu, S.Q., Xu, G.L., Feng, J.L.: Evaluation on comprehensive competitiveness of inland port based on the method of arrtribute theory. J. Pattern Recogn. Image Proress. **4**, 373–380 (2013)
7. Xu, G.L., Min, S.: Research on multi-agent comprehensive evaluation model based on attribute coordinate. In: IEEE International Conference on Granular Computing (GrC), pp. 556–562 (2012)
8. Xu, X.L., Xu, G.L., Feng, J.L.: A kind of synthetic evaluation method based on the attribute computing network. In: IEEE International Conference on Granular Computing (GrC), pp. 644–647 (2009)
9. Xu, X.L., Xu, G.L.: Research on ranking model based on multi-user attribute comprehensive evaluation method. In: Applied Mechanics & Materials, pp. 644–650 (2014)
10. Xu, X.L., Xu, G.L.: A recommendation ranking model based on credit. In: IEEE International Conference on Granular Computing (GrC), pp. 569–572 (2012)
11. Xu, X.L, Feng, J.L.: Research and implementation of image encryption algorithm based on zigzag transformation and inner product polarization vector. In: IEEE International Conference on Granular Computing, vol. 95, no. 1, pp. 556–561 (2010)
12. Xu, G.L., Feng, J.L., Liu, Y.C.: Pattern recognition method based on the attribute computing network. J. Comput. Res. Dev. (S1) (2008)

13. Xu, G.L., Wang, L.F.: Evaluation of aberrant methylation gene forecasting tumor risk value in attribute theory. J. Basic Sci. Eng. **16**(2) (2008)
14. Feng, J.L., Bi, J.Y.: Lung cancer cells' recognition based on attribute theory. J. Guangxi Normal Univ. (Nat. Sci. Ed.) **29**(3) (2011)

The Concept of Applying Lifelong Learning Paradigm to Cybersecurity

Michał Choraś[1(✉)], Rafał Kozik[1], Rafał Renk[2],
and Witold Hołubowicz[1,2]

[1] Faculty of Telecommunications, Computer Science and Electrical Engineering,
UTP University of Science and Technology, Bydgoszcz, Poland
chorasm@utp.edu.pl
[2] Adam Mickiewicz University, Poznań, Poland

Abstract. One of the current challenges in machine learning is to develop intelligent systems that are able to learn consecutive tasks, and to transfer knowledge from previously learnt basis to learn new tasks. Such capability is termed as lifelong learning and, as we believe, it matches very well to counter current problems in cybersecurity domain, where each new cyber attack can be considered as a new task. One of the main motivations for our research is the fact that many cybersecurity solutions adapting machine learning are concerned as STL (Single Task Learning problem), which in our opinion is not the optimal approach (particularly in the area of malware detection) to solve the classification problem. Therefore, in this paper we present the concept applying the lifelong learning approach to cybersecurity (attack detection).

Keywords: Cybersecurity · Machine learning · Lifelong learning intelligent systems · Pattern recognition

1 Introduction

The major goal of this paper is to present the concept of applying the lifelong learning approach to cybersecurity. We want to answer the need of developing the system that does not need to return over to the previous original network data as the knowledge is already encoded and embedded in the trained components. Such capability of intelligent systems (called lifelong learning) is currently needed in cybersecurity, where each new cyber attack type can be considered as a new task.

Our paper focuses on the concept of developing a novel Lifelong Learning Intelligent System (often termed as LLIS) for cybersecurity. We strongly believe that lifelong machine learning systems can overcome limitations of statistical learning algorithms which need large number of training examples and are suitable for isolated single-task learning [1].

Existing lifelong machine learning research is still in its infancy and there are many open challenges. Key functionalities that need to be developed within such systems in order to benefit from past learned knowledge include: knowledge retaining from past learning tasks, knowledge transfer to future learning tasks, prior knowledge update, user feedback.

© Springer International Publishing AG 2017
D.-S. Huang et al. (Eds.): ICIC 2017, Part III, LNAI 10363, pp. 663–671, 2017.
DOI: 10.1007/978-3-319-63315-2_58

Also the concept of "task" that appears in many formal definitions [3] of lifelong machine learning model, seems to be hard to match many real life setups. For example, when considering telecommunication network monitoring for cybersecurity purposes, it is often difficult to distinguish when a particular task finishes and the subsequent one starts i.e. when a different family of attacks has started.

One of the main drawbacks of many cybersecurity solutions adapting machine learning is the fact that the learning process is concerned as STL (Single Task Learning problem). For instance, when developing an anomaly detection system (ADS), researchers usually [2] collect different malicious network traffic samples generated by different malware families, then split the data for training and evaluation (often inspecting manually the data to ensure that the datasets will be similarly distributed), and finally train the model. The following aspects shall be considered:

- The tasks collecting data samples for different malware families are not identical and thus those should not be treated as single task;
- On the other hand, while treating those tasks separately (learning the classifiers independently) the information that could be acquired from other tasks will be missing.

The paper is organized as follows: in Sect. 2 the state of the art in lifelong learning intelligent systems is presented. Section 3 contains the description of the LLIS approach for cybersecurity system, while in Sect. 4 the possible scenarios are discussed. Conclusions are provided afterwards.

2 State of the Art in Lifelong Learning Intelligent Systems (LLIS)

Originally lifelong learning was stated as a sequence of learning tasks that need to be solved using the previous knowledge stored in previously learnt classifiers [4]. According to [3, 5], theoretical considerations on lifelong learning are relatively widely described in the literature, in particular in the perspective of growing popularity of machine learning approaches and applications.

However, scientific communities put more attention to aspects of learning based on well-known knowledge domains and well-labelled training datasets, while approaches to lifelong learning (or "learning to learn") without observed data, e.g. to perform new, future tasks are not yet very popular.

In [6], one of the first attempts to describe model of lifelong learning can be found. The author introduced a formal model called inductive bias learning, that can be applied when the learner is able to learn novel tasks drawn from multiple, related tasks from the same environment. Those considerations focused only on the finite-dimensional output spaces, and mainly on linear machines rather than nonlinear ones, in contrary to [7] work, additionally extending earlier research with algorithmic stability aspects.

In [8] there is proposed an approach to the problem of learning a number of different target functions over time, with assumptions that they are initially unknown for the learning system and that they share commonalities. Different approaches to solve this sequence of tasks include transfer learning [9], multitask learning, supervised, semi-supervised, reinforcement learning [10], and unsupervised techniques.

There are also works defining strong theoretical foundations for lifelong machine learning concept. Particularly, in [5] authors worked on a PAC-Bayesian generalization bound applied for lifelong learning allowing quantification of relation between expected losses in future learning tasks and average losses in already observed (learned) tasks.

Majority of the approaches so far assume that problem representation is not changing, (i.e. feature space). However, recent works increasingly consider that also the underlying feature space can be shifting. To overcome different approaches solutions such as changing kernels for feature extraction [11], changing latent topics [12], or the underlying manifold in manifold learning [13, 14] are proposed.

The HCS (Hybrid Classifiers Systems) paradigm addresses naturally all the challenges of lifelong machine learning such as learning new tasks while preserving knowledge of the previous ones.

In fact, classifier ensemble management resembles some of the algorithms proposed for lifelong learning. For example, critical aspect of the lifelong learning systems is the ability to detect the task shift, which is quite similar to concept drift detection [15], and can be tackled by hyper-heuristics [16]. To deal with debatable cases in ensemble learning and to increase transparency in such debatable decisions, our hypothesis is that argumentation could be more effective than current resolution methods.

Moreover, recent work on hybrid classifiers has demonstrated promising results of using an argumentation-based conflict resolution instead of voting-based methods for debatable cases in ensemble learning [17], showing that the hybridization of ensemble learning and argumentation fits the decision patterns of human agents.

In the next section, we propose the general concept and framework of using lifelong learning paradigm and hybrid classifiers to cybersecurity problems such as intrusion, anomalies and cyber attacks detection.

3   The Concept of Applying Lifelong Learning to Cybersecurity Domain

Network and information security is now one of the most prominent problems of citizens, societies and homeland security [18]. As widely observed, the number of successful attacks on information, citizens and even secure financial systems, as well as critical infrastructures is still growing [19, 20]. One of the problem lies with the inefficiency of signature based approaches to detect cyber attacks. In situations, where new attacks (or even slightly modified families of malware) emerge, those systems are not efficient until the new signatures are created [21]. On the other hand, anomaly based approaches (systems which detect abnormalities in traffic or e.g. requests to databases) tend to produce false positives (false alarms). Such situation and current challenges in network security motivate our research and the concept to apply lifelong learning approach to cybersecurity domain.

The general concept of the lifelong learning for cybersecurity is presented in Fig. 1.

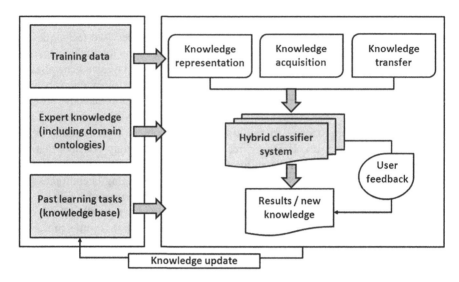

Fig. 1. The general concept diagram of the proposed lifelong learning approach.

Our concept is to implement practical lifelong learning cybersecurity system on the basis of the hybrid classifiers (HCS - Hybrid Classifiers Systems) with strong emphasis on knowledge management. Our implementation should take the advantage of possibly available training data, past learned knowledge as well as the expert knowledge.

In order to achieve such goal, we plan to define:

- how to represent knowledge,
- how to retain knowledge from past learning tasks and transfer it to the future learning tasks,
- what knowledge needs update,
- when to acquire relevant knowledge according to various changes that may occur in the environment, and
- how to incorporate user feedback in order to improve learning.

The knowledge representation model should be far more complex than the solutions already proposed in the literature (e.g. list of model components shared between learning tasks). Therefore, our hybrid intelligent approach will rely both on multi-classifiers and dynamic domain knowledge.

Hybrid Classifiers Systems have already been successfully applied to solve complex machine learning problems in variety of domains. In fact, multi classifier paradigm is akin to some of the algorithms proposed for lifelong learning that build and maintain some kind of reservoir of latent models/modules that may become useful or be somehow reused in the future. Classifiers inside an ensemble can embody this property, but some measure of the relevance of the classifier to the actual task is required, in order to decide which classifiers are alive or latent. The measure of uncertainty of the classifier response may serve this purpose. Therefore, we also look for recent advancements in the area of ensemble learning methods. Particularly, to deal with possible unforeseen situations in lifelong learning, our investigation domain will enact argumentation (agreement) technologies for conflict resolution in multi-classifiers [17].

To facilitate the lifelong learning process, we organise current and past experiences and relevant information about the environment (networks) into a structured and hierarchical model of knowledge. This knowledge will be accumulated and will support the future learning tasks.

One of the key common challenges here addresses the scalability problems in order to allow knowledge-base to grow without impacting the knowledge access (reading) performance. Commonly, the knowledge databases contain information related to learning process, such as final models, learnt patterns, performance statistics, etc.

In that regards, hybrid classifier paradigm seems to be the obvious choice when different lifelong learning schemas are analysed, since HCS systems maintain ensemble of classifiers (that resemble accumulated knowledge) that may be useful or be reused in the future. Therefore, the ensemble will maintain classifiers that will be shared among different tasks. Obviously, there are questions related to how the knowledge (in that case the ensemble) can be transferred from previous tasks to the current one (to learn current task faster) or how to refine existing knowledge. Some solutions already exist, proposing to represent each task parameters as linear combination of latent components shared between tasks.

For the knowledge update, we investigate dialectical argumentation for ontology update and enrichment [22] or reasoning with inconsistent domain knowledge [23, 24]. Argumentation can be seen as a vehicle for knowledge representation [25], but also allows us to explore a new class of "Argument-Based Learning Algorithms for lifelong learning" (in line with explanation-based learning).

When it comes to the cybersecurity and cyber-attacks detection, there is no single classifier or IDS system that will allow recognising all kinds of attacks. Also the same system (even if learnt to detect the same type of attacks) have to be learnt again when changing the monitored network (topology, services, characteristics etc.). In that regards we will need transfer learning mechanism that will allow us to learn to detect attack B from knowledge (e.g. ensembles of classifiers) learnt for attack A.

Another aspect is the overlap of knowledge that our system will need to be aware of. We will leverage this both to facilitate learning of new tasks and improving the effectiveness, when executing the old ones. Using again the cybersecurity example, IDS learnt in one network will use already established knowledge to detect attacks in another new network in a more accurate way (than without lifelong learning approach).

4   Current Use Cases and Future Work

We currently focus on adaptation of lifelong learning mechanism in the area of cybersecurity. In particular, we plan to adapt the above-mentioned ideas to efficiently detect anomalies in the monitored networks.

4.1   Anomaly Detection in the Application Layer

Application layer attacks, such as SQLIA (SQL Injection Attacks) are top-ranked on several threat lists. One of the examples is the "OWASP Top 10" [26] list that has been identified by Open Web Application Security). The list, among others, contains the following items from the application layer: injection flaws (e.g. SQL Injection), broken authentication and session management, Cross-Site Scripting.

The practical implementation of lifelong learning approach to protect the application layer can consider for example the analysis of user requests to web service or database. In such scenario, we can apply sensors and implement complex algorithms to learn the models of normal requests or user behaviour, and detect all the requests that fall outside the model of normality [27].

However, in order to decrease the false positive ratio (indicating anomalies which are not attacks or symptoms of misbehaviour) we suggest implementing lifelong learning solutions. In such scenario, the model of normal requests changes quickly (e.g. due to availability of new services or just new fields in web forms) and therefore anomaly detection approach tends to have high false positive ratio. In our proposed solution, the system will re-learn using the past knowledge to quickly adapt to network changes.

4.2   NetFlows Analysis for Malware Detection

Currently, one of the most challenging problems in the area of cybersecurity is the detection of malicious software in the local area network by means of analysis of the captured traffic. Commonly, these techniques allow the network administrator to detect machines that could be infected and are part of a botnet. In particular passive monitoring (e.g. without direct inspection of the host and content of the exchanged data) allows addressing the privacy issues. Particularly, the information gathered in form of NetFlows[1] can efficiently be used to identify the suspicious behaviour of hosts in the network.

However, as the system can be learnt to identify known botnets, it is commonly more challenging to adapt the system to new situations – e.g. to repeat the learning or adapt the system to different network with different volume of traffic, etc.

Therefore, our ambition in this area is to make the learning system to adapt faster, e.g. using the knowledge from previous learning tasks. We plan to achieve this by decomposing the learning tasks into a set of latent components that can be shared among different learning tasks so that specific learning challenge could be represented by the linear combination of these components.

In order to achieve our goal, we developed the prototype of the platform for distributed processing of network data. The high-level overview is provided in Fig. 2.

The data collected from the network can be stored on distributed file system (DFS) or as a stream transferred to the Apache Spark[2] cluster. On top of the Spark system, there is a machine learning library (MLlib[3]) which facilitates the data

---

[1] https://www.ietf.org/rfc/rfc3954.txt.

[2] http://spark.apache.org/.

[3] http://spark.apache.org/mllib/.

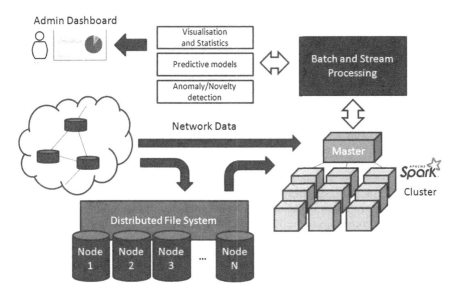

Fig. 2. The general idea for scalable network data processing.

processing and data mining processes. Currently, the system provides the end-user with the graphical interface that enables visualisation of various network characteristics (e.g. a number of connections established by particular IP address in the particular time window).

5 Conclusions

In this paper we have presented the general concept of applying lifelong learning approach to cybersecurity domain. We have proposed the general framework of the system based on the hybrid classifiers and comprising of emerging solutions in machine learning and pattern recognition. We currently work on the implementation of our concept and tests on real life cyber threats (e.g. modern worms and malware).

References

1. Chen, Z., Liu, B.: Lifelong machine learning. Synth. Lect. Artif. Intell. Mach. Learn. **10**(3), 1–145 (2016)
2. García, S., Zunino, A., Campo, M.: Survey on network-based botnet detection methods. Secur. Commun. Netw. **7**(5), 878–903 (2014)
3. Pentina, A., Lampert, C.H.: Lifelong learning with non-i.i.d. tasks. In: Advances in Neural Information Processing Systems (2015)
4. Chen, Z., Lium, B.: Lifelong machine learning in the big data era. In: IJCAI 2015 Tutorial (2015)

5. Pentina, A., Lampert, C.H.: A PAC-Bayesian bound for lifelong learning. In: ICML (2014)
6. Baxter, J.: A model of inductive bias learning. J. Artif. Intell. Res. (JAIR) **12**, 149–198 (2000)
7. Maurer, A.: Algorithmic stability and meta-learning. J. Mach. Learn. Res. **6**, 967–994 (2005)
8. Balcan, M., Blum, A., Vempala, S.: Efficient representations for lifelong learning and autoencoding. In: Workshop on Computational Learning Theory (COLT) (2015)
9. Segev, N., et al.: Learn on source, refine on target: a model transfer learning framework with random forests. IEEE Trans. Pattern Anal. Mach. Intell. **PP**(99), 1 (2015)
10. Ammar, H.B., Tutunov, R., Eaton, E.: Safe policy search for lifelong reinforcement learning with sublinear regret. In: Proceedings of the 32nd International Conference on Machine Learning, vol. 37, pp. 2361–2369. JMLR.org (2015)
11. Qiu, Q., Sapiro, G.: Learning transformations for clustering and classification. J. Mach. Learn. Res. **16**, 187–225 (2015)
12. Chen, Z., Liu, B.: Topic modeling using topics from many domains, lifelong learning and big data. In: Proceedings of the 31st International Conference on Machine Learning (2014)
13. Yang, H.L., Crawford, M.M.: Domain adaptation with preservation of manifold geometry for hyperspectral image classification. IEEE J. Sel. Top. Appl. Earth Obs. Remote Sens. **9**(2), 543–555 (2016)
14. Yang, H.L., Crawford, M.M.: Spectral and spatial proximity-based manifold alignment for multitemporal hyperspectral image classification. IEEE Trans. Geosci. Remote Sens. **54**(1), 51–64 (2016)
15. Widmer, G., Kubat, M.: Learning in the presence of concept drift and hidden contexts. Mach. Learn. **23**(1), 69–101 (1996)
16. Sim, K., Hart, E., Paechter, B.: A lifelong learning hyper-heuristic method for bin packing. Evol. Comput. **23**(1), 37–67 (2015)
17. Conţiu, Ş., Groza, A.: Improving remote sensing crop classification by argumentation-based conflict resolution in ensemble learning. Expert Syst. Appl. **64**, 269–286 (2016)
18. Choraś, M., Kozik, R., Bruna, M.P.T., Yautsiukhin, A., Churchill, A., Maciejewska, I., Eguinoa, I., Jomni, A.: Comprehensive approach to increase cyber security and resielience. In: Proceedings of ARES (International Conference on Availability, Reliability and Security), Touluse, pp. 686–692. IEEE (2015)
19. Kozik, R., Choraś, M., Renk, R., Hołubowicz, W.: Cyber security of the application layer of mission critical industrial systems. In: Saeed, K., Homenda, W. (eds.) CISIM 2016. LNCS, vol. 9842, pp. 342–351. Springer, Cham (2016). doi:10.1007/978-3-319-45378-1_31
20. Choraś, M., Kozik, R., Flizikowski, A., Renk, R., Hołubowicz, W.: Cyber threats impacting critical infrastructures. In: Setola, R., Rosato, V., Kyriakides, E., Rome, E. (eds.) Managing the Complexity of Critical Infrastructures. Studies in Systems, Decision and Control, vol. 90, pp. 139–161. Springer, Heidelberg (2017). doi:10.1007/978-3-319-51043-9_7
21. Choraś, M., Kozik, R., Puchalski, D., Hołubowicz, W.: Correlation approach for SQL injection attacks detection. In: Herrero, Á., et al. (eds.) International Joint Conference CISIS'12-ICEUTE'12-SOCO'12 Special Sessions. AISC, vol. 189, pp. 177–186. Springer, Heidelberg (2012). doi:10.1007/978-3-642-33018-6_18
22. Gómez, S.A., Chesñevar, C.I., Simari, G.R.: ONTOarg: a decision support framework for ontology integration based on argumentation. Expert Syst. Appl. **40**(5), 1858–1870 (2013)
23. Moguillansky, M.O., Simari, G.R.: A generalized abstract argumentation framework for inconsistency-tolerant ontology reasoning. Expert Syst. Appl. **64**, 141–168 (2016)
24. Choraś, M., Kozik, R., Flizikowski, A., Hołubowicz, W.: Ontology applied in decision support system for critical infrastructures protection. In: García-Pedrajas, N., Herrera, F., Fyfe, C., Benítez, J.M., Ali, M. (eds.) IEA/AIE 2010. LNCS, vol. 6096, pp. 671–680. Springer, Heidelberg (2010). doi:10.1007/978-3-642-13022-9_67

25. Bentahar, J., Moulin, B., Bélanger, M.: A taxonomy of argumentation models used for knowledge representation. Artif. Intell. Rev. **33**(3), 211–259 (2010)
26. OWASP: The Open Web Application Project – OWASP Top Ten
27. Kozik, R., Choraś, M., Hołubowicz, W.: Evolutionary-based packets classification for anomaly detection in web layer. Secur. Commun. Netw. **9**(15), 2901–2910 (2016)

Lying Speech Characteristic Extraction Based on SSAE Deep Learning Model

Yan Zhou[1,2(✉)], Heming Zhao[2], and Li Shang[1]

[1] School of Electronics and Information Engineering,
Suzhou Vocational University, Suzhou 215104, Jiangsu, China
zhyan@jssvc.edu.cn
[2] School of Electronics and Information Engineering, Suzhou University,
Suzhou 215100, Jiangsu, China

Abstract. Lie speech detection is a typical psychological calculation problem. As the lie information is hidden in speech flow and cannot be easily found, so lie speech is a complex research object. Lie speech detection is not only need to pay attention to the surface information such as words, symbols and sentence, it is more important to pay attention to the internal essence structure characteristics. Therefore, based on the study of speech signal sparse representation, this paper proposes a Stack Sparse Automatic Encoder (SSAE) deep learning model for lying speech characteristics extraction. The proposed method is an effective one, it can reflect people's deep lying characteristics, and weaken lying person's personality traits. The deep characteristics compensate the lack of lie expression of basic acoustic features. This improved the lying state correct recognition rate. The experimental results show that, due to the introduction of deep learning characteristics, the individual lying recognition rate has increased by 4%–10%. This result suggests that, the lie detection based on speech analysis method is feasible. Furthermore, the proposed lying state detection based on speech characteristic provides a new research way of psychological calculation.

Keywords: Speech analysis · Characteristic exaction · Lie speech detection · SSAE · Deep learning

1 Introduction

Lying characteristic analysis [1, 2] from the speech signal is a novel and complex scientific problems. The extraction of lying characteristic is not only need to pay attention to the surface information, such as pronunciation of words, text symbols, but also should pay more attention to explore the recessive characteristics information which is contained in the lying speech signal. Gaussian Mixture Model (GMM) is a traditional speech processing method and be used a lot. But GMM is essentially a shallow model and cannot fully express the relationship between the speech characteristics and the lying state space. Furthermore the internal class distinction performance of GMM model is also not strong. Therefore, lying speech feature analysis cannot be limited of traditional mode, such as acoustic features, rhythm features. It

© Springer International Publishing AG 2017
D.-S. Huang et al. (Eds.): ICIC 2017, Part III, LNAI 10363, pp. 672–681, 2017.
DOI: 10.1007/978-3-319-63315-2_59

needs to analysis the structure of a large number of speech data through the machine learning algorithm. Machine learning [3–5] is a big data analysis method, and can extract the internal characteristic structure information, that is implied in the lying speech data. Thus, the different characteristics between lying speech and common speech can be significantly expressed.

Deep Learning (DL) algorithm is one of the most effective machine learning methods. It was proposed by Hintonn et al. [6, 7], which is inspired by the brain perceptual system. DL algorithm is a kind of unsupervised data analysis methods. It is been widely used for feature extraction of the data with potential complex structure rules, such as video signal, voice signal and music signal. So DL model [8] is an immense improvement compared with the traditional shallow learning model, such as Boosting algorithm, SVM algorithm and Logistic Regression algorithm, etc. Recently, the research of DL takes more attention of many scholars [9–11], since 2011, the speech researchers of Google and Microsoft have adopted the Deep Neural Network (DNN) technology to reduce the error speech recognition rate by 20%–30%. In 2013, the Baidu CEO robin Li has planned to set up Baidu deep learning institute, which is committed to Baidu web search in the aspects of speech recognition. In 2016, the company of Keda Xunfei has launched speech interaction artificial intelligence open platform. This platform realized the artificial intelligence breakthrough from "listen and speak" to "understand the thinking" through the combination of DL neural network and large training data. Due to the strong ability of data deep learning, some researchers use DL algorithm in the area of speech calculation. For example, Chen [12] proposed the method of Chinese speech recognition based on the deep learning characteristics. Jiang et al. [13] put forward the scheme of binaural classification for reverberant speech segregation using deep neural networks.

Because the perception of lie is a process that starts from the exterior learning, then to the interior learning, it is difficult to reflect people's state of mind only by a word or a phrase. So it is need to set up a layered model to describe the lie characteristics. Here, the process from production to perception of the lie speech can be seen as a speech chain. The lying speech chain approximately can be seen as five layers: the language layer, the physical layer, the acoustic layer, the rhythmic layer and the psychological perception layer. In fact, the essence of the whole speech communication is a dynamic process, and the variables of the layers are closely related. Moreover, they are associated by the potential control mechanism. As be considered above, the SSAE deep learning model is proposed in this paper to learning the lie speech essential characteristics. The number of layer is set for 5, which is corresponding to the five speech chain layers. The diagram of the relations between lie speech characteristics and the layered structure of DL model is shown in the Fig. 1. The nature of lying characteristics can be depicted step by step through using the unsupervised deep learning algorithm, and the obtained characteristics can get the promotional and expressive performance.

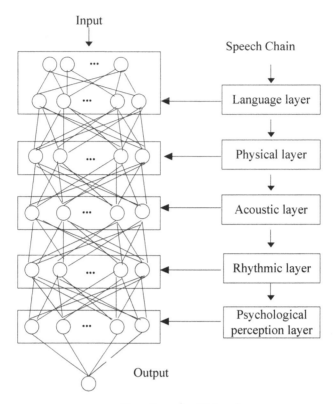

Fig. 1. Deep learning model for characteristic extraction.

2   The SSAE Deep Learning Model

The SSAE deep learning model is proposed by Bengio et al. It was successfully put forward on the basis of Deep Belief Network (DBN). SSAE deep learning model can reconstruct signal through the characteristics extracted by the automatic decoder. It is an unsupervised training method by using the approximate optimal conditions between the reconstruct characteristics and the original characteristics. Moreover, the SSAE deep learning model don't need marked training samples when solves the task of high dimensional speech data classification. SSAE is composed of the Sparse Auto Encoder (SAE) basic unit. SAE is a kind of auto encoder that use sparse coding algorithm. Through speech signal sparse coding, it can not only obtain smaller storage space and more simple calculation, but also can obtain structure characteristic of signal more clearly.

SSAE is the superposition of multiple sparse automatic encoders. It trains sparse automatic encoders step by step based on the input data, finally, it complete the entire network training. As a result, through the comparison of the difference between input

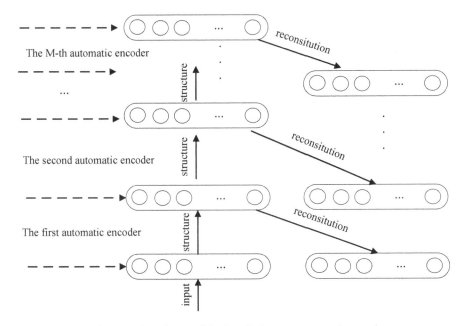

Fig. 2. Deep learning model of stacked sparse automatic encoders.

signal and output signal, it can judge the effectiveness of the characteristics trained by SSAE. The SSAE structure is shown in Fig. 2, which is stacked up by M sparse auto encoders. From the figure, we can see that two layers constitute a sparse automatic encoder, the output of each layer is used to reconstruct signal and also compared with the input signal.

The proposed SSAE deep model is a deep network adding sparse constraint, and directly build model for the hidden layer response. Through setting the constraint of each hidden layer unit, it makes the norm of hidden layer unit response very small. Because response value of each hidden layer is between 0–1, so the sparse response of the hidden layer can be obtained. The training objective function of SSAE can be expressed as following.

$$J_{\min}(W, c, b) = -\sum_{l=1}^{m} \log \sum_{h} \rho(x(l), h(l)) + \lambda \sum_{j=1}^{K} \left| \rho - \frac{1}{m} \sum_{l=1}^{m} E[h_j(l)|x(l)] \right|^2 \quad (1)$$

In the formula above, the $\rho \in (0, 1)$ represents the response expectation of the hidden layer units. Because it needs to make the response of hidden layer unit to be the most sparsest one, so the value ρ is usually relatively small, the $\sum_{h} \rho(x(t), h(t))$ represents the appearance probability of the l-th sample $x(l)$. The $h(l)$ represents the response of hidden layer. The $E[h(l)|x(l)]$ represents the expectation response of the j-th hidden layer, and m represents the number of samples.

In the process of SSAE training, the algorithm trains one layer every time, that is to say, the training network only contains a hidden layer. At the beginning, the first AE is trained by the input data. The training goal is to getting the minimum error between the input signal and the reconstruction of sparse decomposition. After the first layer training, put the output of the first layer as the input of the second AE layer. That is, the output of M-1th layer is as the input for the M-th layer. According to this repeated method, the algorithm executed until all the training AEs are finished.

3   Experimental Results and Analysis

3.1   Experiment Set

In this paper, the experimental samples are taken from the Soochow University Lie Database (SULD). Here, thirty testers are chosen and everyone for 5 different paragraphs respectively. The pretreated processing concluding speech signal segmentation, adding window, and performing the time-domain speech signal converted into the corresponding spectrogram. In order to reduce the high dimension of lying speech, here, the Principal Component Analysis (PCA) algorithm is applied. At last, randomly intercepting small spectrograms form the original spectrograms as the training samples of every tester.

In order to compare the recognition rate between the traditional phonetic characteristics and the characteristics automatically extracted by the proposed SSAE deep learning algorithm. Using the open EAR open source toolkit to calculate traditional phonetic characteristics, including pitch frequency, zero crossing rate, MFCC parameters and $\Delta MFCC$ 24 dimension parameters, and seven statistics of Teager energy operator, in addition, plus two psychoacoustics parameters of roughness and fluctuation strength, so the total traditional phonetic parameters are 36 dimensions. Moreover, the Support Vector Machine (SVM) is applied as the identification model. This paper is aim to implement the exaction of lying characteristic from speech signal that is based on the SSAE deep learning network. The specific steps are as following:

Step 1: Preparing the unsigned lie speech. Firstly, do the endpoint detection, the window size is 3 ms, overlapping is 10 ms, the total training number of everyone is for 10000.

Step 2: Selecting a certain amount of weight matrix randomly in the process of deep learning to realize the visualization and analysis the weight distribution range of the deep learning algorithm.

Step 3: Setting the layer number of deep learning network is 5, and setting the number of hidden layer node is 1024. The size of the sample data block is set as $S = 6 \times 6$. The trained network weight is W and the offset is b, and the Sigmoid function is used to be the activation function of deep network.

Step 4: Using the bottom-up unsupervised greedy algorithm to train the speech samples step by step. Sparse encoder of each layer automatically unsupervised learning feature vectors of this layer, and as the input signal for the top layer.

Step 5: According to the hierarchical structure of speech feature recognition, extracting the language layer, physical layer, acoustic layer, rhythmic layer and psychological perception layer in turn.

Figure 3 shows a training sample of 4 s length and the corresponding spectrogram.

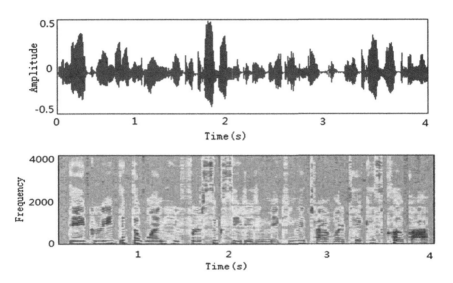

Fig. 3. The randomly selected original training speech signal.

3.2 Experiment Results

After a large amount of data learning, the lie deep learning network model can be obtained. In Tables 1 and 2, there are recognition rate results based on three kinds of feature sets, they are Basic Acoustic Feature Set (BAFS), Deep Feature Set (DFS) and the Fusion of BAF and DF Set (FBDS) respectively. In order to analysis the lying state recognition based on different phonetic features more clearly, Figs. 4 and 5 intuitively express the lying psychological state identification results.

Table 1. Male lie psychological state correct recognition rate (%)

Feature set	Testers														
	1	2	3	4	5	6	7	8	9	10	11	12	13	14	15
BAFS	56.2	55.4	59.3	50.3	58.5	49.7	50.3	60.4	60.9	50.4	55.3	63.7	48.3	49.9	59.4
DFS	59.4	58.3	63.3	54.3	64.4	50.6	51.2	67.4	68.3	50.3	58.5	65.4	54.5	55.6	67.4
FBDS	61.3	59.5	64.7	57.4	66.6	55.4	57.3	68.3	70.9	54.3	59.8	74.6	55.7	58.6	69.5
Feature Set	Testers														
	16	17	18	19	20	21	22	23	24	25	26	27	28	29	30
BAFS	50.3	56.4	59.3	50.6	47.9	51.2	53.8	50.4	60.5	64.3	59.4	55.9	57.8	60.3	65.5
DFS	56.4	65.1	65.5	55.3	54.5	56.4	58.5	58.2	64.3	70.5	67.4	58.5	67.4	69.8	67.9
FBDS	57.7	68.9	68.5	56.4	54.7	61.3	61.5	65.2	70.4	75.8	68.3	66.9	68.4	69.8	76.5

Table 2. Female lie psychological state correct recognition rate (%)

Feature set	Testers														
	1	2	3	4	5	6	7	8	9	10	11	12	13	14	15
BAFS	55.2	53.2	58.3	46.3	58.0	47.9	56.4	59.4	60.4	49.3	57.2	50.2	43.2	56.3	60.3
DFS	56.3	59.2	59.4	55.7	50.4	50.6	60.3	51.3	67.4	50.3	58.6	58.1	49.3	56.5	67.4
FBDS	60.4	63.2	58.4	64.3	56.9	64.5	60.7	65.4	72.3	65.4	63.7	56.4	51.4	60.4	71.4
Feature set	Testers														
	16	17	18	19	20	21	22	23	24	25	26	27	28	29	30
BAFS	41.2	56.4	60.3	50.3	46.8	63.1	55.3	50.2	58.3	50.7	60.6	57.3	66.4	67.3	48.3
DFS	55.8	50.4	67.4	54.3	50.6	67.4	58.3	56.4	59.3	60.3	60.5	58.4	68.4	65.4	50.6
FBDS	60.3	63.2	70.4	64.3	60.5	73.7	61.9	60.5	60.5	65.8	66.1	59.6	74.6	75.3	64.1

3.3   Experiment Analysis

From the experimental data we can see that, the recognition rate by using the fusion of basic characteristic and deep characteristic is higher than the other two feature sets, and the lying state correct recognition rate of men is higher than women. The specific explains are as follows:

(1) The Tables 1 and 2 show that, under the basic acoustic characteristics, the average recognition rate of male is 55.7%, and the average recognition rate of female is 54.8%. Under the deep characteristics, the average recognition rate of male is 60.8%, and the average recognition rate of female is 57.8%. Compared to the basic acoustic characteristics, the average recognition rate based on deep characteristics have increased by 5.1% and 3.0% respectively. These results indicate that, the deep characteristics can essentially excavate the details of the structure characteristics of lying speech, so adopting the deep learning for speech signal is more suitable for lying speech detection.

(2) The test results in Figs. 4 and 5 show that, under the fusion of basic characteristic set and deep characteristic set, the highest correct recognition rate of male is

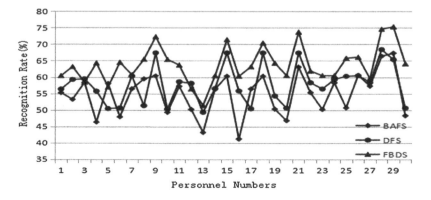

Fig. 4. Lying recognition correct rate of male under different characteristics.

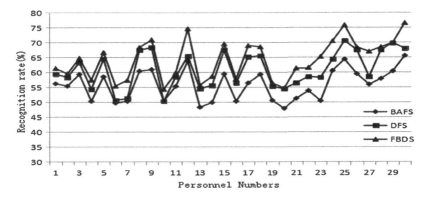

Fig. 5. Lying recognition correct rate of female under different characteristics.

76.5%, and the lowest correct recognition rate is 54.3%, Compared to the deep characteristics set, the highest and lowest correct recognition rate have improved by 11.0% and 6.4% respectively. The highest correct recognition rate of female is 75.3%, and the lowest correct recognition rate is 51.4%, Compared to the deep characteristics, the highest and lowest correct recognition rate had improved by 8.0% and 10.2% respectively. This indicates that, the fusion of deep learning characteristics and the basic acoustic parameters can obviously improve the lying speech recognition rate.

(3) The average correct recognition rate results of male are as follows: the average recognition rate is 55.7% based on the basic acoustic parameters, the average recognition rate is 60.8% based on the deep learning characteristic parameters, and the average recognition rate is 64.2% based on the fusion of basic characteristic and deep characteristic. The average recognition rate of male is 60.2% under the three different conditions. In addition, the average correct recognition rate results of female are as follows: the average recognition rate is 54.8% based on the basic acoustic parameters, the average recognition rate is 57.8% based on the deep learning characteristic parameters, and the average recognition rate is 63.8% based on the fusion of basic characteristic and deep characteristic. The average recognition rate of female is 58.8% under the three different conditions. As we can see that, the average recognition rate of male is higher than female by 1.43%. This suggests that when men tell lies, it is prone to panic, and the nervousness can be reflected through the change of the phonetic features, therefore, male lying state is more easily to be identified.

From the above data analysis, it is feasible to realize the lying psychological state based on the speech analysis. But the lying recognition based on speech is a complexity research, it has a gap between the lying state recognition rate performed in this paper and the practical engineering application, so speech lie detection research is still in the exploratory stage.

4 Conclusion

This paper adopts the method of SSAE deep learning nonlinear network to extract the lie speech characteristics. Through the transformation of characteristics step by step, the psychological characteristics of lying speech can be classified and distributed. Finally, the inner structural information of implied psychological characteristics can be found. The deep lying features can reflect the common characteristics of lying people, and weakening the lying person's personality traits. Moreover, it can improve the lying state recognition rate by combining the deep learning characteristics and the basic acoustic characteristics. The experimental results show that, the lie speech recognition based on the deep learning is an effective analysis method. The SSAE deep learning method proposed in this paper is not only the effective complement to the traditional basic acoustic feature extraction model, but also provides a new research approach for the research of psychological state recognition based on speech analyzing. From the perspective of the existing research results, although the speech feature extraction based on deep learning had achieved some results, but there are still some limitations, for example, it is difficult to estimate the consumption of computing resources such as the parameters size of the training network model, and the number of training samples and computing complexity, et al. These are all the study direction in the future research.

Acknowledgments. This research is supported by: National Natural Science Foundations of China (Grant Nos. 61372146, 61373098). The Youth Found of Natural Science Foundation of Jiangsu Province of China (Grant No. BK20160361). The fund of Qinglan Project Young and Middle-aged Academic Leader of Jiangsu Province.

References

1. Throckmorton, C., Handra, S., William, J.: Financial fraud detection using vocal, linguistic and financial cues. Decis. Support Syst. **74**, 78–87 (2015)
2. Elliott, E., Leach, A.M.: You must be lying because i don't understand you: language proficiency and lie detection. J. Exp. Psychol. Appl. **22**(4), 488–499 (2016)
3. Zhang, X.L., Wu, J.: Deep belief networks based voice activity detection. IEEE Trans. Audio Speech Lang. Process. **21**(4), 697–710 (2013)
4. Srivastava, N., Salakhutdinov, R.: Multimodal learning with deep boltzmann machines. J. Mach. Learn. Res. **15**, 2949–2980 (2014)
5. Brahma, P.P., Wu, D.P., She, Y.Y.: Why deep learning works: a manifold disentanglement perspective. IEEE Trans. Neural Netw. Learn. Syst. **27**(10), 2496–2947 (2015)
6. Hinton, G.E., Li, D., et al.: Deep neural networks for acoustic modeling in speech recognition: the shared views of four research groups. Sig. Process. Mag. **29**(6), 82–97 (2012)
7. Hinton, G.E., Osindero, S., Teh, Y.W.: A fast learning algorithm for deep belief nets. Neural Comput. **18**(7), 1527–1554 (2006)
8. Li, X.G., Yang, Y.N., Pang, Z.H.: A comparative study on selecting acoustic modeling units in deep neural networks based large vocabulary Chinese speech recognition. Neurocomputing **170**, 251–256 (2015)

9. Du, J., Tu, Y.H., Dai, L.R.: A regression approach to single-channel speech separation via high-resolution deep neural networks. IEEE-ACM Trans. Audio Speech Lang. Process. **24** (8), 1424–1437 (2016)

10. Jaitly, N., Hinton, G.E.: Using an auto encoder with deform-able templates to discover features for automated speech recognition. In: 14th Annual Conference of the International Speech Communication Association, Lyon, pp. 25–29 (2013)

11. Cai, M., Shi, Y., Liu, J.: Deep maxout neural networks for speech recognition. In: Workshop on IEEE International Proceedings on the Automatic Speech Recognition and Understanding, pp. 291–296 (2013)

12. Chen, M.M.: Fusion of various characteristics of Chinese speech recognition based on the technology of deep learning. Doctoral dissertation of University of Chinese Academy of Sciences, Beijing (2015)

13. Jiang, Y., Wang, D.L., Liu, R.S.: Binaural classification for reverberant speech segregation using deep neural networks. IEEE-ACM Trans. Audio Speech Lang. Process. **22**(12), 2112–2121 (2014)

A Novel Fire Detection Approach Based on CNN-SVM Using Tensorflow

Zhicheng Wang, Zhiheng Wang$^{(\boxtimes)}$, Hongwei Zhang, and Xiaopeng Guo

Research Center of CAD, Tongji University, 4800 Caoan Road, Jiading District, Shanghai 201804, People's Republic of China
1531707@tongji.edu.cn

Abstract. In this paper, we propose a novel approach to detect fire based on convolutional neural networks (CNN) and support vector machine (SVM) using tensorflow. First of all, we construct a large number of different kinds of fire and non-fire images as the positive and negative sample set. Next we apply Haar feature and AdaBoost cascade classifier to extract the region of interest (ROI). Then, we use CNN-SVM to filter the results of Haar detection and reduce the number of negative ROI. The CNN is constructed to train the dataset with four convolutional layers. Finally, we utilize SVM to replace the fully connected layer and softmax to classify the sample set based on the training model in order. Experimental results show that the method we proposed is better than other methods of fire detection such as CNN or SVM etc.

Keywords: Convolutional neural networks · Support vector machine · Haar feature · AdaBoost · Fire detection

1 Introduction

With the development of computer vision technology, the technology of fire detection by the camera has gradually applied in our life. The original temperature sensors and smoke sensors falls far short of some complex environments needs. Most of the researchers detect the fire by using the color characteristics and motion characteristics and the results got disappointing in the complex environment. It is significant in studying how to extract the characteristics of the weak fire, how to identify the weak fire accurately and how to prevent and control the fire alarm. Therefore, the researchers have proposed various methods of fire detection.

In Twentieth Century, the traditional device of fire detection is mainly composed of smoke sensor, temperature sensor and ion sensor which is based on the physical characteristics of the fire [1, 2]. These devices have the same problem, however, which is the sensors alarmed only after the smoke, temperature and other physical characteristics reached their threshold. And that makes the devices impractical in large space. Besides, the issue of installation cost which from the devices locations and numbers should be considered.

For the past few years, computer vision technology has been expanding rapidly. The fire detection system based on image processing requires no additional cost. The surveillance camera would be installed in public, high street and tunnels. Fire would be

© Springer International Publishing AG 2017
D.-S. Huang et al. (Eds.): ICIC 2017, Part III, LNAI 10363, pp. 682–693, 2017.
DOI: 10.1007/978-3-319-63315-2_60

detected easily through the camera instead of waiting the danger threshold shows up. The early fire detection algorithm was based on a fire color model [3–5]. Çelik and Demirel [6] proposed a method of classifying pixels by fire color models. Noda and Ueda [7] proposed a fire area segmentation technique and analyzed the histogram of the image. Nevertheless, these methods brings error-detecting for lots of the color information such as red clothes and lights are mistaken as fire. There are some other methods of detection. For instance, the color space conversion, which is from RGB to other color space such as HIS [8], HSV [9], LBP [10, 11], YCbCr [12] and YUV [13] etc. Compared with the RGB model, these methods reduce the number of false alarms, but still have some shortcomings and cannot be applied in industry.

As time goes on, the researchers at home and abroad combined the static or dynamic characteristics of fire and machine learning algorithms to improve the precision of fire detection. Toreyin et al. [14] and others separated the fire region from the sequence images by analyzing the time and space wavelet. The time domain wavelet transform is detected around the fire boundary. In addition, the color transform of each sequence images can be detected and the spatial wavelet transform is calculated by the region of the fire color transformation. Although they bring forward some better experimental results, it still be impractically in actual production considering the thresholds under different environment. In the paper [15], they first proposed a study of the fire object by iterative adaptive threshold technique. Then a fire detection algorithm [16] based on Bof in YUV color space was proposed subsequently. A detection algorithm [17] of combined with the characteristics of fire and human activities is proposed. But the fire shape is amorphism and affected by smoke frequently. Even though the above methods achieved certain results, these methods still exist some disadvantage. Usually they have low accuracy of detection, some of them haven't broad applications for the lack of samples, and some algorithms extract features got a weak anti-jamming capability. Hence, the fire detection method still need to conduct the deeper step research.

Owing to the deficiency of the fire dataset on the internet, we collected a large number of images which contain houses, forests, cars, parks and other places of the fire. These images are as shown in Fig. 1. The fire's size, color and flashing changes will be presented in the dataset. Then we collected the same scenes images without fire as the negative sample. Extracting training set and test set after distributing these data randomly and the ratio of them is 3 to 1.

The main steps of our method are as follows: Firstly, Haar feature detection and AdaBoost classifier are used to extract the ROI in each image, which reduce the extraction range of CNN. Secondly, the extraction results are classified as positive samples and negative samples of CNN. Then training CNN and saving the model. Thirdly, we replace the fully connection layer of the CNN by the SVM and keep on training the sample set. Finally, testing the test set with the trained model to obtain the recognition result of the fire.

Fig. 1. A sample of the data set for the collection of the fire

2   System Architecture

The method we proposed consists of two modules, as shown in Fig. 2. First of all, giving a large number of images with resize 32 × 32 size. The fire feature of these images are extracted by Haar feature detection and AdaBoost cascade classifier. The parameters of AdaBoost are obtained by supervised learning. After training the CNN model, we use SVM, which training data are the vector of the output of the fourth convolutional layer from the model, for classification instead of the CNN's fully connected layer.

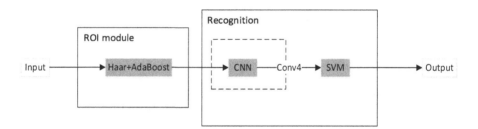

Fig. 2. The architecture of our proposed method

2.1   Reformative Convolution Neural Network

The traditional CNN consists of two convolutional layers and two subsampling layers. The result of the second subsampling layer is fed to the fully connection layers for classification. Compared to the traditional one, we propose an improved CNN architecture which includes an input layer, four convolutional layers, two subsampling layers and two fully connection layers. The input image size of the CNN is 32 × 32, and the convolution kernel size of the four convolutional layers is 3 × 3. The subsampling layer uses the pooling method of max pooling and preserves the edges of the feature map. The extra edges are filled with 0. The activation function is Relu, defined as: Relu $(x) = $ max $(x, 0)$.

In order to train the parameters better in the model and prevent training from overfitting, we bring in the regularization which is a very important technique in the machine learning area. Therefore, we use the L2 regularization method to optimize it. L2 regularization is defined as follows:

$$w^* = \operatorname{argmin} \sum_j \left(t(x_j) - \sum_i w_i h_i(x_j) \right)^2 + \lambda \sum_{i=1}^k w_i^2 \qquad (1)$$

Where $\lambda \sum_{i=1}^k w_i^2$ is regularization parameter. λ controls the balance between the penalty term and the mean square. The larger the λ, the smaller the w.

In order to update the network parameters weights randomly at each iteration, the dropout is introduced to enhance the randomness. When using the optimization algorithm for the cost function, we chose Adam optimizer [18] to optimize the network loss. It is generally recommended to reduce the learning rate in training progress of the model. So we add the Learning Rate Decay algorithm, as defined below:

$$LR = lr^* dr^{(gs/ds)} \qquad (2)$$

Where LR represents the learning rate; lr represents the initial learning rate; dr represents the decay rate; gs represents the global step size, and ds represents the decay step size. According to the above structure, the parameters in the CNN training. Then we save the trained model and pave the way for the SVM.

2.2 CNN and SVM Classifier

The method above preserves the trained CNN model, so the output of the last convolution layer can be obtained directly from the CNN model. The training process of the CNN is basically equal to its parameters in the process of continuous tuning, so we can get a better result as the parameters of the convolution layer get more optimize. Experiments show that if the results of the last convolution layer in the training process are used as inputs to SVM, the classification results will be far less than the above model. So, we propose a CNN-SVM architecture, as shown in Fig. 3. The process is as follow: First, we get the last convolution layer output from the preservation of the CNN model as the SVM classifier input. The radial basis function (RBF) kernel maps a sample to a higher dimension. Compared with the polynomial kernel function, RBF need to determine the parameters to be less, the number of kernel function parameters

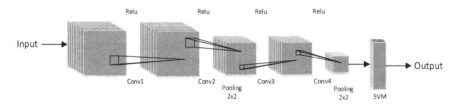

Fig. 3. Structure of the reformative CNN-SVM

directly affect the complexity of the function. Thus, the SVM classifier with the kernel function as the RBF is chosen here. Second, the SVM training and then save the model, for the detection of the fire.

2.3   The Ways of Fire Detection

According to the two steps above, we get the CNN-SVM model. We can detect the fire on the basis of the two models as shown in Fig. 4. Firstly, we get the ROI of the image by the Haar and AdaBoost cascade classifier with the Haar feature model has be trained first. It took us more than a month to train the model and we could omit statements here. The ROI in the image represents the result of the Haar feature, and it also has fire and non-fire region. But it cannot be applied in the actual scene because there exists a large number of non-fire regions. So we use CNN to filter those negative samples, and make it more practical to be used in industry. Secondly, the images of the Haar output will be filtered by using the trained CNN-SVM model. Saving the image with the fire and filtering out the non-fire image, then we can get a higher precision. Finally, the images that showed left over are the fire images. Real-time monitoring video can be used as the input of the architecture. When the camera monitors fire, the system in our method will trigger an alarm and do a certain fire prevention in the practical application of the scene. This specific process as shown in Fig. 5.

Fig. 4. The procedure of fire detection

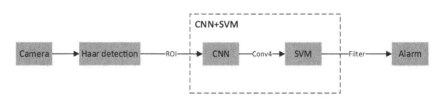

Fig. 5. The process of fire detection in real scene

3   Experiments

The experimental test environment is ubuntu16.04 64-bit operating system and 64 GB of memory. The processor (CPU) is Intel Core i7-6700 k CPU @ 4.00 GHz × 8. The graphics processor (GPU) is GeForce GTX 1080/PCIe/SSE2. The environment of development and experimental is tensorflow framework. And the opencv-2.4.13 library and python will be used in it.

3.1 Experiment Data

We constructed the experimental data and it is consisted of two parts. The first part is the dataset of Haar + AdaBoost feature extraction which including 28137 images. 20923 images of them are negative dataset and the others are positive dataset. Each size of the image is 32 × 32. We divided the training set into two categories and placed them in two folders. Label 0 represents the class without fire and label 1 illustrates the class which contains fire. Then we detect a variety of fire and non-fire video by Haar detection, and get 55197 images as a dataset of the CNN-SVM model. So the second part of the dataset contains 41300 images of the training dataset and the 13897 images of the test dataset. All the data is distributed randomly. Each image is also resized to 32 × 32. In the end, the dataset is generated into tensorflow input format. Through the iterator, the dataset is generated (64, 32, 32, 3) tensor shape. Then we set classification label by one-hot coding method, and get (64, 2) tensor shape for the CNN training. Next, we change the dataset tensor shape to (64, 3072). And then the classification label is set tensor shape (64, 1) for CNN-SVM training. There is no need to label the one-hot coding because we only require the output of the fourth convolution layer.

3.2 Region of Interest Extraction

The trained AdaBoost classifier usually gives a low precision in a ROI module, as shown in Fig. 6.

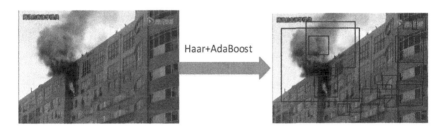

Fig. 6. The result of Haar + AdaBoost classifier

Therefore, the CNN-SVM is applied to improve the precision rates of fire detection. AdaBoost is based on the OpenCV platform. In the part of the training parameter setting, we set the nstages parameter to 20 which means 20 stages were trained. The Maxfalsealarm parameter is set to 0.5 that expect a false alarm rate about $0.5^{20} \approx 9.6e{-}07$. The minhitrate parameter is set to 0.999 and a hit rate about $0.999^{20} \approx 0.98$. And then we extract the ROI by the trained model.

3.3   The Models of Reformative CNN-SVM

This part is mainly to classify the results of the previous step and extract the fire im-ages accurately. The design of the convolution layer as shown in Fig. 7.

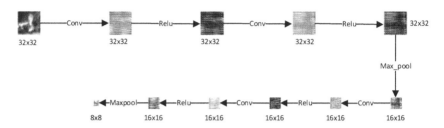

Fig. 7. The structure of convolution layer

First, we operate convolution on the image. Second, we execute the Relu activation function, and then back to the first step. Third, we put the result into max pooling and convolute it again. After that, we repeat the second step and do the last time max pooling on it. After all above, we get the final result of the convolution layer. In the process of training, by the back propagation, and constantly optimize the weights and biases of each layer. And then detect which weight loss is greatest and the loss is reduced, as shown in Fig. 8.

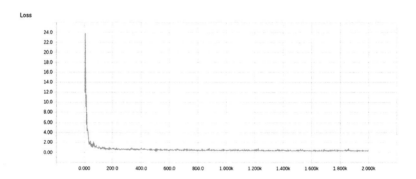

Fig. 8. The process of training loss

After saving the training model, we get the last convolution layer output by the training model and classify them with SVM. Then we save the model again. Finally, we test dataset with the trained model above and get a satisfactory result, as shown in Fig. 9.

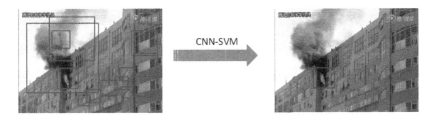

Fig. 9. The result of CNN-SVM

3.4 Running Time

Our method will be used in real-time detection of fire and the video data is processed by frame. Therefore, it is helpful to analyze the time of each module in the method to optimize the algorithm later. In the Haar detection module, the average time per frame in the video is about 0.082 s; In the CNN-SVM module, the average run time per frame in the video is about 0.495 s. The total running time is about 0.553 s. Be-cause each frame contains multiple ROI, and the detection time of CNN-SVM is changed by the number of ROI.

3.5 Detection Results

In this part of the paper, we analyze the results of CNN-SVM module to prove the validity of the method we proposed. Our method will be compared with one reformative CNN and SVM method. We will be using Receiver Operating Characteristic (ROC) metric to evaluate classifier output quality. The X axis of the ROC curve is false positive rate and the Y axis is true positive rate. The larger the area below the ROC curve (AUC) represents the better classification. The other way is by Precision-Recall metric to evaluate classifier output quality. In information retrieval, precision is a measure of the relevance of the results. The larger the area below the curve, both the higher the precision rate and the higher recall rate. Where high precision relates to a low false positive rate, and high recall relates to a low false negative rate. Therefore, the detection method is also the larger the area, the better classification. The experiment used 13897 images as a test set. The ROC curves of the three methods are shown in Fig. 10.

The Precision-Recall curves of the three methods are shown in Fig. 11.

We compare the three methods and the AUC value is 99.72% in our method. The AUC value of the reformative CNN is 95.79%. The AUC value of SVM is 61.68% as shown in Table 1. In order to achieve a better evaluation, we also evaluate the Precision-Recall of these methods. These Precision-Recall values are 99.14%, 90.71% and 65.57% as shown in Table 2. The results of these methods prove the reformative CNN architecture is better than the SVM model. Our method gives a better model than the previous two methods. Then, we show some results of our experiment in Fig. 12. In order to reduce the space complexity of our method, each 60 images as the input of the CNN by the iterator. This approach improves the efficiency of training and reduces space complexity, as compared to putting all the images together as input.

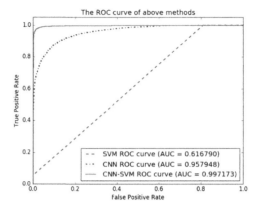

Fig. 10. The ROC curves of the three methods

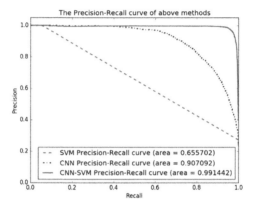

Fig. 11. The Precision-Recall curves of the three methods

Table 1. The AUC values of the three methods.

Method	AUC
Our method (CNN-SVM)	99.72%
One reformative CNN	95.79%
SVM	61.68%

Table 2. The value of the Precision-Recall

Method	Value
Our method (CNN-SVM)	99.14%
One reformative CNN	90.71%
SVM	65.57%

Fig. 12. Some examples of fire detection by our method

4 Conclusions

In this paper, we propose an effective way to detect the fire. First, the Haar feature and the AdaBoost classifier are used to extract the ROI. Second, the trained CNN-SVM model is used to detect ROI and filter out the negative ROI. In addition, a CNN-SVM model is proposed to improve the precision of recognition. The experimental results are based on the dataset that we built. Our method is compared with SVM and one reformative CNN method to prove the precision of our method. Then, we analyzed the various parts of the experiment in order to improve the method and do the relevant work in the future.

Acknowledgments. The authors would like to express their gratitude to the anonymous reviewers for their valuable comments and suggestions to improve the quality of this paper. This work is supported by the National Science and Technology Support Project of China under Grant 2015BAF04B01 and Science and Technology Innovation Project of Shanghai under Grant 16111107602.

References

1. Ryser, P., Pfister, G.: Optical fire and security technology: sensor principles and detection intelligence. In: International Conference on Solid-State Sensors and Actuators, pp. 579–583. IEEE (1991)
2. Kaiser, T.: Fire detection with temperature sensor arrays. In: IEEE International Carnahan Conference on Security Technology, pp. 262–268. IEEE (2000)
3. Chen, T.H., Wu, P.H., Chiou, Y.C.: An early fire-detection method based on image processing. In: International Conference on Image Processing, vol. 3, pp. 1707–1710. IEEE (2004)
4. Horng, W.B., Peng, J.W., Chen, C.Y.: A new image-based real-time flame detection method using color analysis. In: Networking, Sensing and Control, pp. 100–105. IEEE (2005)
5. Liu, C.B., Ahuja, N.: Vision based fire detection. In: International Conference on Pattern Recognition, vol. 4, pp. 134–137. IEEE (2004)
6. Çelik, T., Demirel, H.: Fire detection in video sequences using a generic color model. Fire Saf. J. **44**(2), 147–158 (2009)
7. Noda, S., Ueda, K.: Fire detection in tunnels using an image processing method. In: Vehicle Navigation and Information Systems, pp. 57–62. IEEE (1994)
8. Yunyang, Y.A.N., Xiyin, W.U., Jing, D.U., et al.: Video fire detection based on color and flicker frequency feature. J. Front. Comput. Sci. Technol. **8**(10), 1271–1279 (2014)
9. Yamagishi, H., Yamaguchi, J.: Fire flame detection algorithm using a color camera. In: Proceedings of International Symposium on Micromechatronics and Human Sciene, pp. 255–260. IEEE (1999)
10. Tian, H., Li, W., Ogunbona, P., et al.: Smoke detection in videos using non-redundant local binary pattern-based features. In: IEEE International Workshop on Multimedia Signal Processing (MMSP), pp. 1–4. IEEE (2011)
11. Yuan, F.: A double mapping framework for extraction of shape-invariant features based on multi-scale partitions with AdaBoost for video smoke detection. Pattern Recogn. **45**(12), 4326–4336 (2012)
12. Celik, T., Demirel, H., Ozkaramanli, H.: Automatic fire detection in video sequences. In: European Signal Processing Conference (SIPCO 2006), vol. 6, no. 3, pp. 1–5. IEEE (2006)
13. Marbach, G., Loepfe, M., Brupbacher, T.: An image processing technique for fire detection in video images. Fire Saf. J. **41**(4), 285–289 (2006)
14. Töreyin, B.U., Dedeoğlu, Y., Güdükbay, U., et al.: Computer vision based method for real-time fire and flame detection. Pattern Recogn. Lett. **27**(1), 49–58 (2006)
15. Wang, W., Zhou, H.: Fire detection based on flame color and area. In: IEEE International Conference on Computer Science, vol. 3, pp. 222–226. IEEE (2012)

16. Liu, Z.G., Zhang, X.Y., Yang-Yang, et al.: A flame detection algorithm based on bag-of-features in the YUV color space. In: International Conference on Intelligent Computing and Internet of Things, pp. 64–67. IEEE (2015)
17. Wu, J.: Real-time visual detection of early manmade fire. In: IEEE International Conference on Information, pp. 993–997. IEEE (2015)
18. Kingma, D., Ba, J.: Adam: a method for stochastic optimization. In: The 3rd International Conference for Learning Representations, San Diego (2015)

A Hybrid Learning Algorithm
for the Optimization of Convolutional Neural
Network

Di Zang[1,2(✉)], Jianping Ding[1,2], Jiujun Cheng[1,2],
Dongdong Zhang[1,2], and Keshuang Tang[3]

[1] Department of Computer Science and Technology, Tongji University,
Shanghai, China
zangdi@tongji.edu.cn
[2] The Key Laboratory of Embedded System and Service Computing,
Ministry of Education, Tongji University, Shanghai, China
[3] Department of Transportation Information and Control Engineering,
Tongji University, Shanghai, China

Abstract. The stochastic gradient descend (SGD) is a prevalence algorithm used to optimize Convolutional Neural Network (CNN) by many researchers. However, it has several disadvantages such as occurring in local optimum and vanishing gradient problems that need to be overcome or optimized. In this paper, we propose a hybrid learning algorithm which aims to tackle the above mentioned drawbacks by integrating the methods of particle swarm optimization (PSO) and SGD. To take advantage of the excellent global search capability of PSO, we introduce the velocity update formula which is combined with the gradient descend to overcome the shortcomings. In addition, due to the cooperation of the particles, the proposed algorithm helps the convolutional neural network dampen overfitting and obtain better results. The German traffic sign recognition (GTSRB) benchmark is employed as dataset to evaluate the performance and experimental results demonstrate that proposed method outperforms the standard SGD and conjugate gradient (CG) based approaches.

Keywords: Particle swarm optimization · Convolutional neural network · Hybrid learning algorithm · Stochastic gradient descent

1 Introduction

After years of development, from the initial relatively simple handwriting character recognition applications, the Convolutional Neural Network (CNN) algorithm has gradually extended to some of the more complex areas, such as: pedestrian detection [1], behavior recognition [2], human body posture recognition [3], and so on. Regarding the CNN training, as we all know, the prevalent method is Back Propagation (BP) algorithm which is used for parameters tuning.

However, back propagation algorithm has a number of disadvantages when used alone. For example, BP algorithm is a local search optimization method, when it is used to solve a complex nonlinear problem, the network weight is gradually tuned in the

© Springer International Publishing AG 2017
D.-S. Huang et al. (Eds.): ICIC 2017, Part III, LNAI 10363, pp. 694–705, 2017.
DOI: 10.1007/978-3-319-63315-2_61

direction of loss decrease, which always makes the algorithm occur in local optimum, resulting in network training failure. The algorithm also has vanishing gradient problems, in some deep neural networks, the gradient tends to get smaller as we move backward through the hidden layers. This means that neurons in the earlier layers learn much more slowly than neurons in later layers. Additionally, the same learning rate applies to all parameters tuning in BP algorithm. If our data is sparse and our features have very different frequencies, we might not want to tune all of them to the same extent, but perform a larger tuning for rarely occurring features.

Currently, there have been several algorithms that are widely used by the deep learning community in CNN to deal with the aforementioned disadvantages. Momentum [4] is a method that helps accelerate SGD in the relevant direction and dampens oscillations. It does this by adding a fraction γ of the update vector of the past time step to the current update vector. Nesterov accelerated gradient (NAG) [5] which is based on momentum adds a Nesterov term, making correction when update gradient, to avoid moving too fast and improve the result. Adagrad [6] is an algorithm for gradient-based optimization which adapts the learning rate to the parameters, performing larger updates for infrequent and smaller updates for frequent parameters. Adadelta [7] is an extension of Adagrad that seeks to reduce its aggressive, monotonically decreasing learning rate. RMSprop [8] is an adaptive learning rate method proposed by Geoff Hinton. RMSprop and Adadelta have both been developed independently around the same time stemming from the need to resolve Adagrad's radically diminishing learning rates. Adaptive Moment Estimation (Adam) [9] is another method that computes adaptive learning rates for each parameter. All the above mentioned works use Stochastic Gradient Descent (SGD).

However, in this paper, to address the mentioned disadvantages of BP algorithm, a hybrid algorithm is proposed for CNN training. Particularly, because of the excellent global search capability of Particle Swarm Optimization (PSO), it is introduced to carry on training and combined with Stochastic Gradient Descent (SGD) to achieve better results. The combined algorithm proposed contributes to avoid occurring in local optimum, also it can overcome the premature saturation and sluggishness, which improves experiment results. Furthermore, PSO algorithm has many particles, each of them works out variant results, we can pick out the best particles through specific mechanism to reduce overfitting [10].

The rest of the paper is organized as follows, in Sect. 2 a brief introduction of introducing PSO is presented. Then in Sect. 3, the proposed approach is described in details. In both Sects. 4 and 5, GTSRB datasets is used for model evaluation and conclusions are depicted respectively.

2 Particle Swarm Optimization (PSO)

PSO is a heuristic global optimization which is proposed by Kennedy and Eberhart [11, 12] in 1995. It was inspired by the movement and clustering of organisms in a bird flock. Each solution of optimization problem is a bird in search space, called "the particle". In another word, the position of each particle is a potential solution. PSO randomly initializes the position and velocity of a group of particles, where its amount is called population size m, and usually the amount is set from 20 to 40. The position of the i-th particle in the d-dimensional space is expressed as $x_i = \left(x_i^1, x_i^2, \ldots, x_i^d\right)$ $(i = 1, 2, \ldots, m)$, while the velocity is $v_i = \left(v_i^1, v_i^2, \ldots, v_i^d\right)$ that determines the displacement in search space for a single iteration, and d is the amount of independent variables in the actual problem.

We evaluate the particles by their fitness value. In addition, according to their fitness, particles refresh their own best historical position $pbest$ and the best position of whole swarm, called $gbest$. If the actual problem pursues the minimum, then the smaller fitness is optimal. Each iteration,

$$if\ fitness_i < pbestfitness_i, \quad then\ pbestfitness_i = fitness_i,\ pbest_i = x_i.$$

$$if\ fitness_i < gbestfitness, \quad then\ gbestfitness = fitness_i,\ gbest = x_i.$$

where $fitness_i$ and $pbestfitness_i$ denote the fitness of the particle i and its best historical fitness, respectively, $gbestfitness$ is the best fitness of whole swarm. The following formulas are used to update velocity and position of particles:

$$v_i^{t+1} = v_i^t + c_1 r_1 \left(pbest_i - x_i^t\right) + c_2 r_2 \left(gbest - x_i^t\right) \tag{1}$$

$$x_i^{t+1} = x_i^t + v_i^{t+1} \tag{2}$$

where v_i^t and x_i^t denote the velocity and position of the particle i at moment t, respectively; c_1 and c_2 are accelerating factors, r_1 and r_2 are random numbers between [0,1]. The second formula consists of three parts, the first part is inertia or momentum, which reflects the habit of movement on particle and represents the tendency of the particle to maintain its previous speed; the second part is cognition, representing the memory of their own historical experience; the last part is the social part, reflecting the collaboration and sharing knowledge among the particles and representing the trend to approaching the global best position.

3   Proposed Optimization Algorithm and Model Architecture

Since SGD has several disadvantages, so in this paper, a robust hybrid training algorithm is proposed. The optimization algorithm is combined with PSO and SGD for CNN training, so it is called PSO-CNN, which helps overcome the deficiencies of SGD. In this approach, the unified PSO and SGD algorithms can be critical to achieve superior results and surpass the previous approach, and the PSO is recruited as revival constituent, since the benefits of SGD remain maintained. For example, instead of running one particle representing the entire CNN parameter, multiple particles are used and dispersed within the search space.

In CNN, parameter tuning follows the formula $P = P - \eta \frac{\partial E}{\partial P}$, where η is the learning rate and $\frac{\partial E}{\partial P}$ denotes the gradient of parameters. Inspired by the velocity formula of PSO given by (1), we propose the improved update expression of hybrid algorithm:

$$P_k^{t+1} = P_k^t - \eta \frac{\partial E}{\partial P_k^t} + c_1 r_1 \left(pbest_k - P_k^t\right) + c_2 r_2 \left(gbest - P_k^t\right) \tag{3}$$

where P_k^t denotes the parameters of particle k at moment t, namely, position; similarly, $pbest_k$ is the best position for the particle k, $gbest$ is the best particle in the whole swarm space, c_1, c_2, r_1 and r_2 are the same as Eq. (1). Keeping the gradient term in Eq. (3) is to help the particle move in the right direction towards the optimum. Meanwhile, the introduction of $pbest$ term and $gbest$ term bring into play the advantages of PSO, enhancing the global search ability and reducing oscillation when particles cross the slopes of the ravine. In the proposed algorithm, we use the cross-entropy cost function as fitness $E_k = -\frac{1}{N}\sum_{i=1}^{N}[y \, ln\, a + (1 - y)\, ln\, (1 - a)]$, where N is the total number of items of training data, the sum is over all training inputs, y and a are desired output and real output.

In addition, all particles cooperate with the subtle methods are described in Fig. 1, it shows the procedure that training phase uses the PSO-CNN. First, we initialize network structure mentioned below and parameters as particles, and so on. Then, we model evaluation for all particles. After that, we update parameters according to Eq. (3) and replace $pbestfitness$, $gbestfitness$, $pbest$, $gbest$. Unless condition meet, the training phase is always follow the diagram.

CNN generally consists of alternative two main layers called convolution and max-pooling layer and ends up with fully connected layer. All these layers are connected to each other with weights. However, there are many different other CNN architectures. In this study, the same structure proposed by LeCun et al. [13] is used, as shown in Fig. 2. The CNN structure is 8C-8S-12C-12S-16C-16S-5F, where C stands for Convolution layer, S is for subsampling layer, and F is for full connected layer. The number before C or S denotes the number of feature maps, and 5F is a fully connected layer with 5 outputs.

So how can the CNN parameters be encapsulated into particles? How do the parameters be initialized? How can it justify the best particles and dampens overfitting? These ambiguous steps need to be clarified.

It is obvious that the weights and biases are constituent parameters of CNN. Therefore, in this work, the weights and bias are dismantled and encapsulated into vectors as shown below:

$$W = \{w_1, w_2, \ldots w_l, \ldots, w_L\} \tag{4}$$

$$B = \{b_1, b_2, \ldots b_l, \ldots, b_L\}\, l = 1 \ldots L \tag{5}$$

where l is the layer index, L is the total number of layers, w_l is the weight parameters of layer l, and b_l is the bias parameters of the layer l. Finally, the final total parameters of bias and weights are given by

$$P_k = \left\{w_1^k, b_1^k, w_2^k, b_2^k \ldots w_l^k, b_l^k, \ldots w_L^k, b_L^k\right\} \tag{6}$$

698   D. Zang et al.

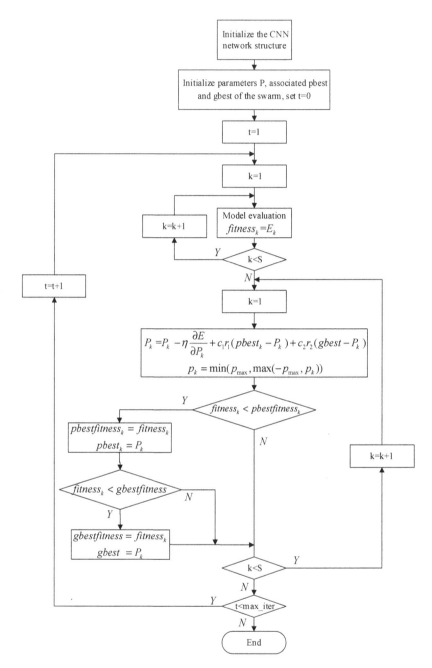

Fig. 1. The diagram of learning using PSO-CNN, where S refers to size of swarm, k is particle's id counter from 1 to S, P_k is k-th particle's parameters, max_iter is the maximum iterations and t is iteration counter from 1 to max_iter.

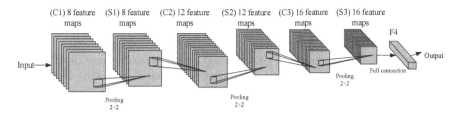

Fig. 2. The used CNN structure ⁻

where P^k is the parameters of particle k, namely, position. In convolution layers of CNN, there are set of filters and each has dimensions n × n and can be vectored to be $1 \times n^2$. Thus, having m filters for l th layer, then the total weight parameters are $w_l^n = mn^2$. In addition, the total bias parameters for the given lth layer are $b_l^n = 1 \times m$.

Although the particles are usually randomly initialized, in this paper they are initialized with Xavier method [14]. The implementation of Xavier initialization is the following uniform distribution:

$$W \sim U\left[-\frac{\sqrt{6}}{\sqrt{fan_{in}+fan_{out}}}, \frac{\sqrt{6}}{\sqrt{fan_{in}+fan_{out}}}\right] \tag{7}$$

where W is the weight parameters, fan_{in}, fan_{out} are the input parameters and output parameters, respectively. And B is initialized by random numbers between [0,1]. Also, in the training phase, the particles follow the boundary law that each parameter is limited in $[-p_{max}, p_{max}]$, where p_{max} is given by the specific situation of the experiment.

Since there are particles that will be trained, each one of them could be the best one among the swarm and can give an optimal solution. In order to justify the best particle among swarm in the training phase, the following notion is used: $P = arg\,min\,E$, where E is the fitness mentioned in Sect. 3. However, usually, the particle with minimum fitness E may not emerge the best performance in the testing phase, and it happens because the entire *pbest*s of swarm are clustered around the optimal particle which may be a little overfitting while some other particle may have the better performance. To obtain the optimal parameters in the experiment, the following equation is introduced: $P = arg\,max\,A$, where A is the classification accuracy of each *pbest* during testing phase: $A = 1 - \frac{bad}{M}$, where *bad* is the number of samples that are misclassified and M is the total number of items of testing data. At the end of experiment, P may be a *pbest* with highest accuracy which contains all parameters of network we want to obtain.

4   Experimental Results

The algorithm is evaluated on part of German traffic sign recognition benchmark (GTSRB) datasets [15]. Traffic sign recognition is a multi-class classification problem with unbalanced class frequencies. Traffic signs can provide a wide range of variations between classes in terms of color, shape, and the presence of pictograms or text.

However, there exist subsets of classes (e.g., speed limit signs) that are very similar to each other. The classier has to cope with large variations in visual appearances due to illumination changes, partial occlusions, rotations, weather conditions, etc.

In this paper, five categories of the most signs were selected, these categories are the most representative for GTSRB and covers the above variations. The dataset consists of 3700 samples, 2700 samples are used for training and each category has 540 pictures, the rest is used for testing and each category has 200 pictures. All the pictures have the various size, so we need to transform them into the same size, which is 28 × 28 pixels, samples for the datasets are shown in Fig. 3. The pixels are scaled to be in [0, 1] before the training. There is no preprocessing or data augmentation utilized in this work. In this dataset, the size of mini-batches is 30 images.

The CNN used in this work consists of alternative convolutional and max pooling layers. Fully connected layer is implemented on the top of the network. The number of particles is 30 and weights of them are randomly initialized with Xavier method. At the

Fig. 3. Samples of scaled GTSRB datasets

beginning, in the feedforward phase, each particle of network uses the above-mentioned structure for model evaluation and cross-entropy cost function is used as fitness assessment for particles. The lowest cross-entropy error is the highest fitness. In the backward phase, PSO-CNN introduces the velocity formula which combines with gradient descend to avoid occurring in local optimum. To evaluate the performance of our approach, the conventional CNNs with standard SGD and conjugate gradient (CG) [16, 17] optimization methods are compared, as Fig. 4 shows, the *gbestfitness* curve of PSO-CNN goes down faster than conventional CNNs with standard SGD and CG. And Table 1 shows the results of the several algorithms, where the training accuracy and training fitness denote the accuracy of training dataset and the loss in the end of training phase, respectively; test accuracy and test fitness are the same meaning as the former in test phase. All the losses are the value of the cross-entropy cost function. The lowest fitness or loss is the best performance. All these indicate that PSO-CNN converges fastest and obtain the best results. At the end of epoch, we can see the curve of conventional CNN drops below the *gbestfitness* curve of PSO-CNN, and the training fitness in Table 1 shows the same situation, however, the final test accuracy is on the contrary, which means the conventional CNNs with standard SGD and CG suffer from serious overfitting while PSO-CNN reduces it. In addition, in the test phase of PSO-CNN, we can further dampen overfitting by making use of the advantage of swarm as described in Sect. 3. So, PSO-CNN obtains the best result and it surpasses conventional CNN and its variants.

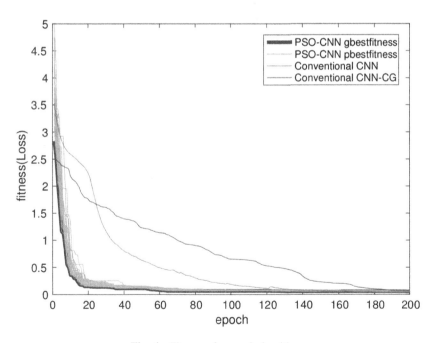

Fig. 4. Fitness of several algorithms

Table 1. The results of GTSRB DATASET

Method	Training accuracy	Training fitness (loss)	Test accuracy	Test fitness (loss)
PSO-CNN	100%	0.059	95.8%	0.078
Conventional CNN	99.9%	0.047	83.4%	1.986
Conventional CNN with CG	100%	0.073	90.5%	0.982

As we can see, Fig. 5 shows the convolutional kernels learned by the convolutional layers of PSO-CNN. The network has learned a variety of frequency and orientation selective kernels to adapt to various conditions in GTSRB datasets. Figure 6 displays the corresponding feature maps generated by the kernels in Fig. 5. In the c1 layer, the feature maps are similar to the original images, and in c2 layer, the feature maps present complex texture and stay the fuzzy outline of input images, while in c3 layer, the feature maps only leave only edges and pixels.

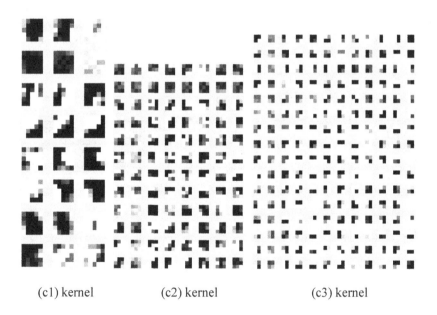

(c1) kernel   (c2) kernel   (c3) kernel

Fig. 5. Convolutional kernel of PSO-CNN

Figure 7 shows the precision recall curves for the several algorithms. Each point of the curves represents the recall and precision of one category. In order to evaluate the performance of the proposed algorithm in detail, area under the precision recall (AUC) [18] is also used as the evaluation criteria, the highest AUC is the best performance. Table 2 shows the AUC values of several algorithms, which proves that the proposed algorithms PSO-CNN is better than conventional CNN and its variants.

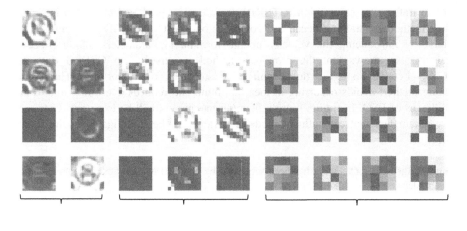

(c1) feature maps   (c2) feature maps   (c3) feature maps

Fig. 6. Feature map of convolutional layers in PSO-CNN

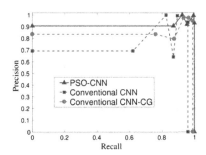

Fig. 7. The precision recall curves of several algorithms

Table 2. The AUC of several algorithms

Method	AUC
PSO-CNN	0.9127
Conventional CNN	0.7232
Conventional CNN with CG	0.8351

5 Conclusion

In this work, to address the shortcomings of stochastic gradient descend (SGD) such as occurring in local optimum and vanishing gradient problems, a new hybrid training process, called Particle Swam Optimization-Convolution neural network (PSO-CNN) algorithm, is proposed and demonstrated for optimizing conventional CNN training. It is established that the algorithm is well suited for achieving nontrivial results on

German traffic sign recognition (GTSRB) datasets. The proposed algorithm is a proficient method for training because it combines both PSO and SGD in an innovative fashion by introducing the velocity update formula to parameters tuning process. Analysis also shows that the proposed method is superior on GTSRB datasets, PSO-CNN converges to the global minimum faster than other algorithms. The hybrid training method avoids occurring in local optimum and reduces overfitting due to the advantages of PSO. In future, more influential parameters will be explored. There are more parameters that can influent model accuracy will be investigated in the future work.

Acknowledgement. This work has been supported by National Natural Science Foundation of China under grant 61472284.

References

1. Sermanet, P., Kavukcuoglu, K., Chintala, S., et al.: Pedestrian detection with unsupervised multi-stage feature learning. In: Proceedings of the IEEE Conference on Computer Vision and Pattern Recognition, pp. 3626–3633 (2013)
2. Karpathy, A., Toderici, G., Shetty, S., et al.: Large-scale video classification with convolutional neural networks. In: Proceedings of the IEEE Conference on Computer Vision and Pattern Recognition, pp. 1725–1732 (2014)
3. Toshev, A., Szegedy, C.: Deeppose: human pose estimation via deep neural networks. In: Proceedings of the IEEE Conference on Computer Vision and Pattern Recognition, pp. 1653–1660 (2014)
4. Qian, N.: On the momentum term in gradient descent learning algorithms. Neural Netw.: Off. J. Int. Neural Netw. Soc. **12**(1), 145–151 (1999)
5. Nesterov, Y.: A method for unconstrained convex minimization problem with the rate of convergence o(1/k2). In: Doklady ANSSSR (translated as Soviet.Math.Docl.), vol. 269, pp. 543–547 (1983)
6. Duchi, J., Hazan, E., Singer, Y.: Adaptive subgradient methods for online learning and stochastic optimization. J. Mach. Learn. Res. **12**, 2121–2159 (2011)
7. Zeiler, M.D.: ADADELTA: an adaptive learning rate method (2012). arXiv preprint arXiv: 1212.5701
8. Tieleman, T., Hinton, G.: Lecture 6.5-rmsprop: divide the gradient by a running average of its recent magnitude. COURSERA: Neural Netw. Mach. Learn. **4**(2), 26–31 (2012)
9. Kingma, D.P., Ba, J.L.: Adam: a method for stochastic optimization. In: International Conference on Learning Representations, pp. 1–13 (2015)
10. Murphy, K.P.: Machine Learning: A Probabilistic Perspective, pp. 82–92. MIT Press, Cambridge (2012)
11. Kennedy, J.: Particle swarm optimization. In: Sammut, C., Webb, G.I. (eds.) Encyclopedia of Machine Learning, pp. 760–766. Springer, Heidelberg (2011)
12. Wang, X., Gao, X.Z., Ovaska, S.J.: A hybrid particle swarm optimization method. In: IEEE International Conference on Systems, Man and Cybernetics, SMC 2006, vol. 5, pp. 4151–4157. IEEE (2006)
13. LeCun, Y., Bottou, L., Bengio, Y., Haffiner, P.: Gradient-based leaning applied to document recognition. Proc. IEEE **86**(11), 2278–2324 (1998)

14. Glorot, X., Bengio, Y.: Understanding the difficulty of training deep feedforward neural networks. In: AISTATS 2010, vol. 9, pp. 249–256 (2010)
15. Stallkamp, J., Schlipsing, M., Salmen, J., Igel, C.: The German traffic sign recognition benchmark: a multi-class classification competition. In: Proceedings of the IEEE International Joint Conference on Neural Networks, pp. 1453–1460 (2011)
16. Fletcher, R., Reeves, C.M.: Function minimization by conjugate gradients. Comput. J. 7(2), 149–154 (1964)
17. Stallkamp, J., Schlipsing, M., Salmen, J., Igel, C.: The German traffic sign recognition benchmark: a multi-class classification competition. In: Proceedings of the IEEE International Joint Conference on Neural Networks, pp. 1453–1460 (2011)
18. Davis, J., Goadrich, M.: The relationship between precision-recall and ROC curves. In: Proceedings of the 23rd International Conference on Machine Learning, pp. 233–240. ACM (2006)

Robust Ranking Model via Bias-Variance Optimization

Jinzhong Li[1,2,3,4(✉)], Guanjun Liu[3,4], and Jiewu Xia[1,2]

[1] Department of Computer Science and Technology, Jinggangshan University,
Ji'an 343009, China
[2] Network and Data Security Key Laboratory of Sichuan Province,
University of Electronic Science and Technology of China,
Chengdu 610054, China
[3] Department of Computer Science and Technology, Tongji University,
Shanghai 201804, China
[4] The Key Laboratory of Embedded System and Service Computing,
Ministry of Education, Tongji University, Shanghai 201804, China
1210510@tongji.edu.cn

Abstract. Improving average effectiveness is an objective of paramount importance of ranking model for the learning to rank task. Another equally important objective is the robustness—a ranking model should minimize the variance of effectiveness across all queries when the ranking model is disturbed. However, most of the existing learning to rank methods are optimizing the average effectiveness over all the queries, and leaving robustness unnoticed. An ideal ranking model is expected to balance the trade-off between effectiveness and robustness by achieving high average effectiveness and low variance of effectiveness. This paper investigates the effectiveness-robustness trade-off in learning to rank from a novel perspective, i.e., the bias-variance trade-off, and presents a unified objective function which captures the trade-off between these two competing measures for jointly optimizing the effectiveness and robustness of ranking model. We modify the gradient based on the unified objective function using LambdaMART which is a state-of-the-art learning to rank algorithm, and demonstrate the strategy of jointly optimizing the combination of bias and variance in a principled learning objective. Experimental results demonstrate that the gradient-modified LambdaMART improves the robustness and normalized effectiveness of ranking model by combining bias and variance.

Keywords: Learning to rank · Ranking model · Effectiveness-robustness tradeoff · Bias-variance tradeoff · LambdaMART algorithm

1 Introduction

Ranking model is one of the important problems in information retrieval (IR) systems. The number of query-document pairs and the relevance judgments between query and documents may be different for each query, it causes the ideal effectiveness of each query may be different for an ideal ranking model. We note that the effectiveness can be measured by some IR metrics, such as Expected Reciprocal Rank (ERR) [1]. For instance, the ideal effectiveness of $ERR@1$ with 5 relevance judgments usually have

© Springer International Publishing AG 2017
D.-S. Huang et al. (Eds.): ICIC 2017, Part III, LNAI 10363, pp. 706–718, 2017.
DOI: 10.1007/978-3-319-63315-2_62

5 different values as 0.9375, 0.4375, 0.1875, 0.0625 and 0 respectively. The unique value of ideal effectiveness for each query causes their corresponding optimization space to be unique even if their actual effectiveness have the same value. Suppose that there are two queries q_1 and q_2, and their values of ideal effectiveness are 0.6 and 0.9 respectively. Assume that the values of actual effectiveness of q_1 and q_2 are both 0.4 for ranking model A, the values of actual effectiveness of q_1 and q_2 are 0.3 and 0.5 for ranking model B, the values of actual effectiveness of q_1 and q_2 are 0.5 and 0.3 for ranking model C, respectively. Furthermore, the values of the optimization space of q_1 and q_2 are 0.2 and 0.5 for A, the values of the optimization space of q_1 and q_2 are 0.3 and 0.4 for B, the values of the optimization space of q_1 and q_2 are 0.1 and 0.6 for C in relation to their ideal effectiveness, respectively. Although the three ranking models have same values of average effectiveness, the ranking models are different from each other and the ranking model B is better from the perspective of robustness. Robustness in a ranking model should minimize the variance of effectiveness across queries when the ranking model is disturbed. Because the normalized effectiveness of each query is closer to each other in ranking model B, the performance of B is relatively stable and exhibit better robustness. Therefore, the actual effectiveness of each query should be normalized by dividing its value by their corresponding ideal effectiveness in order to enhance the robustness of ranking model.

However, existing methods of learning to rank cannot distinguish the three ranking models as stated in the above example. Most of works focused on improving the average effectiveness of retrieval results and leaving the robustness criterion unnoticed, thus causing the ranking model unstable across all queries. Fortunately, recent retrieval system evaluations are insisting more emphasis on robustness, the TREC 2013 and TREC 2014 Web track introduced a risk-sensitive task. Though the effectiveness-robustness trade-off has been detailed in several recent studies [2, 3], the trade-off have not gained sufficiently systematic and quantitative evaluations over the systems participated in the learning to rank tasks.

In brief, the average effectiveness and the robustness are crucial for an effective ranking model and both should be given preference while evaluating the performance of ranking model. To this end, this paper incorporates the effectiveness and robustness of ranking model into the optimization objective for learning to rank algorithm in the training process of ranking model so as to simultaneously optimize the effectiveness and robustness. Further, this paper proposes the effectiveness and robustness trade-off for learning to rank from a new perspective, i.e., the bias-variance trade-off, and also proposes to formulate the bias and variance of effectiveness for the effectiveness and robustness of ranking model respectively. Hereby, reducing the bias and variance of effectiveness directly reflects in improving the effectiveness and robustness of ranking model. We formulate the combination of bias-variance to combine the robustness and effectiveness of ranking model into a unified optimization objective across all queries. This objective can be integrated into many learning to rank algorithms. Further, this paper demonstrates the extension of LambdaMART [4, 5] in order to optimize the new objective. The proposed bias-variance formulation can provide more theoretical insights on the trade-off between effectiveness and robustness of ranking model. To the best of our knowledge, this paper is the first work addressing robust learning to rank via bias-variance optimization.

The main contributions of this work are summarized as follows:

(1) A bias-variance trade-off for training ranking model is investigated in the learning to rank task, the bias and variance of effectiveness are combined into a unified objective for jointly optimizing the effectiveness and robustness of ranking model.
(2) The gradient of LambdaMART algorithm is modified by the unified optimization objective, and a principled strategy of jointly optimizing the combination of bias and variance of effectiveness is demonstrated.

2 Related Work

Learning to rank refers to machine learning techniques for training the model in a ranking task [6]. The studies of learning to rank have gained considerable attention from the IR researches and industry communities in the past decade. A number of machine learning techniques were successfully applied as the learning algorithms for relevance ranking, including multiple additive regression tree [4, 5], perceptron [7], support vector machine [8], neural network [9], extreme learning machine [10], cooperative coevolution [11], Bayesian [12] and random forests [13] and so on.

However most of the previous works typically focused on optimizing average effectiveness using different optimization methods, but they often left the importance of ranking robustness unnoticed. A ranking model is considered to be robust if it minimizes the variance across all the queries when the ranking model is disturbed. This paper considers both effectiveness and robustness of the ranking model in the problem of learning to rank. Effectiveness and robustness are integrated into a unified objective function to optimize the trade-off between them.

Robustness of ranking model has been previously reviewed in some literatures. Dincer et al. [2] examined risk-sensitive evaluation from the perspective of statistical hypothesis testing. Wang et al. [3] explored the robust learning to rank by incorporating the risk measurement and quantitative comparison in the true sense. Further, their works presented a unified framework for jointly optimizing effectiveness and robustness from the perspective of risk and reward using LambdaMART method. However, the key differences between their works and this paper are that: (1) the gain-risk functions are linearly combined into a single objective function using a weighted method in their works. Reasonably assigning weights to the gain-risk functions in advance is involved in several complexities in their method. In this paper, we investigate the effectiveness-robustness trade-off in learning to rank from a new perspective, i.e., the bias-variance trade-off. We present a unified objective function which captures the trade-off between the bias of effectiveness and the variance of effectiveness for jointly optimizing effectiveness and robustness. The objective function of the bias-variance trade-off in our model is different from their objective function of the risk-reward trade-off. The bias-variance functions are not assigned with weights in our method. (2) They take BM25 model as baseline, but BM25 cannot guarantee the effectiveness of each query is the best. Our strategy considers the ideal effectiveness of each query as the corresponding baseline. That is to say, our strategy has a strongest baseline, and it incorporates the most optimal baseline into our unified objective function. In short, our baseline is more reasonable to optimize ranking models.

Zhang et al. [14] introduced the bias-variance into IR evaluation, and [15] formulated the notion of bias-variance in regards to the retrieval performance and estimation quality of query model, and further investigated various query estimation methods using bias-variance analysis, and studied the retrieval effectiveness-stability trade-off in query model estimation. Differing from their works, this paper focuses on investigating the effectiveness-robustness trade-off from the view of bias-variance tradeoff, which is a new perspective in learning to rank. We combine the variance of effectiveness and the square of bias of effectiveness into a unified objective function which captures the bias-variance trade-off for jointly optimizing the effectiveness and robustness of ranking model, and demonstrate the strategy of jointly optimizing the combination of bias and variance of effectiveness in a principled learning objective.

This paper further extends the robust ranking model for learning to rank based on the bias-variance trade-off. The effectiveness and robustness of ranking model are characterized by bias and variance. We combine the bias and variance of effectiveness into a unified objective to jointly optimize the effectiveness and robustness of ranking model, and demonstrate that the unified objective can be optimized by LambdaMART method in a principled learning objective. To our knowledge, this is the first work investigating the performance of learning to rank based on the bias-variance trade-off.

3   Bias-Variance for Learning to Rank

In general, the bias of an estimator is the difference between the estimator's expected value and the true value of the parameter being estimated, while the variance of an estimator is simply the expected value of the squared sampling deviations in probability theory and statistics. The optimization of the bias-variance trade-off is the problem of simultaneously minimizing bias and variance which prevent supervised learning algorithms from generalizing beyond their training set in statistics and machine learning. This motivates us to make an analogy, a ranking model resembles an estimator, and the effectiveness of each query for ranking model resembles the value of sampling for the estimator. Based on this perspective, we introduce bias-variance into the problem of learning to rank to formulate the bias and variance of effectiveness of ranking model which are related to the effectiveness and robustness of ranking model respectively. Specifically, assuming that we have an ideal effectiveness for each query, the bias of effectiveness of ranking model represents the difference between the actual mean effectiveness and the ideal mean effectiveness over all queries, and the smaller bias of effectiveness indicates the smaller expected deviation which implies the higher expected effectiveness of ranking model. The variance of effectiveness of ranking model corresponds to the variance across different individual queries. The smaller variance of effectiveness reflects the stronger robustness of ranking model. In this context, we investigate the problem of improving the effectiveness and robustness of ranking model from the perspective of reducing the bias and variance of effectiveness of ranking model.

3.1 Bias-Variance of Ranking Model

We define the bias-variance of ranking model based on the actual effectiveness and the ideal effectiveness of each query in learning to rank task.

Definition 1. Let M_q^I be the optimal effectiveness of query q in an ideal ranking model I, and M_q be the actual effectiveness of q in ranking model R, the mean of ideal effectiveness of I is defined as $E[M_q^I] = \frac{1}{|Q|} \sum_{q \in Q} M_q^I$, the mean of actual effectiveness of R is defined as $E[M_q] = \frac{1}{|Q|} \sum_{q \in Q} M_q$, where $|Q|$ denotes the number of all queries in query set Q. Effectiveness M can be defined as the *ERR* score or other metrics of interest.

I is a virtual ranking model constructed by sorting all documents w.r.t each query based on their relevance judgments in descending order. The effectiveness of each query is calculated according to the order, thus M_q^I is the optimum value which has the maximum effectiveness for each query q. $E[M_q^I]$ represents the average value of the ideal effectiveness over all the queries in the ideal ranking model I, and $E[M_q]$ represents the average value of the actual effectiveness over all queries in the ranking model R. The bigger value of $E[M_q]$ shows the superior effectiveness of ranking model R.

We introduce an auxiliary variable Y_q to normalize the effectiveness of each query, and assume that $Y_q = \frac{M_q^I - M_q}{M_q^I}$ when $M_q^I > 0$ and $Y_q = 0$ when $M_q^I = 0$.

Definition 2. The bias of effectiveness of ranking model is defined as $B[Y_q] = \frac{1}{|Q|} \sum_{q \in Q} (Y_q - 0) = \frac{1}{|Q|} \sum_{q \in Q} Y_q$, where "0" is a virtual baseline.

The bias of effectiveness represents the average degree of regularized deviation between the actual effectiveness and the ideal effectiveness over all queries. $B[Y_q]$ is the average value of Y_q across all queries in ranking model, and it equals $E[Y_q]$. The smaller value of $B[Y_q]$ shows the normalized effectiveness of ranking model is superior.

Definition 3. The variance of effectiveness of ranking model is defined as $V[Y_q] = \frac{1}{|Q|} \sum_{q \in Q} (Y_q - E[Y_q])^2$.

The variance of effectiveness measures the dispersion degree to which the actually regularized effectiveness across different individual queries vary around their average, i.e. $V[Y_q]$ is the average of the square of the difference between Y_q and $E[Y_q]$ over all queries in ranking model. Smaller the value of $V[Y_q]$, stronger is the robustness of ranking model.

Definition 4. The effectiveness-robustness trade-off of ranking model is defined as

$$Obj = (B[Y_q])^2 + V[Y_q] \qquad (1)$$

In Eq. (1), the variance of effectiveness of ranking model and the squared bias of effectiveness of ranking model are combined to yield a trade-off function *Obj* in an integrated manner. The unified objective function *Obj* considers both effectiveness and robustness, and *Obj* is expected to minimize the value, i.e. the smallest *Obj* is pursued. Equation (1) can be deduced as $Obj = E[Y_q])^2$ according to the following equation:

$$Obj = (B[Y_q])^2 + V[Y_q] = (B[Y_q])^2 + \frac{1}{|Q|}\sum_{q \in Q}(Y_q - E[Y_q])^2$$

$$= (B[Y_q])^2 + \frac{1}{|Q|}\sum_{q \in Q}[(Y_q)^2 - 2 \times E[Y_q] \times Y_q + (E[Y_q])^2]$$

$$= (E[Y_q])^2 + \frac{1}{|Q|}\sum_{q \in Q}(Y_q)^2 - \frac{2 \times E[Y_q]}{|Q|}\sum_{q \in Q}Y_q + (E[Y_q])^2 = E[(Y_q)^2]$$

The effectiveness-robustness trade-off *Obj* is the expected value of the square of the regularized difference between the ideal effectiveness and the actual effectiveness across different individual queries, i.e., it is the second central moment of random variable Y_q, and we conclude that the expected value of the squared difference $E[(Y_q)^2]$ can be expressed as the sum of variance and squared bias of effectiveness.

3.2   Constructing Robust Ranking Model

LambdaMART [4, 5] is the multiple additive regression trees (MART) version of LambdaRank, which is based on RankNet. LambdaMART has been proven to be very successful algorithm for solving real world ranking problems: an ensemble of LambdaMART rankers won the 2010 Yahoo! Learning to Rank Challenge. This paper further extends the LambdaMART for robust ranking model.

As Wang et al. [3] described in their literature, LambdaMART modified the gradient of RankNet, the new gradient captures the absolute change ΔM in IR metric M due to swapping documents d_i and d_j, in addition to capturing the original λ_{ij} value, which the derivative of a cross entropy of RankNet. The new gradient for each document d_i is simply defined as:

$$\lambda_i^{new} = \sum_{j \neq i}\lambda_{ij} \times |\Delta M_{ij}|. \tag{2}$$

The gradient λ_i^{new} defined on each document for each query is optimized by LambdaMART. In principle, LambdaMART can be extended to optimize any standard IR metric by simply replacing ΔM_{ij} in Eq. (2) according to the corresponding change ΔM_{ij}^* in the optimized metric M^*. In general, the consistency property [3–5] is desired since LambdaMART's approximation to the overall gradient is derived from the pairwise gradients. Therefore, it is considered to be suitable for M^* to satisfy the consistency property: when swapping the ranked positions of two documents, d_i and d_j, in a sorted document list where d_i is more relevant than d_j, and ranks before d_j, the objective should decrease.

In this paper, the optimization objective M of the original LambdaMART is modified to *Obj* which defines as a combination of bias and variance of effectiveness in Eq. (1), so $|\Delta M_{ij}|$ will be modified to $|\Delta Obj_{ij}|$ for the gradient λ_i^{new} in Eq. (2). The objective M of the LambdaMART is expected to maximize the value, but the new objective *Obj* of the gradient-modified LambdaMART is expected to minimize the value. We note that M and *Obj* are just opposite optimization objective. Therefore, we now demonstrate that *Obj* satisfies the consistency property: any documents pairwise swap between correctly ranked d_i and d_j for the same q must lead to an increase in *Obj*, any documents pairwise swap between incorrectly ranked d_i and d_j for the same q must lead to a decrease in *Obj*. After swapping ranked positions of a number of pairs of documents for query set Q, the change ΔObj of *Obj* in Eq. (1) can be written as Eq. (3):

$$\Delta Obj = Obj - Obj^*. \tag{3}$$

Equation (3) can be deduced as $\Delta Obj = \frac{1}{|Q|} \sum_{q \in Q} (\frac{\Delta M}{M_q^I} \times \frac{M_q + M_q^* - 2 \times M_q^I}{M_q^I})$ according to the following equation.

$$\Delta Obj = E[(Y_q)^2] - E[(Y_q^*)^2] = \frac{1}{|Q|} \sum_{q \in Q} [(Y_q)^2 - (Y_q^*)^2] = \frac{1}{|Q|} \sum_{q \in Q} [(\frac{M_q^I - M_q}{M_q^I} - \frac{M_q^I - M_q^*}{M_q^I}) \times (\frac{M_q^I - M_q}{M_q^I} + \frac{M_q^I - M_q^*}{M_q^I})]$$

$$= \frac{1}{|Q|} \sum_{q \in Q} (\frac{M_q^* - M_q}{M_q^I} \times \frac{2 \times M_q^I - M_q^* - M_q}{M_q^I}) = \frac{1}{|Q|} \sum_{q \in Q} (\frac{\Delta M}{M_q^I} \times \frac{M_q^* + M_q - 2 \times M_q^I}{M_q^I})$$

Theorem: The effectiveness-robustness trade-off *Obj* in Eq. (1) satisfies the consistency property.

Proof: Let r_i be the relevance grade of document d_i w.r.t query q. Suppose that $i > j$ presents the ith ranked position of the document d_i is ranked higher than the jth ranked position of the document d_j. After swapping ranked positions of d_i and d_j for q, the change ΔObj_{ij} can be written as: $\Delta Obj_{ij} = \frac{\Delta M_{ij}}{M_q^I} \times \frac{M_q^* + M_q - 2 \times M_q^I}{M_q^I}$. Because $0 \le M_q^* \le M_q^I$ and $0 \le M_q \le M_q^I$, thus, $\frac{M_q^* + M_q - 2 \times M_q^I}{M_q^I} \le 0$, then ΔObj_{ij} and ΔM_{ij} are a reverse change.

This can be also proved by considering the following five cases.

Case 1. If $r_i > r_j$ and $i > j$, swap d_i and d_j, then $M_q^* > M_q$, so $\Delta M_{ij} < 0$ and $Obj_{ij} > 0$;
Case 2. If $r_i > r_j$ and $i < j$, swap d_i and d_j, then $M_q^* < M_q$, so $\Delta M_{ij} > 0$ and $Obj_{ij} < 0$;
Case 3. If $r_i = r_j$, swap d_i and d_j, then $M_q^* = M_q$, so $\Delta M_{ij} = 0$ and $Obj_{ij} = 0$;
Case 4. If $r_i < r_j$ and $i > j$, swap d_i and d_j, then $M_q^* < M_q$, so $\Delta M_{ij} > 0$ and $Obj_{ij} < 0$;
Case 5. If $r_i < r_j$ and $i < j$, swap d_i and d_j, then $M_q^* > M_q$, so $\Delta M_{ij} < 0$ and $Obj_{ij} > 0$.

The bigger value of effectiveness metric M indicates that it is more excellent in the original LambdaMART algorithm, while the smaller value of objective *Obj* indicates that it is also more excellent in the gradient-modified LambdaMART algorithm based on bias-variance optimization, thus the optimization directions of M and *Obj* are just opposite. Based on the above five cases, *Obj* satisfies the consistency property. Therefore, the modified gradient via bias-variance optimization can still be adopted in the LambdaMART algorithm to train the ranking model.

4 Experimental Results and Analysis

The original LambdaMART algorithm [4, 5], the modified LambdaMART algorithm via risk-sensitive optimization [3], and our modified LambdaMART algorithm via bias-variance optimization are denoted as ORIG, Alpha, and VARI, respectively. Further, when the value of risk-sensitivity parameter α is equal to 0, it is denoted as Alpha0, i.e., ORIG, and when the value of α is equal to 5 and 10, it is denoted as Alpha5 and Alpha10 in [3], respectively. In order to verify the performance of VARI, we conduct experiments on two large-scale and publicly available learning to rank datasets MSLR-WEB10K and MSLR-WEB30K[1] based on RankLib[2], which is an open-source library of learning to rank algorithms.

4.1 Evaluation Measures of Effectiveness and Robustness

Both the average effectiveness over all the queries and the robustness across different queries are considered in our performance evaluations. We adopt a common IR evaluation metric, i.e., *ERR* [1], to measure the effectiveness of ranking model in the training process of ranking model. $B[Y_q]$ and $V[Y_q]$ in Definitions 2 and 3 are considered to measure the effectiveness and robustness of ranking model, respectively. In addition, the *Distribution* of the effectiveness of each query over 5 intervals are defined as [0.0, 0.2], (0.2, 0.4], (0.4, 0.6], (0.6, 0.8] and (0.8, 1.0], respectively. The *Distribution* counts the number of queries in each interval and reflects the robustness of the ranking model. VARI is compared with ORIG, Alpha5 and Alpha10 on the three measures, including $B[Y_q]$, $V[Y_q]$ and *Distribution* based on *ERR@k*, where $k = 10$ and 20.

4.2 Analysis of Experimental Results

In our experiments, all the evaluations are conducted using the 5-fold cross validation. We report the overall results of the 5-fold test for ORIG, Alpha5, Alpha10 and VARI in terms of their effectiveness and robustness from Figs. 1, 2, 3, 4, 5 and 6.

From Figs. 1 and 2, it can be observed that VARI achieves the smallest $B[Y_q]$ and $V[Y_q]$ in comparison to those of ORIG, Alpha5 and Alpha10 in all cases. Smaller the bias of effectiveness of ranking model, better is its effectiveness. Smaller the variance of effectiveness of ranking model, stronger is its robustness. Thus it can be concluded that VARI is more robust and its normalized effectiveness is more excellent according to $V[Y_q]$ and $B[Y_q]$. In addition, it is evident that the *Distribution* of the number of queries are relatively high in most cases for the middle intervals, but the *Distribution* of the number of queries are relatively low at the ends of the intervals [0.0, 0.2] and (0.8, 1.0] for VARI from Figs. 3, 4, 5 and 6. This reflects the fact that an increased number of queries are closer to the average value of effectiveness of ranking model, thus VARI achieves stronger robustness. The main reason is that the bias and variance of

---

[1] http://research.microsoft.com/en-us/projects/mslr/download.aspx.
[2] http://sourceforge.net/p/lemur/code/HEAD/tree/RankLib/trunk/.

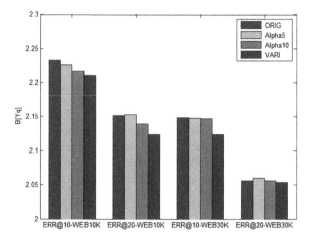

Fig. 1. Performance comparison of $B[Y_q]$

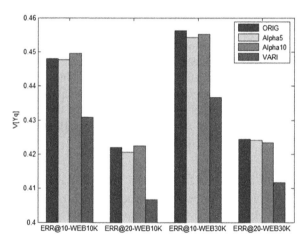

Fig. 2. Performance comparison of $V[Y_q]$

effectiveness of ranking model are integrated into the optimization objective function in the training process of ranking model for VARI. The optimization objective is composed of the variance of effectiveness $V[Y_q]$ and the squared bias of effectiveness $(B[Y_q])^2$. $(B[Y_q])^2$ and $V[Y_q]$ are simultaneously optimized in VARI. Because the values of bias and variance of effectiveness are both between 0 and 1, the value of $(B[Y_q])^2$ will be reduced less than the value of $V[Y_q]$, which makes VARI more focused on the optimization of $V[Y_q]$ in the training process of ranking model, and may affect $B[Y_q]$ when $V[Y_q]$ is minimized. Furthermore, because Y_q is the normalized effectiveness, the optimization of the query with a larger optimized space is enhanced. Thus VARI achieves smaller $V[Y_q]$ and smaller $B[Y_q]$ to some extent.

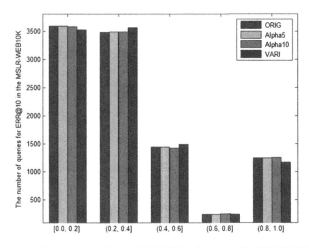

Fig. 3. *Distribution* comparison of *ERR*@10 of queries for MSLR-WEB10K

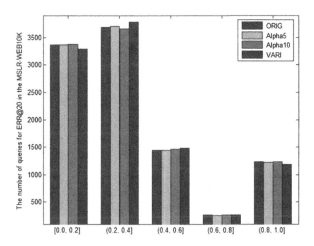

Fig. 4. *Distribution* comparison of *ERR*@20 of queries for MSLR-WEB10K

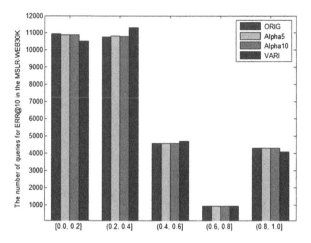

Fig. 5. *Distribution* comparison of *ERR*@10 of queries for MSLR-WEB30K

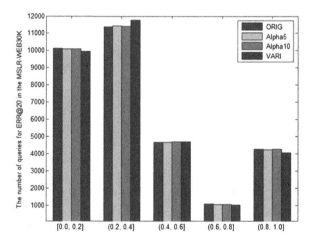

Fig. 6. *Distribution* comparison of *ERR*@20 of queries for MSLR-WEB30K

5   Conclusions and Future Work

The problem of learning to rank increase complexities when robustness of ranking model is considered. This paper studies the effectiveness-robustness trade-off in ranking model for the learning to rank task from the perspectives of bias-variance optimization, and proposes a statistical measure called the bias and variance of effectiveness to quantify the effectiveness and robustness of ranking model. Moreover, we present a unified objective function which captures the trade-off between the bias and variance of effectiveness for jointly optimizing effectiveness and robustness, and demonstrate the strategy of jointly optimizing the combination of bias and variance of effectiveness in a principled learning objective. Experimental results show that VARI is more effective from the perspective of the normalized effectiveness and more robust with regard to the robustness measures than ORIG, Alpha5 and Alpha10. In the future, we plan to further improve the LambdaMART algorithm for achieving better effectiveness and/or robustness of ranking model. Moreover, we will explore and apply the combinational bias-variance to other typical algorithms of learning to rank and analyze the bias-variance optimization for extracting more insights in order to develop new techniques for improving the effectiveness and/or robustness of ranking models.

Acknowledgements. This work was in part supported by the Natural Science Foundation of Jiangxi Province of China (No. 20171BAB202010), the Opening Foundation of Network and Data Security Key Laboratory of Sichuan Province (No. NDSMS201602).

References

1. Chapelle, O., Metlzer, D., Zhang, Y., Grinspan, P.: Expected reciprocal rank for graded relevance. In: Proceedings of the 18th ACM Conference on Information and Knowledge Management, Hong Kong, China, pp. 621–630. ACM (2009)
2. Dinçer, B.T., Macdonald, C., Ounis, I.: Hypothesis testing for the risk-sensitive evaluation of retrieval systems. In: Proceedings of the 37th International ACM SIGIR Conference on Research and Development in Information Retrieval, Gold Coast, Queensland, Australia, pp. 23–32. ACM (2014)
3. Wang, L., Bennett, P.N., Collins-Thompson, K.: Robust ranking models via risk-sensitive optimization. In: Proceedings of the 35th International ACM SIGIR Conference on Research and Development in Information Retrieval, Portland, Oregon, USA, pp. 761–770. ACM (2012)
4. Wu, Q., Burges, C.J.C., Svore, K.M., Gao, J.: Adapting boosting for information retrieval measures. Inf. Retrieval **13**(3), 254–270 (2010)
5. Burges, C.J.C.: From ranknet to lambdarank to lambdamart: an overview. Technical report, Microsoft Research Technical Report MSRTR-2010–82 (2010)
6. Li, H.: A short introduction to learning to rank. IEICE Trans. Inf. Syst. **94**(10), 1854–1862 (2011)
7. Xu, J., Xia, L., Lan, Y., Guo, J., Cheng, X.: Directly optimize diversity evaluation measures: a new approach to search result diversification. ACM Trans. Intell. Syst. Technol. (TIST) **8**(3), 41 (2017)
8. Jung, C., Shen, Y., Jiao, L.: Learning to rank with ensemble ranking SVM. Neural Process. Lett. **42**(3), 703–714 (2015)
9. Rigutini, L., Papini, T., Maggini, M., Scarselli, F.: SortNet: learning to rank by a neural preference function. IEEE Trans. Neural Networks **22**(9), 1368–1380 (2011)
10. Zong, W., Huang, G.B.: Learning to rank with extreme learning machine. Neural Process. Lett. **39**(2), 155–166 (2014)
11. Wang, S., Wu, Y., Gao, B.J., Wang, K., Lauw, H.W., Ma, J.: A cooperative coevolution framework for parallel learning to rank. IEEE Trans. Knowl. Data Eng. **27**(12), 3152–3165 (2015)
12. Guo, H.F., Chu, D.H., Ye, Y.M., Li, X.T., Fan, X.X.: BLM-rank: a Bayesian linear method for learning to rank and its GPU implementation. IEICE Trans. Inf. Syst. **E99-D**(4), 896–905 (2016)

13. Ibrahim, M., Carman, M.: Comparing pointwise and listwise objective functions for random-forest-based learning-to-rank. ACM Trans. Inf. Syst. (TOIS) **34**(4), 20 (2016)
14. Zhang, P., Song, D., Wang, J., Hou,Y.: Bias-variance decomposition of ir evaluation. In: Proceedings of the 36th International ACM SIGIR Conference on Research and Development in Information Retrieval, Dublin, Ireland, pp. 1021–1024. ACM (2013)
15. Zhang, P., Song, D., Wang, J., Hou, Y.: Bias-variance analysis in estimating true query model for information retrieval. Inf. Process. Manage. **50**(1), 199–217 (2014)

A Comparison of Distance Metrics in Semi-supervised Hierarchical Clustering Methods

Abeer Aljohani[1]([✉]), Daphne Teck Ching Lai[2], Paul C. Bell[1], and Eran A. Edirisinghe[1]

[1] Department of Computer Science,
Loughborough University, Loughborough, UK
{A.Aljohani,P.Bell,E.A.Edirisinghe}@lboro.ac.uk
[2] Faculty of Science, Universiti Brunei Darussalam,
Bandar Seri Begawan, Brunei
daphne.lai@ubd.edu.bn

Abstract. The basic idea of ssHC is to leverage domain knowledge in the form of triple-wise constraints to group data into clusters. In this paper, we perform extensive experiments in order to evaluate the effects of different distance metrics, linkages measures and constraints on the performance of two ssHC algorithms: IPoptim and UltraTran. The algorithms are implemented with varying proportions of constraints in the different datasets, ranging from 10% to 60%. We found that both IPoptim and UltraTran performed almost equally across the seven datasets. An interesting observation is that an increase in constraint does not always show an improvement in ssHC performance. It can also be observed that the inclusion of too many classes degrades the performance of clustering. The experimental results show that the ssHC with Canberra distance perform well, apart from ssHC with well-known distances such as Euclidean and Standard Euclidean distances. Together with complete linkages and small amount of constraints of 10%, ssHC can achieve good results of an F-score close to 0.8 and above for four out of the seven datasets. Moreover, the output of non-parametric statistical test shows that using the UltraTran algorithm in combination with the Manhattan distance metric and Ward.D linkage method provides the best results. Furthermore, utilizing IPoptim and UltraTran with the Canberra distance measure performs better for the given datasets.

Keywords: Distance measure · Semi-supervised Hierarchical Clustering · Linkage measure · Ultra-metric

1 Introduction

There has been significant academic interest in semi-supervised learning because it achieves better accuracy and requires less human effort [1]. Standard unsupervised clustering methods have been enhanced by using constraints or class labels as shown in

© Springer International Publishing AG 2017
D.-S. Huang et al. (Eds.): ICIC 2017, Part III, LNAI 10363, pp. 719–731, 2017.
DOI: 10.1007/978-3-319-63315-2_63

various previous studies [2, 3]. The selection of a suitable distance metric is the key component for the success of the application of the clustering technique [4]. The existing comparison of distance metrics has been designed for semi-Supervised partitional clustering. However, no research efforts have been investigated for the distance based ssHC methods.

This paper aims to empirically investigate the effects of different distance metrics on the performance of ssHC in relation to different linkage measures and numbers of constraints in order to understand the internal structure of the different datasets and to recommend the appropriate combination of distance metric, linkage and the number of constraints for better clustering results. The rest of this paper is organized as follows. Section 2 outlines the state-of-the-art research in the field. Section 3 provides a preliminary outline of ssHC, distance metrics and cluster linkages. Section 4 presents the experimental setup and other details, and Sect. 5 presents the results of the experiments and offers a brief discussion of these. Finally, Sect. 6 concludes the paper.

2   Related Work

The important step in a clustering analysis is the choice of an appropriate clustering method, which arises after having determined the variables to be used, the distance measures, and the criterion for clustering [5].

Over recent years, constraints based clustering [6] on semi-supervised data have been used. These constraints can be directly used from the original data (using partially labelled data) or provided by researchers [7]. Wagstaff and Cardie [6] suggest that pairwise constraints have two types based on the data instances provided by the information given: Must-Link (ML) constraints and Cannot-Link (CL) constraints. However, when the linking of data points involves various hierarchy levels in HC methods, CL and ML constraints are unsuitable [2]. In [2], the authors proposed a Must-Link-Before (MLB) constraint that identifies the order in which the clusters are merged, and can be naturally integrated into the HC process. They applied triple-wise constraints based on an ultra-metric dendrogram distance, instead of adopting pairwise constraints. Ultra-metric "is a special kind of tree metric where all elements of the input dataset are leaves in the underlying tree and all leaves are at the same distance from the root" [2].

In [8], the authors compared 5 HC techniques on binary data, and found Ward's method to be the best overall. In a similar study [9] on functional data, Ward's method was usually found to be the best, while average linkage performed well in some special situations, such as when the number of clusters was over specified. In [10], the authors used four agglomerative HC methods and found that the single linkage technique was the least effective, while the group average and Ward's methods gave the best overall recovery.

3   Background

3.1   Semi-supervised Hierarchical Clustering (ssHC)

HC is an unsupervised technique. It has two distinct approaches; data points are grouped into a hierarchical structure using either agglomerative or divisive approaches [11]. ssHC has been shown an important variant of the traditional HC paradigms with knowledge-based constraints [2]. In ssHC, the inclusion of knowledge in the clustering process for a set of instances (X) is done by constructing cluster order constraints (C) based on pairwise dissimilarities (D), which is defined as $X = \{x_1, x_2, ..., x_n\}$, $D = \{d(x_i, x_j) | x_i, x_j \in X\}$, $C = \{(x_i, x_j, x_k) | d(x_i, x_j) < d(x_i, x_k), x_i, x_j, x_k \in X$. The aim of ssHC is to construct tree hierarchies that satisfy as many constraints as possible respecting the order of merge of data points based on their dissimilarities. In constructing the set of constraints, there are two situations which need to be dealt with: handling implicit constraints and removing conflicts. Given two constraints $c_1 = (x_i, x_j, x_k)$ and $c_2 = (x_i, x_j, x_l)$ there exists an additional constraint $c_3 = (x_i, x_j, x_l)$, which is implicit but may not be in the initial set of constraints. To address this, the Floyd-Warshall algorithm is performed to find its transitive closure and to extend the set [6]. Furthermore, within the set of constraints there may be conflicting situations or contradictory constraints, e.g., $c_1 = (x_i, x_j, x_k)$ and $c_2 = (x_j, x_k, x_i)$ and deadlock constraints, e.g., $c_1 = (x_i, x_j, x_k)$, $c_2 = (x_i, x_k, x_l)$ and $c_3 = (x_i, x_l, x_j)$. These are addressed by removing iteratively some of the constraints that form these conflicts.

We have implemented two algorithms based on ultra-metric distance matrices. First, the optimization-based approach which is an optimization of the least square loss function expressed as a distance matrix, subject to ultra-metric and relative constraints known as Iterative Projection optimization (IPoptim) [2]. Second, the transitive dissimilarity-based approach which is a modified form of Floyd-Warshall algorithm. It fits the original objective of minimizing a transitive dissimilarity matrix to an ultra-matrix, subject to relative constraints that are integrated into the process of constructing the transitive dissimilarity and is known as the Ultra Transitive (Ultra-Tran) approach [2].

3.2   Distance Measures

(Dis)similarity of clusters and data points are measured by distance metrics, which must be chosen prior to clustering [4]. Distance metrics used in our experiments is presented in Table 1.

3.3   Linkage Measure Techniques

The procedure of integrating clustering together would be directly influenced by the selection of a linkage measure [17]. The different linkage mechanisms used in our experiments are discussed in Table 2.

Table 1. Different distance measures

Distance metric	Formula	(Dis)advantages
Euclidean	It is defined by [4] as: $D_e(x_i, x_j) = \sqrt{\sum_{k=1}^{d} \left| x_{ik} - x_{jk} \right|^2}$	It often has a negative impact on the performance of clustering in high dimensional datasets having different scales [12]
Mahalanobis	It is defined by [4] as: $D_{MA}(x_i, x_j) = (x_i - x_j)V^{-1}(x_i - x_j)^T$	It is a scale invariant and takes account of correlations of the dataset, it is different from the Euclidean distance. It is also sensitive to sampling fluctuations and violates the triangle inequality [13]
Standardised Euclidean	It is defined by [4] as: $D_{SE}(x_i, x_j) = (x_i - x_j)D^{-1}(x_i - x_j)^T$. where the variance of variable x_j over data points is represented by D as the diagonal matrix with diagonal elements given by ua_j^2	A diagonal Mahalanobis distance measure is produced when the squared standardized Euclidean distance is multiplied by the geometric mean of the variances. In the covariance matrix the diagonal Mahalanobis distance does not use the information of the diagonal [4]
Manhattan	It is defined by [4] as: $D_{Mn}(x_i, x_j) = \sum_{k}^{d} \left| x_{ik} - x_{jk} \right|$	The clusters formed using Manhattan distance tend to give clusters shaped as rectangles, and in high dimension datasets [4]
Cosine	It is defined by [4] as: $D_{\cos}(x_i, x_j) = 1 - \frac{x_i^T x_j}{\|x_i\| \|x_j\|}$	It is not invariant to signal shifts and cannot provide information regarding the magnitude of differences. It violates the triangle inequality and is scale invariant [4]
Spearman	It is defined by [4] as: $D_{Spear}(x_i, x_j) = 1 - S_C(x_i, x_j)$, where $S_C(x_i, x_j) = \frac{\sum_{k=1}^{d} (m_{ik}^r)(m_{jk}^r)}{\sqrt{\sum_{k=1}^{d} (m_{ik}^r)^2 \sum_{k=1}^{d} (m_{jk}^r)^2}}$ $m_{ik}^r = r(x_{ik}) - \bar{r}, m_{Jk}^r = r(x_{Jk}) - \bar{r}$	This measurement adopts non-parametric similarity, and is stronger against outliers than the Pearson correlation, but when data are converted, there is a loss of information, which is a disadvantage [4]
Chebyshev	It if formally defined in [4] as: $D_{Ch}(x_i, x_j) = \max_{1 \le l \le d}\left(\left| x_{li} - x_{jl} \right| \right)$	This measure shortens the time to determine distances between datasets, which is an advantage [14], and for quantitative variables and ordinal variables this metric could be used [15]
Canberra	It is mathematically defined by [4] as: $D_{Can}(x_i, x_j) = \sum_{k=1}^{d} \frac{\left| x_{ik} - x_{jk} \right|}{\|x_{ik}\| + \|x_{jk}\|}$	This distance is normally adopted when data are scattered around the origin [16]. When both coordinates are close to zero, this distance is sensitive to small changes [4]
Bray-Curtis	It is defined by [4] as: $D_{Bc}(x_i, x_j) = \frac{\sum_{k=1}^{d} \left| x_{ik} - x_{jk} \right|}{\sum_{k=1}^{d} (x_{ik} - x_{jk})}$	This distance does not satisfy the triangle inequality. Besides, if two data points are close to zero values, the results are undefined, which is a disadvantage of this distance [4]

Table 2. Different linkage mechanisms

Linkage measure	Definition	(Dis)advantages
Single	It measures the distance between two groups as the minimum distance found between one point from the first cluster and one point from the second cluster	Findings indicate that the single linkage technique is sensitive to outliers and noise, but overall its advantages include effective handling of cluster shapes that are non-elliptical and complicated [4]
Complete	It measures the distance between two groups as the maximum distance found between any pair of points in these two group	The most distant objects determine when two clusters merge in complete linkage clustering, which resolves the issue of large clusters produced by single linkage clustering, as large clusters can be broken, but non-spherical clusters cannot be detected [4]
Average	It measures the distances between each case in the first group and every case in the second group are computed and then averaged	This approach uses average pairwise distances between members to merge clusters, and this criterion forms an effective compromise, but remains susceptible to outliers and noise [18], and is biased towards globular clusters [4]
Centroid	The distance between two groups is computed as the distance between the centroids of the clusters	In contrast to average, complete and single linkage approaches, cluster distances do not demonstrate greater monotony with increased iterations in centroid linkage method, which is considered to be an important characteristic [18]
Median	It calculates the median distance between two clusters	Since this method uses the median instead of the mean that is used in the averaging technique, it reduces the effects of outliers by down-weighting [19]
Ward	It is based on the analysis of variance of the groups, in which the two clusters are joined when there is the lowest increase in the total value of the sum of the squares of the differences, within each cluster, from each data object to the centroid of the cluster	This method tends to produce clusters of approximately equal size. Although Ward's method is the most common technique, it is highly sensitive to outliers [20]

4   Experiments

4.1   Methodology and Evaluation Measures

The experiments were conducted on seven datasets of various dimensions and classes. Among these datasets, six of them were obtained from the UCI Machine Learning Repository [21]. The last one is the NTBC dataset containing 663 observations with 6 classes obtained by [22]. The specifications of the datasets used are summarized in Table 3.

Table 3. Dataset specifications showing number of data patterns (N), number of dimensions (n), number of classes (c) and the missing values (M)

Dataset	N	n	c	M
Iris	150	4	3	No
Wine	178	13	3	NO
Breast Cancer Wisconsin Original (BCWO)	683	10	2	Yes
Breast Cancer Wisconsin Diagnostic (BCWD)	569	32	2	NO
Pima Indian diabetes (Pima)	768	7	2	Yes
Cardiotocography (CTG)	2126	21	3	No
Nottingham Tenovus Breast Cancer (NTBC)	662	25	6	No

We compared the techniques proposed in [2] the iterative projection algorithm (IPoptim) and the transitive dissimilarity transformation algorithm (UltraTran) as described in Sect. 3. We implemented the clustering algorithm using programming language R version 3.3.2. We used the Rcpp library to increase the efficiency with the C++ programming language. As the algorithms make with matrices of high complexity of order $O(n^2)$ and $O(n^3)$, we conducted the experiments under the environment of Linux Ubuntu 14.04 LTS Intel(R) using a machine having a Core 2 Quad 2.50 GHz processor and 14 GB of RAM.

We evaluated the clustering algorithm with different distance metrics using F-score. It is proposed in [23] that the accuracy of HC is evaluated considering the entire hierarchy using the real classes of the data object. Suppose that the hierarchy is cut at a certain level and the group G_i is generated, where D_j is a group of data sharing the same label over C, then

$$F - \text{score}(G_i, D_j) = \frac{2 * \text{Recall}(G_i, C_j) * \text{Precision}(G_i, C_j)}{\text{Recall}(G_i, C_j) + \text{Precision}(G_i, C_j)}.$$

The F-score of group G_i is calculated as the maximum F-score among all classes C To calculate the F-score for the entire hierarchy, we calculate the weighted sum of each group's F-score of the form $F - \text{score}(H) = \sum_{i=1}^{N} \frac{|G_i|}{D} F - \text{score}(G_i)$, where $|D|$ samples are for HC with a total of $N = \frac{(1 + |D|) \cdot |D|}{2}$.

Garcia and Herrera [24] and Derrac et al. [25] considered non-parametric tests for multiple comparisons as well as post-hoc procedures for $N \times N$ comparisons, for classification tasks. The studies illustrate that first the Friedman test should be conducted in order to detect whether statistically significant differences occur among the examined algorithms. Thus, we applied the Friedman test methodology proposed by Demšar [26] for the comparison of several algorithms over multiple datasets. The Friedman test [27] is a non-parametric test (free distribution) used to compare observations repeated on the same subjects. This test determined the rank of the algorithms for each individual dataset.

4.2   Experimental Setup

To reflect the external domain knowledge through in the form of constraints, the actual class of the dataset is used in the construction of the constraints. For example, in a constraint $c = (x_i, x_j, x_k)$, the components x_i, x_j belong to the same class C_1 and the component x_k belongs to a different class C_2, where classes C_1 and C_2 are randomly chosen. The data points composing the constraints are chosen randomly and with replacement, that is, the same data object can appear in different constraints. Due to the randomness in the construction of the constraints 10 runs of the algorithms are performed and the results are averaged for obtaining the final results. The algorithms are run with varying proportions of constraints for all datasets, ranging from 10% to 60%. All constraint sets are preprocessed to eliminate any conflicts.

Varying Distance Metrics: To evaluate the effect of different distance metrics, we used the two most popular linkages, viz. single and complete, and fixed the number of constraints to 10%. The performance of both ssHC algorithms were evaluated for the 10 different distance metrics. The results are presented in Fig. 1.

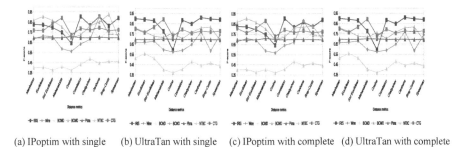

(a) IPoptim with single   (b) UltraTan with single   (c) IPoptim with complete   (d) UltraTan with complete

Fig. 1.   Performance of different distance measures.

Varying Linkages: We evaluated effect of changing linkages on the performance of ssHC algorithms. Two popular distance metrics, viz. Euclidean and Manhattan distance are considered with 10% constraints. The results are presented in Fig. 2.

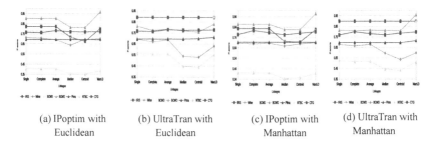

| (a) IPoptim with Euclidean | (b) UltraTran with Euclidean | (c) IPoptim with Manhattan | (d) UltraTran with Manhattan |

Fig. 2. Performance of different linkages.

Varying the Number of Constraints: We evaluated the effect of changing the number of constraints on the two ssHC algorithms. Here we used all ten of the distance metrics, but kept only one linkage, namely the single linkage. The performance of Iris, Wine, NTBC and CTG datasets are evaluated for the two ssHC algorithms. as these results show more interesting trends. The results are presented in Fig. 3.

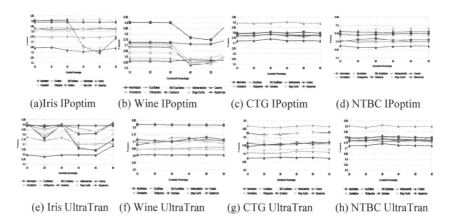

| (a)Iris IPoptim | (b) Wine IPoptim | (c) CTG IPoptim | (d) NTBC IPoptim |

| (e) Iris UltraTran | (f) Wine UltraTran | (g) CTG UltraTran | (h) NTBC UltraTran |

Fig. 3. Performance of varying the number of constraints

Evaluating the Clustering Efficiency: The two types of non-parametric statistical test are conducted on the most favourable 30 combinations of the two ssHC algorithms with the following distance measures: Manhattan, Euclidean, Correlation, Canberra, and BrayCurtis. These were used with the following linkage measures: complete, average and Ward.D with 10% constraints for all combinations. These achieved significant results in terms of F-score when employed to both IPoptim and UltraTran.

First, the non-parametric Friedman test is applied to evaluate the rejection of the hypothesis that all the classifiers perform equally well for a given level. It ranks the algorithms for each dataset separately. The best performing algorithm obtains the higher rank. Then, the Friedman test compares the average ranks of the algorithms and calculates the Friedman statistic. If a statistically significant difference in the performance is detected, this means that the performance of the classifiers have statistically significant difference. Average rankings of the 10 algorithms over seven datasets shown in Table 4. Second, Post-hoc testing is applied after the Friedman statistic in order to find out which pairs of our algorithms are significantly different then each other. It is based on a specific value on the significance level α. We fixed the significance level $\alpha = 0.05$ for all comparisons.

Table 4. Average rank of IPoptim (I) and UltraTran (U) algorithms with distance metric (Man - Manhattan, Eucl - Euclidean, Corr - Correlation, Canb - Canberra, and Bray - Bray-Curtis) and linkage measure, (comp - complete, avg - average and Ward), *Represents the best performing algorithm with distance measure and linkage mechanism.

Algorithm	#Rank	Algorithm	#Rank
1 U-Man-Ward*	7.00	2 U-Man-comp	11.07
3 I-Bray-Ward	11.92	4 U-Eucl-Ward	12.28
5 U-Canb-Ward	12.85	6 I-Canb-Ward	12.85
7 U-Eucl-avg	13.07	8 U-Canb-comp	13.28
9 U-Canb-avg	13.71	10 U-Bray-avg	13.71

5 Results and Discussion

The observations from the experiments performed over several datasets of varying size, class and dimension are as follows.

- Choice of distance measure: It is clearly evident that Cosine distance performs badly, plausibly due to its insensitivity to the scale of the data points (it only measures the angle between two vector points). Euclidean and Manhattan distance measures perform well among a range of datasets. However, Canberra appears to be perform favourably consistently in both IPoptim and UltraTran. It is difficult to tell whether these measures will outperform others for an arbitrary dataset. One could be recommended to use standard distance measures such as Euclidean, Manhattan or Mahalanobis distance first, and then to experiment with Correlation or Canberra (for widely separated clusters) or Chebyshev or Bray-Curtis.
- Choice of clustering algorithm: Both IPoptim and UltraTran perform almost equally for the seven datasets. However, UltraTran performs slightly better than IPoptim in some cases. This seems to be contrary to the observation made in [2] where IPoptim performed slightly better than UltraTran. However, the two algorithms may be experimented on more diverse datasets to evaluate their difference in performance, which appears to be minor. Any one of these two algorithms may be recommended.
- Choice of Linkage: single, complete or average linkages perform consistently better over most datasets. Median and centroid linkages should be avoided, as they are seen to degrade the F-score considerably. Ward.D sometimes proves to be a good measure (even better than the first three), but it possibly works better only when there are no outliers, as suggested in [20]. We would recommend single, complete and average linkages first, and then the Ward.D linkage.

- Choice of constraints: The quality of clustering (as evident from the F-score) with changing number of constraints shows clear trends. For small datasets such as Iris or Wine, the performance seems to be more sensitive to the number of constraints, whereas larger datasets show less sensitivity. The safe threshold would be to use 30% constraints, after which the performance of ssHC decreases in the case of most datasets.

  It may be observed in Figs. 1, 2 and 3, that the performance of F-score on the NTBC dataset with 25 dimensions is considerably lower than the other six datasets, especially with high dimensional data such as BCWD with 2 classes. We considered the issue with importance and decided to perform a new analysis of the same dataset after reducing 6 classes to a 2-class (merging the first three and the last three classes into two broader class) and to a 3-class as discussed in [22]. It was found that the F-score increased from 50–55% to 80–90%. This may suggest that those ssHC algorithms are not able to identify these subgroups directly. It may thus be inferred, not too pre-maturely, that datasets with a higher number of clusters are susceptible to showing a degraded performance in ssHC algorithms.

- Choice of appropriate combination of ssHC algorithm with linkage measure and similarity metric for given dataset: The non-parametric Garcia and Herrera [24] and Derrac et al. [25] statistical tests have been conducted and the output of these tests are presented in Tables 4 and 5.

Table 4 shows the output of Friedman test that determines the average ranking of different combinations of ssHC algorithms and there is any significant difference in performance across the reported observations. The average rank of the algorithms reveal that even for one algorithm (UltraTran) with different linkages and distance metrics, significant difference between their performances is present. For example, there is a difference of 6 ranks among UltraTran Canberra distance with complete linkage, average linkage and Ward.D linkage. Moreover, it is also observed from the Table 4 that among all linkage methods, the Ward.D linkage method performs consistently better in comparison to other linkage measures for given datasets. Additionally, the difference among average ranks of different combinations is a motivation for conducting investigation into which pairs of algorithms differ significantly. Hence, post-hoc analysis has been conducted and the output of the test is presented in Table 5.

Nemenyi [28] Holm [29], and Shaffer [30, 31] tests have been used to demonstrate significant for $N \times N$ comparisons for top 5 algorithm combinations out of 435 combinations produced by the post-hoc test are placed in Table 5 (with p-value < 0.05). These tests have been conducted on F-score values and the output is

Table 5. Adjusted p-VALUES (*Represents the best performing algorithm with distance measure and linkage mechanism).

I	Hypothesis	Unadjusted p	p-Neme	p-Holm	p-Shaf
1	U-Man-Ward vs U-Corr-comp*	0.001434	0.6239	0.6239	0.6239
2	U-Man-Ward vs U-Bray-comp	0.002398	1.0433	1.0409	0.9737
3	I-Corr-avg vs U-Man-Ward	0.005222	2.2716	2.2611	2.1201
4	I-Corr-Ward vs U-Man-Ward	0.006894	2.9988	2.9781	2.7988
5	U-Man-Ward vs U-Corr-avg	0.010768	4.684	4.641	4.3718

p-value for Nemenyi, Holm and Shaffer which is presented in Table 5. The lower the p-value (p-Nemenyi, p-Holm, p-Shaffer), the better the performance of the given algorithm combination. It can be seen from the Table 5 that Ward.D linkage with Manhattan distance is the best combination for the given datasets. These analyses also match the output of the Friedman test and the F-score analysis. Hence, it is advised to use ssHC algorithms with Ward.D and Manhattan distance for given datasets to effectively cluster the data.

6   Conclusion

In the current empirical study, the performance of ssHC algorithms has been evaluated with the ten most commonly used distance measures and six linkage measures. We also incorporated supervised information of benchmark datasets as input from class labels in the form of constraints. The experimental results show that both IPoptim and UltraTran perform almost equally across the seven datasets. However, UltraTran performs slightly better than IPoptim in some cases. This finding is also validated by the non-parametric statistical test namely Friedman test which reveals that UltraTran with Manhattan distance and Ward.D linkage is a top performing combination. Further, the post-hoc analysis called Nemenyi, Holm, and Shaffer tests were conducted and the output of these tests again emphasized that Ward.D linkage with Manhattan distance is better for given datasets. However, we have noticed that it is difficult to identify a single best set of measures that will work best for all datasets with ssHC, but in general, we would recommend that using Euclidean, Std. Euclidean and Canberra measures with single, complete, average or Ward.D linkages and a small number of constraints could serve as a favourable choice for initial exploration when using ssHC. Datasets with a higher number of clusters have been observed to affect the performance of ssHC greatly, especially with a high dimensional dataset. Moreover, it has been observed that increaseing the number of constraints does not always improve ssHC performance. Particularly, the decrease in performance of ssHC with higher number of constraints seems to be even more drastic for small datasets in comparison to with large datasets. Therefore, optimization of the important constraints for ssHC of different datasets needs to be investigated further in our future work.

Acknowledgement.   We would like to thank Professor Jonathan Garibaldi for sharing the NTBC dataset with us.

References

1. Pise, M.N., Kulkarni, P.: A survey of semi-supervised learning methods. In: Proceedings of the International Conference on Computational Intelligence and Security, pp. 30–34 (2008)
2. Zheng, L., Li, T.: Semi-supervised hierarchical clustering. In: 2011 IEEE 11th International Conference on Data Mining (ICDM), pp. 982–991. IEEE (2011)

3. Poria, S., Gelbukh, A., Das, D., Bandyopadhyay, S.: Fuzzy clustering for semi-supervised learning – case study: construction of an emotion lexicon. In: Batyrshin, I., González Mendoza, M. (eds.) MICAI 2012. LNCS, vol. 7629, pp. 73–86. Springer, Heidelberg (2013). doi:10.1007/978-3-642-37807-2_7

4. Kumar, V., Chhabra, J.K., Kumar, D.: Performance evaluation of distance metrics in the clustering algorithms. INFOCOMP J. Comput. Sci. **13**(1), 38–52 (2014)

5. Hair, J.F., Anderson, R.E., Tatham, R.L.: Multivariate Data Analysis. Macmillan, London (1987)

6. Wagstaff, K., Cardie, C.: Clustering with instance-level constraints. In: Proceedings of the 17th International Conference on Machine Learning, pp. 1103–1110 (2000)

7. Dasgupta, S., Ng, V.: Which clustering do you want? Inducing your ideal clustering with minimal feedback. J. Artif. Intell. Res. **39**, 581–632 (2010)

8. Hands, S., Everitt, B.: A monte carlo study of the recovery of cluster structure in binary data by hierarchical clustering technique. Multivar. Behav. Res. **22**, 235–243 (1987)

9. Ferreira, L., Hitchcock, D.B.: A comparison of hierarchical methods for clustering functional data. Commun. Stat. Simul. Comput. **38**(9), 1925–1949 (2009)

10. Milligan, G., Cooper, M.: A study of standardization of variables in cluster analysis. J. Classif. **5**, 181–204 (1988)

11. Bade, K., Nurnberger, A.: Personalized hierarchical clustering. In: Proceedings of the IEEE/WIC/ACM International Conference on Web Intelligence, pp. 181–187 (2006)

12. Jain, A.K., Duin, R.P.W., Mao, J.: Statistical pattern recognition: a review. IEEE Trans. Pattern Anal. Mach. Intell. **22**(1), 4–37 (2000)

13. Cherry, L.M., Case, S.M., Kunkel, J.G., Wyles, J.S., Wilson, A.C.: Body shape metrics and organismal evolution. Evolution **36**(5), 914–933 (1982)

14. Potolea, R., Cacoveanu, S., Lemnaru, C.: Meta-learning framework for prediction strategy evaluation. In: Filipe, J., Cordeiro, J. (eds.) ICEIS 2010. LNBIP, vol. 73, pp. 280–295. Springer, Heidelberg (2011). doi:10.1007/978-3-642-19802-1_20

15. Singh, A., Yadav, A., Rana, A.: K-means with three different distance metrics. Int. J. Comput. Appl. **67**(10), 13–17 (2013)

16. Charulatha, B.S., Rodrigues, P., Chitralekha, T., Rajaraman, A.: A comparative study of different distance metrics that can be used in fuzzy clustering algorithms. Int. J. Emerg. Trends Technol. Comput. Sci. (2013)

17. Mazzocchi, M.: Statistics for Marketing and Consumer Research. Sage Publications, Thousand Oaks (2008)

18. Albalate, A., Minker, W.: Semi-Supervised and Unsupervised Machine Learning: Novel Strategies. Wiley, Hoboken (2013)

19. Yashwant, S., Sananse, S.L.: Comparisons of different methods of cluster analysis with application to rainfall data. Int. J. Innov. Res. Sci. Eng. Technol. **4**(11), 10861–10872 (2015)

20. Gan, G., Ma, C., Wu, J.: Data clustering: theory, algorithms, and applications. In: Proceedings of the ASA-SIAM Series on Statistics and Applied Probability (2007)

21. Lichman, M.: UCI Machine Learning Repository (2013)

22. Soria, D., Garibaldi, J.M., Ambrogi, F., Green, A.R., Powe, D., Rakha, E., Macmillan, R.D., Blamey, R.W., Ball, G., Lisboa, P.J., Etchells, T.A., Boracchi, P., Biganzoli, E., Ellis, I.O.: A methodology to identify consensus classes from clustering algorithms applied to immunohistochemical data from breast cancer patients. Comput. Biol. Med. **40**(3), 318–330 (2010)

23. Larsen, B., Aone, C.: Fast and effective text mining using linear-time document clustering. In: Proceedings of International Conference on Knowledge Discovery and Data Mining, pp. 16–22 (1999)

24. Garcia, S., Herrera, F.: An extension on "statistical comparisons of classifiers over multiple data sets" for all pairwise comparisons. J. Mach. Learn. Res. **9**(Dec), 2677–2694 (2008)
25. Derrac, J., Garcia, S., Molina, D., Herrera, F.: A practical tutorial on the use of nonparametric statistical tests as a methodology for comparing evolutionary and swarm intelligence algorithms. Swarm Evol. Comput. **1**(1), 3–18 (2011)
26. Demšar, J.: Statistical comparisons of classifiers over multiple data sets. J. Mach. Learn. Res. **7**(Jan), 1–30 (2006)
27. Friedman, M.: The use of ranks to avoid the assumption of normality implicit in the analysis of variance. J. Am. Stat. Assoc. **32**(200), 675–701 (1937)
28. Nemenyi, P.B.: Distribution-free multiple comparisons. Ph.D. thesis, Princeton University (1963)
29. Holm, S.: A simple sequentially rejective multiple test procedure. Scand. J. Stat. **6**, 65–70 (1979)
30. Shaffer, J.P.: Modified sequentially rejective multiple test procedures. J. Am. Stat. Assoc. **81** (395), 826–831 (1986)
31. Shaffer, J.P.: Multiple hypothesis testing. Ann. Rev. Psychol. **46**(1), 561–584 (1995)

Classifying Non-linear Gene Expression Data Using a Novel Hybrid Rotation Forest Method

Huijuan Lu[1], Yaqiong Meng[1], Ke Yan[1(✉)], Yu Xue[2],
and Zhigang Gao[3]

[1] College of Information Engineering,
China Jiliang University, Hangzhou 310018, China
yanke@cjlu.edu.cn
[2] Nanjing University of Information Science & Technology,
Nanjing 210044, China
[3] College of Computer Science, Hangzhou Dianzi University,
Hangzhou 310018, China

Abstract. Rotation forest (RoF) is an ensemble classifier based on the combination of linear analysis theories and decision tree algorithms. In existing works, the RoF has demonstrated high classification accuracy and good performance with a reasonable number of base classifiers. However, the classification accuracy drops drastically for linearly inseparable datasets. This paper presents a hybrid algorithm integrating kernel principal component analysis and RoF algorithm (KPCA-RoF) to solve the classification problem in linearly inseparable cases. We choose the radial basis function (RBF) kernel for the PCA algorithm to establish the nonlinear mapping and segmentation for gene data. Moreover, we focus on the determination of suitable parameters in the kernel functions for better performance. Experimental results show that our algorithm solves linearly inseparable problem and improves the classification accuracy.

Keywords: Rotation forest · Kernel principal component analysis · Decision tree · Gene expression data

1 Introduction

Tumor is neoplasm cell, which is a kind of abnormal cell growing with cancer-causing factors [1]. Tumors can be generally categorized into malignant and benign tumor in medical profession, and the malignant tumor is also called as cancer. Cancer as a fatal disease posts great threat to human health. However, the occurrence of cancerous goes through a long incubation period [2]. The tumor can be controlled or killed in the early stage, given that the tumor is found early and the tumor is intervened effectively during this period. The tumor diagnosis and classification of gene expression data are two hot topics recently. However, gene expression data contains thousands of genes with a small number of samples, which makes it difficult to analyze and process [3]. Moreover, the gene expression data is well-known to be linearly inseparable, noisy and imbalanced [4–6]. For cancer/tumor diagnosis, it's important to know the functionalities of differential genes and the consequences of intervention [7]. The focuses of gene

© Springer International Publishing AG 2017
D.-S. Huang et al. (Eds.): ICIC 2017, Part III, LNAI 10363, pp. 732–743, 2017.
DOI: 10.1007/978-3-319-63315-2_64

data classification include classification accuracy, generalization ability, complexity, intelligibility, and the stability [8, 9].

The concept of Rotation forest (RoF) was introduced as an ensemble machine learning technique extending the concept of random forest (RF) based on primary classifiers, e.g. decision trees [10, 11]. The main features of RoF are the disjoint segmentation of the feature space and transformation of training and testing datasets, which result in more favorable classification accuracy comparing with traditional ensemble classifiers. RoF obtains feature subspace by random feature segmentation, transforms the attribute subset using principal component analysis (PCA) and calculates difference in training sample set by replicated sample method. However, the direct application of conventional RoF on gene expression data usually implies low classification accuracy due to the nonlinear property of the gene data.

Hybrid RoF approaches solve the linear inseparable problem of the gene expression data classification by combining RoF with non-linear machine learning methods, such as the kernel principal component analysis (KPCA), support vector machines (SVMs) [12, 13] and multiple-layer neural networks (MNNs) [14]. The KPCA method with a non-linear kernel, e.g. the radial basis function (RBF) or sigmoid function, transforms the original non-linear dataset to a reduced dataset with more separable feature sets [15–17]. To our knowledge, there is no existing work focusing on tuning the parameters for the KPCA mixing with RoF algorithm.

In this study, we present a hybrid RoF algorithm based on kernel principal component analysis (KPCA-RoF). The proposed KPCA-RoF algorithm employs the radial basis function (RBF) kernel for the KPCA part to establish the nonlinear mapping and transformation for gene expression data. In the experiment section, we focus on tuning the optimal solution for parameters of KPCA. The concept of F-measure is utilized to show the improvement of using our method compared to using conventional machine learning techniques. Furthermore, the proposed algorithm shows the ability of separating linearly inseparable datasets as well as improving the classification accuracy.

2 Related Work

Decision tree (DT) algorithm and its extensions, such as EG2, ID3, C4.5, CART and etc., are popular machine learning methods for fast classification of linearly separable datasets [18, 19]. Random forest (RF) is an ensemble DT classifier which splits the feature space using multiple DTs [20]. The rotation forest (RoF) is another extension of RF which assigns different sub-datasets with different DTs [21]. The RoF segments the feature space, eliminates the co-relations in each subset and iteratively obtains the most distinguishable datasets for each DT. It was already an ensemble algorithm when it was first introduced by Rodriguez et al. [10]. In the following year, Kuncheva and Rodríguez [22] proved that the RoF has better performance than traditional ensemble machine learning techniques, such as Adaboost, bagging and RF. They also found that the feature extracted by principal component analysis (PCA) is most suitable for enhancing the classification accuracy of RoF.

At the meanwhile, reports have been made saying that the RoF is not compatible with every dataset, especially for non-linear datasets. A series of hybrid RoF algorithms were proposed to solve the non-linear cases. Zhang and Zhang [23] started to introduce a hybrid algorithm combining RoF with Adaboost to increase the classification accuracy for linearly inseparable datasets. Mousavi et al. [24] proposed a hybrid RoF approach combining ensemble pruning and rotation forest to predict human miRNA target. For ensemble pruning, they utilizes genetic algorithm. In other words, a subset of classifiers from the heterogeneous ensemble is firstly selected by genetic algorithm. Next, the selected classifiers are trained based on the rotation forest method and then combined using weighted majority voting. Wong et al. [25] combined the RoF and local phase quantization (LPQ), which evaluates the performance of the RoF ensemble classifier with the state-of-the-art support vector machine (SVM) classifier. Wong et al. also indicated that the proposed method might play a complementary role for future proteomics research. Ayerdi and Romay [26] introduced a new spectral classifier, which was called anticipative hybrid extreme rotation forest (AHERF). The hybrid method combines learning approach AHERF starts with a model selection phase, uses a small subsample of the training data, and defines a ranking-based selection probability distribution of the classifier architectures which will be used in the ensemble. Thus, the architecture applicable for the data domain will be used more frequently to train individual classifiers in the ensemble. However, the RoF algorithm research and application are still limited for gene expression data classification; there is still quite a large gap to explore.

Recent works also show that the hybrid machine learning techniques integrating KPCA method effectively extract significant information from non-linear datasets. Kuang et al. [27] proposed a novel hybrid method which involves a KPCA with radial basis function (RBF) kernel, a multi-layer support vector machine (SVM) and genetic algorithm (GA) for intrusion detection. Mengqi et al. [28] introduced a hybrid approaches for online fault detection of non-linear data, which include a hybrid method combining KPCA and Hypersphere Support Vector Machine (HSSVM) and another hybrid method mixing Recursive KPCA, Adaptive Control Limit (ACL) and Online Sequential Extreme Learning Machine (OS-ELM) together. Luo et al. [29] adopted the KPCA for Nonlinear Process Fault Diagnosis in Military Barracks. They extended the traditional KPCA by integrating the statistic pattern analysis framework (SPA), and named the extended KPCA as multivariate statistical kernel principal component analysis (MSKPCA). Experimental results showed that the MSKPCA has better performance compared to PCA and KPCA for fault detection and diagnosis in the particular application field.

3   The Proposing Algorithm Based on KPCA

Our proposing KPCA-RoF algorithm divides the training dataset into K subsets, transforms each subset using kernel principal analysis (KPCA) and utilizes multiple DT base classifiers to produce the find classification results. In the testing phase, the testing dataset is again transformed by KPCA and then inserted into the trained KPCA-RoF classifier which is a combination of various DTs.

The overall structure of KPCA-RoF is shown in Fig. 1 and can be described as follows:

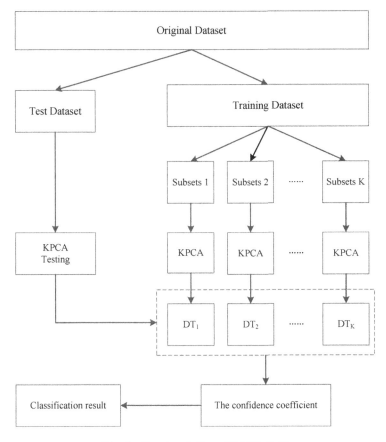

Fig. 1. The description of KPCA-RoF.

(1) For a given n-dimensional dataset $S : \{(x_i, y_i)\}_{i=1}^{L}$, we remove partial class labels and call it H, divide H into K disjoint subsets.

(2) Suppose that $D_1, D_2, \cdots D_L$ are the base classifiers. We randomly select m features to form a feature subset H_{ij} corresponding to D_i, where $m = n/k$, and $1 \leq i \leq L, 1 \leq j \leq k$. From each H_{ij}, KPCA is used to transform the feature subset to a coefficient matrix C_{ij}.

(3) Repeat step (2) K times.

(4) Combine the K coefficient matrixes to form a large sparse matrix. The rotation matrix R_i for the base classifier D_i is defined as:

$$R_i = \begin{bmatrix} C_{i1} & 0 & \cdots & 0 \\ 0 & C_{i2} & \cdots & 0 \\ \vdots & \vdots & \vdots & \vdots \\ 0 & 0 & \cdots & C_{ik} \end{bmatrix} \tag{1}$$

(5) The classifier D_i uses dataset $H_{ij}R_i$ as the training set. In the testing process, each testing sample x is feed into the ensemble classifier after multiplying R_i. The confidence coefficient [30] of determining its class is given as:

$$u_c(x) = \frac{1}{L}\sum_{i=1}^{L} p_i(x) \tag{2}$$

3.1 Kernel Principal Component Analysis (KPCA)

Suppose $x_1, x_2, \cdots, x_m$ are training samples, each of which consists of N features. A kernel function maps input sample x into feature space F by mapping function Φ. The mapping function $\Phi(x)$ from each sample x to a reduced sample is defined as:

$$\Phi : \mathrm{R}^N \rightarrow F, \quad x \mapsto \xi = \Phi(x) \tag{3}$$

The feature space F is generated by $\Phi(x_1), \Phi(x_2), \cdots, \Phi(x_m)$. The training samples after mapping with centralization is processed as follows:

$$\sum_{\mu=1}^{m} \Phi(x_\mu) = 0 \tag{4}$$

The covariance matrix of feature space after mapping is:

$$\bar{C} = \frac{1}{m}\sum_{i=1}^{m} \Phi(x_i)\Phi(x_i)^T \tag{5}$$

The feature equation according to conventional PCA is as follows:

$$\lambda V = \bar{C}V \tag{6}$$

where λ is the eigenvalue and V is the eigenvector both belonging to the space, which is generated by $\Phi(x_i)$ with the following equation:

$$\lambda\{\Phi(x_i) \cdot V\} = \{\Phi(x_i) \cdot \bar{C}V\} \tag{7}$$

A parameter α_i is adopted to make V to be a linear expression with the equation below:

$$V = \sum_{i=1}^{m} \alpha_i \Phi(x_i) \tag{8}$$

By merging Eqs. (7) and (8), the inner product of mapping data is defined as a m-order matrix K. The parameters are calculated by the kernel function that we chose in the early phase, namely:

$$K_{ij} = (\Phi(x_i) \cdot \Phi(x_j)) \tag{9}$$

Next, we obtain Eq. (10), which is equivalent to the Eq. (6):

$$l\lambda\alpha = K\alpha \tag{10}$$

Where $\alpha = (\alpha_1, \alpha_2, \cdots, \alpha_m)^{\mathrm{T}}$, and matrix K is the transformation matrix for the KPCA algorithm.

Suppose that the eigenvalues of matrix K are given as: $\lambda_1 \leq \lambda_2 \leq \cdots \leq \lambda_m$, and the corresponding eigenvectors are given as: $\alpha_1, \alpha_2, \cdots, \alpha_m$. The eigenvalue and eigenvector of matrix K can be calculated.

3.2 Kernel Function and Its Parameters

In order to segment the linearly inseparable dataset, the dataset is required to be transformed into a feature space through a kernel function [31].

Several common kernel functions are shown as follows:

Linear kernel function:

$$K(x_i, x_j) = x_i^T x_j \tag{11}$$

Polynomial kernel function:

$$K(x_i, x_j) = (\gamma x_i^T x_j + r)^p \tag{12}$$

Radial basis function (RBF):

$$K(x_i, x_j) = \exp(-\delta \|x - x_i\|^2) \tag{13}$$

In this study, RBF becomes the best choice among all kernel functions because of the low magnitude complexity, less parameters and strong representation. Moreover, we control the degree of over-fitting by adjusting the size of the parameters of kernel function to get suitable algorithm. Different kinds of kernel functions result in different classification results. The diversity of kernel function can be evaluated by measuring interclass distances [32]. The kernel function is widely used, independent with any particular classification algorithm, and uninfluenced by generalization ability.

The optimization of parameters uses the distances between feature classes as a reference index. The angle and the distance in the feature space of training data after mapping are given as:

$$\cos <\Theta_{i,j}> \; = \frac{K(x_i, x_j)}{\sqrt{K(x_i, x_i) \cdot K(x_j, x_j)}}, \quad 0 \leq \Theta_{i,j} \leq \pi/2 \tag{14}$$

We use $D_{i,j}$ to represent the distance between two vectors:

$$D_{i,j} = \sqrt{K(x_i, x_i) + K(x_j, x_j) - 2K(x_i, x_j)} \tag{15}$$

After replacing a specific kernel function in Eqs. (12) and (13), we have:

$$\cos < \Theta_{i,j} > = \frac{\exp(-\delta \|x_i - x_j\|^2)}{\sqrt{\exp(-\delta \|x_i - x_i\|^2) \cdot \exp(-\delta \|x_j - x_j\|^2)}} = \exp(-\delta \|x_i - x_j\|^2) \tag{16}$$

All $\Theta_{i,j}$ must meet the following condition:

$$\Theta_{i,j} \in [0, \pi/2] \tag{17}$$

The distance between two vectors can also be obtained in a similar way:

$$D_{i,j} = \sqrt{2 - 2\exp(-\delta \|x_i - x_j\|^2)} \tag{18}$$

From the Eqs. (16–18), there is only one parameter δ that influences the class distance and angle, which eventually determines the class distribution in the feature space as well as the RoF classification results.

While the value of the parameter δ approaches to 0, the cosine of the angle approaches to 1, which means the value of two vectors' gap angle tends to 0 after mapping. In this situation, the distance vector tends to be zero, which means all the samples are mapped to one point, leading to unclassifiable samples. While the value of the parameter δ tends to infinity, the two vectors' angle tends to $\pi/2$; the distance of the sample tends to a constant, which means the sample set is mapped into a n-dimensional feature space; and the feature vectors are pairwise orthogonal. While the dimension of the feature space increases to n (n is the number of samples) monotonically with the increment of δ, the vector angle and distance in feature space also increase monotonically, approaching to $\pi/2$ and $\sqrt{2}$ respectively.

Given a training set X containing L samples and C categories:

$$X = \bigcup_{i=1}^{C} x_i, \quad L = \bigcup_{i=1}^{C} l_i \tag{19}$$

The averages of the samples after mapping in the feature space can be calculated as:

$$m_1 = \frac{1}{l_1} \sum_{i=1}^{l_1} \Phi(x_i^{(1)}), \quad m_2 = \frac{1}{l_2} \sum_{i=1}^{l_2} \Phi(x_i^{(2)}) \tag{20}$$

The average distance between the classes in the mapping space is expressed as follow:

$$D(C_1, C_2) = \|m_1 - m_2\|^2 = \frac{1}{l_1^2} \sum_{i=1}^{l_1} \sum_{j=1}^{l_1} \exp(-\delta \|x_i^{(1)} - x_j^{(2)}\|^2)$$

$$+ \frac{1}{l_2^2} \sum_{i=1}^{l_2} \sum_{j=1}^{l_2} \exp(-\delta \|x_i^{(1)} - x_j^{(2)}\|^2) - \frac{2}{l_1 l_2} \sum_{i=1}^{l_1} \sum_{j=1}^{l_2} \exp(-\delta \|x_i^{(1)} - x_j^{(2)}\|^2). \tag{21}$$

The cosine value between classes in the kernel space is:

$$\cos <\Theta_{C1,C2}> \ = \frac{1}{l_1 l_2} \sum_{i=1}^{l_1} \sum_{j=1}^{l_2} \exp(-\delta \|x_i^{(1)} - x_j^{(2)}\|^2) \tag{22}$$

The cosine between classes in the kernel space is:

$$\cos <\Theta_{C_1,C_1}> \ = \frac{1}{l_1^2} \sum_{i=1}^{l_1} \sum_{j=1}^{l_1} \exp(-\delta \|x_i^{(1)} - x_j^{(1)}\|^2)$$

$$\cos <\Theta_{C_2,C_2}> \ = \frac{1}{l_2^2} \sum_{i=1}^{l_2} \sum_{j=1}^{l_2} \exp(-\delta \|x_i^{(2)} - x_j^{(2)}\|^2) \tag{23}$$

Summarizing Eqs. (21), (22) and (23):

$$D(C_1, C_2) = \cos <\Theta_{C_1,C_1}> \ + \cos <\Theta_{C_2,C_2}> \ - 2\cos <\Theta_{C_1,C_2}> \tag{24}$$

Equation (21–24) show that the class separation distance can be calculated by inter class angle and within class angle. When within angle is big and inter class angle is small, the class separation distance is enlarged. According to Eq. (16), inter class angle and within class angle both increase with the increments of parameter δ; therefore there may exist one parameter marking the class separation distance largest.

4   Results

The Acute Lymphocytic Leukemia (ALL), Breast, Central Nervous System (CNS), Colon, Small Round Blue Cell Tumor (SRBCT), Lung datasets are employed in the experiment, all datasets are downloaded from public available website: http://datam.i2r. a-star.edu.sg/datasets/krbd/. We randomly extract features from the datasets 30 times; set the confidence threshold as 0.95. The optimal parameters for the first three datasets are shown in Fig. 1. The range of value δ is given as follows: $\delta = \{10^{-6}, 10^{-5}, \cdots 10^5, 10^6\}$.

We compare our algorithm with the Bagging algorithm and the original RoF algorithm. Our results demonstrate the effectiveness of the proposing algorithm. All algorithms use C4.5 DT as based classifier. The main control variables are the number of base classifiers N and the number of samples M. We adjust one variable at each time to obtain the results.

4.1   Statistical Analysis

In this experiment, we use variance S^2, weighted average $\overline{X}$ and the F-measure to verify the classifier improvement [33].

Table 1 lists the three variables ($\overline{X}$, S^2, F) of different approaches on different gene expression datasets.

Table 1. Statistical analysis

Dataset	Algorithm	$\overline{X}$	S^2	F
Breast	KPCA-RoF	0.8087	0.0059	4.51
	RoF	0.7092	0.0266	
CNS	KPCA-RoF	0.7975	0.0153	2.2
	RoF	0.7523	0.0337	
ALL	KPCA-RoF	0.7787	0.0134	3.65
	RoF	0.7302	0.0489	
Colon	KPCA-RoF	0.8617	0.0040	1.65
	RF	0.7923	0.0066	
SRBCT	KPCA-RoF	0.7852	0.0076	1.81
	RF	0.7010	0.0042	
Lung	KPCA-RoF	0.9136	0.0007	2.71
	Bagging	0.6426	0.0019	

4.2   The Classification Performance Comparison

Table 2 shows the classification accuracy for different base classifier number. From the results, a reasonable accuracy is obtained with 15 base classifiers, and further increment of the number of base classifiers does not increase the accuracy much. We test several other ensemble algorithms for each dataset, i.e.: KPCA-RoF, RoF, RF and Bagging. It is noted that in order for fair comparison, we transform the datasets used in RoF, RF and bagging algorithms to the same number of features as the dataset used in KPCA-RoF algorithm. In general, the classification results can be interpreted as follows: Bagging decision tree (BDT) is the worst among all four methods; our method is the best. The BDT is a simple integration of DT; therefore the accuracy improvement will not be obvious; RF adds the random segmentation of the feature space, which gives better classification results. From the experimental results, RoF and our method produce the best classification accuracy because of the feature space segmentation and transformation. Our method can achieve highest accuracy with a small number of base classifiers. This shows that the nonlinear transformation compared with the linear transformation is more adaptable to various datasets. Meanwhile, nonlinear transformation will increase the computation complexity, and also complicate the algorithm.

Table 2. Classification accuracy and the number of base classifiers

Int. Level	Algorithm	3	7	10	15	20	25	30	35	40
Breast	KPCA-RoF	0.6454	0.7233	0.8524	**0.8576**	0.8575	0.8577	0.8577	0.8576	0.8577
	RoF	0.5921	0.6751	0.7532	**0.7689**	0.7672	0.8119	0.8154	0.8156	0.8163
	RF	0.5827	0.6524	0.6856	**0.7265**	0.7471	0.7521	0.7537	0.7635	0.7613
	Bagging	0.5612	0.6653	0.6748	**0.7167**	0.7425	0.7350	0.7432	0.7498	0.7429
CNS	KPCA-RoF	0.6788	0.7783	0.8742	**0.8745**	0.8743	0.8755	0.8739	0.8735	0.8741
	RoF	0.5823	0.6459	0.7864	**0.7998**	0.8046	0.8076	0.8154	0.8158	0.8164
	RF	0.5299	0.6348	0.6654	**0.7012**	0.7257	0.7659	0.7861	0.7935	0.7956
	Bagging	0.4769	0.5413	0.6557	**0.7068**	0.7163	0.7118	0.7341	0.7323	0.7312
ALL	KPCA-RoF	0.5674	0.6358	0.7618	**0.8416**	0.8505	0.8517	0.8516	0.8516	0.8516
	RoF	0.5465	0.5616	0.7386	**0.7937**	0.8157	0.8162	0.8169	0.8156	0.8160
	RF	0.5412	0.5652	0.6891	**0.7390**	0.7693	0.7971	0.8065	0.8068	0.8071
	Bagging	0.5498	0.5616	0.6256	**0.6689**	0.7006	0.7112	0.7133	0.7135	0.7133
Colon	KPCA-RoF	0.7143	0.8247	0.8759	**0.9091**	0.9093	0.9035	0.8963	0.8986	0.9090
	RoF	0.6339	0.7382	0.8095	**0.8571**	0.8557	0.8512	0.8486	0.8471	0.8457
	RF	0.6429	0.7147	0.7593	**0.7854**	0.7868	0.7884	0.7835	0.7791	0.7818
	Bagging	0.5238	0.5735	0.6465	**0.6858**	0.6867	0.6889	0.6896	0.6863	0.6832
SRBCT	KPCA-RoF	0.6035	0.7243	0.8127	**0.8444**	0.8496	0.8484	0.8455	0.8407	0.8435
	RoF	0.5873	0.6825	0.7907	**0.8205**	0.8274	0.8245	0.8228	0.8165	0.8153
	RF	0.5641	0.6713	0.7264	**0.7409**	0.7425	0.7476	0.7446	0.7425	0.7408
	Bagging	0.4486	0.5614	0.6429	**0.6767**	0.6782	0.6786	0.6772	0.6718	0.6746
Lung	KPCA-RoF	0.8574	0.8994	0.9248	**0.9305**	0.9340	0.9360	0.9349	0.9281	0.9247
	RoF	0.8255	0.8725	0.9060	**0.9195**	0.9195	0.9128	0.9106	0.8993	0.8859
	RF	0.7505	0.8067	0.8434	**0.8718**	0.8735	0.8768	0.8727	0.8693	0.8674
	Bagging	0.5538	0.6191	0.6497	**0.6597**	0.6686	0.6738	0.6746	0.6787	0.6810

5 Conclusion

RoF is a machine learning technique which generates a transformation matrix through the segmentation, sampling and transformation of the dataset. Based on the conventional RoF algorithm, we design a hybrid method to deal with the non-linear gene expression datasets. The original training dataset is randomly segmented into subsets, where customized KPCA transformation is applied to each of the subsets. The RBF kernel is employed to deal with the non-linear property of the data. For each transformed subset, a decision tree based classifier is trained; and the combination of all DTs forms the final ensemble classifier which is named as KPCA-RoF.

In the experiment part, we carefully tune the kernel parameter to obtain the optimal result for classification. With the optimized kernel parameter, the proposed KPCA-RoF algorithm is capable to produce highest classification accuracy over existing classifiers, such as BDT, RF and RoF, which is also verified by the concept of F-measure. For linearly inseparable datasets, such as the gene expression datasets, the KPCA-RoF has demonstrated its clustering power over various datasets downloaded from online resources.

References

1. Wang, W., Wang, Z., Bu, X., Li, R., Zhou, M., Hu, Z.: Discovering of tumor-targeting peptides using bi-functional microarray. Adv. Healthcare Mater. **4**(18), 2802–2808 (2015)
2. Stanbury, J.F., Baade, P.D., Yu, Y., Yu, X.Q.: Impact of geographic area level on measuring socioeconomic disparities in cancer survival in New South Wales, Australia: a period analysis. Cancer Epidemiol. **43**, 56–62 (2016)
3. Liszewski, K.: Exploiting Gene-Expression Data (2012)
4. Liu, Y., Lu, H., Yan, K., Xia, H., An, C.: Applying cost-sensitive extreme learning machine and dissimilarity integration to gene expression data classification. Comput. Intell. Neurosci. **2017**, 1–9 (2016)
5. Wang, Z., Zhang, J.: Impact of gene expression noise on organismal fitness and the efficacy of natural selection. Proc. Natl. Acad. Sci. **108**(16), E67–E76 (2011)
6. Pastinen, T., Sladek, R., Gurd, S., Ge, B., Lepage, P., Lavergne, K., Verner, A.: A survey of genetic and epigenetic variation affecting human gene expression. Physiol. Genomics **16**(2), 184–193 (2004)
7. Lu, H.J., An, C.L., Zheng, E.H., Lu, Y.: Dissimilarity based ensemble of extreme learning machine for gene expression data classification. Neurocomputing **128**, 22–30 (2014)
8. Langdon, W.B., Buxton, B.F.: Genetic programming for mining DNA chip data from cancer patients. Genetic Programm. Evolvable Mach. **5**(3), 251–257 (2004)
9. Furey, T.S., Cristianini, N., Duffy, N., Bednarski, D.W., Schummer, M., Haussler, D.: Support vector machine classification and validation of cancer tissue samples using microarray expression data. Bioinformatics **16**(10), 906–914 (2000)
10. Rodriguez, J.J., Kuncheva, L.I., Alonso, C.J.: Rotation forest: a new classifier ensemble method. IEEE Trans. Pattern Anal. Mach. Intell. **28**(10), 1619–1630 (2006)
11. Su Chong, J., Shenggen, L.Y., et al.: Improving random forest and rotation forest for highly imbalanced datasets. Intell. Data Anal. **19**(6), 1409–1432 (2015)

12. Suykens, J.A., Vandewalle, J.: Least squares support vector machine classifiers. Neural Process. Lett. **9**(3), 293–300 (1999)
13. Gu, B., Sheng, V.S., Tay, K.Y., Romano, W., Li, S.: Incremental support vector learning for ordinal regression. IEEE Trans. Neural Netw. Learn. Syst. **26**(7), 1403–1416 (2015)
14. Krizhevsky, A., Sutskever, I., Hinton, G.E.: Imagenet classification with deep convolutional neural networks. In: Advances in Neural Information Processing Systems, pp. 1097–1105 (2012)
15. Mika, S., Schölkopf, B., Smola, A.J., Müller, K.R., Scholz, M., Rätsch, G.: Kernel PCA and de-noising in feature spaces. In: NIPS, vol. 11, pp. 536–542, December 1998
16. Schölkopf, B., Smola, A., Müller, K.-R.: Kernel principal component analysis. In: Gerstner, W., Germond, A., Hasler, M., Nicoud, J.-D. (eds.) ICANN 1997. LNCS, vol. 1327, pp. 583–588. Springer, Heidelberg (1997). doi:10.1007/BFb0020217
17. Hoffmann, H.: Kernel PCA for novelty detection. Pattern Recogn. **40**(3), 863–874 (2007)
18. Olshen, L.B.J.F.R., Stone, C.J.: Classification and regression trees. Wadsworth Int. Group **93** (99), 101 (1984)
19. Quinlan, J.R.: C4. 5: Programs for Machine Learning. Elsevier, Amsterdam (2014)
20. Breiman, L.: Random forests. Mach. Learn. **45**(1), 5–32 (2001)
21. Rodriguez, J.J., Kuncheva, L.I., Alonso, C.J.: Rotation forest: a new classifier ensemble method. IEEE Trans. Pattern Anal. Mach. Intell. **28**(10), 1619–1630 (2006)
22. Kuncheva, L.I., Rodríguez, J.J.: An experimental study on rotation forest ensembles. In: Haindl, M., Kittler, J., Roli, F. (eds.) MCS 2007, vol. 4472, pp. 459–468. Springer, Heidelberg (2007). doi:10.1007/978-3-540-72523-7_46
23. Zhang, C.X., Zhang, J.S.: RotBoost: a technique for combining rotation forest and AdaBoost. Pattern Recogn. Lett. **29**(10), 1524–1536 (2008)
24. Mousavi, R., Eftekhari, M., Haghighi, M.G.: A new approach to human microRNA target prediction using ensemble pruning and rotation forest. J. Bioinform. Comput. Biol. **13**(06), 1550017 (2015)
25. Wong, L., You, Z.H., Ming, Z., Li, J., Chen, X., Huang, Y.A.: Detection of interactions between proteins through rotation forest and local phase quantization descriptors. Int. J. Mol. Sci. **17**(1), 21 (2015)
26. Ayerdi, B., Romay, M.G.: Hyperspectral image analysis by spectral-spatial processing and anticipative hybrid extreme rotation forest classification. IEEE Trans. Geosci. Remote Sens. **54**(5), 2627–2639 (2016)
27. Kuang, F., Xu, W., Zhang, S.: A novel hybrid KPCA and SVM with GA model for intrusion detection. Appl. Soft Comput. **18**, 178–184 (2014)
28. Mengqi, N., Jingjing, D., Tianzhen, W., Diju, G., Jingang, H., Benbouzid, M.E.H.: A hybrid kernel PCA, hypersphere SVM and extreme learning machine approach for nonlinear process online fault detection. In: IECON 2015-41st Annual Conference of the IEEE Industrial Electronics Society, pp. 002106–002111. IEEE, November 2015
29. Luo, K., Li, S., Ren Deng, W.Z., Cai, H.: Multivariate statistical kernel PCA for nonlinear process fault diagnosis in military barracks. Int. J. Hybrid Inf. Technol. **9**(1), 195–206 (2016)
30. Boujnouni, M.E., Jedra, M., Zahid, N.: Support vector domain description with a new confidence coefficient. In: 2014 9th International Conference on Intelligent Systems: Theories and Applications (SITA-2014), pp. 1–8. IEEE, May 2014
31. Amari, S.I., Wu, S.: Improving support vector machine classifiers by modifying kernel functions. Neural Netw. **12**(6), 783–789 (1999)
32. Devijver, P.A., Kittler, J.: Pattern Recognition: A Statistical Approach. Prentice Hall, Upper Saddle River (1982)
33. Brereton, R.G.: The F distribution and its relationship to the chi squared and t distributions. J. Chemom. **29**(11), 582–586 (2015)

Intelligent Data Analysis and Prediction

A Classification and Predication Framework for Taxi-Hailing Based on Big Data

Changqing Yin$^{(\boxtimes)}$, Yiwei Lin, and Chen Yang

College of Software Engineering, Tongji University, Shanghai 201804, China
yin_cq@qq.com

Abstract. As an important public transportation, Taxi is used for passengers every day, which is one of the primary causes for traffic jams. For passengers, knowing the difficulty degree of taking a taxi at a particular time and place can help us plan the journey effectively. Nevertheless, the existing predication models for traffic are not able to express the difficulty degree of choosing a taxi. In order to solve this problem, we can use historical data of taxi status to analysis and predict the possibility of taxi-hailing at a specific time and place. In this paper, we present a classification and predication framework for taxi-hailing. In this framework, firstly we use K-Means clustering algorithm to divide the taxi data into different clusters. Then we use Echarts to extract the features of each cluster in order to show the different difficulty degree. Next we use neural network to generate the predication result using the result of K-Means. On this basis, we propose a method to make the predication of taxi-hailing at a particular time and place, which can calculate the possibility score of taxi-hailing. Finally, we make a prediction using this framework and compare the predication results with the actual travelling data report. The comparison results verify the reliability of this framework.

Keywords: K-Means · Neural network · Big data · Possibility score · Prediction for taxi-hailing

1 Introduction

The taxicab industry is developing quickly and has been becoming an important way of transport in urban centers. As of the end of June 2016, there are around 52,000 taxies in Shanghai and taxies account for 23% of the total daily traffic. However, the inhomogeneous distribution and scheduling of taxi usually result in increasing difficulty to take a taxi for passengers. To help citizens take a taxi more easily, it is significative to predict the degree of difficulty to take a taxi.

In many existing systems and predication model for taxi predication, they focus on providing real-time peripheral taxi distribution, but not showing the success probability of taking a taxi. In this paper, we proposed a predication framework which can show the success probability of taking a taxi based on a particular place and timestamp. Firstly, our framework processes raw data automatically using clustering algorithm, and then we use processed data to perform the prediction. Finally, our framework will visualize the data to give users a better view of the result. Our work makes the following main contributions:

© Springer International Publishing AG 2017
D.-S. Huang et al. (Eds.): ICIC 2017, Part III, LNAI 10363, pp. 747–758, 2017.
DOI: 10.1007/978-3-319-63315-2_65

- We propose a classification and predication framework to predict the difficulty degree of taxi-hailing at a specific time and place, which is not evolved in the existing traffic predication models.
- We implement our approach in a tool for the predication, which is freely available.
- We evaluate the effectiveness and reliability of our framework in predicting the difficulty using real data.

The rest of this paper is organized as follows. After describing the related work about prediction framework for traffic data, we introduce the workflow and functionality of our program in Sect. 3. In Sect. 3.1, we demo the data preprocessing method and present a new method to estimate the time difference called "Time Ring". Then in Sect. 3.2 we present the pre-experiment and advise the process of clustering and predication. Section 4 is devoted to introduce and evaluate the experiment result for this framework. In Sect. 5, we analyze the advantages and disadvantages of the framework.

2 Related Works

Nowadays, the research in the area of big data becomes very popular. By big data, we can summarize the regularity of something and make a prediction of its trend, such as the prediction of the traffic flow. In the aspect of traffic, the prediction research based on traffic data becomes more and more important for city development. Related works in this field contain nonlinear network traffic prediction, driver destination prediction, regional taxicab service rate prediction and so on. All of these predictions are based on machine learning and there are many prediction algorithms such as decision tree, artificial neural network, and support vector machine in the machine learning field. The commonly applied methods are neural network and k-means in the field of traffic.

In terms of destination prediction, an outstanding research is a competition held in European Conference on Machine Learning and Principles and Practice of Knowledge Discovery in Databases 2016. In this competition, contestants are required to build a predictive framework which is able to infer the final destination of taxi rides on partial trajectories. And the key to first prize [1] is a model architecture based on multilayer perceptron neural network. In this model, one perceptron consists of one input layer (the input size is fixed), one or more hidden layers (input data for calculations), and one output layer (output prediction result). They use the clustering algorithm at the destination of all training trajectories and approximately 3392 popular destination data points were obtained. The penultimate layer of this model is a softmax, which is used to predict the probability of each data point that can become the taxi destinations. To minimize the mistake between the forecast and actual locations, they use a stochastic gradient descent method.

In the field of taxi quantity prediction, Zhang et al. [2] think that the variation of the city taxi quantity is a complex dynamic system with complex features trends: random, nonlinear, and inadequate data. Therefore, the traditional linear prediction model and single nonlinear prediction model can't accurately reflect the complexity of the variance because they have easily resulted in lower prediction accuracy. On this basis, they propose a GM-BP model. In this model, the GM (1, 1) grey prediction model is used to predict the variation trend of the city taxi quantity, and then the residuals of GM (1, 1) model is corrected by BP neural network model in order to enhance the forecast accuracy of city taxi quantity. The result indicates that BP neural network model is suitable for non-linear, volatile data series prediction, and it can predict random values in city taxi quantity variation process accurately.

For traffic abnormity prediction, Shi et al. [3] try to build a basic model based on three-layer neural network. Traffic abnormity can be predicted by neural network in the dimension of time. At the same time, Münz et al. [4] present a novel flow-based anomaly detection scheme based on the K-mean clustering algorithm. In their paper, they divide the datasets into different clusters using the K-means clustering algorithm. Then they deploy K-means clustering to determine whether the cluster is normal traffic cluster or anomalous traffic cluster. The result in their paper shows that the clustering algorithm can improve the detection quality.

3   Workflow and Functionality

In order to make it easier to understand the process, firstly we show the workflow of the program in Fig. 1.

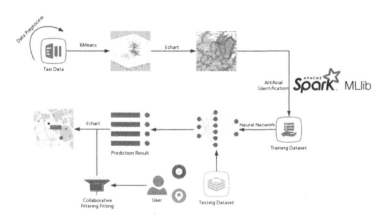

Fig. 1. Program workflow.

In this workflow, firstly we preprocess the raw data and select a training dataset from the processed data for the pre-experiment. In the pre-experiment we practice the training dataset to figure out the best K value of K-Means algorithm. After this experiments, we determine that the best value of K is 5. Using this value, we run the K-Means algorithm and divide the data into five clusters. Based on the divided clusters, we use Echarts to display and extract the features of each cluster in order to show the different difficulty degree. Next, we use neural network to generate the prediction results based on these clusters. Finally, we use a collaborative filtering fitting function to calculate the possibility score of taxi-hailing at a particular time and place. In

Sect. 3.1, we introduce the process of data preprocess, which is mainly applied to process the raw data and improve the accuracy of prediction. In Sect. 3.2, we introduce the process of clustering using K-Means and the process of predication using natural network and possibility score calculation.

3.1 Data Preprocess

In the process of Data Preprocess, it includes two steps: data cleaning and data standardizing. And we will introduce these two steps in this section.

Before introducing the process of data preprocess, we should give an outline of the dataset used in this paper firstly. The Dataset in this report is offered by a taxi company in Shanghai. It takes three different months that come from 2014 to 2016, and we choose some of them for our experimentation. There are more than 110,000,000 lines of data in one day, and each line of data records several attributes for a specific time and taxi. The detailed attributes in the dataset are shown in Table 1.

Table 1. The detailed attributes in the dataset

carID	isAlarm	isEmpty	topLight	isElevated	isBrake
20257	0	1	1	0	0
21868	0	1	5	0	0
GPSTime	longitude	latitude	speed	direction	satellite
2016-04-10 16:25:47	121.50824	31.19750	0.0	155.0	6
2016-04-10 16:08:22	121.45526	31.32458	0.0	145.0	5

In the step of Data Cleaning, firstly we pick out the data that has signals form more than 3 satellites, which are more accurate. Then we check the integrity of data and remove the incomplete data. Lastly, we also should get rid of the useless attributes from these data, such as direction, car ID etc.

In the step of Data Standardizing, we mainly do two tasks. One is to standardize the attribute of speed and another is to combine the speed attribute with the time attribute. In the process of combination, the traditional representation of time is not suitable and we present a new representation for time which is called "Time Ring".

Time Ring is an important procedure in data standardizing. This activity is contrived to work out the inaccuracy problem of linear time difference. In this paper, we divide the data by the attribute of day at first, but this division method will cause a serious problem. Suppose that there are two time values, 0:02 and 23:58. We know the time difference between them is only 4 min. But if a day starts from 0:00 and come to an end at 23:59 in a linear way, the difference should be calculated by subtracting the earlier time from the later time. In this case we subtract 0:02 from 23:58, then we got the time difference, which is 23 h and 56 min.

So which time difference shall we use, 4 min or 24 h? In daily life, people's behaviors are similar between 0:02 and 23:58 because these times are both at midnight. So we should shorten the distance between the two ends of the timeline. One solution is to tie up the two ends of time line together. Obviously, if we tie up the line together, we will get a ring.

Now we can use the unit circle to represent the ring. We take the ring of $x^2 + y^2 = 1$ in a two dimensional coordinate system. So we can represent the points on the circle in Eq. (1).

$$x = \sin\theta, y = \cos\theta, \theta \in [0, 2\pi) \tag{1}$$

Then we use the time circle to represent the whole day. For the convinence of computing, we convert a time in the formation of "hh:mm" into the formation of minutes by formula $t = hh * 60 + mm$, and the value range of t is $[0, 1440)$. At the same time, the value of x also should be converted by using $\theta = \frac{2\pi}{1440} * t$. Finally, the following transformation comes out in Eq. (2).

$$x = \sin\left(\frac{2\pi}{1440} * t\right), y = \cos\left(\frac{2\pi}{1440} * t\right), t \in [0, 1440) \tag{2}$$

In this formula, we use the Euclidean distance to represent the distance between two times t_1 and t_2. The distances between 0:02 and 23:58 is about 0.01745 and the distance between 0:02 and 12:02 is 2. By using the Euclidean distance, we think the distance between times can represent the behavior difference of people in different time. Moreover, the Euclidean distance will be comforted in the distance calculation using K-Means clustering algorithm. At last, the Euclidean distance of two time t_1 and t_2 is shown in Eq. (3).

$$d = \sqrt{(x_1 - x_2)^2 + (y_1 - y_2)^2} \tag{3}$$

3.2   Cluster and Predication

In the process of cluster and predication, firstly we use K-Means clustering to divide the taxi data into different clusters for showing the different degree of difficulty. Then we try to use neural network and possibility score to make the prediction of taxi-hailing at a particular time and place based on the result of K-Means. In this section, we will introduce each procedure in detailed.

K-Means. In the procedure of K-Means, we cluster the data and utilize the cluster result to measure the difficulty degree for taxi-hailing.

Before we use this algorithm, firstly we should find out the optimum value of K for K-Means. For this, we select the data within 30 min and standardize them. The initial value range of k in our experiment is $[2, 8]$. For each value of k, we execute K-Means 5 times and work away the mediocre sum of square error in each bunch. In the end, we set out the variation tendency of line and decide that the optimum value is 5 according to this line chart.

After the procedure of determining parameters, we use the processed data in data preprocess to execute the K-Means clustering algorithm. According to the preliminary

Fig. 2. Cluster 0 and Cluster 4 at 20:00

experiment, the optimum value of K is 5. After we obtain the clusters, we have evaluated the results of these clusters, which are described in more detail in Sect. 4. Based on the evaluation results, we give a possibility score to each cluster. In Fig. 2, we use Echarts to display the taxi distribution of Cluster 0 and Cluster 4 at 20:00. Left one is Cluster 0 and right one is Cluster 4. From this figure, we can indicate that different clusters have different distributions, which are indeed completely different categories.

In the process of analysis, we discover that there are two factors impacting the results sparkly, which are *Speed* and *IsEmpty*. A low empty rate means that the taxi is full of passengers and a low speed means that the taxi is out of service. In either condition it is not easy to find a taxi. Then we try to use the two attributes together to calculate the possibility score of each cluster. Either attribute that is close to zero will result in a low possibility score. So we can define the possibility score a_i in Eq. (4).

$$a_i = \text{Speed}_i * \text{isEmpty}_i (i \in [1, k]) \tag{4}$$

Neural Network. In the process of neural network, we regard the K-Means clusters with different difficulty degree as training dataset and pick some raw data as testing dataset. With these datasets, we use neural network to generate the original predication results.

In the calculation of neural network, we build a neural network model with three layers, three inputs and five outputs. Three inputs are the attributes of latitude, longitude and time for a taxi. Five outputs represent the predicted results of which cluster the taxi should belong to. The value range of outputs is $[0, 1]$. If we use o_i to represent the value of each output, we can define the possibility score of neural network as follows:

$$A_{nn} = \sum_{i=1}^{k} o_i * a_i \tag{5}$$

On the basis of neural network, we can obtain a prediction result for taxi-hailing. However, the prediction result can not give a good indication of difficulty degree for taxi-hailing, which is not convenient and friendly for users. So we propose a new method to evaluate the difficulty degree for taxi-hailing numerically, which is called possibility score in this paper.

Possibility Score Calculation. In order to calculate the possibility score for taxi-hailing at a particular time and place, we try to find the nearby points of the given point. Using these points, we can calculate the possibility score of these points and regard it as the possibility score of the given point.

Firstly, we merge these nearby points together according to the latitude and longitude of these points. The average latitude of Shanghai is $31°$. And the distance of each latitude equals to approximately 111 km. So we can get that the distance of each longitude in Shanghai equals to about 95 km, which is calculated by 111 km $* \cos 31°$. If we divide the distance of each latitude and longitude by 10000, the length and width are 11.1 m and 9.5 m. That means we can use a geographic point to represent an area, in which the length is 11.1 m and the width is 9.5 m.

Now a geographic point may contain several tuples of data points. These data points belongs to the different clusters which are divided by K-Means. In cluster i ($i \in [1, k]$), the number of data points is n_i. And the number of total data points in the geographic point is $S_{one\_point} = \sum_{i=1}^{k} n_i$.

If a user wants to know whether it is easy to find a taxi at point $P(lat_p, lng_p)$, we can do it as follows. Firstly, we select a region near P. Then we can work out the rate of P by calculating the nearby points' influence. In order to do this, we can do this by the following steps.

First, we work out the total number of data points in this region. Supposing that the number of geographic points in this region is m, and we can calculate the total number of data points in this region as follows:

$$S_{region} = \sum_{j=1}^{m} \sum_{i=1}^{k} n_{ij} (i \in [1, k], j \in [1, m]) \qquad (6)$$

In this formula, n_{ij} means the number of data points belonging to the cluster i in geographic point j, and S_{region} means the total number of data points in this region.

Next we should work out the sum of possibility score for each data point in this region. During the process of calculation, we find that the near points contribute more influence and the far ones contribute less to the possibility score. So we can weigh the possibility score by the distance r_j between point j and point p:

$$A_{region} = \sum_{i=1}^{k} \sum_{j=1}^{m} \frac{n_{ij} * a_{ij}}{r_j} \qquad (7)$$

$$r_j = \sqrt{\left((lat_p - lng_p) * \cos(31°)\right)^2} * \frac{1111000m}{1°} \qquad (8)$$

Then we attend to work out the average possibility score. As we can see, the more data points are, the more taxis are in this region and it is easier to find a taxi in this region. But the number of taxis is not linear with the growth trend of possibility score. So what we need is a function that grows fast at first and then slows down with the input increase. After selection and comparison, we think the logarithmic function is a serious pick. So the computing method of possibility score is shown in the Eq. (9).

$$A_p = log\, S_{region} * \frac{A_{region}}{S_{region}} \qquad (9)$$

In this formula, A_p is the possibility score we want. Then we put the point P and the points in the region on a web page with ECharts and display all the point above on the map in Fig. 3.

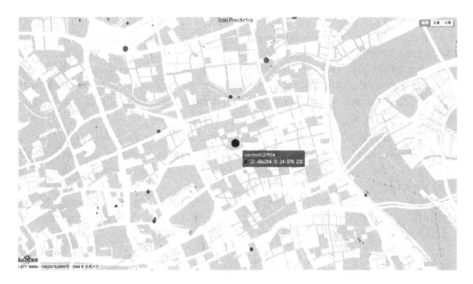

Fig. 3. Sample result.

4   Evaluation

In this section, firstly we analyze the clusters divided by K-Means and find the similarities and differences between the clusters in order to show the different degree of difficulty. After that, we analyze the results of possibility predication using the cluster of K-Means and compare the predication with the actual situation, which verifies the reliability and accuracy of the framework.

4.1   K-Means

In the evaluation of K-Means, We attempt to find out the difference between clusters, firstly we want to utilize the number of taxis as the difference and make the cluster distribution graph as follow in Fig. 4.

In Fig. 4, the X-axis in this coordinate system means the time, and the Y-axis means the number of taxis at a specific time. As we can see, the Cluster 4 and Cluster 0 appear at almost the same time, and Cluster 2 and Cluster 3 appear at almost the same time. So the number of taxis can't be the distinguishable attributes and they should be

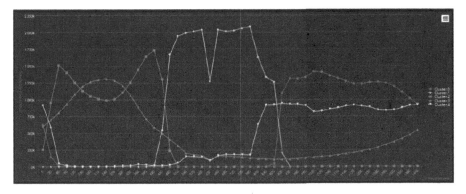

Fig. 4. Distribution of clustered data.

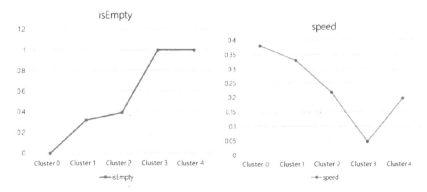

Fig. 5. Distribution of mean value.

distinguished by other attributes. We try to compare with some attributes and finally we focus on the following two attributions in Fig. 5.

About Cluster 4 and Cluster 0, the empty rate of Cluster 4 is almost 1 while the rate of Cluster 0 is nearly 0. Another attribution is speed: the speed of Cluster 4 is half of Cluster 0 and we also find points of Cluster 0 are mostly on highway. We can make a guess that taxis of Cluster 0 are the taxis with passengers that quickly drive to the destination. And Cluster 4 is the taxis which are hunting for passengers.

About Cluster 3, the empty rate is almost 1 but the speed is near 0. The taxis are almost alone the road and the majority data in this Cluster are distributed during the period between 21:00 and 9:00, which is the time for sleep. Thus we guess that Cluster 3 is the taxis out of service for the rest and it is difficult to take a taxi.

On the other hand, Cluster 2 and Cluster 1 have the empty rate of 0.4 and in normal speed. But Cluster 2 is a little slower because it is in the night. We argue that these are the taxis that are running and it is easier for passengers to take a taxi.

In a result, the five clusters can clearly indicate the the different degree of difficulty for taxi-hailing.

4.2   Possibility Prediction

In the process of possibility prediction, firstly we choose some points in downtown area manually, and work out all the possibility of finding a taxi at the point in the whole day. Then we show the possibility values using a line chart in Fig. 6. Finally, we try to check whether their tendency is same with the actual situation.

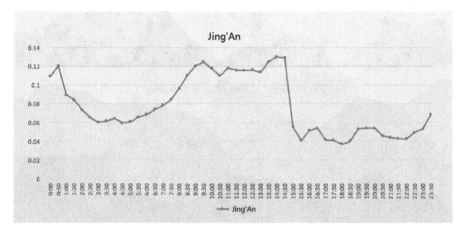

Fig. 6.   Result of Jin'An district.

According to the China Intelligent Travel 2016 big data report issued by Didi and CBNData, the most difficult time to take a taxi is between 2:10 to 5:20, and the easiest time is between 13:00 to 14:20. Then we select different points to make a comparison.

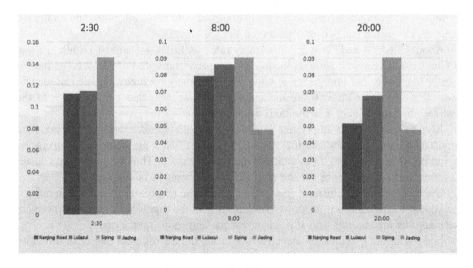

Fig. 7.   Result of four places.

In Fig. 7, we select four points: one is in Siping, one is in Jiading and the other two is in Lujiazui and East Nanjing Road. About the timestamp, we select points in three different times: 2:30, 8:00, and 20:00. Compared with different geographic points, it is overwhelmingly difficult to find a taxi in Jiading at any time but most easy in Siping because Jiading is too far from downtown. Lujiazui keeps the second and East Nanjing Road keeps the third at this time. It does show the ability of our model to distinguish the different possibility to find a taxi in different regions. Comparing with different time, 20:00 is a difficult time because people go home at that time, which raise the requirement of taxis amount. About 8:00, it is easy to find a taxi, maybe because enough empty taxis move on the road at a suitable speed. The empty taxi is enough at early morning because the number of passengers reached bottom at that time.

5   Conclusion and Future Work

In this paper, we present a framework to classify and predicate the situation of taxi-hailing based on big data. In the framework, firstly we simplify the big data because the original data we use is complex and large. After data cleaning and pre-processing, the dataset becomes smaller and improve the accuracy of prediction. In this process, we propose a method called "time ring" to solve the problem of unreliability of the linear time distance. Then we use K-Means clustering algorithm to divide the taxi data into different clusters. Next we use neural network to generate the predication result using the result of K-Means. On this basis, we propose a method to make the predication of taxi-hailing at a particular time and place, which can calculate the possibility score of taxi-hailing. From this model, readers can benefit from the classification and predication framework to predict the difficulty degree of taxi-hailing at a specific time and place, which is not evolved in the existing traffic predication models.

References

1. de Brébisson, A., Simon, É., Auvolat, A., Vincent, P., Bengio, Y.: Artificial neural networks applied to taxi destination prediction. https://arxiv.org/abs/1508.00021
2. Zhang, Z., Wang, L., Jia, L., Li, F., Zhang, L., Zhao, M.: Projective label propagation by label embedding: a deep label prediction framework for representation and classification. Knowl.-Based Syst. **119**, 94–112 (2017)
3. Shi, A., Weiming, K.: Prediction of urban traffic abnormity based on causal network. In: 2015 Sixth International Conference on Intelligent Systems Design and Engineering Applications (ISDEA), pp. 574–577 (2015)
4. Münz, G., Li, S., Carle, G.: Traffic anomaly detection using k-means clustering. In: GI/ITG Workshop MMBnet (2007)
5. Xuewu, Z., Yongjun, L.: The city taxi quantity prediction via GM-BP model. In: 2015 IEEE International Conference on Cyber Technology in Automation, Control, and Intelligent Systems (CYBER), pp. 1594–1598 (2015)

6. Manasseh, C., Sengupta, R.: Predicting driver destination using machine learning techniques. In: 16th International IEEE Conference on Intelligent Transportation Systems (ITSC 2013), pp. 142–147 (2013)

7. Ahmad, B.I., Murphy, J.K., Godsill, S., Langdon, P.M., Hardy, R.: Intelligent interactive displays in vehicles with intent prediction: a Bayesian framework. IEEE Sig. Process. Mag. 34(2), 82–94 (2017)

8. Tak, S., Kim, S., Oh, S., Yeo, H.: Development of a data-driven framework for real-time travel time prediction. Comput.-Aided Civil Infrastruct. Eng. 31(10), 777–793 (2016)

9. Liebig, T., Piatkowski, N., Bockermann, C., Morik, K.: Dynamic route planning with real-time traffic predictions. Inf. Syst. 64, 258–265 (2017)

10. Hu, W., Yan, L., Wang, H., Du, B., Tao, D.: Real-time traffic jams prediction inspired by Biham, Middleton and Levine (BML) model. Inf. Sci. 381, 209–228 (2017)

11. Oliveira, T.P., Barbar, J.S., Soares, A.S.: Computer network traffic prediction: a comparison between traditional and deep learning neural networks. IJBDI 3(1), 28–37 (2016)

12. Li, H.: Research on prediction of traffic flow based on dynamic fuzzy neural networks. Neural Comput. Appl. 27(7), 1969–1980 (2016)

13. Bezuglov, A., Comert, G.: Short-term freeway traffic parameter prediction: application of grey system theory models. Expert Syst. Appl. 62, 284–292 (2016)

14. Li, J., Mei, X., Prokhorov, D., Tao, D.: Deep neural network for structural prediction and lane detection in traffic scene. IEEE Trans. Neural Netw. Learn. Syst. 28(3), 690–703 (2017)

15. Yi, H., Jung, H., Bae, S.: Deep neural networks for traffic flow prediction. In: BigComp, pp. 328–331 (2017)

16. Ma, X., Dai, Z., He, Z., Ma, J., Wang, Y., Wang, Y.: Learning Traffic as Images: A Deep Convolution Neural Network for Large-scale Transportation Network Speed Prediction. CoRR abs/1701.04245 (2017)

17. Mitrovic, N., Asif, M.T., Dauwels, J., Jaillet, P.: Low-dimensional models for compressed sensing and prediction of large-scale traffic data. IEEE Trans. Intell. Transp. Syst. 16(5), 2949–2954 (2015)

18. Park, J., Li, D., Murphey, Y.L.: Real time vehicle speed prediction using a neural network traffic model. In: The 2011 International Joint Conference on Neural Networks (IJCNN), pp. 2991–2996 (2011)

19. Luo, Q.: On discovering regional taxi service disequilibrium with geographical collaborative filtering. In: 2014 International Conference on Informative and Cybernetics for Computational Social Systems (ICCSS) (2014)

20. Akhmetov, D.F., Dote, Y., Ovaska, S.J.: Fuzzy neural network with general parameter adaptation for modeling of nonlinear time-series. IEEE Trans. Neural Netw. 12(1), 148–152 (2001)

PRACE: A Taxi Recommender for Finding Passengers with Deep Learning Approaches

Zhenhua Huang[1], Zhenqi Zhao[1], Shijia E[1], Chang Yu[1],
Guangxu Shan[1], Tienan Li[2], Jiujun Cheng[1], Jian Sun[2(✉)],
and Yang Xiang[1(✉)]

[1] College of Electronics and Information Engineering, Tongji University,
Shanghai 201804, People's Republic of China
shxiangyang@tongji.edu.cn
[2] Department of Transportation Engineering, Tongji University,
Shanghai 201804, People's Republic of China
sunjian@tongji.edu.cn

Abstract. In this paper, we propose a real-time recommender system (PRACE) for taxi drivers to find a next passenger and start a new trip efficiently, based on historical GPS trajectories of taxis. To provide high-quality passenger-seeking advice, PRACE takes passenger prediction, road condition estimation, and earnings into ranking simultaneously. Different from many previous researchers, we not only pay more attention to the driving context of taxis (i.e., driving directions, positions, etc.) but also extract meaningful representations of these attributes, using deep neural networks. To enhance the effect of learning, the result of statistics is added to the input of models. Relying on the map meshing method, we treat the prediction task as a multi-classification problem rather than a regression problem and make comparisons with several state-of-the-art methods. Finally, we evaluate our method through extensive experiments, using GPS trajectories generated by more than 10,000 taxis from the same company over a period of two months. The results verify the effectiveness, efficiency, and availability of our recommender system.

Keywords: Recommender system · Deep learning · Passenger prediction · Trajectory mining

1 Introduction

Now informed driving is becoming more and more important to drivers. It offers great convenience if someone tells them where the next passenger would appear after an occupied trip, which closely relates to their income. Although there are already many taxi recommender systems, most of them use shallow traffic models (i.e. statistical learning) and are still somewhat unsatisfying, such as ignoring context, feature associations, and personalization. It inspires us to rethink this problem based on deep architecture models with such abundant amount of trajectory data.

Research in deep learning attempts to make better representations and create models to learn these representations from large-scale unlabeled data [1]. Applying

© Springer International Publishing AG 2017
D.-S. Huang et al. (Eds.): ICIC 2017, Part III, LNAI 10363, pp. 759–770, 2017.
DOI: 10.1007/978-3-319-63315-2_66

deep learning methods in taxi operations, our work is mainly aimed at passenger recommendation for taxi drivers by deep neural networks. Following this strategy, we also work for the road condition estimation of trips, which contributes to providing advice on passenger searching with high quality. The ultimate goal of our research is proposing a recommender system, named PRACE, i.e., Passenger Recommendation based on Attractiveness, road Condition, and Earnings, to inform drivers where to find a next passenger with high accuracy and profits, as well as reliability. Specifically, our primary contributions are as follows:

- We introduce a novel passenger prediction method based on grids. Conventional researchers usually focus on road segments for passenger prediction. In fact, people's activities are regional in scope [2] and taxi drivers are familiar with the regions they usually cruise, so their activity regions can be demonstrated as spatial grids based on amounts of trajectory data.
- Apply deep learning methods into feature extraction of trajectory mining. For the cruising of taxis, the driving context, such as driving direction, positions, and surroundings, are all meaningful information, which helps with personalized recommendations. We use deep neural networks and t-SNE to extract features from input dimensions, and the results show a more accurate characterization than conventional methods.
- Bring passenger prediction, road condition, and earnings into taxi recommendation, which is a novel estimation method. Passenger prediction helps determine the specific location of potential passengers, and road condition reflects the cost to these places in the way of time and distance. Besides, earnings are what the drivers most concern about and are always obtained through the trip length. As a whole, we can exactly evaluate the quality of cruising recommendation with these factors.

The rest of this paper is arranged as follows. Section 2 shows the literature works of taxi-related recommendation. Section 3 makes an overview of the system, including problem description and framework. Section 4 presents the description of the methodology applied in this paper. Specifically, except for pretreatment, we borrowed knowledge from natural language processing and used different methods to represent data. Section 5 compares experiment results over different methods, and the parameters in various prediction tasks are also discussed in-depth. There is also a subsection presenting the validation result of PRACE for taxi drivers, mainly on profit and cruising time. Finally, Sect. 6 makes a conclusion of our work so far as well as prospecting our future work.

2   Related Work

In the literature, the taxi-related recommendation has been long regarded as a key functional component of location-based service. Yuan's T-finder [3] informs taxi drivers with some passenger locations, using the knowledge of travelers' mobility patterns and taxi drivers' picking-up/dropping-off behaviors learned from the GPS trajectories. Ding et al. [4] propose a dynamic scoring system to evaluate each road segment in different time periods by considering both picking up rate and profit factors.

Qu et al. [5] and Dong et al. [6] are committed to providing profitable routes for drivers to maximize their income. From another aspect, Zhang *et al.* [7] applies graph knowledge into reducing drivers' cruising miles. Xu *et al.* [8] proposes a Taxi-hunting Recommendation System (TaxiRS), to provide passengers with a waiting time to get a taxi ride in a particular location.

The above studies solve the question mainly relying on graph and statistics knowledge. Recently the development of deep learning gives us a chance to take a new look at trajectory data. De Brébisson et al. [9] applied different kinds of artificial neural networks to make taxi destination prediction in Portugal. Lv *et al.* [10] apply a deep learning approach to make traffic flow prediction with big data. Zhang et al. [11] applies a deep ST-ResNet into forecasting the flow of crowds. However, research on taxi-related recommendation with deep learning methods is rarely seen. Deep learning allows computational models that are composed of multiple processing layers to learn representations of data with multiple levels of abstraction [12], widely used in speech recognition and images classification and achieved dramatical improvement. In this paper, we apply deep neural networks into feature extraction of trajectory mining and explore new lands for the taxi-related recommendation.

3   Overview

3.1   Problem Description

The problem we are committed to solving is illustrated as follows. After finishing an occupied trip, where to find the next passenger is a question for the driver. Passenger prediction is to forecast where the driver would roll for a next passenger, according to people's movement patterns and taxi drivers' cruising experience learned from amounts of trajectory data. Like,

$$R(['110', 'old\,street', '13:10', \ldots]) = 'new\,street'$$

which means a taxi (id is '110') is in 'old street' at '13:10' (with other information), and 'R' thinks it would roll for 'new street' for the next passenger. 'R' is just the mapping relationship we'd like to learn from data. Assuming that there are several candidate places after prediction, the main challenge is choosing a most appropriate one. Similarly, road condition prediction helps decide how long it takes if a taxi rolls from a place to another. Here we would take the attractiveness, the road condition, and the earnings into consideration comprehensively, and give each of them a score to evaluate their quality. Eventually, the place of the best evaluation will be selected and recommended to the corresponding taxis.

3.2   Framework

Figure 1 illustrates the overall framework of our system. From bottom to top, the initial task is data processing, including map meshing and trajectories segmenting. Rather than road networks, we can get a grid map from map meshing. The next step of trajectories segmenting is map matching, which attaches GPS points to corresponding roads, and then we can get whole idle or occupied trips. Next, apply statistical methods to process grid map and trips and generate information about hot pick-up spots, road condition, and grid earnings. Another branch is preparing for the input of models. These trips cannot be the direct input of prediction models, and the pretreatment is extracting useful features from them, such as **taxi id**, **latitude**, **longitude** and so on. Then these data would go through cleaning and filtering operations to ensure availability. Next step is embedding these features (including discrete values, e.g., taxi id) into space vectors. The final task is taxi recommendation and consists of three parts: passenger prediction, road condition prediction, and grid earnings, which are undertaken by prediction models and statistics, respectively. At this point the system structure is complete. If the taxi makes a request, PRACE will return a place with potential passengers and high profit.

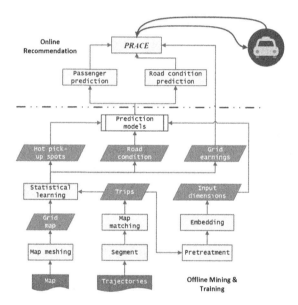

Fig. 1. Framework overview

4 Methodology Description

4.1 Map Meshing

To reduce the difficulty of processing large amounts of trajectory data, map meshing method is applied to process map and GPS positions to set up a road network of grids. Compared with longitude and latitude alone, grid map could merge more information, such as grid index and visiting frequency. The size of the grid is critical to the accuracy of prediction and data processing time. In this paper, the size of map grid is **400 m × 400 m**, which is calculated based on average cruising distance (1.57 km per trip). As for taxi drivers, pick-up positions closely relate to their cruising strategy. However, nearby pick-up positions often form a region, in which drivers have a higher chance of meeting potential passengers.

4.2 Sudoku

With the map meshing method, **Sudoku** is used to mark the driving context of taxis. This operation partitions the surroundings into discrete grids, which is commonly used in spatial processing [11]. We'd like to count the boarding of other taxis in each neighbor grid during a time window, which reflects the demand for taxis in this region and may affect the driver's cruising strategy. Thus we collect and dump the information into a database to facilitate the query, indexed by (grid id, time window). Also, discretization of continuous features could reduce the amount of calculation, enhance the robustness of the model, and lower the risk of over-fitting in machine learning [17].

4.3 Prediction Models

The core architecture of this model is based on a MLP (multi-layer perceptrons), which is one of the basic neural network architectures and trains using backpropagation. Specifically, a MLP consists of an input layer, one or more hidden layers, and an output layer. The input layer is always with fixed size representation of input variables, and the hidden layers are used to calculate the intermediate representation of the input variables. Finally, the output layer is used to give the prediction of the output value. As shown in Fig. 2, we use embedding to process data with discrete dimensions as the input layer of the neural network. Then we add three hidden layers, and the last layer is output layer based on softmax which is a generalization of the logistic function to fit the multi-label classification.

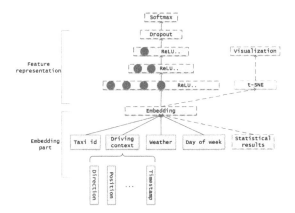

Fig. 2. Deep neural networks with multi-hidden layers structure

Especially, we add one dropout layer before the output layer to prevent overfitting. Finally, the output layer returns the prediction for the corresponding inputs. In our case, the input layer receives a representation of the taxi's profile with associated metadata and the output layer predicts the destination or road condition (labels). We used standard hidden layers consisting of a matrix multiplication followed by a bias and nonlinearity. The nonlinearity we choose to use is the Rectifier Linear Unit (ReLU) [13]. Compared to traditional sigmoid-shaped activation functions, the ReLU limits the gradient vanishing problem so that we could train a network with deep architecture.

The objective function used in our neural networks is multi-class log loss, also known as the categorical cross-entropy. It is a commonly used loss function in the field of multi-label classification, and can be optimized by Adam [14], which is a variant of the typical stochastic gradient descent (SGD). The overall model is just like a linear stack of layers, simple but effective. Moreover, this approach could easily be adapted to other similar applications in which the goal is to predict an output from an input with continuous and discrete features.

5   Experiments

5.1   Data Set

The trajectory data came from 10882 taxis equipped with GPS sensors, recorded from March 1st, 2016 to April 30th, 2016 in Shanghai. The trajectories were segmented into idle and occupied trips according to research needs. In general, trajectory data usually has offset from real road networks caused by internal and external factors. Therefore, a map matching algorithm was applied to calibrate trajectories as well as attaching GPS positions to road segments. Finally, over 10 million trajectories of taxis were applied in our experiment. After data cleaning and filtering, we extracted useful features from the trajectory data set. After shuffling, we selected training set with all the possible prefixes of the complete trajectories. Then we chose data which had the same distribution over prefixes with the training set as the testing set. In this paper, the proportion of training set and testing test were 85% and 15%, respectively.

5.2   Passenger Prediction

Different from traditional predicting models, neural networks use neurons to fit data by adjusting weights and mining inner principals. Our research takes trajectory data as input, aimed at predicting the positions where potential passengers would appear during a day. In this case, we select totally 2823 hot pick-up grids as candidate places, of which the visiting frequency is above the average.

As shown in Table 1, both NN and non-NN methods are applied in classification, and the comparison of performance covers accuracy and running time. All the algorithms are running on a server with 24 Intel Xeon CPUs E5-2620, and 4 T K20c GPUs. The MLP model used in this task is similar to De Brébisson's model applied to taxi destination prediction in kaggle's competition [9], which is with a single hidden layer of 500 ReLUs for the encoder. When it comes to DNN, first we add an embedding

Table 1. Performance of evaluation methods on passenger prediction

Method	Parameters	Time per loop	Accuracy
SVM	C: 1.0, kernel: rbf, gamma: 1	>4 h	–
RNN	Single hidden layer with 500 neurons; training with GPU	4.25 min	0.066
XGBoost	max_depth: 4, eta: 0.1; training with 16 CPU threads	>4 h	0.117
Random forest	max_depth: 30, n_estimators: 10, min_samples_split: 2	42.8 min	0.211
MLP	Single hidden layer with 500 neurons; training with GPU	14.7 min	**0.498**
DNN	Three hidden layers, [2048, 1024, 512]; training with GPU	21.3 min	**0.628**

layer to transforms lists of feature-value mappings to vectors. Then besides three hidden layers, we add one dropout layer before the output layer to prevent the model from overfitting. Also, though the training set of DNN could be well fitted (loss → 0.6), we adopt the **early stop** technique to get the best test result (0.628). However, this task fails many state-of-the-art models such as XGBoost [15] and Random Forest, for a long time of calculation, high memory usage, and inaccuracy. When it comes to SVM, the reason why it did not even give a result is that the multiclass support is handled according to a one-vs-one scheme, which makes it hard to scale to, a dataset with more than a couple of 10000 samples with quadratic time complexity. RNN has weak performance, which means less dependency between attributes of input, and the output category is not an input sequence item in the provided samples.

When the training of DNN is finished or passes by a checkpoint, we can save the model into an .hdf5 file, which contains the weights of the neural networks. Thus the training step could go on by loading .hdf5 when new data comes rather than training a model from the very beginning.

5.3 Road Condition Prediction

According to the speed range of taxis, the congestion mode of roads is divided into three categories, as Table 2 shows. Road condition is of great importance when taxi drivers choose whether to go this way or not. Therefore, after predicting where the next passenger would appear, telling drivers how long they will take to get the destination is crucial. It can be expressed as

$$R(['110', 'old\ street'', 'new\ street', \ldots]) = smooth$$

which means the taxi (id is '110') would get a smooth trip between 'old street' and 'new street' to find the next passenger (with other information). Especially, 'id' is a dimension of input so that the model could fit driving habits of different drivers.

Table 2. Categories of road condition based on speed

Category	Speed range	State
0	<20 km/h	Severe congestion
1	20 km/h ~ 35 km/h	Light congestion
2	>35 km/h	Smooth

The number of the category in road condition is far fewer than passenger prediction. It is part of the reason why non-NN methods can get better performance. As shown in Table 3, DNN achieves the highest accuracy of 0.753 with a running time of 72 s per iteration. Compared to MLP, the deep structure of DNN could get more implicit information for feature representation. However, RNN performs the worst in this task, with the lowest accuracy and the longest time. When it comes to non-NN methods, XGBoost has a rather good performance. Though the running time per iteration of SVM is less, it needs thousands of iterations, which is more than any other methods. Besides, due to the characteristics of the algorithm itself, Random Forest just runs a single iteration of 107.2 s and gets the accuracy of 0.732, proving it has the time advantage in solving this problem.

Table 3. Performance of evaluation methods on road condition prediction

Method	Parameters	Time per loop	Accuracy
RNN	Single hidden layer with 500 neurons; training with GPU	108 s	0.438
MLP	Single hidden layer with 500 neurons; training with GPU	28 s	0.649
SVM	C: 1.0, kernel: rbf, gamma: 1	5 s	0.705
Random forest	max_depth: None, n_estimators: 10, min_samples_split: 2	107.2 s	**0.723**
XGBoost	max_depth: 40, eta: 0.1; training with 32 CPU threads	24 s	**0.739**
DNN	Three hidden layers, [512, 256, 128]; training with GPU	72 s	**0.753**

To enhance learning effect and improve accuracy, we add more meaningful features into input dimension of models. Still taking road condition prediction as an example, some researchers apply statistical methods to solve the problem, such as using similar trips and road network data [16]. That inspired us to combine the result of statistics with current input, and then we use t-SNE to visualize the data, which also helps with feature selection. Figure 3 reflects that the new dimension of feature added from statistics would make the feature extraction more accurate. Moreover, for the case of classifications, there is still a significant intraclass distance, which means categories could be refined in detail.

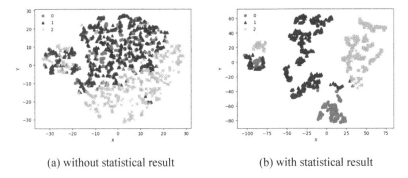

(a) without statistical result   (b) with statistical result

Fig. 3. Data visualization on road condition prediction

5.4 Validation on PRACE

In this section, we will discuss the recommender system PRACE, which could evaluate the score for each grid. Synthesize the following indicators: passenger prediction, road condition estimation and the average earnings of grids to sum up a formula, which reflects the quality of a searching trip. After obtaining the possible appearance places of passengers, we propose a choice of them based on the formula to decide their final ranks. The ranking formula for grids is shown as follows:

$$\zeta(O, D, t) = \alpha H_{D,t'} \cdot \beta C_{O,D,t} \cdot \frac{\alpha \gamma L_{D,t'}}{Dis(O, D)} \tag{1}$$

In Eq. (1), O represents the current grid where the taxi drives in, while D is the potential grid of next passengers. $H_{D,t'}$ is the attractiveness of D at time t', which is evaluated by visiting frequency per unit time. t' is the expected time when the taxi arrives in D. For the second part, $C_{O,D,t}$ is an estimation of traffic speed from O to D at t, and this reflects the congestion status of road networks. As for the last part, the numerator $L_{D,t'}$ is the potential trip length if the taxi meet a passenger in D at t', which is the average value of trips obtained by statistics and indexed by (grid id, time window), while the denominator $Dis(O, D)$ represents the Manhattan Distance from O to D and is just the cost of passenger searching.

$$Dis(O, D) = (|D.x - O.x| + |D.y - O.y|) \cdot s \tag{2}$$

where s is the size of the grid (square), and $(D.x, D.y)$ is coordinate of D. In fact, Eq. (2) is an estimation of searching trip length. Then the arriving time t' could be generated by Eq. (3):

$$t' = t + \frac{Dis(O, D)}{C_{O,D,t}} \tag{3}$$

α, β and γ are the weight of each part. In this case, we set β, $\gamma \leftarrow 1$ to highlight the importance of different Ds. Softmax could give out the probability of every candidate so that we can sort the results. At this time α is calculated by softmax, which is also the weight of the corresponding D. Equation (1) would rate the potential pick-up spots for taxi drivers, and then we rank them by scores. Considering the navigation system (e.g. Bing map and Amap) is becoming more and more advanced, the recommender system only provides pick-up spots and road condition from origin to destination rather than specific paths.

$$g(l) = \begin{cases} 14 & l \leq 3km \\ 14' + 2.5 * (l - 3) & 3km < l \leq 15km \\ 44 + 3.6 * (l - 15) & 15km < l \end{cases} \tag{4}$$

The test model took the ground truth data as input, and then compared the result of prediction with real data. ***Top-3*** results of prediction were picked as candidate solutions for the recommendation, and we selected the one scoring most in as the final answer. Furthermore, this operation is under the assumption that if the grid of passenger prediction (top-1 answer) is consistent with the ground truth, the prefix will dominate the rest. The **domination** means the rest results are also in keeping with reality and this trip does not get a promotion. To demonstrate the validation result, 1000 trips were selected randomly from 5:00 to 23:00 as the test set, and we apply Eq. (4) to calculate earnings of a trip, where l is the trip length. Figure 4 (a) presents the transformation vectors between ground truth and recommendation. Upper left is the best direction, because it means taxis not only shorten the time of finding passengers but also increase the earnings, while **dot** means the trip does not get a promotion and is consistent with the ground truth. Eventually, a total of 306 trips got promotions, shortening 1.3 min of cruising and increasing 2.3 RMB per trip in average. Also, there is a phenomenon that revenue decreases as searching time decreases according to the rules, like the lower left vector shows, but it also improves the efficiency of cruising. From an overall view, the

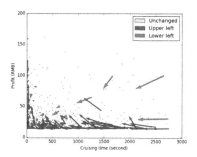 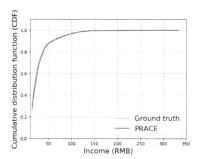

(a) Transformation vectors after apply-   (b) Cumulative distribution function of
ing PRACE   earnings

Fig. 4. Performance on applying PRACE

comparison of cumulative distribution functions about earnings is shown in Fig. 4 (b), proving that our method is effective in passenger recommendation.

6   Conclusion and Future Work

In this paper, we propose a real-time recommender model, called PRACE, to recommend suitable passengers for taxi drivers, using the GPS trajectories from a sample of vehicles (e.g. taxicabs). A good recommendation system for drivers should not only lay stress on the availability, i.e., accuracy, but also pay more attention to the quality of the complete customer search procedures. Hence we divided the question into three parts, i.e., passenger prediction, road condition prediction and grid earnings, and we ruled them separately.

For the two subtasks of prediction, different methods were applied to learn good representations of the input data. DNN could fully understand the linkages between the attributes, while t-SNE focuses more on dimensionality reduction and data visualization. When using neural networks to solve classification problems, different optimization methods were applied to facilitate training. Secondly, the statistical method was applied to add more information to the input attributes of training models. GPS trajectories generated by more than 10,000 taxis over a period of two months from the same company were used to train with GPUs acceleration. We also made comparisons between neural networks and several state-of-the-art non-NN methods. Finally, we evaluated the result of PRACE with ground truth data, proving the effectiveness, efficiency, and availability of our method.

In the future, we would like to study the impact of taxis' earnings and take grid earnings prediction into consideration. Just like traffic flow prediction, we want the model could understand the relation between adjacent journeys in the same grids. Moreover, models such as RNN would be tried to solve this question.

Acknowledgement. This work is supported by the Natural Science Foundation of Shanghai (17ZR1445900, 71571136), the Shanghai Rising-Star Program (15QA1403900). The Fok Ying-Tong Education Foundation (142002) and the Fundamental Research Funds for the Central Universities.

References

1. Szegedy, C.: An overview of deep learning. In: AITP 2016 (2016)
2. Gonzalez, M.C., Hidalgo, C.A., Barabasi, A.L.: Understanding individual human mobility patterns. Nature **453**(7196), 779–782 (2008)
3. Yuan, N.J., Zheng, Y., Zhang, L., et al.: T-finder: a recommender system for finding passengers and vacant taxis. IEEE Trans. Knowl. Data Eng. **25**(10), 2390–2403 (2013)
4. Ding, Y., Liu, S., Pu, J., et al.: Hunts: a trajectory recommendation system for effective and efficient hunting of taxi passengers. In: 2013 IEEE 14th International Conference on Mobile Data Management (MDM), vol. 1, pp. 107–116. IEEE (2013)

5. Qu, M., Zhu, H., Liu, J., et al.: A cost-effective recommender system for taxi drivers. In: Proceedings of the 20th ACM SIGKDD International Conference on Knowledge Discovery and Data Mining, pp. 45–54. ACM (2014)

6. Dong, H., Zhang, X., Dong, Y., et al.: Recommend a profitable cruising route for taxi drivers. In: 2014 IEEE 17th International Conference on Intelligent Transportation Systems (ITSC), pp. 2003–2008. IEEE (2014)

7. Zhang, D., He, T., Lin, S., et al.: Online cruising mile reduction in large-scale taxicab networks. IEEE Trans. Parallel Distrib. Syst. **26**(11), 3122–3135 (2015)

8. Xu, X., Zhou, J., Liu, Y., et al.: Taxi-RS: taxi-hunting recommendation system based on taxi GPS data. IEEE Trans. Intell. Transp. Syst. **16**(4), 1716–1727 (2015)

9. de Brébisson, A., Simon, É., Auvolat, A., et al.: Artificial neural networks applied to taxi destination prediction. arXiv preprint arXiv:1508.00021 (2015)

10. Lv, Y., Duan, Y., Kang, W., et al.: Traffic flow prediction with big data: a deep learning approach. IEEE Trans. Intell. Transp. Syst. **16**(2), 865–873 (2015)

11. Zhang, J., Zheng, Y., Qi, D.: Deep spatio-temporal residual networks for citywide crowd flows prediction. arXiv preprint arXiv:1610.00081 (2016)

12. LeCun, Y., Bengio, Y., Hinton, G.: Deep learning. Nature **521**(7553), 436–444 (2015)

13. Nair, V., Hinton, G.E.: Rectified linear units improve restricted Boltzmann machines. In: Proceedings of the 27th International Conference on Machine Learning (ICML 2010), pp. 807–814 (2010)

14. Kingma, D., Ba, J.: Adam: a method for stochastic optimization. arXiv preprint arXiv:1412. 6980 (2014)

15. Chen, T., Guestrin, C.: XGBoost: a scalable tree boosting system. In: Proceedings of the 22nd ACM SIGKDD International Conference on Knowledge Discovery and Data Mining, pp. 785–794. ACM (2016)

16. Singh, A.D., Wu, W., Xiang, S., et al.: Taxi trip time prediction using similar trips and road network data. In: 2015 IEEE International Conference on Big Data (Big Data), pp. 2892–2894. IEEE (2015)

17. Fayyad, U., Irani, K.: Multi-interval discretization of continuous-valued attributes for classification learning (1993)

Local Sensitive Low Rank Matrix Approximation via Nonconvex Optimization

Chong-Ya Li[1(✉)], Wenzheng Bao[1], Zhipeng Li[1], Youhua Zhang[2],
Yong-Li Jiang[3], and Chang-An Yuan[4]

[1] School of Electronics and Information Engineering, Tongji University,
Shanghai 201804, China
li_chongya2015@163.com
[2] School of Information and Computer,
Anhui Agricultural University, Hefei, Anhui, China
zhangyh@ahau.edu.cn
[3] Ningbo Haisvision Intelligence System Co., Ltd.,
Ningbo 315000, Zhejiang, China
[4] Science Computing and Intelligent Information Processing of GuangXi Higher
Education Key Laboratory, Guangxi Teachers Education University Nanning,
Guangxi, China
yca@gxtc.edu.cn

Abstract. The problem of matrix approximation appears ubiquitously in recommendation systems, computer vision and text mining. The prevailing assumption is that the partially observed matrix has a low-rank or can be well approximated by a low-rank matrix. However, this assumption is strictly that the partially observed matrix is globally low rank. In this paper, we propose a local sensitive formulation of matrix approximation which relaxes the global low-rank assumption, leading to a representation of the observed matrix as a weighted sum of low-rank matrices. We solve the problem by nonconvex optimization which exhibits superior performance of low rank matrix estimation when compared with convex relaxation. Our experiments show improvements in prediction accuracy over classical approaches for recommendation tasks.

Keywords: Matrix completion · Low rank · Nonconvex optimization · Recommender systems

1 Introduction

Matrix approximation is an important topic in machine learning [1], numerical analysis and signal processing. Given a matrix with m rows and n columns, we get to observe k entries which is comparably much smaller than mn, the total number of entries. It is impossible to recover the Matrix M without additional assumptions, there are a great number of matrixes accord with the observed entries of M. The prevailing assumption is that the target matrix M has a low-rank or can be well approximated by a low-rank matrix, which help us complete the matrix by the observed entries accurately [3].

In the field of recommender systems, users submit ratings on a subset of entries, and the vendor recommends new items based on the user's preferences [31]. The

© Springer International Publishing AG 2017
D.-S. Huang et al. (Eds.): ICIC 2017, Part III, LNAI 10363, pp. 771–781, 2017.
DOI: 10.1007/978-3-319-63315-2_67

famous example of recommender systems is the Netflix problem [2], users rate movies, but only a few movies are rated so that there are very few observed entries of the data matrix.

To investigate this problem, firstly, we introduce the singular value decomposition (SVD) [14]. Formally, let $M \in R^{m \times n}$, then the reconstructed matrix $\hat{M}$ rank is r, that is:

$$\hat{M} = U \sum V^T \tag{1}$$

Where U has dimensions $m \times r$, V has dimensions $n \times r$, and $\sum$ is a diagonal $r \times r$ matrix, $r \ll \min(m,n)$. The low-rank assumption not only allows us use a small amount of observed entries to predict a large number of unobserved entries, but also enhance the effectiveness of the machine learning model in many practical problems.

In matrix completion, we can recover the data by matrix factorization or nuclear norm method. However, existing algorithms all assume that the recommender matrix data M can be globally approximated by a low-rank matrix. In the area of recommender systems, there are a few key latent factors which determines whether a user would like a product or not. To different groups of users and items, key latent factors always are different. For example, in movie recommender systems, factors of one group of users and items are genre, actor and released year, but factors of another group of users and items are actress, time and director. This proves the rationality of assuming that M behaves as a low-rank matrix in the vicinity of certain row-column combinations.

In this paper, we assumed local sensitiveness on data matrix, which took both local consistency and global correlation into considered. The local sensitive low rank matrix approximation is proposed in this paper. Thus, we construct several local matrices by kernel smoothing, and then we use matrix factorization approach to reach the optimization solution of our problem. Moreover, not only should the frobenius norm of every local matrix be minimized when estimating missing values, local matrices should also overlap to make sure the value of every missing point can be constraint by information from different local matrices. Our experiments show that, by evaluating RMSE on three popular recommendation systems datasets, our method outperforms most recent works in the context of recommendation systems.

2 Related Works

In this section, we introduce several popular low-rank matrix approximation methods. First, we establish notations that used in this paper. The matrix which we need to complete is M, and the approximation of M is $\hat{M} = UV^T$, where U has dimensions $m \times r$, V has dimensions $n \times r$. The set of columns is denoted $\{1, 2 \ldots, n\}$ by $[n]$, similarly the set of rows $\{1, 2 \ldots m\}$ is denoted by $[m]$.

And $\Omega \overset{def}{=} \{(a_1, b_1), \ldots, (a_k, b_k)\} \in [m] \times [n]$ denotes the indices of all observed entries. Therefore, the training set is $\{M_{a,b} : (a, b) \in \Omega\}$. We let $\prod_\Omega$ be the matrix that contains the observed entries Ω, and $0'$ is filled with in the other positions.

$$\left[\prod_{\Omega}(M)\right]_{a,b} = \begin{cases} M_{a,b} & (a,b) \in \Omega \\ 0 & otherwise \end{cases} \tag{2}$$

We use two matrix norms in this paper, the Nuclear (trace) norm is

$$\|X\|_* \overset{def}{=} \sum_{i=1}^{r} \sigma(X) \tag{3}$$

the Frobenius norm is

$$\|X\|_F \overset{def}{=} \sqrt{\sum_i \sum_j X_{i,j}^2} \tag{4}$$

There are two popular low-rank matrix approximation approaches: the approach based on nuclear norm and the approach based on matrix factorization. In the first approach, the problem of matrix completion relies on the rank minimization, therefore there is a unique low-rank matrix which is consistent with the observed entries. And one could, in principle, recover the unknown matrix by solving

$$\min_X \ rank(X) \tag{5}$$
$$s.t. \ X_\Omega = M_\Omega$$

Unfortunately, this algorithm is NP-hard, and all known algorithms for solving it are doubly exponential in theory and in practice. A popular alternative of (5) is nuclear norm, which is the convex relaxation. Nuclear-norm minimization is the tightest convex relaxation of the NP-hard rank minimization problem [3, 4].

$$\min_X \ \|X\|_* \tag{6}$$
$$s.t. \ X_\Omega = M_\Omega$$

In the second approach, the unknown rank r matrix is expressed as the product of two much smaller matrices UV^T, where $U \in R^{m \times r}$, $V \in R^{n \times r}$, so that the low-rank requirement is automatically fulfilled. Such a matrix factorization model has long been used in PCA(principle component analysis) and many other applications

$$F(U,V) = \min_{U,V} \frac{1}{2} \sum_{(a,b) \in \Omega} \left(M_{a,b} - \left(UV^T\right)_{ab}\right)^2 \tag{7}$$

The convex optimization algorithms for solving (6) are computationally inefficient, in the sense that they incur high per-iteration computational cost, and only attain sublinear rates of convergence to the global optimum [29]. Instead, in the nonconvex optimization method (7), the reparametrization of $M = UV^T$, though making the optimization problem in nonconvex, significantly improves the computational efficiency. Existing literature has established convincing empirical evidence (7) that can be effectively solved by a board variety of gradient-based nonconvex optimization algorithms, including gradient descent, alternating exact minimization, as well as alternating gradient descent.

3 Local Sensitive Low Rank Matrix Approximation

In this section, we introduce a local sensitive low rank matrix learning method to solve the matrix completion problem. In recommender systems, there are some key latent factors that determine whether a user like the item. For example, when some users choose a movie, they always pay attention to the genre and director. These key latent factors can be learned automatically and we don't need to know what they are actually. For instance, if the rank is 6, whether the user prefer to the movie depends on 6 item characteristics (the columns of U, V).

We would introduce the method of constructing several local low-rank matrices. First, we assume that there exists metric structure in matrix M. We use $d((a,b),(a',b'))$ to describe the similarity between the users a and a' and items b and b'. The distance between two users and two items is not related to the indices of those rows or columns. In our experiment, the distance d is computed based on the observed entries of M. In the paper [6], Lee et al. found arc-cosine performed better than 2-norm and cosine similarity.

Therefore, we use SVD method $(M = UV^T)$ to factorize the partially observed matrix M. We then compute the similarity between the rows of factor matrix U or V. For example, we can use the following formula to get the similarity between users a and a' or items b and b'.

$$d(\text{a,a}') = \arccos\left(\frac{\langle U_a, U_{a'}\rangle}{\|U_a\|\|U_{a'}\|}\right)$$
$$d(b,b') = \arccos\left(\frac{\langle U_b, U_{b'}\rangle}{\|U_b\|\|U_{b'}\|}\right) \tag{8}$$

Apart from the distance metric, the anchor points $(s_1, \ldots, s_q)$ decide the number of local matrices. Anchor points are chosen from the observed entries and determine the local models by smoothing kernel method. While the number of anchor points increases, the accuracy of matrix completion enhances. There are three popular smoothing kernels:
Uniform kernel:

$$K_h(s_1, s_2) \Leftrightarrow \mathbf{1}[\text{d}(s_1, s_2) < h] \tag{9}$$

Triangular kernel:

$$K_h(s_1, s_2) \Leftrightarrow \left(1 - h^{-1}d(s_1, s_2)\right)\mathbf{1}[\text{d}(s_1, s_2) < h] \tag{10}$$

Epanechnikov kernel:

$$K_h(s_1, s_2) \Leftrightarrow \left(1 - d(s_1, s_2)^2\right) \mathbf{1}[d(s_1, s_2) < h] \tag{11}$$

Where $s_1, s_2 \in [m] \times [n]$, and parameter h denotes the bandwidth and $h > 0$. $K(s_1, s_2)$ has wide spread when h is large, and $K(s_1, s_2)$ has narrow spread when h is small.
The kernel function is defined as:

$$K_h((a, b), (a', b')) = K_{h_1}(a, a') K_{h_2}(b, b') \tag{12}$$

Where K_{h_1} and K_{h_2} are kernels on the spaces $[m]$ and $[n]$. In our paper, we use uniform kernel for K_{h_1} and K_{h_2}.

Figure 1 shows the locally low-rank linear assumption that the neighborhood $\{s' : d(s, s') < h\}$ in the original matrix M for all $s \in [m] \times [n]$ is approximately described by the corresponding entries of the low-rank matrix $T(s)$. The number of anchor points plays a significant role in our algorithm. There are serval ways to choose the anchor points. We choose anchor points from the observed entries of M since the row and column indices of the test data is provided. If the indices is not provided, the way of choosing uniformly from the entry set or selecting anchor points meet the requirement that no entry in $[m] \times [n]$ is far away from the other anchor points.

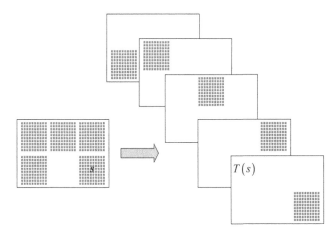

Fig. 1. For all, the neighborhood of in the original matrix is approximately described by the corresponding low rank matrix. In the figure, adjacent rows (or columns) are semantically similar just for illustration purpose. In real data, similar users (or items) may scattered over the entry space.

According to the matrix factorization, we propose to solve the following optimization problem:

$$\min \sum_{i=1}^{q} \left\| \prod_{\Omega_i} (X - M) \right\|_F \tag{13}$$

$$s.t.\, rank(X) = r,\, 1 \leq i \leq q$$

The original matrix is divided into q local matrices, Ω_i presents the index set of elements in the $i'th$ local matrix, and $X = UV^T$.

We summarize our algorithm in Algorithm 1. Our algorithm bases on local low rank matrices, we use kernel smoothing to get local matrices, which be mentioned before.

Algorithm 1.Local Sensitive Low Rank Matrix Learning

Input: whose entries are defined over Ω

Parameters: kernel function K of bandwidths h_1 and h_2

rank r and the number of local models q

for $t = 1, ..., q$ do

  select (a_t, b_t) at random from Ω

  for all $i = 1, ..., m$ do

    construct entry i : $\left[K^{a_t} \right]_i := K_{h_1} (a_t, i)$

  end for

  for all $j = 1, ..., n$ do

    construct entry j : $\left[K^{b_t} \right]_j := K_{h_2} (b_t, j)$

  end for

  for all $i = 1, ..., m$ do

    for all $j = 1, ..., n$ do

     $T_t = \left[K^{a_t} \right] \left[K^{b_t} \right]_j M$

    end for

  end for

end for

$\min \sum_{i=1}^{q} \left\| \prod_{\Omega_i} (X - M) \right\|_F$

$s.t.\, rank(X) = r, 1 \leq i \leq q$

Output

---

4   Results

In this section, we present numerical experiments on popular recommendation systems datasets for matrix completion and compare the performance with SVD to support our theoretical analysis.

In our experiments, we use three popular recommendation systems datasets, MovieLens 1 M(6 K × 4 K with 106 observations), MovieLens 10 M(70 K × 10 K with 107 observations), and the Netflix(480 K × 18 K with 108 observations). We used the Uniform kernel to construct several local matrices which help us deal with unobserved entries easily. The bandwidth is $h_1 = h_2 = 0.8$. The maximum number of iterations is $T = 100$, and the number of anchor points is $q = 50$.

The prediction accuracy is evaluated by the root mean squared error (RMSE) on each experiment, which is mathematically defined as follow

$$RMSE = \sqrt{\frac{\sum_{i=1}^{N} (r_i - \widehat{r}_i)^2}{N}} \tag{14}$$

RMSE is a measure that puts more emphasis on large errors compared with the alternative of mean absolute error.

Table 1 shows the performance of our method with 50 anchor points and SVD. The fixed rank we choose 1, 5, 7, 10, 15, 20, our method always performs better than SVD. On these three recommendation systems datasets, SVD with 10 rank performs better than when rank is 15. The RMSE of Our method decreases with the increase of the rank, the increasing amplitude is reduced.

Table 1. RMSE of different algorithms on three datasets. Results on SVD and our method

	MovieLens 1 M		MovieLens 10 M		Netflix	
Rank	SVD	Our method	SVD	Our method	SVD	Our method
1	0.9201	0.9116	0.8723	0.8541	0.9388	0.9155
5	0.8737	0.8540	0.8348	0.8132	0.8836	0.8647
7	0.8678	0.8412	0.8234	0.7947	0.8788	0.8516
10	0.8650	0.8380	0.8219	0.7910	0.8765	0.8445
15	0.8652	0.8317	0.8225	0.7818	0.8758	0.8376
20	0.8647	0.8305	0.8220	0.7803	0.8742	0.8322

From Fig. 2, we can clearly see the performance of SVD and our method on MovieLens 1 M, Movielens 10 M and Netflix. The improvement of SVD's performance is rather mirror when rank is greater than 7 while our method's improvement is still evident until the rank is 20. With the rank increasing, our method improve evidently, and our method performs better than SVD when the rank is same. We can also find our method of rank 5 outperforms SVD of rank 20.

Figure 3 graphs the RMSE of our method and SVD, we choose several values of r (5, 10, 15) as a function of the number of anchor points. As in the case of Table 1, our method improves as r increase, but our method always outperforms SVD. Figure 3 shows the RMSE of our method, SVD on MovieLens 1 M(a), MovieLens 10 M(b), Netflix(c) datasets. The results for our method are depicted by solid lines, while for SVD with dotted lines. Models of the same rank are same colors. Our method's performance improves with the increment of anchor points while SVD's performance keeps unchanged.

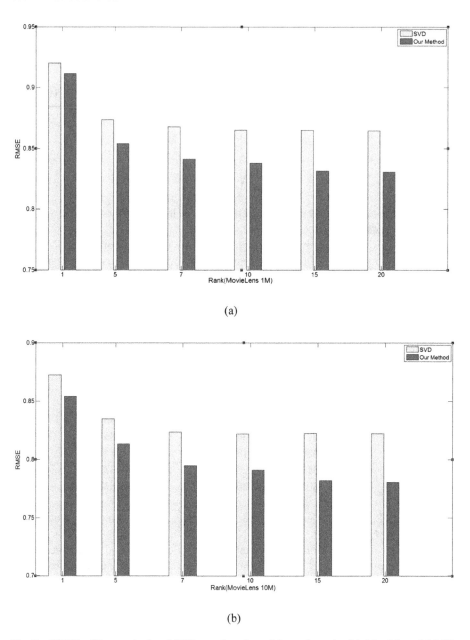

(a)

(b)

Fig. 2. RMSE of Our method and SVD as a function of the rank on the (a). Movielens 1 M (b). Movielens 10 M.

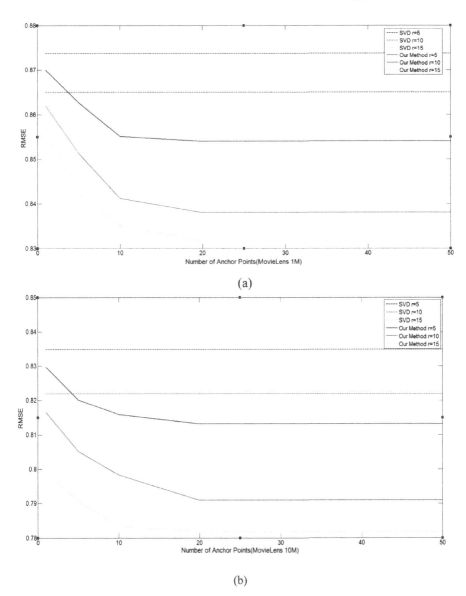

(a)

(b)

Fig. 3. RMSE of Our method, SVD on MovieLens 1 M (a), MovieLens 10 M (b) datasets. The results for our method are depicted by solid lines, while for SVD with dotted lines. Models of the same rank are same colors. (Color figure online)

In summary, we get the conclusion that our method outperforms better than SVD, our method achieve high performance with a low rank and only few anchor points.

5 Conclusions

This paper introduced a novel low rank matrix approximation model for matrix completion problem. Different to existing related methods, we take local consistency of recommendation systems data into considered, and use kernel smoothing to construct several local matrices, to achieve a naturally and precisely approximation of recommendation data matrix. We use a nonconvex optimization algorithm to tackle the model. Our experiments show that our method outperforms the related work by evaluating the accuracy in completion of recommendation systems data matrix. Our method is not only applicable for recommendation systems, but also applicable for signal and image denoising so long as the locality assumption holds.

Acknowledgments. This work was supported by the grants of the National Science Foundation of China, Nos. 61472280, 61472173, 61572447, 61373098, 61672382, 61672203, 61402334, 61520106006, 31571364, U1611265, and 61532008, China Postdoctoral Science Foundation Grant, No. 2016M601646.

References

1. Abernethy, J., Bach, F., Evgeniou, T., et al.: Low-rank matrix factorization with attributes. arXiv preprint cs/0611124 (2006)
2. ACM SIGKDD, Netflix, Proceedings of KDD Cup and Workshop (2007). http://www.cs. uic.edu/∼liub/KDD-cup-2007/proceedings.html
3. Candès, E.J., Recht, B.: Exact matrix completion via convex optimization. Found. Comput. Math. **9**(6), 717–772 (2009)
4. Candès, E.J., Tao, T.: The power of convex relaxation: near-optimal matrix completion. IEEE Trans. Inf. Theory **56**(5), 2053–2080 (2010)
5. Argyriou, A., Evgeniou, T., Pontil, M.: Multi-task feature learning. Adv. Neural. Inf. Process. Syst. **19**, 41 (2007)
6. Lee, J., Kim, S., Lebanon, G., et al.: Local low-rank matrix approximation. ICML **28**(2), 82–90 (2013)
7. Recht, B., Fazel, M., Parrilo, P.A.: Guaranteed minimum-rank solutions of linear matrix equations via nuclear norm minimization. SIAM Rev. **52**(3), 471–501 (2010)
8. Osher, S., Burger, M., Goldfarb, D., et al.: An iterative regularization method for total variation-based image restoration. Multiscale Model. Simul. **4**(2), 460–489 (2005)
9. Ji, H., Liu, C., Shen, Z., et al.: Robust video denoising using low rank matrix completion. In: CVPR, pp. 1791–1798 (2010)
10. Candes, E.J., Eldar, Y.C., Strohmer, T., et al.: Phase retrieval via matrix completion. SIAM Rev. **57**(2), 225–251 (2015)
11. Lee, D.D., Seung, H.S.: Learning the parts of objects by non-negative matrix factorization. Nature **401**(6755), 788–791 (1999)
12. Deerwester, S., Dumais, S.T., Furnas, G.W., et al.: Indexing by latent semantic analysis. J. Am. Soc. Inf. Sci. **41**(6), 391 (1990)
13. Buckland, M.: Annual review of information science and technology. J. Documentation (2013)

14. Sarwar, B., Karypis, G., Konstan, J., et al.: Application of dimensionality reduction in recommender system-a case study. Minnesota Univ Minneapolis Dept of Computer Science (2000)
15. Koren, Y., Bell, R., Volinsky, C.: Matrix factorization techniques for recommender systems. Computer **42**(8), 30–37 (2009)
16. Koren, Y.: Collaborative filtering with temporal dynamics. Commun. ACM **53**(4), 89–97 (2010)
17. Wang, X.-F., Huang, D.S., Xu, H.: An efficient local Chan-Vese model for image segmentation. Pattern Recogn. **43**(3), 603–618 (2010)
18. Li, B., Huang, D.S.: Locally linear discriminant embedding: an efficient method for face recognition. Pattern Recogn. **41**(12), 3813–3821 (2008)
19. Huang, D.S., Du, J.-X.: A constructive hybrid structure optimization methodology for radial basis probabilistic neural networks. IEEE Trans. Neural Netw. **19**(12), 2099–2115 (2008)
20. Singer, A., Cucuringu, M.: Uniqueness of low-rank matrix completion by rigidity theory. SIAM J. Matrix Anal. Appl. **31**(4), 1621–1641 (2010)
21. Cai, J.F., Candès, E.J., Shen, Z.: A singular value thresholding algorithm for matrix completion. SIAM J. Optim. **20**(4), 1956–1982 (2010)
22. Ma, S., Goldfarb, D., Chen, L.: Fixed point and Bregman iterative methods for matrix rank minimization. Math. Program. **128**(1–2), 321–353 (2011)
23. Wen, Z., Yin, W., Zhang, Y.: Solving a low-rank factorization model for matrix completion by a nonlinear successive over-relaxation algorithm. Math. Program. Comput. **4**(4), 333–361 (2012)
24. Toh, K.C., Yun, S.: An accelerated proximal gradient algorithm for nuclear norm regularized linear least squares problems. Pac. J. Optim. **6**(615–640), 15 (2010)
25. Beck, A., Teboulle, M.: A fast iterative shrinkage-thresholding algorithm for linear inverse problems. SIAM J. Imaging Sci. **2**(1), 183–202 (2009)
26. Huang, D.S.: Systematic theory of neural networks for pattern recognition. Publishing House of Electronic Industry of China, May 1996. (in Chinese)
27. Huang, D.S., Jiang, W.: A general CPL-AdS methodology for fixing dynamic parameters in dual environments. IEEE Trans. Syst. Man Cybern. - Part **42**(5), 1489–1500 (2012)
28. Lin, Z., Chen, M., Ma, Y.: The augmented lagrange multiplier method for exact recovery of corrupted low-rank matrices. arXiv preprint arXiv:1009.5055 (2010)
29. Huang, D.S.: Radial basis probabilistic neural networks: model and application. Int. J. Pattern Recogn. Artif. Intell. **13**(7), 1083–1101 (1999)

Author Index